02

DISCARD

D0889875

EARTH SCIENCE

EARTH SCIENCE
VOLUME III

EARTH MATERIALS
AND
EARTH RESOURCES

EDITOR
JAMES A. WOODHEAD
Occidental College

EDITORIAL BOARD

CHARLES W. ROGERS, VOLUMES III & V
Southwestern Oklahoma State University

DENNIS G. BAKER, VOLUME IV
University of Michigan

ANITA BAKER-BLOCKER, VOLUME IV
Applied Meteorological Services

DAVID K. ELLIOTT, VOLUME II
Northern Arizona University

RENÉ DE HON, VOLUME I
University of Louisiana at Monroe

SALEM PRESS, INC.
Pasadena, California Hackensack, New Jersey

Managing Editor: Christina J. Moose
Project Development: Robert McClenaghan
Manuscript Editors: Doug Long, Amy Allison
Acquisitions Editor: Mark Rehn
Research Supervisor: Jeffry Jensen
Photograph Editor: Philip Bader
Assistant Editor: Andrea E. Miller
Indexers: Melanie Watkins, Lois Smith
Research Assistant: Jeffrey Stephens
Production Editor: Cynthia Beres
Page Design and Layout: James Hutson
Additional Layout: William Zimmerman
Graphics: Electronic Illustrators Group

Library of Congress Cataloging-Publication Data

Earth science / editor, James A. Woodhead.
    p. cm.
Expands and updates Magill's survey of science: earth science series.
Includes bibliographical references and indexes.
Contents: v. 1. The physics and chemistry of earth — v. 2. The earth's surface and history — v. 3. Earth materials and earth resources — v. 4. Weather, water, and the atmosphere — v. 5. Planetology and earth from space.
ISBN 0-89356-000-6 (set : alk. paper) — ISBN 0-89356-001-4 (v. 1 : alk paper) —
ISBN 0-89356-002-2 (v. 2 : alk. paper) — ISBN 0-89356-003-0 (v. 3 : alk. paper) —
ISBN 0-89356-004-9 (v. 4 : alk. paper) — ISBN 0-89356-005-7 (v. 5 : alk. paper)
1. Earth sciences. I. Woodhead, James A. II. Magill's survey of science. Earth science series.

QE28 .E12 2001
550—dc 21

00-059567

First Printing

PRINTED IN THE UNITED STATES OF AMERICA

CONTENTS

EARTH SCIENCE

1
MINERALS AND CRYSTALS

BIOPYRIBOLES

Biopiriboles are important rock-forming minerals, third in abundance only to feldspars and quartz. They are especially abundant in igneous and metamorphic rocks. Important groups of biopyriboles include micas, pyroxenes, and amphiboles.

PRINCIPAL TERMS

AMPHIBOLES: a group of generally dark-colored, double-chain silicates crystallizing largely in the orthorhombic or monoclinic systems and possessing good cleavage in two directions intersecting at angles of about 56 and 124 degrees

CHAIN SILICATES: a group of silicates characterized by joining of silica tetrahedra into linear single or double chains alternating with chains of other structures; also known as "inosilicate"

CLEAVAGE: the tendency of a mineral or chemical compound to break along smooth surfaces parallel to each other and across atomic or molecular bonds of weaker strength

CRYSTAL SYSTEM: one of any of six crystal groups defined on the basis of length and angular relationship of the associated axes

MICAS: a group of complex, hydrous sheet silicates crystallizing largely in the monoclinic system and possessing pearly, elastic sheets with perfect one-directional cleavage

MONOCLINIC: a crystal system possessing three axes of symmetry, generally of unequal length; two axes are inclined to each other obliquely, and the third is at right angles to the plane formed by the other two

ORTHORHOMBIC: referring to a crystal system possessing three axes of symmetry that are of unequal length and that intersect at right angles

PYROXENES: a group of generally dark-colored, single-chain silicates crystallizing largely in the orthorhombic or monoclinic systems and possessing good cleavage in two directions intersecting at angles of about 87 and 93 degrees

SHEET SILICATES: a group of silicates characterized by the sharing of three of the four oxygen atoms in each silica tetrahedron with neighboring tetrahedra and the fourth with other atoms in adjacent structures to form flat sheets; also known as "phyllosilicate" or "layer" silicate

SILICATE: a chemical compound or mineral whose crystal structure possesses silica tetrahedra (a structure formed by four charged oxygen atoms surrounding a charged silicon atom)

CHEMICAL STRUCTURE AND COMPOSITION

Biopiriboles include numerous but related groups of minerals. They are important constituents of both igneous rocks, whose minerals largely form as a result of cooling and crystallization of a melt (liquid), and metamorphic rocks, whose minerals are largely crystallized in a solid state at elevated temperatures or pressures. Some biopyriboles also occur as fragments in sedimentary rocks and sediments. The three most important mineral groups in biopyriboles are micas, pyroxenes, and amphiboles.

Biopiriboles are silicate minerals, meaning that they are composed of atoms of the chemical elements silicon and oxygen as well as atoms of other chemical elements. The silicon atom is surrounded by four oxygen atoms attached or bonded to it. This forms a structure known as a silica tetrahedron, which can be represented as a four-sided solid with triangular faces. This silicon-oxygen bond is very strong. These silica tetrahedra may be repeated in various ways; biopyriboles are expressed by the chain (line) and sheet (plane) structures. Pyroxenes and amphiboles are examples of chain silicates and micas of sheet silicates.

Other atoms with attached or bonded oxygen (or in some cases oxygen attached to hydrogen, or hydroxyl) may also occur as chains or sheets alternating with the chains or sheets of silica tetrahedra. In sheet silicate groups, the sheet or plane of silica tetrahedra might alternate with aluminum bonded to its oxygens. For some sheet silicate minerals, other layers (potassium atoms bonded to oxygen or hydroxyl, for example) may be layered within. In chains, silica tetrahedra may alternate with structures of iron atoms with attached oxygens. In some groups (amphiboles), two chains of silica tetrahedron alternate with one or more other chains.

CRYSTALS AND CRYSTALLINE MASSES

Biopiriboles are found as crystals or as crystalline masses. Crystals may be defined as inorganic or organic solids which are chemical elements or compounds. These are formed by growth and are bounded by faces or surfaces with a definite geometric relationship to one another. This relationship is expressive of an orderly internal arrangement of atoms and molecules. Crystalline masses are intergrowths of crystals that show incomplete expression of external faces, generally because there was either not enough space or some substance in the solution or melt inhibited crystal growth.

Crystals belong to systems, based on the relationship of three lines or axes to each other. Most micas, pyroxenes, and amphiboles fall into the orthorhombic and monoclinic systems under normal geologic conditions in the Earth's crust (a very few are hexagonal or triclinic).

MICAS AND BRITTLE MICAS

The micas are one of the most common groups of rock-forming minerals. They are characterized by the following properties: the formation of thin, sheetlike crystals, which are stronger within the sheets than between them; a tendency to break in one smooth direction parallel to the sheets, a property called "cleavage"; crystallization in the monoclinic system for most species; considerable elasticity; and a luster that is vitreous (glasslike) to pearly. Two of the most common members of this group are biotite and muscovite mica.

Biotite mica occurs in black, brown, or dark green flakes and is a potassium-magnesium-iron-aluminum hydroxyl silicate. The magnesium and iron atoms can substitute for each other in the oc-

---

MAJOR BIOPIRYPOLE GROUPS

A. Sheet silicates
  1. Mica group (examples: biotite and muscovite)
  2. Chlorite group
  3. Serpentine group (examples: antigorite and chrysotile asbestos)
  4. Talc group (examples: talc and pyrophyllite)
  5. Clay mineral group (examples: kaolinite, smectite, and illite)
  6. Brittle mica group

B. Chain silicates
  1. Single silica tetrahedron chains
    a. Orthorhombic pyroxenes (examples: enstatite and hypersthene)
    b. Monoclinic calcic pyroxenes (examples: diopside and augite)
    c. Monoclinic alkali pyroxenes (example: jadeite)
    d. Pyroxenoids (example: wollastonite)
  2. Double silica tetrahedra chains
    a. Orthorhombic amphiboles (example: anthophyllite)
    b. Monoclinic magnesium-iron amphiboles (example: cummingtonite)
    c. Monoclinic calcic amphiboles (examples: tremolite and hornblende)
    d. Monoclinic alkali amphiboles (examples: glaucophane, riebeckite—form "blueschists")

C. Complex biopyriboles

Sheets of muscovite mica. (© William E. Ferguson)

of hydrous calcium-bearing, iron-rich, or aluminous sheet silicates. Samples are characteristically easily broken across, as well as between, sheets, hence the term "brittle mica." One aluminous variety, margarite, occurs in attractive lilac or yellow crystals associated with corundum (aluminum oxide) or as veins in chlorite schist.

CHLORITES

The chlorite group is similar to the micas in that these minerals also tend to have sheetlike crystals that show cleavage parallel to the sheets and possess a pearly luster. The color of chlorite flakes is commonly dark to bright green, although it can also occur in brown, pink, purple, and colorless crystals. The potassium, sodium, and lithium that are present in mica are absent in chlorite. Its crystals are flexible but not elastic.

Chlorite may occur as fine-grained masses in clay or claystone. The largest crystals occur in altered ultramafic rocks, such as serpentinite, and in the metamorphic rock chlorite schist. Varieties

tahedral layers (which alternate with a silica tetrahedron and potassium-hydroxyl layer), because charged atoms (ions) of magnesium and iron have nearly the same size as well as the same charge. The mineral occurs widely in most igneous and metamorphic rocks but is most abundant in granite, which is composed of interlocking crystals of feldspar, quartz, and, commonly, mica. It is also abundant in the coarsely layered (foliated) metamorphic rock, mica schist.

Muscovite mica occurs in clear to smoky yellowish, greenish, or reddish flakes and is a potassium-aluminum hydroxyl silicate. Aluminum occurs both in the octahedral layer and (substituting for silica) in the tetrahedral layer. Muscovite mica is especially abundant in granite and occurs in large sheets and crystals in the very coarse rock granite pegmatite. It is also prominent in mica schist and occurs microscopically in slate. Flakes of muscovite can occur as fragments in some sandstones, which are sedimentary rocks.

The brittle mica group is less common, but consists largely

Steam and pressure change mafic rocks to serpentinite, one of the chlorites. (© William E. Ferguson)

of this mineral group also may occur associated with metallic ore deposits. Chlorite differs in the arrangement and number of octahedral layers from the related serpentine and talc groups.

SERPENTINES

The serpentine mineral group is hydrous and somewhat complex. It includes magnesium-rich sheet silicates, which commonly occur as an alteration of the magnesium-iron silicates olivine and pyroxene in altered ultramafic rocks known as metaperidotites. Serpentines may also form as an alteration of olivine (forsterite) pyroxene or other minerals in marble. The various kinds vary from rather soft to moderately hard. The most common are antigorite, with a platy-massive structure, and chrysotile asbestos, which is fibrous. Serpentines are most commonly green, but brown, red, blue and black varieties are known. They are economically important in some areas and occur in mountain belts such as the Alps and the Appalachian Piedmont.

TALCS

The talc group consists of sheet silicates that are rich in magnesium, aluminum, or iron. They are characterized by a very pearly luster, great softness, so they can be scratched with a fingernail, and a soapy or greasy feel; they occur in thin sheets or scaly or radiated masses. Talc is a hydrous magnesium silicate that can form through the alteration of olivine, pyroxene, or serpentine. It is common in altered ultramafic rocks; it also occurs as talc schist and in some marbles. Pyrophyllite, the aluminum-rich analogue of talc, occurs in schists or through metamorphic alteration of aluminous rocks. The iron-rich analogue minnasotaite occurs in metamorphosed iron formations.

CLAY MINERALS

Clay minerals are of the most widespread groups of sheet silicates. They include chlorite, kaolinite (china clay), smectite (also known as montmorillonite) and illite. Clays can occur as mixed sheets, for example, of chlorite and illite or illite and smectite.

Kaolinite is usually white and soft and is a hydrous aluminum silicate, prized for use in china. It consists of silica tetrahedron layers alternating with aluminum hydroxide octahedral layers.

Smectite is unusual: It has a layer that takes up water or liquid organic molecules, and therefore its mineral structure is expandable. Illite has a structure similar to that of muscovite mica, but some of the potassiums are replaced by hydroxyl. It is sometimes included in the mica group.

All these minerals may be important constituents of sedimentary rocks (especially chlorite, illite, and smectite) and of soils. Claystones and mudstones are largely made up of clay minerals. Other members of this group occur associated with metallic ore deposits and hot springs or geyser areas.

PYROXENES

The chain silicate group includes orthopyroxenes, clinopyroxenes, pyroxenoids, orthoamphiboles, and clinoamphiboles. The pyroxene group is characterized by single silica tetrahedron chains alternating with octahedron chains and other chains with cubic structures. The octahedron chains tend to have smaller internal atoms than those of the cubic chains. Octahedron chain atoms include magnesium, iron, or aluminum, and cubic chain atoms include calcium and sodium. Sometimes the larger atoms such as calcium are called X-type, and the smaller atoms such as magnesium are called Y-type. Other letter classifications may be used by crystallographers. Pyroxenes have two directions of smooth breakage (cleavage), nearly at right angles to each other. Most pyroxenes tend to be green, brown, or green-black in color and are moderately hard.

Orthorhombic pyroxenes (also called orthopyroxenes) occur mostly in dark-colored high-temperature igneous rocks such as pyroxenites and gabbros. They range from light bronze-brown to dark green-brown in color and may have a bronzy luster. In this group, ferrous iron atoms can substitute freely for the magnesium atoms in the structure. Two important orthopyroxenes are enstatite and hypersthene.

Monoclinic pyroxenes (also called clinopyroxenes) occur in dark-colored igneous rocks and in siliceous or aluminous marbles metamorphosed at high temperatures. There are two main groups, the calcium-rich or calcic types and the sodium-rich or alkali types. Substitution of sodium and calcium atoms can occur to a certain extent between the two types. The calcic pyroxenes are usually green or brown; sodic pyroxenes may be light or bright green or, if iron-rich, blue-black to black. Common examples of calcic clinopyroxenes are diopside and augite. Magnesium and ferrous iron atoms substitute freely. Two important sodium-alkali pyroxenes are jadeite, an important carving and gem material, and aegirine.

Pyroxenoids are similar in some respects to pyroxenes; the former are also high-temperature minerals, but differ in that octahedral and cubic chains both have Y-type atoms, so that monoclinic or even triclinic structures result. They also tend to be more tabular and less blocky than are pyroxenes in their structure. Wollastonite, a calcium silicate found in siliceous marbles, is an important example of this group.

AMPHIBOLES

Minerals having double silica tetrahedral chains are called amphiboles. Amphiboles are distinguished from pyroxenes, which are closely similar in color, hardness, and occurrence, by the two directions of cleavage, 124 and 56 degrees instead of close to 90 degrees, as is the case for pyroxenes. Orthorhombic amphiboles (orthoamphiboles) include the magnesium-iron amphibole anthophyllite and the aluminum-magnesium-iron amphibole gedrite. These types are restricted largely to magnesium-rich, calcium-poor metamorphosed ultramafic and mafic plutonic and volcanic rocks. Anthophyllite-gedrite varies from purple or clove-brown to yellow-brown or gray in color and may be columnar or fibrous in structure.

One group of clinoamphiboles (monoclinic amphiboles) is a magnesium-iron silicate similar to the anthophyllite group. The magnesium-rich end member is called cummingtonite, and the iron-rich member is known as grunerite. Cummingtonite usually occurs as fibrous or radiating crystals and is brown to gray in color. The largest group of clinoamphiboles is the calcic amphiboles. In all the calcic amphiboles, the amount of calcium exceeds that of alkalis (sodium and potassium). In the tremolite group, which is analogous to the diopside group in the clinopyroxenes, calcium and magnesium or ferrous iron are the major constituents of the cubic and octahedral chains, respectively, although sodium-rich tremolites occur. These amphiboles vary from colorless to green and occur mostly in marble and meta-

morphosed dark igneous rocks. The hornblende group of calcic amphiboles, which is somewhat analogous to the augite group of clinopyroxenes, can have a much more varied composition: Some sodium may substitute for calcium, and aluminum and ferric iron may substitute for magnesium and ferrous iron in the octahedral chains and aluminum for silica in the tetrahedral chains. Hornblendes are commonly black to dark green and occur in a wide array of igneous and metamorphic rocks.

The alkali amphiboles are rich in sodium (or, very rarely, potassium), and most range in color from dark to light blue or violet to blue-black. Important members are the sodium-magnesium-iron-aluminum amphiboles riebeckite (blue) and glaucophane (blue to violet). Glaucophane is an important constituent of blueschist, which is formed at high pressures and is especially common in some parts of California and Japan.

Complex biopyriboles consist of combined anthophyllite-talc structures. First described from Vermont, they occur elsewhere as well. Other combinations of amphibole-talc structures are possible. Multiple chain units are characteristic of this group.

METHODS OF STUDY AND ANALYSIS

Many methods have been used to study and analyze biopyriboles. Simple physical techniques can determine color, cleavage directions, crystal form, hardness (resistance to abrasion), density, and other properties. The major biopyribole groups and the more common or distinctive kinds of micas, amphiboles, and pyroxenes can be identified with such techniques. Examination under a binocular or compound microscope can extend this process to smaller grains or crystals. Association of other minerals rich in certain elements may also be helpful in identification.

Polarizing microscopes are more powerful tools that force light to travel in a certain direction through a sample by means of polarizers and producing interference and refraction (bending) effects in the light. The chemical makeup of many pyroxenes can be studied in this way, but the more complex amphiboles and sheet silicates require more sophisticated methods.

In X-ray diffraction, X rays are generated by electron bombardment and produce multiple reflections off atomic planes in crystals, allowing determination of the dimensions of spacing of these planes and hence identification of the mineral. Micas, clays, serpentines, and other sheet silicates are often readily differentiated and analyzed by X-ray diffraction. Special cameras for X-ray diffraction permit the study of structure, mineral unit cell dimensions, and atomic position of the elements.

Scanning electron microscopy produces an electron photograph of the surface of a fine-grained material or small crystals. This method is necessary, in conjunction with X-ray diffraction, for unequivocal identifications of fine-crystalline clay and serpentine group minerals. Transmission electron microscopy permits resolution to a few angstroms (atomic dimensions), thus allowing direct studies of mineral structure. This method is necessary for studying detailed molecular structure of the chain silicates and complex biopyriboles.

The electron microprobe, useful for chemical analysis, focuses an intense beam of electrons on some coated material (usually gold or carbon); the material then emits characteristic X rays, whose wavelength and intensity can be examined with an X-ray spectroscope. Through calculations and with adequate corrections applied, an analysis can be produced, provided there is a mineral standard for comparison.

Differential thermal analysis uses a thermocouple method for measuring temperature differences between the material being tested and a standard material. A useful method to detail heat-absorbing (endothermic) dehydration reactions for minerals, especially for clay minerals and sheet silicates, thermal analysis aids in the identification and structural analysis of these minerals.

INDUSTRIAL USES

Biopiriboles are important constituents of rocks in both the crust and the mantle of the Earth. Some of the rocks containing pyroxenes and amphiboles, such as traprock and diorite, are used for road and railroad gravels, building stone, and monuments. Clay minerals, especially kaolinite and a hydrated type called halloysite, are used to make fine china and pottery and are a constituent of ceramics, brick, drain tile, and sewer pipe. Kaolinite is also used as a filter in medical research and as a filler in paper. Bentonite (smectite or montmorillonite) is used in drilling muds that support the bit and drilling apparatus in oil exploration.

Muscovite mica has been used as an electric insulating material and as a material for wallpaper, lubricants, and nonconductors. Lepidolite is a source of lithium and is used in the manufacture of heat-resistant glass. Talc is highly important in the cosmetics industry. As the massive variety, soapstone, it is used for tabletops and in paint, ceramics, paper, and insecticides. Pyrophyllite, the aluminum analogue, is used for the same purposes. Serpentine has been used as an ornamental and building stone. The variety chrysotile is the main source of asbestos, used in the past for fireproof fabrics and construction material. Fibrous varieties of anthophyllite (also called amosite), tremolite, and riebeckite (also called crocidolite) were also used in the past as sources of asbestos. Health considerations have largely forced the discontinuance of its manufacture and use.

Pyroxenes are not so widely used, but clear and transparent colored varieties of diopside and spodumene have been used as gemstones, and both jadeite and rhodonite are prized gem materials for carving. Spodumene is also a major source of lithium for ceramics, batteries, welding flux, fuels, and the compound lithium carbonate, used to treat manic-depressives.

David F. Hess

CROSS-REFERENCES

Basaltic Rocks, 1274; Carbonates, 1182; Clays and Clay Minerals, 1187; Earth's Age, 511; Evolution of Life, 999; Feldspars, 1193; Fossil Record, 1009; Gem Minerals, 1199; Geologic Time Scale, 1105; Granitic Rocks, 1292; Hydrothermal Mineralization, 1205; Ionic Substitution, 1209; Magnetite, 1214; Metamictization, 1219; Minerals: Physical Properties, 1225; Minerals: Structure, 1232; Nonsilicates, 1238; Orthosilicates, 1244; Oxides, 1249; Radioactive Minerals, 1255; River Valleys, 943.

BIBLIOGRAPHY

Cepeda, Joseph C. *Introduction to Minerals and Rocks*. New York: Macmillan, 1994. Cepeda's text provides a good introduction to biopyriboles for students just beginning their studies in the Earth sciences. Includes illustrations and maps.

Deer, W. A., R. A. Howie, and J. Zussman. *An Introduction to the Rock-Forming Minerals*. 2d ed. New York: Longman, 1992. This volume discusses in detail the crystallography, properties, chemistry, occurrence, and origin of the pyroxenes and amphiboles, as well as the micas and other sheet silicates. The work's quality is excellent; it is very well written and well organized. Suitable for college-level students.

Klein, Cornelis, and C. S. Hurlbut, Jr. *Manual of Mineralogy*. New York: John Wiley & Sons, 1999. A general text of mineralogy, revised many times since James D. Dana first published the prototype in 1862. Extremely useful, it lists properties, occurrence, and uses of the amphiboles, pyroxenes, and sheet silicates and gives background information in crystallography and descriptive mineralogy. Suitable for the high school and college reader.

Leake, Bernard E. "Nomenclature of Amphiboles." *The American Mineralogist* 63 (November, 1978): 1023-1052. A complete, but rather technical, description of amphiboles and their classification. Appropriate for the college-level reader.

Oldershaw, Cally. *Rocks and Minerals*. New York: DK, 1999. This small, 53-page volume is filled with color illustrations and is therefore of great use to new students who may be unfamiliar with the rock and mineral types discussed in classes or textbooks.

Robinson, George. "Amphiboles: A Closer Look." *Rocks and Minerals* (November/December, 1981). An excellent and readable summary of the properties and occurrence of this complex group. Suitable for both the high school and college reader.

CARBONATES

Carbonate minerals are characterized by having the carbonate ion in their composition. The common carbonate minerals are divisible into the calcite, aragonite, and dolomite groups.

CARBONATE MINERAL FORMATION

Carbonates are one among several classes of mineral. Minerals are the stuff of which rocks are made. They are natural substances with a definite chemical composition and an ordered internal arrangement. Minerals can be divided into chemical groups based on their atoms and into structural groups based on the atoms' ordered arrangement.

All carbonates contain the carbonate ion as their defining anionic group. An ion is an atom that has lost or gained electrons and so has become chemically reactive. When an ion has lost an electron, it has a positive charge and is called a cation; when an ion has gained an electron, it has a negative charge and is called an anion. The number of electrons lost or gained is the charge of the ion. The charge and the radius (size) of an ion determine how it is chemically bonded to another ion, how strong the bonding is, and the number of ions that can be coordinated to it or surround it.

Scientists have found that the carbonate ion contains one carbon ion in the middle of a triangle formed by three oxygen ions, which occupy the corners of the triangle. Two-thirds of the charge on each oxygen ion is used for bonding with the central carbon ion. As a result, only one-third of the charge on each oxygen ion is available for bonding with cations, and the carbonate ion group acts as if it were a single ion with an overall double negative charge. To form a carbonate mineral, the negative charge on the carbonate ion must be neutralized by cations such as calcium, magnesium, strontium, manganese, and barium.

CARBONATE MINERAL STRUCTURE

Atoms in a mineral lie on imaginary planes, which are called atomic planes. These planes cut across other planes, forming definite and knowable intersection angles. Those atomic planes that contain many atoms tend to develop into crystal faces. Minerals are crystalline, and the crystal faces of a mineral are related to the internal arrangement of atoms. The bonds between atoms of the same atomic plane tend to be stronger than the bonds across the atomic planes. Consequently, minerals break or cleave along preferred planes, or cleavage planes, when struck by a hammer. A set of parallel cleavage planes yields one cleavage direction. Some minerals have more than one cleavage direction; others have none, since the bond strength between atoms is the same in all directions. Common carbonate minerals have three cleavage directions.

The atoms of a mineral are symmetrically related to one another. Since crystal faces are related to the internal arrangement of the atoms, these crystal faces are also symmetrically related. Scientists have found that all minerals can be grouped into thirty-two crystal classes based on their symmetry relations. A small group of atoms forms the basic building block of crystals. This basic building block is called a unit cell; it is an arrangement of the smallest number of atoms which, as a unit, may be repeated over and over again to form a visible crystal. The volume of a unit cell can be determined from the lengths of its lines and the angles subtended by them. Depending on the cell's shape, these imaginary lines may be parallel to its edges or may pass through opposite corners, sides, or edges. The lines are called crystallographic axes. The various kinds of unit cell have led scientists to regroup the thirty-two crystal classes of minerals into six crystal systems.

The common carbonates belong to two of the six: the orthorhombic and hexagonal systems. The orthorhombic crystal system is characterized by having three crystallographic axes which are unequal in length and which are perpendicular to one another. The hexagonal system has four axes, all of which pass through a common center. Three of these axes lie on the same plane, are of equal length, and are separated from one another by 120 degrees. The fourth axis is different in length from the others and is perpendicular to them.

The three common carbonates, calcite, aragonite, and dolomite, belong to three structural types. Calcite and aragonite are polymorphs of the same compound; that is, the same chemical compound occurs in different structures. Aragonite is orthorhombic and less symmetrical than calcite, which is hexagonal. Calcite and dolomite are both hexagonal, but they do not have identical structures. In calcite, calcium atomic planes lie between carbonate ion planes. In dolomite, alternating calcium planes are occupied by magnesium atoms. As a result, calcite has a higher symmetry content than dolomite, although both belong to the same crystal system. Also, dolomite shows more internal order than calcite, because it requires the positioning of more different atoms in specific atomic sites.

All carbonates that contain divalent cations whose ionic radii are less than or equal to that of the calcium ion are isostructural to calcite, which has a hexagonal structure. Different chemical compounds which have identical structures are said to be isostructural. Siderite, magnesite, and rhodochrosite are isostructural to calcite.

CALCITE GROUP

Calcite is by far the most common of all carbonate minerals. It is commonly off-white, colorless, or transparent. It may exhibit different crystal shapes, but in all cases, a careful examination will reveal the hexagonal crystal structure. It is fairly resistant to abrasion, although it can be scratched by a steel knife. Its resistance to scratching is partly why marble, which is composed of calcite, is used as a building stone. Calcite fizzes and dissolves in

Sinkholes that form by the collapse of roof rocks of near-surface caves in the limestones of warm and humid regions are a problem not only to farmers but also to homeowners. It is not unusual for a house to sink suddenly into a depression caused by a collapse into an underground cave, as this one did in 1967. (U.S. Geological Survey)

acid. Clear calcite crystals exhibit sets of parallel cleavage lines that intersect with other sets in such a way as to give the impression that calcite is rhombus-shaped.

In its most abundant form, calcite is synthesized by aquatic organisms that are common in shallow marine environments of warm latitudes, such as the Gulf Coast. As the marine organisms die, their shells, which are made of calcite, settle at the bottom. These shells may be broken into smaller fragments by browsing organisms or wave action. The fragments may become cemented together by calcite and form a rock called limestone. Many caves and some sinkholes are found in regions where the rocks are limestone. Caves are formed because calcite is dissolved by groundwater. Sinkholes are formed when the roofs of caves collapse.

DOLOMITE GROUP

In the dolomite group, alternating calcium atomic planes are occupied by other divalent ions, such as magnesium in dolomite, iron in ankerite, and manganese in kutnohorite. Since the magnesium, iron, and manganese ions are smaller than the calcium ion, the dolomite group structure is not as highly symmetrical as the calcite structure.

Consequently, although both the calcite and the dolomite group minerals are hexagonal, they belong to different classes among the thirty-two classes of crystals. Of the dolomite group minerals, the mineral dolomite is by far the most common. It is fairly common in ancient carbonate rocks called dolostones. Dolomite is white to pinkish and is similar to calcite in many ways; however, it does not fizz readily when diluted acid is dropped on it.

ARAGONITE GROUP

In the aragonite group, carbonate ions do not lie on simple atomic planes. Adjacent carbonate ions are slightly out of line; also, adjacent carbonate ions face in opposite directions. Divalent cations whose ionic radii are larger than or equal to the calcium ion form carbonate minerals which are isostructural to aragonite. Of the minerals in the aragonite group, aragonite is the most common. It is generally white and elongate. It is the carbonate mineral that readily precipitates from marine water, but it is not a stable mineral, and in time it changes to the more symmetrical and hexagonal structure of calcite. Aragonite is the least symmetrical polymorph of calcium carbonate and is found in metamorphic rocks which were formed under high pressure and comparatively low temperature.

IDENTIFICATION OF CARBONATE MINERALS

Carbonates are generally light-colored, soft, and easy to scratch with a knife as compared with most other common rock-forming minerals. They form a white powder when scratched. They are harder than fingernails, however, and cannot be scratched by them. When acid is poured on carbonate powder, it fizzes, liberating carbon dioxide. Carbonates such as calcite do not even have to be powdered for the acid test, because they fizz readily. The cleavage planes and the angles between cleavages are another physical

A travertine formation, the Liberty Cap, at Mammoth Hot Springs, Yellowstone National Park. Travertine, or tufa, is a banded calcite rock that precipitates at the mouths of hot springs. (U.S. Geological Survey)

method by which carbonate minerals can be distinguished. Clear crystals such as those of calcite can produce double refraction of objects, another property that identifies carbonates without the aid of instruments.

Better mineral identification is done by scientists after a rock is cut to a small size, mounted on glass, and ground to a very thin section of rock (0.03 millimeter thick), which is then capable of transmitting light. The thin section is placed on a stage of a transmitted-light polarizing microscope. A lens below the stage polarizes light; it allows the transmission only of light which vibrates in one direction, for example, east to west. Another lens, above the stage, allows the passage only of light that vibrates in the other direction, north to south. When glass is placed on a stage and the lower polarizer is inserted across the light source, the color of the glass can be seen. When the upper polarizer is also inserted, however, the glass appears dark, because no light is transmitted. The optical properties of most minerals, including the carbonates, are different from those of glass. Other accessories are used in addition to the polarizing lenses in order to determine minerals' optical properties. Magnification by microscope permits the better determination of minerals' physical properties, such as their shape and cleavage.

DETERMINATION OF CRYSTAL STRUCTURE

The crystal structure of carbonate crystals can be determined with the aid of a contact goniometer. Its simplest version is a protractor with a straight edge fastened at its middle. The goniometer is used to measure the angles between crystal faces, from which scientists can determine the crystal structure.

X-ray diffraction can ascertain the crystal structure of any substance, including carbonates. Diffraction peaks characteristic of each mineral can be displayed on a chart recorder when X rays bombard a sample. Each diffraction peak results from the reinforcement of X-ray reflections from mutually parallel atomic planes within the sample. Several diffraction peaks from one mineral indicate equivalent numbers of sets of atomic planes within the minerals. The difference in the peak heights corresponds to the density of atoms in the pertinent atomic planes. The detection device does not have to be a chart recorder; it can be a

photographic paper or a digital recorder that can be appropriately interfaced to a computer for the quick identification of minerals.

PRACTICAL APPLICATIONS

Carbonates are a fairly common group of minerals which form in environments that range from arid lands to shallow seas. Hot springs are one place where calcite precipitates. Travertine, or tufa, is a banded rock that precipitates at the mouth of springs. Caliche, deposits of carbonate that precipitate from groundwater in arid climates, is a source of serious problems to irrigation farmers. Sinkholes that form by the collapse of roof rocks of near-surface caves in the limestones of warm and humid regions are a problem not only to farmers but also to homeowners. It is not unusual for part of a house, or the whole of it, to sink suddenly into a depression caused by a collapse into an underground cave.

Carbonates are important for their regulation of the pH, or acidity content, of ocean waters. Carbonate minerals dissolve when the acid content of water is raised and precipitate when the acid content is reduced. In this way, the pH of ocean water is regulated to a steady value of 8.1.

Carbonates are also known for their industrial applications, which range from dolomite tablets to building materials such as cement and mortar. One of the finest building rocks is marble, which is composed of carbonate minerals. Marble often is delicately banded with different colors. The banding arises because the minerals are lined up in directions perpendicular to the natural pressure under which an impure limestone was metamorphosed and converted to marble. If the original limestone was pure and composed entirely of calcite, the marble that is metamorphosed from it would be white and not banded. Polished marble is used as a building material or often as decorative stone for doors or exteriors. Polished travertine is also used as building stone, but it is placed in the interiors of buildings because of banded porous zones that can accumulate rainwater. Regular limestone is used in buildings and, most commonly, in retaining walls alongside houses and roads. Most limestone is used for cement. Cement is up to 75 percent limestone; the rest is silica and aluminum.

Habte Giorgis Churnet

CROSS-REFERENCES

Biopyriboles, 1175; Clays and Clay Minerals, 1187; Earth Resources, 1741; Electron Microscopy, 463; Experimental Petrology, 468; Feldspars, 1193; Gem Minerals, 1199; Hydrothermal Mineralization, 1205; Ionic Substitution, 1209; Magmas, 1326; Magnetite, 1214; Metamictization, 1219; Metamorphic Rock Classification, 1394; Minerals: Physical Properties, 1225; Minerals: Structure, 1232; Nonsilicates, 1238; Orthosilicates, 1244; Oxides, 1249; Radioactive Decay, 532; Radioactive Minerals, 1255; X-ray Powder Diffraction, 504.

BIBLIOGRAPHY

Bathurst, Robin G. C. *Carbonate Sediments and Their Diagensis.* 2d ed. New York: Elsevier, 1975. This is an excellent book on carbonate rocks. Chapter 6 discusses the chemistry and structure of the more common carbonate minerals.

Butler, James Newton. *Carbon Dioxide Equilibria and Their Applications.* Chelsea, Mich.: Lewis, 1991. Butler discusses in great detail the role of carbonates in the chemical equilibria of carbon dioxide. Includes a short but useful bibliography and an index.

Deer, W. A., R. A. Howie, and J. Zussman. *An Introduction to the Rock-Forming Minerals.* 2d ed. New York: Longman, 1992. A work of reference useful for Earth science students. Carbonates are discussed in detail.

Klein, Cornelis, and C. S. Hurlbut, Jr. *Manual of Mineralogy.* 21st ed. New York: John Wiley & Sons, 1999. An excellent book on the study of minerals. The details of carbonates are treated in chapter 10. Suitable for college-level students.

Loucks, Robert G., and J. Frederick Sarg, eds. *Carbonate Sequence Stratigraphy: Recent Developments and Applications.* Tulsa, Okla.: American Association of Petroleum Geologists, 1993. This book contains several essays that address advances in the study of carbonates as they relate to stratigraphy, as well as their relevance to the petroleum industry. Bibliography and index.

Mason, Brian, and L. G. Berry. *Elements of Mineralogy.* San Francisco: W. H. Freeman, 1968. An excellent and easy-to-read book on the study of minerals. Used by many colleges. Carbonates are discussed in chapter 7.

Parker, Sybil P., ed. *McGraw-Hill Encyclopedia of the Geological Sciences.* 2d ed. New York: McGraw-Hill, 1988. This reference has complete entries on all the common carbonate minerals, including aragonite, dolomite, limestone, and calcite. Written at a college level. Illustrated.

Prinz, Martin, George Harlow, and Joseph Peters, eds. *Simon and Schuster's Guide to Rocks and Minerals.* New York: Simon & Schuster, 1978. Rocks and minerals are described and illustrated with color photographs in this easy-to-read book.

CLAYS AND CLAY MINERALS

Clays are fine-grained materials with unique properties, such as plastic behavior when wet. They form by weathering of silicate rocks at the Earth's surface, by diagenetic reactions, and by hydrothermal alteration. An understanding of clays is important in solving problems in petroleum geology, engineering, and environmental science.

PRINCIPAL TERMS

AUTHIGENIC MINERALS: minerals which formed in place, usually by diagenetic processes

CATION EXCHANGE CAPACITY: the ability of a clay to adsorb and exchange cations, or positively charged ions, with its environment

CHEMICAL WEATHERING: a change in the chemical and mineralogical composition of rocks by means of reaction with water at the Earth's surface

DETRITAL MINERALS: minerals which have been eroded, transported, and deposited as sediments

DIAGENESIS: the conversion of unconsolidated sediment into consolidated rock after burial by the processes of compaction, cementation, recrystallization, and replacement

HYDROLYSIS: a chemical weathering process which produces clays by the reaction of carbonic acid with aluminosilicate minerals

PHYLLOSILICATE: a mineral with silica tetrahedra arranged in a sheet structure

SHALE: a sedimentary rock with a high concentration of clays

DEFINITION AND PROPERTIES

The definition of clays varies depending on the scientific discipline or application. An engineer's definition, which is based on particle size, differs from a mineralogist's definition, which is based on crystal structure. In the broadest sense, clays are materials that have a very fine grain size (less than 0.002 millimeter) and behave plastically when wet. A more specific definition of a clay mineral is a hydrous aluminum phyllosilicate, or, more simply stated, a mineral that contains water, aluminum, and silicon and has a layered structure. The term "clays" will be used for the broad definition, "clay mineral" for the specific definition. Rock flour, or material that was ground to a fine powder by glaciers, would fit the definition of clays; however, it may contain minerals such as quartz and feldspars that do not fit the definition of a clay mineral. Mica is a hydrous aluminum phyllosilicate, but it often occurs as large crystals, so it does not fit the definition of clays. Certain minerals such as zeolites and hydroxides (goethite and gibbsite) have a very fine grain size and physical properties similar to clay minerals, so they are often included with the clay minerals. Unique properties of clays, including their plastic behavior when wet and ability to adsorb water and ions in solution, can be attributed to their small crystal size, high surface area, and unique crystal structure.

CLAY MINERAL STRUCTURE

There are two basic elements of the clay mineral structure: a tetrahedral sheet and an octahedral sheet. A silica atom surrounded by four oxygen atoms forms the basic building block of all silicate minerals, the four-sided silica tetrahedron. In phyllosilicate minerals, the silica tetrahedra are linked together by sharing the three oxygen atoms at the corners of the tetrahedra, forming a continuous sheet. The tetrahedral sheet has a negative charge and a general chemical formula of $Si_4O_{10}^{4-}$. An octahedron is an eight-sided figure that consists of a cation, or a positively charged ion, surrounded by six hydroxyl anions, or negatively charged ions (OH^-). The octahedra are linked together to form a sheet by sharing the hydroxyl anions on the edges of the octahedra. The octahedral sheets are neutral, and there are two types: trioctahedral and dioctahedral. The trioctahedral sheet is composed of divalent cations such as Mg^{2+} and Fe^{2+}. For every six hydroxyl anions, the trioctahedral sheet contains three cations, resulting in

Vertisol, a clay soil, cracks when dry. (© William E. Ferguson)

a sheet where all the available octahedral sites contain a cation. The dioctahedral sheet contains trivalent cations such as Al^{3+} and Fe^{3+}. The dioctahedral sheet has only two cations per six hydroxyl anions, resulting in only two-thirds of the available octahedral sites being filled with cations. The general chemical formulas for the trioctahedral and dioctahedral sheets are Mg_3OH_6 and Al_2OH_6, respectively. Layers in clay minerals are made of combinations of tetrahedral and octahedral sheets. The unshared oxygen of the silica tetrahedra take the place of some of the hydroxyl anions in the octahedral sheet, resulting in neutralization of the negative charges on the tetrahedral sheet. There are two types of layers: a 1:1 or T-O layer made up of one tetrahedral sheet and one octahedral sheet and a 2:1 or T-O-T layer which contains one octahedral sheet sandwiched between two tetrahedral sheets.

The prototype clays have neutral layer charges, with the layers held together by weak van der Waals bonds. Kaolinite and serpentine are 1:1 prototype clay minerals. Kaolinite is dioctahedral and serpentine is trioctahedral. Kaolinite, a pure white clay, is the major constituent of fine porcelain. Serpentine can occur as chrysotile asbestos, which was widely used as insulation material before it was recognized as a health hazard. Talc and pyrophyllite are the trioctahedral and dioctahedral 2:1 prototype clay minerals. The waxy or slippery feel of talc is the result of cleavage along the weak van der Waals bonds between the layers. Because of the various types of ionic substitutions in the tetrahedral and octahedral sheets, the layers may develop a negative charge, which needs to be balanced by interlayer materials. A lower-valence cation may substitute for a higher-valence cation in the tetrahedral or octahedral sheets. Layers can also develop a negative charge if some of the sites in a trioctahedral sheet are left vacant. In the mica group of clay minerals, the charge on the 2:1 layer is −1 and is balanced by a positively charged potassium ion that occurs between the layers. Muscovite is a dioctahedral 2:1 phyllosilicate, and biotite is trioctahedral. The perfect cleavage of micas is in the direction parallel to the tetrahedral sheets and allows them to be peeled into paper-thin sheets. Micas usually occur as larger crystals and therefore are not considered true clay minerals; however, illite, a common clay mineral, has a structure and chemical formula similar to muscovite, a mica. The chlorite clay minerals have an extra octahedral sheet between the 2:1 layers to balance the excess negative charges. The vermiculite and smectite groups of clay minerals have layer charges which are less than 1 and are balanced by hydrated cations between the layers; a hydrated cation is a positively charged ion, such as sodium, that is surrounded by water. In smectites, this water is held very loosely between the 2:1 layers and can be easily lost or gained depending on the humidity of the environment, causing these clays to shrink or swell.

FORMATION AND DEPOSITION

Clay minerals occur in soils, sediments, sedimentary rocks, and some metamorphic rocks. Sedimentary rocks cover approximately 80 percent of the Earth's surface, and shales are the most common type of sedimentary rock. Because shales are composed predominantly of clays, their abundance makes clays one of the most important constituents of the Earth's surface.

Most clays form by the breakdown and weathering of minerals rich in aluminum and silicon at the Earth's surface. Physical weathering is the breaking and fragmentation of rocks with no change in the mineralogical or chemical composition. This process can form clay-sized particles, but it does not form clay minerals. Physical weathering, however, increases the surface area of minerals, which favors chemical weathering. A change in the chemical and mineralogical composition of rocks by reaction with water at the Earth's surface is called chemical weathering. The process called hydrolysis is a chemical weathering reaction that results in the formation of clays. A weak acid and water react with aluminosilicate minerals, resulting in the production of a clay mineral plus ions in solution. (The weak acid is called carbonic acid; it forms when carbon dioxide, a common gas in the atmosphere, is dissolved in rainwater.) The type and amount of clay that forms by this reaction depends on the nature of the rock being weathered (the parent material) and the intensity of weathering.

Clays produced by weathering are eventually eroded, transported, and deposited as sediments. Most clays are transported in suspension. The brown, muddy waters of rivers are a reflection of the clays being carried in suspension. Clays are deposited and accumulate in quiet water environments, where the energy is low enough to allow the clays to settle out of suspension. Several processes enhance the deposition of clays. When fresh river water mixes with salty ocean water, the negative charges on clay surfaces are neutralized, causing them to flocculate. Clay floccules, or aggregates of clay-sized particles that behave like larger silt- or sand-sized grains, rapidly settle out of suspension. Biodeposition is a process whereby organisms ingest clays with their food; the resultant fecal pellets settle to the bottom as sand-sized particles.

TYPES OF CLAY MINERALS

There are two types of clays in sedimentary rocks: detrital and authigenic. Minerals which are transported and deposited as sediments are called detrital. Authigenic minerals form within the rocks during diagenesis, a process whereby sediments buried within the Earth's crust undergo increases in temperature and pressure, resulting in compaction and cementation of loose sediments and the formation of sedimentary rocks. Water expelled from the sediments during this process

---

USES OF COMMON CLAY AND SHALE IN THE U.S.
IN METRIC TONS

Use	1993	1994
Ceramics and glass	159,000	136,000
Civil engineering and sealing	84,200	177,000
Floor and wall tile:		
Ceramic	274,000	283,000
Other	100,000	45,100
Heavy clay products:		
Brick, extruded	10,900,000	11,400,000
Brick, other	1,450,000	1,620,000
Drain tile and sewer pipe	142,000	140,000
Flowerpots	35,700	35,800
Flue linings	55,300	60,600
Structural tile	35,000	37,100
Other	582,000	710,000
Lightweight aggregate:		
Concrete block	2,210,000	2,400,000
Highway surfacing	243,000	247,000
Structural concrete	787,000	801,000
Miscellaneous	286,000	305,000
Portland and other cements	7,540,000	6,920,000
Refractories	153,000	303,000
Miscellaneous	231,000	311,000
Total	**25,300,000**	**25,900,000**

SOURCE: U.S. Bureau of Mines, *Minerals Yearbook, 1994.* U.S. Government Printing Office, 1996.
NOTE: "Ceramics and glass" includes earthenware, pottery, and roofing granules. "Other" under "Heavy clay products" includes roofing tiles, sewer pipe, and terra cotta. Due to rounding, figures do not sum exactly.

may react with other minerals in the sediment to form clay minerals. Kaolinite, chlorite, illite, and smectite formed by diagenesis have been observed in sandstones. Diagenesis may also result in one clay mineral being converted into another clay mineral. A reaction which commonly occurs is the alteration of smectite to form illite as a result of an increase in temperature. This reaction is important to petroleum geologists because intermediate mixed-layered illite/smectite clays form at different temperatures. The clay mineralogy of a shale can be used to determine the maximum burial temperature of a rock.

Clay minerals may also form in metamorphic rocks as a result of hydrothermal alteration. Hydrothermal fluids are hot, chemically active fluids that accompany igneous intrusions. In addition to forming clay minerals as they pass through a host rock, hydrothermal fluids are responsible for producing important ore deposits such as copper ores. Economic geologists may use the distribution of hydrothermal clay minerals to locate valuable ore deposits.

IDENTIFICATION AND ANALYSIS

Clay minerals are difficult to identify and analyze because of their small crystal size. Because they can be observed neither with the naked eye nor by standard petrographic microscopes, they require the use of sophisticated equipment for their identification. Analysis is further complicated by the fact that it is difficult to obtain pure samples of the clay minerals because they often occur as mixtures with other minerals. Before a clay mineral can be analyzed, it must be isolated from the sample by means of special physical and chemical techniques.

The tool a clay mineralogist uses most often is an X-ray diffractometer (XRD). This instrument focuses a beam of X rays onto the sample. The crystal structure of the minerals acts as a diffraction grating, and the instrument records X rays that are diffracted from the mineral. The geometry of the XRD unit is such that mineralogists can determine the spacing between planes of atoms in the crystal structure by measuring the position of "reflections" produced by X rays diffracted by the mineral. Clay mineralogists prepare specially oriented samples which enhance the reflections between the layers of clay minerals, called basal reflections. The basal reflections are used in determining the type of clay mineral present.

Other instruments that are used to investigate clay minerals include the scanning electron microscope (SEM) and the transmission electron microscope (TEM). The very high resolution of these microscopes allows the scientist to observe clay minerals at a very great magnification. The scientist is thus able to observe the outward crystal form of clay minerals and their texture or orientation with respect to other grains in the sample. The SEM is very helpful in distinguishing between detrital and authigenic clays in sedimentary rocks.

A property that is helpful in identifying clays and understanding their behavior is the cation exchange capacity (CEC). Because of their small size and unique crystal structure, clay surfaces are negatively charged and have the ability to adsorb positive ions on their surfaces and within clays between the layers. These cations are easily exchanged with solutions. If a clay that is saturated with cations is placed in a solu-

Kaolinite clay on the floor of an open pit mine; cracks are due to shrinkage when water evaporates. (© William E. Ferguson)

tion saturated with sodium ions, it will exchange its cations for the sodium ions. The ability of a clay to adsorb and exchange cations is called the "cation exchange capacity" and depends on the type of clay mineral. Kaolinite, chlorite, and illite have relatively low CECs; smectite has a relatively high CEC. This property is especially important to soil scientists, because it controls the availability of nutrients necessary for plant growth.

INDUSTRIAL APPLICATIONS

Clays have a variety of applications in industry. They are a readily available natural resource and are relatively inexpensive. Approximately 50 million tons of clay materials worth more than $1 billion are used industrially each year.

Shales, formed predominantly of clays, are the most common form of sedimentary rock. (© William E. Ferguson)

Sedimentary rocks that produce oil and gas by the heating of organic matter after it is buried are called source rocks. The source rock for most oil and gas is shale. By determining the type of diagenetic clay minerals present, the petroleum geologist can determine if the source rock has been heated to a temperature that is high enough to produce oil or gas. Rocks that contain oil or gas which can be easily extracted are called reservoir rocks. The best reservoir rocks are sandstones with a high porosity and permeability. Some sandstones contain clay minerals that occur between the sand grains as a cement or within the pores. Clay minerals have the potential to reduce the porosity and permeability of a reservoir. It is important to know the type and amount of clay minerals in a reservoir in order to evaluate its quality. A knowledge of clays is important to engineering because the concentration of clays in a soil determines it stability. Soils which contain high percentages of smectites could cause damage to the foundations of buildings because smectites swell when saturated with water and subsequently shrink when dried out.

Clays may be used as liners for sanitary landfills because the small grain size of clays allows them to be packed together very closely, forming an impermeable layer. The liner prevents toxic leachate, which forms when rainwater reacts with solid waste, from moving out of the landfill and contaminating surface water and groundwater supplies. Chemical engineers have developed what are called designer clays by altering the properties of naturally occurring clays. These clays act as catalysts in the breakdown of toxic substances to form less toxic products. Designer clays are helpful in the destruction and disposal of toxic wastes such as dioxin and in the cleaning of existing toxic waste sites.

Clays are the basic raw material of the ceramics industry. When clays are mixed with water, they become plastic and are easily molded. A hard ceramic material is produced by firing the molded clay. In addition to the familiar pottery and dinnerware, fired clays are used in the production of brick, tiles, sewer pipes, sanitaryware pottery, kiln furniture, cement, and lightweight aggregates. Kaolinite is used as a coating on fine paper, in paints, and as a filler in plastics and rubber. The petroleum industry uses kaolinite as a cracking catalyst in the refinement of petroleum. Swelling clays such as smectite are used as binders in animal feed and iron ore pellets (taconite), as drilling muds, as industrial absorbents, and as pet litter.

Annabelle M. Foos

CROSS-REFERENCES

Biopyriboles, 1175; Building Stone, 1545; Carbonates, 1182; Carbonatites, 1287; Cement, 1550; Feldspars, 1193; Gem Minerals, 1199; Hydrothermal Mineralization, 1205; Igneous Rock Classification, 1303; Ionic Substitution, 1209; Karst Topography, 929; Magnetite, 1214; Metamictization, 1219; Minerals: Physical Properties, 1225; Minerals: Structure, 1232; Non-silicates, 1238; Orthosilicates, 1244; Oxides, 1249; Radioactive Minerals, 1255; Rocks: Physical Properties, 1348; X-ray Powder Diffraction, 504.

BIBLIOGRAPHY

Blatt, Harvey, Gerard Middleton, and R. Murray. *Origin of Sedimentary Rocks.* 2d ed. Englewood Cliffs, N.J.: Prentice-Hall, 1980. This comprehensive textbook covers the classification, origin, and interpretation of sedimentary rocks. The formation and classification of clays are discussed in detail in the chapter on weathering. The chapter on mudrocks gives an excellent discussion of the distribution of clay minerals. Suitable for college-level students.

Eslinger, Eric, and David Pevear. *Clay Minerals for Petroleum Geologists and Engineers.* Tulsa, Okla.: Society of Economic Paleontologists and Mineralogists, 1988. Covers the major geologic aspects of clay mineralogy. The crystal structure, classification of clay minerals, origin by weathering and diagenesis, and distribution of clays are discussed in detail. Application of clay mineralogy to exploration and production of petroleum is also covered. The appendix contains a summary of sample preparation and X-ray diffraction analysis of clays. A suitable text for college-level students with a science background.

Klein, Cornelis, and Cornelius S. Hurlbut, Jr. *Manual of Mineralogy.* 21st ed. New York: John Wiley & Sons, 1999. A general textbook on mineralogy, covering crystallography, physical properties of minerals, and systematic mineralogy. A very good discussion on X-ray diffraction is given in chapter 6. Phyllosilicate minerals are discussed in chapter 11. Suitable for college-level students.

Longstaffe, F. J. *Short Course in Clays and the Resource Geologist.* Toronto: Mineralogical Association of Canada, 1981. A collection of papers by well-recognized experts in the field of clay mineralogy. The first three chapters cover the crystal structures of clay minerals and their identification by X-ray diffraction.

Subsequent chapters give case histories and specific examples of applications of clay mineralogy to petroleum geology. Suitable for college-level students with a science background.

Paquet, Haelaene, and Norbert Clauer, eds. *Soils and Sediments: Mineralogy and Geochemistry.* New York: Springer, 1997. This detailed overview of the processes of weathering and soil sedimentation and deposition includes a vast amount of information on clay minerals. Bibliography and index.

Velde, B. *Introduction to Clay Minerals: Chemistry, Origins, Uses, and Environmental Significance.* New York: Chapman and Hall, 1992. This useful introductory text includes sections on the structure and classification of clay, sedimentation, and the absorption of organic pollutants into clays. Extensive bibliography and index.

Velde, B., ed. *Origin and Mineralogy of Clays: Clays and the Environment.* London: Springer-Verlag, 1995. This advanced text includes essays on such subjects as the composition and mineralogy of clay minerals, the origin of clays by weathering and soil formation, compaction and diagenesis, and the formation of clay minerals in hydrothermal environments.

Welton, J. E. *SEM Petrology Atlas.* Tulsa, Okla.: American Association of Petroleum Geologists, 1984. An atlas of scanning electron microscope (SEM) graphs of authigenic minerals that occur in sandstones. In addition to the SEM images, the chemical composition of the minerals is given. Text is kept to a minimum; however, a summary of sample preparation and scanning electron microscope analysis is given in the introduction. Suitable for all levels.

FELDSPARS

Feldspars are the most abundant group of minerals within the Earth's crust. There are many varieties of feldspar, distinguished by variations in chemistry and crystal structure. Although feldspars have some economic uses, their principal importance lies in their role as rock-forming minerals.

PRINCIPAL TERMS

CRYSTAL: a material with a regular, repeating atomic structure

IGNEOUS ROCKS: rocks formed from the molten state; they may be erupted on the surface (volcanic) or harden within the Earth's crust (plutonic or intrusive)

ION: an atom that has gained or lost electrons and thereby acquired an electric charge; most atoms in minerals are ions

METAMORPHIC ROCKS: rocks formed by the effects of heat, pressure, or chemical reactions on other rocks

POLARIZED LIGHT: light whose waves vibrate or oscillate in a single plane

SEDIMENTARY ROCKS: rocks formed on the Earth's surface from materials derived by the breakdown of previously existing rocks

SILICA TETRAHEDRON: the fundamental molecular unit of silica; a silicon atom bonded to four adjacent oxygen atoms in a three-sided pyramid arrangement

SILICATE: a mineral containing silica tetrahedra, which may be separate from one another or joined into larger units by sharing their corner oxygen atoms

COMPOSITION AND PROPERTIES

Considered as a group, the feldspars are the most abundant minerals in the Earth's crust. They form in igneous, metamorphic, and sedimentary rocks and are among the principal repositories for sodium, calcium, potassium, and aluminum in the crust. The feldspars are thus compounds of aluminum, oxygen, and silicon, together with one or more of the elements sodium, potassium, and calcium. They form two principal series: potassium feldspars and plagioclase feldspars. There are a few rare barium feldspars as well. The feldspars are a more complex group than this summary suggests at first glance, because they undergo subtle but important changes in crystal structure depending on temperature and pressure. Also, there are several distinctive mixtures of potassium and plagioclase feldspars.

The physical properties of all the feldspars are similar. They all have a hardness of 6 on Mohs scale (they can scratch most glass, but cannot scratch quartz). Their densities are all in the range 2.6-2.75 grams per cubic centimeter, or about the average density for most common rock-forming minerals. Feldspars are usually, though not always, light in color. They are usually translucent in thin splinters but in rare cases can be transparent. All the feldspars have good cleavage, that is, a tendency to split easily along smooth planes dictated by the atomic structure of the mineral. They cleave along two perpendicular or nearly perpendicular planes.

CRYSTAL STRUCTURE

Most of the feldspars crystallize in the triclinic crystal system; a few feldspars are monoclinic. Crystals are classified according to their atomic arrangements. The fundamental atomic unit that makes up any crystalline material can be pictured as fitting inside an imaginary box, or unit cell, with parallel sides. Unit cells stack together to form a crystal, and the shape of the crystal reflects the shape of its unit cell. In triclinic crystals, the angles between faces or edges of the unit cell are never right angles. In monoclinic crystals, two pairs of faces of the unit cell are perpendicular, but the third is not. A monoclinic feldspar unit cell looks like a carton with no top or bottom, sheared slightly out of shape so that its outline is an oblique parallelogram instead of a rectangle.

One feature of the crystal structures of the feldspars is especially notable. The feldspar minerals have a pronounced tendency to exhibit distinctive kinds of crystal twinning, or abrupt changes in crystal growth patterns. The growth of a crystal can be pictured as stacking planes of atoms on one another in a specific pattern. There are often many equally possible ways to stack one plane on the next. When a crystal has been built up according to one stacking pattern and then begins following a different pattern, there is an abrupt change in the atomic structure of the mineral. This changeover of atomic structure is called twinning. Sometimes the results are visible to the unaided eye, and the mineral appears to consist of two crystals stuck to one another or penetrating each other. In other cases, the results of twinning may only be visible through the microscope. Twinning of feldspars is a valuable clue to the geologist, because twinning often makes it easy to distinguish feldspars that are otherwise very similar.

Feldspars belong to the tektosilicate group: silicate minerals in which silica tetrahedra link to form three-dimensional networks. The silica tetrahedra in feldspars link to form zigzag chains, and the chains in turn are linked to adjacent chains to create a continuous network. The aluminum in feldspars occupies the centers of some of the tetrahedra in place of silicon, and the potassium, sodium, or calcium occupy the open spaces between the chains.

POTASSIUM FELDSPARS

There are three important potassium feldspars: microcline, orthoclase, and sanidine. Microcline is perhaps the most familiar potassium feldspar, because it is the normal potassium feldspar in granitic rocks and is the principal potassium feldspar in metamorphic rocks. Microcline can be any light color but is most often white or pink. The familiar pink color of many granites is caused by microcline, which is colored pink by microscopic plates of the iron oxide mineral hematite within the feldspar. Amazonite is a distinctive variety of microcline, unusual for its bright green color. Amazonite is a minor gem stone. The green color is of uncertain origin. It has been attributed to small amounts of the elements rubidium or lead or to changes in the crystal structure of the

microcline because of natural radioactivity in the surrounding rocks. Colors in minerals frequently have complex causes and are often the result of very tiny amounts of impurities. It is common for a given color to have several different causes. Orthoclase is a variety of potassium feldspar that forms at somewhat lower pressure than does microcline. It occurs chiefly in granitic rocks that form near the Earth's surface and cool quickly and also in volcanic rocks. Sanidine is a high-temperature potassium feldspar found in some volcanic rocks. All three varieties of potassium feldspar are distinguished by differences in their crystal structure, particularly as seen under the microscope.

PLAGIOCLASE FELDSPARS

The plagioclase series of feldspars is one of the best natural illustrations of a solid solution. A solid solution is a blend of two or more distinct materials on the atomic scale, just as solutions exist in liquids. Metallic alloys are other familiar examples of solid solutions. Solid solutions differ from chemical compounds in that the components can have variable proportions. They differ from simple mixtures (for example, salt and pepper) in that the components are interchangeable on the atomic level. These concepts are important to understand because feldspars include both solid solutions and mixtures.

The plagioclase series consists of a solid solution of two components, or end members: albite and anorthite. The proportions of aluminum and silicon are different in anorthite because calcium ions in minerals normally have a +2 electric charge, compared with the +1 of sodium. Therefore, an aluminum ion (+3) must substitute for silicon (+4) to compensate for the extra charge on the calcium ion. The plagioclase series is subdivided into six members, depending on the relative amounts of albite and anorthite. In increasing order of anorthite content, the plagioclase feldspars are albite, oligoclase, andesine, labradorite, bytownite, and anorthite. The plagioclase feldspars become somewhat denser and usually darker in color with increasing anorthite content.

Albite contains less than 10 percent of the anorthite component and forms in sodium-rich environments. Albite forms in marine sedimentary rocks as a cementing mineral and forms in marine volcanic rocks when sodium from seawater replaces calcium in their plagioclase feldspars. Albite can also form during metamorphism of sodium-rich rocks. Albite is rare in igneous rocks because igneous rocks are rarely so rich in sodium and poor in calcium that the mineral would form. Oligoclase contains 10-30 percent of the anorthite component and is very common because it is the normal plagioclase feldspar in granitic rocks. Andesine contains 30-50 percent anorthite and is common in igneous rocks less silica-rich than granite, such as diorite. Labradorite contains 50-70 percent anorthite and is the principal plagioclase feldspar in silica-poor igneous rocks such as basalt or gabbro. Labradorite is also the principal feldspar in rare igneous rocks called anorthosite, which consist mostly of plagioclase. Anorthosite is a common rock type of the Moon. Terrestrial anorthosites, rare as they are, are commonly used as ornamental building stones. They generally are dark gray, with large feldspar crystals a centimeter or more across, and show attractive bright bluish reflections from cleavage planes within the feldspar. Bytownite, with 70-90 percent anorthite, is perhaps the rarest plagioclase. It occurs most often in very silica-poor igneous rocks. Anorthite is any plagioclase with 90 percent or more of the anorthite end member (or less than 10 percent albite). Anorthite most often forms by meta-morphism of rocks rich in calcium and aluminum but very poor in sodium, such as clay-rich limestones.

OTHER FELDSPARS

To some extent, the potassium feldspars and the sodium-rich plagioclases also form a solid solution. Feldspars containing roughly equal parts of potassium feldspar and albite are called anorthoclase. In addition to the pure feldspar minerals, there does exist a wide variety of mixtures of feldspars. At high temperatures, the potassium feldspars and plagioclases coexist in solid solution much more readily than at low temperatures. Crystals that formed a stable solid solution when an igneous or metamorphic rock formed often become unstable when the rock cools. As the feldspar cools, the plagioclase and potassium feldspar often separate. The final result is a patched or streaked network of plagioclase and potassium feldspar filaments interlocked with one another. This texture is easily visible to the unaided eye. When the feldspar consists mostly of potassium feldspar enclosing small amounts of plagioclase, the mixture is called perthite. Perthite is very common; almost any large microcline crystal will exhibit perthitic texture, with the pink microcline enclosing milky filaments of plagioclase. Less often, plagioclase is the dominant feldspar, enclosing small inclusions of potassium feldspar. Such a mixture is called antiperthite.

There are very few feldspar minerals other than the potassium and plagioclase feldspars. The openings in the atomic structure of feldspar are so large that only very large ions can be held there. Magnesium and iron ions are too small, so there are no magnesium or iron feldspars. Lithium, although chemically similar to sodium and potassium, is also too small to form feldspars. Cesium and rubidium feldspars have been made artificially; some feldspars are rich in cesium and rubidium, but no special names have been as-

Feldspar is one of the two most abundant mineral types in the Earth's crust; here its typical two-directional, 90-degree cleavage can be seen. (U.S. Geological Survey)

signed to these minerals. Barium feldspars, such as celsian, do exist in nature. Hyalophane can be considered a solid solution of celsian, albite, and potassium feldspar. Banalsite is another barium feldspar. The rare mineral buddingtonite forms when ammonia-rich volcanic solutions alter plagioclase, replacing sodium and calcium with ammonia.

FELDSPAR-LIKE MINERALS

A few feldspars and feldspar-like minerals form when other small ions substitute for aluminum or silicon within the silica tetrahedra. Feldspar-like minerals that form in this manner include reedmergnerite, eudidymite, danburite, and hurlbutite. All are uncommon.

A few minerals that are geologically akin to the feldspars deserve mention. The feldspathoids have feldspar-like chemical compositions and form when rocks are too poor in silica to form feldspars. They are usually softer than feldspars and with quite different crystal forms. The scapolite minerals are essentially plagioclase feldspars that include molecules of sodium chloride (halite or table salt), calcium sulfate (gypsum), or calcium carbonate (calcite) within their atomic structures. It has been suggested that scapolite may be a common mineral on Mars, formed when plagioclase absorbed carbon dioxide from the planet's atmosphere, and that much of the planet's original carbon dioxide may now be locked up in scapolite.

TECHNIQUES FOR IDENTIFYING FELDSPARS

A wide range of techniques has been developed to probe the structure of minerals with polarized light. The two most obvious features of minerals in polarized light are interference color and extinction. When light enters a mineral, it splits into two beams polarized in perpendicular planes. The orientation of the planes is closely related to the crystal structure of the mineral. The two light beams travel at different speeds through the mineral. When the beams emerge, they recombine into a single light beam whose direction of vibration is usually different from the original direction. Some light passes through the second polarizer. The resulting interference color is determined strictly by the amount the two beams of light separated within the mineral. If the specimen is rotated (petrographic microscopes are normally equipped with rotating specimen stages), it will black out, or undergo extinction, at intervals of 90 degrees. Extinction occurs when the vibration directions of light in the mineral match those of the polarizing filters. In these positions, light leaving the mineral experiences no change in vibration direction and is blocked by the second polarizing filter.

Under the microscope, in normal illumination, feldspars are colorless and similar in appearance to quartz. They can often be distinguished from quartz by cleavage (which appears as straight, parallel cracks) and by a dusty appearance caused by chemical alteration. Quartz is almost immune to chemical alteration. Between crossed polarizers, quartz and feldspar display similar interference colors, but the twinning habits of the feldspars usually make it easy to distinguish them. Because the crystal structure changes abruptly when twinning occurs, twinned crystals are obvious as a result of abrupt changes in the optical properties. A crystal that looks like a single entity in normal illumination appears as distinct regions with different interference color or extinction between crossed polarizers.

Geologists sometimes use staining techniques to identify feldspars. The commonest method involves etching the specimen with hydrofluoric acid (an extremely hazardous material) and applying a series of chemicals that stain potassium feldspars yellow and plagioclase pink.

For probing the atomic structure of feldspars in detail, many highly sophisticated techniques are in use. The chemical composition of feldspars can be determined for even tiny specimens by bombarding the specimen with electrons (electron microprobe) or with X rays (X-ray fluorescence) and measuring the energies of radiation given off by the specimen. The atomic arrangement of feldspars is determined by X-ray diffraction, in which X rays are reflected off atoms within the specimen. The amount of X rays reflected in different directions can be used to determine the geometric arrangement of atoms within the crystal.

GEOLOGIC SIGNIFICANCE

Feldspars have some economic value, but their principal importance lies in their role as major building blocks of the Earth. Because they are such tremendously important reservoirs of common elements, feldspars are key minerals in classifying rocks, and the composition of feldspar minerals in a rock is a powerful clue to its origin and history.

Feldspars are also geologically significant for their role in radiometric dating. Both the potassium-argon and rubidium-strontium dating methods rely on elements that are found in feldspar, either as principal ingredients (potassium) or as common trace elements (rubidium and strontium).

ECONOMIC VALUE

Feldspars have minor use as gemstones. Amazonite is a green variety sometimes used as a gem, and moonstone is translucent feldspar with microscopic inclusions that give it a milky appearance. Aventurine is a clear feldspar with tiny included plates of other minerals that impart a sparkly appearance. Some feldspar-rich rocks called anorthosite are used as an ornamental building stone.

The most important uses of feldspar are less glamorous. Feldspar is the principal ingredient of porcelain. Indirectly, feldspar is the source of aluminum. Rocks that are rich in feldspar and poor in iron are broken down by tropical weathering so that all but the aluminum is dissolved away. The final result is a mixture of aluminum minerals called bauxite, the principal ore of aluminum. In temperate climates, weathering of feldspar releases potassium, an essential plant nutrient.

Steven I. Dutch

CROSS-REFERENCES

Biopyriboles, 1175; Carbonates, 1182; Cement, 1550; Clays and Clay Minerals, 1187; Diagenesis, 1445; Electron Microscopy, 463; Expansive Soils, 1479; Gem Minerals, 1199; Hydrothermal Mineralization, 1205; Industrial Nonmetals, 1596; Ionic Substitution, 1209; Landfills, 1774; Magnetite, 1214; Metamictization, 1219; Minerals: Physical Properties, 1225; Minerals: Structure, 1232; Nonsilicates, 1238; Oil and Gas Origins, 1704; Orthosilicates, 1244; Oxides, 1249; Radioactive Minerals, 1255; Silica Minerals, 1354; Soil Chemistry, 1509; Soil Formation, 1519; Soil Liquefaction, 334; Weathering and Erosion, 2380; X-ray Powder Diffraction, 504.

BIBLIOGRAPHY

Blackburn, William H., and William H. Dennen. *Principles of Mineralogy.* Dubuque, Iowa: Wm. C. Brown, 1988. A college-level mineralogy text with chapters on mineralogical theory and methods, plus descriptions of common minerals. As in most such texts, there is a lengthy section on the feldspars. Chapter 12 is useful for its good coverage of modern analytical techniques used in the study of minerals.

Deer, W. A., R. A. Howie, and J. Zussman. *An Introduction to the Rock-Forming Minerals.* 2d ed. New York: Longman, 1992. The chapters on the feldspars contain detailed descriptions of chemistry, crystal structure, and identification techniques.

Klein, Cornelis, and Cornelius S. Hurlbut, Jr. *Manual of Mineralogy.* 21st ed. New York: John Wiley & Sons, 1999. One of the most widely used college mineralogy texts. Contains chapters on most mineralogical methods, plus descriptions of common minerals. Particularly good for its attractive illustrations of atomic structure of minerals, its description of optical properties and the petrographic microscope, and its survey of research on the causes of color in minerals.

O'Donoghue, Michael. *The Encyclopedia of Minerals and Gemstones.* London: Orbis, 1976. The first half of this book is a summary of basic geological, crystallographic, and optical concepts that relate to gem minerals. The language is aimed at a general audience. The second half is devoted to a description of about 1,000 minerals. Liberally illustrated with large, attractive color photographs.

Parsons, Ian, ed. *Feldspars and Their Reactions.* Dordrecht, the Netherlands: Kluwer, 1994. This book is a collection of papers presented at the North Atlantic Treaty Organization (NATO) Advanced Study Institute on Feldspars and Their Reactions in Edinburgh, Scotland, in 1993. Although many of the discussions may be too technically advanced for the high school student, the bibliography is useful for anyone seeking information on feldspars and geochemistry.

Roberts, Willard Lincoln, George Robert Rapp, Jr., and Julius Weber. *Encyclopedia of Minerals.* New York: Van Nostrand Reinhold, 1974. A summary of the physical properties of twenty-two hundred minerals. Notable for its almost complete collection of color photographs. Its alphabetic arrangement simplifies locating mineral descriptions, but the lack of cross-reference lists of minerals by chemistry, crystal structure, or other properties is a serious drawback for students who wish to compare minerals.

Wood, Elizabeth A. *Crystals and Light.* Princeton, N.J.: Van Nostrand Reinhold, 1964. A short book, aimed at beginning college students, that describes many of the optical phenomena encountered in the study of minerals. Describes basic concepts of crystallography and the behavior of polarized light in simple language.

GEM MINERALS

A gem mineral is any mineral species which yields a gem upon cutting and/or polishing. Gem minerals have value based upon their potential to produce gems. Although gems have had various uses in the past, including special powers attributed by folklore, their principal uses have been personal adornment (as jewels), a mode of investment, and a symbol of wealth and power.

PRINCIPAL TERMS

GEM: a cut and polished stone that possesses durability, rarity, and beauty necessary for use in jewelry and therefore of value

GEMSTONE: any rock, mineral, or natural material that has the potential for use as personal adornment or ornament

INCLUSION: a foreign substance enclosed within a mineral; often very small mineral grains and cavities filled with liquid or gas; a cavity with liquid, a gas bubble, and a crystal is called a three-phase inclusion

MINERAL: a naturally occurring inorganic substance with a characteristic chemical composition and atomic structure, manifested in its external geometry and other physical properties

MINERAL SPECIES: a mineralogic division in which all the varieties in any one species have the same basic physical and chemical properties

MINERAL VARIETY: a division of a mineral species based upon color, type of optical phenomenon, or other distinguishing characteristics of appearance

ROUGH: gem mineral material of suitable quality to be used for fashioning gemstones

SYNTHETIC MINERAL: a human-made reproduction of the structure, composition, and properties of a particular mineral

PHYSICAL PROPERTIES

Gem minerals are those mineral species that have yielded the material from which specimens have been fashioned into gems. Gems and therefore gem mineral varieties must have the same properties. These properties are beauty (color, phenomenon, or clarity), durability (hardness and toughness), and value (rarity, demand, and tradition). Of the approximately three thousand mineral species known, only about ninety of these produce material in a quality and quantity suitable for gems, and this quantity of suitable material makes up generally only a very small portion of the gem mineral found. Often the quality of this gem material is such that the material is given a gem varietal name or mineral varietal name based upon color or phenomenon. A different name is given for each color variation; hence, there is a large number of gem names compared with the gem mineral names. Two or more different gem minerals may have the same gem name, as in the case of jade and the minerals nephrite (actinolite and tremolite) and jadeite. Finally, new gem varieties are found from time to time. For example, tanzanite was found in the 1960's and tasvorite during the 1970's.

BEAUTY

The beauty of a gem may be inherent in the gem mineral species—such as clarity, color, or phenomena (stars, eyes, and the like)—or it may be brought out as it is fashioned by cutting and polishing. Faceting is the cutting of a stone to add faces (facets) or flat surfaces, generally with a regular geometric form. Because faceting enhances the brilliance of a stone by causing reflections of both front and back facets, those mineral varieties having superior clarity became important gem varieties. The faceting of a clear gem mineral produces a color phenomenon called dispersion. This is the ability of a particular gem mineral to break light up into a rainbow of component colors, as does a prism. This property is commonly known as "fire" and is well known in diamonds.

Color is a very important property of gems. There are many gem varieties of gem minerals based solely on this property. For example, the named color varieties of beryl include aquamarine (blue-green), chrysolite (yellow-green), emerald (intense green), heliodor (brownish-green), and morganite (pink or orange). If one combines clarity with color, something very beautiful happens, such as with the sparkling of a group of cut sapphires, rubies, and emeralds. A "play of color" is produced by various combinations of reflection, refraction, diffraction, and interference phenomena; it is well known from such gems as opal (opalescence) and labradorite (labradorescence). Some stones are selected entirely for their color. These may be opaque or nearly so and may or may not take a high polish. They include such stones as turquoise, jade, and many varieties of quartz, such as jasper, bloodstone, agate, and carnelian.

Some stones have inclusions that produce aesthetic effects, such as stars, eyes, spangling, or even pictures. Stars and eyes are produced by minute aligned voids or needles of minerals like rutile. If they are in one direction, they produce eyes, as in chryoberyl and tourmaline. If they are in a hexagonal or in other star-shaped patterns, they produce stars, as in star rubies, sapphires, diopside, and garnet. Spangling results from variously oriented small inclusions of reflective minerals within the gem mineral, such as mica, rutile, or tourmaline. Picture agate has variously distributed clumps of pyrolusite dendrites and masses which, like clouds, one forms into pictures, or landscapes in the mind. It is apparently why mystical powers have been ascribed to such stones as these.

DURABILITY

Durability of a gem mineral is its resistance to scratching or breaking, a necessity for an owner of jewelry because one does not want to damage a gem simply by wearing it and thus markedly reduce its value.

Hardness is the property of a gem mineral by which it resists scratching. Mineralogists and gemologists test unknown minerals against a hardness scale of common minerals, called Mohs hardness scale. It is as follows: (10) diamond (9) corundum (8) topaz (7) quartz (6) orthoclase (feldspar) (5) apatite (4) fluorite (3) calcite (2) gypsum, and (1) talc. Number 10 is the hardest mineral, and number 1 is the softest, most easily scratched mineral. The hardness of 7, the hardness of quartz, a very

common mineral, is very important in the selection of gems or gem minerals. Quartz is a very common mineral, which means a stone with a hardness of less than 7 becomes easily scratched in our environment and therefore lacks durability. Most of our well-known gem minerals have a hardness greater than or equal to 7.

Toughness is the resistance to breaking and is referred to by mineralogists and gemologists as a quality of tenacity. Some other qualities associated with tenacity are brittleness, tenderness, elasticity, flexibility, ductility, and malleability. Most minerals are brittle, but native gold is malleable, ductile, and sectile, while muscovite mica is flexible and elastic.

Minerals break in two major ways, called cleavage and fracture. Cleavage is the breaking of minerals parallel to more weakly bonded atomic planes found within the mineral and produces flat reflecting surfaces. Fracture is the random breaking of a mineral across these atomic planes. It may be in the form of curved surfaces called conchoidal fracture or simply of irregular surfaces. Toughness is directly related to the bond strength between adjacent atoms. Cleavage thus reduces toughness, as in easily cleaved topaz, while an intergrown mat of fine crystals increases toughness of a specimen, as in the case of the two jades, nephrite and jadeite. Cleaving was probably the first way to facet gems and still is very important in the shaping of diamond rough before cutting and polishing.

VALUE

The value of a gemstone or a particular gem is based upon the interaction of many factors that can be divided into four major groups: beauty, rarity, quality of fashioning, and demand/tradition/folklore. Beauty includes such properties as color, quality of color, and clarity (freedom of inclusions, fractures, or cloudiness); it also includes phenomena such as quality of stars, eyes, or play of color. Rarity involves not only the rarity of the variety of the gem mineral but also the rarity of that quality of stone being used. For example, size influences the value because large stones are rarer than small stones; perfection influences the value because high-quality stones are less common than flawed stones. Rarity of the gem mineral can increase value or decrease value. Very common varieties of quartz, for example, may have less value than similar qualities and sizes of corundum because corundum is less common. Some gems, however, are so rare that people have not heard of them, and there is no demand except by connoisseurs and hobbyists, which in turn decreases value. Quality of fashioning includes degrees of perfection of symmetry and customizing of facet junction angles to the properties of the gem mineral species, which maximizes brilliance, color, and dispersion, and fineness or degree of polishing. Demand/ tradition/folklore includes such issues as fashion, fads, and engagement ring and birthstone traditions. These associations, in general, increase value. Good

Garnet, one of many gemstones, forms equidimensional crystals with twelve to thirty-six faces. (U.S. Geological Survey)

or bad luck attributed to gemstones may increase or decrease their value, respectively.

The terms "precious" and "semiprecious" used with a gemstone name are misleading in assessing the value of the stone. "Precious" has been historically applied to gems such as diamonds, rubies, sapphires, opals, and emeralds, while "semiprecious" has been applied to such gems as aquamarine, tourmaline, chrysoberyl, citrine, topaz, and amethyst. Almost any gem mineral is found in a wide variety of qualities and therefore so are gems; thus, there are high-quality stones called semiprecious that are often more valuable than low-quality stones called precious. Unfortunately, the term "precious" has also been applied all too often to low-quality diamond, ruby, sapphire, opal, or emerald in order to delude the buyer.

FORMATION OF GEM MINERALS

Gem minerals originate in the same way as all other minerals. They occur in four broad categories: igneous deposits (formed from molten rock—lava and magma), hydrothermal and pneumatolitic deposits (formed from hot water and steam, respectively), metamorphic deposits (formed by heat, pressure, and chemically active fluids with crystallization in the solid state), and sedimentary

deposits (formed by processes of weathering, erosion, and deposition). Corundum is an example of a gem mineral that may be found under several conditions. A well-known igneous gem mineral formed from the cooling of molten rock is olivine (peridot), which is found in lava. Gem minerals found in plutonic rock (molten rock cooled within the earth) are diamond, various garnets, zircon, corundum, spinel, labradorite, orthoclase, and albite-oligoclase. Well-known gem minerals formed by hydrothermal and pneumatolitic processes are varieties of quartz (amethyst, citrine, crystal, smoky), varieties of beryl (aquamarine, morganite, emerald), various kinds of tourmaline, topaz, spodumene, and microcline. These minerals are commonly found in bodies or rock called pegmatites, or veins. Metamorphic rocks containing gem minerals form in the roots of mountain chains and/or adjacent to igneous bodies, where there are high pressures and temperatures. Well-known gem minerals formed here include corundum, kyanite, lazurite (lapis lazuli), various garnets, jade (both nephrite and jadeite), spinel, beryl, and chrysoberyl. Gem minerals found in sedimentary deposits may be divided into two major divisions: placer deposits (gravels which contain heavy minerals resistant to weathering and left by erosion and deposition of streams) and primary sedimentary minerals grown in the sediments. Placer gem minerals include diamond, corundum, spinel, tourmaline, topaz, various garnets, zircon, and chrysoberyl and may be associated with placer gold. Primary sedimentary minerals are mainly varieties of quartz (agate, chalcedony, onyx, sard, carnelian, petrified wood, heliotrope, bloodstone, jasper). Gem minerals may also result from weathering processes. A fine example is turquoise, which is produced by the weathering of copper-bearing hydrothermal veins under arid conditions.

TOOLS OF GEMOLOGIST

The binocular microscope is one of the most useful tools to the gemologist, whether making identifications, grading, or preparing a stone for lapidary work. Other magnifying instruments used include the hand lens and loupe, which require considerable skill for competent use and are very limited in their magnification abilities. Properties of minerals determinable for identification purposes with the binocular microscope include hardness, by noting polish quality or wear characteristics, and crystallography, from traces of cleavage on the unpolished girdle of the stone, from noting single or double refraction, from orientation of inclusions, from birefringence by the doubling of facets on the back side of the stone, from the shape of gas or liquid inclusions, and from banding of color. Properties of minerals determinable for grading of gems include clarity, quality of cutting and polishing, symmetry, and color. Properties of minerals determinable for lapidary purposes include crystallographic orientation, cleavage directions, and location of color variations and inclusions. Most gems are considered flawless when no inclusions, fractures, surface blemishes, or cutting flaws are visible to the naked eye.

The refractometer is also one of the most useful tools of the gemologist for purposes of gem identification. This technique makes use of the fact that each mineral species has a distinctive and characteristic index of refraction, or indices of refraction. Different minerals may have one, two, or three indices of refraction. A refractometer such as the Rayner Refractometer or Duplex Refractometer can measure the indices of refraction of a gem by use of the critical angle of light for that mineral, which is dependent upon the mineral's refractive index or indices. The index or indices of refraction are measured from flat facets of cut gems or crystal faces; an average of the indices may be obtained from a curved face or from any polished surface of a fine-grained aggregate, such as jade or chalcedony. Tables of refractive indices allow gemologists to compare results of their testing with those of known gem materials.

The polariscope is a simple yet very valuable instrument. It is a light source with two Polaroid plates mounted above and below the gem. The polariscope is used to distinguish between crystalline double refractive material and singly refractive crystalline material or amorphous materials, and to identify doubly refractive fine-grained aggregates, such as jades and chalcedony. Some skill in interpretation of results is needed to distinguish anomalous double refraction found in some garnets and strain effects found in some gems and glass. Strain may be an indicator to the jeweler that the stone may be easily broken and that care should therefore be taken when mounting the stone in jewelry.

The spectroscope may on occasion be a useful instrument. The hand-held types are difficult to use, and the table models are very expensive. They are generally not necessary for identification of stones but are useful in confirming or determining the presence of metallic ions like iron or chromium. Light traveling through the mineral is separated into its color components in a rainbow of color by prism or by diffraction. There will be dark lines on this rainbow, corresponding to which cations are in the mineral. Genuine ruby will show absorption lines for chromium; red garnet does not.

ROLE IN ECONOMICS AND TECHNOLOGY

The study of gems or gem minerals interests the connoisseur, hobbyist, investor, and those whose business is the gem trade. Two reasons for a large public ownership of gemstones in Western society are the tradition of the diamond engagement ring in North America and the use of the diamond as an investment in Europe. It is particularly this market where most of the fraudulent practices in the gem and gem mineral trade exist. Investing in gems and gem minerals has been common for several reasons, including short-term gains, long-term gains, and as a hedge against inflation. Gemstones have also been valued for their portability, especially in unstable political situations. As with any investment, however, there is an element of risk. The degree of risk may vary with the gem mineral; for example, finding a large new deposit drops the price. Cornering of the market by a particular group forces the price up or down for some political or economic reason. Diamond pricing is probably the most structured and stable because it has been controlled by the De Beers Central Selling Organization (CSO) in London for the last hundred years. Nevertheless, prices for a D-colored flawless one-carat diamond (round brilliant) fluctuated from more than $7,000 in 1977 through $50,000 to $60,000 in 1980 down to about $12,000 in 1985 to more than $17,000 in 1988 in a very volatile world market.

Because of their perfection, gems are used by scientists for research concerned with crystals and crystallization processes. They also have many technological applications, from ruby (corundum) lasers to quartz crystals in citizen-band radios and telephone communications equipment. Because of the large demand for these minerals, synthetic substitutes are generally utilized today for these technological applications. Mineral species, of which the gem-quality varieties may make up only a very small quantity, may also have other major non-gem-related industrial uses. The mineral diamond, for example, has use as a gem not only in transparent, attractively colored, and colorless varieties but also in translucent, nearly opaque, or highly included stones, which are not pleasing to the eye yet have important industrial cutting and abrasive uses. Much high-technology

grinding, cutting, and polishing is based upon diamonds, from the bits that drill oil wells and cut metals and the saws that cut stone to the fine abrasives used to polish gems and lenses. The mineral corundum, the second hardest mineral known, not only produces specimens of ruby and sapphire but also is extensively used as an abrasive. Some of the abrasive uses, such as coarse grinding with emery, have decreased because of the introduction of relatively inexpensive artificial diamonds, but other uses, such as polishing with alumina, have increased. The mineral hematite, perhaps not so well known for its black metallic gems, is best known as the chief ore of iron. The jewelry trade knows it best for yet another use, jewelers' rouge, a reddish material used to polish metals and some gems. The gem mineral quartz, with many gem varieties—amethyst, citrine, smoky, rose, crystal, agate, carnelian, bloodstone—has many nongem uses, including glassmaking and the production of abrasives (sandpaper) and sand for sandboxes, beaches, and concrete.

Charles I. Frye

CROSS-REFERENCES

Aluminum Deposits, 1539; Biopyriboles, 1175; Building Stone, 1545; Carbonates, 1182; Clays and Clay Minerals, 1187; Diagenesis, 1445; Feldspars, 1193; Hydrothermal Mineralization, 1205; Igneous Rock Classification, 1303; Ionic Substitution, 1209; Magmas, 1326; Magnetite, 1214; Metamictization, 1219; Minerals: Physical Properties, 1225; Minerals: Structure, 1232; Non-silicates, 1238; Orthosilicates, 1244; Oxides, 1249; Pegmatites, 1620; Petrographic Microscopes, 493; Radioactive Minerals, 1255.

BIBLIOGRAPHY

Baurer, Max. *Precious Stones.* 2 vols. Mineola, N.Y.: Dover, 1968. An older book that has been reprinted. The original was published in German in 1896, went through several editions, and was translated by L. J. Spencer of the British Museum in 1904. Most of the information is still up-to-date, though some noteworthy gems have been found since publication and new localities and new technologies developed. Written for the hobbyist as well as the gemologist. The original German and English editions are collector's items.

Cavey, Christopher. *Gems and Jewels.* London: Studio, 1992. Cavey sets out not only to provide information about gems and jewels but also to correct many of the misconceptions concerning their study. Although the book includes many illustrations, only a few of them are in color.

Desautels, Paul E. *The Gem Kingdom.* New York: Random House, 1970. This book is not intended to be an exhaustive study of gems or gem minerals but instead touches on most aspects of the gem world using the Smithsonian Institution Gem and Mineral Collections for illustration. Might be considered more useful for gem appreciation than gemology. Excellent photographs and a lively text. For anyone, such as collectors, interested in gems, particularly those found at the Smithsonian.

Gubelin, E. J. *Internal World of Gemstones.* 3d ed. Santa Monica, Calif.: Gemological Institute of America, 1983. A lovely book of exquisite photomicrographs. Good information on the internal nature of gemstones and a classification of inclusions. Very useful to the gemologist or the jeweler wishing information concerning inclusions in gemstones. Also excellent for the hobbyist and collector.

Hall, Cally. *Gemstones.* London: Darling Kindersley, 1994. This introductory overview of gemstones and their properties incudes numerous color illustrations and an index.

Hurlbut, Cornelius S., Jr., and George S. Switzer. *Gemology.* New York: John Wiley & Sons, 1979. A commonly available textbook on gemology, written for college-level students.

Klein, Cornelis, and Cornelius S. Hurlbut, Jr. *Manual of Mineralogy.* 21st ed. New York: John Wiley & Sons, 1999. An introductory college-level text on mineralogy. Discusses the physical and chemical properties of minerals and describes the most common minerals and their varieties, including gem mineral varieties. Contains more than 500 mineral name entries in the mineral index and describes in detail about 150 mineral species.

Kunz, George. *The Curious Love of Precious Stones.* Mineola, N.Y.: Dover, 1971. A delightful book about the mythology surrounding gems. Written at the high school level, this volume provides a wealth of information on the folklore of gems and was first published in 1913 by J. B. Lippincott. The original edition is a collector's item.

Liddicoat, Richard T., Jr. *Handbook of Gem Identification.* 11th ed. Santa Monica, Calif.: Gemological Institute of America, 1981. The best all-around text written for the gemologist, jeweler, or layperson who wishes to identify gemstones. Gems are listed in alphabetical order. Many tables of gem properties are included, and gem-testing equipment is described. Synthetics, imitations, and fakes are discussed, and information is frequently updated with new editions. High school level.

Shipley, Robert M. *Dictionary of Gems and Gemology.* 6th ed. Santa Monica, Calif.: Gemological Institute of America, 1974. This dictionary is an international reference source for the gemological profession. Contains, with definitions, the vocabulary associated with the gem trade, including gem and gem mineral names, equipment names, and names associated with synthetics and other imitations. The glossary has more than 4,000 entries.

Sinkankas, John. *Gemstones of North America.* New York: Van Nostrand Reinhold, 1959. Recommended for those who wish to know more about North American gem mineral locations. Written for the hobbyist, at the high school reading level.

Webster, Robert F. G. A. *Gems: Their Sources, Descriptions, and Identification.* 5th ed. Boston: Butterworth-Heinemann, 1994. A technical college-level text that is an important reference for the gemologist or the senior jeweler. Reflects the British attention to detail. Better for sources and descriptions than for identification. Essential for the gemologist.

HYDROTHERMAL MINERALIZATION

Hydrothermal mineralization refers to the formation of minerals from the interaction of hot aqueous (water) solutions with ordinary rocks. The minerals could form by precipitating directly out of the solutions, or they could result from the replacement of minerals by other minerals.

PRINCIPAL TERMS

DIFFUSION: the movement of ions or molecules through a medium (solid, liquid, or gaseous) from a location of high concentration to that of a lower concentration

FLUID INCLUSION: a bubble, within a mineral, filled partly with liquid and partly with gas

ION PAIR: a loose combination of two or more ions, such that the combination acts as a single ion having a neutral, negative, or positive charge

IONS: atoms that have either gained or lost electrons

ISOTOPES: atoms of the same element with different numbers of neutrons in their nuclei

MINERAL: a natural substance with a definite chemical composition and with an ordered internal arrangement of atoms

REDUCED SULFUR: a negatively charged sulfur ion

SULFIDE: a mineral type containing reduced sulfur, as in zinc sulfide

ECONOMIC MINERAL DEPOSITS

Hydrothermal mineralization is the formation of minerals from naturally hot aqueous solutions. Minerals that form at the mouths of hot springs or geysers are hydrothermal minerals. Such minerals are usually banded and colorful.

A number of economic mineral deposits are formed by hydrothermal mineralization processes. Economic mineral deposits are commonly exploited for their metallic components. The properties of metals are such that they have many applications: furniture; materials for electrical wiring, cables, and plumbing systems; kitchen utensils; and the frameworks of buildings. These can be partly or completely made of metals. Automobiles, trucks, trains, ships, submarines, airplanes, rockets, guns, and munitions are almost entirely made of metals. Jewelry, too, is commonly metallic. Metallic coins serve as currencies. Clearly, metals have profound impact on civilization. Economic mineral deposits are sought not only for their metals. Certain economic minerals, because of their industrial applications, are called industrial minerals. Industrial minerals may be used as insulators of heat and electric current, as cement and concrete for buildings, or as fertilizers.

Commonly, economic hydrothermal minerals are dominated by the mineral group called sulfides. Sulfides are a combination of metallic ions such as those of copper, zinc, and lead with the sulfur ion. Ions are atoms that have lost electrons and are thus positively charged or that have gained electrons and are characterized by negative charges. The negatively charged sulfur ion is also called reduced sulfur. Examples of sulfides include chalcopyrite, a golden-colored copper-iron sulfide; sphalerite, a greenish-brown zinc sulfide; and galena, a shiny, silver-colored and heavy lead sulfide.

FORMATION OF ECONOMIC HYDROTHERMAL MINERALS

Many economic deposits exhibit textures indicating that the minerals formed by precipitation from solutions. In some places, hydrothermal solutions may have dissolved certain rock layers and caused the breakup of overlying rocks. The ensuing rock fragments (breccia) may be cemented together by minerals, including economic minerals, that precipitated from a later phase of hot aqueous solution processes.

In the circulating hydrothermal solutions, the atoms necessary for the formation of economic minerals are dissolved in the form of ions. Metallic ions and reduced sulfur combine readily, however, to form sulfide minerals, which should settle out from the solutions. Thus, it is not easy to transport metallic ions and reduced sulfur in the same solution. Some scientists believe that metallic ions combine with other ions to form pairs that act as single ions. In such ion pairs, the metals are kept in solution and then transported. There is no agreement among scientists, however, about the nature of these ion pairs or whether metallic ions and reduced sulfur are transported in the same or different solutions.

From the circulating hydrothermal solutions, minerals precipitate out and form economic deposits at certain places. Appropriate geologic settings for viable economic deposits include the following: a plumbing system that is continuous up to the source areas from which the mineralizing ions enter the hydrothermal solution; a rock sequence that slows down the circulation of hydrothermal solution so that minerals, some of which are extremely coarse-grained, can precipitate out; or a place where different conditions combine to cause the precipitation of minerals.

Scientists have suggested that the following conditions are possibilities that lead to the precipitation of minerals: the mixing of a metallic ion-carrying solution with a reduced sulfur-carrying solution; a hot solution encountering cold conditions; solutions encountering conditions where their hydrogen ion content (pH) must change; or solutions encountering conditions that lead to the breaking of ion-pairing mechanisms that had kept metallic ions dissolved. In some localities, the common rocks might serve as essentially passive places where the conditions that cause mineral precipitation are mingling. In other places, the rocks are reactive, and the formation of economic deposits results from solution-rock interaction. Scientists can infer the geologic conditions and processes that led to the precipitation of economic mineral deposits from stable isotope ratios in the minerals.

IDENTIFICATION OF HYDROTHERMAL MINERALIZATION

Hydrothermal mineralization may be inferred from the analysis of minerals by using a variety of methods and instruments: petrographic microscopes for distinguishing mineral types, heating- and freezing-stage microscopes for estimating temperature and salinity of fluids, mass spectrometers for estimating geologic conditions and pro-

cesses, and radiant-energy emission-measuring instruments such as electron microprobes, X-ray fluorescence spectrometers, and instrumental neutron activation analysis equipment for determining elemental composition.

Rocks can be sliced into extremely thin sections (0.03 millimeter thick) after they have been mounted on transparent glass. In these thin sections, some minerals allow the transmission of light through them; they are called transparent minerals. Others, such as sulfides, reflect light and are opaque. Such optical properties can be used for identifying minerals under appropriate microscopes. Reflected-light microscopes are used to identify opaque minerals, and transmitted-light microscopes are used for transparent minerals.

In some transparent minerals, spherical and ellipsoidal bubbles, filled partly by a liquid phase and partly by gas, are observed. These are fluid inclusions. A thin section containing fluid inclusions is placed on a heating and cooling stage, a specially designed stage, of a transmitted-light microscope. On the heating stage, the heat is increased until the bubbles in a thin section are filled with a homogeneous phase—that is, until there is no more separation between liquid and gas in the fluid inclusions. The temperature of homogenization of the fluids in the bubbles is then recorded and is deduced to be the minimum temperature of the fluid from which the enclosing mineral was formed. The fluid may be examined in place or extracted for further chemical analysis—for example, to determine the type and amount of dissolved ions. Other samples from the same rock, containing identical bubbles, may be frozen until the fluids in the bubbles become solids. The freezing temperature is recorded, and scientists use this temperature to infer the amount of salts that were dissolved in the solution.

STUDY OF HYDROTHERMAL MINERALIZATION

Geologic conditions and processes that led to the formation of minerals may be determined from isotopes in minerals. For this analysis, selected samples are powdered and dissolved in acids, producing gases. The gas container is then attached to a mass spectrometer, and the gases are made to enter a chamber. In this chamber, the gases are bombarded with electrons so that atoms (including isotopes) in the gases lose electrons and are converted into positive ions. The ionized isotopes are accelerated and made to pass through a magnetic field before they arrive at a detector. Electrical signals from the detector can be digitized or displayed on a strip-chart recorder. A series of peaks, each representing an isotope (the peak height being proportional to the abundance of the isotope), can be recorded by adjusting the accelerating voltage or the magnetic field. This technique permits the identification of isotopes and determination of their relative abundance. If the isotopes are stable isotopes (isotopes that do not change into other ones), then isotope ratios in rocks yield information on geologic conditions. Geologic conditions such as oxidation-reduction reactions, temperature, or diffusion, as well as the type of solution that led to the formation of hydrothermal minerals, can be extrapolated from stable isotope ratios in rocks. If the isotopes are naturally radioactive, they can be used to determine the age of formation of hydrothermal minerals.

Change in the composition of a hydrothermal solution during the growth of a mineral may be estimated by using an electron microprobe (EMP). The EMP is a powerful tool for determining the variation of composition within a single mineral, because several analyses can be conducted on one surface of a mineral. For an EMP analysis, polished rock samples are placed on the stage of a special microscope, which is fitted with an X-ray spectrometer for the determination of elemental composition of a sample. The polished sample is carbon-coated and then bombarded with electrons, causing electrons from the sample to be excited to higher energy levels. Upon deexcitation, irradiated X rays are guided to a crystal that will cause separation of the rays so that they may be detected by X-ray counting devices or displayed on strip-chart records as peaks and valleys. The peak positions on the charts correspond to the elements in the rock sample, and the peak heights are proportional to the concentration of the elements.

An analytical technique that is similar to the EMP is the use of the X-ray fluorescence spectrometer (XRF). For an XRF analysis, selected samples are powdered and then bombarded with X rays. The elements in the sample then irradiate X rays, which can be analyzed as in the EMP method. Major and trace elements in minerals can be identified by this method, which will provide comprehensive information about the hydrothermal solution from which the minerals were formed.

Detailed information on hydrothermal mineralization may be determined from elements that occur in extremely small amounts and cannot be determined by the methods discussed above. Instrumental neutron activation analysis (INAA) can circumvent this problem. For the INAA, selected samples are powdered and placed in small containers. The samples are then placed in a chamber where they are bombarded by neutrons supplied by a nuclear reactor or a particle accelerator. The atoms in the sample incorporate these neutrons and become artificially induced heavy and unstable (radioactive) isotopes. The samples are removed and placed in a chamber where the activity of gamma rays emitted by the decay of the artificial radioactive isotopes can be detected by a recording device. The elements are identified from the decay constants and half-lives of the isotopes.

Habte Giorgis Churnet

CROSS-REFERENCES

BIBLIOGRAPHY

Barnes, H. L., ed. *Geochemistry of Hydrothermal Ore Deposits.* 2d ed. New York: Wiley-Interscience, 1979. An excellent collection of basic articles on ore-forming processes.

Deer, W. A., R. A. Howie, and J. Zussman. *An Introduction to the Rock-Forming Minerals.* 2d ed. New York: Longman, 1992. This work includes descriptions of nonsilicate minerals such as sulfides, oxides, and carbonates. Useful for advanced students and research workers.

Guilbert, John M., and Charles F. Park, Jr. *The Geology of Ore Deposits*. New York: W. H. Freeman, 1985. One of the textbooks used in colleges for the study of ore deposits. Provides an excellent description of hydrothermal fluids, their migration, and other processes that lead to the formation of deposits. Chapters 2 through 5 describe various aspects of hydrothermal processes.

Sawkins, F. J., ed. *Metal Deposits in Relation to Plate Tectonics*. 2d rev. ed. New York: Springer-Verlag, 1990. No specific chapter is devoted to hydrothermal solutions in this book. Yet discussions of mineral deposit types in the book address hydrothermal fluid movement and deposition. Provides an excellent survey of the geologic settings in which mineral deposits occur.

Skinner, B. J., ed. *Economic Geology: Seventy-fifth Anniversary Volume, 1905-1980*. Lancaster, Pa.: Economic Geology, 1981. An excellent collection of articles. Useful for advanced students and researchers in the geological sciences.

Stanton, R. L. *Ore Petrology*. New York: McGraw-Hill, 1972. A very helpful college-level book on ore deposits. Stanton emphasizes the association of mineral deposits with ordinary rocks. Hydrothermal solutions are discussed in chapter 6.

Thompson, J. F. H. *Magmas, Fluids, and Ore Deposits*. Nepean, Ontario: Mineralogical Association of Canada, 1995. This anthology contains papers presented at a Mineralogical Association of Canada symposium in 1995 on such subjects as ore deposits, magmatism, hydrothermal alteration, and geological intrusions. Although somewhat technical, the bibliography will be useful to anyone interested in hydrothermal mineralization.

IONIC SUBSTITUTION

Ionic substitution varies the chemical composition of minerals by replacing atoms with other atoms of similar size, electrical charge, and chemical properties. Because of ionic substitution, many minerals are not distinct substances but are members of series whose composition varies over a wide range.

PRINCIPAL TERMS

ANGSTROM UNIT: a unit of size often used for describing the dimensions of atoms, equal to one one-hundred-millionth (10^{-8}) centimeter

ANION: an atom that has gained electrons and thus acquired a negative electrical charge

CATION: an atom that has lost electrons and therefore has a positive charge

END MEMBER: any of the pure materials that make up a solid solution; a mineral intermediate in composition between two end members is called intermediate

ION: an atom that is electrically charged because it has gained or lost electrons

IONIC RADIUS: the effective radius of an ion in a mineral; ionic radius for a given ion can vary depending on the number and type of ions around it

OXIDATION STATE: the number of electrons gained or lost by an ion, usually specified as the electric charge on the ion; also called valence

SOLID SOLUTION: a mixture of two or more solid materials on the atomic scale, which can occur if the materials have very similar atomic structures

BONDING BEHAVIOR OF IONS

Ionic substitution is the substitution of one atom for another in minerals. Although minerals are naturally occurring, inorganic chemical compounds, which ideally should have definite, fixed compositions, the compositions of minerals are often far from definite. Much of the variability of mineral compositions results from ionic substitution. Ionic substitution is governed chiefly by the size, electric charge, and bonding behavior of the ions in minerals.

Most of the bonding in minerals is ionic bonding: If some atoms in a mineral can lose electrons easily to become cations, and others can acquire electrons to become anions, their opposing electrical charges hold them together. Table salt, or halite, is a common material with ionic bonding: Each sodium atom loses an electron, while each chlorine atom acquires an electron. The sodium cations and chlorine anions are then bonded together by their opposite electric charges. In covalent bonding, electrons orbit adjacent atoms and fill out the electron shells of both atoms simultaneously. The most important example in geology is the bond between silicon and oxygen. This bond is partially ionic and partially covalent. That is, electrons spend most of the time attached to the oxygen anion, but there is also some sharing of electrons between oxygen and silicon. In metallic bonding, electrons wander freely through the metal. In effect, the metal consists of cations held together by a negatively charged electron "gas." The free electrons cause metals to be such good conductors of heat and electricity. Gold is a mineral with metallic bonding.

IONIC RADIUS

In minerals an ion generally behaves as a rigid sphere of definite size, or ionic radius. Anions are generally larger than cations for two reasons. First, cations often form when atoms lose their outermost shell of electrons, whereas anions form by adding atoms to their outermost shell. Second, cations have more protons than electrons, so the positively charged nucleus of the atom pulls the remaining electrons in closer. Anions, on the other hand, have more electrons than protons. The extra electrons repel one another, and the electrons are pushed farther apart. In effect, the electron shells around an anion become slightly inflated. Ionic radius is also closely related to the position of an element in the periodic table. Cations become larger as their atomic number increases, simply because each row in the periodic table has one more electron shell than does the row above.

The size difference between the cations and anions has some interesting consequences. The most abundant element in the Earth's crust, oxygen, is an anion in minerals. Its ionic radius is two to three times that of the commonest cations and its volume about twenty-five times as great. Oxygen accounts for fully 95 percent of the volume of all atoms in the Earth's crust. The structure of minerals can often be pictured in terms of stacking spheres of various sizes in the right proportions. Because anions are so large, minerals often consist of anions packed in various ways, with the smaller cations filling the voids between. Frequently, the atomic arrangement of a mineral remains unchanged if one cation is substituted for another. Thus, it is usually more useful to classify minerals, and chemical compounds in general, according to their anions rather than their cations. For example, magnesium sulfate (epsomite) is very soft and soluble in water; magnesium carbonate (magnesite) is harder and insoluble in water but soluble in acid; and magnesium silicate (forsterite) is hard, glassy, and insoluble in water or acid. Magnesium minerals, as a group, have no properties in common. On the other hand, the carbonates of magnesium, calcium (calcite), iron (siderite), and manganese (rhodochrosite) are all quite similar in density, hardness, crystal form, and many other physical properties.

ELECTRIC CHARGE

In addition to size, the other important property of ions that governs the structure of minerals is electric charge. Size and electric charge dictate the degree of substitution of elements in minerals. Sometimes, elements with very different chemistries are similar enough in size and charge to substitute for each other. For example, nickel and magnesium are very different chemically, but they have nearly identical ionic radii and valences and substitute easily for each other in minerals.

The more different two ions are in radius, the less likely they are to substitute for each other. The difficulty is compounded if the ions have different charges. For an ion of one charge to substitute for another, some additional ionic substitution must be made to restore electrical charge balance. For example, in the plagioclase feldspars, sodium (+1) and calcium (+2) substitute, but the extra charge on the calcium ion is balanced by substituting aluminum (+3) for silicon (+4). Finally, the bonding behavior of some elements affects substitution. For example, silicon is bonded to oxygen partly by electrical attraction (ionic bonding) but also partly by sharing electrons (covalent bonding). Silicon usually bonds to four neighboring oxygen atoms in a tetrahedral (triangular pyramid) arrangement. Other elements can and do substitute for silicon but only in a limited way, because they lack silicon's distinctive bonding behavior. In some cases, ions are different enough in size, charge, and bonding behavior that they can substitute only to a limited degree. In other cases, ions substitute freely. A mineral series that varies over a wide range of composition because of free ionic substitution is called a solid solution. The ideal pure minerals at the extremes of the composition range are called end members of the series.

COMMON SUBSTITUTIONS

Elements that substitute for one another generally occupy distinct portions of the periodic table. In the periodic table, elements with similar electron structures fall into vertical columns, called groups. These groups are usually atoms of similar size and valence. One important group includes the alkali metals: potassium, rubidium, and cesium. These ions all have charges of +1 and large ionic radii (greater than 1.0 angstrom unit). Another important group is the alkaline Earth elements: calcium, strontium, and barium. These ions also have large ionic radii, but they have charges of +2. Sodium is often too small to substitute for the other alkali metals but commonly substitutes for calcium. Another group of large cations that substitute readily for one another are the lanthanide and actinide rare-earth elements, mostly with valences of +3 or +4.

There are many medium-sized cations (ionic radii 0.6-0.8 angstrom unit) that substitute for one another. One large group consists of cations with +2 valence, mostly from the middle of the periodic table: iron, nickel, cobalt, manganese, magnesium, and sometimes copper and zinc. Iron and magnesium are perhaps the most important ionic substitution in minerals. If there is enough room in the mineral structure, calcium will sometimes substitute for some of these elements. Among the smaller cations, substitution is common among valence +3 cations of small to medium size: aluminum, chromium, and ferric iron. Another important group of elements are silicon and other small cations (with ionic radii of less than 0.6 angstrom unit) that substitute for it in mineral structures: aluminum, beryllium boron, and titanium. These ions have charges of +3 to +5.

Among the anions, ionic substitution occurs most commonly among the halogens: fluorine, chlorine, bromine, and iodine. These have charges of −1 and large ionic radii. The hydroxyl radical (OH) also has a charge of −1 (-2 for oxygen, plus +1 for hydrogen) and a large ionic radius. Fluorine often substitutes for the hydroxyl radical.

In some minerals, cations are attached so weakly that they can be easily removed and replaced by other cations. This process, called ion exchange, occurs commonly among a group of silicates called the zeolites as well as among clay minerals. The ion-exchange capacity of zeolites was once widely exploited in water softeners.

IMPURITIES

If ionic substitution occurs on a small scale, the substitute ion is considered an impurity rather than a major chemical component. Nevertheless, impurities can be geologically and economically significant. One of the most conspicuous effects of impurities in minerals is to create or modify their color. In their pure state, the gem minerals quartz, corundum, beryl, and diamond are all colorless. Colored gem varieties of these minerals are the result of small ionic substitutions. For example, the green of emerald (beryl) and the red of ruby (corundum) both result from tiny amounts of chromium substituting for aluminum. Amethyst (purple quartz) owes its color to ferric iron, rose quartz (pink) to manganese or titanium. Only certain atoms are capable of producing color in minerals. Most electrons in atoms are paired, attracted to each other by the weak nuclear force so strongly that visible light lacks enough energy to affect them. Only those atoms with unpaired electrons can absorb visible light, and of the most

abundant elements in the Earth's crust (oxygen, silicon, aluminum, iron, magnesium, sodium, potassium, and calcium), only iron has unpaired electrons. Iron is the dominant coloring agent in most rocks and many minerals. Ferric iron imparts various shades of red, yellow, or brown; ferrous iron tends to color minerals green or black.

For some elements, ionic substitution is so effective that the element occurs only as impurities in other minerals. Rubidium is much more abundant in the Earth's crust than are vital metals such as copper or zinc, yet there are no rubidium minerals. Instead, rubidium substitutes for potassium. Similarly, gallium substitutes for aluminum, and hafnium for zirconium. Such elements are said to be dispersed.

STUDY OF MINERAL CHEMISTRY

The study of ionic substitution in minerals requires an understanding of both the chemical composition of minerals and the geometric arrangement of atoms within them. The basic understanding of mineral chemistry was acquired by methods that are now considered primitive: Minerals were dissolved in solvents, and the individual components were separated by chemical reactions and weighed. From this information, the chemical formula of the mineral could be calculated. Ionic substitution in most common minerals was first detected by such means. These techniques require great care, are time-consuming, and are limited to an accuracy of about 0.1 percent.

Modern methods of chemical analysis include the electron microprobe and X-ray fluorescence. Both methods rely on bombarding a specimen with radiation (electrons or X rays, respectively), and analyzing the wavelengths of X rays emitted by the specimen. Each chemical element emits characteristic X-ray wavelengths, so that the X rays emitted by a specimen are a direct indicator of its composition. These methods can be easily automated, do not damage the specimen, and can be used on very tiny samples.

STUDY OF MINERAL ATOMIC STRUCTURE

The atomic structure of a mineral dictates how large or small an ion can be to fit in a particular site. For some simple minerals, it has been possible to infer their atomic structure using geometrical reasoning and such clues as the crystal form of the mineral. For example, table salt (halite) always cleaves into cubic fragments, and it has a simple formula: equal numbers of sodium and chlorine atoms. It is reasonable (and correct) to surmise that sodium and chlorine atoms alternate in a simple cubic pattern.

Complex minerals, including many of the silicates, are far too complex for such methods. Atomic structures of complex minerals are usually determined by X-ray diffraction. When a specimen is bombarded by X rays, X rays scatter or diffract off atoms in the specimen. When the wavelength of the X rays, the angle at which the X rays strike, and the arrangements of atoms are just right, the diffracted X rays emerge in a particularly strong beam. The intensity of the X rays depends both on the geometry of the diffraction and on the chemical elements involved. There are many ways to make use of X-ray diffraction to determine the atomic structure of minerals. These methods can show not only the locations of atoms but also subtle details such as distortions that result from the substitution of unusually large or small ions. A computer is required to handle the complex computations needed for such studies.

GEOLOGIC SIGNIFICANCE

Ionic substitution in minerals enriches the variety of the mineral world and has played a significant role in the chemical evolution of the Earth. Although the crust of the Earth makes up only one three-hundredth of the total mass of the Earth, it contains more than 1 percent of the Earth's total potassium and half or more of the entire Earth's rubidium, uranium, thorium, and some rare-earth elements. These elements are enriched in the crust because they do not substitute easily for the iron and magnesium that are principal components of minerals in the Earth's mantle, the layer beneath the crust.

Finally, without ionic substitution, radiometric dating of rocks and minerals would be far more difficult. Radiometric dating relies on the decay of radioactive atoms at a known rate and the accumulation of the resulting decay products. Two of the major dating systems are based on elements that occur as impurities. The rubidium-strontium method is based on the decay of rubidium 87, which occurs as an impurity in potassium minerals. The uranium-lead dating method can make use of uranium minerals but is most frequently applied to zircon, in which the uranium occurs as an impurity substituting for zirconium.

Steven I. Dutch

CROSS-REFERENCES

Biopyriboles, 1175; Carbonates, 1182; Clays and Clay Minerals, 1187; Feldspars, 1193; Fluid Inclusions, 394; Gem Minerals, 1199; Geothermal Phenomena and Heat Transport, 43; Geysers and Hot Springs, 694; Groundwater Movement, 2030; Heat Sources and Heat Flow, 49; Hydrothermal Mineralization, 1205; Magmas, 1326; Magnetite, 1214; Metamictization, 1219; Metasomatism, 1406; Minerals: Physical Properties, 1225; Minerals: Structure, 1232; Non-silicates, 1238; Orthosilicates, 1244; Oxides, 1249; Pegmatites, 1620; Radioactive Minerals, 1255; Sub-Seafloor Metamorphism, 1423; Water-Rock Interactions, 449.

BIBLIOGRAPHY

Blackburn, William H., and William H. Dennen. *Principles of Mineralogy.* Dubuque, Iowa: Wm. C. Brown, 1988. A college-level mineralogy text, with chapters on mineralogical theory and methods, as well as descriptions of common minerals. Chapters 4, 5, and 6 describe the chemistry and atomic structure of minerals, with considerable detail on atomic bonding. Chapter 12 is useful for its good coverage of modern analytical techniques used in the study of minerals.

Bloss, F. D. *Crystallography and Crystal Chemistry.* New York: Holt, Rinehart and Winston, 1971. A text for advanced college students that covers crystal form, atomic structure, physical properties of minerals, and X-ray methods. Although the mathematics can be complex, there are also many passages where important concepts are explained in clear and simple terms. Illustrations are abundant and highly informative.

Chen, Philip S. *A New Handbook of Chemistry.*

Camarillo, Calif.: Chemical Elements Publishing, 1975. A concise and simplified chemical reference book for introductory college chemistry courses. Although a general chemical reference, it contains much useful information including ionic radii, abundances of isotopes, and oxidation states of the elements for Earth scientists.

Day, Frank H. *The Chemical Elements in Nature.* New York: Van Nostrand Reinhold, 1964. A survey of the occurrence of the elements in the Earth and atmosphere as well as their biological and economic significance. The treatment is simple and nonmathematical. Some remarks on the chemistry of other planets are dated and a few are now known to be incorrect. On the whole, however, the book is a good survey for the layperson.

Deer, William A., R. A. Howie, and J. Zussman. *Rock-Forming Minerals.* 5 vols. New York: John Wiley & Sons, 1988.

_____. *An Introduction to Rock-Forming Minerals.* 2d ed. New York: Longman, 1992. Standard references on mineralogy for advanced college students and above. Each chapter contains detailed descriptions of chemistry and crystal structure, usually with chemical analyses. Discussions of chemical variations in minerals are extensive. *An Introduction to Rock-Forming Minerals* is a condensation of a five-volume set originally published in the 1960's.

Frye, Keith. *Mineral Science: An Introductory Survey.* New York: Macmillan, 1993. This basic text, intended for the college-level reader, provides an easily understood overview of mineralogy, petrology, and geochemistry, including descriptions of specific minerals. Illustrations, bibliography, and index.

Hammond, Christopher. *The Basics of Crystallography and Diffraction.* New York: Oxford University Press, 1997. Hammond's book covers crystal form, atomic structure, physical properties of minerals, and X-ray methods. Illustrations help clarify some of the more mathematically complex concepts. Includes bibliography and index.

Klein, Cornelis, and Cornelius S. Hurlbut, Jr. *Manual of Mineralogy.* 21st ed. New York: John Wiley & Sons, 1999. One of the most widely used college mineralogy texts. Contains chapters on most mineralogical methods, as well as descriptions of common minerals. Particularly good for its attractive illustrations of atomic structure of minerals and its survey of research on the causes of color in minerals. Chapter 4 contains many useful chemical tables and a good description of the way chemical analyses are converted into chemical formulas.

Muecke, Gunter K., and Peter Möller. "The Not-So-Rare Earths." *Scientific American* 258 (January, 1988): 72-77. A survey of one important group of elements that commonly substitute in minerals. The article explores the variations in their natural occurrence and shows how variations in ionic radius and valence allow the rare-earths to be used as tracers for many geologic processes. *Scientific American* is written for nonspecialists at a college reading level.

O'Nions, R. K., P. J. Hamilton, and N. M. Evensen. "Chemical Evolution of the Earth's Mantle." *Scientific American* 242 (May, 1980): 120-135. This article, written for nonspecialists at a college reading level, shows how trace elements and isotopes in minerals can be used to deduce the chemical history of the Earth's interior. In particular, elements with large ionic radii have become concentrated in the Earth's crust.

Turekian, Karl K. *Chemistry of the Earth.* New York: Holt, Rinehart and Winston, 1972. A short but information-packed summary of geochemistry, written for college students at an introductory level. The description of mineral structure, bonding, and ionic substitution is well illustrated.

MAGNETITE

Magnetite is an iron oxide found in nature. Its dominant characteristic is its natural magnetism, and it is used as a raw material in the production of iron and steel. Magnetite occurs in many parts of the world, but it exists in unusually high concentrations in a few areas that support large mining operations.

PRINCIPAL TERMS

FLOTATION: a method of separating pulverized ores by placing them in a solution in which some particles float and others sink

IRON OXIDE: a compound formed of the elements iron (Fe) and oxygen (O) in chemically stable combinations

MAGNETISM: the specific properties of a magnet, regarded as an effect of molecular interaction

MINERAL: a naturally occurring, homogeneous substance formed by inorganic processes and having a characteristic set of physical proper-

ties, a definite and limited range of chemical composition, and a molecular structure usually expressed in crystalline form

PELLETIZING: a method of rolling fine particles of iron ore under damp conditions into balls that are heated to become concentrated and durable

SINTERING: to make or become cohesive by the combined action of heat and pressure

TACONITE: an inferior grade of iron ore found in the Mesabi district in Minnesota, consisting of a very hard chert

MAGNETISM

Magnetite has several physical "signatures" that make it readily identifiable. To the eye, it is black and opaque, with a dull or metallic luster. Its most notable feature, however, is its natural magnetism, which distinguishes it among other metal minerals. A fine sliver of magnetite placed on water will float because of surface tension and will invariably move so that it has a north-south alignment, like a compass needle. It derives its magnetic properties from an internal flow of electrons, which produces magnetism just as magnetism, in reverse, produces electricity in electric generators. A molecule of magnetite contains two different ions of iron, which are located in two specific sites; this arrangement causes a transfer of electrons between the different irons in a structured path, or vector. This electric vector generates the magnetic field.

Magnetism can be induced by a magnetic field such that particular minerals are more susceptible than others. A few minerals have much stronger positive susceptibilites than others and may also exhibit a magnetism that remains after the magnetic field has been removed. These minerals are termed "magnetic" or, more correctly, "ferromag-

netic." This characteristic has strong application in the concentration of magnetite iron ores.

IRON AND IRON-OXIDE FORMATIONS

Magnetite is an iron-oxide mineral. Iron is the world's most widely used metal and constitutes about 5 percent of the Earth's crust. Precambrian iron-bearing deposits are distributed worldwide. They are found in the Lake Superior region of North America, in South America, in India, and in Sweden. The definition of an "iron formation," as deposits are known, is typically a chemical sediment, usually thin-bedded or laminated, containing 15 percent or more iron of sedimentary origin, commonly but not necessarily containing layers of chert. However, few iron-ore deposits containing less than 25 percent iron would be considered as commercial unless they existed in large amounts and could be concentrated to higher percentages of iron very inexpensively.

Two major sources of iron have been postulated. The first is the decomposition of rocks during weathering. The second is volcanic activity, from which the iron is obtained from magmatic emissions or contact of lava and seawater. For some time, it was assumed that a combination of these activities would be necessary for the very large accumulations of iron-bearing deposits observed. Later investigations of weathering under certain tropical or subtropical conditions have shown that there would be sufficient volumes of materials produced by weathering alone, without the necessity for the assistance of igneous activities.

Iron in solution is precipitated as various compounds, or facies, including oxide, silicate, carbonate, and sulfide. The oxide facies are subdivided into two parts: hematite-banded (nonmagnetic) and magnetite-banded (magnetic). The deposits of magnetite in North America are thought by some scientists to have been metamorphosed to a certain extent, although other opinions maintain that they are entirely sedimentary in origin. However, most authorities believe that a combination of metamorphism and sedimentation is responsible for magnetite and other iron-oxide formations that occur in nature.

Magnetite is the dominant iron-oxide mineral in what is referred to as "taconite" in the Mesabi Range, or district, of Minnesota. The Mesabi is about 175 kilometers long and about 2.4 kilometers wide, having iron ore thicknesses between 120 and 230 meters. Characteristics of individual ore bodies in the Mesabi are lengths of up to 5.6 kilometers, widths from 0.8 to 1.6 kilometers, and thicknesses of 150 meters. Most of the world's magnetite is produced in the Mesabi and is thought to have been formed as a result of metamorphism and sedimentation. In mining parlance, "taconite" denotes an unenriched iron-oxide formation; it is characteristically a very hard material because of its high silica content.

DETERMINING MAGNETITE RESERVES

There is a fundamental difference in a magnetite "reserve" and a magnetite "resource." Every deposit is classified as a resource, while reserves are only those deposits that can be extracted and utilized using current guidelines of technology, price, and cost. The commitment of funds for the development of magnetite reserves depends on the reliable estimation of recoverable deposits under current economic conditions. These estimates are used to predict the life of the mine and its cash flow and to project the manner and time in which the reserve will be extracted. The low-grade magnetic taconites of the Mesabi range in northeastern Minnesota are near the surface and typical of gently dipping ore bodies that can be assessed from ground and aerial surveys. These deposits have broad areal extent and usually have long productive lives.

Both conventional and computer methods are used in determining these magnetite reserves. In the conventional method, topographic maps are made from ground or aerial surveys. Cross sections are drawn to accommodate existing drill-hole patterns and extensions. Geologic information and test data are added to the cross sections, and the configuration and stratigraphic layering of the magnetite ore body are determined. For each drill hole, the minable taconite layer or layers are determined from core analyses or well logs (electrocardiograms for rocks), and the ore-to-waste ratio is calculated and compared to the cutoff-stripping ra-

Carbonatitic lapilli tuff with lapilli particles cored by magnetite crystals. (Geological Survey of Canada)

tio. This second ratio is the maximum amount of waste material that can be excavated and handled while still permitting economical operation. Incomplete data usually dictate the drilling of additional exploratory holes, which becomes a tug-of-war in maximizing data while minimizing expenses.

Using data from mining, metallurgical, and economic studies, a mining plan is developed that includes an approximation of the open-pit outline. Cross sections are prepared to include as much of the pit as possible. The information in these cross sections is weighted for influence, and an overall stripping ratio for the pit is determined. The pit outline is then adjusted to accommodate the overall stripping ratio, and refined economic data using the current pit design are generated to determine profitability and directions for proceeding with the mine.

The computer method uses a series of computer programs to transform topographic elevations, drill-hole data, and geologic information into a three-dimensional model of the deposit and then performs calculations, as was done manually in the conventional method. In either method, these data are used to make formal proposals regarding funding sources, property leasing, equipment selection, and manpower requirements. In practice, both conventional and computer methods are used in determining these magnetite reserves.

MINING MAGNETITE ORE

The taconite deposits of the Mesabi region of Lake Superior offer a typical example of the method of mining the dense magnetite ore. Very large open-pit excavations are deepened and widened from the surface through the use of blasting and bulk materials handling. The dense and abrasive nature of the deposit requires the use of wear-resistant drills and even flame drilling, a process called "jet piercing." Jet piercing is similar to using a blowtorch to burn a hole into the deposit through the differential expansion (spalling) of the silica and iron and the use of compressed air to expel the spalled material to the surface. The high cost associated with jet piercing, together with improvements in percussive drilling, has tended to render jet-piercing equipment obsolete.

The blast-hole drills are set up on benches, or flat areas, around the periphery of the open pit so that the blasted material will be expelled into the bottom of the pit. After drilling is performed to a depth just below the height of the bench, explosives are loaded into the hole for blasting. Either dynamite or a mixture of ammonium nitrate and fuel oil can be used as an explosive. The blasted material is loaded by power shovel or mobile loaders into large trucks having capacities between 80 and 200 tons for transportation out of the pit. Operations involving the mining of the raw magnetite are simple and straightforward. The problems center on equipment wear, operating cost, and economy of scale.

CONVERSION OF RAW ORE

Considering that typical Mesabi taconite has an iron concentration less than 50 percent, the conversion of these raw ores to a product greater than 50 percent iron and without objectionable impurities is the greatest problem facing iron-ore producers. Typically, most iron ores are priced on a basis that they contain at least 51.50 percent iron at a specified "natural" moisture content.

Magnetite ore in its natural form is associated with waste materials, often silicates. The ore must be subdivided in order to liberate the magnetite and thus increase the iron concentration of the ore. Several concentration techniques are available, depending on the nature of the association of magnetite and its "gangue," or waste material, as well as the nature of the waste materials themselves. Common methods make use of physical and chemical properties of magnetite.

Common magnetite ore is sufficiently heavy that the magnetite, with a specific gravity of 5.1, can be easily separated from the relatively less heavy waste materials, which have specific gravities around 2.6. Heavy media separators using a ferrosilicon suspension in a rotary drum are most common for coarse particles. For medium-sized particles, heavy media separators using a magnetite suspension in hydraulic cyclones are used. For fine particles, machines known as "Humphreys spirals" are normally used; these machines use the tangential velocity differences between materials of different specific gravities to separate them.

Magnetite can also be separated from its waste material by using a low-intensity magnetic field. Such separation is inexpensive, because the process is very simple. Either electromagnets or permanent magnets can be used with equal efficiency. Rotary-drum separators utilizing permanent magnets are favored in this process, as permanent magnets provide freedom from electrical maintenance that is inherent in electromagnets.

Magnetite can also be concentrated by froth flotation, a process whereby the fine magnetite ore clings to bubbles formed by mixing a chemical reagent into water or a hydrocarbon, then frothing it. Froth flotation can be used to separate magnetite ores from waste material or to upgrade low-quality ore. Magnetite is also separated by cationic flotation, a process in which cationic reagents, often amines, are used as a medium for separation. In this process, the free quartz and locked quartz-magnetic particles are carried away from the free magnetite grains. The floated product can then be reground and reconcentrated magnetically. This process is carried out on a large scale by large batteries of flotation cells.

Pelletizing and Sintering

The process of blasting, loading, transporting, crushing, and grinding necessary to remove magnetite ore from its natural habitat and concentrate it for sale produces a considerable amount of dust-sized fine material, or "fines." Fines, if not utilized as sale material, can severely affect the profitability of a magnetite-concentration process. Blast furnaces are the usual endpoint for the magnetite concentrate. Such furnaces do not tolerate fine material very well, and thus a uniform-sized material without fines is necessary as feed material. It is necessary to utilize fines in a cost-effective manner, and this is done by agglomerating, whereby the fine material is pressed into various shapes by a variety of equipment.

The term "pelletizing" in the minerals industry refers to processes whereby the fines concentrate is rolled in a damp condition into balls that can be fired or indurated until they become hard. Fine-size ore concentrates are usually pelletized at the mine and shipped as pellets to the steel plant, although pelletizing processes are frequently constructed near transportation points, particularly in the Great Lakes region. Pellets are usually greater than 1 centimeter in diameter. A common practice in some areas producing natural high-grade ores is to screen the ore at the mine to about 1 centimeter, with the oversized material used for blast-furnace feed. The undersized material may be shipped for use as fines, pelletized at the mill site, or pelletized at the mine and shipped as pellets by the ore carrier. Pelletizing virtually revolutionized iron mining in North America. Although pelletizing had been known and pilot plants constructed before 1920, it was the exhaustion of the higher grades of ore that caused it to be reevaluated in the late 1940's. A decade later, pelletizing was firmly established as a normal operating practice in the Lake Superior iron ranges. Not only did pelletizing permit the economic use of ore fines, it also prevented these fines from becoming stream sediments and therefore detrimental to the environment.

Magnetite ore fines may also be concentrated by the sintering process. In this process, the ore fines are mixed with fine carbonaceous material called "coke breeze," then fed on a moving bed grate across a heat source so that the coke breeze combusts and partially fuses the ore fines. The sintered product traveling off the end of the grate is crushed and sized for shipment to the blast furnace.

Ore Transportation

Since magnetite ore is a relatively inexpensive commodity, its transportation can frequently exceed the cost of mining and processing the ore. For effective commercial exploitation, therefore, it is necessary to find low-cost transportation, and this usually means that some form of bulk shipment on water is made. Typical are oceangoing vessels of 200,000 tons capacity that effect economies of scale over smaller vessels requiring about the same number of crew members.

Recent developments in ore transportation include the shipping of magnetite fines with enough moisture to bind them in the ship's hold but not enough to cause them to be fluid. After arrival at the destination, the mass is re-pulped by the addition of moisture and agitation, then pumped from the vessel to a dewatering facility onshore for pelletizing or sintering. Besides saving a costly dewatering operation at the shipping point, the cargo is shipped dust-free and can be unloaded at a reduced cost as compared to dry unloading operations.

Charles D. Haynes

CROSS-REFERENCES

BIBLIOGRAPHY

Bateman, Alan M., ed. *Economic Geology.* Urbana, Ill.: Economic Geology Publishing, 1955. A classic text, often referenced by more recent writers. Discusses many of the principles of economic geology used today.

Evans, Anthony M. *An Introduction to Economic Geology and Its Environmental Impact.* Malden, Mass.: Blackwell Science, 1997. Although Alan M. Bateman's classic text remains useful, Evans provides information on economic geology that is relevant to the current student, particularly concerning the environmental aspects of mines and mineral resources. Illustrations, maps, bibliography, and index.

Kennedy, Bruce A. *Surface Mining.* Littleton, Colo.: Society of Mining, Metallurgical, and Exploration Engineers, 1990. A comprehensive treatment of the surface mining profession, including detailed descriptions of all phases of this endeavor. Case studies are mixed with topical descriptions. Extensive references are cited in the book.

Lamey, Carl L. *Metallic and Industrial Mineral Deposits.* New York: McGraw-Hill, 1966. A comprehensive treatment of mineral formation and utilization. Includes a description of the formation of iron minerals, including magnetite, as well as all other known industrial minerals.

Tarling, D. H., and F. Hrouda. *The Magnetic Anisotropy of Rocks.* New York: Chapman & Hall, 1993. A complex book intended for use as a textbook at the college level or for reference by the practicing professional. An exhaustive treatment of a complicated and little-known phenomenon. Contains an extensive reference list.

Weiss, Norman L. *SME Mineral Processing Handbook.* New York: Society of Mining, Metallurgical, and Petroleum Engineers, 1985. This two-volume set provides a detailed description of mineral processing systems, equipment, and case studies. This is an advanced treatment of the subject of mineral preparation, but it is well illustrated and can be understood by the layperson.

METAMICTIZATION

Metamictization breaks down the original crystal structure of certain rare minerals to a glassy state by radioactive decay of uranium and thorium. It is accompanied by marked changes in properties and often water content. Many minerals start to lose their radioactive components upon metamictization. Those that do not may be candidates for synthetic rocks grown from high-level nuclear wastes in order to isolate them until they are safe.

PRINCIPAL TERMS

ALPHA PARTICLE: a helium nucleus emitted during the radioactive decay of uranium, thorium, or other unstable nuclei

CRYSTAL: a solid made up of a regular periodic arrangement of atoms; its form and physical properties express the repeat units of the structure

GLASS: a solid with no regular periodic arrangement of atoms; that is, an amorphous solid

GRANITE: a light-colored igneous rock made up mainly of three minerals, two feldspars, and quartz, with variable amounts of darker minerals

ISOTOPES: atoms of the same element with identical numbers of protons but different numbers of neutrons in their nuclei

ISOTROPIC: having properties that are the same in all directions—the opposite of anisotropic, having properties that vary with direction

PEGMATITE: a very coarse-grained granitic rock, often enriched in rare minerals

SILICATE: a substance whose structure includes silicon surrounded by four oxygen atoms in the shape of a tetrahedron

SPONTANEOUS FISSION: uninduced splitting of unstable atomic nuclei into two smaller nuclei, an energetic form of radioactive decay

X RAY: a photon with much higher energy than light and a much shorter wavelength; its wavelength is about the same as the spacing between atoms in crystal structures

METAMICT MINERAL PROPERTIES

Metamict minerals are an anomaly. They have the form of crystals, but in all other ways they resemble glasses. They fracture like glass, are optically isotropic like glass, and to all appearances are noncrystalline. Yet minerals cannot grow without an ordered crystalline arrangement of their constituent atoms—even metamict ones. The term "metamict" is from the Greek roots *meta* + *miktos*, meaning "after + mixed." It aptly describes minerals that must originally have grown as crystals and subsequently have been rendered glassy by some process that has destroyed their original crystallinity. The discovery that all metamict minerals are at least slightly radioactive and that metamict grains have more uranium or thorium than do their nonmetamict equivalents led to the realization that radiation damage resulting from the decay of uranium and thorium causes metamictization. Although controversial at first, the concept of radiation damage is now so easy for scientists to accept that the puzzle is not why some minerals become metamict, but why others with relatively large concentrations of uranium and thorium never do.

The varieties and chemical compositions of minerals that undergo metamictization are quite diverse. Yet all metamict minerals share several common properties: they are all radioactive, with measurable contents of uranium, thorium, or both; they are glassy, brittle, and fracture like glass; they are usually optically isotropic for both visible and infrared light; they are amorphous to X-ray diffraction; and they often have nonmetamict equivalents with the same form and essentially the same composition. Finally, compared to their nonmetamict equivalents, they have lower indices of refraction, are more darkly colored, are softer, are less dense, are more soluble in acids, and contain more water. In those minerals that exhibit a range of metamictization from crystalline to metamict, partially metamict samples have intermediate properties.

For example, X-ray diffraction patterns of zircon show that radiation damage causes it to swell markedly in proportion to accumulated radiation damage up to the point of total metamictization. Beyond that point, the structure is so disordered that it can no longer diffract X rays, even though a continued decrease in density indicates further expansion. Changes in other properties parallel the decrease in density. Partially metamict zircons that are heated at temperatures well below the melting point recrystallize readily to grains that are aligned with their original form, while those that are completely metamict recrystallize just as readily, but not to the original alignment. Several other metamict minerals recrystallize in the same way, but many produce mixtures of different minerals on heating because either they are outside their fields of stability or their compositions have changed subsequent to metamictization.

METAMICT MINERAL GROUPS

Metamict minerals belong to only a few broad groups of chemical compounds. In each group, some have uranium or thorium as the dominant metal ion present, but most have only small amounts substituting in the crystal structure for other ions, such as zirconium, yttrium, and the lanthanide rare-earth elements (REE), which just happen to have similar sizes and charges. This is known as isomorphous substitution and is common in most mineral groups.

The largest group of metamict minerals are yttrium-, REE-, uranium-, and thorium-bearing complex multiple oxides of niobium, tantalum, and titanium. All the metal-oxygen bonds in these minerals have about equal strength and about equal susceptibility to radiation damage. As it happens, most of these minerals have little resistance to metamictization and are commonly found in the metamict state.

Samarskite, brannerite, and columbite are examples of complex multiple oxide minerals. Samarskite is always totally metamict and compositionally altered with added water. It usually recrystallizes to a mixture of different minerals on heating, making its original structure and chemical formula hard to determine. Brannerite can range from partially to totally metamict. It has been found to be the primary site for uranium in some 1,400-million-year-old granites from the

southwestern United States. In those rocks, it was recrystallized as recently as 80 million years ago, but even in that geologically short period of time, it has become metamict again. Columbite is never more than partially metamict. Apparently, the presence of iron in its formula helps prevent metamictization from progressing as far as it does in other niobates. These minerals are generally found in coarse-grained pegmatites associated with granites. They may also be found as accessory minerals within the granites themselves, but even when they account for most of the rock's uranium or thorium, they are so rare that it takes special concentrating techniques simply to find a few grains.

Silicates are the largest group of minerals in the Earth's crust, but they account for only the second largest number of metamict minerals. They are characterized by having very strongly bonded silicon-oxygen groups in their crystal structures. Each silicon ion is surrounded by four oxygen ions in a tetrahedral (three-sided pyramid) shape. The chemical bonding between silicon and oxygen in the tetrahedral group is considerably stronger than the bonding between oxygen and any other metal ions in the structure and is more resistant to radiation damage. The only silicates susceptible to metamictization are those in which the tetrahedral groups are not linked to form strong chains, sheets, or networks that make the structure resistant to radiation damage. The most commonly occurring metamict mineral is zircon, a zirconium silicate with isolated tetrahedral groups. Only a small amount of uranium can substitute for zirconium in the structure, up to about 1.5 percent, but it is usually less than 0.5 percent. Up to about 0.1 percent of thorium can also be present. These small amounts are sufficient to cause zircon to occur in a wide range of degrees of metamictization. Zircon is only one member of a group of silicates that grow with essentially the same crystal structure and subsequently become metamict. Thorite (thorium silicate) and coffinite (uranium silicate) also belong to that group. They each require about the same amount of radiation damage as does zircon to become metamict, but because their concentrations of radioactive elements are much higher, metamictization occurs much more quickly.

Phosphates are the smallest group of metamict minerals, and they are usually found only partially metamict. One of particular interest is xenotime (yttrium phosphate), which has the same crystal structure as zircon. It commonly occurs in the same rocks as zircon, with equal or even greater contents of uranium and thorium, but is usually much less radiation damaged. Similarly, monazite (a rare-earth phosphate) has the same structure as huttonite, a mineral with the same composition as thorite, and takes up about the same amount of uranium and considerably more thorium than does zircon but is seldom more than slightly damaged. Apparently, the substitution of phosphorus for silicon in these structures makes the phosphates less susceptible to metamictization than are the equivalent silicates.

RADIATION DAMAGE TO MINERAL STRUCTURE

In order to understand metamictization fully, one must also understand how the radioactive decay of uranium and thorium damages crystal structures. In nature, uranium and thorium are both made up of a number of isotopes. Their most common isotopes are uranium 238, uranium 235, and thorium 232. Each of these isotopes decays, through a series of emissions of alpha particles, into an isotope of lead—lead 206, lead 207, and lead 208, respectively. The alpha particle is emitted from the decaying nucleus with great energy, and the emitting nucleus recoils simultaneously in the opposite direction. The energy transmitted to the mineral structure by the alpha particle and the recoiling nucleus is the major cause of radiation damage. The alpha particle has a very short range in the mineral structure (only about 20 wavelengths of light) and imparts most of its energy to it by ionizing the atoms it passes. Near the end of its path, when it has slowed down enough, it can collide with hundreds of atoms. The much larger recoil nucleus has a path that is about one thousandth as long, but it collides with ten times as many atoms. Thus, the greatest amount of radiation damage is caused by the recoil nucleus rather than by the alpha particle itself. Both particles introduce an intense amount of heat in a very small region of the structure, disrupting it but also increasing the rate at which the damage is spontaneously repaired. The accumulation of radiation damage depends on a balance between the damage and self-repair processes. In simple terms, radioactive minerals that remain crystalline have

high rates of self-repair, while those that become metamict do not.

Another contribution to radiation damage in uranium-bearing minerals is the spontaneous fission of uranium 238, in which its nucleus splits into two separate nuclei of lighter elements. That process is much more energetic than is alpha emission, but it happens at a much lower rate. It probably produces only about one-tenth to one-fifteenth the damage that alpha emission and alpha recoil cause. The radiation damage done by the decay of uranium and thorium is not always confined to the metamict mineral itself. The decay process also involves the emission of gamma rays that penetrate the surrounding rock, often causing the development of dark halos in the host minerals. Metamict grains are also often surrounded by a thin, rust-colored, iron-rich rim at their contact with other minerals. These rims are enriched in uranium and lead that have leaked out of the damaged grain as well as the iron from the surrounding rock. These phenomena draw dark outlines around radioactive grains and make them easy to spot in most granites and pegmatites.

X-RAY STUDIES

X-ray diffraction results from the reinforcement of X-ray reflections by repeated planes of atoms in a crystal. The diffraction pattern is characteristic of each crystal structure and depends on the spacing of the planes. The strength of diffraction peaks depends on the atomic density in the diffracting planes and on the regularity of the crystal structure. If the structure is damaged, both the intensity and the sharpness of the diffraction peaks decrease. Glasses produce no X-ray diffraction pattern, because there is no long-range order or regularity in their structures and thus no planes of atoms exist on which X rays can diffract. They are said to be "X-ray amorphous."

X-ray diffraction studies of metamict minerals have shown them to resemble glasses in being devoid of a regular crystal structure. In partially metamict samples, the spacing between planes increases and the regularity of the structure decreases with progressive radiation damage. For example, in some 570-million-year-old zircons, 0.25 percent uranium is enough to make them X-ray amorphous—completely metamict. Much older zircons from other localities with higher uranium contents, however, yield X-ray patterns that reveal them to be only slightly damaged. In those areas, a relatively recent geologic event caused the zircons to be completely recrystallized with partial loss of lead, produced by radioactive decay of uranium and thorium, but little or no loss of uranium or thorium themselves.

Once a mineral is X-ray amorphous, X-ray diffraction is not as effective as is X-ray absorption spectroscopy for studying its structure. The way that a material absorbs X rays depends on the spacing of the atoms in it. X-ray absorption spectroscopy has shown that the spacing between adjacent ions is little changed in metamict minerals, but the regularity at greater distances has been lost. It has also shown that among complex multiple oxides, their crystal structures are highly regular and easily distinguished, but their metamict structures are nearly indistinguishable. Thus, the oxides all approach the same glassy state on metamictization.

ELECTRON MICROSCOPY AND INFRARED SPECTROSCOPY

Electron diffraction from single grains of metamict minerals yields results very similar to those of X-ray diffraction techniques. For example, it has been shown that partially metamict zircons are made up of misaligned crystalline domains that are destroyed with progressive metamictization.

High-resolution electron microscopy allows the scientist to examine a mineral's structure directly. While actual atoms cannot be seen, the regularity of the array of atoms making up the structure can. Applied to zircons, this technique has shown that fission particles from the spontaneous fission of uranium 238 leave long tracks of disruption behind them as they pass through the crystal structure. In partially metamict zircons, highly disordered damaged patches (called domains) are interspersed with undamaged domains. As metamictization proceeds, the damaged domains begin to overlap and eventually wipe out all remaining crystalline domains.

Infrared spectroscopy is a powerful tool for studying the bonds between atoms in minerals. The absorption of infrared light by a crystal structure depends on the strength and regularity of bonds within the structure. Infrared study has shown that the silicon-oxygen bonds of the tetra-

hedral groups in zircon remain intact to a large extent even in metamict samples, while the regularity of virtually every zirconium-oxygen bond has been disturbed. Thus, silicon remains surrounded by four oxygen atoms even in the most metamict zircons when the regularity of the formerly crystalline structure has been destroyed. Infrared spectroscopy is also often used to study the occurrence of water in minerals. Because most metamict minerals are enriched in water to some extent, its role in producing or stabilizing the metamict state is a question of some interest. Infrared studies of zircons in a wide range of metamict states show that while there is no water in their structures, small amounts of hydroxyl are common. The existence of "dry" zircons in all states from crystalline to metamict, however, shows that neither water nor hydroxyl is necessary to the metamictization process. The tendency for metamict zircons to have more hydroxyl than do less damaged samples suggests that hydroxyl enters zircons only after metamictization has opened up their structures sufficiently for it to diffuse in.

ROLE IN ECONOMY AND PUBLIC SAFETY

Metamict minerals are important to people for two principal reasons. First, some gemstones, such as zircon, occur in the metamict state, and several others, such as topaz, can be enhanced in appearance by irradiation. Metamict gemstones are often considerably more valuable than the crystalline varieties, because the anisotropic optical properties of the crystalline gems make them look "fuzzy" inside and less beautiful than the isotropic metamict stones. Radiation damage also often imparts color to the gemstone, increasing its value. (Artificial means of imparting color to gemstones by irradiation have been developed, but the gem often fades back to its original color with time. Consequently, it is best to know the entire history of a colored gem, or have it examined by a knowledgeable and trustworthy expert, before investing large amounts of money in it.)

Second, and ultimately more important, the understanding of metamictization may someday be invaluable to the safe disposal of high-level nuclear wastes. Ways must be found to keep high-level wastes from leaking into the environment and poisoning groundwater for a period of at least ten thousand years—after which they will have lost most of their radioactivity. The only way that scientists can determine what could conceivably lock up hazardous, highly radioactive atoms for that length of time is to examine results of long-term experiments on potentially leak-proof containers—solid materials that could be grown from the radioactive wastes that would be impervious to alteration, breakdown, or damage. The "experiments" of that duration are those that have already occurred in nature: radioactive minerals that have been damaged internally for millennia. Geochemists have found that some minerals retain their radioactive elements over millions of years despite metamictization, while others do not. Synthetic analogues of the ones that do may be grown from a mixture of the radioactive elements and added compounds to produce artificial rocks that are resistant to leaching and that have the potential to trap the hazardous substances for the required time.

James A. Woodhead

Cross-References

Aluminum Deposits, 1539; Biopyriboles, 1175; Carbonates, 1182; Clays and Clay Minerals, 1187; Feldspars, 1193; Gem Minerals, 1199; Hydrothermal Mineralization, 1205; Ionic Substitution, 1209; Iron Deposits, 1602; Magnetite, 1214; Minerals: Physical Properties, 1225; Minerals: Structure, 1232; Mining Processes, 1780; Mining Wastes, 1786; Nickel-Irons, 2718; Non-silicates, 1238; Orthosilicates, 1244; Oxides, 1249; Radioactive Minerals, 1255; Rock Magnetism, 177; Sedimentary Mineral Deposits, 1629; Stony Irons, 2724.

Bibliography

Cavey, Christopher. *Gems and Jewels*. London: Studio, 1992. Cavey sets out not only to provide information about gems and jewels but also to correct many of the misconceptions concerning their study. Although the book includes many illustrations, only a few of them are color.

Deer, W. A., R. A. Howie, and J. Zussman. *Rock-Forming Minerals*. Vol. 1A, *Orthosilicates*. 2d ed. New York: John Wiley & Sons, 1988.

_____. *Rock-Forming Minerals*. Vol. 1B, *Disilicates and Ring Silicates*. 2d ed. New York: John Wiley & Sons, 1988.

_____. *Rock-Forming Minerals*. Vol. 5, *Non-Silicates*. 2d ed. New York: John Wiley & Sons, 1988. These are part of a five-volume set that describes the most important rock-forming minerals. Metamict minerals are covered in the three volumes listed: oxides and phosphates in volume 5, silicates in the expanded volumes 1A and 1B. The discussion of metamictization in the volumes for zircon and allanite is especially lucid and complete. As a whole, the work is well written, very well illustrated, and well indexed—an indispensable reference for geologists. Suitable for college-level students.

Fleischer, M. *Glossary of Mineral Species*. 5th ed. Tucson, Ariz.: The Mineralogical Record, 1987. This glossary is an alphabetical listing of all currently accepted mineral names and formulas, along with the now-accepted assignments of discredited names. Includes separate lists of important mineral groups as well as the chemical word-formulas for several minerals. Does not include a list of minerals by chemical contents. Not indexed or illustrated and has a minimum of text. Suitable for anyone needing the formula of a particular mineral, but less suitable for identifying a mineral from its composition.

Hurlbut, C. S., Jr., and G. S. Switzer. *Gemology*. New York: John Wiley & Sons, 1979. A well-illustrated and complete introduction to gemstones, describing the methods of study of gems. A good introduction to mineralogy for the nonscientist, this volume includes a section with descriptions of minerals and other materials prized as gems. Suitable for high-school-level readers.

Mitchell, R. S. "Metamict Minerals: A Review, Part 1. Chemical and Physical Characteristics, Occurrence." *The Mineralogical Record* 4 (July/August, 1973): 177.

_____. "Metamict Minerals: A Review, Part 2. Origin of Metamictization, Methods of Analysis, Miscellaneous Topics." *The Mineralogical Record* 4 (September/October, 1973):

214. This pair of articles on metamict minerals covers the process of metamictization in great depth without becoming overly technical. *The Mineralogical Record* is written for lay mineral and gem collectors as well as for Earth scientists. Its editorial aim is to cover the famous mineral and gem localities of the world with state-of-the-art photography. Several issues have become collector's items. Suitable for anyone interested in gems or minerals.

Smith, David G., ed. *The Cambridge Encyclopedia of Earth Sciences.* New York: Crown Publishers, 1981. Chapter 5, "Earth Materials: Minerals and Rocks," gives a good description of the areas of mineralogy essential to an understanding of metamictization, although there is no discussion of metamict minerals. The text is suitable for college-level readers not intimidated by somewhat technical language. A well-illustrated and carefully indexed reference volume.

Webster, Robert F. G. A. *Gems: Their Sources, Descriptions, and Identification.* 5th ed. Boston: Butterworth-Heinemann, 1994. A technical college-level text that is an important reference for the gemologist or the senior jeweler. Reflects the British attention to detail. Better for sources and descriptions than for identification. Essential for the gemologist.

MINERALS: PHYSICAL PROPERTIES

Many minerals are readily identified by their physical properties, but identification of other minerals may require instruments designed to examine details of their chemical composition or crystal structure. The characteristic physical properties of some minerals, such as hardness, malleability, and ductility, make them commercially useful.

PRINCIPAL TERMS

CLEAVAGE: the tendency for minerals to break in smooth, flat planes along zones of weaker bonds in their crystal structure

CRYSTAL: a solid bounded by smooth planar surfaces that are the outward expression of the internal arrangement of atoms; crystal faces on a mineral result from precipitation in a favorable environment

DENSITY: in an informal sense, the relative weight of mineral samples of equal size; it is defined as mass per unit volume

LUMINESCENCE: the emission of light by a mineral

LUSTER: the reflectivity of the mineral surface; there are two major categories of luster: metallic and nonmetallic

MOHS HARDNESS SCALE: a series of ten minerals arranged in order of increasing hardness, with talc as the softest mineral known (1) and diamond as the hardest (10)

TENACITY: the resistance of a mineral to bending, breakage, crushing, or tearing

COLOR, STREAK, LUSTER

Minerals have diagnostic physical properties resulting from their chemistry and crystal structure. Physical properties of minerals include color, streak, luster, crystal shape, cleavage, fracture, hardness, and density, or specific gravity. Several minerals have additional diagnostic physical properties, including tenacity, magnetism, luminescence, and radioactivity. Some minerals, notably halite (common table salt), are easily identified by their taste. Other minerals, such as calcite, effervesce or fizz when they come into contact with hydrochloric acid.

Color is an obvious physical property, but it is one of the least diagnostic for mineral identification. In some minerals, color results from the presence of major elements in the chemical formula; in these minerals, color is a diagnostic property. For example, malachite is always green, azurite is always blue, and rhodochrosite and rhodonite are always pink. In other minerals, color is the result of trace amounts of chemical impurities or of defects in the crystal lattice structure. Depending on the impurities, a particular species of mineral can have many different colors. For example, pure quartz is colorless, but quartz may be white (milky quartz), pink (rose quartz), purple (amethyst), yellow (citrine), brown (smoky quartz), green, blue, or black.

Streak is the color of the mineral in powdered form. Streak is more definitive than mineral color, because although a mineral may have several color varieties, the streak will be the same for all. Streak is best viewed after rubbing the mineral across an unglazed porcelain tile. The tile has a hardness of approximately 7, so minerals with a hardness of greater than 7 will not leave a streak, although their powdered color may be studied by crushing a small piece.

Luster refers to the reflectivity of the mineral surface. There are two major categories of luster: metallic and nonmetallic. Metallic minerals include metals (such as native copper and gold), as well as many metal sulfides, such as pyrite and galena. Nonmetallic lusters can be described as vitreous or glassy (characteristic of quartz and olivine), resinous (resembling resin or amber, characteristic of sulfur and some samples of sphalerite), adamantine or brilliant (diamond), greasy (appearing as if covered by a thin film of oil, including nepheline and some samples of massive quartz), silky (in minerals with parallel fibers, such as malachite, chrysotile asbestos, or fibrous gypsum), pearly (similar to an iridescent pearl-like shell,

PHYSICAL PROPERTIES OF MINERALS

Property	Explanation
Chemical composition	Chemical formula that defines the mineral
Cleavage	Tendency to break in smooth, flat planes along zones of weak bonding; depends on structure
Color	Depends on presence of major elements in the chemical composition; may be altered by trace elements or defects in structure; often not definitive
Crystal shape	Outward expression of the atomic crystal structure
Crystal structure	Three-dimensional ordering of the atoms that form the mineral
Density	Mass per unit volume (grams per cubic centimeter)
Electrical properties	Properties having to do with electric charge; quartz, for example, is piezoelectric (emits charge when squeezed)
Fracture	Tendency for irregular breakage (not along zones of weak bonding)
Hardness	Resistance of mineral to scratching or abrasion; measured on a scale of 1-10 (Mohs hardness scale)
Luminescence	Emission of electromagnetic waves from mineral; some minerals are fluorescent, some thermoluminescent
Luster	Reflectivity of the surface; may be either metallic or nonmetallic
Magnetism	Degree to which mineral is attracted to a magnet
Radioactivity	Instability of mineral; radioactive minerals are always isotopes
Specific gravity	Relative density: ratio of weight of substance to weight of equal volume of water at 4 degrees Celsius
Streak	Color of powdered form; more definitive than color
Taste	Salty, bitter, etc.; applies only to some minerals
Tenacity	Resistance to bending, breakage, crushing, tearing: termed as brittle, malleable, ductile, sectile, flexible, or elastic

such as talc), and earthy or dull (as in clays).

CRYSTAL SHAPE, CLEAVAGE, FRACTURE

Crystal shape is the outward expression of the internal three-dimensional arrangement of atoms in the crystal lattice. Crystals are formed in a cooling or evaporating fluid as atoms begin to slow down, move closer, and bond together in a particular geometric gridwork. If minerals are unconfined and free to grow, they will form well-shaped, regular crystals. On the other hand, if growing minerals are confined by other, surrounding minerals, they may have irregular shapes. Some of the common shapes or growth habits of crystals include acicular (or needlelike, such as natrolite), bladed (elongated and flat like a knife blade, such as kyanite), blocky (equidimensional and cubelike, such as galena and fluorite), and columnar or prismatic (elongated or pencil-like, such as quartz and tourmaline). Other crystal shapes are described as pyramidal, stubby, tabular, barrel-shaped, or capillary.

Cleavage is the tendency for minerals to break in smooth, flat planes along zones of weaker bonds in their crystal structure. Cleavage is one of the most important physical properties in identifying minerals because it is so closely related to the internal crystal structure. Cleavage is best developed in minerals that have particularly weak chemical bonds in a given direction. In other minerals, differences in bond strength are less pronounced, so cleavage is less well developed. Some minerals have no planes of weakness in their crystal structure; they lack cleavage and do not break along planes. Cleavage can occur in one direction (as in the micas, muscovite, and biotite) or in more than one direction. The number and orientation of the cleavage planes are always the same for a particular mineral. For example, orthoclase feldspar has two directions of cleavage at right angles to each other.

Fracture is irregular breakage that is not controlled by planes of weakness in minerals. Conchoidal fracture is a smooth, curved breakage surface, commonly marked by fine concentric lines, resembling the surface of a shell. Conchoidal fracture is common in broken glass and quartz. Fibrous or splintery fracture occurs in asbestos and sometimes in gypsum. Hackly facture is jagged with sharp edges and occurs in native copper. Uneven or irregular fracture produces rough, irregular breakage surfaces.

Hardness, Density, Specific Gravity

Hardness is the resistance of a mineral to scratching or abrasion. Hardness is a result of crystal structure; the stronger the bonding forces between the atoms, the harder the mineral. The Mohs hardness scale, devised by a German mineralogist, Friedrich Mohs, in 1822, is a series of ten minerals arranged in order of increasing hardness. The minerals on the Mohs hardness scale are talc (the softest mineral known), gypsum, calcite, fluorite, apatite, potassium feldspar (orthoclase), quartz, topaz, corundum, and diamond (the hardest mineral known). A mineral with higher hardness number can scratch any mineral of equal or lower hardness number. The relative hardness of a mineral is easily tested using a number of common materials, including the fingernail (a little over 2), a copper coin (about 3), a steel nail or pocket knife (a little over 5), a piece of glass (about 5.5), and a steel file (6.5).

Density is defined as mass per unit volume (typically measured in terms of grams per cubic centimeter). In a very informal sense, density refers to the relative weight of samples of equal size. Quartz has a density of 2.6 grams per cubic centimeter, whereas a "heavy" mineral such as galena has a density of 7.4 grams per cubic centimeter (about three times as heavy). Specific gravity (or relative density) is the ratio of the weight of a substance to the weight of an equal volume of water at 4 degrees Celsius.

The terms "density" and "specific gravity" are sometimes used interchangeably, but density requires the use of units of measure (such as grams per cubic centimeter), whereas specific gravity is unitless. Specific gravity is an important aid in mineral identification, particularly when studying valuable minerals or gemstones, which might be damaged by other tests of physical properties. The specific gravity of a mineral depends on the chemical composition (type and weight of atoms) as well as the manner in which the atoms are packed together.

Tenacity

Tenacity is the resistance of a mineral to bending, breakage, crushing, or tearing. A mineral may be brittle (breaks or powders easily), malleable (may be hammered out into thin sheets), ductile (may be drawn out into a thin wire), sectile (may be cut into thin shavings with a knife), flexible (bendable, and stays bent), and elastic (bendable, but returns to its original form). Minerals with ionic bonding, such as halite, tend to be brittle. Malleability, ductility, and sectility are diagnostic of minerals with metallic bonding, such as gold. Chlorite and talc are flexible, and muscovite is elastic.

Magnetism and Luminescence

Magnetism causes minerals to be attracted to magnets. Magnetite and pyrrhotite are the only common magnetic minerals, and they are called ferromagnetic. Lodestone, a type of magnetite, is a natural magnet. When in a strong magnetic field, some minerals become weakly magnetic and are attracted to the magnet; these minerals are called paramagnetic. Examples of paramagnetic minerals include garnet, biotite, and tourmaline. Other minerals are repelled by a magnetic field and are called diamagnetic minerals. Examples of diamagnetic minerals include gypsum, halite, and quartz.

Luminescence is the term for emission of light by a mineral. Minerals that glow or luminesce in ultraviolet light, X rays, or cathode rays are fluorescent minerals. The glow is the result of the mineral changing invisible radiation to visible light, which happens when the radiation is absorbed by the crystal lattice and then reemitted by the mineral at lower energy and longer wavelength. Fluorescence occurs in some specimens of a mineral but not all. Examples of minerals which may fluoresce include fluorite, calcite, diamond, scheelite, willemite, and scapolite. Some minerals will continue to glow or emit light after the radiation source is turned off; these minerals are phosphorescent. Some minerals glow when heated, a property called thermoluminescence, present in some specimens of fluorite, calcite, apatite, scapolite, lepidolite, and feldspar. Other minerals luminesce when crushed, scratched, or rubbed, a property called triboluminescence, present in some specimens of sphalerite, corundum, fluorite, and lepidolite and, less commonly, in feldspar and calcite.

RADIOACTIVITY AND ELECTRICAL CHARACTERISTICS

Radioactive minerals contain unstable elements that alter spontaneously to other kinds of elements, releasing subatomic particles and energy. Some elements come in several different forms, differing by the number of neutrons present in the nucleus. These different forms are called isotopes, and one isotope of an element may be unstable (radioactive), whereas another isotope may be stable (not radioactive). Radioactive isotopes include potassium 40, rubidium 87, thorium 232, uranium 235, and uranium 238. Ex-

amples of radioactive minerals include urananite and thorianite.

Some minerals also have interesting electrical characteristics. Quartz is a piezo-electric mineral, meaning that when squeezed, it produces electrical charges. Conversely, if an electrical charge is applied to a quartz crystal, it will change shape and vibrate as internal stresses develop. The oscillation of quartz is the basis for its use in digital quartz watches.

STUDY TECHNIQUES

In many cases, the physical properties of minerals can be studied using relatively common, inexpensive tools. The relative hardness of a mineral may be determined by attempting to scratch one mineral with another, thereby bracketing the unknown mineral's hardness between that of other minerals on the Mohs hardness scale. Streak may be determined by rubbing the mineral across an unglazed porcelain tile to observe the color (and sometimes odor) of the streak, if any. Color and luster are determined simply by observing the mineral.

The angles between adjacent crystal faces may be measured using a goniometer. There are several types of goniometers, but the simplest is a protractor with a pivoting bar, which is held against a large crystal so that the angles between faces can be measured. There are also reflecting goniometers, which operate by measuring the angles between light beams reflected from crystal faces.

MINERAL COMPOSITION OF COMMON ROCKS

Mineral (percentages)	Granite	Basalt	Amphibolite	Schist	Shale	Sandstone	Limestone
Quartz	30	—	—	32	17	97	3
Alkali feldspar	60	5	—	—	—	1	1
Plagioclase	5	45	42	18	—	—	—
Pyroxene	—	40	—	—	—	—	—
Amphibole	—	—	50	—	—	—	—
Olivine	—	5	—	—	—	—	—
Biotite	4	—	5	7	—	—	—
Muscovite	—	—	—	38	1	1	—
Magnetite	1	5	3	3	1	1	1
Staurolite	—	—	—	2	—	—	—
Clay minerals	—	—	—	—	80	—	1
Calcite	—	—	—	—	1	94	—

SOURCE: Data are from *McGraw-Hill Encyclopedia of Earth Sciences*, ed. Sybil P. Parker (2d ed., McGraw-Hill, Inc., 1987).

Density can be determined by measuring the mass of a mineral and determining its volume (perhaps by measuring the amount of water it displaces in a graduated cylinder), then dividing these two measurements. Specific gravity (SG) is usually determined by first weighing the mineral in air and then weighing it while it is immersed in water. When immersed in water, it weighs less because it is buoyed up by a force equivalent to the weight of the water displaced. A Jolly balance, which works by stretching a spiral spring, can measure specific gravity. For tiny mineral specimens weighing less than 25 milligrams, a torsion balance (or Berman balance) is useful for accurate determinations. Heavy liquids, such as bromoform and methylene iodide, are also used to determine the specific gravity of small mineral grains. The mineral grain is placed into the heavy liquid and then acetone is added to the liquid until the mineral grain neither floats nor sinks (that is, until the specific gravity of the mineral and the liquid are the same). Then, the specific gravity of the liquid is determined using a Westphal balance.

An ultraviolet light source (with both long and short wavelengths) is used to determine whether minerals are fluorescent or phosphorescent. A portable ultraviolet light can be used to prospect for fluorescent minerals. Thermoluminescence can be triggered by heating a mineral to 50-100 degrees Celsius. Radioactivity is measured using a Geiger counter or scintillometer.

COMMERCIAL APPLICATIONS

The physical properties of minerals affect their usefulness for commercial applications. Minerals with great hardness, such as diamonds, corundum (sapphire and ruby), garnets, and quartz, are useful as abrasives and in cutting and drilling equipment. Other minerals are useful because of their softness, such as calcite, which is used in cleansers because it will not scratch the surface being cleaned. Also, calcite in the form of marble is commonly used for sculpture because it is relatively soft and easy to carve. Talc (hardness 1) is used in talcum powder because of its softness.

Some metals, such as copper, are ductile, which makes them useful for the manufacture of wire. Copper is one of the best electrical conductors. Copper is also a good conductor of heat and is often used in cookware. Gold is the most malleable and ductile mineral. Because of its malleability, gold can be hammered into sheets so thin that 300,000 of them would be required to make a stack 1 inch high. Because of its ductility, 1 gram of gold (about the weight of a raisin) can be drawn into a wire more than a mile and a half long. Gold is the best conductor of heat and electricity known, but it is generally too expensive to use as a conductor.

Other minerals are valuable because they do not conduct heat or electricity. They are used as electrical insulators or for products subjected to high temperatures. For example, kyanite, andalusite, and sillimanite are used in the manufacture of spark plugs and other high-temperature porcelains. Muscovite is also useful because of its electrical and heat-insulating properties; sheets of muscovite are often used as an insulating material in electrical devices.

Cleavage is the property responsible for the use of graphite as a dry lubricant and in pencils. Graphite has perfect cleavage in one direction, and is slippery because microscopic sheets of graphite slide easily over one another. A "lead" pencil is actually a mixture of graphite and clay; it writes by leaving tiny cleavage flakes of graphite on the paper.

The color of the streak (or crushed powder) of many minerals makes them valuable as pigments. Hematite has a red streak and is used in paints and cosmetics. Silver is commonly used in photographic films and papers because in the form of silver halide, it is light-sensitive and turns black. After developing and fixing, metallic silver remains on the film to form the negative.

The uranium-bearing minerals (urananite, carnotite, torbernite, and autunite) are used as sources of uranium, which is important because its nucleus is susceptible to fission (splitting or radioactive disintegration), producing tremendous amounts of energy. This energy is used in nuclear power plants for generating electricity. Pitchblende, a variety of urananite, is a source of radium, which is used as a source of radioactivity in industry and medicine. The high specific gravity of barite makes it a useful additive to drilling muds to prevent oil well gushers or blowouts. It is also opaque to X rays and is used in medicine for "barium milkshakes" before patients are X-rayed so that the digestive tract will show up clearly.

Pamela J. W. Gore

CROSS-REFERENCES

Biopyriboles, 1175; Carbonates, 1182; Clays and Clay Minerals, 1187; Earth's Oldest Rocks, 516; Electron Microscopy, 463; Feldspars, 1193; Fission Track Dating, 522; Gem Minerals, 1199; Groundwater Pollution and Remediation, 2037; Hydrothermal Mineralization, 1205; Infrared Spectra, 478; Ionic Substitution, 1209; Magnetite, 1214; Metamictization, 1219; Minerals: Structure, 1232; Non-silicates, 1238; Nuclear Waste Disposal, 1791; Orthosilicates, 1244; Oxides, 1249; Radioactive Decay, 532; Radioactive Minerals, 1255; Water-Rock Interactions, 449; X-ray Powder Diffraction, 504.

BIBLIOGRAPHY

Blackburn, W. H., and W. H. Dennen. *Principles of Mineralogy.* Dubuque, Iowa: Wm. C. Brown, 1988. This book is divided into three parts. The first part is theoretical and includes crystallography and crystal chemistry, along with a section on the mineralogy of major types of rocks. The second part is practical and includes chapters on physical properties of minerals, crystal geometry, optical properties, and methods of analysis. The third part contains systematic mineral descriptions. Designed for an introductory college course in mineralogy, but should be useful for amateurs as well.

Bloss, F. D. *An Introduction to the Methods of Optical Crystallography.* New York: Holt, Rinehart and Winston, 1961. This book is for the advanced student of mineralogy who is interested in the ways in which the crystal structure of a mineral changes the characteristics of a beam of light passing through it, as studied with the petrographic microscope. Theoretical, it may be of interest to persons with a background in physics or geology.

Cepeda, Joseph C. *Introduction to Minerals and Rocks.* New York: Macmillan, 1994. Cepeda's text provides a good introduction to the structure of minerals for students just beginning their studies in the Earth sciences. Includes illustrations and maps.

Chesterman, C. W., and K. E. Lowe. *The Audubon Society Field Guide to North American Rocks and Minerals.* New York: Alfred A. Knopf, 1988. This book contains 702 color photographs of minerals, grouped by color, as well as nearly a hundred color photographs of rocks. All the mineral photographs are placed at the beginning of the book, and descriptive information follows, with the minerals grouped by chemistry. Distinctive features and physical properties are listed for each of the minerals, and information on collecting localities is also given. A section at the back of the book discusses various types of rocks. A glossary is also included. Suitable for the layperson.

Deer, William A., R. A. Howie, and J. Zussman. *An Introduction to Rock-Forming Minerals.* 2d ed. New York: Longman, 1992. A standard reference on mineralogy for advanced college students and above. Each chapter contains detailed descriptions of chemistry and crystal structure, usually with chemical analyses. Discussions of chemical variations in minerals are extensive.

Desautels, P. E. *The Mineral Kingdom.* New York: Ridge Press, 1968. An oversize, lavishly illustrated coffee-table book with useful text supplementing the color photographs. It covers how minerals are formed, found, and used and includes legends about minerals and gems as well as scientific data. Provides a broad introduction to the field of mineralogy. Intended for the amateur, it also makes fascinating reading for the professional geologist.

Frye, Keith. *Mineral Science: An Introductory Survey.* New York: Macmillan, 1993. This basic text, intended for the college-level reader, provides an easily understood overview of mineralogy, petrology, and geochemistry, including descriptions of specific minerals. Illustrations, bibliography, and index.

_____. *Modern Mineralogy.* Englewood Cliffs, N.J.: Prentice-Hall, 1974. This book addresses minerals from a chemical standpoint and includes chapters on crystal chemistry, structure, symmetry, physical properties, radiant energy and crystalline matter, the phase rule, and mineral genesis. Designed as an advanced college textbook for a student who already has some familiarity with mineralogy. Includes short descriptions of minerals in a table in the appendix.

Hammond, Christopher. *The Basics of Crystallography and Diffraction.* New York: Oxford University Press, 1997. Hammond's book covers crystal form, atomic structure, physical properties of minerals, and X-ray methods. Illustrations help clarify some of the more mathematically complex concepts. Includes bibliography and index.

Kerr, P. F. *Optical Mineralogy.* New York: McGraw-Hill, 1977. This book is designed to instruct advanced mineralogy students in the study and identification of minerals using a petrographic microscope. The first part concerns the basic principles of optical mineralogy, and the second part details the optical properties of a long list of minerals.

Klein, C., and C. S. Hurlbut, Jr. *Manual of Mineralogy.* 21st ed. New York: John Wiley & Sons, 1999. One of a series of revisions of the original mineralogy textbook written by James D. Dana in 1848. The first part of the book is dedicated to crystallography and crystal chemistry, with shorter chapters on the physical and optical properties of minerals. The second part of the book provides a classification and detailed, systematic description of various types of minerals, with sections on gem minerals and mineral associations. Considered to be the premier mineralogy textbook for college-level geology students; many parts of it will be useful for amateurs.

Lima-de-Faria, Josae. *Structural Minerology: An Introduction.* Dordrecht, the Netherlands: Kluwer, 1994. This book provides a good college-level introduction to the basic concepts of crystal structure and the classification of minerals. Illustrations, extensive bibliography, index, and a table of minerals on a folded leaf.

Mottana, A., R. Crespi, and G. Liborio. *Simon & Schuster's Guide to Rocks and Minerals.* New York: Simon & Schuster, 1978. This book is fully illustrated with color photographs of 276 minerals. It provides background information on physical properties, environment of formation, occurrences, and uses of each mineral. A sixty-page introduction to minerals provides sophisticated technical coverage that will be of interest to both mineralogy students and amateurs. The last part of the book illustrates and describes one hundred types of rocks. A glossary is also included.

Pough, F. H. *A Field Guide to Rocks and Minerals.* Boston: Houghton Mifflin, 1976. This well-written and well-illustrated book is suitable for readers of nearly any age and background. One of the most readable and accessible sources, it provides a fairly complete coverage of the minerals. Designed for amateurs.

Zussman, J., ed. *Physical Methods in Determinative Mineralogy.* New York: Academic Press, 1967. A reference book that describes technical methods used in the study of rocks and minerals, including transmitted and reflected light microscopy, electron microscopy, X-ray fluorescence spectroscopy, X-ray diffraction, electron microprobe microanalysis, and atomic absorption spectroscopy. Written for the professional geologist or advanced student.

MINERALS: STRUCTURE

The discovery of the internal structures of minerals by the use of X-ray diffraction was pivotal in the history of mineralogy and crystallography. X-ray analysis revealed that the physical properties and chemical behavior of minerals are directly related to the highly organized arrangements of their atoms, and this knowledge has had important scientific and industrial applications.

PRINCIPAL TERMS

CLEAVAGE: the capacity of crystals to split readily in certain directions

CRYSTAL: externally, a solid material of regular form bounded by flat surfaces called faces; internally, a substance whose orderly structure results from a periodic three-dimensional arrangement of atoms

ION: an electrically charged atom or group of atoms

IONIC BOND: the strong electrical forces holding together positively and negatively charged atoms

MINERAL: a naturally formed inorganic substance with characteristic physical properties, a definite chemical composition, and, in most cases, a regular crystal structure

X RAY: radiation interpretable in terms of either very short electromagnetic waves or highly energetic photons (light particles)

THE FATHER OF CRYSTALLOGRAPHY

During the seventeenth and eighteenth centuries, natural philosophers used two basic ideas to explain the external structures of minerals: particles, in the form of spheres, ellipsoids, or various polyhedra; and an innate attractive force, or the emanating "glue" needed to hold particles together. These attempts to rationalize mineral structures still left a basic problem unanswered: How does one explain the heterogeneous physical and chemical properties of minerals with homogeneous particles? This problem was not answered satisfactorily until the twentieth century, but the modern answer grew out of the work of scientists in the eighteenth and nineteenth centuries. The most important of these scientists was René-Just Haüy, often called the Father of Crystallography.

Haüy, a priest who worked at the Museum of Natural History in Paris, helped to make crystallography a science. Before Haüy, the science of crystals had been a part of biology, geology, or chemistry; after Haüy, the science of crystals was an independent discipline. His speculations on the nature of the crystalline state were stimulated when he accidentally dropped a calcite specimen, which shattered into fragments. He noticed that the fragments split along straight planes that met at constant angles. No matter what the shape of the original piece of calcite, he found that broken fragments were rhombohedra (slanted cubes). He reasoned that a rhombohedron, similar to the ones he obtained by cleaving the crystal, must be preformed in the inner structure of the crystal. For Haüy, then, the cleavage planes existed in the crystal like the mortar joints in a brick wall. When he discovered similar types of cleavage in a variety of substances, he proposed that all crystal forms could be constructed from submicroscopic building blocks. He showed that there were several basic building blocks, which he called primitive forms or "integral molecules," and they represented the last term in the mechanical division of a crystal. With these uniform polyhedra, he could rationalize the many mineral forms observed in nature. Haüy's building block was not, however, the same as what later crystallographers came to call the unit cell, the smallest group of atoms in a mineral that can be repeated in three directions to form a crystal. The unit cell is not a physically separable entity, such as a molecule; it simply describes the repeat pattern of the structure. On the other hand, for Haüy, the crystal was a periodic arrangement of equal molecular polyhedra, each of which might have an independent existence.

GEOMETRICAL AND SYMMETRICAL ANALYSES

In the nineteenth century, Haüy's ideas had many perceptive critics. For example, Eilhardt Mitscherlich, a German chemist, discovered in 1819 that different mineral substances could have the same crystal form, whereas Haüy insisted that each substance had a specific crystal structure. Some crystallographers shunned the concrete study of crystals (leaving it to mineralogists), and they defined their science as the study of ordered space. This mathematical analysis bore fruit, for crystallographers were able to show that, despite great variety of possible mineral structures, all forms could be classified into six crystal systems on the basis of certain geometrical features, usually axes. The cube is the basis of one of these systems, the isometric, in which three identical axes intersect at right angles. Symmetry was another factor in describing these crystal systems. For example, a cube has fourfold symmetry around an axis passing at right angles through the center of any of its faces. As some crystallographers were establishing the symmetry relationships in crystal systems, others were working on a way to describe the position of crystal faces. In 1839, William H. Miller, a professor of mineralogy at the University of Cambridge, found a way of describing how faces were oriented about a crystal, similar to the way a navigator uses latitude and longitude to tell where his ship is on the Earth. Using numbers derived from axial proportions, Miller was able to characterize the position of any crystal face.

Friedrich Mohs, a German mineralogist best remembered for his scale of the hardness of minerals, was famous in his lifetime for his system of mineral classification, in which he divided minerals into genera and species, similar to the way biologists organized living things. His system was based on geometrical relationships that he derived from natural mineral forms. He wanted to transform crystallography into a purely geometrical science, and he showed that crystal analysis involved establishing certain symmetrical groups of points by the rotation of axes. When these point groups were enclosed by plane surfaces, crystal forms were generated. The crystallographer's task, then, was to analyze the symmetry operations characterizing the various classes of a crystal system.

BRAVAIS LATTICES

Beginning in 1848, Auguste Bravais, a French physicist, took the same sort of mathematical approach in a series of papers dealing with the kinds of geometric figures formed by the regular grouping of points in space, called lattices. Bravais applied the results of his geometric analysis to crystals, with the points interpreted either as the centers of gravity of the chemical molecules or as the poles of interatomic electrical forces. With this approach, he demonstrated that there is a maximum of fourteen kinds of lattices, which differ in symmetry and geometry, such that the environment around any one point is the same as that around every other point. These fourteen Bravais lattices are distributed among the six crystal systems. For example, the three isometric Bravais lattices are the simple (with points at the vertices of a cube), body-centered (with points at the corners along with a point at the center of a cube), and face-centered (with corner points and points at the centers of the faces of the cube). With the work of Bravais, the external symmetry of a mineral became firmly grounded on the idea of the space lattice, but just how actual atoms or molecules were arranged within unit cells remained a matter of speculation.

In the latter part of the nineteenth century, various European scientists independently advanced crystallography beyond the point groups of the Bravais lattices by recognizing that the condition of translational equivalence was a restriction justified only by an external consideration of points. The condition of translational equivalence means that if one found oneself within a lattice and could move from point to point, one would find the same view of one's surroundings from each position. In the late 1880's, the Russian mineralogist Evgraf Federov introduced the glide plane, in which a reflection in a mirror plan is combined with a translation without rotation along an axis. Using various symmetry elements, Federov derived the 230 space groups, which represent all possible distributions that atoms can assume in minerals.

PAULING'S RULES

Shortly after the work of Federov, William Barlow, an English chemist, began to consider the problem of crystal symmetry from a more concrete point of view. He visualized crystals not in terms of points but in terms of closely packed spherical atoms with characteristic diameters. In considering atoms to be specifically sized spheres, he found that there are certain geometric arrangements for packing them efficiently. One can appreciate his insight by thinking about arranging coins in two dimensions. For example, six quarters will fit around a central quarter, but only five quarters will fit around a dime. Barlow showed that similar constraints hold for the three-dimensional packing of spherical atoms of different diameters.

As scientists determined more and more mineral structures, they became convinced that minerals are basically composed of spherical atoms or ions, each of characteristic size, packed closely together. For example, the silicate minerals were of central concern to William Lawrence Bragg in England and Linus Pauling in the United States. The basic unit in these minerals is the tetrahedral arrangement of four oxygen atoms around a central silicon atom. Each tetrahedral unit has four negative charges, and so one would expect that electric repulsion would force these tetrahedral building blocks to fly apart. In actual silicate minerals, however, these units are linked, in chains or rings or sheets, in ways that bring about charge neutralization and stability. These tetrahedra may also be held together by such positively charged metal ions as aluminum, magnesium, and iron. These constraints lead to a fascinating series of structures. Pauling devised an enlightening and useful way of thinking both about these silicate structures and about complex inorganic substances in general. In the late 1920's, he proposed a set of principles (now known as Pauling's Rules) that govern the structures of ionic crystals, that is, crystals in which ionic bonding predominates. The silicate minerals provide striking examples of his principles. One of his rules deals with how a positive ion's electrical influence is spread among neighboring negative ions; another rule states that highly charged positive ions tend to be as far apart as possible in a structure. Pauling's Rules allowed him to explain why certain silicate minerals exist in nature and why others do not.

TEMPERATURE AND PRESSURE STUDIES

In the 1960's the structure determination of minerals became an important activity in some large geology departments. By this time, through computerized X-ray crystallography, it was possible to determine, quickly and elegantly, the exact atomic positions of highly complex minerals. In the 1970's and 1980's, scientific interest shifted to the study of minerals at elevated temperatures and pressures. These studies often showed that temperature and pressure changes cause complex internal structural modifications of the mineral, including shifts in distances between certain ions and in their orientation to others. New minerals continue to be discovered and their structures determined. Structural chemistry has played an important role in deepening understanding both of these new minerals and of old minerals under stressful conditions. This knowledge of mineral structure has benefited not only mineralogists, crystallographers, and structural chemists but also inorganic chemists, solid-state physicists, and many Earth scientists.

STUDY OF EXTERNAL STRUCTURE

The first methods of examining the external structure of minerals were quite primitive. In the seventeenth century, Nicolaus Steno cut sections from crystals and traced their outlines on paper. A century later, Arnould Carangeot invented the contact goniometer. This device, which enabled crystallographers to make systematic measurements of interfacial crystal angles, was basically a flat, pivoted metal arm with a pointer that could move over a semicircular protractor. William Hyde Wollaston invented a more precise instrument, the reflecting goniometer, in 1809. This device used a narrow beam of light reflected from a mirror and directed against a crystal to make very accurate measurements of the angles between crystal faces. The reflecting goniometer ushered in a period of quantitative mineralogy that led to the multiplication of vast amounts of information about the external structure of minerals.

The discovery of the polarization of light in the nineteenth century led to another method of mineral investigation. Ordinary light consists of electromagnetic waves oscillating in all directions at right angles to the direction of travel, but a suitable material can split such light into two rays, each vibrating in a single direction (this light is then said to be plane polarized). Various inventors perfected the polarizing microscope, a versatile instrument using plane-polarized light to identify minerals and to study their fine structure. Even the darkest minerals could be made transparent if sliced thin enough. These transparent slices produced complex but characteristic colors because of absorption and interference when polarized light passed through them.

STUDY OF INTERNAL STRUCTURE

In the twentieth century, X-ray diffraction provided scientists with a tool vastly more powerful than anything previously available for the investigation of internal mineral structures. Before the development of X-ray methods in 1912, the internal structure of a mineral could be deduced only by reasoning from its physical and chemical properties. After X-ray analysis, the determination of the detailed internal structures of minerals moved from speculation to precise measurement. The phenomenon of diffraction had been known since the seventeenth century. It can be readily observed when a distant street light is viewed through the regularly spaced threads of a nylon umbrella, causing spots of light to be seen. In a similar way, Max von Laue reasoned that the closely spaced sheets of atoms in a crystal should diffract X rays, with closely spaced sheets diffracting X rays at larger angles than more widely spaced ones. William Lawrence Bragg then showed how this technique could be used to provide detailed information about the atomic structure of minerals.

The powder method of X-ray diffraction consists of grinding a mineral specimen into a powder that is then formed into a rod by gluing it to a thin glass fiber. As X rays impinge on it, this rod is rotated in the center of a cylindrical photographic film. The diffraction pattern on the film can then be interpreted in terms of the arrangement of atoms in the mineral's unit cell.

Although X rays have been the most important type of radiation used in determining mineral structures, other types of radiation, in particular infrared (with wavelengths greater than those of visible light), have also been effective. Infrared radiation causes vibrational changes in the ions or molecules of a particular mineral structure, and that permits scientists to map its very detailed atomic arrangement. The technique of neutron diffraction makes use of relatively slow neutrons

from reactors to determine the locations of the light elements in mineral structures (the efficiency of light elements in scattering neutrons is generally quite high).

In recent decades, scientists have continued to develop sophisticated techniques for exploring the structure of minerals. Each of these methods has its strengths and limitations. For example, the electron microprobe employs a high-energy beam of electrons to study the microstructure of minerals. This technique can be used to study very small amounts of minerals as well as minerals in situ, but the strong interaction between the electron beam and the crystalline material produces anomalous intensities, and thus electron-microprobe studies are seldom used for a complete structure determination. Many new techniques have helped scientists to perform structural studies of minerals in special states—for example, at high pressures or at temperatures near the melting point—but the most substantial advancements in determining mineral structures continue to involve X-ray analysis.

SCIENTIFIC AND ECONOMIC APPLICATIONS

A central theme of modern mineralogy has been the dependence of a mineral's external form and basic properties on its internal structure. Because the arrangement of atoms in a mineral provides a deeper understanding of its mechanical, thermal, optical, electrical, and chemical properties, scientists have determined the atomic arrangements of many hundreds of minerals by using the X-ray diffraction technique. This great amount of structural information has proved to be extremely valuable to mineralogists, geologists, physicists, and chemists. Through this information, mineralogists have gained an understanding of the forces that hold minerals together and have even used crystal-structure data to verify and correct the formulas of some minerals. Geologists have been able to use the knowledge of mineral structures at high temperatures and pressures to gain a better understanding of the eruption of volcanoes and other geologic processes. Physicists have used this structural information to deepen their knowledge of the solid state. Through crystal-structure data, chemists have been able to expand their understanding of the chemical bond, the structures of molecules, and the chemical behavior of a variety of substances.

Because minerals often have economic importance, many people besides scientists have been interested in their structures. Rocks, bricks, concrete, plaster, ceramics, and many other materials contain minerals. In fact, almost all solids except glass and organic materials are crystalline. That is why a knowledge of the structure and behavior of crystals is important in nearly all industrial, technical, and scientific enterprises. This knowledge has, in turn, enabled scientists to synthesize crystalline compounds to fill special needs: for example, high-temperature ceramics, electrical insulators, semiconductors, and many other materials.

Robert J. Paradowski

CROSS-REFERENCES

Biopyriboles, 1175; Carbonates, 1182; Clays and Clay Minerals, 1187; Diamonds, 1561; Earth Resources, 1741; Feldspars, 1193; Gem Minerals, 1199; Gold and Silver, 1578; Hydrothermal Mineralization, 1205; Industrial Metals, 1589; Industrial Nonmetals, 1596; Ionic Substitution, 1209; Magnetite, 1214; Metamictization, 1219; Minerals: Physical Properties, 1225; Non-silicates, 1238; Orthosilicates, 1244; Oxides, 1249; Petrographic Microscopes, 493; Radioactive Minerals, 1255; Silica Minerals, 1354; X-ray Powder Diffraction, 504.

BIBLIOGRAPHY

Bragg, William Lawrence. *Atomic Structure of Minerals*. Ithaca, N.Y.: Cornell University Press, 1937. Bragg wrote this book while he was Baker Professor at Cornell University in the spring semester of 1934. Primarily a discussion of mineralogy from the perspective of the vast amount of new data generated by the successful application of X-ray diffraction analysis to crystalline minerals. Because of its provenance in a series of general lectures, the text is highly readable, though a knowledge of elementary physics and chemistry is presupposed. Of use to mineralogists, physicists, chemists, and all other scientists interested in the physical and chemical properties of minerals.

Bragg, William Lawrence, G. F. Claringbull, and W. H. Taylor. *The Crystalline State*. Vol. 4, *The Crystal Structures of Minerals*. Ithaca, N.Y.: Cornell University Press, 1965. The X-ray analysis of crystals generated so much data that proved to be of interest to workers in so many branches of science that Bragg needed several collaborators and volumes to survey the subject; this volume is a comprehensive compilation of crystal-structure information on minerals. Because each collaborator wrote on that aspect of the subject of which he had expert knowledge, the analyses of structures are authoritative. Can be appreciated and used by anyone with a basic knowledge of minerals, as crystallographic notation is kept to a minimum and the actual structures take center stage.

Cepeda, Joseph C. *Introduction to Minerals and Rocks*. New York: Macmillan, 1994. Cepeda's text provides a good introduction to the structure of minerals for students just beginning their studies in the Earth sciences. Includes illustrations and maps.

Deer, William A., R. A. Howie, and J. Zussman. *An Introduction to Rock-Forming Minerals*. 2d ed. New York: Longman, 1992. Standard references on mineralogy for advanced college students and above. Each chapter contains detailed descriptions of chemistry and crystal structure, usually with chemical analyses. Discussions of chemical variations in minerals are extensive.

Evans, Robert Crispin. *An Introduction to Crystal Chemistry*. 2d ed. New York: Cambridge University Press, 1964. In this book, Evans, a Cambridge chemist, analyzes crystal structures in terms of their correlation with physical and chemical properties. His approach is not comprehensive; rather, he discusses only those structures that are capable of illustrating basic principles that govern the behavior of these crystals. Though the author's approach demands some knowledge of elementary chemistry and physics on the part of the reader, there is no need for detailed crystallographic knowledge.

Hammond, Christopher. *The Basics of Crystallog-*

raphy and Diffraction. New York: Oxford University Press, 1997. Hammond's book covers crystal form, atomic structure, physical properties of minerals, and X-ray methods. Illustrations help clarify some of the more mathematically complex concepts. Includes bibliography and index.

Lima-de-Faria, Josae. *Structural Minerology: An Introduction.* Dordrecht, the Netherlands: Kluwer, 1994. This book provides a good college-level introduction to the basic concepts of crystal structure and the classification of minerals. Illustrations, extensive bibliography, index, and a table of minerals on a folded leaf.

Lipson, Henry S. *Crystals and X-Rays.* New York: Springer-Verlag, 1970. Lipson wrote this book, which is part of the Wykeham Science series, to give advanced high school students and college undergraduates an inspiring introduction to the present state of X-ray crystallography. Though many scientists treat crystallography as a mathematical subject, Lipson stresses the observational and experimental, for example, by showing how the X-ray diffraction technique was used to determine the structures of some simple minerals.

Pauling, Linus. *The Nature of the Chemical Bond and the Structure of Molecules and Crystals: An Introduction to Modern Structural Chemistry.* 3d ed. Ithaca, N.Y.: Cornell University Press, 1960. Pauling's first scientific paper was the X-ray analysis of a mineral, molybdenite, and he went on to determine many other mineral structures. He used both crystal structures and quantum mechanics to develop a classic theory of the chemical bond. This book grew out of his own work and his tenure as Baker Professor at Cornell University during the fall semester of 1937. The beginner will encounter difficulties, but readers with a good knowledge of chemistry will find this book informative and inspiring.

Sinkankas, John. *Mineralogy for Amateurs.* New York: Van Nostrand Reinhold, 1966. As the title suggests, Sinkankas intended his book primarily for the amateur mineralogist. Because of its simplified presentation of many complex ideas, it has become popular with nonprofessionals. Includes a good chapter on the geometry of crystals, in which the basic ideas of mineral structure are cogently explained. Very well illustrated with photographs and drawings (many of the latter done by the author).

Smith, David G., ed. *The Cambridge Encyclopedia of Earth Sciences.* New York: Crown, 1981. This volume is part of a Cambridge series of reference works dedicated to the sciences. The various sections of this encyclopedia (on geology, mineralogy, oceanography, seismology, and the physics and chemistry of the earth) were written by authorities from England and the United States. Though primarily a reference work, this book is both readable and informative in most sections; for example, part 2 contains a good analysis of the internal structure of minerals. Some knowledge of elementary physics and chemistry is needed for a full understanding of most sections. Profusely illustrated with helpful diagrams and photographs.

NON-SILICATES

Non-silicate minerals (exclusive of the carbonates and oxides), although not as abundant as the silicates in the part of the Earth that is accessible to humankind, are important because they are the major sources of many of the critical elements and compounds upon which civilized society is based.

PRINCIPAL TERMS

CRYSTALLINE: a property of a chemical compound to have an orderly internal atomic arrangement that may or may not have well-developed external faces

ION: an atom that has a positive or negative charge

METAL: an element with a metallic luster, high electrical and thermal conductivity, ductility, and malleability

MINERAL: a naturally occurring, solid chemical compound with a definite composition and an orderly internal atomic arrangement

ORE: a mineral or minerals present in large enough amounts in a given deposit to be profitably mined for the metal(s)

ROCK-FORMING MINERAL: the common minerals that compose the bulk of the Earth's crust (outer layer)

SEMIMETAL: elements that have some properties of metals but are distinct because they are not malleable or ductile

CLASSIFICATION OF MINERALS

The Earth is divided into several distinct layers (crust, mantle, and core), each with unique physical and chemical properties. Since only the crust is readily accessible to scientists, this discussion is largely restricted to this outermost layer. Estimates suggest that the crust is dominated by only eight elements (oxygen, silicon, aluminum, iron, magnesium, calcium, sodium, and potassium).

One of the most important elements is silicon; geologists commonly divide minerals into two broad categories, silicates and non-silicates. Silicates are those minerals that contain silicon as an essential part of the composition; non-silicates lack silicon in their formulas. The silicates make up the vast majority of the minerals in the crust, with all the other classes accounting for only 3 percent of the total.

A common classification scheme for minerals recognizes eleven mineral classes. One class is silicates. The other classes (or non-silicates) include the native elements; sulfides and sulfosalts; oxides and hydroxides; carbonates; halides; nitrates; borates; phosphates, arsenates, and vanadates; sulfates and chromates; and tungstates and molybdates.

NATIVE ELEMENTS

Of all the elements, only 20, the native elements, are known to occur in the Earth in the free state. These elements can be separated into metals, semimetals, and nonmetals. Within the metals, based on atomic structure, three groups are recognized: the gold group (gold, silver, copper, and lead), the platinum group (platinum, palladium, iridium, and osmium), and the iron group (iron and nickel-iron). Mercury, tantalum, tin, and zinc are metals that have also been identified. Within the semimetals, two groups are commonly recognized: the arsenic group (arsenic, antimony, and bismuth) and the tellurium group (selenium and tellurium). Sulfur and two forms of carbon (graphite and diamond) are the nonmetal minerals.

SULFIDES AND SULFOSALTS

The sulfides include a great number of minerals, many of which are important economically as sources of metals. Although sulfur is the dominant anion (negatively charged ion), this group also includes compounds with arsenic (arsenides), tellurium (tellurides), selenium (selenides), antimony, and bismuth as the anion. Commonly included with the sulfides are the sulfosalts. These sulfosalts are generally distinct because they contain the semimetals arsenic and antimony in the metal site. Only four sulfides are considered to be rock-forming minerals: pyrite, marcasite, chalcopyrite, and pyrrhotite.

HALIDES, NITRATES, AND BORATES

The halides are minerals that contain one of four anions—fluorine, chlorine, bromine, and iodine. The chlorine-bearing chlorides are the most abundant halides, with the fluorides second in abundance, but the bromides and iodides are very rare. The only two halides that are considered to be rock-forming minerals are halite and fluorite.

The nitrates include minerals made up of one nitrogen ion surrounded by three oxygen ions. None of the nitrates is considered common, and they are relatively few in number. Most occur only in deposits in very arid regions.

The borates are minerals that have boron strongly bound to either three or four oxygen ions. All the borates are restricted to dry lake deposits in extremely arid regions. None is considered a common rock-forming mineral, but borax is probably the most readily recognized.

PHOSPHATES AND SULFATES

In the phosphate class (which includes the arsenates and vanadates), the phosphorus cation (positively charged ion) is surrounded by four oxygens. Although this class includes many minerals, most are extremely rare; only one, apatite, is common.

The sulfate minerals are another large class. The sulfur cation is surrounded by four oxygens. The chromates are included with the sulfates. Two main subgroups of sulfates are recognized: the anhydrous sulfates and the hydrous sulfates. Although this class includes many minerals, very few are considered to be common. Examples of anhydrous sulfates are barite and anhydrite. The most common hydrous sulfate is gypsum.

TUNGSTATE AND MOLYBDATE

The tungstate and molybdate minerals are very similar to the sulfates, with either the tungsten or molybdenum cations surrounded by four oxygens in a pattern slightly different from that in the sulfates. None of the tungstates or molybdates is a common rock-forming mineral, and all are relatively rare.

STUDY OF PHYSICAL PROPERTIES

A number of physical properties are studied not only because they help identify particular minerals but also because the physical properties dictate whether minerals have any commercial use. Measurements of the angular relationships between faces with the optical goniometer is helpful in describing minerals. The systematic study of the external form of minerals is commonly called morphological crystallography. Other properties that are studied include the tendency of minerals to fracture (break along irregular surfaces) or to cleave (break along straight planar surfaces that represent planes of internal weakness) and tenacity (resistance or response to attempts to break, bend, or cut). Hardness, another essential property, refers to the resistance of a substance to abrasion. Hardness is commonly evaluated on a scale of relative hardness, called the

Barite, an anhydrous sulfate, is a non-silicate of the sulfates group. It is used in drilling fluids and as an additive to rubber, glass, and plastics. (U.S. Geological Survey)

Mohs hardness scale, which uses a set of common minerals of different hardnesses for comparison. Other properties that may have industrial applications include magnetism and electrical properties. Minerals containing uranium and thorium exhibit another property, radioactivity, that is measurable with a Geiger counter or scintillation counter.

A variety of thermal properties are commonly determined. Differential thermal analysis (DTA) measures the temperatures at which compositional and structural changes take place in minerals. The DTA curve, which is a graphic recording of the thermal changes in a mineral, is a characteristic of many minerals. Density (mass per unit of volume) is another key property. Density is normally determined as specific gravity (the ratio of the density of the substance to that of an equal volume of water at 4 degrees Celsius and 1 atmosphere of pressure). Depending on the amount and size of the mineral sample available, three methods may be used to determine density: Jolly Balance, Berman Balance, or pycnometer.

The petrographic microscope is an important analytical tool used to study the optical properties of minerals in polarized light transmitted through the specimen. Both crushed samples and thin sections (0.030 millimeter thick) are commonly utilized. Many of the metal-bearing minerals, particularly the native metals and sulfides, do not readily allow polarized light to pass through them, so microscopic analysis is conducted on highly polished samples in reflected light.

STRUCTURAL AND CHEMICAL ANALYSIS

The study of the orderly internal atomic arrangement within minerals is commonly called structural crystallography. The most common method used to study this internal morphology is the X-ray diffraction powder method, in which a small, finely powdered sample is bombarded with X rays. Like the human fingerprint, every mineral has unique diffraction patterns (caused by X rays interacting with atomic planes). Several other important X-ray diffraction methods involve single crystals (oscillations, rotation, Weissenberg, and precession methods) and are primarily used not for identification purposes, but for refining the complex internal geometries of minerals.

In addition to the physical properties of minerals, a variety of chemical methods can be used to identify and study non-silicate minerals. Until the 1960's, quantitative chemical analysis methods such as colorimetric tests, X-ray fluorescence spectrography, and atomic absorption spectrography were commonly used for mineral analysis. Since that time, however, most mineral analyses have been produced by electron microprobe analysis.

ECONOMIC IMPORTANCE OF NATIVE ELEMENTS

A great number of the non-silicates under consideration touch people's lives in a multitude of ways. Of the native elements, one only needs to look over the commodity reports in the daily newspaper to understand the importance of the likes of gold, silver, and platinum to world economies. Most of the gold in the world is owned by national governments, which commonly use it to settle international monetary accounts. It, like silver and platinum, is also becoming increasingly popular as a form of investment. In addition, gold is used in jewelry, scientific instruments, and electroplating. Silver is used in the photographic, electronics, refrigeration, jewelry, and tableware industries. Most copper is used in the production of electrical wire and in a variety of alloys, brass (copper and zinc) and bronze (copper, tin, and zinc) being the most common. Platinum is a very important metal because of its high melting point, hardness, and chemical inertness. It is primarily used by the chemical industry as a catalyst in the production of chemicals but is also used in jewelry and surgical and dental tools. Sulfur has a wide variety of uses in the chemical industry (production of insecticides, fertilizers, fabrics, paper, and soaps). In addition to their use as gems, diamonds are used in a variety of ways as cutting, grinding, and drilling agents.

USES OF OTHER NON-SILICATES

Like the native elements, the sulfides are also the source of a number of metals. The most important ores of silver, copper, lead, zinc, nickel, mercury, arsenic, antimony, molybdenum, and arsenic are sulfides. The government of the United States considers most of these commodities to be critical in time of war and has huge stockpiles of them in reserve throughout the country, since most are mined in other countries.

The two most important halides, halite and fluorite, are widely used. Halite, or ordinary table salt, is a source of sodium and sodium compounds, chlorine, and hydrochloric acid, which are all important in the chemical industry. It is also used as a de-icing compound, in fertilizers, in livestock feeds, and in the processing of hides. Most fluorite is used to make hydrofluoric acid for the chemical industry or as a flux in the production of steel and aluminum.

In the borates, borax is used to produce glass, insulation, and fabrics. It is also used in medicines, in detergents, in soaps, and as a preservative.

Of the phosphates, only apatite is a common mineral, but at least three others are important for the elements that they contain. Monazite is the primary source of thorium, which is a radioactive element with considerable potential as a source of nuclear energy. Autunite, a complex phosphate, and carnotite, a complex vanadate, are both important sources of uranium. Apatite, which is a source of phosphorus, is most widely used in fertilizers and detergents.

Of the sulfates, barite, anhydrite, and gypsum are the three most commonly used. The bulk of the barite is used as a drilling mud in the minerals and energy exploration industry. Anhydrite and gypsum occur in similar geological conditions and have similar compositions and uses. Both are used as soil conditioners. Gypsum is mainly used for the manufacture of plaster of Paris, which is used in wallboard.

In the tungstates, wolframite and scheelite are

Fluorite, one of the two major members of the non-silicate subgroup known as the halides, is found worldwide in hydrothermal deposits. Its crystals have an octahedral cleavage pattern. (U.S. Geological Survey)

the main ores of tungsten, which is used as a hardening alloy, in lamp filaments, and in tungsten carbide for cutting tools.

Ronald D. Tyler

CROSS-REFERENCES

Biopyriboles, 1175; Carbonates, 1182; Clays and Clay Minerals, 1187; Electron Microprobes, 459; Electron Microscopy, 463; Feldspars, 1193; Gem Minerals, 1199; Hydrothermal Mineralization, 1205; Infrared Spectra, 478; Ionic Substitution, 1209; Magnetite, 1214; Mass Spectrometry, 483; Metamictization, 1219; Minerals: Physical Properties, 1225; Minerals: Structure, 1232; Neutron Activation Analysis, 488; Orthosilicates, 1244; Oxides, 1249; Radioactive Minerals, 1255; X-ray Powder Diffraction, 504.

BIBLIOGRAPHY

Berry, L. G., B. Mason, and R. V. Dietrich. *Mineralogy: Concepts, Descriptions, Determinations.* 2d ed. San Francisco: W. H. Freeman, 1983. A college-level introduction to the study of minerals that focuses on the traditional themes necessary to understand minerals: how they are formed and what makes each chemically, crystallographically, and physically distinct from others. Descriptions and determinative tables include almost two hundred minerals (more than one-half of which are non-silicates).

Cepeda, Joseph C. *Introduction to Minerals and Rocks.* New York: Macmillan, 1994. Cepeda's text provides a good introduction to the structure of minerals for students just beginning their studies in the Earth sciences. Includes illustrations and maps.

Chesterman, C. W., and K. E. Lowe. *The Audubon Society Field Guide to North American Rocks and Minerals.* New York: Alfred A. Knopf, 1988. This book contains 702 color photographs of minerals, grouped by color, as well as nearly a hundred color photographs of rocks. All the mineral photographs are placed at the beginning of the book, and descriptive information follows, with the minerals grouped by chemistry. Distinctive features and physical properties are listed for each of the minerals, and information on collecting localities is also given. A section at the back of the book discusses various types of rocks. A glossary is also included. Suitable for the layperson.

Deer, William A., R. A. Howie, and J. Zussman. *An Introduction to Rock-Forming Minerals.* 2d ed. New York: Longman, 1992. A standard reference on mineralogy for advanced college students and above. Each chapter contains detailed descriptions of chemistry and crystal structure, usually with chemical analyses. Discussions of chemical variations in minerals are extensive.

Dietrich, Richard V., and B. J. Skinner. *Rocks and Rock Minerals.* New York: John Wiley & Sons, 1979. This short, readable college-level text provides a relatively brief but excellent treatment of crystallography and the properties of minerals. Although the descriptions of minerals focus on the silicates, the important rock-forming non-silicates are also considered. The book is very well illustrated and includes a subject index and modest bibliography.

Ernst, W. G. *Earth Materials.* Englewood Cliffs, N.J.: Prentice-Hall, 1969. The first four chapters of this compact introductory text deal with minerals and the principles necessary to understand their physical and chemical properties, as well as with their origins. Chapter 3 specifically deals with a number of the important rock-forming non-silicate minerals. The text includes a subject index and short bibliography.

Frye, Keith. *Mineral Science: An Introductory Survey.* New York: Macmillan, 1993. This basic text, intended for the college-level reader, provides an easily understood overview of mineralogy, petrology, and geochemistry, including descriptions of specific minerals. Illustrations, bibliography, and index.

Hammond, Christopher. *The Basics of Crystallography and Diffraction.* New York: Oxford University Press, 1997. Hammond's book covers crystal form, atomic structure, physical properties of minerals, and X-ray methods. Illustrations help clarify some of the more mathematically complex concepts. Includes bibliography and index.

Hurlbut, C. S., Jr., and G. S. Switzer. *Gemology.* New York: John Wiley & Sons, 1979. A well-illustrated introductory textbook for the reader with little scientific background. Its coverage includes the physical and chemical properties of gems, their origins, and the instruments used to study them. Later chapters treat methods of synthesis, cutting and polishing, and descriptions of gemstones.

Klein, C., and C. S. Hurlbut, Jr. *Manual of Mineralogy.* 21st ed. New York: John Wiley & Sons, 1999. An excellent second-year college-level text for use as an introduction to the study of minerals. The topics discussed include external and internal crystallography, crystal chemistry, properties of minerals, X-ray crystallography, and optical properties. The book also systematically describes more than one hundred non-silicate minerals.

Lima-de-Faria, Josae. *Structural Minerology: An Introduction.* Dordrecht, the Netherlands: Kluwer, 1994. This book provides a good college-level introduction to the basic concepts of crystal structure and the classification of minerals. Illustrations, extensive bibliography, index, and a table of minerals on a folded leaf.

Pough, Frederick H. *A Field Guide to Rocks and Minerals.* 4th ed. Boston: Houghton Mifflin, 1976. One of the most popular and easily accessible books dealing with non-silicate minerals. Intended for the reader with no scientific background, it includes chapters on collecting and testing minerals, descriptions of environments of formation, physical properties, classification schemes, and mineral descriptions.

Tennissen, A. C. *Nature of Earth Materials*. 2d ed. Englewood Cliffs, N.J.: Prentice-Hall, 1983. This text is written for the non-science student and treats minerals from the perspective of both the internal relationships (atomic structure, size, and bonding) and external crystallography. It includes an excellent overview of the physical properties of minerals and classification and description of 110 important minerals.

ORTHOSILICATES

The orthosilicates are a large and diverse group of rock-forming minerals. The group contains a number of minerals of geologic importance, among them olivine, which may be the most abundant mineral of the inner solar system. Orthosilicates have a few special but limited industrial uses and include a number of important gemstones.

PRINCIPAL TERMS

ANION: an atom that has gained electrons to become a negatively charged ion

CATION: an atom that has lost electrons to become a positively charged ion

CRYSTAL STRUCTURE: the regular arrangement of atoms in a crystalline solid

IGNEOUS ROCK: a rock formed by the solidification of molten, or partially molten, rock

METAMORPHIC ROCK: a rock formed when another rock undergoes changes in mineralogy, chemistry, or structure owing to changes in temperature, pressure, or chemical environment at depth within a planet

MINERAL: a naturally occurring, solid, inorganic compound with a definite composition and an orderly internal arrangement of atoms

SILICATES: minerals containing both silicon and oxygen, usually in combination with one or more other elements

SOLID SOLUTION: a solid that shows a continuous variation in composition in which two or more elements substitute for each other on the same position in the crystal structure

TETRAHEDRON: a four-sided pyramid made out of equilateral triangles

CHEMICAL STRUCTURE

The orthosilicates are one of the major groups of silicate minerals. Oxygen and silicon are the two most common elements in the rocky outer layers of the Earthlike inner planets. Thus, it is not surprising that the silicates are the most abundant rock-forming minerals on these planets. They are not only very abundant, but also very diverse: It has been estimated that almost one-third of the roughly 3,000 known minerals are silicates.

In all but a few of these minerals, each silicon atom is surrounded by a cluster of four oxygen atoms distributed around the silicon in the same way that the corners of a tetrahedron are distributed around its center. Silicate tetrahedra can join together by sharing one oxygen at a corner. By this means, two or more silicate tetrahedra can link together to form pairs, rings, chains, sheets, and three-dimensional frameworks. This linking provides the basis of the classification of silicate structures: Sorosilicates contain pairs of tetrahedra, cyclosilicates contain rings, inosilicates contain chains, phyllosilicates contain sheets, and tectosilicates contain three-dimensional frameworks. In the orthosilicates, however, silicate tetra-hedra do not link to each other. Each silicate tetrahedra is isolated from the others as if it were an island, and hence these minerals are sometimes known as island silicates. In the structure of some minerals, isolated silicate tetrahedra are mixed with silicate pairs formed when two tetrahedra share an oxygen at a common corner. Many scientists classify these structures as orthosilicates and use the term nesosilicates to refer to a subdivision of the orthosilicates containing minerals in which all the silicon occurs in isolated tetrahedra.

CHEMICAL BONDING

The electrostatic attraction between negatively charged ions (anions) and positively charged ions (cations) is the basis of the ionic chemical bond. The chemical bonding in orthosilicates is predominantly ionic. Under normal geological conditions, silicon loses four electrons to become a cation with a charge of +4, while oxygen gains two electrons to become an anion with a charge of −2. The ionic bond between oxygen and silicon is usually the strongest bond in orthosilicate structures. For a mineral to be stable it must be electrically neutral; in other words, the total number of negative charges on the anions must be equal to the total number of positive charges on the cations. It would take two oxygen anions to balance the charge on one silicon cation. In the silicate tetrahedra, however, the silicon cation is surrounded by four oxygen anions, so in the group as a whole there are four excess negative charges. In the orthosilicate structures this excess negative charge is balanced by the presence of cations outside the silicate tetrahedra. It is the bond between the oxygen in the tetrahedra and these other cations that holds the tetrahedra together to form a coherent, three-dimensional structure. The most common cations to play this role are aluminum with a charge of +3, iron with a charge of +3 or +2, calcium with a charge of +2, and magnesium with a charge of +2. Aluminum, iron, and calcium are, respectively, the third, fourth, and fifth most abundant elements in the Earth's crust, while magnesium is the seventh most abundant. Conspicuously absent from the common orthosilicate structures are the two alkali cations sodium and potassium, which are the sixth and eighth most abundant elements in the Earth's crust.

Of all the different kinds of silicates, orthosilicates have the lowest ratio of silicon to oxygen. Thus, they often form in environments relatively low in silicon. The atoms in orthosilicates tend to be packed closer together than the atoms in many other silicates, causing them to be somewhat denser. Greater densities are favored at higher pressure, hence a number of the more important minerals stable at high pressure are orthosilicates. Orthosilicates also tend to be harder (resistant to being scratched) than the average silicate. This property helps give many orthosilicates high durability (resistance to wear), which contributes to their use as gemstones. The orthosilicates are a large and diverse group, and a description of more than the most common of them is well beyond the scope of this article.

OLIVINE

The mineral olivine is the most common and widespread of the orthosilicates: Indeed, it is probably the most abundant mineral in the inner solar system. Olivine contains magnesium (Mg) and iron (Fe) in addition to silicon (Si) and oxygen (O). The chemical formula of olivine is generally written as $(Mg,Fe)_2SiO_4$. The parentheses in this formula indicate that olivine is a solid solution; in other words, magnesium and iron can substitute for each other in the olivine structure. Pure magnesium olivines (Mg_2SiO_4) are known as forsterite, and pure iron olivines (Fe_2SiO_4) are known as fayalite. Most olivines have both magnesium and iron and hence have compositions that lie between these two extremes. Each iron and magnesium atom is surrounded by six oxygen atoms, while each oxygen atom is bonded to three iron or magnesium atoms and one silicon atom, thereby creating an extended three-dimensional structure. Olivine is usually a green mineral with a glassy luster and a granular shape.

Rocks made up mostly of olivine are known as peridotites. Although peridotites are relatively rare in the Earth's crust (that layer which begins at the surface and extends to depths of between 5 and 80 kilometers), it is the rock that makes up most of the Earth's uppermost mantle. This olivine-rich layer begins at depths ranging from between 5 and 80 kilometers and extends downward to depths of roughly 400 kilo-

The orthosilicates include olivine, which may be the most abundant mineral of the inner solar system. Olivine is also the dominant mineral in Earth's mantle. With its beautiful green color, olivine is sometimes used as a gem mineral, peridot. (© William E. Ferguson)

meters. Evidence indicates that the Moon and the other inner planets (Mercury, Venus, and Mars) also have similar olivine-rich layers. Olivine can also be an important mineral in basalts and gabbros, which are the most abundant igneous rocks in the crusts of the inner planets. In addition, it is an abundant mineral in many different kinds of meteorites and some kinds of metamorphic rocks.

GARNETS

The garnets are a group of closely related orthosilicate minerals, each of which is a solid solution. The chemistry of the common garnets can be fairly well represented by the somewhat idealized general formula $A_3B_2Si_3O_{12}$ where the A stands for either magnesium, iron (with a charge of +2), manganese (Mn), or calcium (Ca), and the B stands for either aluminum (Al), iron (with a charge of +3), or chromium (Cr). In the crystal structures of these minerals, the cations in the A site are surrounded by eight oxygen anions, while the cations in the B site are surrounded by six oxygens. Most garnets can be described as a mixture of two or more of the following molecules: pyrope ($Mg_3Al_2Si_3O_{12}$), almandine ($Fe_3Al_2Si_3O_{12}$), spessartine ($Mn_3Al_2Si_3O_{12}$), grossular ($Ca_3Al_2Si_3O_{12}$), andradite ($Ca_3Fe_2Si_3O_{12}$), and uvarovite ($Ca_3Cr_2Si_3O_{12}$).

The most abundant and widespread garnets are almandine-rich garnets, which form during the metamorphism of some igneous rocks and of sediments rich in clay minerals. Grossular-rich and andradite-rich garnets are found in marbles formed through the metamorphism of limestone. The formation of spessartine-rich or uvarovite-rich garnets occurs during the metamorphism of rocks with high concentrations of manganese or chrome. Rocks with these compositions are relatively unusual, and hence these garnets are fairly rare. Pyrope-rich garnets are widespread in the Earth's mantle, although they typically do not occur in abundance (more than 5 or 10 percent of the rock). Garnets that have weathered out of other rocks are sometimes found in sands and sandstones. Garnets may also occur in small amounts in some igneous rocks.

ALUMINOSILICATES

Two or more minerals are called polymorphs if they are made up of the same kinds of atoms in the same proportions but in different arrangements. Polymorphs are minerals with the same compositions but different crystal structures. The aluminosilicates are a group of orthosilicate minerals containing three polymorphs: kyanite, sillimanite, and andalusite. Each of these minerals has the chemical formula Al_2SiO_5. The differences in the structures of these minerals are best illustrated by considering the aluminum atoms; in kyanite, all the aluminum atoms are surrounded by six oxygen atoms; in sillimanite, one-half of the aluminum atoms are surrounded by six oxygens, while the other half are surrounded by four oxygens; in andalusite, half the aluminum atoms are surrounded by four oxygens, and half are surrounded by five oxygens. Kyanite usually forms elongated rectangular crystals with a blue color. Sillimanite typically occurs as white, thin, often fibrous crystals. Andalusite is most commonly found in elongated crystals with a square cross section and a red to brown color. The aluminosilicates typically form during the metamorphism of clay-rich sediments; such rocks have the relatively high ratios of aluminum to silicon necessary for the formation of these minerals. The identity of the aluminosilicate formed depends upon the temperature and pressure of metamorphism; kyanite forms at relatively high pressures, sillimanite forms at relatively high temperatures, and andalusite forms at low to moderate temperatures and pressures.

TOPAZ, ZIRCON, TITANATE, AND EPIDOTE

Topaz is another aluminum-rich orthosilicate; although this mineral also contains fluorine (F) and/or the hydroxyl molecule (OH). The chemical formula of topaz is $Al_2SiO_4(F,OH)_2$. As the parentheses indicate, this mineral is a solid solution in which fluorine and hydroxyl can substitute for each other. Topaz is formed during the late stages in the solidification of a granite liquid.

The mineral zircon contains the relatively rare element zirconium (Zr). Zircon has the chemical formula $ZrSiO_4$; it generally also contains small amounts of uranium and thorium. The decay of these radioactive elements can be used to obtain the age of a rock, making zircon particularly important to geologists. It is most commonly found as brown rectangular crystals with a pyramid on either end. It is a widespread mineral in igneous rocks, although it generally occurs in relatively minor amounts.

Titanite ($CaTiSiO_5$), sometimes known as sphene, is one of the most common minerals bearing titanium (Ti). It is a fairly widespread mineral, occurring in many different kinds of igneous and metamorphic rocks, but is rarely present in abundance.

Epidote ($Ca_2(Al,Fe)Al_2Si_3O_{12}(OH)$) contains both isolated silicate tetrahedra and tetrahedral pairs; hence, it would be classified as an orthosilicate but not a nesosilicate. Epidote is a fairly common mineral most typically formed during low-temperature metamorphism in the presence of water. It most often occurs as masses of fine-grained, pistachio-green crystals.

ANALYTICAL TECHNIQUES

To characterize a mineral requires both its chemical composition and crystal structure. There are many different analytical techniques that will give chemical compositions; probably the most popular to use on orthosilicates is electron microprobe analysis. In this technique, part of the mineral is bombarded by a high-energy beam of electrons, which causes it to give off X rays. Different elements in the mineral give off X rays of different wavelengths, and the intensities of these different X rays depend on the abundance of these elements. By measuring the wavelengths and intensities of the X rays, the composition of the mineral can be obtained. This technique has the advantages of being nondestructive (the mineral is still available and undamaged after the analysis) and being applicable to very small spots on a mineral: The typical electron microprobe analysis gives the composition of a volume of mineral only a few tens to hundreds of cubic microns (millionths of a meter) in size.

The crystal structures of orthosilicates are generally obtained using the technique of X-ray diffraction. In this technique an X-ray beam is passed through the mineral. As the beam interacts with the atoms in the mineral, it breaks up into many smaller, diffracted beams traveling in different directions. The intensities and directions of these diffracted beams depend on the positions of atoms in the mineral. By analyzing the diffraction pattern, a scientist can discover the crystal structure of a mineral.

SCIENTIFIC AND ECONOMIC VALUE

Orthosilicates are one of the important building blocks of the planets of the inner solar system, and this reason alone provides an important scientific rationale for studying them. Despite their importance in nature, these minerals have had only a limited technological use. The relatively high hardness of garnet makes it suitable as an abrasive, and it is used in some sandpapers and abrasive-coated cloths. The aluminosilicates are used in the manufacture of a variety of porcelain that is noted for its high melting point, resistance to shock, and low electrical conductivity. This material is used in spark plugs and brick for high-temperature furnaces and kilns. Zircon is mined in order to obtain zirconium oxide and zirconium metal. Zirconium oxide has one of the highest known melting points and is used in the manufacture of items that have to withstand exceptionally high temperatures. Zirconium metal is used extensively in the construction of nuclear reactors. Titanite is mined as a source of titanium oxide. Titanium oxide has a number of uses but is most familiar as a white pigment in paint.

Probably the most widespread uses of the orthosilicates are as gemstones. Especially fine, transparent crystals of olivine make a beautiful green gem generally known as peridot, although the names chrysolite and evening emerald are sometimes used instead. Relatively transparent garnets also make very beautiful gems. The most common garnets are a deep red, and this is the color usually associated with the stone. Yet gem-quality garnets can also be yellow, yellow-brown, orange-brown, orange-yellow, rose, purple, or green. Garnet is a relatively common mineral and therefore is typically among the least valuable of gemstones. The major exception is the green variety of andradite garnet: Known in the gem trade as demantoid, this relatively rare material is one of the more valuable gems. When properly cut, zircon has a brilliancy (ability to reflect light) and fire (the ability to break white light up into different colors) second only to diamond and is a popular gemstone. Topaz is also widely used as a gem. The most valuable topaz is orange-yellow to orange-brown in color; unfortunately, all yellow gems are sometimes incorrectly referred to as topaz: When this practice is followed, true topaz is generally known as precious topaz or oriental topaz. Gem-quality topaz may also be colorless, faintly green, pink, red, blue, and brown.

Edward C. Hansen

CROSS-REFERENCES
Biopyriboles, 1175; Carbonates, 1182; Clays and Clay Minerals, 1187; Feldspars, 1193; Gem Minerals, 1199; Hydrothermal Mineralization, 1205; Ionic Substitution, 1209; Magnetite, 1214; Meta- mictization, 1219; Minerals: Physical Properties, 1225; Minerals: Structure, 1232; Non-silicates, 1238; Oxides, 1249; Radioactive Minerals, 1255.

BIBLIOGRAPHY

Deer, W. A., R. A. Howie, and J. Zussman. *Rock-Forming Minerals*. Vol. 1A, *Orthosilicates*. New York: John Wiley & Sons, 1988. This is one of the most complete treatments of the orthosilicates in English. A very well-written book, it is an excellent place to go for detailed information on individual minerals. Most suitable for college-level audiences.

Frye, Keith. *Mineral Science: An Introductory Survey*. New York: Macmillan, 1993. This basic text, intended for the college-level reader, provides an easily understood overview of mineralogy, petrology, and geochemistry, including descriptions of specific minerals. Illustrations, bibliography, and index.

Gait, R. I. *Exploring Minerals and Crystals*. Toronto: McGraw-Hill Ryerson, 1972. This is a lower-level introduction to the science of mineralogy. Well written and illustrated, it is more scientific than the typical rock-hound manual but less detailed than the introductory textbook. Suitable for high school students.

Hammond, Christopher. *The Basics of Crystallography and Diffraction*. New York: Oxford University Press, 1997. Hammond's book covers crystal form, atomic structure, physical properties of minerals, and X-ray methods. Illustrations help clarify some of the more mathematically complex concepts. Includes bibliography and index.

Hurlbut, Cornelius S., Jr., and G. S. Switzer. *Gemology*. New York: John Wiley & Sons, 1979. Written in the style of an introductory textbook, it is a very good starting point for someone whose primary interest in mineralogy is gems. It includes an introduction to the chemistry, physics, and mineralogy needed for the study of gems; a discussion of the general technical aspects of gemology; and a description of individual gemstones. Suitable for upper-level high school students.

Klein, Cornelis, and Cornelius S. Hurlbut, Jr. *Manual of Mineralogy*. 21st ed. New York: John Wiley & Sons, 1999. A popular textbook for a first course in mineralogy, this is an excellent book for someone interested in an introduction to the scientific study of minerals. Contains a very good sixteen-page section on the neosilicates. Suitable for upper-level high school students.

Lima-de-Faria, Josae. *Structural Minerology: An Introduction*. Dordrecht, the Netherlands: Kluwer, 1994. This book provides a good college-level introduction to the basic concepts of crystal structure and the classification of minerals. Illustrations, extensive bibliography, index, and a table of minerals on a folded leaf.

Ribbe, P. H., ed. *Orthosilicates: Reviews in Mineralogy*. Vol. 5. Washington, D.C.: Mineralogical Society of America, 1980. This book consists of eleven articles, all but one of which specialize in a specific mineral or mineral group. The articles tend to be fairly technical, and some previous exposure to mineralogy or chemistry would be very helpful to the reader. Suitable for college-level students.

Sinkankas, J. *Mineralogy for Amateurs*. New York: Van Nostrand Reinhold, 1964. Not quite as rigorous as an introductory text, this book still contains much more scientific detail than the average book on minerals written for amateurs. Suitable for high school students.

Webster, Robert F. G. A. *Gems: Their Sources, Descriptions, and Identification*. 5th ed. Boston: Butterworth-Heinemann, 1994. A technical college-level text that is an important reference for the gemologist or the senior jeweler. Reflects the British attention to detail. Better for sources and descriptions rather than for identification. Essential for the gemologist.

OXIDES

The oxides represent one of the most important classes of minerals. They are the source of several key metals upon which the world is dependent; these metals include iron, aluminum, titanium, manganese, uranium, and chromium. Products manufactured from these metals touch virtually every aspect of modern living.

PRINCIPAL TERMS

CRUST: the outer layer of the Earth; it extends to depths of from 5 kilometers to at least 70 kilometers and is the only layer of the Earth directly accessible to scientists

IGNEOUS ROCK: a major group of rocks formed from the cooling of molten material on or beneath the Earth's surface

METAL: an element with a metallic luster and high electrical and thermal conductivity; it is ductile, malleable, and of high density

METAMORPHIC ROCK: a major group of rocks that are formed from the modification of sedimentary or igneous rocks by elevated temperatures and/or pressures beneath the Earth's surface

MINERAL: a naturally occurring, solid chemical compound with a definite composition and an orderly internal atomic arrangement

ORE: a mineral or minerals with a valuable constituent present in large enough amounts in a given deposit to be minable for the metal(s) at a profit

QUARTZ: a very common silicate mineral

ROCK: an aggregate of one or more minerals

SEDIMENTARY ROCK: a major group of rocks formed from the breakdown of preexisting rock material or from the precipitation of minerals by organic or inorganic processes

MINERAL CLASSIFICATION

Compositionally, the Earth contains eighty-eight elements, but only eight of these comprise more than 99 percent (by weight) of the crust. These eight elements combine to form some two dozen minerals that make up more than 90 percent of the crust. Within the rocks and minerals of the crust, the element oxygen accounts for almost 94 percent of the total volume. On the atomic scale, that implies that most minerals are virtually all atoms of oxygen, with the other elements filling in the intervening spaces in orderly arrangements. Most minerals form because ions (atoms that have gained or lost one or more electrons) become mutually attractive. More precisely, ions with positive charges (cations), which have lost electrons, become attracted to ions with negative charges (anions), and if the charges are balanced and several rules of crystal chemistry are satisfied, a mineral will form. A simple example is the combination of the sodium cation and the chlorine anion to form the mineral halite.

Ionic combinations include complex anions (radicals) that are strongly bound cation and anion groupings. These radicals take on a negative charge and will attract more weakly bound cations. Silica and carbonate are examples of these complex anions. These anion radicals and the simple anions form the basis for one of the most widely used classification systems of minerals, with the following classes of minerals recognized: the native elements; sulfides and sulfosalts; oxides and hydroxides; carbonates; halides; nitrates; borates; phosphates, arsenates, and vanadates; sulfates and chromates; tungstates and molybdates; and silicates.

Although this classification is based entirely on chemical composition, subdivisions within these classes are based on both structural and additional chemical criteria. Of these eleven classes, the silicates dominate the crust, forming approximately 97 percent of this layer. Although all the other classes represent only 3 percent of the crust, these classes include the majority of the minerals that society has come to depend upon. In economic value alone, the oxides undoubtedly rank at or near the top of all the classes of minerals, including the silicates.

AO, A$_2$O, AND A$_2$O$_3$ OXIDES

The oxides are those minerals that have oxygen combined with one or more metals. Generally, the oxides are subdivided according to the ratio of the number of metals to the number of oxygens in the formula. Many of the oxides have relatively simple metal (A) to oxygen (O) ratios, so the following categories are recognized: AO, A$_2$O, AO$_2$, and A$_2$O$_3$. Some oxides have atomic structures in which different metals occupy different atomic (structural) sites. These minerals are commonly referred to as complex or multiple oxides, and most have the general formula AB$_2$O$_4$, where A and B are separate atomic sites.

In both the AO and A$_2$O oxides, there are no minerals that are considered common. In the AO$_2$ oxides, however, several minerals are important, with two subdivisions recognized: the rutile group and the individual mineral uraninite. The three important minerals that occur in the rutile group are rutile (TiO$_2$), cassiterite (SnO$_2$), and pyrolusite (MnO$_2$). Rutile is a common minor mineral in a wide variety of quartz-rich igneous and metamorphic rocks. It is also found in black sands along with several other oxides. Pyrolusite is a very widespread mineral found in manganese-rich nodules on the floors of the oceans, seas, lakes, and bogs. Cassiterite is a common minor constituent in quartz-rich igneous rocks. Uraninite (UO$_2$) is a separate AO$_2$ oxide that, like cassiterite and rutile, is characteristically associated with quartz-rich igneous rocks. In several places it also occurs with gold in stream deposits modified by metamorphism.

Even more common than the AO$_2$ oxides are the A$_2$O$_3$ oxides, which include the common minerals hematite (Fe$_2$O$_3$), corundum (Al$_2$O$_3$), and ilmenite (FeTiO$_3$). Hematite, the most widespread iron oxide mineral in the crust, occurs in a wide variety of conditions, such as metamorphic deposits, quartz-rich igneous rocks, and sedimentary rocks. Corundum is a widespread minor constituent in metamorphic rocks low in silicon and relatively rich in aluminum. It is also found in igneous rocks that have low silicon contents. Ilmenite is another mineral that typically occurs in small amounts in many types of igneous rocks and in black sands with several other oxides.

COMPLEX OXIDES AND HYDROXIDES

In the more complex oxides (AB$_2$O$_4$), the most common minerals occur in the spinel group, but the individual mineral columbite-tantalite (Fe,Mn)(Nb,Ta)$_2$O$_4$ is also commonly given consideration. The spinel group contains many minerals that have complex interrelationships. Several, however, are more abundant than others. Spinel (MgAl$_2$O$_4$) is common in some metamorphic rocks formed at high temperatures and in igneous rocks rich in calcium, iron, and magnesium. Magnetite (Fe$_3$O$_4$) is a common minor mineral in many igneous rocks. Relatively resistant to weathering, it also occurs in black sands and in very large sedimentary banded iron deposits. Chromite (FeCr$_2$O$_4$), another important mineral in the spinel group, is found only in calcium-poor and iron- and magnesium-rich igneous rocks. Unlike members of the spinel group, columbite-tantalite is a separate subdivision of the AB$_2$O$_4$ oxides. Like so many of the other oxides, it is characteristically associated with quartz-rich igneous rocks.

The hydroxides are a group of minerals that includes a hydroxyl group (OH)$^{-1}$ or the water (H$_2$O) molecule in their formulas. These minerals tend to have very weak bonding and, as a consequence, are relatively soft. Five hydroxides are briefly considered here: brucite (Mg(OH)$_2$), manganite (MnO(OH)), and three minerals in the goethite group (diaspore, AlO(OH); goethite, FeO(OH); and bauxite, a combination of several hydrous aluminum-rich oxides including diaspore). Brucite occurs as a product of the chemical modification of magnesium-rich igneous rocks and is associated with limestones. Manganite tends to occur in association with the oxide pyrolusite. Of the minerals in the goethite group, diaspore commonly is associated with corundum and occurs in aluminum-rich tropical soils. Goethite, an extremely common mineral, occurs in highly weathered tropical soils, in bogs, and in by-products of the chemical weathering (breakdown) of other iron oxide minerals. Bauxite is also found in highly weathered tropical soils.

STUDY OF HAND SPECIMENS

There are several general approaches to the study and identification of the oxide minerals. They include the study of hand specimens, optical properties, the internal atomic arrangements and their external manifestations (crystal faces), chemical compositions, and the synthesis of minerals. In the study of hand specimens, minerals possess a variety of properties that are easily determined or measured. These properties both aid in identification and may make the minerals commercially useful. Several properties are important for the oxides and hydroxides. Luster describes the way the surface of the mineral reflects light. Many minerals have the appearance of bright smooth metals, and this sheen is referred to as a metallic luster. Some minerals, like hematite and goethite, do not readily transmit light but may exhibit this type of luster. Other minerals that are able to transmit light have nonmetallic lusters, regardless of how shiny the outer surface of the mineral may be. Some common nonmetallic lusters include glassy, vitreous, and resinous. Minerals that have metallic lusters also characteristically produce powders that have diagnostic colors. The color of the powdered mineral is called the streak. Hematite, for example, may be silvery gray in color, but its powder is red. The shapes of minerals can also be important. Corundum, for example, is typically hexagonal in outline; magnetite forms octahedrons; and hematite has thin plates that grow together in rosettes. Specific gravity, the ratio of the weight of a substance to the weight of an equal volume of water, is also an important property of many of these minerals. Oxides tend to have much higher than average specific gravities. Heavy liquids, the Jolly balance, the pyncometer, and the Berman balance are all means by which specific gravity may be determined. Hardness is another property that is used for identification purposes and makes some of the oxides useful. Hardness is simply the measure of a substance's resistance to abrasion. Some minerals like corundum have particularly high hardnesses and therefore can be used commercially as abrasives.

Microscopic Study and Crystallography

The study of the optical properties of the oxides is conducted either in polarized light transmitted with a petrographic microscope or in reflected light. One important optical property of minerals in transmitted light is refractive index, which is a measure of the velocity of light passing through them. Minerals have up to three unique refractive indices, because light may travel at different velocities in different directions. Minerals that are nontransparent, such as many of the oxides, are studied in reflected light with an ore microscope. Such

properties as reflectivity, color, hardness, and reactivity to different chemicals are all considered.

The study of the orderly internal atomic arrangements within minerals and the associated external morphologies is called crystallography. The primary methods used to study these internal geometries are a variety of X-ray techniques that look at single crystals or powered samples of minerals. The most commonly used procedure is the X-ray diffraction powder method. This technique takes advantage of the principle that the internal atomic arrangement in every mineral is different from all others. X rays striking the powdered sample are reflected or diffracted at only specific angles (dictated by internal geometries). Thus, the measurement of the specific angles of diffraction provides enough information to identify the substance. Single crystals, however, are normally used in the more detailed studies with X rays. With respect to the study of crystal faces on mineral specimens, a reflecting goniometer allows for the precise measurement of the angular relationships between these faces.

Chemical and Synthesis Studies

Until the latter part of the twentieth century, most chemical analyses of minerals were conducted by methods generally referred to as wet-chemical analyses. In these analyses, the mineral is first dissolved in solution. The amounts of the individual elements in the solution are then determined either by separation and weighing of precipitates or by measurement via spectroscopic methods of elemental concentrations in solution. Over the years, the accuracy and speed of completing these analyses have been improved, but several problems have never been completely solved. The invention of the electron microprobe in the 1950's has revolutionized mineral analyses and has largely eliminated many of the problems and greatly decreased the time necessary to conduct most analyses. The beam of electrons that is used can be precisely focused on samples or areas of samples as small as 1 micron (10^{-3} millimeter) in diameter. Thus, not only small samples may be analyzed, but also individual crystals may be evaluated in several places to check for compositional variations.

Another important method used to study and understand the oxides and other minerals in the laboratory involves the synthesis of minerals. The primary purpose of these studies is to determine the temperature and pressure conditions at which individual minerals form. In addition, the roles of fluids and interactions with other minerals are also evaluated. High-temperature studies are typically conducted in furnaces containing platinum or tungsten heating elements. High-pressure studies are produced in large hydraulic apparatus or presses.

INDUSTRIAL USES

The oxides are one of the most important mineral groups because they have provided civilization with some very important metals. Iron, titanium, chromium, manganese, aluminum, and uranium are several of these key metals extracted from the ores of the oxides. Iron is the second most common metal in the crust. Along with aluminum, manganese, magnesium, and titanium, iron is considered an abundant metal because it exceeds 0.1 percent of the average composition of the crust. The steelmaking industry uses virtually all the iron mined. Steel, an alloy in which iron is the main ingredient, also includes one or more of several other metals (manganese, chromium, cobalt, nickel, silicon, tungsten, and vanadium) that impart special properties to steel. Hematite, magnetite, and goethite are three of the most important ore minerals of iron.

Unlike iron, which has been utilized for more than three thousand years, aluminum is a metal that did not gain prominence until the twentieth century. Since aluminum is light in weight and exhibits great strength, it is widely used in the automobile, aircraft, and shipbuilding industries. It is also utilized in cookware and food and beverage containers.

A third abundant metal that is primarily extracted from oxide and hydroxide minerals is manganese. Manganese is used in the production of steel. The other abundant metal in the crust that is produced from oxides is titanium. Titanium is important as a metal and as an alloy because of its great strength and light weight. It is a principal metal in the engines and essential structural components of modern aircraft and space vehicles.

Several scarce metals, representing less than 0.1 percent of the average composition of the crust, that occur as oxides and have important uses include chromium, uranium, tin, tantalium, and niobium. Of these metals, chromium and uranium are the most prominent. Chromium is utilized in the steel industry, where it is a principal component in stainless steel. In addition, because of its high melting temperature, chromium is used in bricks for metallurgical furnaces. Uranium is an important metal because it spontaneously undergoes nuclear fission and gives off large amounts of energy. Uranium is utilized in nuclear reactors to generate electricity. Tin is another scarce metal that for tens of centuries was utilized as an alloy in bronze. It has lost many of its older uses, but remains a prominent metal as new applications are developed. Tantalium is highly resistant to acids, so it is used in equipment in the chemical industry, in surgical inserts and sutures, and in specialized steels and electronic equipment. Niobium is used in the production of stainless steels and refractory alloys (alloys resistant to high temperatures) used in gas turbine blades in aircraft engines.

Emery, a combination primarily of black corundum, magnetite, and hematite, is used as an abrasive. Several oxides also form gemstones. Rubies are the red gem variety of corundum, and sapphire is also a gemstone of corundum that can have any other color. Spinel, if transparent, is a gem of lesser importance. If red, the gem is called a ruby spinel.

Ronald D. Tyler

CROSS-REFERENCES

BIBLIOGRAPHY

Berry, L. G., Brian Mason, and R. V. Dietrich. *Mineralogy.* 2d ed. San Francisco: W. H. Freeman, 1983. A college-level introduction to the study of minerals that focuses on the traditional themes necessary to understand minerals: how they are formed and what makes each chemically, crystallographically, and physically distinct from others. Descriptions and determinative tables include almost two hundred minerals (with twenty-eight oxides and hydroxides).

Deer, William A., R. A. Howie, and J. Zussman. *An Introduction to Rock-Forming Minerals.* 2d ed. New York: Longman, 1992. A standard reference on mineralogy for advanced college students and above. Each chapter contains detailed descriptions of chemistry and crystal structure, usually with chemical analyses. Discussions of chemical variations in minerals are extensive.

Dietrich, R. V., and B. J. Skinner. *Rocks and Rock Minerals.* New York: John Wiley & Sons, 1979. This short, readable college-level text provides a relatively brief but excellent treatment of crystallography and the properties of minerals. Although the descriptions of minerals focus on the silicates, several important oxides are considered. Very well illustrated and includes a subject index and modest bibliography.

Frye, Keith. *Mineral Science: An Introductory Survey.* New York: Macmillan, 1993. This basic text, intended for the college-level reader, provides an easily understood overview of mineralogy, petrology, and geochemistry, including descriptions of specific minerals. Illustrations, bibliography, and index.

Hammond, Christopher. *The Basics of Crystallography and Diffraction.* New York: Oxford University Press, 1997. Hammond's book covers crystal form, atomic structure, physical properties of minerals, and X-ray methods. Illustrations help clarify some of the more mathematically complex concepts. Includes bibliography and index.

Hurlbut, C. S., Jr., and G. S. Switzer. *Gemology.* New York: John Wiley & Sons, 1979. A well-illustrated introductory textbook for the reader with little scientific background. Coverage includes the physical and chemical properties of gems, their origins, and the instruments used to study them. Later chapters treat methods of synthesis, cutting and polishing, and descriptions of gemstones.

Klein, Cornelis, and C. S. Hurlbut, Jr. *Manual of Mineralogy.* 21st ed. New York: John Wiley & Sons, 1999. An excellent second-year college-level text that is an introduction to the study of minerals. Topics discussed include external and internal crystallography, crystal chemistry, properties of minerals, X-ray crystallography, and optical properties. Also systematically describes twenty-two oxide minerals.

Lindsley, Donald H, ed. *Oxide Minerals: Petrologic and Magnetic Significance.* Washington, D.C.: Mineralogical Society of America, 1991. The essays in this text, part of the Mineralogical Society of America's Reviews in Mineralogy series, cover topics relevant to oxide minerals and their magnetic properties, petrogenesis, and petrology. Extensive bibliography.

Ransom, Jay E. *Gems and Minerals of America.* New York: Harper & Row, 1974. A readily available book intended for the nonscientist who is interested in rock and mineral collecting. Introductory chapters introduce basic mineral characteristics and their environments of formation. Later chapters focus on the locations and collection of gems and minerals throughout the United States. A number of oxides are considered.

Skinner, Brian J. *Earth Resources.* 3d ed. Englewood Cliffs, N.J.: Prentice-Hall, 1986. A compact, well-illustrated introductory text that discusses the distribution of and rates of use of a variety of mineral and energy resources. Chapters 5 and 6 cover a number of important metals that are derived from oxides and hydroxides. Indexed and includes a modest bibliography.

Tennissen, A. C. *Nature of Earth Materials.* 2d ed. Englewood Cliffs, N.J.: Prentice-Hall, 1983.

A text written for the nonscience student that treats minerals from the perspective of both the internal relationships (atomic structure, size, and bonding) and external crystallography. Includes an excellent overview of the physical properties of minerals and the classification and description of 110 important minerals.

RADIOACTIVE MINERALS

Radioactive minerals combine uranium, thorium, and radium with other elements. Useful for nuclear technology, these minerals furnish the basic isotopes necessary not only for nuclear reactors but also for advanced medical treatments, metallurgical analysis, and chemicophysical research.

PRINCIPAL TERMS

AUTORADIOGRAPHY: a method by which a photograph of radioactive minerals is taken, using the emissions of the minerals themselves

CARNOTITE: an important ore of uranium, formed by reaction of groundwater with minerals already present

HALF-LIFE: the amount of time required for exactly one-half of an element's original material to decay into a daughter product

ISOTOPES: atoms of the same element with identical numbers of protons but different numbers of neutrons in their nuclei

MONAZITE: a rare-earth phosphate widely disseminated over the crust; the major ore source for thorium

PEGMATITE: a very coarse-grained granitic rock, often enriched in rare minerals, including uranium and thorium

RADIOACTIVE EMISSIONS: particles and radiation thrown off during the breakdown of a nucleus, including alpha and beta particles and gamma rays

RADIOACTIVITY: the spontaneous breakdown of elements in nature into other elements with differing nuclear properties

URANINITE: the chief ore of uranium and radium, formed of uranium oxide in combination with rare-earth elements

AVAILABILITY OF RADIOACTIVE MINERALS

Of the more than twenty-five hundred species of minerals known to exist on the Earth, only a small fraction are characterized by the presence of radioactive elements, primarily uranium, thorium, and radium. These elements are in a continuous state of radioactive breakdown, over time changing the chemical and structural composition of the minerals in which they occur. Approximately 150 of these compounds have uranium or thorium as an essential element in their crystallographic structure. Many of these minerals are quite rare; others, more important economically, can be characterized as present in small and variable amounts, paired with other elements in a solid solution of variable composition. Many such compounds, whether abundant or sparse, have current or future economic potential as sources for usable uranium and thorium.

URANIUM

Although in the Earth's crust uranium possesses an average concentration of only 4 parts per million, its abundance is much higher in several minerals that have become of great economic interest. Uranium itself is a dense, very electropositive, reactive metal that forms many alloys that are usable in nuclear technology. Uranium reacts with almost all nonmetallic elements and their binary compounds. Several hundred uranium-containing minerals have been identified, including such rare ones as tyuyamunite, torbernite, autunite, eudialyte, pyrochlore, and rinkite.

The chief ore of uranium is uraninite, which is basically uranium oxide (UO_2). Numerous other elements are often present, particularly thorium, rare-earths (mostly cerium), and lead, the latter element formed as a radioactive decay by-product of the uranium breakdown. A complete solid-solution series covers the range between uranium oxide, thorianite (ThO_2), and cerianite (CeO_2). The material has been found in pegmatites with zircons, in hydrothermal deposits with arsenides (compounds of arsenic and a metal such as cobalt, iron, or silver), and in some sedimentary deposits associated with coal. Because uraninite is the chief ore source of uranium and radium, the basic raw material of nuclear energy, it is extremely important. Its chemical variants include bröggerite and cleveite.

Another important ore of uranium and radium is carnotite, a secondary mineral resulting from the reaction of groundwater with preexisting uranium-bearing minerals. Existing as a hydrous vanadate of potassium and uranium (the water varying from one to three molecules depending on the temperature), the largest deposits are in the Colorado Plateau, in Australia, and in Zaire.

Minerals containing uranium possess the element in two ionic forms: as a quadrivalent ion and as a hexavalent one known as the uranyl ion. Minerals with the positive four-valence ion are often black, do not fluoresce in ultraviolet light, and usually occur as primary deposits. Such minerals include uraninite and coffinite. Most uranium-containing minerals, however, are in the uranyl ion form, characterized by a bright lemon-yellow to green color and fluorescing in ultraviolet as a lemon-yellow color. All these minerals are secondary in origin, deriving from preexisting sediments, and are formed from solutions at relatively low temperature and pressure. Uranium 238 has a half-life of 4.5×10^9 years and decays in ten steps to lead 206; uranium 235 has a half-life of 7.1×10^8 years and decays in eleven steps to lead 207. This means that when the Earth was formed about 4.5 $\times 10^9$ years ago, uranium 238 was twice as abundant, and uranium 235 was eight times as abundant as it is now.

THORIUM

With an average concentration of 12 parts per million, thorium is much more abundant in the crust than is uranium. The major source for thorium is the mineral monazite, a rare-earth phosphate containing 3-10 percent thorium as thorium oxide (ThO_2). Thorium substitutes for cerium (Ce) and lanthanum (La), giving a series of monazite minerals that may range all the way to 30 percent thorium oxide. Monazite is very widely disseminated in granites, pegmatites, and gneiss rocks as accessory grains and crystals. The weathering products of such materials, found in the form of beach sands or fluvial deposits, often include large amounts of thorium.

Thorium has a half-life of 1.4×10^{10} years. Its decay series, going through ten successive disintegra-

tions, terminates at lead 208. It is useful as an alloying agent in some structural metals and in magnesium technology; in a nuclear reactor, it can be converted to uranium 233, an atomic fuel.

RADIUM

Radium, with a decay half-life of 75,400 years, is another radioactive element usually found in uranium minerals but only in concentrations of 1 part to 3 million parts uranium. Thirteen radium isotopes are known, all radioactive, but only radium 226 is technologically important. The best mineral source at present is pitchblende. Radium salts tend to ionize the surrounding atmosphere, causing a bluish glow to appear. Radium compounds also cause other compounds in mineral form, such as zinc sulfide, to phosphoresce and fluoresce.

CONCENTRATIONS OF RADIOACTIVE ELEMENTS

All three elements occur in high concentrations in the continental crust compared to the oceanic material. The reason relates to the large ionic radii of the isotopic elements compared to those of silicon, aluminum, magnesium, and iron, which combine with oxygen to make up the bulk of the Earth's oxide minerals. The larger ions do not fit as readily into the crystal lattice formed with oxygen, so that in magmatic and metamorphic events, they are freely mobile and tend to follow magmatic products and volatiles upward into the crust. There, they incorporate themselves more easily into the more open crystal structures found in the lower-pressure environments of the surface crustal material.

Uranium, in one mineral form or another, has been found in every geologic environment of the United States, with the exception of ultramafic igneous rocks. Concentrations of radioactive elements have been located in igneous rocks, veins, clastic sediments, precipitates, evaporites, coals, marine black shales, and limestones. The principal sources of uranium in the United States are in terrestrial sandstones and limestones, principally in areas characterized by simple folded mountains alternating with broad basins, such as the Black Hills of South Dakota. The ore bodies themselves originate by a number of means, including deposits from molten magma, hot fluids, groundwater, surface solutions, volcanic ash, and seawater.

Three major types of uraniferous veins have

been recognized by their mineral associations: nickel-cobalt-native silver veins, silica-iron-lead veins, and iron-titanium veins. The first group is characterized by pitchblende, as is the second (with different associated minerals), while the third type of vein is characterized by uranium titanites, such as davidite, occurring mainly in igneous intrusive rocks, such as at Radium Hill, South Australia. The geologic processes that lead to the formation of mineral deposits include differentiation by fractional crystallization in magmas, the escape of hot liquids and gases from cooling melts into surrounding rocks, concentration by underground waters, concentration by sedimentation (the principal process for the formation of monazite), and fixation by biochemical processes such as plant rooting. All these mechanisms have been shown to be active in depositing uranium and thorium in the surface rocks.

Because of the means by which uranium is produced and concentrated, uranium and thorium can be shown to cluster within broad, poorly defined areas of the globe. Initial concentration probably occurred during the formation of the Earth's crust. Geologic analyses and histories provide detailed information on where to look, what minerals to expect, and the feasibility of extracting the radioactive minerals located in a particular area.

Concentration Studies

Because concentrations of the radioactive elements in rocks are usually in trace amounts, measured in parts per million, very precise chemical and geological analysis is required. The isotopic abundance is most often obtained through multi-channel gamma-ray spectrometry, a technique by which gamma-ray emissions from the various elements are separated and counted on the basis of energy. For each el-

emental decay, gamma radiation has very characteristic energies, which the decay counter can detect. The counts at specific levels of the energy spectrum are indicative of the particular isotopic abundance in the sample.

Concentration studies can also be done using atomic absorption spectrometry. This involves heating the sample to a vapor, then passing a light through the gas to see which wavelengths are absorbed. The elements present and their abundances can be easily determined in this way. Three different methods of physically separating the isotopes are common. Gaseous diffusion involves passing the gases through a porous material; the light elements pass through more swiftly than do the heavier ones. In a mass spectrograph, the particles are forced to follow a curved path in a vac-

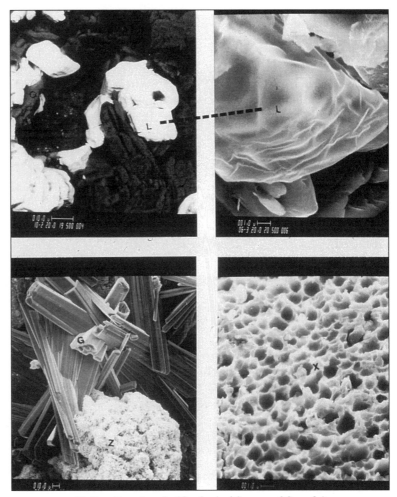

Radioactive minerals. (Geological Survey of Canada)

uum chamber under the influence of a magnetic field, the amount of bending of the path depending on the mass of the element. In thermal diffusion, the sample is heated to 500 degrees Celsius and allowed to move in a convection gradient, separating because of the ability of lighter particles to move faster than heavier ones.

Minerals with radioactive elements can be detected by many techniques. Autoradiography uses the interaction of the radiation emitted by radioactive elements with a photographic plate, causing an ion trail that essentially breaks down the photographic emulsion. The density of the image depends on the concentration of radioactive elements present. In this very direct method, results can be quantified if the tracks are clearly defined, for the numbers of trails can simply be counted. The fission track method, used for uranium and thorium, is similar, in that a polished sample is broken apart by the emitted radiation, leaving tracks which can be seen once the sample is etched. Generally, the material is embedded in plastics that record the tracks upon themselves. Such a technique has been used to detect concentrations of uranium from 4,000 parts per million in zircons to 10^{-3} parts per million in hypersthene and anorthite.

Identification can also be done by four somewhat similar methods. With the polarizing microscope, particular minerals can be identified by their color changes under the action of polarized light. In X-ray diffraction, the mineral is powdered and bombarded with X rays. The radiation will be diffracted by the planes of atoms present, forming a diffractogram, which reveals the crystal structure. In electron microscopy, a small sample, as small as 1 micron, is bombarded with an electron beam, so that X rays with wavelengths characteristic of the particular radioactive element are produced. The intensity of radiation also reveals the concentration of the element in the sample. Finally, with the scanning electron microscope, the sample is coated with a gold-palladium layer and bombarded with electrons, revealing the detailed topography of the mineral. Magnifications of up to 100,000 are possible.

EXTRACTION OF MINERALS

To obtain purified materials, the major technique used with all the minerals is the extraction

method, which permits high selectivity and complete separation. For uranium, the best method is extraction of uranyl nitrate with ether from a nitrate acid solution saturated with ammonium nitrate. Tributyl phosphate may replace the ether, while adding salts, such as trivalent aluminum or iron, or divalent zinc, causes a salting-out effect to occur, precipitating even more uranium oxide. Ion exchange is used to extract the other radioactive elements (and works for uranium as well). In this process, complex anions on various exchange media, most commonly sulfuric acid or carbonate solutions, are used for their sorbable qualities. The material is passed through a column of basic anion (negatively charged) exchangers. Under the appropriate conditions, uranium, for example, in a nitrate acid in 80 percent alcohol solution, separates completely from other rare-earth metals.

SCIENTIFIC AND ECONOMIC USES

Though radioactive minerals are rare, they have become extremely important scientifically and economically. Uranium, thorium, and radium are useful in many fields: medicine, industrial processes, and physics and chemical theoretical studies. The study of radioisotopes, formed from the radioactive elements separated from the Earth's minerals, has led to immense increases in knowledge of how the human body works, how plants employ photosynthesis, and how individual cells manage to maintain homeostatic equilibrium. In chemistry, such isotopes provide a picture of intermediaries in chemical reactions, allowing better understanding of the workings of chemical compounds such as enzymes. In physics, a theoretical understanding of the nuclear structure itself— down to the fundamental level of quarks, the supposed elementary particles of the universe—has been gained through watching the breakdown of radioactive nuclei in nature. In the iron and steel industries, their use has helped to shed light on the thermodynamics and kinetics of metal-slag reactions, the hydrodynamics of molten steel, the movement of charged materials in a blast furnace, and crystallization in metallurgical products.

In the field of geochronology, the only means of setting absolute dates on events in the distant past is to examine the breakdown, at a precise decay rate, of radioactive elements. Not only can geoscientists investigate the origin of the Earth and the rest of the solar system in this way, but they can also study the evolution of the Earth's surface, oceans, and atmosphere—the collage of geological events, such as mountain building or wandering continents, that have shaped the globe.

More important for the average person, however, is the fact that radioactive elements provided the basic materials necessary for the development of nuclear fission as an energy source. Nuclear technology is dependent on a continuous supply of such minerals to create the energy required to meet the demands of industrialized countries at present; such demands are likely to increase. Unfortunately, the hazards posed by the manufacture of nuclear weapons, radiation fall-out from accidents at energy plants, and nuclear waste disposal loom large. It is to be hoped that as these rare minerals continue to be mined and used, the attendant problems will be addressed and eventually solved.

Arthur L. Alt

CROSS-REFERENCES

Biopyriboles, 1175; Carbonates, 1182; Clays and Clay Minerals, 1187; Feldspars, 1193; Gem Minerals, 1199; Hydrothermal Mineralization, 1205; Ionic Substitution, 1209; Magnetite, 1214; Metamictization, 1219; Minerals: Physical Properties, 1225; Minerals: Structure, 1232; Non-silicates, 1238; Orthosilicates, 1244; Oxides, 1249.

BIBLIOGRAPHY

Blatt, Harvey, and Robert J. Tracy. *Petrology: Igneous, Sedimentary, and Metamorphic.* New York: W. H. Freeman, 1996. A study of the major rock types, tracing the processes by which igneous, sedimentary, and metamorphic materials are formed. Discusses mineral concentration factors and relates minerals, including radioactive ones, to global tectonics. Contains a section on the use of isotopes. Well written but detailed; includes many references to other works.

Faul, Henry, ed. *Nuclear Geology.* New York: John Wiley & Sons, 1954. After a readable introduction into the realm of nuclear physics, which defines terms and describes instruments used, the work details the distribution of uranium and thorium in rocks and the ocean. Discusses types of minerals, how they are found geographically, and how they are used geochemically. Contains extensive references.

Faure, Gunter. *Principles of Isotope Geology.* 2d ed. New York: John Wiley & Sons, 1986. This well-written work explains how isotopic data are used to interpret and solve geologic problems, showing how radioactive isotopes make important contributions to virtually every area of science. Includes discussion of minerals, geochronology, and the uses of natural radioactivity. Features helpful references.

Finch, Warren Irvin. *Principal Radioactive Minerals Encountered in Mining and Associated Environmental Concerns.* Reston, Va.: U.S. Geological Survey, 1994. This open-file report by the U.S. Geological Survey contains information on the mining of radioactive substances such as uranium and thorium ores, and the environmental hazards that often accompany such ventures.

Glasstone, Samuel, et al. *Principles of Nuclear Reactor Engineering.* New York: Van Nostrand Reinhold, 1955. Dealing with the entire scope of reactor engineering, this work has two important chapters on reactor materials and the processing of radioactive minerals. Provides a detailed overview of how reactors work, how energy is produced from radioactive minerals, the types of radiation emitted, and the nature of the reactions. Rather extensive mathematics but very interesting.

Gurinsky, David, and G. J. Dienes. *Nuclear Fuels.* New York: Van Nostrand Reinhold, 1956. A guide to the metallurgy of uranium and thorium, including means of preparation and production for use. Describes the metals' growth and formation in nature and discusses alloys and properties useful for technology. A well-written presentation of radiation effects. Includes numerous references and explanatory diagrams.

Heaman, L., and J. N. Ludden, eds. *Applications of Radiogenic Isotope Systems to Problems in Geology*. Nepean, Ontario: Mineralogical Association of Canada, 1991. This short-course handbook has articles by a variety of authors on isotope geology, radioisotopes in geology, and radioactive dating techniques. Bibliography.

Lavrukhina, Augusta K., et al. *Chemical Analysis of Radioactive Materials*. Cleveland: CRC Press, 1967. This work discusses the states and behaviors of radioactive elements and the theoretical bases for methods of analysis. Topics include detecting radioisotopes, concentrating elements, and means of isolating each one. The references are helpful. Features detailed analysis and chemistry.

Mason, Brian, and Carleton B. Moore. *Principles of Geochemistry*. 4th ed. New York: John Wiley & Sons, 1982. A well-written work on elements that are significant in the Earth's structures. Covers their physical and chemical evolution within the system. Contains chapters on isotope geochemistry for stable isotopes, radiation emitters, and their usefulness in geochronology. A very competent discussion of the important radioactive minerals. Good references are provided.

Nininger, Robert D. *Formation of Uranium Ore Deposits*. Vienna: International Atomic Energy Agency, 1974. This work appraises the world supply of uranium minerals. Mechanisms for the formation and concentration of minerals are examined closely, and the isotope geology and mineralogy of diverse deposits around the world are surveyed. Includes numerous maps and illustrations. Some of the papers are not in English. Well equipped with references.

_____. *Minerals for Atomic Energy*. New York: Van Nostrand Reinhold, 1954. A comprehensive work on the analysis of the radioactive minerals. Covers ore deposits and formations and the types of minerals containing uranium, thorium, and radium in association with other elements. Describes where such minerals occur in nature. Informative identification guide to the majority of important minerals, with good illustrations.

Scheid, Werner, and Aurel Sandulescu. *Frontier Topics in Nuclear Physics*. New York: Plenum, 1994. The essays in this book, published in cooperation with the North Atlantic Treaty Organization (NATO) Scientific Affairs Division, cover both basic and advanced concepts in nuclear physics, nuclear structure, heavy elements, heavy ion collisions, and nuclear astrophysics. Bibliography and indexes.

2
IGNEOUS ROCKS

ANDESITIC ROCKS

Andesite is an intermediate extrusive igneous rock. Active volcanoes on the Earth erupt andesite more than any other rock type. Andesites are primarily associated with subduction zones along convergent tectonic plate boundaries.

INTEREST TO GEOLOGISTS

Andesite takes its name from lavas in the Andes mountains of South America. To most geologists, andesites are light gray porphyritic volcanic rocks containing phenocrysts of plagioclase but little or no quartz and no sanidine or feldspathoid. Despite their lackluster appearance, andesites are of great interest to geologists for several reasons. First, active volcanoes on the Earth erupt andesite more than any other rock type; andesite is the main rock type at 61 percent of the world's active volcanoes. Second, andesites have a distinctive tectonic setting. They are primarily associated with convergent plate boundaries and occur elsewhere only in limited amounts. Of the active volcanoes that occur within 500 kilometers of a subduction zone, 78 percent include andesite; only three active volcanoes not near a destructive plate boundary include it. Third, andesites have bulk compositions similar to estimates of the composition of continental crust. This similarity, in association with the tectonic setting of andesites, suggests that they may play an important role in the development of terrestrial crust. Fourth, the development and movement of andesitic magma seem to be closely related to the formation of many ore deposits, including such economically important ores as molybdenum and porphyry copper deposits.

VISCOSITY

Rocks are classified chemically according to how much silica they contain; rocks rich in silica (more than 64 percent) are called silicic. They consist mostly of quartz and feldspars, with minor amounts of mica and amphibole. Rocks low in silica (less than 54 percent), with no free quartz but high in feldspar, pyroxene, olivine, and oxides, are called basic. Basic rocks, free of quartz, tend to be dark, while silicic rocks are lighter and contain only isolated flecks of dark minerals. Basalts are examples of silicic volcanic rocks. Andesites, having a silica content of about 60 percent, are volcanic rocks termed intermediate. Andesite's plutonic equivalent is diorite. There is no cut-and-dried difference between basalt and andesite or between andesite and rhyolite. Instead, there is a broad transitional group of rocks that carry names such as "basaltic andesite" or "andesitic rhyolite."

Nevertheless, some generalizations can be applied to the lavas and magmas that form these rocks. One generalization has to do with viscosity. Andesite lavas are more viscous than are basalt lavas and less viscous than are rhyolite lavas. This difference is primarily an effect of the lava's composition, and to a certain extent it is a result of the high portion of phenocrysts present in the more viscous lavas.

Different minerals crystallize at different temperatures. As a basalt magma cools, a sequence of minerals appears. The first mineral to crystallize is usually olivine, which continues to crystallize as the magma cools until a temperature is reached at which a second mineral, pyroxene, begins to crystallize. As the temperature continues to drop, these two continue to crystallize. Cooling continues until a temperature

Andesite lava from Mount Lassen, California. (© William E. Ferguson)

is reached when a third mineral, feldspar, crystallizes. This chain of cooling and mineral crystallization is known as Bowen's reaction principle. Often, olivine and pyroxene crystallize out early in the process, so they may be present in the final rock as large crystals, up to a centimeter across. These crystals are called phenocrysts. The size of these phenocrysts is in direct contrast to the fine-grained crystals of the groundmass. Igneous rocks with phenocrysts in a fine-grained groundmass are known as porphyries. Most volcanic rocks contain some phenocrysts. The groundmass crystals form when the lava cools on reaching the surface. If the lava has a low viscosity, reaches the surface, spreads out, and cools quickly, individual crystals do not have enough time to grow. The overall rock remains fine-grained. The phenocrysts crystallized out much earlier, while the magma was still underground. There, they had plenty of time to grow and were then carried to the surface with the magma during eruption.

ANDESITE FLOWS

Basalts, as a result of their low viscosity, tend to produce thin lava flows that readily spread over large areas. They rarely exceed 30 meters in thickness. Andesite flows, by contrast, are massive and may be as much as 55 meters thick. The largest single andesite flow described, which is in northern Chile, has an approximate volume of 24 cubic

kilometers. Because of their low viscosity, basalt flows can advance at considerable rates; speeds up to 8 kilometers per hour have been measured. Andesite flows often move only a matter of tens of meters over several hours. As a result of higher viscosity, andesite flows show none of the surface features of more "liquid" lavas. Flow features such as wave forms, swirls, or ropy textures often associated with basalt flows never occur in andesite flows. Andesite flows tend to be blocky, with large, angular, smooth-sided chunks of solid lava. The flow tends to behave as a plastic rather than a liquid. An outer, chilled surface develops on the slow-moving flow, with the interior still molten. Plasticity within the flow increases toward the still-molten inner portion. As the flow slowly shifts, moves, and cools, the hard, brittle outer layer breaks into the large angular blocks characteristic of andesite lava flows. As the flow slowly moves, the blocks collide and override one another to form piles of angular andesite blocks.

The viscous nature of andesite lavas is responsible for many classic volcanic features. Most notable is the symmetrical cone shape of the stratovolcanoes of the Circum-Pacific region. Short, viscous andesitic flows pouring down the flanks of these volcanoes are alternately covered by pyroclastic material and work to build the steep central cone characteristic of composite stratovolcanoes. When an andesite's silica content rises to a point that it approaches rhyolite composition, it is termed a dacite. Dacite is often so viscous that it cannot flow and blocks the vent of the volcano. This dacite plug is called a lava dome. Such a dome can be seen in pictures of Mount St. Helens. If the plugged volcano becomes active again, the lava dome does not allow for a release of accumulating pressure and explosive gases. Pressure builds until the volcano finally erupts with great force and violence.

ANDESITE LINE

The differences between basalts and andesites reflect their differences in composition, which is a function of the environment in which their source lavas occur. Basalts are typically formed at mid-ocean ridges and form oceanic crust. The generation of andesite magma is characteristic of de-

structive plate margins. Here, oceanic plates are being subducted below continental plates. Destructive plate boundaries tend to produce a greater variety of lavas than do spreading zones (ocean ridges). The close association of andesite with convergent plate boundaries is the significance of the "Andesite Line" often drawn around the Pacific Ocean basin. This fairly well-defined line separates two major petrographic regions. Inside this line and inside the main ocean basin, no andesites occur. All active volcanoes inside the line erupt basaltic magma, and all volcanic rocks associated with dormant volcanoes within this region are basaltic. Outside the line, andesite is common. The Andesite Line is the western and northern boundary of the Pacific plate and the eastern boundary of the Juan de Fuca, Cocos, and Nazca plates. The Andesite Line parallels the ma-

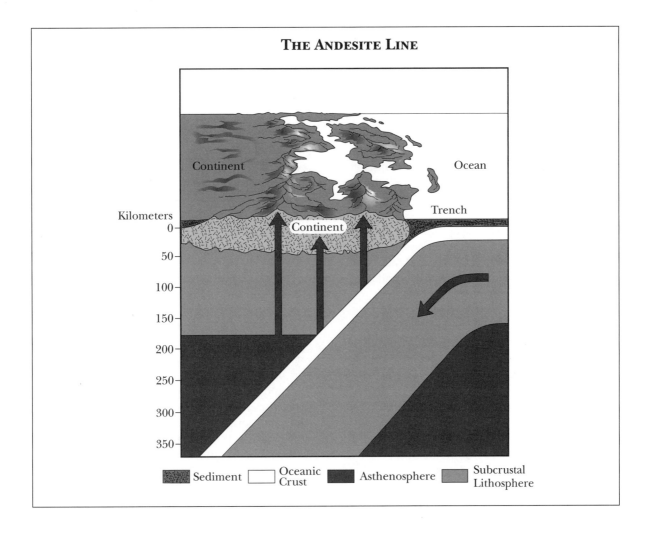

THE ANDESITE LINE

jor island arc systems, the subducting edges of the tectonic plates listed above, and a chain of prominent and infamous stratovolcanoes known as the Ring of Fire. These stratovolcanoes are exemplified by Mount Rainier and other volcanoes of the western United States, El Chichón of Mexico, San Pedro of Chile, Mounts Egmont and Taupo of New Zealand, Krakatau and Tambora of the Indonesian Arc, Fujiyama of Japan, Bezymianny of Kamchatka, and the Valley of Ten Thousand Smokes in Alaska.

BENIOFF ZONE

The Benioff zone is a plane dipping at an angle of 45 degrees below a continent, marking the path of a subducting oceanic plate at the destructive plate margin. It is a region of intense earthquake activity. At the Benioff zone, the subducting plate melts, producing a chemically complex, destructive margin magma (andesitic). As andesitic magma is formed at the Benioff zone and begins its slow rise toward the surface, it passes through and comes in contact with regions of the asthenosphere, lithosphere, and continental crust. During its rise, the magma also comes in contact with circulating meteoric water, and new magmatically derived fluids are formed. Because andesitic magma rises through a variety of host rocks, the variety of minerals that enrich the new fluids is increased. The rising magma also stresses the surrounding host rocks, causing them to fracture. These fractures and similar open spaces are filled with mineral-rich water and magmatically derived fluids. As the fluids cool, mineral ores precipitate from the solutions and fill the open spaces. These filled spaces often become mineral-rich dikes and sills.

FIELD STUDY

The study and interpretation of andesitic rocks is accomplished in three basic ways: fieldwork, in which researchers travel to locations to assess and interpret a specific region of the Earth that is known or suspected of being andesitic terrain; petrological studies, which use all available methods of study to ascertain the history, origin, conditions, alterations, and decay of the rocks collected during fieldwork; and mineralogical studies, which through intensive laboratory investigations identify the specific mineral characteristic of a sample.

If a scientist is interested in andesitic rocks and knows, or suspects, that a region of the Earth is andesitic terrain, travel to this location to do fieldwork is likely. There, the researcher will most likely collect a large number of samples. The location from which each sample is taken is carefully recorded on a map. Additional data describing the stratigraphic interval from which the sample is taken are recorded. These data include thickness, areal extent, weathering, and strike and dip of the location. Photographs are also commonly used to record the surrounding environment of the sample location. Samples are usually given preliminary study at the researcher's field camp.

PETROLOGICAL AND MINERALOGICAL STUDIES

When the fieldwork is over, the researcher returns with the samples to a laboratory setting. In the laboratory, petrological and mineralogical investigations begin. A petrological study of an andesitic sample will include both a petrographic and petrogenetic analysis. The petrographic analysis will describe the sample and attempt to place it within the standard systematic classification of igneous rocks. Identification of the sample is accomplished by means of examining a thinly sliced and polished portion of the sample under a petrographic microscope—a process known as thin-section analysis. The information obtained by microscopic examination gives a breakdown of the type and amount of mineral composition within the sample. Knowing the conditions under which these minerals form allows the researcher to make a petrogenetic assessment.

Petrogenesis deals with the origin and formation of rocks. If a mineral is known to form only at certain depths, temperatures, or pressures, its presence within a sample allows the researcher to draw some specific conclusions concerning the rock's formation and history. Mineralogical studies of field-gathered samples aid in the petrogenetic portion of the analysis. Mineralogy involves the study of how a mineral forms, its physical properties, chemical composition, and occurrence. Minerals can exist in a stable form over only a narrow range of pressure and temperature. Experimental confirmation of this range enables the researcher to make a correlation between the occurrence of a mineral in a rock and the conditions under which the rock was formed. In this sense, mineralogical and petrological analyses complement each other and work to formulate a concise history of a given sample.

When data gathered from the field and laboratory are combined, the geological history of a given field area can begin to be interpreted. When areas of similar igneous rock types, ages, and mineral compositions are plotted on a map, they form a petrographic province. A petrographic province indicates an area of similar rocks that formed during the same period of igneous activity. On a global scale, the Andesite Line marks the boundary between two great provinces: the basaltic oceanic crust and the andesitic continental crust. Both crustal forms have distinctly different geological histories and mineral compositions.

Randall L. Milstein

BIBLIOGRAPHY

Bowen, N. L. *The Evolution of Igneous Rocks.* Mineola, N.Y.: Dover, 1956. An unmatched source and reference book on igneous rocks, written by the father of modern petrology. This book is the basis of Bowen's reaction principle. Written for graduate students and professional scientists.

Carmichael, I. S. E., F. J. Turner, and J. Verhoogen. *Igneous Petrology.* New York: McGraw-Hill, 1974. A college-level textbook on the formation and development of igneous rocks. Very detailed and complete.

Gill, J. B. *Orogenic Andesite and Plate Tectonics.* New York: Springer-Verlag, 1981. A well-documented summary of the entire field of andesite genesis. Written for graduate students and professional Earth scientists.

Klein, Cornelis, and Cornelius S. Hurlbut, Jr. *Manual of Mineralogy.* 21st ed. New York: John Wiley & Sons, 1999. An introductory college-level text on mineralogy. Discusses the physical and chemical properties of minerals and describes the most common minerals and their varieties, including gem mineral varieties. Contains more than 500 mineral name entries in the mineral index and describes in detail about 150 mineral species.

Perch, L. L., ed. *Progress in Metamorphic and Magmatic Petrology.* New York: Cambridge University Press, 1991. Although intended for the advanced reader, several of the essays in this multiauthored volume will serve to familiarize the new student with the study of igneous rocks. In addition, the bibliography will lead the reader to other useful material.

Science Mate: Plate Tectonic Cycle. Fremont, Calif.: Math/Science Nucleus, 1990. This book is part of the Integrating Science, Math, and Technology series of manuals intended to help teachers explain basic scientific concepts to elementary students. As such, it provides a basic understanding of plate tectonics, earthquakes, volcanoes, and general geology for the student with no background in the Earth sciences.

Sutherland, Lin. *The Volcanic Earth: Volcanoes and Plate Tectonics, Past, Present, and Future.* Sydney, Australia: University of New South Wales Press, 1995. Although Sutherland focuses on volcanic activity in Australia, the book provides an easily understood overview of volcanic and tectonic processes, including the role of igneous rocks. Includes color maps and illustrations, as well as a bibliography.

Williams, H., and A. R. McBirney. *Volcanology.* San Francisco: Freeman, Cooper, 1979. A classic textbook on volcanoes and volcanology. Well illustrated and very descriptive. Written for the undergraduate or graduate student.

Windley, B. F. *The Evolving Continents.* New York: John Wiley & Sons, 1977. An excellent reference book on plate tectonics and tectonic processes. Written for the college-level reader.

ANORTHOSITES

Anorthosites are coarse-grained, intrusive igneous rocks composed principally of plagioclase feldspar. They are useful for what they reveal about the early crustal evolution of the Earth, and they are the source of several economic commodities.

PRINCIPAL TERMS

CRUST: the upper layer of the Earth and other "rocky" planets; it is composed mostly of relatively low-density silicate rocks

GABBRO: coarse-grained, iron-magnesium-rich, plutonic igneous rock; anorthosite is an unusual variety of gabbro

HYPERSTHENE: a low-calcium pyroxene mineral

MAGMA: molten silicate liquid plus any crystals, rock inclusions, or gases trapped in that liquid

MANTLE: a layer in the Earth extending from about 5 to 50 kilometers below the crust

MASSIF: a French term used in geology to describe very large, usually igneous intrusive bodies

NORITE: gabbro in which hypersthene is the principal pyroxene; it is commonly associated with anorthosites

PERIDOTITE: the most common rock type in the upper mantle, where basalt magma is produced

PLAGIOCLASE: a silicate mineral found in many rocks; it is a member of the feldspar group

PLUTONIC: formed by solidification of magma deep within the Earth and crystalline throughout

PRECAMBRIAN EON: a large span of geologic time extending from early planetary origins to about 600 million years ago

ANORTHOSITE COMPOSITION

Anorthosites are igneous rocks that are composed primarily of plagioclase feldspar (calcium-sodium-aluminum silicate). Minor minerals in these rocks may include pyroxene minerals (calcium-iron-magnesium silicates), iron-titanium oxides, and, in metamorphosed varieties, garnet. All anorthosites are coarse-grained, plutonic rocks (they crystallize at depth), and they may have plagioclase crystals 10 centimeters or more in length. Because the color of plagioclase changes with minor changes in chemical composition, anorthosites come in a variety of colors. Light gray is the most common, but dark gray, black, light blue, green, and brown varieties are known.

ANORTHOSITE TYPES

In the Earth, there are two types of anorthosite, massif-type anorthosites and layered intrusive anorthosites. The former occur as large, lens-shaped or dome-shaped intrusions that may be exposed by erosion over an area of several square kilometers. The latter variety is associated with intrusions of gabbroic (iron-rich, low-silica) magma that has segregated into mineralogically distinct layers, anorthosite occurring in the uppermost layers. The lighter areas on the Moon, known as the "lunar highlands," are composed of anorthosite that originated in a similar manner to terrestrial, layered intrusive anorthosites.

Massif-type anorthosites are unusual among igneous rocks in their restricted distribution in time and space and their composition. They contain at least 85 percent plagioclase, the other silicate minerals being either augite (high-calcium pyroxene), hypersthene (low-calcium pyroxene), or both. Ilmenite and apatite commonly occur as minor accessory minerals. Massif anorthosites are found almost exclusively in a wide belt from the southwestern United States through Labrador and on the other side of the Atlantic through Sweden and Norway. Another belt extends from Brazil, through Africa (Angola, Tanzania, Malagasy), through Queen Maud Land in Antarctica, and across Bengal, India, to Australia (a pre-continental drift reconstruction). The best examples of massif anorthosites in the United States are in the Adirondack Mountains in northeastern New York

Anorthosite that is 90 percent labradorite feldspar. (© William E. Ferguson)

State. In Canada, the Nain and Kiglapait intrusions are notable for their excellent surface exposures of anorthosite and associated rocks.

ANORTHOSITE FORMATION

Ages determined with radiometric dating techniques demonstrate that nearly all anorthosites, including the layered-intrusion varieties, are Precambrian; most are between about 1,700 million and 1,100 million years old. No lava flows of anorthositic composition are known; they are exclusively plutonic. In addition, experimental evidence shows that liquids of anorthositic composition cannot be produced by any known process of melting at depth in the Earth. Although specific mechanisms are not fully understood, it is generally accepted that anorthosites form from concentrated plagioclase crystals that had previously crystallized from gabbroic or other magmas. Massif anorthosites are commonly associated with gabbroic rocks called "norites" (gabbro in which the pyroxene mineral is hypersthene), and it can be demonstrated in most cases that anorthosite is produced by adding progressively more plagioclase to norite magma.

Anorthosites in layered intrusives clearly result from the same processes that produce the other layered rocks associated with the anorthosite layers. Layered intrusives, commonly termed "layered complexes," are generally tabular bodies that vary in thickness from a few hundred to thousands of meters thick. They consist of extremely iron- and magnesium-rich rocks called peridotites in their lower reaches but grade upward into gabbros and, in some cases, anorthosites in their upper extents. This layering of different rock types is attributed to the gravitationally controlled settling of dense minerals to the bottom of a large body of originally gabbroic magma. Rock layers of different compositions are built up over time as different minerals crystallize and are deposited on the chamber floor. Plagioclase generally crystallizes relatively late in this process and, depending on its composition and that of the enclosing silicate liquid, may actually float to the top of the magma chamber instead of settling to the bottom. Anorthosite deposits are known to have been formed by both settled and floated plagioclase. In both cases, anorthosite results from the concentration of plagioclase by mechanical processes after crystallization.

ANORTHOSITE DISTRIBUTION

Layered intrusive anorthosites occur in the early Precambrian Stillwater Complex of western Montana and late Precambrian Duluth Complex that parallels the north shore of Lake Superior in northern Minnesota. The Stillwater contains three anorthosite units, each about 400 to 500 meters thick. In contrast, the Duluth Complex, about 40 kilometers thick, is mostly anorthosites and "gabbroic anorthosites" (somewhat richer in iron-magnesium silicates than anorthosites), with a mostly unexposed, relatively thin peridotite unit at its base. Norites and troctolites (olivine plus plagioclase) occur as minor associated rocks. Other layered anorthosites occur in the Bushveld complex in South Africa, the Fiskenaesset Complex in Greenland, and the Dore Lake and Bell River complexes in Ontario, Canada, among others. All the layered intrusions in the United States are Precambrian.

Without question, the most obvious (but least accessible) exposures of anorthosite occur on the Moon. Relatively rare on Earth, anorthosite is the dominant rock type on the Moon, where it forms the bulk of the rocks in the lunar highlands. The lunar highlands are the Moon's ancient, highly cratered, light-colored areas. The dark areas are basalt (iron-rich silicate rock) lava flows that fill huge craters blasted in the highlands by large meteorite collisions. Ever since the first Moon rocks were returned to Earth by the Apollo 11 astronauts in 1969, scientists have studied lunar anorthosites and associated rocks for clues to how the Moon's crust and interior originated and evolved. The generally accepted model postulates that early in its history, more than 4 billion years ago, the entire surface of the Moon became molten to a depth of several kilometers. This "magma ocean" then cooled, and as it cooled, it crystallized various silicate and oxide minerals. The heavier, iron-rich minerals, being denser than the silicate magma, sank to form deep-seated layers, similar to the peridotite layers in the Stillwater and Bushveld complexes noted above. Plagioclase, on the other hand, was less dense than the iron-rich liquid, so it floated to the top, solidifying to form the early lunar highland anorthosites. Over time, the highlands became increasingly cratered by the meteorite impacts, until they acquired their present appearance about 3 billion years ago. Since that time, meteorite impacts have been sporadic. Interestingly, lunar highland anorthosite occurs with minor quantities of norite and troctolite, two rock types that also are commonly associated with terrestrial anorthosites. To lunar scientists, this peculiar group of associated rocks is the "ANT suite"; ANT is an acronym for anorthosite, norite, troctolite.

ASSESSING ANORTHOSITE ORIGIN AND SIGNIFICANCE

Paradoxically, the origin of terrestrial anorthosites is not as clear-cut as the origin of lunar anorthosites, and many models and hypotheses have been offered to explain their unique composition and space-time relationships. No explanation is readily accepted as applying to all or most anorthosite occurrences, but enough is known about anorthosite bodies to allow some good, educated guesses. Scientific work on anorthosites has included numerous field-mapping projects to determine their spatial extent and structure. Laboratory work has included radiochemical dating studies to determine the absolute ages of anorthosites and associated rocks and experimental studies that have explored possible parent materials and melting environments of anorthosite magmas.

Several factors must be considered in assessing the origin and significance of anorthosite bodies, specifically the massif varieties. First, anorthosites are restricted, for the most part, to the very narrow time band between about 1.7 and 1.1 billion years before the present, as determined mostly by potassium-argon, uranium-lead, and rubidium-strontium dating. Second, they occur in belts where orogeny (mountain building) may have occurred before or after their emplacement but where they intruded during anorogenic times or during rifting. They are unequivocally igneous as opposed to metamorphic; that is, they did not result from some other type of rock that changed in the solid state but rather from molten magmas. This fact is determined in the field by examining anorthosite contacts with older rocks to see whether these rocks show evidence of thermal heating caused by intrusion or hot magma. Anorthosites lack minerals that contain water in their structures, so water was not an important constituent of anorthosite parent magmas. Finally, many anorthosite complexes, such as those in New York's Adirondack Mountains, show an association not only with norites and troctolites but also with pyroxene-bearing granitic rocks called "charnockites."

Although experimental evidence shows that a liquid of plagioclase composition cannot be generated under high pressures from any known Earth materials, magmas can form from rocks in the mantle or lower crust that—under anhydrous conditions—are capable of generating the plagioclase "mush" (crystals plus a small amount of liquid) by plagioclase flotation. Laboratory melting experiments show that one type of magma, called "quartz diorite" (relatively siliceous magma of the type erupted in 1980 by Mount St. Helens), could produce a plagioclase mush at great pressures. Intrusion of this mush at higher levels in the crust could squeeze out some of the interstitial liquid, which would crystallize as charnockites. Norites and gabbroic anorthosites represent cases where the plagioclase mush has trapped greater or smaller amounts of more iron-rich silicate liquid.

Other scenarios are possible, and others have been proposed to explain anorthosite bodies. Whatever the precise mechanism might have been, it produced large volumes of a very unusual suite of rocks over a geologically brief time interval. No doubt this extensive melting event required higher heat flow in the crust and mantle than is now present in areas of anorthosite occurrences or in most other areas of the Earth. The explanation for anorthosites therefore requires an evaluation of what unique thermal or cataclysmic event they record in Earth history. About that, geologists can only conjecture.

DECORATIVE AND PRACTICAL APPLICATIONS

Anorthosites are not very common in the Earth's crust, and most people are not as familiar with this stone as they are with the more common building stones, granite and marble. Ironically, anyone who has ever seen the Moon has seen huge spans of anorthosite; it is the principal rock type composing the lunar highlands, the dominant light-colored areas of the Moon.

Anorthosite is used as a building stone or for decorative building facings. A beautiful variety composed mostly of the high-calcium plagioclase known as labradorite is in particular demand. Labradorite exhibits blue, violet, or green iridescence that varies in color with the angle of incidence of light. Anorthosite composed of labradorite plagioclase is sometimes polished and used in tabletops and floor or wall panels. It has other decorative and practical applications as well. Like other feldspar-rich rocks, anorthosite can be used to make ceramic products. Porcelain bath fixtures, insulators, and dining ware are made from finely pulverized feldspar that is heated to very high temperatures. The resistance to heat and electricity and the general durability of porcelain come from the same properties inherent in feldspar.

Anorthosites and associated rocks are also the sites of economically exploitable iron-titanium deposits in some localities. The ore occurs mostly as ilmenite (iron-titanium oxide) and magnetite (iron oxide), mostly concentrated in associated rocks such as norite. Some notable deposits occur at Lofoten, in northern Norway; Allard Lake, in Quebec; Iron Mountain, in the Laramie Range of Wyoming; Duluth, Minnesota; and Sanford Lake (near Tahawus), in the Adirondacks of New York State.

John L. Berkley

CROSS-REFERENCES

Andesitic Rocks, 1263; Basaltic Rocks, 1274; Batholiths, 1280; Carbonatites, 1287; Continental Crust, 561; Earthquake Distribution, 277; Eruptions, 739; Granitic Rocks, 1292; Hot Spots, Island Chains, and Intraplate Volcanism, 706; Igneous Rock Bodies, 1298; Igneous Rock Classification, 1303; Island Arcs, 712; Kimberlites, 1310; Komatiites, 1316; Lamproites, 1321; Magmas, 1326; Orbicular Rocks, 1332; Plate Margins, 73; Plate Motions, 80; Plate Tectonics, 86; Plutonic Rocks, 1336; Pyroclastic Rocks, 1343; Rocks: Physical Properties, 1348; Silica Minerals, 1354; Stratovolcanoes, 792; Ultramafic Rocks, 1360; Xenoliths, 1366.

BIBLIOGRAPHY

Carmichael, Ian S. E., F. J. Turner, and John Verhoogan. *Igneous Petrology.* New York: McGraw-Hill, 1974. This textbook is designed for upper-level undergraduates and graduate students; the section on anorthosites does an excellent job of summarizing their occurrences and the problems associated with their interpretation and geologic significance. Much of this section can be comprehended by anyone with a minimal background in geology. Critical references from the scientific literature are listed in full near the end of the book.

Isachsen, Yngvar W., ed. *Origin of Anorthosite and Related Rocks.* Memoir 18. Albany, N.Y.: State Department of Education of New York, 1968. A massive compendium of articles presented at the Second Annual George H. Hudson Symposium, held at Plattsburg, New York, in 1966. The conference was devoted to massif anorthosite bodies and problems of their origins and was attended by experts on anorthosites as well as those with general interest in their study. It is without question one of the most extensive and comprehensive works on anorthosites ever prepared. It includes a very detailed appendix, and each article has an extensive reference list at the end. For students who are interested in the details of anorthosites, this book is the best place to start.

Klein, Cornelis, and Cornelius S. Hurlbut, Jr. *Manual of Mineralogy.* 21st ed. New York: John Wiley & Sons, 1999. An introductory college-level text on mineralogy. Discusses the physical and chemical properties of minerals and describes the most common minerals and their varieties, including gem mineral varieties. Contains more than 500 mineral name entries in the mineral index and describes in detail about 150 mineral species.

Perch, L. L., ed. *Progress in Metamorphic and Magmatic Petrology.* New York: Cambridge University Press, 1991. Although intended for the advanced reader, several of the essays in this multiauthored volume will serve to familiarize the new student with the study of igneous rocks. In addition, the bibliography will lead the reader to other useful material.

Science Mate: Plate Tectonic Cycle. Fremont, Calif.: Math/Science Nucleus, 1990. This book is part of the Integrating Science, Math, and Technology series of manuals intended to help teachers explain basic scientific concepts to elementary students. It provides a basic understanding of plate tectonics, earthquakes, volcanoes, and general geology.

Sutherland, Lin. *The Volcanic Earth: Volcanoes and Plate Tectonics, Past, Present, and Future.* Sydney, Australia: University of New South Wales Press, 1995. Although Sutherland focuses on volcanic activity in Australia, the book provides an easily understood overview of volcanic and tectonic processes, including the role of igneous rocks. Includes color maps and illustrations, as well as a bibliography.

Taylor, G. Jeffrey. *A Close Look at the Moon.* New York: Dodd, Mead, 1980. This book is designed for junior and senior high school readers. It has a very good section on lunar rocks by an author who is one of the world's authorities on lunar rocks and lunar crustal evolution. Excellent cross-sectional diagrams trace the history of the lunar highlands (anorthosite-rich areas) and the lunar crust in general. The book contains photographs of Moon rocks, including some taken through a microscope. A brief volume, but one packed with useful information about one of the most familiar anorthosite bodies, the Moon.

Windley, Brian F. *The Evolving Continents.* New York: John Wiley & Sons, 1977. An advanced text that focuses on the evolution of the Earth's continental crust as revealed by its complex geology. The section on anorthosite bodies is very descriptive and should be comprehensible to most nonspecialists. The book also contains a brief description of economic commodities associated with anorthosites, and other references to various anorthosite occurrences and problems may be found in the index. Includes a complete listing of technical references for those who wish to dig deeper.

BASALTIC ROCKS

"Basalt" is the term applied to dark, iron-rich volcanic rocks that occur everywhere on the ocean floors, as oceanic islands, and in certain areas on continents. It is the parent material from which nearly any other igneous rock can be generated by various natural processes.

PRINCIPAL TERMS

AUGITE: an essential mineral in most basalts, a member of the pyroxene group of silicates

CRUST: the upper layer of the Earth and the other "rocky" planets; it is composed mostly of relatively low-density silicate rocks

LITHOSPHERIC PLATES: giant slabs composed of crust and upper mantle; they move about laterally to produce volcanism, mountain building, and earthquakes

MAGMA: molten silicate liquid, including any crystals, rock inclusions, or gases trapped in that liquid

MANTLE: a layer beginning at about 5 to 50 kilometers below the crust and extending to the Earth's metallic core

OCEANIC RIDGES: a system of mostly underwater rift mountains that bisect all the ocean basins; basalt is extruded along their central axes

OLIVINE: a silicate mineral found in mantle and some basalts, particularly the alkaline varieties

PERIDOTITE: the most common rock type in the upper mantle, where basalt magma is produced

PLAGIOCLASE: one of the principal silicate minerals in basalt, a member of the feldspar group

SUBDUCTION ZONES: areas marginal to continents where lithospheric plates collide

BASALT COMPOSITION

Basalt is a dark, commonly black, volcanic rock. It is sometimes called "trap" or "traprock," from the Swedish term *trapp*, which means "steplike." On cooling, basalt tends to form hexagonal columns, which in turn form steplike structures after erosion. Excellent examples can be seen at Devils Postpile, in the Sierra Nevada in central California, and the Giant's Causeway, in County Antrim, Ireland. The term "trap," however, is used mostly by miners and nonspecialists; scientists prefer the word "basalt" for the fine-grained, volcanic rock that forms by the solidification of lava flows. Basaltic magma that crystallizes more slowly below the Earth's surface, thus making larger mineral grains, is called "gabbro."

The importance of basalt to the evolution of planets like the Earth cannot be overemphasized. Basalt is considered to be a "primary silicate liquid," in that the first liquids to form by the melting of the original minerals that made up all the so-called rocky planets (those composed mostly of silicates) were basaltic in composition. In turn, basalt contains all the necessary ingredients to make all the other rocks that may eventually form in a planet's crust. Furthermore, many meteorites (which are believed to represent fragments of planetoids or asteroids) are basaltic or contain basalt fragments, and the surfaces of the Moon and the planets Mercury, Venus, and Mars are known to be covered to various degrees by basalt lava flows.

Like most rocks in the Earth's crust, basalt is composed of silicate minerals, substances whose principal component is the silica molecule. Compared with other silicate rocks, basalts contain large amounts of iron and magnesium and small amounts of silicon. This characteristic is reflected in the minerals in basalts, which are mostly pyroxene minerals (dark-colored, calcium-iron-magnesium silicates, such as augite) and certain feldspar minerals called plagioclase (light-colored, calcium-sodium-aluminum silicates). Pyroxenes and plagioclase are essential minerals in basalts, but some types of basalt also contain the mineral olivine (a green, iron-magnesium silicate). The high abundance of dark green or black pyroxene and, in some cases, olivine gives basalts their characteristic dark color. This color is mainly attributable to the high iron content in pyroxene and olivine.

OCEANIC BASALT DISTRIBUTION

Basalt is the most common type of igneous rock (rock formed by the crystallization of magma) on or near the surface of the Earth. Relatively rare on continents, it is the principal rock in the ocean floors. Drilling into the sea floor by specially designed oceanographic research ships reveals that basalt invariably lies just below a thin cover of fine, sedimentary mud. Ocean floor basalt flows out of mid-ocean ridges, a system of underwater mountain ranges that spans the globe. These ridges commonly trend roughly down the middle of ocean basins, and they represent places where the Earth's lithospheric plates are being literally split apart. In this process, basalt magma is generated below the "rift mountains"; it flows onto the cold ocean floor and solidifies. Although oceanic ridges are normally hidden from view under the oceans, a segment of the Mid-Atlantic Ridge emerges above the waves as the island of Iceland.

Oceanic islands are also composed of basalt. The islands of Hawaii, Fiji, Mariana, Tonga, and Samoa, among others, are large volcanoes or groups of volcanoes that rise above water from the ocean floor. Unlike the basalts that cover the ocean floors, however, these volcanoes do not occur at oceanic ridges but instead rise directly from the sea floor. Basalt is also a fairly common rock type on island arcs, volcanic islands that occur near continental margins. These curvilinear island chains arise from melting along subduction zones, areas where the Earth's lithospheric plates are colliding. This process generally involves material from the ocean basins diving under the more massive continents; andesite (a light-colored rock) volcanoes are the main result, but some basalt erupts there as well. The Japanese, Philippine, and Aleutian island chains are examples of island arcs, as are the island countries of New Zealand and Indonesia.

CONTINENTAL BASALT DISTRIBUTION

Basalt lava flows are not nearly as common on continents as in oceanic areas. Andesites, rhyolites, and related igneous rocks are far more abundant than basalt in continental settings. In North America, basalt lava flows and volcanoes are best exposed to view in the western United States and Canadian provinces, western Mexico, Central America, and western South America. The greatest accumulations of basalt lava flows in the United States are in the Columbia Plateau of Washington, Oregon, and Idaho. In this large area, a series of basalt lava flows have built up hundreds of feet of nearly flat-lying basalt flows over a few million years. These "fissure flows" result from lava's pouring out of long cracks, or fissures, in the crust. They are similar in many respects to the basalt flows produced at oceanic ridges, because no actual volcanic cones are produced—only layer after layer of black basalt. The Snake River Plain in southern Idaho has a similar origin, and other extensive basalt plateaus occur in the Deccan area of southern India, the Karroo area of South Africa, and Paraná State in Brazil.

A basaltic flow that was underlying sedimentary rock. (Geological Survey of Canada)

Not all basalt is erupted as lava flows. If the lava is particularly rich in volatiles such as water and carbon dioxide, it will be explosively ejected from the volcano as glowing fountains of incandescent particles that rain down on the surrounding area. Conical volcanoes composed almost exclusively of basalt ejecta particles are called cinder cones. Good examples of cinder cones are Sunset Crater in northern Arizona, the numerous cinder cones in Hawaii and Iceland, and the very active volcano in Italy known as Stromboli. In fact, the rather violent eruptions that produce cinder cones are called "strombolian eruptions."

BASALT CLASSIFICATION

Although basaltic rocks may all look alike to the nonspecialist, there are actually many different kinds of basalt. They are arranged by scientists into a generally accepted classification scheme based on chemistry and, to some extent, mineralogy. To begin with, basalt can be distinguished from the other major silicate igneous rocks by its relatively low (about 50 percent) silica content. Within the basalt clan itself, however, other means of classification are used. Basalts are divided into two major groups, the alkaline basalts and the subalkaline basalts. Alkaline basalts contain high amounts of the alkali metal ions potassium and sodium but relatively low amounts of silica. In con-

trast, subalkaline basalts contain less potassium and sodium and more silica. As might be expected, this chemical difference translates into differences in the mineral content of the basalt types as well. For example, all alkaline basalts contain one or more minerals called "feldspathoids" in addition to plagioclase feldspar. They also commonly contain significant olivine. Subalkaline basalts, on the other hand, do not contain feldspathoids, although some contain olivine, and they may be capable of crystallizing very tiny amounts of the mineral quartz. The presence of this very silica-rich mineral reflects the relatively high silica content of subalkaline basalt magmas versus alkaline basalt magmas. Within these two major groups are many subtypes, too numerous to discuss here.

STUDY TECHNIQUES

Like other igneous rocks, basalts are analyzed and studied by many techniques. Individual studies may include extensive field mapping, in which the distribution of various types of basalt is plotted on maps. Especially if geologic maps of basalt types are correlated with absolute ages determined through radiometric dating techniques, the history of magma generation and its relation to tectonic history (earth movements) can be reconstructed for a particular area. Good examples of such studies are those conducted in recent years on the Hawaiian Islands. These studies indicate that the alkaline basalts on any given island are generally older than the subalkaline basalts, showing that magma production has moved upward, to lower pressure areas, in the mantle with time. This finding supports the idea that oceanic island basalts such as those in Hawaii are generated within so-called mantle plumes, roughly balloon-shaped, slowly rising masses of mantle material made buoyant by localized "hot spots."

Samples of basalt are also analyzed in the laboratory. The age of crystallization of basalt is obtained by means of radiometric dating techniques that

Columnar jointing in basaltic rocks in Yellowstone National Park. (© William E. Ferguson)

involve the use of mass spectrometers to determine the abundances of critical isotopes, such as potassium 40 and argon 40 or rubidium 87 and strontium 86 and 87. To obtain information on how basalt magma is generated and how it subsequently changes in composition before extrusion as a lava flow, scientists place finely powdered samples in metallic, graphite, or ceramic capsules and subject them to heating and cooling under various conditions of pressure. Such procedures are known as experimental petrology. It has been proved that nearly any other igneous magma composition can be derived from basalt magma by the process of crystal fractionation. Widely believed to be the major factor influencing chemical variation among igneous rocks, this process results in ever-changing liquid compositions as the various silicate minerals crystallize, and are thus removed from the liquid, over time. Basalt's parental role gives it enormous importance in the discipline of igneous petrology.

TRACE ELEMENT ANALYSIS

Another fruitful avenue of research is trace element analysis of basalts. Trace elements occur in such low abundances in rocks that their concentrations must usually be expressed in terms of parts per million or even parts per billion. Among the most useful substances for tracing the history of basalt are the rare-earth elements. Chromium, vanadium, nickel, phosphorus, strontium, zirconium, scandium, and hafnium are also used. There are many methods for measuring these elements, but the most common, and most accurate, is neutron activation. This method involves irradiating samples in a small nuclear reactor and then electronically counting the gamma-ray pulses generated by the samples. Since different elements tend to emit gamma rays at characteristic energies, these specific energies can be measured and the intensity of gamma pulses translated into elemental concentrations.

Once determined for a particular basalt sample, trace element abundances are sensitive indicators of events that have transpired during the evolution of the basalt. There are two reasons for this sensitivity. First, trace elements are present in such low concentrations as compared with major elements (iron, aluminum, calcium, silicon, and the like) that any small change in abundance

caused by changes in the environment of basalt production will be readily noticed. Second, different minerals, including those crystallizing in the magma and those in the source peridotite, incorporate a given trace element into their structures or reject it to the surrounding liquid to widely varying degrees. Therefore, trace element concentrations can be used to show which minerals were involved in producing certain observed chemical signatures in basalts and which were likely not involved.

For example, it is well known that the mineral garnet readily accepts the rare-earth element lutecium into its structure but tends to reject most lanthanum to any adjacent liquid. Basalts with very little lutecium but much lanthanum were therefore probably derived by the melting of garnet-bearing mantle rocks. Since garnet-bearing mantle rocks can exist only at great depth, basalts with such trace element patterns must have originated by melting at these depths in the mantle. In fact, that is one of the most important lines of evidence that alkaline basalts originate at high-pressure regions in the mantle.

ECONOMIC APPLICATIONS

Basaltic islands, particularly in the Pacific basin, are some of the most popular tourist stops in the world. More important, however, basalt magma contains low concentrations of valuable metals that, when concentrated by various natural processes, provide the source for many important ores. Copper, nickel, lead, zinc, gold, silver, and other metals have been recovered from ore bodies centered in basaltic terrains. Some of the richest mines of metallic ores in the world are located in Canada, where ores are found associated with extremely old basaltic rocks, called "green-stones," from long-vanished oceans. The richest of these mines is Kidd Creek, in northern Ontario. These ore-bearing basalts were first extruded more than 2 billion years ago, during what geologists call Precambrian times (the period from 4.6 billion to about 600 million years ago). Other notable ore deposits include the native, or metallic, copper in late Precambrian basalts that was mined for many years in the Keweenaw Peninsula of northern Michigan. In addition, the island of Crete in the Mediterranean Sea has copper mines that were mined thousands of years ago during the "copper" and "bronze" ages of human history. The basalt enclosing these ores is believed to have erupted from an ancient mid-ocean ridge trending between Africa and Europe.

Basalt can also be used as a building stone or raw material for sculptures, but its high iron content makes it susceptible to rust stains. It is also ground up to make road gravel, especially in the western United States, and it is used as decorative stone in yards and gardens.

John L. Berkley

CROSS-REFERENCES

Andesitic Rocks, 1263; Anorthosites, 1269; Batholiths, 1280; Building Stone, 1545; Carbonatites, 1287; Feldspars, 1193; Granitic Rocks, 1292; Igneous Rock Bodies, 1298; Igneous Rock Classification, 1303; Industrial Metals, 1589; Iron Deposits, 1602; Kimberlites, 1310; Komatiites, 1316; Lamproites, 1321; Lunar History, 2539; Lunar Rocks, 2561; Magmas, 1326; Metamorphic Mineral Deposits, 1614; Mountain Belts, 841; Orbicular Rocks, 1332; Plutonic Rocks, 1336; Pyroclastic Rocks, 1343; Rocks: Physical Properties, 1348; Silica Minerals, 1354; Ultramafic Rocks, 1360; Xenoliths, 1366.

BIBLIOGRAPHY

Ballard, Robert D. *Exploring Our Living Planet.* Washington, D.C.: National Geographic Society, 1983. This book covers every aspect of the Earth's volcanic and tectonic features and is lavishly illustrated with color photographs, illustrations, and diagrams. The sections on "spreading" and "hotspots" largely deal with basalt volcanism and its relationship to plate tectonic theory. Well written and indexed, the text will be easily understood and appreciated by specialists and laypersons alike.

Decker, Robert, and Barbara Decker. *Volcanoes.* New York: W. H. Freeman, 1997. This brief book gives a comprehensive treatment of volcanic phenomena. It is illustrated with numerous black-and-white photographs and diagrams. Chapters 1, 2, 3, and 6 deal almost exclusively with basalt volcanism. The last four chapters deal with human aspects of volcanic phenomena, such as the obtaining of energy and raw materials, and the effect of volcanic eruptions on weather. Includes an excellent chapter-by-chapter bibliography. Suitable for high school and college students.

Lewis, Thomas A., ed. *Volcano.* Alexandria, Va.: Time-Life Books, 1982. One of the volumes of the Planet Earth series, this book is written with the nonspecialist in mind. Wonderful color photographs, well-conceived color diagrams, and a readable narrative guide the reader through the world of volcanism. The book is especially good for its descriptions of past eruptions and their effects on humankind. Basalt is covered mainly in the chapter on Hawaii and the chapter on Heimaey, Iceland. Has a surprisingly extensive bibliography and index for a book of this kind.

Lutgens, Frederick K., and Edward J. Tarbuck. *Earth: An Introduction to Physical Geology.* 6th ed. Upper Saddle River, N.J.: Prentice Hall, 1999. This college text provides a clear picture of the Earth's systems and processes that is suitable for the high school or college reader. In addition to its illustrations and graphics, it has an accompanying computer disc that is compatible with either Macintosh or Windows. Bibliography and index.

Macdonald, Gordon A. *Volcanoes.* Englewood Cliffs, N.J.: Prentice-Hall, 1972. Written by one of the premier volcanologists in the world, this book is ideal for those desiring a serious but not overly technical treatment. Every conceivable aspect of volcanic phenomena is covered, but the sections on basalt (particularly as it occurs in Hawaii) are particularly good. Includes suggested readings, a comprehensive list of references, a very good index, and an appendix that lists the active volcanoes of the world. Somewhat lengthy.

Perch, L. L., ed. *Progress in Metamorphic and Magmatic Petrology.* New York: Cambridge University Press, 1991. Although intended for the advanced reader, several of the essays in this multiauthored volume will serve to familiarize the new student with the study of igneous rocks. In addition, the bibliography will lead the reader to other useful material.

Putnam, William C. *Geology.* 2d ed. New York: Oxford University Press, 1971. A comprehensive and accessible text. Chapter 4, "Igneous Rocks and Igneous Processes," uses a vivid description of the 1883 eruption of Krakatau as a way of introducing the formation processes of igneous rocks. Other famous volcanic eruptions are also discussed. The rocks' classification and composition are described in detail. The chapter concludes with a list of references. Illustrated.

Science Mate: Plate Tectonic Cycle. Fremont, Calif.: Math/Science Nucleus, 1990. This book is part of the Integrating Science, Math, and Technology series of manuals intended to help teachers explain basic scientific concepts to elementary students. As such, it provides a basic understanding of plate tectonics, earthquakes, volcanoes, and general geology for the student with no background in the Earth sciences.

Sutherland, Lin. *The Volcanic Earth: Volcanoes and Plate Tectonics, Past, Present, and Future.* Sydney, Australia: University of New South Wales Press, 1995. Although Sutherland focuses on volcanic activity in Australia, the book provides an easily understood overview of volcanic and tectonic processes, including the role of igneous rocks. Includes color maps and illustrations, as well as a bibliography.

Tarbuck, Edward J., and Frederick K. Lutgens. *The Earth: An Introduction to Physical Geology.* 2d ed. Columbus, Ohio: Merrill, 1987. Aimed at the reader with little or no college-level science experience, this textbook includes a chapter devoted to igneous rocks and their textures, mineral compositions, classification, and formation. Illustrated with photographs and diagrams. Includes review questions and list of key terms.

BATHOLITHS

Batholiths are gigantic bodies of granitic rock located in mobile belts surrounding the ancient cores of the continents. The growth of continental crust during the past 2.5 billion years is intimately related to the origin and emplacement of major volumes of granitic magma that solidify as batholiths.

PRINCIPAL TERMS

CRYSTALLIZATION: the solidification of molten rock as a result of heat loss; slow heat loss results in the growth of crystals, but rapid heat loss can cause glass to form

GRANITIC/GRANITOID: descriptive terms for plutonic rock types having quartz and feldspar as major mineral phases

I-TYPE GRANITOID: granitic rock formed from magma generated by partial melting of igneous rocks in the upper mantle or lowermost crust

MAGMA: molten rock material that crystallizes to form igneous rocks

MIGMATITE: a rock exhibiting both metamorphic and plutonic textural traits

MOBILE BELT: a linear belt of igneous and de-

formed metamorphic rocks produced by plate collision at a continental margin; relatively young mobile belts form major mountain ranges

PARTIAL MELTING: a process undergone by rocks as their temperature rises and metamorphism occurs; magmas are derived by the partial melting of preexisting rock; also known as ultrametamorphism or anatexis

PLUTON: a generic term for an igneous body that solidifies well below the Earth's surface; plutonic rocks are coarse-grained because they cool slowly

S-TYPE GRANITOID: granitic rock formed from magma generated by partial melting of sedimentary rocks within the crust

BATHOLITH FORMATION

Batholiths are large composite masses of granitoid rock formed by numerous individual bodies of magma that have risen from deep source areas in molten form and solidified near enough to the surface to be exposed by erosion. The resulting rocks are relatively coarse-grained in texture and markedly heterogeneous in chemical and mineralogical composition. A well-studied example is that of the coastal batholith of Peru, which forms an almost continuous outcrop 1,100 kilometers long and 50 kilometers wide along the western flank of the Andes. This enormous body has steep walls and a flat roof. It is composed of more than one thousand individual plutons emplaced along a narrow belt parallel to the present coastline during a volcanic-plutonic event that extended over a period of 70 million years. Many such batholiths are known in the mountainous areas of the world, but few are as large or as magnificently exposed to view as that in Peru. Geological glossaries often define a batholith as a "coarse-tex-

tured igneous mass with an exposed surface area in excess of 100 square kilometers"; this description has the virtue of simplicity but is misleading because it encompasses granitic plutons, and even nongranitic plutons, which form under conditions quite removed from those associated with the world's major batholiths.

In most instances, there is evidence to indicate that the individual plutons of a batholith were emplaced as hot, viscous melts containing suspended crystals. This molten material is called magma. Cooling and crystallization occur during the surfaceward ascent of magma and gradually transform it to solid rock, which prevents further upward movement. The depth at which total solidification occurs varies and is strongly dependent upon the initial temperature and water content of the magma. Extreme levels of ascent, within 3 to 5 kilometers of the surface, are possible only for very hot magmas with very low initial water contents. Most granitic plutons complete their crystallization at depths in the range of 8 to 20 kilometers. The characteristic coarse textures observed in most granitic plutons are the result of slow cooling, which, in turn, implies that the rate of magma ascent is also slow. These traits distinguish plutonic rocks from their volcanic counterparts. As would be expected, the formation of a batholith is a complex and lengthy event that is the sum of the processes responsible for the emplacement of each member pluton. Each member pluton has an individual history involving the generation of magma in the source region, ascent and partial crystallization, physical displacement of overlying solid rock, chemical interaction with the solid rocks encountered during ascent, and the terminal crystallization phase. Consequently, each member pluton of a batholith can be expected to exhibit a combination of textural, mineralogical, and chemical variations that are peculiar to itself.

MOBILE BELTS

It has long been recognized that major batholiths are confined to narrow zones elongated parallel to the margins of older continental crust. In such zones, granitic melts intrude either thick sequences of chemically related volcanic rocks or highly deformed and metamorphosed sedimentary rocks. These granite-dominated zones are called mobile belts. The ancient cores of continents are all older than 2.5 billion years. They are surrounded by mobile belts that become successively younger away from the core. The most recent mobile belts form major mountain chains along continental margins. The resulting age pattern clearly shows that continents grow larger with time by the marginal accretion of mobile belts. In the late 1960's, the emergence of plate tectonic theory provided a basis for understanding how mobile belts form and are accreted to preexisting continent margins. The impetus provided by this theory sparked intensive study of the world's mobile belts. These studies amply show that logical time-space relationships exist between plate collisions, deformation styles, and rock types that occur in mobile belts. Two distinct types of mobile belts are now recognized, and each is dominated by granitic batholiths. These batholiths, however, are very different in terms of granitic rock types, modes of pluton emplacement, rock associations, metamorphic effects, and the metallic ores they host. The batholiths of the two mobile belt types

are called I-type and S-type batholiths.

Mobile belts along the eastern margin of the Pacific Ocean contain I-type batholiths exclusively. Their size and collective volume is staggering. The Peruvian batholith is an example already mentioned. Others of this type include the Sierra Nevada batholith, the Idaho batholith, and the tremendous Coast Range batholith, which extends from northern Washington to the Alaska-Yukon border. In contrast, the western margin of the Pacific Ocean is dominated by mobile belts with S-type batholiths, although some I-types are also present. The batholiths of Western Europe are also mainly of the S-type. In southeastern Australia, where the two types of batholiths were first recognized, S-type and I-type granitoids form a paired belt parallel to the coastline. Although their geographical distribution is uneven, both I-type and S-type batholiths occur worldwide.

I-TYPE BATHOLITHS

The most distinctive trait of I-type batholiths is the broad range of granitic rock types they contain. In these batholiths, the rock types gabbro-diorite, quartz diorite-granodiorite, and granite occur in the approximate proportions of 15:50:35. This means that quartz diorite (also called tonalite) and granodiorite jointly compose 50 percent of I-type batholiths, and true granite is a subordinant component in them. This wide compositional spectrum not only characterizes an entire I-type batholith but also is typical of the individual member plutons. Usually the major plutons are concentrically zoned with small central cores of true granite enveloped by extensive zones of granodiorite that grade outward into margins of quartz diorite. Small plutons in the compositional range of gabbro-diorite are common but subordinate to the zoned granitic bodies. Most member plutons of I-type batholiths have domal or cylindrical shapes and very steep contacts with the surrounding rock. Others may have a steeply tilted sheetlike form, but, regardless of shape, most I-type plutons cut through the preexisting rock layers at a steep angle. The emplacement of these plutons appears to be controlled by near-vertical fractures that may extend downward to the base of the crust. In younger I-type batholiths such as that in Peru, the granitoids have intruded into a roof of chemically related volcanic rocks that show the same compositional spectrum as the granitic plutons. This volcanic pile, dominated by andesite, may be 3 to 5 kilometers thick at the time of pluton emplacement. Gradually, this volcanic roof is stripped away by erosion so that in older, deeply eroded batholiths, volcanic rocks are generally absent. The grade of regional metamorphism in the rocks enclosing I-type batholiths is relatively low, and there is little evidence of large-scale horizontal compression or crustal shortening. Structural displacements and the movements of rising plutons are dominantly vertical and typically occur over a time span of 50 to 100 million years.

S-TYPE BATHOLITHS

S-type batholiths contrast with I-types in almost every respect. To begin with, the ratio of gabbro-diorite to quartz diorite-granodiorite to granite is 2:18:80 in S-type batholiths. These plutonic complexes are very much dominated by true granite, and gabbro-diorite plutons are rare or absent. In many cases, S-type granites are the distinctive "two mica granites," which contain both biotite and muscovite and are frequently associated with major tin and tungsten ore deposits. The batholiths of northern Portugal are typical examples of this association. S-type batholiths, as well as their member plutons, lack the concentric zoning that characterizes I-type plutons. Compositional homogeneity is their trademark. S-type plutons are intruded into thick sequences of regionally metamorphosed sedimentary rocks. The metamorphic grade ranges from moderate to very high, and, frequently, the granites are located within the zone of highest metamorphic grade. In such cases, migmatites are often present. The enclosing metamorphosed rocks are intensely folded in response to marked crustal compression, and volcanic rocks are conspicuously absent. S-type batholiths are smaller in volume and form over a shorter period of time (usually less than 20 million years) than their I-type counterparts.

Half Dome and the upper Yosemite Valley, carved and smoothed by glaciers in a great granite batholith. (© William E. Ferguson)

DIFFERING ORIGINS OF BATHOLITH TYPES

The many contrasting traits of I-type and S-type batholiths are an indication that the conditions of magma generation and emplacement are very different in the mobile belts in which they are found. In the case of I-type batholiths, it appears that magmas are generated at relatively great depths and above the subduction zones formed at destructive plate boundaries. The magmas are derived by partial melting of upper-mantle basic igneous rocks and, perhaps, lower crustal igneous rocks. The melts rise along the steep fractures produced by crustal tension over the subduction zone. The igneous ancestry of these melts is the reason for calling the resulting plutons "I-type." The hottest and driest of these I-type magmas will reach the surface to produce extensive fields of volcanic rocks and large calderas. In some cases, like the Peruvian batholith, the rise of magma was "passive" in the sense that room was provided for the rising plutons by gravitational subsidence of the overlying roof rock. This is the process of cauldron subsidence. In the case of the Sierra Nevada batholith, however, it appears that I-type magmas were emplaced by "forceful injection." In this process, rising magma makes room for itself by shouldering aside the surrounding solid rock. On the other hand, the evidence suggests that S-type magmas originate by partial melting of metamorphosed sedimentary rocks. This sedimentary parentage of the magmas is the reason for designating them as "S-type." Melting is made possible by dehydration of water-bearing minerals under conditions of intense metamorphism. The frequent presence of migmatites is evidence for this transition from metamorphic conditions to magmatic conditions. The essential requirements for relatively high-level crustal melting are high temperatures, intense horizontal compression to produce deep sedimentary basins, and a thick pile of sediments to fill these basins. Such conditions are best met in back-arc basins, which also form at destructive plate margins but considerably inland from the volcanic-plutonic environment of I-type batholiths. This may explain why some I-type and S-type batholiths occur in paired belts parallel to a continental margin, as in southeastern Australia. The

Fry Creek Batholith in British Columbia. (Geological Survey of Canada)

collision environment that arises when continent meets continent in the terminal stage of subduction may also provide suitable conditions for S-type magma production. The magmas that result will be relatively cool and wet and will not be able to rise far above their zone of melting. During this limited ascent, the S-type magmas tend to assume the shape of a light bulb, with a neck tapering down to the zone of melting. Because of their limited capacity for vertical movement, S-type plutons require no special mechanisms to provide additional space for them.

FIELD STUDY

The study of a batholith begins with the study of its individual member plutons. This always involves fieldwork, laboratory analysis of rock samples returned from the field, and comparison of the resulting data with those obtained from other batholiths. Because of their great size, batholiths present special problems for field study. The most informative studies are those in mountainous areas, such as the Peruvian Andes, where erosion has exposed the batholith roof contact and cut steep canyons between and through individual plutons. High topographic relief is essential if the geologist is to learn anything about the variation in shape and composition with depth in the plutons. Most studies are in remote mountain ranges where climatic factors and the absence of roads are obstacles to fieldwork. Often, the summer, snow-free period is as short as four weeks, and, even during this period, winter snow and glaciers may cover much of the study area. Helicopters may be employed to supply remote base camps and transport geologists to sites that are virtually inaccessible by other means.

Study of even a small portion of a major batholith requires several well-trained geologists working intensively during the short field seasons over a period of several years. The geologists make traverses on foot across and around the individual plutons as topography permits, and they record the textural, mineralogical, and structural features observed on maps or aerial photographs of the area. These maps eventually reveal the overall shape of plutons and the patterns of concentric zoning within them. Special maps are prepared to show the distribution of fractures and flow structures within individual plutons. These features indicate how fluid the magma was at the time of emplacement. Contacts between the plutons and older enclosing rocks are closely examined for deformation effects and evidence of thermal and chemical interaction with the magma. Fragments of older rock engulfed by magma are often preserved in a recrystallized state, and these are scrutinized carefully, since they provide clues as to whether the emplacement process was passive or forceful. As the end of the field season draws near, the mapped plutons are sampled. Large, fresh samples must be collected from each recognized zone of each pluton for subsequent laboratory study. The number of samples collected from a single pluton depends upon its size and homogeneity but is frequently in the range of one hundred to five hundred samples. A smaller number of samples is collected from the host rocks at varying distances from the plutonic contact in order to study the thermal effects produced by the

pluton. The field description, identifying number, and exact location of each sample site must be meticulously recorded. If, at the end of the field season, several plutons have been studied and sampled, there may be several thousand rock samples to label, pack securely, and ship to the laboratory, where they will receive further study.

LABORATORY STUDY

At the laboratory, the samples are usually cut in half and labeled in a permanent fashion. One half of each sample is stored for future reference, and the other is prepared for the laboratory procedures. Paper-thin slices of each sample are glued to glass slides for examination under a petrographic microscope. The microscopist identifies the mineral phases present in each slide and determines the abundance of each. The texture of each rock, as revealed under the microscope, is carefully described and interpreted in terms of crystallization sequence and deformation history. When the microscopic study is complete, certain samples, perhaps fifty to one hundred, are chosen for chemical analysis. Most will be analyzed because they are judged to be representative of major zones of a pluton; a few may be analyzed because they exhibit unusual minerals or some peculiar trait not explained by the microscopic study. If the age of a pluton is not known, a few samples (one to ten) will be shipped to a laboratory that specializes in age determinations by radioisotope methods.

DETERMINING ORIGIN AND EMPLACEMENT

Finally, on the basis of the field observations, microscopic examinations, and chemical data, the investigators will assemble rival hypotheses, or scenarios, for the origin and emplacement of the plutons that have been studied. Any scenario that conflicts seriously with known facts is discarded. Those remaining are compared with well-known laboratory melting-crystallization experiments on synthetic and natural rock systems. The size and shape of the plutons, as determined by the field mapping, can be compared with those of "model plutons" derived through sophisticated, but idealized, centrifuge experiments in laboratory settings. The investigators will compare their data, in detail, with data reported in the geological literature by workers in other parts of the world. They will also compare their results with earlier studies of the same plutons, or studies in the same region, if they exist. Ideally, this approach leads to elimination of all hypotheses for the origin and emplacement of the plutons but one. In the majority of cases, particularly those where the study of a major batholith is still in its infancy, it is found that two or even several rival hypotheses explain the known facts equally well. This is an indication that the scope of the study must be enlarged to include even more member plutons of the batholith. As additional plutons are studied in detail, more constraints on the mode of origin and emplacement of the batholith are obtained.

ROLE IN CRUSTAL GROWTH

Mobile belts, dominated by immense granitic batholiths, have been systematically accreted to the ancient continental cores for the last 2.5 billion years of Earth history. Modern plate tectonic theory has provided the basis for understanding the periodic nature of mobile belt accretion and the ways in which crustal and mantle materials are recycled. It is evident that the emplacement of batholiths is at present, and has been for at least 2.5 billion years, the major cause for progressive crustal growth. It is also clear that the rate at which batholiths formed during this lengthy period has far exceeded the rate of continental re-

duction by erosion. The generation of large volumes of granitic magma and its subsequent rise to form batholiths a few kilometers below the crustal surface must be viewed as fundamental to crustal growth. Batholiths play a major role in the formation of mountain systems and are the most important element in the complex rock and metallic ore associations of mobile belts. The very existence of continents is, in fact, a result of the long-standing process of batholith emplacement.

Gary R. Lowell

CROSS-REFERENCES

Achondrites, 2621; Andesitic Rocks, 1263; Anorthosites, 1269; Asteroids, 2640; Basaltic Rocks, 1274; Carbonatites, 1287; Eruptions, 739; Flood Basalts, 689; Granitic Rocks, 1292; Hawaiian Islands, 701; Hot Spots, Island Chains, and Intraplate Volcanism, 706; Igneous Rock Bodies, 1298; Igneous Rock Classification, 1303; Island Arcs, 712; Kimberlites, 1310; Komatiites, 1316; Lamproites, 1321; Magmas, 1326; Mars's Volcanoes, 2457; Ocean Ridge System, 670; Orbicular Rocks, 1332; Plutonic Rocks, 1336; Pyroclastic Rocks, 1343; Recent Eruptions, 780; Rocks: Physical Properties, 1348; Shield Volcanoes, 787; Silica Minerals, 1354; Spreading Centers, 727; Ultramafic Rocks, 1360; Volcanic Hazards, 798; Xenoliths, 1366.

BIBLIOGRAPHY

Atherton, Michael P., and J. Tarney, eds. *Origin of Granite Batholiths: Geochemical Evidence*. Orpington, England: Shiva Publishing, 1979. A summary of the views of major authorities on the origin of batholiths. Suitable for advanced students of geology.

Best, Myron G. *Igneous and Metamorphic Petrology*. 2d ed. New York: W. H. Freeman, 1995. A popular university text for undergraduate majors in geology. A well-illustrated and fairly detailed treatment of the origin, distribution, and characteristics of igneous and metamorphic rocks. Chapter 4 covers granite plutons and batholiths.

Hamilton, Warren B., and W. Bradley Meyers. *The Nature of Batholiths*. Professional Paper 554-C. Denver, Colo.: U.S. Geological Sur-

vey, 1967. In this classic paper, the authors propose their controversial "shallow batholith model." The account is short and very descriptive and can be followed by college-level readers. This and the following article influenced many geologists to abandon the traditional view of batholiths.

_____. "Nature of the Boulder Batholith of Montana." *Geological Society of America Bulletin* 85 (1974): 365-378. The authors apply their 1967 model to the Boulder batholith and reply to the heated criticism of colleagues who did the fieldwork on this batholith. This work is aimed at professionals but can be understood by those with moderate knowledge of plutonic processes.

Judson, S., M. E. Kauffman, and L. D. Leet.

Physical Geology. 7th ed. Englewood Cliffs, N.J.: Prentice-Hall, 1987. A traditional text for beginning geology courses. Simplified but suitable for high school readers. Contains a good index, illustrations, and an extensive glossary. Chapter 3 treats igneous processes and rocks. Chapters 4, 7, 9, and 11 examine fundamental processes related to mountain building, metamorphism, volcanism, and plate tectonics.

Klein, Cornelis, and Cornelius S. Hurlbut, Jr. *Manual of Mineralogy.* 21st ed. New York: John Wiley & Sons, 1999. An introductory college-level text on mineralogy. Discusses the physical and chemical properties of minerals and describes the most common minerals and their varieties, including gem mineral varieties. Contains more than 500 mineral name entries in the mineral index and describes in detail about 150 mineral species.

Meyers, J. S. "Cauldron Subsidence and Fluidization: Mechanisms of Intrusion of the Coastal Batholith of Peru into Its Own Volcanic Ejecta." *Geological Society of America Bulletin* 86 (1975): 1209-1220. Possibly the best available account of a major batholith. The excellent cross-section diagrams clearly indicate that the author was influenced by the model of Hamilton and Myers. Aimed at professionals but can be understood by college-level readers with some background in geology.

Perch, L. L., ed. *Progress in Metamorphic and Magmatic Petrology.* New York: Cambridge University Press, 1991. Although intended for the advanced reader, several of the essays in this multiauthored volume will serve to familiarize the new student with the study of igneous rocks. In addition, the bibliography will lead the reader to other useful material.

Press, F., and R. Siever. *Earth.* 4th ed. New York: W. H. Freeman, 1986. Chapter 5, "Plutonism," is a more thorough treatment of the subject than that in Judson et al. but requires a slightly higher level of reading. This is one of the best university-level texts for a first course in geology.

Press, Frank, and Raymond Siever. *Understanding Earth.* 2d ed. New York: W. H. Freeman, 1998. This comprehensive physical geology text covers the formation and development of the Earth. Readable by high school students, as well as by general readers. Includes an index and a glossary of terms.

Smith, David G., ed. *The Cambridge Encyclopedia of Earth Sciences.* New York: Crown, 1981. More of a super-text than an encyclopedia. The authors skillfully place their fields of expertise in the plate tectonic context and provide a modern overview of the entire field of Earth science. Includes comprehensive index and glossary as well as high-quality maps, tables, and photographs. For both general and college-level readers.

Sutherland, Lin. *The Volcanic Earth: Volcanoes and Plate Tectonics, Past, Present, and Future.* Sydney, Australia: University of New South Wales Press, 1995. Although Sutherland focuses on volcanic activity in Australia, the book provides an easily understood overview of volcanic and tectonic processes, including the role of igneous rocks. Includes color maps and illustrations, as well as a bibliography.

CARBONATITES

Carbonatites are composed of carbonate minerals that appear to have formed from carbonate liquids. They typically contain many minerals of unusual composition that are seldom found elsewhere. Carbonatites have been mined for a variety of elements, including the rare-earth elements, niobium, and thorium. The rare-earths are used as phosphors in television picture tubes and in high-quality magnets in stereo systems; niobium is used in steel to make it resist high temperature.

PRINCIPAL TERMS

CALCITE: a mineral composed of calcium carbonate

DIKE: a tabular igneous rock formed by the injection of molten rock material through another solid rock

DOLOMITE: a mineral composed of calcium magnesium carbonate

IGNEOUS ROCKS: rocks formed from liquid or molten rock material

IJOLITE: a dark-colored silicate rock containing the minerals nepheline (sodium aluminum silicate) and pyroxene (calcium, magnesium, and iron silicate)

ISOTOPES: different atoms of the same element that have different numbers of neutrons (neutral particles) in their nuclei

LIMESTONE: a sedimentary rock composed mostly of calcium carbonate formed by organisms or by calcite precipitation in warm, shallow seas

MINERAL: a naturally occurring element or compound with a more or less definite chemical composition

ROCK: a naturally occurring, consolidated material that usually consists of two or more minerals; sometimes, as in carbonatites, the rock may consist mainly of one mineral

SILICATE MINERAL: a mineral composed of silicon, oxygen, and other metals, such as iron, magnesium, potassium, and sodium

COMPOSITION OF CARBONATITES

Carbonatites are unusual igneous rocks because they are not primarily composed of silicate minerals, as are most other rocks formed from molten rock material. Instead, carbonatites are composed mostly of carbonate minerals and of minor amounts of other minerals that are rare in other rocks. The carbonate minerals composing carbonatites are usually calcite (calcium carbonate) or dolomite (calcium magnesium carbonate). The other minerals found in carbonatites often contain large concentrations of elements that rarely become concentrated enough to form these minerals in other rocks.

Carbonatites are often associated with rare silicate rocks injected at about the same time as the carbonatites. These silicate rocks are unusual, as they normally lack feldspars (calcium, sodium, and potassium aluminum silicate minerals), which are abundant in most igneous rocks formed within the Earth. Instead of feldspar, many of these silicate rocks contain varied amounts of the minerals nepheline (sodium aluminum silicate) and clinopyroxene (calcium, magnesium, iron silicate) and minor amounts of minerals not commonly found in other igneous rocks. The silicate rock consisting of more or less equal amounts of nepheline and clinopyroxene is called ijolite.

OCCURRENCE OF CARBONATITES

Carbonatites often occur as small bodies of various size and shape that cut across the surrounding rocks. Often, they occur as dikes that may be only a few feet to tens of feet wide, but they may be greater in length. An example of carbonatites occurring as dikes is found at Gem Park near Westcliffe, Colorado. Carbonatites sometimes occur as somewhat equidimensional bodies that are larger than those forming dikes. The Sulfide Queen carbonatite at Mountain Pass, California, for example, is about 800 meters long and 230 meters wide. Often, carbonatites contain foreign rock fragments. The cross-cutting relations and foreign rocks found within the carbonatites suggest to many geologists that these carbonatites form by the injection or intrusion of molten carbonate into the surrounding solid rock. The foreign rock fragments within the carbonatite could be solid rocks ripped off the walls by the moving molten carbonate material as it was injected through solid rock.

The occurrence of molten carbonate at the active volcano at Oldoinyo Lengai in northern Tanzania, Africa, confirms that such material exists. The abundance of recently formed volcanic material composed of carbonatite at Ol Doinyo Lengai and at other locations indicates that considerable carbonatite can form at the surface. The carbonatite liquids at the surface may flow as a lava out of the volcano or be ejected explosively into the air, similar to that of other volcanoes formed mostly from silicate liquids. Carbonatites of volcanic origin are especially abundant in Africa along a portion of the continent called the East African rift zone. There, Africa is slowly being ripped apart by forces within the Earth. Some carbonatites that intruded below the surface as dikes may also have fed volcanoes at the surface at one time. Erosion of the extinct volcano and associated silicate rocks may have exposed the dikes composed of solidified carbonatite.

COMPLEXES

The carbonatites, ijolites, and other associated igneous rocks that formed at the same place and were injected at about the same time are called complexes. These complexes have small areas of exposure at the surface; most are exposed over a surface area of only about 1-35 square kilometers. The Magnet Cove complex in Arkansas, for example, has about a 16-square-kilometer exposure. The Gem Park complex in Colorado is only about 6 square kilometers in area. The associated silicate rocks usually compose most of the area of the exposed complex; only a small portion is carbonatite. The overall shape of these complexes is often circular, elliptical, or oval, but departures from these shapes are common. Often, the different rock types within a complex are built of concentric zones much like the layers of an onion.

Examples of these complexes include Seabrook Lake, Canada, where the complex is roughly circular and is about 0.8 kilometer across. Its "tail," however, extends from the main circular body to about 1.2 kilometers to the south. The central core of carbonatite, composed mostly of calcite, is about 0.3 kilometer across. Smaller carbonatite dikes can be found within other rocks. The largest carbonatite body is surrounded by a dark rock containing angular blocks composed mostly of carbonate, clinopyroxene, or biotite (a dark, shiny potassium, iron, magnesium, and aluminum silicate). This latter body is surrounded by a mixture of ijolite and pyroxenite (a rock containing only pyroxene). The ijolite and pyroxenite also compose much of the tail to the south.

Another complex is located at Gem Park in Colorado. Gem Park is oval in shape, roughly 2 by 3.3 kilometers. There is no central carbonatite there. Instead, scores of small, dolomite-rich carbonatites have intruded as dikes across the silicate rocks of the complex. The main silicate rocks of the complex are pyroxenite and a feldspar-clinopyroxene rock called gabbro. The pyroxenite and gabbro form concentric rings with one another. The large amount of gabbro makes this complex unusual, as gabbros are seldom present with carbonatite in the same complex.

The complexes at Gem Park and Seabrook Lake are rather simple, as they contain very few rock types. A wide variety of minerals can be found in some complexes, resulting in a large number of rock types. The Magnet Cove complex in Arkansas, for example, has twenty-eight major rock types listed on the geologic map. (The reading accompanying the map extends the total rock types to an even greater number.) A large proportion of the rock names in geology are generated by the wide variation of minerals found in these complexes that compose merely a tiny portion of the Earth's surface.

HYPOTHESES OF CARBONATITE FORMATION

Up until the 1950's, geologists believed that carbonatites were limestones that melted and were intruded as molten rock material or that circulating waters formed carbonatites by replacing carbonate minerals with silicate minerals. Some geologists thought that carbonatites could have formed from limestones that were remobilized by a solid, plastic flow—much like the flow of toothpaste squeezed out of a tube.

A major problem with the suggestion that carbonate material could melt, however, was the apparently high melting point of pure calcite or dolomite. Few geologists could believe that the temperature within the Earth was high enough to melt calcite or dolomite. Another problem concerning the belief that limestones were the source of carbonatite was that in some areas, no limestones could be found anywhere near the occurrences of carbonatites. Also, the concept of the intrusion of limestones by plastic flow of a solid was difficult to reconcile with many observations of carbonatites, including the occurrence of foreign igneous rock fragments composed of silicate minerals within them. The absence of fossils in the "limestone" also was noted as unusual, as most limestones have abundant fossils. The lack of fossils could be explained, however, by the melting hypothesis: The fossil evidence would have been destroyed during the melting.

Experiments in furnaces at temperatures and pressures similar to those expected deep within the Earth have done much to support the melting hypothesis for the igneous formation of carbonatites. Several experiments in the late 1950's showed that carbonate minerals could melt at reasonably low temperatures (about 600 degrees Celsius) if abundant carbon dioxide and water vapor coexisted with the carbonate minerals. This dispelled the notion that molten carbonate could not exist within the Earth. Also, the discovery of a volcano in Africa in 1960 that was extruding carbonate lavas confirmed that carbonate liquids could exist within the Earth. Similar experiments at high temperature and pressure on the composition of carbon dioxide or water vapor suggested that their composition could not produce carbonatites by replacement of silicate minerals with carbonate minerals.

These experiments, combined with field observations (including the way carbonatites cross-cut surrounding rocks and the presence of foreign rock fragments), confirmed that most carbonatites formed by the intrusion of carbonate liquids. Even so, the experiments fell short of dispelling the notion that carbonate liquids could have been derived from melted limestones. The melted limestone hypothesis met objections, because isotopic and element concentrations in carbonatites were much different from those observed in limestones. For example, the elements lanthanum and niobium are hundreds of times more concentrated in carbonatites than they are in limestones. No way known can produce this magnitude of enrichment of the elements by melting or by leaching of the carbonate liquid from the solid rock through which it moved. Such observations have caused the limestone origin of carbonatites to be rejected by most geologists.

CONTINUING STUDY OF CARBONATITE FORMATION

Scientists continue to try to understand how carbonatites form. Experiments in furnaces suggest that some carbonate liquids may separate from some silicate liquids similar to those occurring with carbonatites. This process would be like the separation of oil and water as immiscible liquids. Other experiments in furnaces suggest that rocks more than 80 kilometers deep within the Earth may melt in small amounts and produce the carbonate liquids and associated silicate liquids similar in composition to those observed in the natural rocks. Although these experiments fail to prove that carbonate liquids form in these ways, several other lines of evidence have convinced geologists that either of these possibilities could produce carbonatites in nature. For example, some possible source rocks that could melt and produce carbonatites or associated silicate rocks are sometimes carried up with lava from deep within the Earth. The strontium 87-strontium 86 ratios and element contents of these possible source rocks have been measured and are similar to those expected for rocks that would melt and produce carbonate and silicate liquids.

Geologists want to understand how carbonatites form partly because of their economic importance. Thus, they can design better strategies to find carbonatites or the associated silicate rocks that are as yet undiscovered. For example, carbonatites are very small targets to find on the surface of the Earth; therefore, if geologists know that the silicate rocks associated with carbonatites contain abundant magnetic minerals, they can fly an airplane systematically across the surface of the area and detect high magnetic fields by using a device designed to detect the magnetism. Once geologists find areas with magnetic anomalies, they can collect soil or stream samples in the area to see if any unusual minerals associated with carbonatites or the associated silicate rocks are present. They could also drill the area to see if any carbonatites were below the surface.

ECONOMIC VALUE OF CARBONATITES

Their unusually high concentrations of some elements in certain minerals make carbonatites potential ores for these elements. Carbonatites have high concentrations of niobium, thorium, and the rare-earth elements of lower atomic number (such as lanthanum and cerium). Iron, titanium, copper, and manganese also have been mined from carbonatites.

Niobium has been economically extracted from the mineral pyrochlore at Fen, Norway, and at Kaiserstuhl, Germany. Niobium is used as an alloy in steel to resist high temperature, used in gas turbines, rockets, and atomic power plants. Rare-earth elements have been mined from the large carbonatite at Mountain Pass, California. The reserves of rare-earths are enormous at Mountain Pass—averaging about 7 percent—compared to other carbonatites. Other carbonatites enriched with rare-earths occur in Malawi in Africa. The rare-earths are concentrated in many minerals, including perovskite, monazite, xenotime, and a variety of rare-earth carbonate minerals. The rare-earths are used as color phosphors in television picture tubes and as components in high-quality magnets used in stereo speakers and headphones. Thorium, often enriched along with the rare-earths in many carbonatites, tends to concentrate in the same minerals and deposits as do the rare-earths. Thorium is radioactive and has been used as a source of atomic energy. It has also been used for the manufacture of mantles for incandescent gas lights.

Robert L. Cullers

CROSS-REFERENCES

Andesitic Rocks, 1263; Anorthosites, 1269; Basaltic Rocks, 1274; Batholiths, 1280; Continental Crust, 561; Continental Growth, 573; Granitic Rocks, 1292; Igneous Rock Bodies, 1298; Igneous Rock Classification, 1303; Kimberlites, 1310; Komatiites, 1316; Lamproites, 1321; Magmas, 1326; Mountain Belts, 841; Orbicular Rocks, 1332; Plate Tectonics, 86; Plutonic Rocks, 1336; Pyroclastic Rocks, 1343; Rocks: Physical Properties, 1348; Silica Minerals, 1354; Subduction and Orogeny, 92; Ultramafic Rocks, 1360; Xenoliths, 1366.

BIBLIOGRAPHY

Bell, Keith, ed. *Carbonatites: Genesis and Evolution.* Boston: Unwin Hyman, 1989. These papers, collected from a meeting of Geological Association of Canada, cover the formation and evolution of carbonatites. Several essays also cover research trends and recent findings. Illustrations, maps.

Bell, Keith, and J. Keller, eds. *Carbonatite Volcanism: Oldoinyo Lengai and the Petrogenesis of Natrocarbonatites.* New York: Springer-Verlag, 1995. The articles in this collection of essays consitute an in-depth study of volcanism and carbonatites in the Oldoinyo Lengai region of of Tanzania in Africa. Among the illustrations are several useful maps of the area. Bibliography and index.

Deer, W. A., R. A. Howie, and J. Zussman. *An Introduction to Rock-Forming Minerals.* 2d ed. New York: Longman, 1992. A standard reference on mineralogy for advanced college students and above. Each chapter contains detailed descriptions of chemistry and crystal structure, usually with chemical analyses. Discussions of chemical variations in minerals are extensive. *An Introduction to Rock-Forming Minerals* is a condensation of a five-volume set originally published in the 1960's.

Heinrich, E. William. *The Geology of Carbonatites.* Skokie, Ill.: Rand McNally, 1966. A layperson can gain useful information from this volume designed for specialists. Of special interest are sections on the history of carbonatite studies (chapter 1), the economic aspects of carbonatites (chapter 9), and descriptions and locations of carbonatites of the world (chapters 11-16).

Kapustin, Yuri L. *Mineralogy of Carbonatites.* Washington, D.C.: Amerind, 1980. This book summarizes the vast variety of minerals that have been found in carbonatites. Much geologic jargon, but a reader with some geologic background or a mineral collector will find it useful.

Larsen, Esper Signius. *Alkalic Rocks of Iron Hill, Gunnison County, Colorado.* Geological Survey Professional Paper 197-A. Washington, D.C.: Government Printing Office, 1942. This publication describes the wide variety of minerals and rocks at this complex containing carbonatite. The rocks are located on a geologic map. The area is easy to visit to collect the minerals, so a mineral collector would find this publication useful.

Menzies, L. A. D., and J. M. Martins. "The Jacupiranga Mine, São Paulo, Brazil." *The Mineralogical Record* 15 (1984): 261-270. *The Mineralogical Record* is a journal for the lay reader or mineral collector that summarizes the geologic occurrence of minerals with a minimum of technical language. It often contains beautiful photographs of minerals. This article is an example of one on carbonatite minerals at a specific location.

Olsen, J. C., D. R. Shawe, L. C. Pray, and W. N. Sharp. *Rare Earth Mineral Deposits of the Mountain Pass District, San Bernardino County, California.* U.S. Geological Survey Professional Paper 261. Washington, D.C.: Government Printing Office, 1954. An excellent description of the rare-earth carbonatite at Mountain Pass. Many detailed rock and mineral names, but a geologic map of the rock locations is included, so a mineral collector might find this source useful.

Parker, Raymond L., and William N. Sharp. *Mafic-Ultramafic Igneous Rocks and Associated Carbonatites of the Gem Park Complex, Custer and Fremont Counties, Colorado.* U.S. Geological Survey Professional Paper 649. Washington, D.C.: Government Printing Office, 1970. This publication gives a detailed description of the large number of minerals and the rocks found at Gem Park. Useful for a mineral collector. A layperson could read it with a dictionary of mineral and rock names. Photographs of some of the minerals are included.

Press, Frank, and Raymond Siever. *Understanding Earth.* 2d ed. New York: W. H. Freeman, 1998. This comprehensive physical geology text covers the formation and development of the Earth. Readable by high school students, as well as by general readers. Includes an index and a glossary of terms.

Roberts, W. L., G. R. Rapp, and J. Weber. *Encyclopedia of Minerals.* New York: Van Nostrand Reinhold, 1974. One of a variety of mineral references available to the layperson that gives common properties, composition, and color photographs of many minerals, including some found in carbonatites.

Sutherland, Lin. *The Volcanic Earth: Volcanoes and Plate Tectonics, Past, Present, and Future.* Sydney, Australia: University of New South Wales Press, 1995. Although Sutherland focuses on volcanic activity in Australia, the book provides an easily understood overview of volcanic and tectonic processes, including the role of igneous rocks. Includes color maps and illustrations, as well as a bibliography.

Tuttle, O. F., and J. Gittins. *Carbonatites.* New York: Wiley-Interscience, 1966. Another fairly technical book, but it could be useful to a layperson who is not intimidated by mineral and rock names. One interesting section describes the only active volcano extruding carbonate lava by the person who first descended into the volcanic vent. A section on the location and description of carbonatites around the world is also included.

GRANITIC ROCKS

Granitic rocks are coarse-grained igneous rocks consisting mainly of quartz, sodic plagioclase, and alkali feldspar, with various accessory minerals. These rock types occur primarily as large intrusive bodies that have solidified from magma at great depths. Granitic rocks can also occur to a lesser degree as a result of metamorphism, a process referred to as granitization.

PRINCIPAL TERMS

APHANITIC: a textural term that applies to an igneous rock composed of crystals that are microscopic in size

CRYSTAL: a solid made up of a regular periodic arrangement of atoms

CRYSTALLIZATION: the formation and growth of a crystalline solid from a liquid or gas

GRANITIZATION: the process of converting rock into granite; it is thought to occur when hot, ion-rich fluids migrate through a rock and chemically alter its composition

ISOTOPES: atoms of the same element with identical numbers of protons but different numbers of neutrons, thus giving them a different mass

MAGMA: a body of molten rock typically found at depth, including any dissolved gases and crystals

MIGMATITE: a rock exhibiting both igneous and metamorphic characteristics, which forms when light-colored silicate minerals melt and then crystallize, while the dark silicate minerals remain solid

PHANERITIC: a textural term that applies to an igneous rock composed of crystals that are macroscopic in size, ranging from about 1 millimeter to more than 5 millimeters in diameter

PLUTON: a structure that results from the emplacement and crystallization of magma beneath the surface of the Earth

PORPHYRITIC: a texture characteristic of an igneous rock in which macroscopic crystals are embedded in a fine phaneritic or even aphanitic matrix

---

GRANITIC ROCK OCCURRENCE

The term "granitic rocks" generally refers to the whole range of plutonic rocks that contain at least 10 percent quartz. They are the main component of continental shields and also occur as great compound batholiths in folded geosynclinal belts. Granitic rocks are so widespread, and their occurrence and relation to the tectonic environment are so varied, that generalizations often obscure their complexity. Basically, major granitic complexes are a continental phenomenon occurring in the form of batholiths and migmatite complexes.

When large masses of magma solidify deep below the ground surface, they form igneous rocks that exhibit a coarse-grained texture described as phaneritic. These rocks have the appearance of being composed of a mass of intergrown crystals large enough to be identified with the unaided eye. A large mass of magma situated at depth may require tens of thousands, or even millions, of years to solidify. Because phaneritic rocks form deep within the crust, their exposure at ground surface reflects regional uplift and erosion, which has removed the overlying rocks that once surrounded the now-solidified magma chamber.

Along continental margins, belts of granitic rocks developed as batholiths composed of hundreds of individual plutons. Formation of batholithic volumes of granitic magmas generally appears to require continental settings. Some of the more prominent batholiths in North America are the Coast Range, Boulder-Idaho, Sierra Nevada, and Baja California. The largest are more than 1,500 kilometers long and 200 kilometers wide, and have a composite structure. The Sierra Nevada, for example, is composed of about 200 plutons separated by many smaller plutons, some only a few kilometers wide.

GRANITIC ROCK CLASSIFICATION

As with other rock types, granitic rocks are classified on the basis of both mineral composition and fabric or texture. The mineral makeup of an igneous rock is ultimately determined by the chemical composition of the magma from which it crystallized. Feldspar-bearing phaneritic rocks containing conspicuous quartz (greater than 10 percent in total volume) in addition to large amounts of feldspar can be designated as granitic rocks. This nonspecific term is useful where the type of feldspar is not recognizable because of alteration or weathering, for purposes of quick reconnaissance field studies, or for general discussion.

Granitic rocks consist of two general groups of minerals: essential minerals and accessory minerals. Essential minerals are those required to be present for the rock to be assigned a specific name based on a classification scheme. Essential minerals in most granitic rocks are quartz, sodic plagioclase, and potassium-rich alkali plagioclase (either orthoclase or microcline). Accessory minerals include biotite, muscovite, hornblende, and pyroxene.

When an initial phase of slow cooling and crystallization at great depths is followed by more rapid cooling at shallower depths or at the surface, porphyritic texture develops, as is evident in the presence of large crystals enveloped in a finer-grained matrix or groundmass. The presence of porphyritic texture is evidence that crystallization occurs over a range of temperatures, and magmas are commonly emplaced or erupted as mixtures of liquid and early-formed crystals.

Classification of granitic rocks can be based either on the bulk chemical composition or on the mineral composition. Chemical analysis units are in the weight percentages of oxides, whereas mineral composition units are in approximate percentages in total volume. The mineral composition of granitic rocks, unlike that of volcanic rocks, provides a reliable basis for classification. Because the two primary feldspars may be difficult to distinguish as a result of extensive solid solution and unmixing, the chemical composition of the rock in terms of the normalized proportions of quartz, plagioclase, and alkali feldspar is recast. Thus, specific rock types are defined on the basis of their ratios. Accessory minerals may or may not be present in a rock of a given type, but the presence of certain accessory minerals may be indicated in the form of a modifier (such as biotite granite).

ORE MINERALS ASSOCIATED WITH GRANITIC PLUTONS

Mineral	Metal	Comments
Chalcopyrite [$CuFeS_2$]	Copper	"Porphyry copper," finely disseminated grains
Native gold [Au]	Gold	Hydrothermal quartz veins, placer deposits
Molybdenite [MoS_2]	Molybdenum	Used in steel alloys
Cassiterite [SnO_2]	Tin	Mostly mined as placer deposits
Urananinite [UO_2]	Uranium	Mostly mined in granite pegmatites
Rutile [TiO_2]	Titanium	Granite pegmatites, quartz veins
Scheelite [$CaWO_4$]	Tungsten	Granite pegmatites
Wolframite [$(Fe,Mn)WO_4$]	Tungsten	Granite pegmatites
Beryl [$Be_3Al_2Si_6O_{18}$]	Beryllium	Pegmatites; gem crystals are emerald and aquamarine
Lepidolite mica [$K(Li,Al)_{2-3}(Al,Si_3O_{10})(OH)_2$]	Lithium	Pegmatites; used in lubricants and antipsychotic drugs
Spodumene [$LiAlSi_2O_6$]	Lithium	Pegmatites, with lepidolite
Columbite-tantalite	Niobium, tantalum	Used in heat-resistant alloys, stainless steel

GRANITIC ROCK GROUPS

Granitic rocks include granodiorite, quartz monzonite granite, soda granite, and vein rocks of pegmatite and aplite. The mineralogy of these vary, and the distinction between different granitic rocks can be gradational. Granodiorite is composed predominantly of andesine-oligoclase feldspar, with subordinate potassium feldspar and biotite, hornblende, or both as accessory minerals. Quartz monzonite is composed of subequal amounts of potassium and oligoclase-andesine feldspars, with biotite, hornblende, or both as accessory minerals. Granite is composed predominantly of potassium feldspar with subordinate oligoclase feldspar and biotite alone or with hornblende or muscovite as accessory minerals. Soda granite is composed predominantly of albite or albite-oligoclase feldspar, with small amounts of algerine pyroxene or sodic amphibole.

Pegmatite is a very coarse-grained and mineralogically complex rock. Structurally, pegmatites occur as dikes associated with large plutonic rock masses. Dikes are tabular-shaped intrusive features that cut through the surrounding rock. The large crystals are inferred to reflect crystallization in a water-rich environment. Aplite is a very fine-grained, light-colored granitic rock that also occurs as a dike and consists of quartz, albite, potassium feldspar, and muscovite, with almandine garnet as an occasional accessory mineral. Most pegmatites can be mineralogically simple, consisting primarily of quartz and alkali feldspar, with lesser amounts of muscovite, tourmaline, and garnet; they are referred to as simple pegmatites. Other pegmatites can be very complex and contain other elements that slowly crystallize from residual, deeply seated magma bodies. High concentrations of these residual elements can result in the formation of minerals such as topaz, beryl, and rare-earth elements, in addition to quartz and feldspar.

Other rocks that are occasionally grouped with granitic rocks are migmatites. Migmatites, meaning mixed rocks, are heterogeneous granitic rocks which, on a large scale, occur within regions of high-grade metamorphism or as broad migmatitic zones bordering major plutons. Migmatites appear as alternating light and dark bands. The light-colored bands are broadly granitic in mineralogy and chemistry, while the darker bands are clearly metamorphic.

Geochemically, granitic rocks vary in several ways, including isotopic composition; proportion of low-melting constituents, such as quartz and alkali feldspars, to high-melting constituents, such as biotite, hornblende, and calcic plagioclase; relative proportion of the low-melting constituents; alumina saturation; and accessory mineral content. Granitic rocks can further be divided into two groups, S-types and I-types, according to whether they were derived from predominantly sedimentary or igneous sources, respectively. Enormous volumes of I-type granites, which constitute most plutons, occur along continental margins overriding subducting oceanic material.

INCLUSIONS

A common feature of granitic rocks, notably in granodiorite to dioritic plutons, is the presence of inclusions or rock fragments that differ in fabric and/or composition from the main pluton itself. The term "inclusions" indicates that they

A large granite boulder shaped by exfoliation, in the Kaweah Basin, Sequoia National Park, California. (U.S. Geological Survey)

Early Mississippian granitic rocks of the Simpson Allochthon Thrust onto oceanic rocks of the Anvil Allochthon, in the Simpson ranges northeast of the Tintina Trench. (An allochthon is an overthrust block of rocks that have been moved along a fault far from their point of origin.) (Geological Survey of Canada)

originated in different ways. Foreign rock inclusions, called xenoliths, include blocks of wall rocks that have been mechanically incorporated into the magma body. This process is referred to as stopping. Some mafic inclusions are early-formed crystals precipitated from the magma itself after segregation along the margin of the pluton, which cools first. These inclusions are called antoliths. Other inclusions may reflect clots of solidified mantle-derived magma that ascended into the granitic source region or residual material that accompanied the magma during its ascent.

PETROGRAPHIC STUDY

Granitic magmas can be derived from a number of sources, notably the melting of continental crust, the melting of subducted oceanic crust or mantle, and differentiation. The problem facing the geologist is to decide on the importance of these sources in relation to their tectonic environment. To accomplish this objective, both petrographic and geochemical information is used.

The standard method of mineral identification and study of textural features and crystal relationships is by the use of rock thin sections. A thin section is an oriented wafer-thin portion of rock 0.03 millimeter in thickness that is mounted on a glass slide. The rock thin section shows mineral content, abundance and association, grain size, structure, and texture. The thin section also provides a permanent record of a given rock that may be filed for future reference.

Thin sections are studied with the use of a petrographic microscope, which is a modification of the conventional compound microscope commonly used in laboratories. The modifications that render the petrographic microscope suitable to study the optical behavior of transparent crystalline substances are a rotating stage, an upper polarizer or lower polarizer, and a Bertrand lens, used to observe light patterns formed on the upper surface of the objective lens. With a magnification that ranges from about 30 to 500 times, the petrographic microscope allows one to examine the optical behavior of transparent crystalline substances or, in this case, crystals that make up granitic rocks.

The study of granitic rocks may be greatly facilitated by various staining techniques of both hand specimens and thin sections. Staining is employed occasionally to distinguish potassium feldspar from plagioclase and quartz, the three main mineral constituents of granitic rocks. A flat surface on the rock is produced by sawing and then polishing. The rock surface is etched using hydrofluoric acid. This step is followed by a water rinse and immersion in a solution of sodium cobaltinitrate. The potassium feldspars will then turn bright yellow. After rinsing with water and covering the surface with rhodizonate reagent, the plagioclase becomes brick red in color. Staining techniques are available for other minerals, including certain accessory minerals such as cordierite, anorthoclase, and feldspathoids.

Measurement of the relative amounts of various mineral components of a rock is called modal analysis. The relative area occupied by the individual minerals is estimated or measured on a flat surface (on a flat-sawed surface, on a flat outcrop surface, or in thin section) and then related to the relative volume. Caution must be used, because the relative area occupied by any mineral species on a particular planal surface is not always equal to the modal (volume) percentage of that mineral on the rock mass.

GEOCHEMICAL STUDY

When a rock specimen is crushed to a homogeneous powder and chemically analyzed, the bulk chemical composition of the rock is derived. Chemical analyses are normally expressed as oxides of the respective elements, which reflects the overwhelming abundance of oxygen. Analysis of granitic rocks shows them to be typically rich in silica potassium and sodium, with lesser amounts of basic oxides such as magnesium, iron, and calcium oxides. Magnesium, iron, and calcium oxides are present in higher abundance in basalts, which contain plagioclase feldspar, pyroxene, and olivine.

Isotopes such as strontium, oxygen, and lead can also be used as tools in evaluating granitic magma sources such as a mantle origin and crustal melting of certain rock types. Some accessory minerals reflect the trace element content of the magma and thus the possible nature of their source. Some, such as garnet or topaz, are products of contamination, while others, such as andalusite, magnetite, or limenite, are products of late hydrothermal alteration. Much research is focused on chemical tracers. Tracers help distinguish the source region of a granitic magma, such as lower-crustal igneous or sedimentary rock or mantle material.

INDUSTRIAL APPLICATIONS

Granitic rocks have been used as dimension stones for many years. Dimension stones are blocks of rock with roughly even surfaces of specified shape and size used for the foundation and facing of expensive buildings. When crushed, granitic rocks can be used as aggregate in the cement and lime industry. In addition to these uses, granitic rocks are valued because of their geographic association with gold. Gold ores are found in close proximity to the contacts of the granitic bodies within both the granitic rocks and surrounding rocks.

Pegmatites can also be very valuable. Simple pegmatites are exploited for large volumes of quartz and feldspar, used in the glass and ceramic industries. Complex pegmatites can also be a source of gem minerals, including tourmaline, beryl, topaz, and chrysoberyl. In spite of varied mineral composition, relatively small size, and unpredictable occurrence, pegmatites constitute the world's main source of high-grade feldspar, electrical-grade mica, certain metals (including beryllium, lithium, niobium, and tantalum), and some piezoelectric quartz.

Stephen M. Testa

CROSS-REFERENCES

Andesitic Rocks, 1263; Anorthosites, 1269; Basaltic Rocks, 1274; Batholiths, 1280; Carbonates, 1182; Carbonatites, 1287; Igneous Rock Bodies, 1298; Igneous Rock Classification, 1303; Kimberlites, 1310; Komatiites, 1316; Lamproites, 1321; Magmas, 1326; Metamorphic Mineral Deposits, 1614; Minerals: Structure, 1232; Orbicular Rocks, 1332; Plutonic Rocks, 1336; Pyroclastic Rocks, 1343; Rocks: Physical Properties, 1348; Silica Minerals, 1354; Ultramafic Rocks, 1360; Xenoliths, 1366.

BIBLIOGRAPHY

Carmichael, Ian S., Francis J. Turner, and John Verhoogen. *Igneous Petrology*. New York: McGraw-Hill, 1974. A well-known reference presenting both the mineralogical and geochemical diversity of granitic rocks and their respective geologic settings. Chapter 2, "Classification and Variety of Igneous Rocks," and chapter 12, "Rocks of Continental Plutonic Provinces," are particularly recommended.

Deer, W. A., R. A. Howie, and J. Zussman. *An Introduction to Rock-Forming Minerals*. 2d ed. New York: Longman, 1992. A standard reference on mineralogy for advanced college students and above. Each chapter contains detailed descriptions of chemistry and crystal structure, usually with chemical analyses. Discussions of chemical variations in minerals are extensive. *An Introduction to Rock-Forming Minerals* is a condensation of a five-volume set originally published in the 1960's.

Hutchison, Charles S. *Laboratory Handbook of Petrographic Techniques*. New York: John Wiley & Sons, 1974. Stresses the practical aspects of laboratory and petrographic methods and techniques.

Klein, Cornelis, and Cornelius S. Hurlbut, Jr. *Manual of Mineralogy*. 21st ed. New York: John Wiley & Sons, 1999. An introductory college-level text on mineralogy. Discusses the physical and chemical properties of minerals and describes the most common minerals and their varieties, including gem mineral varieties. Contains more than 500 mineral name entries in the mineral index and describes in detail about 150 mineral species.

Perch, L. L., ed. *Progress in Metamorphic and Magmatic Petrology*. New York: Cambridge University Press, 1991. Although intended for the advanced reader, several of the essays in this multiauthored volume will serve to familiarize the new student with the study of igneous rocks. In addition, the bibliography will lead the reader to other useful material.

Phillips, William Revell. *Mineral Optics: Principles and Techniques*. San Francisco: W. H. Freeman, 1971. A standard textbook discussing mineral optic theory and the petrographic microscope and its use in the study of minerals and rocks.

Press, Frank, and Raymond Siever. *Understanding Earth*. 2d ed. New York: W. H. Freeman, 1998. This comprehensive physical geology text covers the formation and development of the Earth. Readable by high school students, as well as by general readers. Incudes an index and a glossary of terms.

Smith, David G., ed. *The Cambridge Encyclopedia of Earth Sciences*. New York: Crown, 1981. Chapter 5, "Earth Materials: Minerals and Rocks," gives a good discussion of the processes involved in the formation of granitic rocks and description of the mineral and chemical composition of granitic rocks. A well-illustrated and carefully indexed reference volume.

Sutherland, Lin. *The Volcanic Earth: Volcanoes and Plate Tectonics, Past, Present, and Future*. Sydney, Australia: University of New South Wales Press, 1995. Although Sutherland focuses on volcanic activity in Australia, the book provides an easily understood overview of volcanic and tectonic processes, including the role of igneous rocks. Includes color maps and illustrations, as well as a bibliography.

IGNEOUS ROCK BODIES

The geometry of the bodies in which igneous rocks are found can provide useful information about the physical characteristics of the magmas or lavas that produced them and about the stress field that was active in the region at the time of their formation.

PRINCIPAL TERMS

ASH: solid particles from an erupting volcano, usually formed as ejected molten material cools during its flight through the atmosphere

CONCORDANT: having sides (contacts) that are nearly parallel to the layering in the country rock

COUNTRY ROCK: the rock into which magma is injected to form an intrusion

DIKE: a discordant sheet intrusion

DISCORDANT: having sides (contacts) that are at a substantial angle to the layering in the country rock

LACCOLITH: a concordant, nearly horizontal intrusion that has lifted the country rock above it into a dome-shaped geometry

LAVA: molten rock at or above the surface of the Earth

MAGMA: molten rock, still beneath the surface of the Earth

SHEET INTRUSION: an intrusion that is tabular, or sheetlike, in shape

SILL: a concordant sheet intrusion

LAVA FLOWS AND ASH FALL DEPOSITS

Igneous rocks form as molten rock cools and solidifies. If this happens at or above the surface of the Earth—during a volcanic eruption, for example—the extrusive rock bodies that result are lava flows and ash fall deposits. If it happens beneath the surface of the Earth, the resulting rock bodies are called intrusions or plutons. The igneous rocks involved can have a wide range of compositions and textures, and hence a great variety of names. Common ones include basalt and granite; less common ones are monchiquite and lamprophyre. Igneous petrologists study the rock itself, seeking information about how, where, and from what it melted and how it might have been modified prior to its final solidification. Their emphasis is on chemical and mineralogical composition, and they employ phase diagrams and chemical reactions in their work. Information obtained from the body itself—its size, shape, and orientation—is also worth studying. Extrusive rock bodies can also be studied to learn about former volcanoes. A volcano's location, some of the characteristics of its lava, and even details of the topography at the time it was active can be reconstructed long after the volcano has stopped erupting, and even after it has eroded completely away.

Lava flows develop as fluid lava moves downhill, but lava is not a simple fluid. Its behavior is complex. How easily a fluid flows (how "thick" it is) is a function of its strength and its viscosity. A sensitive function of composition, lava viscosities range from those similar to motor oil to those more like asphalt. Within a single flow, strength and viscosity will vary with temperature, gas content, and flow rate—all of which change as the lava is moving. At the edges of flows, cooler lava may form natural levees, confining the flow. Within the flow itself, tubes may develop beneath the surface through which quickly flowing lava can move with little cooling. At the surface, the loss of heat and gases can result in a nearly solid rind that deforms by cracking and breaking.

An ash fall deposit is thickest near the site of the eruption. Its areal extent will be influenced by the winds prevailing at the time of the eruption and the height reached by the ash before its descent. Being blown by the wind, ash may accumulate beneath areas of stagnant air. If still sufficiently hot, particles of ash may weld together as they settle, producing a hard rock that resists the forces of erosion.

INTRUSIONS

Molten rock that has not reached the surface is called magma. As complex as lava, magma moves from regions of higher pressure to areas where the pressure is lower, rather than flowing downhill. During this process, magma may wedge apart solid rock and intrude into the crack it produces. In this way, magma can forge its own subsurface path for many kilometers. Eventually, this forging may bring it to the surface, where an eruption will ensue. An eruption will permit much greater flow through the crack. Flow will be easiest and fastest where the crack is widest. The flowing magma erodes the walls of the crack; the walls near the fastest-flowing magma will erode most quickly. In the narrow parts of the crack, flow will be much less, the magma will be harder to deform, and it will cool and solidify. Thus, the magma conduit, which initially was a fluid-filled crack, transforms during the eruption into a nearly cylindrical form. At any point in this process, the flow of magma may be interrupted, permitting the magma within the conduit system to solidify. The form taken by this frozen magma will be inherited from the walls of the conduits, just as a casting takes its form from a mold.

There are many names for different intrusive forms, and their classification is not very systematic. If the sides of the intrusion are generally parallel to the layering in the rock into which it intrudes (the country rock), it is said to be concordant. If its sides are at a significant angle to that layering, the intrusion is said to be discordant. The names for irregularly shaped discordant intrusions are based on size. Batholiths are huge; stocks are much smaller. Perhaps more important than the sizes which define them, though, are the differences in depths of formation. Batholiths form at great depths, where the temperatures are so high that the country rock behaves in a fairly ductile fashion, whereas many stocks are thought to be remains of the subsurface cylindrical conduits which fed volcanoes.

SHEET INTRUSIONS

Commonly, igneous intrusions will form tabular bodies, with one dimension much smaller than the other two. These sheet intrusions are called dikes if they are discordant and sills if they are concordant. It is now understood that the orientation of a sheet intrusion is more likely to be controlled by the direction of least compression than by attitude of existing layers.

When molten rock is forced into the crust, it forms dikes or sills depending on which direction is under the least compression. At the surface of the Earth, the direction of least compression must be either horizontal or vertical. Sheet intrusions that form near the surface, then, will be emplaced in either a vertical or a horizontal position. Because sedimentary strata are commonly nearly horizontal, a horizontal sheet intruded into such strata is concordant, while a vertical one is discordant. As magma pushes the sides of a sheet intrusion away from each other, stress is concentrated at the tip of the crack. When this stress exceeds the strength of the rock, the rock splits apart and the crack grows longer. The longer crack fills with fluid under pressure, forcing the sides farther apart. Now, with even greater lever-

Because igneous rock is more resistant to weathering and erosion than the sedimentary rock into which it may have been emplaced, igneous rock bodies often form impressive scenic features, such as Shiprock in New Mexico. (© William E. Ferguson)

age acting on it, the crack tip fails again. The process is repeated as long as there is enough fluid pressure.

The country rock ahead of a sheet intrusion fails in extension. The fracture produced is much like a joint and as such will have many of the surface decorations common to joints. One of these decorations, called plumose structure or twist hackle, occurs when the crack breaks into a number of smaller cracks, each slightly rotated from the parent crack to produce a series of parallel offset cracks. These are called *en echelon* cracks. If they become filled with magma, a set of *en echelon* dikes will result. Continued propagation of the crack and filling with magma will cause the individual segments to coalesce. The resulting intrusion will be a dike with a series of matching offsets along its edges, showing where individual *en echelon* segments existed earlier. Such offsets can be used to infer the direction in which the crack initially grew.

As the area over which the pressure acts increases, the force produced by that pressure increases also. This is the principle behind pneumatic jacks. In the case of a horizontal sill not too far beneath the surface of the Earth, the force pushing up increases with area, while the resisting forces increase with the perimeter. Because area grows faster than does perimeter, there may come a time when the rock lying above the sill will be domed up, producing a laccolith: a concordant intrusion, with a flat floor and a domed roof. Many other names have been assigned to intrusions with different geometries. Lopoliths, sphenoliths, bysmaliths, phacoliths, ductoliths, harpoliths, akmoliths, and even cactoliths have been described. It is not clear whether such nomenclature has useful general applicability, and the use of these terms has fallen out of favor. Indeed, even the distinction between dikes and sills is often no longer made, the phrase "sheet intrusions" being preferred by many geologists.

STUDY OF LAVA FLOWS AND ASH FALL DEPOSITS

Lava flows and ash fall deposits are usually studied in the field. If the outcrops are favorable, detailed maps are constructed, often on top of aerial photographs. These might show variations in thickness, locations, and orientations of lava tubes, levees, and surface fracture patterns. From

the patterns that emerge and by radiometric dating of the deposits, geologists seek to understand the history of the volcanoes involved. Which vents were active when? How large was the largest flow? The smallest? What is the average rate at which lava has been produced by this volcano?

Even if the deposits are buried under thousands of feet of sedimentary rock, they may still be susceptible to study where they are exposed in canyons or encountered in wells. Most of the detail will have been obliterated, but trends in the variation of their thickness can still be used to indicate where the eruptions that produced them occurred. This study has been conducted on rocks hundreds of millions of years old, enhancing the understanding of the Earth's tectonic history.

STUDY OF INTRUSIONS

Intrusions are also studied in the field. Measurements are made of the size and shape of the body being studied. Because much of the intrusion is usually buried and much of it has been removed by erosion, reconstructing the original shape may not be easy. Yet, even a thin slice of data through an intrusion can provide important information. When examined carefully, many dikes show an asymmetric cross section, something like a long, thin teardrop. This cross section has been interpreted to indicate gradients in magma pressure or regional stresses active during emplacement. The edges of sheet intrusions frequently display offsets and occasionally grooves, which are thought to indicate the direction in which the crack occupied by the intrusion initially grew. Such information can be utilized to help reconstruct the three-dimensional form of the intrusion and to define the sequence of events that produced it. The transition from sill to laccolith represents a particularly opportune situation. Estimates of the depth of overburden, strength, and resistance to bending of the overlying rock, and the pressure and mechanical behavior of the magma may all be derived if sufficient data are available for a substantial number of laccoliths and sills in an area.

Dikes often occur in groups, called dike swarms. By mapping such swarms, geologists may find that they reveal systematic patterns that can be interpreted as images of the stress field in the region. The next step is to use computers and the theory of elasticity to analyze such stress fields, and then try to find causes for the stresses discovered. Because the igneous rock making up the intrusions can often be dated, a stress history of the region may be developed.

To understand field exposures, geologists may construct models representing the intrusions they wish to study. Some of these models are physical, with motor oil or petroleum jelly acting as magma and being intruded under pressure into gelatin, plaster, or clay. Conditions can be controlled, and the experiments can be halted to make measurements, photographs, and even films of the process. Another way to model intrusions is on a computer. Using a system of equations derived from the theory of elasticity, geologists can predict the stresses and displacements in the vicinity of an intrusion with a given shape, containing a magma with known properties, and surrounded by rock with known elastic behavior subjected to a known stress field. By letting each of these variables change, their effects can be studied independently.

Laboratory simulations can suggest which field measurements are most likely to be significant in learning about the conditions at the time of intrusion. Armed with an intuition developed in the lab, the field geologist is better prepared to understand field exposures. Field data, in turn, often pose dilemmas that yield to analysis in the laboratory.

GEOLOGIC SIGNIFICANCE

The sizes, shapes, orientations, and distribution of igneous rock bodies have been interpreted to reveal important physical characteristics of the molten rocks that produced them, as well as the conditions prevailing at the time of their forma-tion. Because they are useful pressure gauges, igneous rock bodies can also provide information about the stresses active at the time of their emplacement. Patterns produced by swarms of dikes have been interpreted in terms of the horizontal stresses acting throughout the region. Such information has direct bearing on questions concerning the mechanism of mountain building, plate tectonics, and the history of the Earth in general.

Finally, owing to the fact that igneous rock is frequently more resistant to weathering and erosion than is the sedimentary rock into which it may have been emplaced, igneous rock bodies often form impressive scenic features. Devils Postpile in California, the Henry Mountains of Utah, Shiprock in New Mexico, and the Palisades Sill across the Hudson from New York City are some of the scenic attractions produced by igneous rock bodies. Appreciation of these and similar features is enhanced when one understands something about their formation.

Otto H. Muller

CROSS-REFERENCES

Andesitic Rocks, 1263; Anorthosites, 1269; Basaltic Rocks, 1274; Batholiths, 1280; Building Stone, 1545; Carbonatites, 1287; Continental Crust, 561; Feldspars, 1193; Fold Belts, 620; Granitic Rocks, 1292; Igneous Rock Classification, 1303; Kimberlites, 1310; Komatiites, 1316; Lamproites, 1321; Magmas, 1326; Metamorphic Mineral Deposits, 1614; Minerals: Physical Properties, 1225; Mountain Belts, 841; Orbicular Rocks, 1332; Pegmatites, 1620; Petrographic Microscopes, 493; Plutonic Rocks, 1336; Pyroclastic Rocks, 1343; Rocks: Physical Properties, 1348; Silica Minerals, 1354; Ultramafic Rocks, 1360; Xenoliths, 1366.

BIBLIOGRAPHY

Billings, Marland P. *Structural Geology*. 3d ed. Englewood Cliffs, N.J.: Prentice-Hall, 1972. Chapter 16, "Intrusive Igneous Rocks," presents classic descriptions of many different intrusive rock bodies. There is little emphasis on the mechanisms of formation or on the interpretation of the bodies described, but each is defined and most are illustrated. De-scriptive, with little prior knowledge assumed, this book is suitable for the general reader.

Decker, Robert, and Barbara Decker. *Volcanoes*. San Francisco: W. H. Freeman, 1997. Suitable for the general reader, this book provides useful background for the understanding of igneous rock bodies. Of particular interest are chapter 10, "Roots of Volcanoes," and chapter 8, "Lava, Ash, and Bombs."

Hargraves, R. B., ed. *Physics of Magmatic Processes.* Princeton, N.J.: Princeton University Press, 1980. This book is a fairly technical review of much of the work in progress at the time of its publication. The first seventeen pages of chapter 6, "The Fracture Mechanisms of Magma Transport from the Mantle to the Surface," by Herbert R. Shaw, are very informative and easily understood by the general reader. Most readers will also learn much from chapter 7, "Aspects of Magma Transport," by Frank J. Spera, which describes some of the complexities of magma behavior.

Hunt, Charles B., Paul Averitt, and Ralph L. Miller. *Geology and Geography of the Henry Mountain Region, Utah.* U.S. Geological Survey Professional Paper 228. Washington, D.C.: Government Printing Office, 1953. A classic paper, this 234-page report describes the laccoliths of this region in detail. Profusely illustrated and accompanied by maps and cross sections, the information presented will provide the reader with an excellent sense of what these laccoliths look like and the map patterns they produce. There is little, however, in the way of a convincing discussion concerning how they formed or what inferences may be drawn from their locations or geometries. Suitable for the general reader.

Johnson, Arvid M. *Physical Processes in Geology.* San Francisco: Freeman, Cooper, 1970. This book has strongly influenced the way igneous rock bodies have been studied ever since its publication. Building on an approach developed to study the formation of laccoliths, Johnson leads the reader through discussions of elasticity and viscosity and proceeds to apply some of the results obtained to problems of dike intrusion and the flow of magma. Although differential and integral calculus are needed to follow the derivations, much of the general approach can be appreciated by those with less mathematical training. Suitable for the technically oriented college student.

Perch, L. L., ed. *Progress in Metamorphic and Magmatic Petrology.* New York: Cambridge University Press, 1991. Although intended for the advanced reader, several of the essays in this multiauthored volume will serve to familiarize the new student with the study of igneous rocks. In addition, the bibliography will lead the reader to other useful material.

Press, Frank, and Raymond Siever. *Understanding Earth.* 2d ed. New York: W. H. Freeman, 1998. This comprehensive physical geology text covers the formation and development of the Earth. Readable by high school students, as well as by general readers. Includes an index and a glossary of terms.

Suppe, John. *Principles of Structural Geology.* Englewood Cliffs, N.J.: Prentice-Hall, 1985. Chapter 7, "Intrusive and Extrusive Structures," shows how the emphasis in the study of igneous rock bodies shifted during the 1960's and 1970's. In this text, there is little in the way of categorizing bodies in terms of their shapes and much more on their interpretation in terms of the physical conditions existing during the time of their emplacement. Although mechanically sound, the math used is not intimidating. Suitable for the general reader.

Sutherland, Lin. *The Volcanic Earth: Volcanoes and Plate Tectonics, Past, Present, and Future.* Sydney, Australia: University of New South Wales Press, 1995. Although Sutherland focuses on volcanic activity in Australia, the book provides an easily understood overview of volcanic and tectonic processes, including the role of igneous rocks. Includes color maps and illustrations, as well as a bibliography.

Williams, Howel, and Alexander R. McBirney. *Volcanology.* San Francisco: Freeman, Cooper, 1979. This book is a general text on volcanoes. Chapter 5, "Lava Flows," and chapter 6, "Airfall and Intrusive Pyroclastic Deposits," are most relevant to igneous rock bodies. The writing seems to put an unnecessary stress on terminology. Suitable for college students with some background in geology.

IGNEOUS ROCK CLASSIFICATION

The classification of igneous rocks depends on their texture and composition. "Texture" refers to the grain size of the constituent minerals of the rock and depends on how slowly or quickly a magma cooled to form the igneous rock. "Composition" refers to both chemical and mineralogical features.

PRINCIPAL TERMS

COLOR INDEX: the percentage by volume of dark minerals in a rock; it is used for quick identification of rocks

EXTRUSIVE ROCK: a fine-grained, or glassy, rock which was formed from a magma that cooled on the surface of the Earth

FELSIC: characterized by a light-colored mineral such as feldspar or quartz, or a light-colored rock dominantly composed of such minerals

INTRUSIVE ROCK: an igneous rock which was formed from a magma that cooled below the surface of the Earth; it is commonly coarse-grained

MAFIC: characterized by a dark-colored mineral such as olivine, pyroxene, amphibole, or biotite, or a rock composed of such minerals

MAGMA: a molten rock material largely composed of silicate ions

MINERAL: a natural substance with a definite chemical composition and an ordered internal arrangement of atoms

MODE: the type and amount of minerals actually observed in a rock

NORM: the type and amount of minerals derived by a set of calculations from a chemical analysis of a rock

SILICA: silicon dioxide, or quartz

CLASSIFICATION ACCORDING TO TEXTURE

Igneous rocks are classified according to their texture and composition. With regard to texture, there are two groups: fine-grained, or glassy, and coarse-grained. With regard to composition, the rocks can be classed chemically or mineralogically. Chemically, they can be grouped according to their silicon dioxide content or their combination of oxides; mineralogically, they can be grouped according to mode, norm, or color index.

Rocks that are formed from magmas that cool beneath the surface of the Earth are called "intrusives." Depending on grain size, intrusives are grouped into plutonics, hypabyssal rocks, and pegmatites. Plutonics are coarse-grained igneous rocks that are formed from magmas that cooled slowly deep beneath the surface of the Earth. Hypabyssal rocks are fine-grained, because of a magma's comparatively rapid rate of cooling at a shallow depth. A variety of intrusive which commonly occurs as a tabular body that cuts across other rocks is a pegmatite. A pegmatite is a light-colored and extremely coarse-grained rock.

Igneous rocks which are formed from a magma that has extruded onto the surface of the Earth are known as either "extrusives" or "volcanics." In such environments, the magmas cool rapidly, so the minerals are fine-grained. In some cases, a magma may cool so rapidly that minerals may not form at all. Instead, the magmas turn into rocks with no ordered arrangement of atoms, and as such with no crystalline structure. Such an amorphous and unorganized arrangement of atoms characterizes glass. A rock composed of glass is called "obsidian" and generally is black.

Extrusives are subdivided into "pyroclastics" and "lavas." Pyroclastics are formed by the forcible extrusion of highly gas-charged magmas. The top portion of such magma is full of gas and frothy. The melt that surrounds the spherical gas bubbles cools into glass shards, yielding a "pumice" rock while, or soon after, the magma is extruded. Glass shards, pumice fragments, early formed crystals, fragmented magma, and rocks that surround the conduit may be ejected into the atmosphere or flow laterally. An accumulation of such fireborne fragmentary material is called a "pyroclastic rock." Commonly, the fragments are flattened and welded together because of heat and overlying weight, which produces a "welded tuff." When very fine-grained glass shards which were explosively ejected into the atmosphere rain down and settle on the ground, they produce volcanic ash, or ash fall tuff. Occasionally, a rising magma may encounter a mass of water which creates steam and causes a powerful phreatomagmatic eruption. In such a case, a ring-shaped cloud of steam with minor solid particles moves swiftly away from a vertical eruption column above the vent. Rocks that are formed from the ring of cloud are called "base surge deposits" and show many structures similar to sedimentary rocks.

Lavas are volcanic rocks formed from magmas that flow gently. "Lava" is also a name for the magma itself. Such magmas do not contain much gas. The gas that may be derived from dewatering vegetation and soil over which the magma moves ascends to the top surface of the cooling magma body, so the top part of a lava rock may contain air bubbles, or vesicles. When there are abundant vesicles in the rock, the lava is called a "scoria." In some lavas, the vesicles may be filled by secondary minerals which precipitated from groundwater. Such rocks are called "amygdaloidal lavas." Some magmas erupt and flow beneath a body of water.

The ensuing lavas have a texture that resembles pillows and are therefore called "pillow lavas." Depending on their composition, they are pillow basalts (fine-grained, dark rocks) or spilites (greenish rocks). In the latter, the basalt is altered by interaction with the surrounding body of water.

Some igneous rocks are porphyritic, in that a few minerals in the rock are coarser-grained than the majority of the minerals. The coarse-grained minerals formed first and probably at greater depths than the fine-grained ones; commonly, the term "porphyritic" is used as a prefix or suffix of a rock name, as in porphyritic lava or lava porphyry.

CLASSIFICATION ACCORDING TO CHEMICAL COMPOSITION

Although a few igneous rocks are identified only by their textural characteristics, most are classified on the basis of their composition. The composition of an igneous rock depends on the magma from which the rock was derived. A magma is itself derived by the melting of some portion of the top part of the Earth. This part of the Earth is composed primarily of the following eight major elements: oxygen, silicon, aluminum, iron, calcium, sodium, potassium, and magnesium. Oxygen and silicon together comprise about 75 percent of the top part of the Earth. Thus, the combination of silicon and oxygen dominates rocks which are formed from this part. When igneous rocks are chemically analyzed, the results are dominated by seven different oxides, including silicon dioxide. These oxides are used to classify igneous rocks.

The most common parameter used for classifying igneous rocks is the silicon dioxide, or silica, content of a rock. Based on its weight percentage of silica, a rock could be placed in one of the following four major classes of igneous rocks: felsic (greater than 66 percent), intermediate (52 to 66 percent), mafic (45 to 52 percent), and ultramafic (less than 45 percent). These four major

Pillow lavas, which form only under water, are a form of igneous volcanic rock that appear where lava formerly erupted into, or flowed into, water. These pillow-lava rocks, found near the Yuba River in California, were probably once on the ocean floor. (© William E. Ferguson)

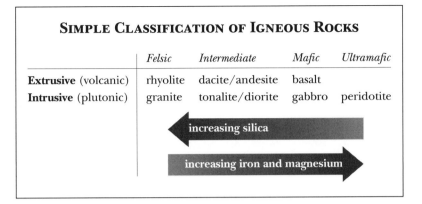

SIMPLE CLASSIFICATION OF IGNEOUS ROCKS

	Felsic	Intermediate	Mafic	Ultramafic
Extrusive (volcanic)	rhyolite	dacite/andesite	basalt	
Intrusive (plutonic)	granite	tonalite/diorite	gabbro	peridotite

← increasing silica

increasing iron and magnesium →

groups of igneous rocks can also be identified by the percentage of dark minerals, or the color index (CI): Felsics have a CI of less than 30 percent; intermediates, 30 to 60 percent; mafics, 60 to 90 percent; and ultramafics, more than 90 percent. An uncommon group of rocks called "carbonatites" are dominantly composed of carbonate minerals instead of silica minerals.

The combination of the seven oxides in igneous rocks is used as another chemical classification scheme. The exact definition of the types of igneous rock distinguished by such a classification scheme requires using graphs that show the variation of the selected oxides. Such a graph is augmented by trace element and isotope ratios of elements in the rocks. Essentially, however, the method classes igneous rocks into alkaline, peralkaline, and subalkaline rocks. The subalkaline rocks are divisible into tholeiitic and calc-alkali rocks. Generally, the tholeiitic rocks have mafic to ultramafic rock associations as a predominant component. In calc-alkali rocks, felsic to intermediate rock associations are most abundant. Alkaline and subalkaline rocks can form at any place where a deep-seated magma source, or a hot spot, penetrates the surface of the Earth, such as in Hawaii. More commonly, though, igneous rock associations are formed at plate boundaries, zones that separate adjacent shifting plates.

Scientists believe that the top part of the Earth is compartmentalized into tectonic plates that are in constant motion with respect to one another. The plate boundaries are the geologic settings in which igneous rock associations are formed. Adjacent tectonic plates may move toward each other and converge at their mutual boundary. This geologic setting is called a convergent plate boundary, and the igneous rocks that form in such a setting are in the calc-alkali class. The group of igneous rocks found in the Andes of South America or the mountains of the Caribbean islands are examples of such igneous rocks. In another geologic setting, adjacent plates move away from each other and a rising molten rock material pushes upward at their mutual boundary. This mutual boundary is called a divergent plate boundary, and there are two types of igneous rock association that can form in such a geologic setting. In continental areas, where the plate motion has not succeeded in tearing apart the continent and where there is a continental bulge with a depression at its center (called a rift valley), the igneous rocks at the boundary are of the alkaline class. The East African Rift Valley is an example of a place where alkaline rocks are found. In contrast, where plate motion has succeeded in tearing apart a continent and has created an intervening oceanic floor, the igneous rocks are of the tholeiitic class, particularly the mafic-ultramafic associations.

CLASSIFICATION ACCORDING TO MINERALOGIC COMPOSITION

Another compositional parameter used for classifying igneous rocks is mineralogic composition. The mineral composition is dependent on the chemical composition of the parental magma. A magma is commonly called a silicate melt, because it is composed mainly of silicon and oxygen that are combined to form the silicate ion. This ion has a four-sided configuration, with oxygens at the corners and silicon at the center. The charge on the corner oxygens of a silica tetrahedron is such that the oxygen atoms can bond with equal strength to adjacent tetrahedra. Silicates are a class of minerals that contain the silicate ion in their composition. The resulting silicates are divisible into groups based on how many of their oxygen atoms are shared among adjacent tetrahedra.

Felsic minerals are composed of oxygen, silicon, aluminum, sodium, potassium, and calcium.

TYPICAL ORE MINERALS ASSOCIATED WITH IGNEOUS ROCKS

Rock Type	Mineral	Metal or Other Commodity Obtained
Felsic—Intermediate		
Granite	Feldspar	Porcelain, scouring powder
	Native gold	Gold
Pegmatite	Cassiterite	Tin
	Beryl	Berryllium, gemstones (emerald; aquamarine
	Tourmaline	Gemstone
	Spodumene	Lithium
	Lepidolite	Lithium
	Scheelite	Tungsten
	Rutile	Titanium
	Apatite	Phosphorus
	Samarskite	Uranium, niobium, tantalium, rare-earth elements
	Columbite, Tantalite	Niobium, tantalium, used in electronics
	Thorianite	Uranium, thorium
	Uraninite	Uranium
	Amazonite (microcline feldspar)	Gemstone
	Rose quartz	Gemstone
	Topaz	Gemstone
	Sphene (titanite)	Titanium, gemstone
	Muscovite mica	Electrical insulation
	Zircon	Zirconium
Rhyolite	Chalcopyrite	Cooper
	Molybdenite	Molybdenum
Mafic—Ultramafic		
Gabbro and Anorthosite	Ilmenite	Titanium
	Labradorite (plagioclase feldspar)	Gemstone
	Chalcopyrite	Copper
	Bornite	Copper
	Pentlandite	Nickel
Peridotite	Chromite	Chromium
	Native plantinum	Platinum
	Sperrylite	Platinum
	Serpentine	Nickel (from weathered soils)

They are silicates in which all oxygen atoms are shared among adjacent tetrahedra. Felsic minerals include quartz (silicon dioxide), which is commonly colorless to transparent; feldspars, such as potassic feldspar, sodic feldspar, and calcic feldspar; and feldspathoids, such as nepheline and leucite. In general, the felsic minerals are light-colored, because they may not contain transition metals. Transition metals are elements whose atomic numbers range from 21 to 30 and whose outer electrons can be excited by light to the same energy levels that correspond to the colors of visible light. Thus, felsic igneous rocks that are dominantly composed of felsic minerals are light-colored, or leucocratic, with a color index of less than 30 percent. A felsic rock is formed from the cooling of a felsic magma, a highly polymerized and viscous magma. Typically, a magma is such that it either does or does not have sufficient silicon dioxide to form quartz; in the latter case, feldspathoids form. In other words, feldspathoids and quartz are incompatible minerals and are not to be found in the same igneous rock.

In contrast to felsic minerals, mafic minerals are dominantly composed of magnesium and iron. These minerals have a structure in which a maximum of three oxygens are shared among adjacent silica tetrahedra; the unshared oxygens are bonded with magnesium and iron. The mafic minerals include olivine, green and equidimensional; pyroxene, dark green or black, and stout; amphibole, black and elongate; and biotite, black and tabular. In most cases, the mafic minerals are dark-colored as a result of the presence of transition metals such as iron. A mafic igneous rock which is dominantly composed of these silicates is therefore also dark-colored.

The visible mineralogic composition, or the mode, is used to classify a rock if it consists of minerals of sizes that can be identified and amounts that can be counted; however, some rocks are very fine-grained, or they contain glass, so their mineral composition cannot be estimated even after magnification under a microscope. In such cases, the rocks are analyzed for their major element content, and the minerals that would have formed had the magma cooled slowly are obtained by calculation. The minerals and their amounts determined by calculation give the norm of a rock, and the individual minerals are normative minerals. The normative mineral composition can be used for classification purposes in the same way as the modal composition.

DETERMINING ROCK MODE

Division into coarse-grained intrusives and fine-grained, or glassy, extrusives is done by visual examination of the texture of the igneous rock. Further classification of these is undertaken by determining the mode or the norm of the rock and applying a classification scheme.

The mode may be determined by trained persons who can identify the minerals in a rock with the naked eye or after magnifying the minerals ten to thirty times with a magnifying glass, such as a pocket-sized hand lens or a binocular microscope. Better mineral identification is done by scientists after a rock is cut to a small size, mounted

on glass, and then ground until a very thin section (0.03 millimeter) of the rock, capable of transmitting light, is prepared. The thin section is placed on a stage of a transmitted light polarizing microscope. A lens below the stage polarizes light by allowing the transmission only of light which vibrates in one direction—for example, east to west. A lens above the stage allows the passage of light vibrating in a north-south direction. When glass is placed on the stage and the lower polarizer is inserted across the transmission of light, the color of the glass is seen. When the upper polarizer is also inserted, the glass appears dark, because no light is transmitted. The properties of most minerals with respect to the transmission of light (optical properties) are different from those of glass.

Other accessories are used in addition to the polarizing lenses in order to characterize accurately the optical properties of minerals. Magnification by the microscope permits better determination of the physical properties of minerals, such as their shape and cleavage. Cleavage refers to a set of planes along which minerals break, and it is related to the ordered internal arrangement of the atoms of a mineral. The felsic minerals are used to classify the non-ultramafic rocks. A rock sample or thin section of such a rock may be inserted into a dye that stains one of the felsics, usually the alkali feldspars. This method simplifies the counting of minerals for classification.

DETERMINING ROCK NORM

The norm of a rock is calculated from the major elements obtained by chemical analysis. Modern methods of chemical analysis use emission or absorption of radiant energy that is unique to each atom and therefore can lead to the identification of that atom. There are four such methods: X-ray fluorescence (XRF) spectrometry, electron microprobe (EMP) analysis, instrumental neutron activation analysis (INAA), and atomic absorption spectrometry (AAS).

In the XRF method, radiant energy is focused on a sample of a rock. This energy removes electrons from the lower electron-energy levels of the atoms. The place of the removed electrons may be taken by other electrons, which fall from higher energy levels by emitting radiant energy. Depending on the energy levels from which the electrons fall, X rays of different energy are released from one element, and the spectrum of these is unique to that element. The X rays from many elements are guided to a crystal, which diffracts and disperses them for easy detection by an X-ray counter. The counter triggers electronic signals, which may be either recorded digitally and interfaced to computers or displayed on strip-chart recorders as separate peaks. The peak positions on the chart correspond to the elements in the rock sample. The peak heights are related to the concentration of the elements, the exact amount being determined by a comparison to that of a standard element admixed to the rock sample.

In EMP analysis, electrons are used as the radiant energy source, and this energy is focused on a rock sample that is polished and coated with carbon. In INAA, the sample is bombarded by neutrons supplied by either a nuclear reactor or an accelerator. Atoms in a sample acquire neutrons and become heavy and unstable (radioactive) isotopes. The samples are removed and placed in a chamber, where the emitted radiant energy (gamma rays) resulting from the decay of the artificially induced radioactive isotopes is guided to a crystal which separates the sample rays for easy detection by a recording device. This method is suitable for detecting the type and concentration of even those elements which may be found only in trace amounts.

In AAS, one uses the radiant energy emitted from a known element in a cathode-ray tube to detect the presence of an element in a sample by noting whether the energy is reduced and by how much. A rock sample is dissolved in a solution and then heated until chemical bonds are broken and the solution contains individual atoms. When radiant energy is made to pass through the atomized sample, the energy will be absorbed by the sample if there are elements in the sample that are identical to those in the cathode. This absorption of energy by the sample causes a reduction in the detection of the source radiant energy, and the reduction corresponds to the amount of the element in the sample.

The XRF method is by far the most widely used technique for analyzing chemical composition. The other techniques have their application in the determination of trace elements that are used to facilitate the classification of igneous rocks by igneous rock association, such as alkaline or subalkaline.

ROLE IN LOCATING ECONOMIC MINERAL DEPOSITS

The classification of igneous rocks has its most useful application in the search for economic metal deposits. Ultramafic and mafic rocks are associated with chromium, platinum, palladium, iridium, osmium, rhodium, ruthium, nickel, iron, titanium, and gold deposits. Intermediate and felsic rocks are associated with copper, molybdenum, tin, tungsten, silver, lead, and zinc deposits. Alkaline rocks, particularly the kimberlites, are associated with diamonds. These different igneous associations are found in different geologic environments. Knowing the geologic environment, then, helps both in classifying the rocks and in anticipating the types of metallic deposit that could be sought in such regions.

Habte Giorgis Churnet

CROSS-REFERENCES

BIBLIOGRAPHY

Best, Myron G. *Igneous and Metamorphic Petrology.* New York: W. H. Freeman, 1995. An easy-to-read book that provides classification as well as descriptions of igneous rocks.

Carmichael, Ian S., Francis J. Turner, and Joan Verhoogen. *Igneous Petrology.* New York: McGraw-Hill, 1974. An excellent book on the classification and description of igneous rocks for college-level students.

Decker, Robert, and Barbara Decker. *Volcanoes.* San Francisco: W. H. Freeman, 1997. Suitable for the general reader, this book provides useful background for the understanding of igneous rock bodies. Of particular interest are chapter 10, "Roots of Volcanoes," and chapter 8, "Lava, Ash, and Bombs."

Hutchison, Charles S. *Laboratory Handbook of Petrographic Techniques.* New York: John Wiley & Sons, 1974. This book provides a discussion of techniques used in the laboratory for identifying minerals and rocks.

International Union of Geological Sciences. Subcommission on the Systematics of Igneous Rocks. *A Classification of Igneous Rocks and Glossary of Terms: Recommendations of the International Union of Geological Sciences Subcommission on the Systematics of Igneous Rocks.* Boston: Blackwell, 1989. This is an extremely useful tool for the student seeking information on igneous rock classification. The glossary and bibliography will be of particular interest.

Klein, Cornelis, and C. S. Hurlbut, Jr. *Manual of Mineralogy.* 21st ed. New York: John Wiley & Sons, 1999. A useful book for study of minerals. Chapter 12 provides a succinct treatment of igneous rocks.

Perch, L. L., ed. *Progress in Metamorphic and Magmatic Petrology.* New York: Cambridge University Press, 1991. Although intended for the advanced reader, several of the essays in this multiauthored volume will serve to familiarize the new student with the study of igneous rocks. In addition, the bibliography will lead the reader to other useful material.

Press, Frank, and Raymond Siever. *Understanding Earth.* 2d ed. New York: W. H. Freeman, 1998. This comprehensive physical geology text covers the formation and development of the Earth. Readable by high school students, as well as by general readers. Includes an index and a glossary of terms.

Prinz, Martin, et al., eds. *Simon & Schuster's Guide to Rocks and Minerals.* New York: Simon & Schuster, 1978. Rocks and minerals are described in this easy-to-read book. Illustrated in color.

Shelley, David. *Igneous and Metamorphic Rocks Under the Microscope: Classification, Textures, Microstructures, and Mineral-Preferred Orientations.* New York: Chapman and Hall, 1993. This well-illustrated text provides an abundance of information on igneous and metamorphic rock classification and nomenclature. Includes an index and a useful bibliography that is nearly twenty-five pages long.

Suppe, John. *Principles of Structural Geology.* Englewood Cliffs, N.J.: Prentice-Hall, 1985. Chapter 7, "Intrusive and Extrusive Structures," shows how the emphasis in the study of igneous rock bodies shifted during the 1960's and 1970's. In this text, there is little in the way of categorizing bodies in terms of their shapes and much more on their interpretation in terms of the physical conditions existing during the time of their emplacement. Although mechanically sound, the math used is not intimidating. Suitable for the general reader.

Sutherland, Lin. *The Volcanic Earth: Volcanoes and Plate Tectonics, Past, Present, and Future.* Sydney, Australia: University of New South Wales Press, 1995. Although Sutherland focuses on volcanic activity in Australia, the book provides an easily understood overview of volcanic and tectonic processes, including the role of igneous rocks. Includes color maps and illustrations, as well as a bibliography.

Zussman, J., ed. *Physical Methods in Determinative Mineralogy.* 2d ed. London: Academic Press, 1978. An excellent collection of articles on microscopy and instrumental analytical methods. Suitable for advanced students and research geologists.

KIMBERLITES

Kimberlite is a variety of ultramafic rock that is fine- to medium-grained, with a dull gray-green to bluish color. Often referred to as a mica peridotite, kimberlite originates in the upper mantle under high temperature and pressure conditions. It frequently occurs at the Earth's surface as old volcanic diatremes and dikes. Kimberlite can be the source rock for diamonds.

PRINCIPAL TERMS

BLUE GROUND: the slaty blue or blue-green kimberlite breccia of the South African diamond pipes

DIAMOND: a high-pressure, high-temperature mineral consisting of the element carbon; it is the hardest naturally occurring substance and is valued for its brilliant luster

DIATREME: a volcanic vent or pipe formed as the explosive energy of gas-charged magmas breaks through crustal rocks

DIKE: a tabular body of igneous rock that intrudes vertically through the structure of the existing rock layers above

MAGMA: a semiliquid, semisolid rock material that exists at high temperatures and pressures and that is mobile and capable of intrusion and extrusion; igneous rocks are formed from magma as it cools

MANTLE: the layer of the Earth's interior that lies between the crust and the core; it is believed to consist of ultramafic material and is the source of magma

PERIDOTITE: any of a group of plutonic rocks that essentially consist of olivine and other mafic minerals, such as pyroxenes and amphiboles

ULTRAMAFIC: a term used to describe certain igneous rocks and most meteorites that contain less than 45 percent silica; they contain virtually no quartz or feldspar and are mainly of ferromagnesian silicates, metallic oxides and sulfides, and native metals

XENOCRYSTS: minerals found as either crystals or fragments in some volcanic rocks; they are foreign to the body of the rock in which they occur

XENOLITHS: various rock fragments that are foreign to the igneous body in which they are present

KIMBERLITE COMPOSITION AND EMPLACEMENT

The rock known as kimberlite is a variety of mica-bearing peridotite and is characterized by the minerals olivine, phlogopite (mica), and the accessory minerals pyroxene, apatite, perovskite, and various opaque oxides such as chromite and ilmenite. Chemically, kimberlite is recognized by its extraordinarily low silicon dioxide content (25-30 percent), high magnesia content (30-35 percent), high titanium dioxide content (3-4 percent), and the presence of up to 10 percent carbon dioxide. It is a dark, heavy rock that often exhibits numerous crystals of olivine within a serpentinized groundmass. Upon weathering, kimberlite is commonly altered to a mixture of chlorite, talc, and various carbonates and is known as blue ground by diamond miners. Occasionally,

kimberlites contain large quantities of diamonds, which makes them important economically.

"Kimberlite" was first used in 1887 by Carvill Lewis to describe the diamond-bearing rock found at Kimberly, South Africa. There, it primarily occurs as a breccia found in several deeply eroded volcanic pipes and also in an occasional dike. Unfortunately, the kimberlite is so thoroughly brecciated and chemically altered that it does not lend itself to detailed petrographic study. Instead, the kimberlite of Kimberly, like kimberlite elsewhere, is more often noted for the varied assortment of exotic xenoliths and megacrysts it contains. The explosive nature of a kimberlite pipe leads to the removal of country rock (rock surrounding an igneous intrusion) as the magma passes thorough the crust and thus can provide scientists with samples of material that originated at great depths. Among the many xenoliths found in the kimberlite breccia pipes are ultramafic rocks such as garnet-peridotites and eclogites, along with a variety of high-grade metamorphic rocks. These specimens provide scientists with an excellent vertical profile of the rock strata at various locations, serve to construct a model of the Earth's crust at various depths, and provide information on the chemical variations in magma.

The emplacement of kimberlite—seen as a calcite-rich kimberlite magma that has intruded rapidly up through a network of deep-seated fractures—clearly attests its volcanic nature. The magma's rate of upward mobility must have been rapid, as evidenced by the positioning of high-density xenoliths such as eclogite and peridotite within the kimberlite pipes. The final breakthrough of the kimberlite magma may have taken place from a depth of 2-3 kilometers, where contact with groundwater contributed to its propulsion and explosive nature. Brecciation rapidly followed, along with vent enlargement by hydrologic fracturing of the country rock. Fragments of deep-seated rock, along with other country rocks, were then incorporated within the kimberlite magma.

KIMBERLITE MINERALOGY AND TEXTURE

As a rock, kimberlite is very complex. Not only does it contain its own principal mineral phases,

but also it has multicrystalline fragments or single crystals derived from the various fragmented xenoliths that it collected along the way. These fragments represent upper mantle and deep crustal origins and are further complicated by the intermixing with the mineralogy of a highly volatile fluid. As a result, no two kimberlite pipes have the same mineralogy. The continued alteration of the high-temperature phases after crystallization can produce a third mineralogy that can affect the interpretation of a particular kimberlite's occurrence.

The characteristic texture of kimberlite is inequigranular because of the presence of xenoliths and megacrysts within an otherwise fine-grained matrix. In relation to kimberlite, the term "megacryst" refers to both large xenocrysts and phenocrysts, with no genetic distinctions. Among the more common megacrysts present are olivine (often altered to serpentine), picro-ilmenite, mica (commonly phlogopite), pyroxene, and garnet. These megacrysts are usually contained in a finer-grained matrix of carbonate and serpentine-group minerals that crystallized at considerably lower temperatures. Among the more common matrix minerals are phlogopite, perovskite, apatite, calcite, and a very characteristic spinel. Found within the textures of these matrix minerals are examples of both rapid and protracted cooling, with the latter evidenced by zoning. Zoning indicates that the matrix liquid cooled after emplacement and that there was sufficient time for crystals present to react with the remaining liquid, which is common to the megacrysts as well. In addition, the megacrysts exhibit an unusual, generally rounded shape that is believed to be a result of their rapid transport to the surface during the eruptive phase.

KIMBERLITE OCCURRENCE

The geological occurrence of kimberlite, clearly volcanic in origin, takes the form of diatremes, dikes, and sills of relatively small size. In shape, kimberlite diatremes usually have a rounded or oval appearance but can oc-

Kimberlite. (Geological Survey of Canada)

cur in a variety of forms. Quite often, diatremes occur in clusters or as individuals scattered along an elongated zone. They rarely attain a surface area greater than a kilometer but in profile will resemble an inverted cone that descends to a great depth. When it occurs as a dike, kimberlite is quite small—often not more than a few meters wide. It may occur as a simple ring dike or in swarms of parallel dikes. Kimberlite's occurrence as a sill is quite rare and will have a wide variability in thickness.

As compared with other types of igneous rocks, the occurrence of kimberlite is considered to be quite rare. Based on factors such as specific matrix color, density, mineral content, and xenolith size, shape, and number, distinction between a certain kimberlite and a specific occurrence can be made. Aside from their scientific value, most kimberlites are economically worthless, except when they contain diamonds.

DIAMOND-BEARING KIMBERLITE

Of the many minerals that constitute kimberlite, diamond is the most noteworthy because of its great economic value. Not all kimberlite contains diamonds, and when diamonds are present, they are not always of gem quality. Even in the most diamondiferous kimberlites, the crystals and cleavage fragments are rare and highly dispersed. Diamond mining from kimberlite requires the removal and processing of enormous amounts of rock to produce relatively few diamonds. This method is expensive and can be dangerous. A much better rate of return can be gained from the placer mining of riverbeds and shorelines where nature has weathered out the diamonds and concentrated them in more readily accessible areas. The discovery of a large diamond in a riverbed in 1866 led to the eventual search for the source rock that produced diamonds. Up to that time, diamonds were recovered only from alluvial deposits and not from the host rock. In 1872, as miners were removing a diamond-bearing gravel, they uncovered a hard bluish-green rock that also contained diamonds. Further mining revealed the now familiar circular structure of the diatreme that continued downward to an undetermined depth. In 1887, the first petrographic description was made of the rock now known as kimberlite, and it was then recognized to be similar to other

volcanic breccias from around the world.

Diamond-bearing kimberlite locations can be found around the world, the most famous being in South Africa. The diamond pipes of South Africa have been the leading producer of diamonds for well over a hundred years and are still unrivaled in terms of world production. Other locations of kimberlites that produce diamonds include the pipes of Siberia, western Australia, and Arkansas.

KIMBERLITE ORIGIN HYPOTHESES

The chemistry and mineralogy characteristic of kimberlite indicate a very complex set of conditions that existed during their emplacement and subsequent crystallization. This fact, combined with kimberlite's limited occurrence and the altered nature of its matrix, makes kimberlite a puzzle to scientists. Three hypotheses have been proposed to explain its origin. A zone-refining hypothesis describes a liquid generated at great depth (600 kilometers) that is dynamically unstable and, as a result, rises toward the surface. As it reaches specific lower pressures, its composition is altered through partial crystallization and fractionation. A second hypothesis envisions a residual process: Partial melting of a garnet-peridotite parent, at depths of 80-100 kilometers, produces fractional crystallization of a picritic basalt (olivine-rich, or more than 50 percent olivine) at high pressure, which could lead to the formation of ecologite cumulates and a residual liquid of kimberlite composition. The third hypothesis describes kimberlite as either the residual end product of a long fractionation process or the product of a limited amount of partial melting. Each of the three hypotheses has merit but falls short of providing a definitive answer, partly because the near-surface environment where kimberlites are found is one of complex chemical reactions, which makes interpretation difficult.

Although kimberlites tantalize scientists with their complexity, the xenoliths and megacrysts that arise with the kimberlites provide substantial data on the relationship between the lower crust and the upper mantle. Kimberlites show the Earth's upper mantle to be petrographically very complex and define both large- and small-scale areas of chemical and textural heterogeneities. Pressure and temperature conditions have been accurately established for depths down to 200 kilometers through studies based on the speciments brought up during kimberlite eruptions.

KIMBERLITE SPECIMEN ANALYSIS

Analysis of a rock such as kimberlite begins with the collection of a fresh specimen (as unaltered by weathering as possible) and continues with the preparation of a series of thin sections and microprobe sections. In this process, a slice of rock is cut with a diamond saw and glued to a glass plate. A second cut is then made to reduce the specimen's thickness to nearly 0.03 millimeter. A final polishing will achieve this thickness, which will permit light to pass through the specimen and provide the opportunity for microscopic examination. A similar procedure will produce a microprobe-thin section, but it requires extra polishing to ensure a uniform surface.

Once the thin sections have been prepared, a geologist uses a petrographic microscope, which employs polarized light, to make proper identification of the mineral phases present, along with their specific optical parameters. Opaque minerals require reflective light study. Microscopic examination is usually the first step in classifying a rock, and it may include an actual point count of the minerals present to determine their individual ratios. Afterward, an electron microprobe or electron microscope is used to give specific chemical compositions for individual mineral phases. With these devices, the scientist can analyze mineral grains as small as a few microns with a high degree of accuracy. Mineral grains or crystals that are smaller require an analyzing electron microscope to reveal their composition and fine detail. If the individual minerals are large enough and can be extracted, then X-ray diffraction can be used for a definitive identification of a particular mineral phase.

It is also important to understand the bulk chemistry of a rock specimen. Here, the scientist can select several different methods, including neutron activation analysis (NAA) and atomic absorption spectrometry (AAS). Both techniques provide excellent sensitivity and precision for a wide range of elements. A third method, X-ray fluorescence (XRF), is also commonly used and provides an accurate and quick means for analysis.

In NAA, a specimen is activated by thermal neutrons generated in a nuclear reactor. In this process, radioisotopes formed by neutron bombardment decay with characteristic half-lives and are measured by gamma-ray spectrometry. In AAS, a specimen is first dissolved in solution and then identified element by element. XRF, which provides sensitivity and precision for quantitative determination of a wide range of elements (best for those above atomic number 12), is the principal method used by most geologists to gather element amounts in both rocks and minerals.

After all the bulk chemical analyses have been gathered and run through various computer programs to determine mineral compositions, the data are compared to the microscopy results for evaluation and classification of the rock. This evaluation is compared to similar data from other locations. A final check and comparison to other rock types and strata in the field collection area complete the analysis.

SCIENTIFIC AND INDUSTRIAL APPLICATIONS

Although some kimberlites have economic value in their diamond content, most do not. No dramatic breakthroughs from the study of kimberlites will make diamonds more abundant or cheaper in price. The result of these studies is a better understanding of the processes and conditions under which diamonds and semiprecious minerals such as garnet are formed. This knowledge has led to the development of synthetic gems that can be used for industrial applications.

The study of kimberlites also has a direct bearing on scientists' understanding of volcanic activity. The prediction of eruptions is an aim of geology, as it affects the hundreds of thousands of people who live near potentially dangerous volcanoes. Kimberlite magma, by the nature of its chemical composition, is a very explosive material and can produce violent eruptions. Even so, it is true that most kimberlite magmas do not reach the surface, and when they do, they affect only a small area. Kimberlite magmas are more effective at depth, where rock is being fractured by the rising magma and by the hydraulic pressures exerted by the various gases moving through the magma. By examining the results of the movement of a kimberlite magma, scientists can gain a better understanding of volcanic behavior.

Paul P. Sipiera

CROSS-REFERENCES

Andesitic Rocks, 1263; Anorthosites, 1269; Basaltic Rocks, 1274; Batholiths, 1280; Carbonatites, 1287; Feldspars, 1193; Granitic Rocks, 1292; Igneous Rock Bodies, 1298; Igneous Rock Classification, 1303; Komatiites, 1316; Lamproites, 1321; Magmas, 1326; Metamorphic Mineral Deposits, 1614; Minerals: Physical Properties, 1225; Orbicular Rocks, 1332; Plutonic Rocks, 1336; Pyroclastic Rocks, 1343; Rocks: Physical Properties, 1348; Silica Minerals, 1354; Ultramafic Rocks, 1360; Xenoliths, 1366.

BIBLIOGRAPHY

Basaltic Volcanism Study Project. *Basaltic Volcanism on the Terrestrial Planets.* Elmsford, N.Y.: Pergamon Press, 1981. This work represents the efforts of a team of scientists to provide the most up-to-date review of the subject of basalt and its relationship to planetary structure. Included are good references to kimberlites. Best suited for undergraduate and graduate students.

Blackburn, William H., and William H. Dennen. *Principles of Mineralogy.* Dubuque, Iowa: Wm. C. Brown, 1988. A basic textbook on the subject of mineralogy. Provides an excellent review of the principles of mineralogy and crystallography, along with the various techniques used in their study. Offers the reader both a broad overview and specific references to mineral families and their relationship to rock groups. Suitable for undergraduate and graduate students.

Boyd, F. R., and H. O. A. Meyer, eds. *Kimberlites, Diatremes, and Diamonds: Their Geology, Petrology, and Geochemistry.* Washington, D.C.: American Geophysical Union, 1979.

_____. *The Mantle Sample: Inclusions in Kimberlites and Other Volcanics.* Washington, D.C.: American Geophysical Union, 1979. Each is part of a two-volume set that examines the many aspects of kimberlites and their formation conditions. Specialized articles cover a wide range of related geochemical, petrological, and mineralogical topics. Volume 2 also provides an in-depth study of the minerals and foreign rocks brought up by the kimberlite pipes. An excellent reference set suitable for undergraduate and graduate students.

Dawson, J. Barry. *Kimberlites and Their Xenoliths.* New York: Springer-Verlag, 1980. A very technical work that examines the geochemistry and mineralogy of kimberlites, along with their relationship to the other rock types. Complete and detailed in its approach. Bibliography provides a wealth of journal article references. An excellent review source on kimberlites for the undergraduate and graduate student.

Decker, Robert, and Barbara Decker. *Volcanoes.* San Francisco: W. H. Freeman, 1997. Suitable for the general reader, this book provides useful background for the understanding of igneous rock bodies.

Mitchell, Roger H. *Kimberlites, Orangeites, and Related Rocks.* New York: Plenum Press, 1995. Mitchell provides a good introduction to the study of kimberlites and related rocks, including an extensive bibliography that will lead the reader to additional information.

_____. *Kimberlites, Orangeites, Lamproites, Melilitites, and Minettes: A Petrographic Atlas.* Thunder Bay, Ontario: Almaz, 1997. This book covers the worldwide distribution of kimberlites and other related rock types. Color illustrations and bibliography.

Nixon, Peter H., ed. *Mantle Xenoliths.* New York: John Wiley & Sons, 1987. This collection of highly technical articles covers a wide range of topics that are related to kimberlites. Considerable attention is paid to regional kimberlite occurrences and to the foreign rocks that are brought up by the kimberlite diatremes. Several of the articles are general enough to suit a beginning reader, but the work is best suited for the undergraduate and graduate student.

Sutherland, Lin. *The Volcanic Earth: Volcanoes and Plate Tectonics, Past, Present, and Future.* Sydney, Australia: University of New South Wales Press, 1995. Although Sutherland focuses on volcanic activity in Australia, the book provides an easily understood overview of volcanic and tectonic processes, including the role of igneous rocks. Includes color maps and illustrations, as well as a bibliography.

Wyllie, P. J., ed. *Ultramafic and Related Rocks.* New York: John Wiley & Sons, 1967. A collection of specialized articles that deal with the various types of rock that are derived from mantle material. Several of the articles focus specifically on kimberlites, especially as an overall review of the subject. Best suited as a reference work for undergraduate and graduate students.

KOMATIITES

Komatiites are volcanic rocks with abundant olivine and pyroxene and little or no feldspar. They also contain large magnesium-oxide concentrations (greater than 18 percent). They are most abundant in the lower portion of exceedingly old piles of volcanic rocks. Economic deposits of nickel, copper, platinum group minerals, antimony, and gold have been found in some komatiites.

PRINCIPAL TERMS

BASALT: a dark rock containing olivine, pyroxene, and feldspar, in which the minerals often are very small

CRUST: the veneer of rocks on the surface of the Earth

FELDSPAR: calcium, potassium, and sodium aluminum silicate minerals

IGNEOUS ROCK: a rock solidified from molten rock material

METAMORPHISM: the process in which a rock is buried in high temperature and pressure so that the original minerals are transformed into new minerals in the solid state

OLIVINE: a magnesium and iron silicate mineral

PYROXENE: a calcium, magnesium, and iron silicate mineral

SILICATE MINERAL: a naturally occurring element or compound composed of silicon and oxygen with other positive ions to maintain charge balance

SKELETAL CRYSTALS: elongate mineral grains that may resemble chains, plates, or feathers

UPPER MANTLE: the region of Earth immediately below the crust, believed to be composed largely of periodotite (olivine and pyroxene rock), which is thought to melt to form basaltic liquids

KOMATIITE COMPOSITION

Komatiites are unusual volcanic rocks. They contain mostly large grains of olivine (magnesium and iron silicate mineral) and pyroxene (calcium, iron, and magnesium silicate mineral) scattered among fine mineral grains that at one time were mostly glass. The glass was unstable and in time slowly converted into individual minerals. The larger minerals often form elongate grains composed of olivine, pyroxene, or chromite (a magnesium, chromium oxide mineral) called skeletal grains at the top of some lava flows. No other rocks form these skeletal grains. The skeletal grains are believed to have formed by quick cooling at the top of the lava flow. The glass must have also formed by quick cooling of the lava, often in contact with water. Because most igneous rocks contain feldspar (calcium, sodium, and potassium aluminum silicate minerals), the lack of feldspar in komatiites also makes them unusual. They also contain magnesium concentrations (magnesium-oxide greater than 18 percent) larger than most other igneous rocks.

KOMATIITE LAVA FLOWS

Each individual lava flow of a komatiite may be a meter to tens of meters thick. A given lava flow can often be traced for a long distance without much variation in thickness. Some of the lava flows contain rounded or bulbous portions called pillows. Pillows can be formed only by extrusion of the lava into water. As lava breaks out of the solid front of a flow, it is quickly cooled into a rounded pillow. Since this process continually takes place, numerous pillows form as the lava advances. The more magnesium-rich komatiites occur as lava flows in the lower portions of these volcanic piles. They gradually become less magnesium-rich in the younger or upper portions of these volcanic piles.

The mineralogy and observed mineral shapes can vary vertically through a given lava flow. The most spectacular and beautiful komatiite lava flows are the ones in which the upper portions contain the skeletal olivine and pyroxene grains. Skeletal grains may resemble chains, plates, or feathers. Some grains grow as large as 3 centimeters. The thickness of these lava flows varies from about 1 meter to 20 meters. In one type of vertical variation, there are abundant fine minerals formed from the original glass at the top of these flows, along with irregular fractures caused by quick cooling at the top of the lava. Underlying this quickly cooled zone is a layer of the larger skeletal grains of olivine and pyroxene that increase in grain size downward. These skeletal crystals probably form very rapidly in the lava by growth from the top downward.

The lower portion of these komatiite flows often contains abundant olivine grains that are much more rounded and less elongated than those of the upper zone. The more rounded grains are believed to have formed by slow crystallization of the olivine from the liquid lava. As the olivine formed from the liquid, the grains slowly sank and piled up on the bottom of the flow until the lava solidified. The base of the lava also may contain very fine-grained, skeletal grains of olivine with irregular cracks, much like the top of the flow. Presumably, the base of the flow formed by quick cooling, as did the top portion.

Observing several cross sections shows how komatiite flows can vary. There is, for example, a considerable difference in the amount of the flow that contains the skeletal minerals from that containing the more rounded minerals. Most komatiite flows do not contain any skeletal minerals. Instead, they consist mostly of the more rounded mineral grains and have irregular fractures extending throughout the lava flow. These differences could be caused by different cooling rates or by the lava's viscosity (how easily it flows). Other flows are composed mostly of rounded pillows.

Komatiite lava flows are often interbedded with rocks composed of angular fragments that came from all parts of the lavas. Some of these rocks show features that might be found along the edge of a body of water, such as ripples formed by wave action. These rocks have formed by the action of waves breaking up some of the solidified lavas and reworking them after they were deposited. (Rocks that have been reworked by water are sedimentary rocks.)

KOMATIITE OCCURRENCE

Komatiites were discovered in the late 1960's in exceedingly old Archean rocks (more than 2.5 billion years old) in Zimbabwe and South Africa. Only a few komatiites occur in younger rocks. One unusual occurrence of a young komatiite is located on Gorgona Island off the coast of Colombia. Komatiites typically occur in the lower or older portions of vast piles of volcanic rocks containing many layers of different lava flows. Komatiites gradually become less abundant further up the volcanic pile and in younger volcanic rocks. Other dark, feldspar-rich, volcanic rocks called basalt (calcium-rich feldspar and pyroxene rock) are interlayered with komatiites. Basalts gradually become more abundant in the younger

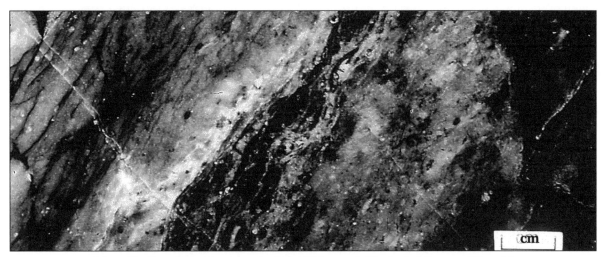

A komatiite flow adjacent to a large rhyolite lapilli fragment. (Geological Survey of Canada)

volcanic flows, along with more light-colored and more silica-rich rocks such as andesite. Small amounts of sedimentary rocks may be interlayered with the volcanic rocks. Sedimentary rocks are derived by water reworking the volcanic rocks. Komatiites eventually disappear in the upper portions of these volcanic piles.

Many komatiites are located in rather inaccessible regions. One area that is accessible is the Vermilion district of northeastern Minnesota in the vicinity of Little Long, Bass, Low, Cedar, Shajawa, and Fall lakes. The age of the district is 2.7 billion years old. Here, there are no true komatiites with magnesium-oxide concentrations greater than 18 percent. There are, however, very magnesium-rich basalts that contain the skeletal crystals found in true komatiites. All the rocks have been buried deep enough that the original minerals were changed or metamorphosed to new minerals in response to the high temperature and pressure. The temperature of metamorphism was still low enough that the original igneous relations may still be observed.

A second example of a komatiite sequence is located at Brett's Cove in Newfoundland. The komatiites here are much younger than most komatiites (formed during the Ordovician, about 450 million years ago). These komatiites formed within layers of rocks called ophiolites, which are believed to be sections of ruptured and tilted oceanic crust and part of the upper mantle. The lower part of the ophiolites contains an olivine and/or pyroxene mineralogy thought to compose much of the upper mantle of the Earth. Overlying the upper-mantle rocks are rocks of basaltic composition containing coarse crystals of olivine, pyroxene, and feldspar. Above these rocks are numerous basaltic lavas that were extruded at the surface and that built up large piles of lavas on the ocean floor. Some komatiite lavas are interbedded in the lower portion of these mainly basaltic lavas. These komatiite flows have a lower zone rich in pyroxene and an upper pillow lava that contains skeletal crystals of pyroxene.

STUDY OF FIELD AND CHEMICAL CHARACTERISTICS

The characteristics of komatiites that are exposed at the surface are described carefully during a field study. These characteristics can suggest how komatiites form at the surface. For example, features such as abundant pillows indicate that the komatiite lava was extruded into water. A small amount of reworking of the lavas by moving water suggests that there was little time between eruptions. A sign of a rapid eruption is the spread of lava over a great distance and at a gentle flow rate.

In addition to the field characteristics of komatiites, geologists study their chemical characteristics in order to understand how they form and evolve. Komatiites are igneous rocks, so they form by the melting of another rock. Experiments using furnaces suggest that much of the rock called peridotite (olivine-pyroxene rock) must melt to form high-magnesium magmas. The magma then may evolve or change in composition by processes such as the crystallization of minerals from the magma, by dissolving some of the solid rock through which it moves, or by mixing with magma of a different composition. The komatiite may even change composition after the magma solidifies because of water vapor or carbon dioxide-rich solutions moving through the solid. It is difficult to assess the relative importance of these processes to modify the composition of a given komatiite.

ANALYZING ELEMENTAL CONCENTRATIONS

One way to test for mineral crystallization of the magma is to plot the elemental concentrations of several analyzed rocks from the same general area against another element, such as magnesium. If the concentration of all elements systematically increases or decreases with increasing magnesium concentrations, then the lavas were probably related by fractional crystallization. The formation and settling of olivine from lava appear to control the concentration of most elements. For example, magnesium, chromium, nickel, and cobalt all concentrate in lava-related olivine; thus, a plot of chromium, nickel, and cobalt shows smooth and systematic decreases when compared to magnesium. Elements such as calcium, titanium, aluminum, silicon, iron, and scandium are rejected from lava-related olivine; they gradually increase with decreasing magnesium. Some elements, such as sodium, potassium, barium, rubidium, and strontium, should also systematically increase with decreasing magnesium, as they are also rejected from lava-related olivine. Instead, these elements are greatly scattered when they are plotted relative to magnesium. These elements are notorious for being moved by carbon dioxide or by water-vapor-rich fluids. Thus, it is assumed that the scatter of these elements is a result of movement by these fluids. The fractionation of these elements because of olivine crystallization, therefore, is obscured.

A few changes in elemental concentrations in komatiite lavas cannot be explained by crystallization or alteration processes. For example, a rare-earth element with a lower atomic number, such as lanthanum, should not systematically increase more than a rare-earth element with a higher atomic number, such as lutetium, with olivine crystallization, as olivine should not fractionate these elements. (The atomic number of an element is the number of protons in its nucleus.) Either the theory is incorrect or some other process produces this variation. Some geologists believe that melting processes control the lanthanum concentration relative to that of lutetium. This melting control may be attributable to a combination of differences in composition of the rock that melts to form komatiites and by the degree and nature of the rock's melting process.

ECONOMIC VALUE

Komatiites contain several types of economic deposits. They may contain important deposits of nickel sulfides, along with large concentrations of platinum group elements (such as platinum and palladium) and copper. The nickel sulfide was probably formed from a sulfide liquid that separated as an immiscible liquid from the komatiite liquid. This is similar to the way oil and water separate when they are mixed together. Nickel sulfide ores have been found in Western Australia, Canada, and Zimbabwe. The most important deposits are found in the komatiite lava flows in the lower portion of a lava pile. The nickel sulfide ore is

concentrated in a portion of the base of a lava flow where it is thicker than other portions of the flow. The immiscible and dense nickel-sulfide liquid may have settled in a thick portion of the flow that was not stirred as much as other portions of the flow. The platinum group metals have a stronger affinity for the sulfide liquid than for the komatiite liquid; consequently, they concentrate in the sulfide liquid.

Gold, antimony, and a few other elements are concentrated in some komatiite flows. Running water may alter and rework some komatiite lavas and form sedimentary rocks. Examples of these deposits occur northeast of Johannesburg in South Africa. There, the lower portion of the rock pile is mostly layers of successive komatiite or basaltic lava flows. This portion is overlain by sedimentary rocks composed of mudrocks changed, at high temperature and pressure, into metamorphic rocks. Some quartz-carbonate rocks within the sedimentary rocks probably formed by the alteration of komatiites. The quartz-carbonate rocks contain high concentrations of antimony in the mineral stibnite and small particles of gold. Solutions moving through the komatiites probably altered the komatiites, leaving the high concentrations of these elements.

Robert L. Cullers

BIBLIOGRAPHY
Arndt, N. T., D. Frances, and A. J. Hynes. "The Field Characteristics and Petrology of Archean and Proterozoic Komatiites." *Canadian Mineralogist* 17 (1985): 147-163. An advanced article summarizing the way komatiite lavas occur. A number of diagrammatic representations of these lavas make it easier to visualize what they look like. A college student with a petrology course could read the discussion; someone with a course in introduc-

tory geology could read it if he or she were willing to look up terminology in a geologic dictionary.

Arndt, N. T., and E. G. Nisbet. *Komatiites*. Winchester, Mass.: Allen & Unwin, 1982. An advanced book reviewing many aspects of komatiites. Much of the book could be read by someone who has had a petrology course. Someone taking an introductory geology course could read chapters 1 and 2 on the history and definition of komatiites. A mineral collector can find references to the location of komatiites. The skeletal crystals developed by some komatiites may be especially beautiful; there are black-and-white photographs of some of the skeletal crystals in the book.

Best, M. G. *Igneous and Metamorphic Petrology*. San Francisco: W. H. Freeman, 1995. This petrology book has a short section on komatiites. Several photographs of the skeletal olivine and pyroxene crystals and a diagrammatic cross section of a komatiite lava flow are included. A person taking an introductory geology course could read the text with the help of a geologic dictionary.

Decker, Robert, and Barbara Decker. *Volcanoes*. San Francisco: W. H. Freeman, 1997. Suitable for the general reader, this book provides useful background for the understanding of igneous rock bodies.

Hyndman, D. W. *Petrology of Igneous and Metamorphic Rocks*. New York: McGraw-Hill, 1985. This petrology book contains a short section on komatiites, including a section on the economic importance of komatiites. A black-and-white photo of the skeletal crystals is provided. A person taking an introductory geology course could read this section with the help of a geologic dictionary.

Mitchell, Roger H. *Kimberlites, Orangeites, and Related Rocks*. New York: Plenum Press, 1995. Mitchell provides a good introduction to the study of kimberlites and related rocks, including an extensive bibliography that will lead the reader to additional information.

Press, Frank, and Raymond Siever. *Understanding Earth*. 2d ed. New York: W. H. Freeman, 1998. This comprehensive physical geology text covers the formation and development of the Earth. Readable by high school students, as well as by general readers. Includes an index and a glossary of terms.

LAMPROITES

Lamproites are potassium-rich and magnesium-rich rocks. They contain the highest ratios of potassium to sodium oxides and potassium to aluminum oxides of any igneous rocks. Some lamproites in Australia have the world's largest concentrations of diamonds.

PRINCIPAL TERMS

AMPHIBOLE: a calcium, magnesium, iron, and aluminum silicate and hydroxide mineral, which may have a variety of other elements

CLINOPYROXENE: a calcium, magnesium, and iron silicate mineral

DIKE: a tabular igneous rock body that cuts across the fabric of the solid rocks

IGNEOUS ROCK: a rock formed from molten rock material or magma

MAGMA: molten rock material usually of silicate composition

MICA: a potassium, magnesium, iron, and aluminum silicate mineral

MINERAL: a naturally occurring element or compound that has a more or less constant composition and arrangement of atoms

OLIVINE: a magnesium and iron silicate mineral

ROCK: a material usually composed of two or more minerals

SILL: a tabular igneous rock body that is parallel to the fabric of the solid rocks; the fabric is commonly the sedimentary layering in these kinds of rocks

---

LAMPROITE COMPOSITION

Lamproites are unusual igneous rocks (formed from molten rock material) because they contain little or no feldspar (potassium, calcium, and sodium silicate). In this respect, lamproites are like kimberlites, carbonatites, or komatiites. Instead of feldspar, they contain potassium- and magnesium-rich minerals such as magnesium-rich olivine and clinopyroxene, magnesium- and titanium-rich mica, and potassium- and magnesium-rich amphiboles. A variety of other, normally scarce, minerals may also occur in lamproites. Often, their original minerals are transformed by water-vapor- and carbon-dioxide-rich fluids into a wide variety of hydrous or carbonate minerals. Lamproites, besides being enriched in potassium and magnesium, are depleted in aluminum and sodium compared to other igneous rocks. They have the highest ratios of potassium to sodium oxide (greater than 3:1) and potassium to aluminum oxide of any igneous rocks.

Lamproites are frequently confused with lamprophyres, kimberlites, and other sodium- and potassium-rich igneous rocks. Lamprophyres are dark-colored, igneous rocks found in dikes that have large, dark minerals with many fine, dark minerals of similar composition. The dark minerals in lamprophyres also are found as well-developed geometric solids with smooth, planar surfaces. Unlike lamproites, the finer minerals often contain feldspars. Kimberlites contain large minerals such as magnesium-rich ilmenite (magnesium, iron, and titanium oxide), titanium-rich pyrope (magnesium and aluminum silicate), olivine, clinopyroxene, and magnesium-rich mica surrounded by abundant smaller minerals of olivine, magnesium-rich mica, clinopyroxene, or minerals formed when the original minerals were transformed by later fluids. The lamproites contain much higher potassium-to-aluminum and potassium-to-sodium ratios than those of lamprophyres or kimberlites. Lamproites also have much higher silicon contents than do kimberlites. Compared with lamprophyres and kimberlites, lamproites contain a higher average concentration of the trace elements rubidium, strontium, uranium, zirconium, and the rare-earths. (Trace elements are those elements that substitute in small amounts for other elements composing the bulk of the minerals in a rock.) The ranges of concentrations of some trace elements, however, often overlap among these rock types.

1321

LAMPOITE OCCURRENCE

Geologists have known about lamproites for some time. Prior to the late 1970's, however, lamproites were believed to be rare rocks without any economic importance. The discovery of large concentrations of diamonds in lamproites from northwestern Australia has elevated them to an important economic position. Thus, the study of lamproites, and the technical literature written about them, has increased. Lamproites occur in more than twenty regions in the world, including the United States, Greenland, Canada, Australia, Europe, Africa, Antarctica, Asia, and Brazil. Although not restricted to any given time period, they appear to have formed most abundantly during the Cretaceous (about 70 million years ago).

Lamproites often occur as lavas. The lava flows are often associated with rocks with other typical volcanic features, such as pyroclastic material thrown out of the volcano into the air. Some pyroclastic material may build up high enough around the vent to create a volcano called a cinder cone. These types of lamproites occur at Leucite Hills, Wyoming, and in the Antarctic. Pyroclastic material formed by the settling of fine volcanic ash is present in some deposits, including those at Prairie Creek, Arkansas, and with lavas and cinder cones at Leucite Hills.

Lamproites commonly occur as tabular igneous rock bodies called dikes. Dikes cut across the layering of other solid rocks (sedimentary rock layers) into which the magma was injected below the land surface. Some lamproites occur as tabular bodies, called sills, that are parallel to the layering into which the magma was injected. The lamproites in Woodson County, Kansas, are horizontal sills parallel to the sedimentary rock layers.

Lamproites also occur as vertical pipelike bodies below the land surface. Lamproite pipes occur at Prairie Creek and at Argyle and the Fitzroy Basin in Western Australia. Lamproite pipes differ from kimberlite pipes as they are shaped like inverted cones, with approximately 0.5 kilometer of flaring outward near the surface. In contrast, kimberlite pipes are more funnel-shaped, with several kilometers of flaring near the surface. The flaring is probably produced by the magma as it begins to boil and explode near the surface when carbon-dioxide- and water-vapor-rich gases are rapidly lost. The explosive activity probably occurs much like shaking a capped soft drink container and suddenly taking off the cap. The kimberlites probably contain more carbon dioxide or water vapor than do the lamproites, and they thus lose these fluids at a greater depth than do the lamproites.

Lamproites vary in size, but they are generally small in surface outcrop area compared to many bodies of igneous rock. The smaller bodies are sills and dikes and may be only a few meters wide. The pipes sometimes flare outward as much as 100 meters. Some volcanic cinder cones rise several hundreds of meters around the surroundings and are hundreds to several thousand meters wide.

OCCURRENCE IN THE UNITED STATES

All known lamproites are found on continents. Most of them occur nearer the edge of continents than do most kimberlites, but it is impossible to relate them to plate tectonic processes such as subduction (the movement of one crustal plate below another) or rifting (the ripping apart of the crust by tension). The lamproite occurrences in the United States and Australia illustrate the range of field relations in these rocks.

Lamproite is exposed over a 400-square-kilometer area in Leucite Hills, Wyoming, where it occurs as volcanic rocks and as shallow magma bodies that crystallized only a million years ago—making them among the youngest of any lamproites. The volcanic rocks originated as lava flows and pyroclastic rocks. In some places, erosion has exposed magma bodies that had solidified below the surface. These intrusive rocks occur as dikes or pipes. The common minerals in these rocks consist of a combination of magnesium-rich mica, magnesium-rich clinopyroxene, magnesium-rich olivine, and leucite (potassium and aluminum silicate).

Lamproite sills that were intruded horizontally about 70 million years ago into limestones and shales are found in Woodson County, Kansas. The sills—a few meters to approximately 30 meters thick—mostly have been located by drilling 330 meters below the surface. One near-surface exposure is being mined for alleged "growth nutrients," which could be added to animal feed. Large granite (quartz and feldspar rock) boulders are exposed at the surface at another location where lamproite is very close to the surface. Apparently, the granite was carried up as a solid by the molten lamproite magma. The granite was probably ripped off the sides of the magma conduit in the lower crust and carried upward by the lamproite magma to become accidental foreign rock fragments in the lamproite. The largest minerals found in the lamproite are magnesium-rich mica, magnesium-rich clinopyroxene, potassium-rich amphibole, and altered magnesium-rich olivine located in an altered mass of much smaller minerals. The overlying and underlying shales and limestones show evidence of having been "baked" or metamorphosed by the heat emitted by the hot magma. For example, minerals in the shale have been changed to sanidine, a potassium-rich feldspar formed only at very high temperatures. Movement of water vapor-rich fluids has apparently carried barium, potassium, and rubidium from the magma into the overlying shales.

Lamproite also is found in the Prairie Creek area and in the Crater of Diamonds Park close to Murfreesboro, Arkansas. The lamproite at Prairie Creek occurs as pipes, dikes, pyroclastic deposits, and ash deposits. The intrusion at the Crater of Diamonds is a pipelike body. Diamonds have been found in both areas. The pipes and dikes are composed of large, altered olivine grains with other small minerals of magnesium-rich mica, potassium-rich amphibole, chromium-rich spinel (magnesium and chromium oxide), magnesium-rich clinopyroxene, magnetite (iron oxide), and other altered minerals. The ash deposit contains magnesium-rich mica, sanidine, and quartz (silicon oxide), with small minerals of carbonate and clays (aluminum silicate minerals with a wide variety of metallic elements).

OCCURRENCE IN AUSTRALIA

Lamproites also exist in the northern portion of Western Australia at more than 100 locations. Diamonds are found at approximately 30 sites. Most of these lamproites are quite young (formed 17-25 million years ago), but some may be Precambrian (older than 600 million years). The intrusive lamproites occur as dikes, sills, pipes, and larger bodies of various shapes. The extrusive lamproites occur as fine-to-coarse volcanic material which was thrown into the air rather than extruded as lava. The diamond-bearing lamproites contain large olivine grains with smaller minerals of olivine, magnesium-rich mica, magnesium-rich clinopyroxene, potassium-rich amphibole, and various minor minerals. The lamproites that do not contain diamonds have the mineral leucite along with many of the minerals found in diamond-bearing lamproites.

One of the olivine-rich lamproites at Argyle contains incredible concentrations and amounts of diamonds. Erosion has also concentrated diamonds in stream deposits adjacent to this lamproite. The rocks here are a mixed ash and sand deposit formed by ejection of material into the air and by dikes formed by injection of lamproite into the other rocks. The ash-sand deposit consists mostly of sand-sized pieces of lamproite and volcanic ash. The lamproite contains large olivine and magnesium-rich mica grains with fine grains of magnesium-rich mica, anatase (titanium oxide), sphene (calcium and titanium silicate), perovskite (calcium and titanium oxide), and other minor minerals.

STUDY OF ORIGIN AND EVOLUTION

The field relationships and chemical composition of lamproites may be used to infer something about their origin. Experiments using furnaces also may help to explain how lamproites form and evolve.

Lamproites in the form of pipes, sills, and dikes are thought to have intruded as magma into solidified rocks, especially if the lamproite is younger than the rocks above or below it. Lamproites were probably injected while they were very hot and contained fluids. These fluids bubbled out and altered the overlying and underlying rocks as observed in the sills at Woodson County, Kansas. The shales into which the lamproite was intruded were "baked" so that the original minerals were transformed into minerals such as sanidine.

The presence of lamproite material located at or near a volcano may be used to infer its volcanic origin. In older rocks, the volcanic cone may have eroded long ago; in this case, the volcanic origin of the rocks may be obscure. Other features must be used to infer their volcanic origin. For example, the presence of fine ash or angular fragments (bombs) that may have been blown out of a volcano suggest a volcanic origin. The presence of glass or altered glass suggests quick cooling of the magma, which is also consistent with a volcanic origin. Other rocks deposited over the volcanic lamproites would not be baked, as they would have been in sills.

The distinctive chemical composition of lamproites helps a geologist to understand what kind of rock melts to form a lamproite magma. The high potassium and magnesium combined with low aluminum and sodium are simply not found in other igneous rocks, so there must be something distinctive about the rocks that melt to form lamproites. Only peridotite, a rock deep within the Earth that contains olivine, pyroxene, garnet (magnesium and aluminum silicate), and magnesium-rich mica, may melt to form the major constituents in the lamproite magma. For example, about 60 percent of the magnesium-rich mica and 40 percent of the pyroxene must have melted to form the major element content of the lamproite at Woodson County, Kansas. This is also consistent with experiments in furnaces in which rocks of similar composition have been melted to produce such magma. In addition, the lamproite magmas contain exceedingly large concentrations of certain trace constituents, such as lanthanum, rubidium, barium, and strontium. It is impossible to melt most peridotites (peridotite is carried up as foreign rock fragments in some magmas) and produce such large concentrations of these trace constituents. Mica peridotites have been found, however, that contain enriched amounts of these trace constituents. If these mica peridotites were to melt, they would produce the large concentrations of those trace constituents found in the lamproite magmas. The presence of diamonds in some lamproites proves that the magma must have formed very deep within the Earth. Diamond is stable only at depths greater than 150 kilometers below the Earth's surface. The magma must have ascended rapidly; otherwise, the diamonds would have had time to decompose as they rose to the surface.

ROLE IN DIAMOND INDUSTRY

The world's largest concentrations of diamonds have been found in lamproites from the northern portion of Western Australia. Diamonds have also been found in lamproites at Prairie Creek, Arkansas; the Luangwa graben, Zambia; and at Bobi and Séguéla, Ivory Coast. There is little information on lamproites in Zambia and the Ivory Coast, but the others have been well studied. The grade of diamonds at the more than twenty locations known to contain diamonds in Australia varies from about 5 to 680 carats per ton. The largest known reserves are at Argyle, where there are at least 61 million tons of known reserves at 680 carats per ton: These reserves are seven to thirty times more enriched in diamonds than the kimberlites in South Africa. Other reserves at Argyle contain lower concentrations of diamonds.

Diamonds have been valued for their beauty for at least twenty-four centuries. More recently, diamonds have become important for industrial purposes, as they are exceedingly hard. Indeed, this elemental form of carbon is the hardest substance in nature. Diamonds are also one of the most expensive natural materials. Even industrial diamonds typically cost $10,000 per pound. Until the 1970's, most of the world's diamond supply and the price of diamonds were controlled by the De Beers Corporation. In the 1980's, the selling of investor diamonds on the world market and the opening of large new diamond mines in Australia led to a new supply of diamonds. De Beers, while trying to maintain its high prices by buying many of these diamonds, considerably strained its resources.

About 20 percent of diamonds are used for gemstones. Prices vary depending on the demand for diamonds of differing size, clarity (lack of impurities), and color. The other 80 percent of diamonds are used for industrial purposes. Diamond is used as an abrasive to cut, polish, or grind a variety of materials. Diamonds also are used in tools such as saws, glass cutters, grinding wheels, and drilling bits.

Robert L. Cullers

CROSS-REFERENCES
Andesitic Rocks, 1263; Anorthosites, 1269; Basaltic Rocks, 1274; Batholiths, 1280; Carbonatites, 1287; Gold and Silver, 1578; Granitic Rocks, 1292; Igneous Rock Bodies, 1298; Igneous Rock Classification, 1303; Kimberlites, 1310; Komatiites, 1316; Magmas, 1326; Orbicular Rocks, 1332; Phase Changes, 463; Platinum Group Metals, 1626; Plutonic Rocks, 1336; Pyroclastic Rocks, 1343; Regional Metamorphism, 1418; Rocks: Physical Properties, 1348; Silica Minerals, 1354; Ultramafic Rocks, 1360; Xenoliths, 1366.

BIBLIOGRAPHY

Bates, R. L. *Geology of the Industrial Rocks and Minerals.* Mineola, N.Y.: Dover, 1969. The author discusses many types of economic deposits, including diamonds. The reader should have had an introductory course in geology, as the author assumes the reader is familiar with common geologic terms. No glossary; few illustrations.

Bergman, S. C. "Lamproites and Other Potassium-Rich Igneous Rocks: A Review of Their Occurrence, Mineralogy, and Geochemistry." In *Alkaline Igneous Rocks*, edited by J. G. Fitton. Boston: Blackwell Scientific for the Geological Society, 1987. This technical discussion of lamproites is suited for the advanced student or the mineral collector who wishes to locate known lamproites: There is often a geologic map of the lamproite and many references.

Craig, J. R., D. J. Vaughan, and B. J. Skinner. *Resources of the Earth.* Englewood Cliffs, N.J.: Prentice-Hall, 1988. A well-written book on economic deposits, for persons with little technical background. Numerous figures and pictures. Includes a glossary and several sections on diamonds.

Cross, C. W. "The Igneous Rocks of the Leucite Hills and Pilot Butte, Wyoming." *American Journal of Science* 4 (1987): 115-141. This article gives the main rock types of this area and locates them on a map. Suitable for a mineral collector who is not intimidated by complex rock names.

Cullers, R. L., et al. "Geochemistry and Petro-genesis of Lamproites, Late Cretaceous Age, Woodson County, Kansas, U.S.A." *Geochimica et Cosmochimica Acta* 49 (1985): 1383-1402. A technical paper suitable for an advanced student. A mineral collector can find the location and the minerals found in this lamproite.

Mitchell, Roger H. *Kimberlites, Orangeites, and Related Rocks.* New York: Plenum Press, 1995. Mitchell provides a good introduction to the study of kimberlites and related rocks, including an extensive bibliography will lead the reader to additional information.

_____. *Kimberlites, Orangeites, Lamroites, Melilitites, and Minettes: A Petrographic Atlas.* Thunder Bay, Ontario: Almaz, 1997. This book covers the worldwide distribution of kimberlites and other related rock types. Color illustrations and bibliography.

Mitchell, Roger H., and Steven C. Bergman. *Petrology of Lamproites.* New York: Platinum Press, 1991. This book gives a thorough account of the origins and evolution of lamproites. Suitable for the high school or college reader. Bibliography and index.

Steele, K., and G. H. Wagner. "Relationship of the Murfreesboro Kimberlite and Other Igneous Rocks of Arkansas, U.S.A." In *Kimberlites, Diatremes, and Diamonds: Their Geology, Petrology, and Geochemistry*, edited by H. R. Boyd and H. O. A. Myers. Washington, D.C.: American Geophysical Union, 1979. The reading is appropriate for a college student with a course in petrology. A mineral collector can find the locations of the lamproites.

MAGMAS

Magma is a naturally occurring liquid usually composed of silicate material with suspended minerals that either concentrate or reject elements, thereby either depleting or enriching the liquid in a given element. This process gives geologists a means to understand how some elements are concentrated in the minerals or liquids to form potential economic deposits. In addition, recognition of the origin of magmas is essential in understanding the complex relationship between volcanism and regional deformation processes, or plate tectonics.

PRINCIPAL TERMS

BASALT: a dark igneous rock containing plagioclase feldspar, pyroxene, or olivine; it contains mineral grains too small to see with the naked eye as well as various amounts of larger minerals

CRYSTALLIZATION: the formation and growth of a crystalline solid from a liquid or gas

DIFFERENTIATION: the process of developing more than one rock type from a common magma

FELDSPAR: a silicate mineral categorized as either plagioclase or alkali

FRACTIONAL CRYSTALLIZATION: the process by which minerals form in a magma and either float or settle from the magma, depending on whether they are lighter or heavier than the magma

GRANITE: an even-textured, light-colored igneous rock made up of quartz and two feldspars, with variable amounts of micas and dark minerals

IGNEOUS ROCK: a rock formed from molten rock material or magma

LITHOSPHERE: that portion of the Earth that comprises the crust and part of the upper mantle where deformation at geologic rates occurs

MINERAL: a naturally occurring substance with a more or less constant composition and thus constant physical properties that reflect that composition

OLIVINE: a silicate mineral containing iron and magnesium

QUARTZ: a silicate mineral containing only silica

SILICATE MINERALS: minerals composed of silicon, oxygen, and other metal ions, such as potassium, sodium, calcium, magnesium, iron, and aluminum

VISCOSITY: a measure of how readily a fluid flows; the more viscous a fluid is, the less easily it flows

LIQUID COMPOSITION

Magmas are composed of a liquid, derived from the melting of silicate rocks at great depth, as well as suspended minerals. The minerals slowly form and grow larger. In contrast to water crystallizing to ice at a constant temperature, magmas crystallize several kinds of minerals over a range of temperatures. The kinds and sequences of minerals crystallizing from the magma depend on the liquid composition, the total pressure, the kinds of vapor (for example, carbon dioxide or water vapor), and the pressure of the vapor. The kinds of minerals crystallizing will in turn influence the composition of the gradually decreasing amount of magma.

Suppose a magma exists at a shallow depth of about 5 kilometers below the surface, or crust, of the Earth. Further suppose that this magma is similar in composition to lavas erupted in the Hawaiian Islands called basalt. A mineral called olivine (magnesium and iron silicate) is often one of the first minerals to crystallize out of these basaltic magmas. Olivine contains more magnesium and less silicon than does the basaltic magma; therefore, the lesser amount of remaining liquid becomes more depleted in magnesium and more enriched in silicon. Another mineral called plagioclase (calcium, sodium, and aluminum silicate) may also begin to crystallize with the olivine. The plagioclase is more enriched in calcium and aluminum and depleted in silicon than the magma. Thus, crystallization and separation of plagioclase will result in the depletion of calcium and aluminum and enrichment of silicon in the remaining magma. Plagioclase has a density similar to that of the magma, and it may float or sink depending on whether it is less or more dense than the surrounding magma. Many other elements are not incorporated into the olivine and plagioclase as they undergo fractional crystallization and gradually become more concentrated in the magma.

As the magma changes composition, it may begin to crystallize other minerals, and olivine may stop crystallizing. For example, quartz (all silica) may crystallize from the last bits of the magma left after most of the liquid has solidified. The magma may not increase much in silicon if minerals enriched in silicon, such as quartz, are crystallizing. In some magmas, the last small amounts of liquid left after crystallization may contain a considerable amount of water vapor and unusually large concentrations of some normally rare elements. These rare elements may sometimes form minerals not normally found in most rocks, and they may be economic to mine.

PRESSURE

The pressure at which a magma crystallizes may also influence the kinds of minerals that form. Suppose the basaltic magma discussed above crystallized at an exceedingly high pressure, corresponding to a depth of about 60 kilometers below the surface of the Earth. Instead of olivine, garnet (magnesium, iron, and aluminum silicate mineral) may be the first mineral to crystallize. Clinopyroxene (calcium, magnesium, and iron silicate mineral) may then shortly join the garnet crystallizing from the magma. No plagioclase or olivine would ever crystallize, as would have occurred at lower pressure. The crystallization and removal of garnet and clinopyroxene would result in a much different change in the composition of the remaining magma from that which would occur at lower pressure. For example, the garnet and clinopyroxene are more concentrated in silicon than the basaltic liquid, and thus the liquid would decrease in silica during fractional crystallization rather than increase, as it would at lower pressure.

The starting composition of the magma will also influence the kinds of minerals that crystallize and in turn influence how the composition of the remaining magma will crystallize. For example, a silicate-rich magma of granitic composition will crystallize different minerals at lower pressures from those of the basalt. The granitic magmas also contain much lower magnesium and iron concentrations and higher potassium concentrations than those of basaltic magmas. Correspondingly, the granitic magmas crystallize more silica-rich and potassium-rich minerals, such as quartz and the potassium-rich feldspars, than do those of basaltic magmas.

WATER VAPOR

The crystallization of minerals from granitic magmas may result in much different changes in composition of the remaining granitic magmas from those of the basaltic magmas. For example, some of the last fractions from the granitic magmas may contain much water vapor and produce exceedingly large crystals of quartz, feldspar, and many rare minerals, such as beryl (a beryllium and aluminum silicate), tourmaline (a complex silicate mineral with boron, sodium, lithium, and aluminum among the many elements), or spodumene (a lithium and aluminum silicate). At other times, gases rich in water vapor may bubble out of these last bits of granitic magmas, much like the gases in a soft drink, and carry out many rare elements not normally concentrated in most rocks. These hydrothermal, or hot water, deposits may form a variety of mineral deposits with minerals such as molybdenite (a molybdenum ore) or wolframite (a tungsten ore).

The presence of water vapor in the magma may also affect the temperature of crystallization, influence the kinds of minerals that crystallize from the magma, and influence the viscosity and settling rates of minerals. The presence of water vapor drastically lowers the melting point of the silicate rocks or crystallization temperature of the silicate magmas. Indeed, some magmas would not exist in nature if water vapor were not present, as the temperature in the Earth would not be high enough to keep the silicate magma liquid. Also, if water vapor suddenly boils out of a magma with suspended crystals, the magma may suddenly freeze to such a fine material that individual crystals may be too small to be seen. The resultant rock would contain the large, early-formed crystals in a fine-grained rock. Some minerals, such as biotite (a potassium, magnesium, iron, and aluminum silicate), have water as an essential part of the mineral. Thus, biotite cannot form without the presence of water vapor in the magma.

Water vapor also breaks down the long chains of silicate atoms in the magma and makes them shorter. This in turn allows the magma to flow more easily, or become less viscous. Less viscous magmas may flow more readily to the surface to form volcanic rocks, or minerals may settle faster in a less viscous magma.

MAGMA GENERATION

The ultimate source of all magmas is a result of partial melting of the solid preexisting rock, notably the mantle, which underlies the Earth's crust. Partial melting can be induced in several ways, most clearly by the increase in temperature. Melting is, however, rarely complete. Most melts probably coalesce to form discrete magma bodies that migrate away from the source area, leaving some residue behind. Partial melting of the originally solid crust or mantle allows for the diverse variety of igneous rock types observed. Only a very small fraction of igneous rocks are representative of primary magmas.

Magma generation is closely associated with tectonic processes. Basaltic magmas, for example, are generated by direct partial melting of the rock comprising the mantle. This magma type is most abundant beneath active oceanic ridges, which are the sites where lithosphere plates are being formed. In fact, the oceanic crust is basically basaltic in composition and composed of volcanic rocks solidified from magma erupted on the ocean floor. These rocks are underlain by basaltic dikes, which reflect fissures or conduits for molten magma. These dikes are in turn underlain by gabbros and cumulate layers, which are evolved rocks that formed when large amounts of crystals settled out of a cooling magma.

Andesitic magmas are generated above subduction zones, where lithospheric plates converge. Subduction zones are long, narrow belts where one lithospheric plate descends beneath another. In this case, the melting of rock within the mantle is complicated by the presence of wet sediments and basalts that comprise the descending oceanic or lithospheric plate. These magmas thus tend to be more silica-rich and are responsible for the formation of new continental crust.

Granitic magmas to a large extent are generated above subduction zones. The processes at work here include both partial melting at the base of the andesitic crust and differentiation, the process of developing more than one rock type from a common magma of andesitic or even basaltic magmas that have been modified by some mixing with crustal material.

True primary magmas are of basaltic composition, as they can only be generated by partial melting of rock within the mantle. In contrast, andesites and granites can be formed either by differentiation of basaltic magmas or by direct partial melting within the mantle or crust. In fact, almost all other igneous rocks evolve in this manner, reflecting the complex interrelationship of these two fundamental processes. The study of both terrestrial meteorite impact craters and lunar rocks returned by the Apollo space program has demonstrated that igneous-looking rocks can also be produced by the melting of rocks by meteorite impact.

STUDY OF BASALTIC MAGMAS

Some geologists conduct experiments in furnaces at the high temperatures and pressures corresponding to various depths within the Earth to see how natural basalts crystallize or to see which solid rocks melt to produce the basaltic magmas. The process might involve placing a natural basalt in a furnace at a known temperature and pressure until all the basalt melts, for example, at 1,200 degrees Celsius. Then the basaltic liquid could be cooled to a slightly lower temperature and suddenly cooled to a very low temperature. The kinds, amounts, and composition of the minerals and the composition of the glass (the liquid forms a glass when rapidly cooled) could then be determined.

The composition of minerals and glass may be determined using an electron microprobe. This instrument shoots a narrow beam of electrons onto a small portion of the mineral or glass. Each element gives off a characteristic energy of radiation that identifies that element. The intensity of the radiation depends on the amount of the element in the sample so that the concentration of the element may be determined. Thus, the min-

The throat of an ancient volcano, Devil's Tower National Monument in Wyoming formerly occupied a magma chamber. The solidified magmatic columns remain standing long after the volcano has eroded away. (© William E. Ferguson)

eral and glass compositions may be measured. The experiment may continue at a series of lower temperatures until all the magma crystallizes. The changing composition and kinds of minerals forming from the basaltic magma can thus be precisely related to the changing composition of the basaltic magma.

Similar experiments for the same basalt can then be done at different pressure or composition of vapor coexisting with the basaltic liquid to determine the effect of changing these parameters on the composition and kinds of minerals forming from the liquid. Also, natural rocks that differ in composition from that of the first basalt may be crystallized over a range of temperature, pressure, and fluid composition to determine their effect on the course of the crystallization.

If the compositional differences among a series of lavas piled on top of one another are similar to the changing composition of the glass during the experimental crystallization of a basalt, then the compositional changes in the series of lava flows could be a result of fractional crystallization. The experiments do not prove that the series of lavas formed by fractional crystallization; they merely show that this is one possible way that the lavas may have formed. The lavas, for example, might have formed by different degrees of melting of the same rock. Sometimes, geologists have to prove that all but one of the possible hypotheses for the formation of a series of rocks cannot be true. The possibility left after all other hypotheses have been rejected then becomes the most attractive hypothesis for the formation of a series of lavas.

SCIENTIFIC AND INDUSTRIAL VALUE

Knowledge concerning the processes involved in the origin of magmas and the subsequent rock types formed is essential in understanding the formation and evolution of the Earth's crust. By understanding these processes on Earth, scientists are able to apply this information to the study of other planets within the solar system.

In addition, geologists believe that the crystallization of minerals from magmas creates rich mineral deposits that can be found in many parts of the world. For example, the Muskox Intrusion, a body of rock formed in Canada by solidification of magma below the surface of the Earth, contains high nickel contents in portions of the intrusion that contain a considerable amount of the mineral olivine (concentrated nickel). The olivine and the mineral chromite settled from the magma at different rates, leaving the chromite concentrated in other layers. Chromite is the source of chromium, which is used with steel to give it hardness and toughness and make it resistant to chemical attack.

Other elements such as tungsten, tin, beryllium, the rare-earth elements, molybdenum, boron, lithium, niobium, tantalum, and uranium may eventually concentrate in the last portions of the crystallizing granitic magma and may form economic deposits of minerals enriched in these elements. Beryllium is used as an alloy with copper to increase the hardness, strength, and fatigue resistance of copper. The mineral beryl may be many different colors, and it has been used as a gemstone. Tantalum and niobium are used for alloys in steel. Tantalum makes steel resistant to chemical corrosion, and niobium helps steel resist high temperatures. Lithium is used in grease to help it keep its lubricant properties over a range of temperatures.

Sometimes, fluids rich in water vapor may bubble out of the magmas and carry out elements—such as tin, copper, tungsten, antimony, gold, silver, and bismuth—that form hydrothermal deposits. Tungsten is used to harden steel used at high speeds in tools, valves, and similar apparatuses. Gold and silver are used for ornamental purposes. Gold is also a monetary standard and is used in dental and scientific applications. Silver is used in photographic film and electronic equipment.

Robert L. Cullers, Stephen M. Testa, and
William C. Sidle

CROSS-REFERENCES

Andesitic Rocks, 1263; Anorthosites, 1269; Basaltic Rocks, 1274; Batholiths, 1280; Carbonatites, 1287; Contact Metamorphism, 1385; Diamonds, 1561; Eruptions, 739; Granitic Rocks, 1292; Igneous Rock Bodies, 1298; Igneous Rock Classification, 1303; Kimberlites, 1310; Komatiites, 1316; Lamproites, 1321; Orbicular Rocks, 1332; Phase Changes, 463; Phase Equilibria, 442; Plutonic Rocks, 1336; Pyroclastic Rocks, 1343; Rocks: Physical Properties, 1348; Silica Minerals, 1354; Ultramafic Rocks, 1360; Xenoliths, 1366.

BIBLIOGRAPHY

Best, M. G. *Igneous and Metamorphic Petrology.* San Francisco: W. H. Freeman, 1995. This college-level geology text provides a brief, general overview of processes affecting the formation of magmas. A useful, well-written introduction for the general reader.

Carmichael, Ian S., et al. *Igneous Petrology.* New York: McGraw-Hill, 1974. Chapter 1, "Magma and Igneous Rocks," and chapter 7, "Sources of Magma: Geophysical-Chemical Constraints," are recommended. A well-known reference presenting in-depth discussion of both the diversity of igneous rock types and their evolution and magma evolution.

Craig, J. R., D. J. Vaughan, and B. J. Skinner. *Resources of the Earth.* Englewood Cliffs, N.J.: Prentice-Hall, 1988. An excellent book for the layperson who has had some high school science and is interested in ore deposits. Chapters 6 and 7 contain a discussion of metals concentrated by magmatic crystallization. For each element, there are sections on its history and uses, its geologic occurrence and ore minerals, and its production and reserves. A glossary is included.

Decker, Robert, and Barbara Decker. *Volcanoes.* San Francisco: W. H. Freeman, 1997. Suitable for the general reader, this book provides useful background for the understanding of igneous rock bodies.

Dietrich, R. V., and B. J. Skinner. *Rocks and Rock Minerals.* New York: John Wiley & Sons, 1979.

One of many books that give information on the identification and origin of minerals and rocks for readers with a minimal science background. A reader with a high school course in chemistry can better appreciate the chemical information. There is a section at the beginning of chapter 4 on the crystallization of magmas.

Hall, Anthony. *Igneous Petrology.* 2d ed. New York: John Wiley & Sons, 1996. A standard text discussing the general aspects of the occurrences, composition, and evolution of igneous rocks and magmas. Each chapter discusses a different rock group. The role of water in magmas is emphasized in the sections on granites. Text is college-level, but the illustrations are useful to the general reader.

Jensen, Mead L., and Alan M. Bateman. *Economic Mineral Deposits.* 3d ed. New York: John Wiley & Sons, 1981. Chapter 4, "Petrology of Mineral Deposits: Magmas, Solutions, and Sediments," and chapter 5, "Magmatic Concentrations," are recommended. A standard and easy-to-read, well-illustrated text discussing the occurrence and formation of mineral deposits.

Middlemost, Eric A. K. *Magmas, Rocks, and Planetary Development: A Survey of Magma/Igneous Rock Systems.* Harlow: Longman, 1997. Middlemost writes for the person with some background in the Earth sciences, covering such topics as igneous rocks, magmas, geodynamics, geochemistry, and volcanism. Illustrations and maps help clarify difficult concepts. Bibliography and index.

Perchuk, Leonid L., and Ikuo Kushiro, eds. *Physical Chemistry of Magmas.* New York: Springer-Verlag, 1991. A college-level text that examines the geochemistry and physical properties of magma. Sections also focus on silicates and silicate minerals.

ORBICULAR ROCKS

Orbicular rocks are igneous or metamorphic stones that contain spherical objects, known as orbicules, consisting of layers of minerals that form onionlike shells around a core. This unusual formation has assisted greatly in the study of how rocks are formed from molten magma.

PRINCIPAL TERMS

AUTOLITH: a fragment of previously crystallized rock that is enclosed within material from the same magma that crystallized later

EUTECTIC: an alloy or solution having its components present in such proportions that the melting point is the minimum possible with those components

FERROMAGNESIAN: containing iron and magnesium

HYDROSILICATE: a compound of silicon and oxygen that also contains water

LIESEGANG RINGS: a series of concentric bands of a precipitated substance that are separated by clear bands and that are often formed in gels by repeated cycles of precipitation

MAFIC: relating to a group of minerals containing magnesium and iron, usually dark in color

METASOMATISM: a metamorphic process that involves important changes in the chemical composition of rock, as well as changes in mineral content and texture

ORTHOCLASE: a mineral consisting of potassium feldspar, with sodium in place of some of the potassium, which may be colorless, white, red, or cream-yellow

PLAGIOCLASE: a feldspar, commonly called moonstone, that is transparent or translucent and pearllike in luster

XENOLITH: a fragment of rock contained in rock of a different kind

COMPOSITION OF ORBICULAR ROCKS

Orbicular rocks tend to be grainy, with grains so large that they can be seen by the unaided eye. The composition of such rock ranges from those that have high amounts of silica (silicon dioxide), such as granite, to rocks containing relatively little silica. Orbicules form in igneous rocks around a core, or nucleus, which may consist of a foreign rock called a xenolith. Cores in igneous rocks can also be formed from a piece of rock, called an autolith, that was segregated early on from the original molten rock from which the orbicular rock was formed. In metamorphic rocks, cores can take shape as recrystallized portions of the original rock, as a different kind of rock portion that was shaped out of the same magma as the surrounding rock, or as a fragment of a rock that existed before metamorphosis occurred. Some orbicules do not have any cores.

The orbicules themselves are spherical shapes within the rock that are onionlike; that is, they consist of several layers of spheres that form concentric circles when cut through the core. The layers within individual orbicules are typically thin, irregular, and sharply defined, and each layer differs from its immediate neighbors in composition or texture. Some layers are composed of thin, platelike mineral grains or mineral grains that are prism-shaped. The minerals of most orbicules are the same as those of the enclosing rock, but they are not necessarily present in the same proportions. Orbicules range in size from less than 3 centimeters to more than 30 centimeters in diameter. The number of shells around a core can number from one to more than twenty. The spheres may be spaced irregularly from each other or in perfect geometrical order. Each orbicule is usually spheroidal or ellipsoidal, but some may be of a more irregular shape or even fragmented because of deformation of the surrounding rock by pressure.

The shells typically form, when cut, a pattern of circles that alternate light and dark. In granitic orbicular rocks, the center is usually biotite (a form of mica), while the dark bands are hornblende or biotite. The light bands are feldspar, a type of quartz. In all the various types of orbicular rocks, the orbicule shells usually consist of various oxides of aluminum, sodium, or calcium. The surrounding matrix usually consists of various oxides of silicon, titanium, iron, or magnesium.

Orbicular rocks have been found on all the continents at more than one hundred different sites. Most of the samples have been reported in Europe and the United States, but this may be because most Western scientists reside in these areas. When orbicular rocks are discovered, they tend to exist in small regions, the greatest dimension of which is no more than several hundred meters in length.

DISCOVERY OF ORBICULAR ROCKS

In 1798, Christian Leopold von Buch, a German geologist noted for his writings about his research and travels, disputed the current theory of Neptunism, in which it was believed that all rocks were sedimentary. He visited Vesuvius, the famous Italian volcano, to see if perhaps some rocks could be formed from molten lava. In 1802, he visited the Auvergne Mountains in France. There he confirmed his belief that granite and many other rocks are formed by volcanic action. Buch at this time first saw orbicular rocks and mentioned them in his works.

The curious pattern of light and dark rings made by different minerals puzzled geologists for many years as they attempted to determine how orbicular rocks could form. In 1913, German physicist and chemist Raphael Eduard Liesegang devised an experiment to re-create concentric rings in the laboratory. Using a gelatin substance infused with a metallic salt such as potassium dichromate, he placed another metallic salt, such as silver nitrate, in the center of the gelatin. The silver nitrate acted as the core or nucleus of a resultant pattern of rings. The potassium dichromate interacted with the silver nitrate to form silver chromate, which formed a band when the reaction was completed locally. The concentration of the mixture became lower in the neighborhood of the bands that had already formed, so the reaction would later tend to take place in regions farther from the nucleus. Eventually, an entire series of concentric rings would appear. The pattern formed was similar to that of orbicular rocks, and the process used to make Liesegang rings is considered similar to that which forms orbicules.

By the late nineteenth century, a theory had already been established concerning orbicular rock formation. During the early stages of the process in which slowly moving magma hardens to form igneous rock, the concentric structure begins to form. In a series of stages that are regularly repeated, a crystallization about a specific center takes place. Early researchers also acknowledge the similarity between orbicules and spherulitic layers in volcanic rocks. Spherulites are usually spherical, crystalline bodies made up of radiating crystal fibers. They are often found in glassy volcanic rocks, such as obsidian or perlite, and they are composed of an intergrowth of quartz and feldspar. Another closely related rock formation is called rapakivi texture. Rapakivi consists of large, salmon-pink alkali feldspars in the shape of rounded crystals a few centimeters in diameter. These crystals are usually surrounded by a final coating of plagioclase—a white, pearly feldspar—about 1 to 2 millimeters thick.

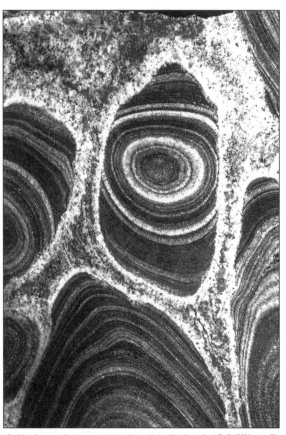

Orbicular rocks in granite, found in Finland. (© William E. Ferguson)

Occasionally, the crystals of alkali feldspar alternate in layers with orthoclase, an alkali feldspar with some of the potassium being replaced by sodium, making rapakivi quite similar to orbicular rocks. The rapakivi are usually found embedded in quartz, feldspar, biotite, or amphibole and are most commonly found in granite, as are orbicular rocks. As early as the 1920's, geologists believed that rapakivi were primitive orbicules that had the same origin as orbicular rocks.

THEORIES OF ORBICULAR ROCK FORMATION

Hypotheses related to orbicule formation in igneous rock date from the 1890's, at which time geologists speculated that parts of the magma could not blend to form a homogeneous molten mass. These parts would then crystallize from their margins inward to form the characteristic orbicular shapes. Later, the crystallization was discovered to act from the inside core outward. In the early twentieth century, many other theories were postulated by petrological researchers. One theory explained that eutectics—alloys that combine in such a way as to reduce the melting point to a minimum—form inside the magma, which adjusts to new conditions by melting and solidifying into concentric rings. The more accepted theory that orbicules form around xenoliths was presented around the same time as a less-accepted theory that suggested that the nucleus or core of the orbicule traveled through different regions of the magma to create different shells.

Another less-accepted theory is that crystallization occurs in a magma that is very thick. This viscosity results in the very slow diffusion of crystallized minerals through the thick magma, so rings form from envelopes of different composition. A theory concerning the formation of orbicules around eutectics explains that crystallization of feldspar causes the surrounding area to increase its level of ferromagnesian components (minerals containing iron and magnesium). These components then begin to crystallize, causing the level of feldspar to increase around them. This process continues until a series of concentric alternating rings is produced. A final theory about orbicular rock formation in igneous rocks explains that it is similar to Liesegang ring formation. This was the generally accepted theory during the late twentieth century.

Theories concerning the formation of orbicules in metamorphic rocks came to the fore by the middle of the twentieth century. One theory merely repeated that the process was similar to that which occurs in Liesegang rings. Another theory concerns dark minerals largely consisting of iron and magnesium known as mafic rocks. During a metamorphic process called sodium metasomatism, sodium ions invade the interior of the core, while calcium, magnesium, and iron ions diffuse outward from the core, to produce a pattern of rings.

Later theorists looked back at early descriptions of the reaction of orbicular rocks under laboratory conditions of induced deformation. When applying high pressure to the orbicules, researchers observed a certain plasticity to be present. Some described the state of the shells as slimy, while others described the initial state of formation as an emulsion of crystals within a gelatinous mass of rock. One researcher described the action of orbicules when squeezed together as similar to the action of soft-boiled eggs when put together in the same cup and pressed. This colorful metaphor captured exactly an image of slippery orbs sliding against each other with little friction.

The significance of the purported plasticity of orbicular rocks led researchers to study more fully the environment in which orbicular rocks form. Since, in a large region of magma, only certain small sections are conducive to orbicular rock formation, petrologists began to study what was different about these sections. Orbicular rocks were found to form in pipes or pockets of formerly wet parts of the magma. In the 1970's, geologists postulated that a trap of surrounding solid rock channeled an amount of upwardly migrating water rich in sulfides and minerals in solution into a small space. This theory was refined in the 1980's by certain researchers who claimed that the mixture from which orbicular rocks formed was actually a molten rock in a gelatinous state composed of silicates combined with water, also called hydrosilicates. In order to crystallize, these hydrosilicates lose the water and attract each other to form larger aggregates of minerals. Another theory proposed in the 1980's stated that hotter magma or water intruded upon a section of magma, causing superheating. As a result of this superheating, all the present silicates would melt. A period of supercooling would then follow, and crystallization would occur around pieces of rock that had not melted during the superheating. In the 1990's, geologists proposed, in the case of volcanic rocks, that these conditions are most likely to take place during volcanic eruptions.

SIGNIFICANCE

To the interested layperson, an orbicular rock may only seem to be a rare and beautiful natural formation. To the geologist, however, an orbicular rock presents a mysterious phenomenon that cannot be easily explained by simple theories of rock formation.

Igneous rocks form when magma, which is molten rock, cools. Slow cooling tends to produce mineral crystals large enough to see, as in orbicular rocks. Quick cooling tends to produce rocks with microscopic crystals or glassy rocks that have no crystals. Slow cooling tends to occur deep within the Earth, while quick cooling tends to occur at the Earth's surface. For this reason, orbicular rocks tend to be formed at a high temperature and pressure within the Earth.

Metamorphic rocks form when sedimentary rocks are subjected to increased temperature or pressure. In general, this occurs when sedimentary rocks come into contact with magma or when the movement of sections of the Earth causes sedimentary rocks to be pressed together. For these reasons, metamorphic orbicular rocks tend to be formed deep within the Earth or in areas of volcanic activity. However, volcanic orbicular rocks are extremely rare. Thus, scientists are generally limited to studying orbicular rocks that were formed far within the interior of the Earth.

Because the formation of orbicular rocks cannot be observed directly, geologists will continue to develop theories of formation that will explain their observed characteristics. These theories could lead to a more sophisticated understanding of the Earth's interior.

Rose Secrest

CROSS-REFERENCES

Andesitic Rocks, 1263; Anorthosites, 1269; Basaltic Rocks, 1274; Batholiths, 1280; Carbonatites, 1287; Eruptions, 739; Experimental Petrology, 468; Feldspars, 1193; Flood Basalts, 689; Fluid Inclusions, 394; Granitic Rocks, 1292; Hot Spots, Island Chains, and Intraplate Volcanism, 706; Igneous Rock Bodies, 1298; Igneous Rock Classification, 1303; Kimberlites, 1310; Komatiites, 1316; Lamproites, 1321; Magmas, 1326; Orthosilicates, 1244; Pegmatites, 1620; Phase Changes, 463; Phase Equilibria, 442; Plutonic Rocks, 1336; Pyroclastic Rocks, 1343; Rocks: Physical Properties, 1348; Silica Minerals, 1354; Spreading Centers, 727; Ultramafic Rocks, 1360; Xenoliths, 1366.

BIBLIOGRAPHY

Augustithus, S. S. "Orbicular Structures." In *Atlas of the Sphaeroidal Textures and Structures and Their General Significance*. Athens: Theophrastus, 1982. A highly technical, extended discussion of recent research into the possible origins of orbicular rocks, with illustrations supporting each theory. For the advanced, highly curious student.

Elliston, John N. "Orbicules: An Indication of the Crystallization of Hydrosilicates." *Earth-Science Reviews* 20 (August, 1984): 265-344. A relatively easy-to-read discussion of the different kinds of orbicular rocks followed by a description of the circumstances surrounding their formation. Contains a glossary, a table of the features of orbicular rocks, and numerous photographs.

Leveson, David J. "Orbicular Rocks: A Review." *Geological Society of America Bulletin* 77 (April, 1966): 409-426. Although out of date for not having a list of viable theories about orbicular rock formation, this is actually a good introduction to the subject. The history of the discovery of orbicular rocks is covered, and a map shows where orbicular rocks have been discovered.

Ort, Michael H. "Orbicular Volcanic Rocks of Cerro Panizos: Their Origin and Implications for Orb Formation." *Geological Society of America Bulletin* 104 (August, 1992): 1048-1058. An overview of theories concerning the origin of orbicular rocks, followed by a detailed description of the location and possible method of formation of magmatic orbicular rocks.

PLUTONIC ROCKS

Plutonic rocks crystallize from molten magma that is intruded deep below the Earth's surface. Exposed at the surface by erosion, they occur mainly in mountain belts and ancient "shield" areas of continents. Many of the world's principal ore deposits are associated with plutonic rock bodies; the rocks themselves may be also exploited economically, as in the production of granite for building stone.

PRINCIPAL TERMS

ANORTHOSITE: a light-colored, coarse-grained plutonic rock composed mostly of plagioclase feldspar

BATHOLITH: the largest type of granite/diorite pluton, with an exposure area in excess of 100 square kilometers

FELDSPAR: an essential aluminum-rich mineral in most igneous rocks; two types are plagioclase feldspar and alkali feldspar

GABBRO: a coarse-grained, dark-colored plutonic igneous rock composed of plagioclase feldspar and pyroxene

GRANITE: a coarse-grained, commonly light-colored plutonic igneous rock composed primarily of two feldspars (plagioclase and orthoclase) and quartz, with variable amounts of dark minerals

MAGMA: a molten silicate liquid that upon cooling crystallizes to make igneous rocks

STOCK: a granite or diorite intrusion, smaller than a batholith, with an exposure area between 10 and 100 square kilometers

PLUTONIC ROCK OCCURRENCE

Plutonic rocks crystallize from molten silicate magmas that intrude deep into the Earth's crust. This same magma (molten silicate liquid) may eventually flow out onto the surface of the Earth as lava that crystallizes to produce volcanic rocks. Volcanic rocks can be distinguished from plutonic rocks by the relative size of their mineral grains. The rapidly cooled volcanic rocks have nearly invisible crystals (or glass), while the slowly cooled plutonic rocks have larger crystals clearly visible to the naked eye.

Because they form deep underground, plutonic rocks require special circumstances to become exposed at the surface. Uplift of the crust in a mountain range or high plateau accompanied by nearly constant erosion by streams or glaciers may eventually uncover once-buried plutonic rocks. Therefore, the best places to see excellent exposures of these rocks are in mountain ranges such as the Rockies, Sierra Nevada, and Appalachian ranges of North America, the Alps of Europe, and the Himalayas of Asia, among many others. Another major plutonic rock terrain exists in "shield" areas found on all the world's major con-

tinents. These areas comprise the ancient cores of the continents and consist of rocks that are billions of years old, most of the rocks having once existed in ancient mountain ranges now eroded down to relatively flat plains. In North America, this area is called the "Canadian Shield" and covers most of Canada and the northern portions of the states of Minnesota, Wisconsin, Michigan, and New York. Comparably ancient plutonic rocks, mostly granites, also occur in the center of the Ozark Plateau of southeastern Missouri.

PLUTONIC ROCK VARIETIES

Plutonic rocks come in many varieties depending upon the chemistry of the parent magma and their mode of emplacement in the crust. Igneous magmas vary chemically between two major extremes: "felsic" magma, in which the concentration of dissolved silica (silicon dioxide) is high and the concentrations of iron and magnesium are relatively low, and "mafic" magma, in which the concentration of silica is low and the concentrations of iron and magnesium are relatively high. Granite and related (generally light-colored) rocks are produced by crystallization of felsic magmas. Their light color and the presence of the mineral quartz (silicon dioxide) distinguishes granitic rocks from the dark plutonic rock gabbro, which crystallizes from mafic magmas. Other rocks, such as diorite, crystallize from magma that is intermediate in composition between felsic and mafic extremes. Diorite and its relatives are generally gray-colored and are commonly mistaken for granite. For example, much of the Sierra Nevada range in eastern California is composed of diorite, although it is popularly known as a "granite" mountain range.

GRANITIC OR DIORITE BODIES

Plutonic rocks are emplaced in the crust as a variety of geometric forms, collectively called "plutons." By far the largest plutons are batholiths, huge masses of granite, diorite, or both with surface exposures exceeding 100 square kilometers.

In western North America, some batholiths are exposed over a considerable portion of whole states, such as the Boulder and Idaho batholiths of Montana and Idaho and the Sierra Nevada and Southern California batholiths of California. Batholiths also occur in the Appalachian Mountains, particularly in the White Mountains of New Hampshire, and in parts of Maine. Most batholiths attain their large size by the successive addition of smaller plutons called "stocks." Stocks are generally exposed over tens of square kilometers but range between 10 and 100 square kilometers in surface area.

Minor plutonic bodies include dikes and sills, tabular intrusions that commonly represent magma that has filled in fractures that either cut across layers in country rock (dikes) or that intruded parallel to rock layers (sills). Some sills fill up with so much magma that they expand and

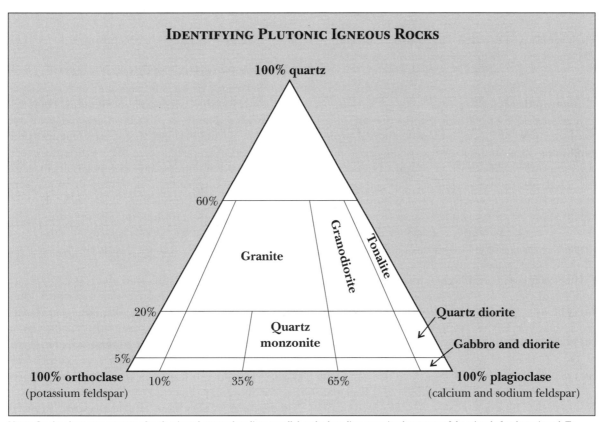

IDENTIFYING PLUTONIC IGNEOUS ROCKS

NOTE: On the chart, percentages of each mineral are read on lines parallel to the base line opposite the corner of the triangle for that mineral. To identify the rock, one must compare the amounts of plagioclase, feldspar, potassium feldspar, and quartz and plot the result on the chart. For example, a rock that is coarse-grained (therefore plutonic or intrusive) and is 40 percent quartz, 30 percent orthoclase (potassium feldspar), and 30 percent plagioclase (feldspar) is called granite.

SOURCE: U.S. Geological Survey; URL: http://wrgis.wr.usgs.gov/docs/parks/rxmin/igclass.html.

Granite, one of the most common of igneous plutonic rocks, a coarse-grained, often light-colored rock composed primarily of two feldspars. (© William E. Ferguson)

Sudbury "nickel irruptive" in Ontario, Canada, is a well-known example of a rich sulfide ore body associated with gabbroic magma. Melting to produce this magma has been attributed to the ancient impact of a large meteoroid. Gravitational segregation of metallic oxides is also known from gabbroic magmas. The rich iron deposits in Kiruna, Sweden, are believed to form from the settling of large blobs of liquid iron oxide that later crystallized to the mineral magnetite. Titanium deposits in anorthosites (feldspar-rich gabbro) at Allard Lake, Quebec, may have formed by gravitational concentration of iron-titanium-rich fluids within the gabbroic magmas that also produced the associated anorthosite rock.

force the overlying layers of rock to bow upward. Such plutons are called "laccoliths," and some, like the Henry Mountains of southeastern Utah, have attained the scale of small mountains.

GABBROIC BODIES

Batholiths, stocks, and laccoliths are predominantly granitic or diorite bodies. Mafic (gabbroic) intrusions occur in their own particular geometric forms, mostly as dikes and sills. Some of these bodies reach enormous size and are exposed over areas that rival large stocks or even granitic batholiths. Many of these bodies show evidence of having concentrated layers of crystals that gravitationally settled after crystallization. Known variously as "gravity-stratified complexes" or "layered mafic-ultramafic complexes," these bodies are commonly rich sources of economically important metallic ores, particularly those of chromium and platinum. The best examples of these bodies are in South Africa (the Bushveld and the Great Dyke), Greenland (the Skaergaard Complex), and North America (the Stillwater Complex in Montana and the Muskox Complex and Kiglapait Complex in Canada).

Other mafic-ultramafic complexes may produce massive layers of metallic sulfides rich in copper, nickel, and variable amounts of gold and silver, among other metals. In North America, the

PLUTON-PRODUCING MAGMAS

The origin of the magmas that create plutons depends in large part on their chemical compositions. Most granite-composition magma probably arises by partial melting of siliceous metamorphic rocks in the deeper parts of the continental crust. Rocks in these regions have compositions that are already close to granitic, so melting them leads inevitably to the production of granitic liquids. These liquids (magma), being less dense than the surrounding, cooler rocks, rise through the crust and join with other bodies to make stocks and, possibly, batholiths.

Mafic magmas of the kind that crystallize gabbro originate in the upper mantle, where they arise by the partial melting of mantle peridotite. Peridotite is the major constituent of the upper mantle and consists of the mineral olivine (iron-magnesium silicate) with minor pyroxene (calcium iron-magnesium silicate) and other minerals. Laboratory experiments have shown that partially melting peridotite at pressures like those in the mantle produces mafic liquids capable of crystallizing gabbro (or basalt at the surface).

A special kind of gabbro, anorthosite, is produced by the separation and concentration of plagioclase feldspar from the gabbroic magma subsequent to its production in the mantle. Because plagioclase commonly is less dense than the surrounding iron-rich gabbroic liquid from which it crystallized, plagioclase crystals may literally float to the top of the magma chamber to form a concentrated mass of feldspar. Anorthosite makes up most of the light-colored regions of the Moon (the lunar highlands) but also occurs as large intrusions on Earth, as in the Duluth Gabbro Complex of northern Minnesota, in the Adirondack Mountains of New York State, and the Kiglapait, Nain, and Allard Lake Complexes in Labrador. Mysteriously, most of Earth's anorthosite bodies were generated during only one restricted period in geologic time, between about 1.1 billion and 1.3 billion years ago.

The origin of diorite magma is complicated by the fact that it can arise in a variety of ways. Diorite plutons, like those in the Sierra Nevada range, represent magma reservoirs under now-vanished volcanoes. The magma that produced these intrusions was generated by melting along a subduction zone, where the Pacific Ocean lithospheric plate (crust plus uppermost mantle) was diving under the North American plate in a zone roughly parallel to the West Coast. A similar process is currently producing the active volcanoes in the Cascade Range of the Pacific Northwest. Any number of parent materials being pulled down to great depth (and, thus, greater temperature) in a subduction zone may mix with subducted ocean water and subsequently melt to produce diorite-type magma. These materials include hydrated (water-saturated) oceanic basalt lava, and even some hydrated mantle materials (serpentine). Ample evidence suggests that some diorite is also produced by the physical mixing of granitic magma produced by melting of continental crustal materials, and gabbroic magma produced in the mantle. These contrasting liquids intermingle on their way upward, finally intruding as intermediate-composition diorite magmas.

ALKALINE PLUTONS

One other group of plutonic rock is worthy of mention, even though surface exposures are relatively rare. These are the "alkaline" intrusives, so named because they tend to be enriched in the alkali elements potassium and sodium and relatively depleted in silica. Alkaline magmas originate by partial melting of mantle peridotite at extreme depths and high pressures (20 kilobars or greater). Because high pressure acts to discourage melting (even at high temperatures), special conditions are required—commonly tectonic rifting—to generate alkaline magmas, making them among the rarest of igneous rocks.

A fairly familiar variety of alkaline-type volcano-intrusive body is the kimberlite "pipe" that in some areas contains diamond. Diamond-bearing kimberlites are especially common in South Africa and other African countries, India, and Russia. North American localities include Murfeesboro, Arkansas, and the Arctic of Canada. Kimberlite rock itself is a complex mixture of mantle and crustal fragments, carbonate minerals, and silicate minerals crystallized from the alkaline magma. Diamonds, if present, were swept up by the magma from high-pressure areas of the mantle and propelled by expanding carbon-dioxide gas to the surface at speeds estimated to exceed supersonic velocities in some cases. Surface deposits (diatremes) of kimberlite consist of volcanic ash ejected during powerful volcanic explosions.

Other important alkaline plutons include the nepheline syenites, light-colored, coarse-grained rocks consisting mostly of the mineral nepheline (sodium aluminum silicate). Important North American localities include Magnet Cove, Arkansas, and Bancroft, Ontario. Depending on the particular locality, these rocks are potential repositories of rare elements and, thus, unusual minerals. Commercial exploitation of nepheline syenites has produced the metals beryllium, cesium, thorium, uranium, niobium, tantalum, and zirconium, to name just a few. These rock bodies may also be a fertile source of apatite (hydrated calcium phosphate), an ore of phosphorous, and the dark-blue mineral sodalite (sodium aluminum silicate chloride), used as a semiprecious gemstone and for sculpted carvings. Another alkaline plutonic rock rich in rare elements and minerals is carbonatite, a magmatic rock composed mostly of calcium, magnesium, and sodium carbonates. Like nepheline syenites, carbonatites are highly prized for their mineral treasures, although they are relatively rare, and their individual surface exposures are limited in size. The best examples are in the East African rift and in South Africa.

STUDY METHODS

Plutonic rocks are igneous rocks; thus, their study entails the same methods that might also apply to volcanic rocks. Field studies of plutons include the construction of geologic maps showing spatial distribution and structure of the various rock types in the pluton. Most plutons contain more than one type of igneous rock or at least show some chemical and mineralogical variation from one locality to the next. During mapping or other field surveys, samples are normally collected to be chemically analyzed by various techniques, including atomic absorption analysis, X-ray fluorescence, or neutron activation analysis. Individual minerals in the rocks may be chemically analyzed using an electron microprobe, a machine that gives a full chemical analysis of a 1-micron spot on a single mineral in a matter of minutes. More sophisticated analyses include isotopic abundance ratios and radiometric ages determined from mass spectrometry. Collected chemical data are normally plotted on diagrams that help show how and why the plutons may have changed over time, and may lead to an understanding of the ultimate origin of their parent magmas.

The identities, textures, and spatial orientations of minerals in rocks are assessed by preparing microscopic slides of thin slices of the rocks called "thin sections." A skilled igneous petrologist (geologist who specializes in igneous rocks) can determine much about the history of a plutonic rock merely by studying a thin section under a microscope. For example, the identification of constituent minerals under the microscope serves to classify the rock, and the relative volume of individual minerals provides hints about the rock's chemistry. For example, a rock with a high volume of dark, mafic (iron and magnesium-rich) minerals would suggest a high concentration of iron and magnesium in that rock compared to one with fewer dark minerals.

ECONOMIC VALUE

Plutonic rocks are the source of many of the raw materials that are used in industrial society. Granite and diorite are used as construction stones in buildings and monuments and as crushed stone for roadways and concrete. Gabbro, however, is generally shunned for decorative purposes because its high iron content causes it to oxidize (rust) over time. On the other hand, the special kind of light-colored gabbro called "anorthosite" (mostly plagioclase feldspar) is prized as a polished building facing stone; it is also used to grace floors or countertops in banks and office buildings.

In addition, many of the richest metallic ore deposits originate in or adjacent to plutonic bodies. The extensive list of metals from granitic deposits includes copper, gold, silver, lead, zinc, molybdenum, tin, boron, beryllium, lithium, and uranium. Gabbroic and related deposits contribute nickel, iron, titanium, chromium, and platinum as well as copper, gold, and silver.

Plutons are also the source of some of the most precious and semiprecious gemstones. For example, the precious gems topaz and emerald occur in ultra-coarse-grained granitic deposits called "pegmatites." Pegmatites also provide the semiprecious gems aquamarine (a blue-green form of emerald), tourmaline (elbaite), rose quartz, citrine (yellow quartz), amethyst, amazonite (aqua-colored microcline feldspar), and zircon (zirconium silicate). Also, large mica crystals from pegmatites are used as electrical and thermal insulators, and feldspar minerals are powdered to make porcelain products and potassium-rich fertilizer. The principal source of the element lithium, used in lubricants and psychoactive drugs, is granitic pegmatite. Lithium is obtained from spodumene (a lithium pyroxene) and lepidolite (a lithium mica), minerals that occur exclusively in pegmatites.

In addition to providing a highly desirable construction stone, anorthosite plutons may also be the source of the semiprecious gemstone labradorite. Labradorite is a high-calcium form of plagioclase feldspar (the major mineral in anorthosites) that displays a green, blue, and violet iridescence similar to the play of colors in the tail of a male peacock. Gem-quality labradorite crystals may be made into jewelry, and polished slabs of labradorite anorthosite are used as counter and desk tops and as building facings.

GEOLOGIC VALUE

In a larger sense, plutonic rocks (particularly granitic and diorite plutons) form the bulk of the world's continents. During the early evolution of the Earth, low-density granitic plutons rising from the early, primitive mantle coalesced to form the cores of the first continents. In later eras, continents have continued to grow, as plutons and overlying volcanic rocks have added new material to the margins. This process is well illustrated by the plutonic terrain of the Sierra Nevada range, the andesite (volcanic equivalent of diorite) volcanoes of the Cascades of the Pacific Northwest, and the Andes Mountains of South America. Were it not for the formation and expansion of continents by the continuing addition of plutons over geologic time, life on Earth would be considerably different. Without continents and the dry land they provide, life would be confined to the oceans, with obvious implications for the evolution of human beings.

Finally, plutonic rocks in natural settings enhance the beauty and general aesthetic value of the landscape. Deeply eroded plutons have produced some of the most striking landscapes in North America, Europe, and Asia, many of which have been set aside as parks and recreation areas. In North America, especially magnificent landscapes in eroded plutons occur at Yosemite National Park in the Sierra Nevada range of California, the Boulder batholith area of Montana, the high peaks area of the Adirondack Dome of New York State, and the White Mountains of New Hampshire.

John L. Berkley

CROSS-REFERENCES

Andesitic Rocks, 1263; Anorthosites, 1269; Basaltic Rocks, 1274; Batholiths, 1280; Carbonatites, 1287; Feldspars, 1193; Granitic Rocks, 1292; Igneous Rock Bodies, 1298; Igneous Rock Classification, 1303; Kimberlites, 1310; Komatiites, 1316; Lamproites, 1321; Magmas, 1326; Metamorphic Rock Classification, 1394; Orbicular Rocks, 1332; Pyroclastic Rocks, 1343; Rocks: Physical Properties, 1348; Silica Minerals, 1354; Ultramafic Rocks, 1360; Xenoliths, 1366.

BIBLIOGRAPHY

Ballard, Robert D. *Exploring Our Living Earth.* Washington, D.C.: National Geographic, 1983. Aimed at the general reader and should be available in most public libraries. Covers the Earth as an "energy machine," with sections on volcanism, mountain-building, earthquakes, and related phenomena. Richly illustrated with colored photographs and drawings. An excellent source of information on plate tectonics, the unifying theory that seeks to explain the origin of major Earth forces and surface features. Because plutonic rocks are involved in both mountain-building and volcanic activity, an understanding of plate tectonics is essential to understanding the origin of plutons. Glossary, bibliography, and comprehensive index.

Best, Myron G. *Igneous and Metamorphic Petrology.* 2d ed. Cambridge, Mass.: Blackwell, 1995. This is a college-level textbook, but it should be accessible to most general readers. Part 1 contains comprehensive treatments of all major plutonic rock bodies, complete with drawings, diagrams, and photographs detailing the essential features of each pluton type. Chapter 1 contains a section on how petrologists study rocks. The appendix contains chemical analyses of plutonic rocks and descriptions of important rock-forming minerals. One of the best books available for the serious student of igneous and metamorphic rocks.

Beus, S. S., ed. *Rocky Mountain Section of the Geological Society of America: Centennial Field Guide.* Vol. 2. Denver, Colo.: Geological Society of America, 1987. This field guide was produced for professional geologists but should also be of use to the general reader. It describes automobile field trips to geological localities in the Rocky Mountain states. Many of the stops on these tours are at plutonic localities, including the Idaho batholith and the Stillwater Complex of Montana. Illustrated with photographs, geological cross sections, various diagrams, and maps. Descriptions and explanations of the various localities allow an appreciation for the details of plutons and their relationships to associated rocks.

Decker, Robert, and Barbara Decker. *Volcanoes.* San Francisco: W. H. Freeman, 1997. Suitable for the general reader, this book provides useful background for the understanding of igneous rock bodies.

Hacker, Bradley R., Ju G. Liou, et al., eds. *When Continents Collide: Geodynamics and Geochemistry of Ultra-High Pressure Rocks.* Boston: Kluwer, 1998. An excellent introduction to the study of geologic metamorphism, this collection of essays examines the geochemical and geophysical effects of high pressures on metamorphic rocks. Intended for readers with some background in Earth sciences. Bibliography and index.

Jensen, Mead L., and Alan M. Bateman. *Economic Mineral Deposits.* New York: John Wiley & Sons, 1981. One of the most comprehensive single-volume treatments of world ore deposits available. Illustrated with abundant monochrome diagrams and photographs; includes tables listing statistics about major ore-bearing regions. All important ore deposits associated with plutonic intrusions are described in detail. Extensive index. This text should be available in most college libraries or in metropolitan public libraries.

Press, Frank, and Raymond Siever. *Understanding Earth.* 2d ed. New York: W. H. Freeman, 1998. This is one of many introductory textbooks in geology for college students.

Prinz, Martin, George Harlow, and Joseph Peters. *Simon & Schuster's Guide to Rocks and Minerals.* New York: Simon & Schuster, 1978. This field guide to rocks and minerals is readily available in most bookstores. It contains beautiful color photographs of museum-quality minerals and rocks, along with comprehensive descriptions. The introduction to the section on rocks contains a concise description of igneous processes, including plutonic bodies and intrusive processes. Explains how magmas are generated and emplaced in the crust and how they become differentiated. Includes a glossary and index. Highly recommended both for "rock hounds" and for professionals interested in exploring plutonic rocks in the field.

Smith, David G., ed. *The Cambridge Encyclopedia of Earth Sciences.* New York: Cambridge University Press, 1981. One of the best and most comprehensive resources on Earth processes available for the nonspecialist. Richly illustrated with colored photographs, drawings, and diagrams. Pertinent sections to plutonic rocks include chapters 10, 13, 14, and 16. Plate-tectonic concepts relevant to plutonic and volcanic processes are explored in depth, as are topics about plutonic rocks and minerals. Includes an extensive glossary and "further reading" section.

Will, Thomas M. *Phase Equilibria in Metamorphic Rocks: Thermodynamic Background and Petrological Applications.* New York: Springer, 1998. The author offers a clear description of phase rule and its effects on the equilibrium of metamorphic rocks. There is an in-depth examination of the geochemical makeup of metamorphic rocks and their properties. Bibliographic references.

PYROCLASTIC ROCKS

Pyroclastic rocks form from the accumulation of fragmental debris ejected during explosive volcanic eruptions. Pyroclastic debris may accumulate either on land or under water. Volcanic eruptions that generate pyroclastic debris are extremely high-energy events and are potentially dangerous if they occur near populated areas.

PRINCIPAL TERMS

ASH: fine-grained pyroclastic material less than 2 millimeters in diameter

IGNIMBRITE: pyroclastic rock formed from the consolidation of pyroclastic-flow deposits

LAPILLI: pyroclastic fragments between 2 and 64 millimeters in diameter

PUMICE: a vesicular glassy rock commonly having the composition of rhyolite; a common constituent of silica-rich explosive volcanic eruptions

PYROCLASTIC FALL: the settling of debris under the influence of gravity from an explosively produced plume of material

PYROCLASTIC FLOW: a highly heated mixture of volcanic gases and ash that travels down the flanks of a volcano; the relative concentration of particles is high

PYROCLASTIC SURGE: a turbulent, low-particle-concentration mixture of volcanic gases and ash that travels down the flanks of a volcano

STRATOVOLCANO: a volcanic cone consisting of both lava and pyroclastic rocks

TEPHRA: fragmentary volcanic rock materials ejected into the air during an eruption; also called pyroclasts

TUFF: a general term for all consolidated pyroclastic rocks

VOLATILES: fluid components, either liquid or gas, dissolved in a magma that, upon rapid expansion, may contribute to explosive fragmentation

FORMATION OF PYROCLASTIC ROCKS

Pyroclastic rocks form as a result of violent volcanic eruptions such as that of Mount St. Helens in 1980 or Mount Vesuvius in A.D. 79. Molten rock, or magma, within the Earth sometimes makes its way to the surface in the form of volcanic eruptions. These eruptions may produce lava or, if the eruption is highly explosive, fragmental debris called tephra or pyroclasts. The term "pyroclastic" is from the Greek roots *pyros* (fire) and *klastos* (broken). Dissolved water and gases (volatiles) are the source of energy for these explosive eruptions. All molten rock contains dissolved fluids such as water and carbon dioxide. When the molten rock is still deep within the Earth, the confining pressure of the overlying rock keeps these volatiles from being released. When the magma rises to the surface during an eruption, the pressure is lowered, and the gases and water may be violently released, causing fragmentation of the molten rock and some of the rock surrounding the magma. This type of explosive eruption is more common in rocks rich in silica, which are more viscous (flow more thickly) than those that are silica-poor. External sources of water, such as a lake or groundwater reservoir, may also provide the necessary volatiles for an explosive eruption.

Pyroclastic debris can be produced from any of three different types of volcanic eruptions: magmatic explosions; phreatic, or steam, explosions; and phreatomagmatic explosions. Magmatic explosions occur when magma rich in dissolved volatiles undergoes a decrease in pressure such that the volatiles are rapidly released or exsolved. The solubility of volatiles in magma is partially controlled by confining pressure, which is a function of depth. Solubility decreases as the magma rises toward the surface. At a certain depth, carbon dioxide and water begin exsolving and become separate fluid phases. At this point, the magma may undergo explosive fragmentation either through an open vent or by destroying the overlying rock in a major eruptive event.

As a magma rises toward the surface, it may encounter a groundwater reservoir or, in a subaqueous vent, interact with surface water. In both cases, the superheating and boiling of water followed by its explosive expansion to gas may fragment the magma and the surrounding country rock. The ratio between the mass of water and the mass of magma controls the type of eruption. If there is little water in relation to magma, the explosive activity may be confined to the eruption of steam and is called a phreatic explosion. If the magma contains significant quantities of dissolved volatiles and encounters a large amount of water, the resulting explosion is termed phreatomagmatic.

COMPOSITION OF PYROCLASTIC DEPOSITS

Pyroclastic deposits are composed of tephra, or pyroclasts. These fragments can have a wide range of sizes. Particles less than 2 millimeters in diameter are termed ash, those between 2 and 64 millimeters are called lapilli, and those greater than 64 millimeters are called blocks, or bombs. There are three principal components that make up pyroclastic debris: lithic fragments, crystals, and vitric, or juvenile, fragments. Lithic fragments can be subdivided into pieces of the surrounding rock explosively fragmented during an eruption (accessory lithics), pieces of already solidified magma (cognate lithics), and particles picked up during transport of eruptive clouds down the flanks of a

volcano (accidental lithics). Crystal fragments are whole or fragmented crystals that had solidified in the magma before eruption. Vitric or juvenile fragments represent samples of the erupting, still molten, magma. They may be either partly crystallized or uncrystallized (glass). Pumice is a type of juvenile fragment that contains many vesicles, or holes, as a result of the rapid exsolution of gases during eruption. Small, very angular glass fragments are called shards.

TYPES OF PYROCLASTIC DEPOSITS

Three types of pyroclastic deposit can be distinguished based on the type of process that forms the deposit: pyroclastic-fall deposits, pyroclastic-flow deposits, and pyroclastic-surge deposits. These types can all be formed by any of the previously described different types of volcanic eruption. Any of these deposits may be termed a tuff if the grain size is predominantly less than 2 millimeters.

Pyroclastic-fall deposits form from the settling, under the influence of gravity, of particles out of a plume of volcanic ash and gases erupted into the atmosphere forming an eruption column. Tuffs formed in this way are coarsest near the eruption center and become progressively finer farther away. Ash falls can also be derived from the tops of more dense pyroclastic flows, as the finer-grained material is turbulently removed from the upper portion of the pyroclastic flow and then settles to the ground.

Two types of pyroclastic deposits result from the formation of dense clouds of ash during an eruption and the subsequent transport of debris in the form of a hot cloud of ash, lapilli, and gases. Pyroclastic flows have a relatively high particle concentration and, in some areas, the western United States for example, form enormous deposits with volumes as large as 3,000 cubic kilometers. Pyroclastic-flow deposits rich in pumice are termed ignimbrites. Pyroclastic surges are expanded, low-particle-concentration density currents that are

A pyroclastic flow deposit. (U.S. Geological Survey)

generally very turbulent. Surge deposits are volumetrically less important than those of pyroclastic flows but can be very destructive. Both flows and surges may have emplacement temperatures of up to 800 degrees Celsius. Pyroclastic surges, however, because of their lower density and turbulent nature, may attain velocities up to 700 kilometers per hour. It is this combination of speed and temperature that makes pyroclastic surges so dangerous.

STUDY OF PYROCLASTIC ROCKS

Geologists study pyroclastic rocks using field techniques, laboratory analyses, and theoretical co nsiderations of eruption processes. Observations of deposits in the field remain the cornerstone of much of geologic interpretation. Pyroclastic deposits form essentially as sedimentary material, that is, as fragments or clasts moved in air or water and deposited in layers. As such, many of the techniques used by sedimentologists (geologists interested in the formation and history of sedimentary rocks) are employed in the study of pyroclastic deposits. Careful examination of a variety of different sedimentary features within pyroclastic deposits can aid in the interpretation of the processes of transport and deposition. This information, studied over as wide a geographic area as possible to ascertain systematic changes in the deposits, will assist in understanding the geologic history of the region.

Much can be learned through analysis of the composition of pyroclastic rocks, which is generally done using a variety of laboratory techniques. The use of specialized microscopes allows geologists to examine very thin sections of rock in order to observe textures and to discern mineral composition. Geochemical techniques have become very popular and powerful in the study of all kinds of rocks, including pyroclastic rocks. By looking at the amounts of certain elements that occur in extremely low abundances and also at relative proportions of certain types of isotopes (naturally occurring forms of the same element that differ only in the number of neutrons in the nucleus), scientists can understand more about the processes taking place deep within the Earth that lead to the formation of magma and eventually to the eruption of pyroclastic debris. Scanning electron microscopes have been used to study in detail the surface features and textures of fine volcanic ash

particles. This information can lead to better understanding of eruptive and transport processes that formed and deposited the pyroclastic particles.

Theoretical studies associated with pyroclastic rocks revolve primarily around considerations of the mechanics of high-temperature, high-velocity eruption clouds and their transport and deposition. This type of reasoning allows a geologist to infer certain conditions of eruption from an analysis of the deposits. The geologic rock record has abundant pyroclastic deposits, and it is through this type of inference that geologists interpret the geologic history of a region. A comprehensive understanding of pyroclastic deposits must include a thorough understanding of the processes by which the deposit forms.

ASSOCIATION WITH VIOLENT ERUPTIONS

Most pyroclastic rocks are associated with stratovolcanoes, also called composite volcanoes. These volcanic edifices are built by a combination of extrusive and explosive processes and thus are formed of both pyroclastic debris and lava. Well-known examples of stratovolcanoes include Mount St. Helens, Mount Fuji, and Mount Vesuvius. Volcanoes such as those found on the Hawaiian Islands are of a less energetic variety called shield volcanoes and produce insignificant quantities of pyroclastic material. Stratovolcanoes are located around the world and are associated with the global process of plate tectonics, which explains the movement of the continents; the generation of new "plates," which include oceanic crust; and the destruction of old plates. It is at the zone of plate destruction that rock is melted and makes its way to the surface, sometimes producing pyroclastic eruptions.

Violent volcanic eruptions that may produce pyroclastic deposits are among the most powerful events occurring on the Earth. Historically, many of the most destructive volcanic eruptions have involved pyroclastic surges. The eruption in A.D. 79 of Mount Vesuvius generated pyroclastic debris that buried the towns of Pompeii and Herculaneum, killing most of the inhabitants. In 1902, on the island of Martinique in the Caribbean, the violent eruption of Mount Pelée produced a pyroclastic surge that swept down on the city of St. Pierre, killing all but a handful of a population of

Phreatic vents in Mount St. Helens debris flows, May 29, 1980. (U.S. Geological Survey)

about thirty thousand. The eruption of Mount St. Helens in 1980 and of El Chichón in Mexico in 1982 both produced pyroclastic surges. Two thousand people were killed as a result of the El Chichón eruption. Pyroclastic surges and flows generally do not present a hazard beyond about a 20-kilometer radius.

Pyroclastic deposits form a major portion of some volcanic terrains. Some of these deposits are enormously extensive, indicating that the eruptions that produced them were much larger than any witnessed in modern times. It is not clear whether these deposits reflect an overall increase in volcanic activity in the Earth's past or whether this type of titanic eruption occurs sporadically throughout geologic time. Titanic eruptions inject so much debris into the upper atmosphere that

global weather can be affected. The Earth experienced brilliant red sunsets and lowered temperatures because dust blocked the Sun for several years following the 1883 eruption of Krakatau in the strait between Java and Sumatra, which completely destroyed an island and discharged nearly 20 cubic kilometers of debris into the air. The explosion was heard nearly 5,000 kilometers away in Australia, and darkness fell over Jakarta, 150 miles away. Some geologists speculate that enormous volcanic eruptions in the Earth's past have even led to extinctions, including that of the dinosaurs, by producing so much ash that the amount of sunlight received on the Earth's surface is reduced, and the entire food chain is disrupted.

Bruce W. Nocita

CROSS-REFERENCES

Andesitic Rocks, 1263; Anorthosites, 1269; Basaltic Rocks, 1274; Batholiths, 1280; Building Stone, 1545; Carbonatites, 1287; Continental Crust, 561; Earth Resources, 1741; Granitic Rocks, 1292; Igneous Rock Bodies, 1298; Igneous Rock Classification, 1303; Kimberlites, 1310; Komatiites, 1316; Lamproites, 1321; Magmas, 1326; Mountain Belts, 841; Orbicular Rocks, 1332; Pegmatites, 1620; Plutonic Rocks, 1336; Rocks: Physical Properties, 1348; Silica Minerals, 1354; Ultramafic Rocks, 1360; Xenoliths, 1366.

BIBLIOGRAPHY

Blong, R. J. *Volcanic Hazards: A Sourcebook on the Effects of Eruptions.* Sydney, Australia: Academic Press, 1984. Discusses the nature of volcanic hazards with case histories. Suitable for college-level students.

Cas, R. A. F., and J. V. Wright. *Volcanic Successions: Modern and Ancient.* Winchester, Mass.: Unwin Hyman, 1987. Eleven of the fifteen chapters in this excellent book deal wholly

or in part with pyroclastic rocks. Takes a sedimentological approach to the study and interpretation of pyroclastic deposits. Indispensable for geologists interested in pyroclastic rocks. Suitable for college-level students.

Cattermole, Peter John. *Planetary Volcanism: A Study of Volcanic Activity in the Solar System.* 2d ed. New York: Wiley, 1996. Cattermole examines volcanism and geology on Earth and other planets in the solar system. A good introduction for the layperson without much background in astronomy or Earth sciences.

Decker, Robert, and Barbara Decker. *Volcanoes.* San Francisco: W. H. Freeman, 1997. This book provides the reader with a good overview of different types of volcanoes and volcanic processes, including those that produce pyroclastic debris. Suitable for high school students.

Fisher, R. V., and H. U. Schmincke. *Pyroclastic Rocks.* New York: Springer-Verlag, 1984. This book, along with that of Cas and Wright, provides the most comprehensive treatment of pyroclastic deposits and the processes which form them. Suitable for college-level students.

Press, Frank, and Raymond Siever. *Understanding Earth.* 2d ed. New York: W. H. Freeman, 1998. This is one of many introductory textbooks in geology for college students. It has a good chapter on igneous rocks.

Sigurdsson, Haraldur, ed. *Encyclopedia of Volcanoes.* San Diego, Calif.: Academic Press, 2000. This book contains a complete summary of the scientific knowledge of volcanoes. It contains eighty-two well-illustrated overview articles, each of which is accompanied by a glossary of key terms. Although this is a college-level text, it is written in a clear style that makes it generally accessible. Cross-references and index.

Simkin, Tom, L. Siebert, L. McClelland, D. Bridge, C. Newhall, and J. H. Latter. *Volcanoes of the World: A Regional Directory, Gazetteer, and Chronology of Volcanism During the Last Ten Thousand Years.* Stroudsbourg, Pa.: Hutchinson & Ross, 1981. Suitable for all readers.

Sutherland, Lin. *The Volcanic Earth: Volcanoes and Plate Tectonics, Past, Present, and Future.* Sydney, Australia: University of New South Wales Press, 1995. Although Sutherland focuses on volcanic activity in Australia, the book provides an easily understood overview of volcanic and tectonic processes, including the role of igneous rocks. Includes color maps and illustrations, as well as a bibliography.

ROCKS: PHYSICAL PROPERTIES

Rocks and rock products are used so widely in everyday life that physical properties of rocks affect everyone. Major properties of rocks fall into two categories: those properties that compose the exterior nature of the rock, such as color and texture, and those properties that make up the rock's internal nature, such as strength, resistance to waves, and toughness.

PRINCIPAL TERMS

COEFFICIENT OF THERMAL EXPANSION: the linear expansion ratio (per unit length) for any particular material as the temperature is increased

COMPRESSIVE STRENGTH: the ability to withstand a pushing stress or pressure, usually given in pounds per square inch

DENSITY: mass per unit volume

ELASTICITY: the maximum stress that can be sustained without suffering permanent deformation

HARDNESS: the resistance to abrasion or surface deformation

SHEAR, or SHEARING, STRENGTH: the ability to withstand a lateral or tangential stress

TOUGHNESS: the degree of resistance to fragmentation or resistance to plastic deformation

COLOR

Physical properties of rocks are important not only to geologists and geophysicists but also to construction engineers, technicians, architects, builders, and highway planners. In fact, rock properties affect everyone. Major physical properties include color, hardness, toughness, density, compressive strength, shear strength, tensional and bending (transverse) strength, elasticity, coefficient of thermal expansion, electrical resistivity, absorption rate of fluids, rates of weathering, chemical activity, response to freeze-thaw tests, texture, spacing between fractures, ability to propagate waves, radioactivity, and melting point.

Color is one of the most interesting physical properties of rocks. It is determined by the colors of the component minerals, arrangement of these minerals, and weathering pigments. Color and patterns of colors are important in assessing the aesthetic qualities of ornamental and building stone. The unweathered colors must be pleasing to the eye, and staining detrimental to the appearance must not occur in appreciable amounts. The quality and beauty of many marbles are renowned when used in ornamental stone and sculpture. Building limestone must have a pleasing gray or tan color and not contain iron or manganese minerals, which would stain the rock brown or black during weathering. The pink-and-black or white-and-black mottling of attractive granites, diorites, and syenites is a familiar sight in business, school, university, and other public buildings. It is not, therefore, necessary or even always desirable that a color be uniform throughout.

HARDNESS, TOUGHNESS, DENSITY

Hardness, or resistance to abrasion, can be found by using the Mohs scale, which was derived by Friedrich Mohs in 1822 to measure relative resistance to abrasion in minerals and rocks. Any substance will be able to scratch any substance softer than or as hard as itself. Ten standard minerals are used: Talc is the softest and designated as hardness equal to 1, and diamond is the hardest and designated as hardness equal to 10. Rock hardness is of particular importance to those people who use rocks or rock products in building façades, monuments, tombstones, patios, and other structures. A variety of tests are used to determine resistance to abrasion over a time interval.

Toughness differs from hardness. The property is defined as the degree of resistance to brittleness or plastic deformation of a particular substance. Toughness is determined by a test using repeated impacts of a heavy object. A hammer is dropped from a specified height upon a sample, and this height is continually increased. The height of the fall in centimeters upon breakage is then defined as toughness. An example of a very tough material is jade, whereas rock salt is very brittle. The so-called French coefficient of wear, which is 40 divided by percentage wear in a test known as the Deval test, is also used to measure toughness. Toughness is a very important property in rocks that are used in building roads and airstrips, which are subject to repeated stresses.

The density of a rock is its mass or weight divided by gravitational acceleration, divided by unit volume (amount of space). More frequently, the rock is compared with an equal volume of water that has a density of almost exactly 1 gram per cubic centimeter. This number, which is dimensionless, is known as specific gravity. There is a considerable variation in specific gravity among rock types and even within them. Most limestone and dolomite rocks range from about 2.2 to 2.7 in specific gravity, whereas basalt and traprock are considerably heavier, at 2.8 to 3.0. Density is not synonymous with strength and toughness. Some low-density materials are strong, and some very dense materials cut by fractures are weak.

THE MOHS HARDNESS SCALE

The Mohs scale measures the hardness of a rock or mineral in terms of the ease with which it can be scratched by other substances. The scale is based on the following minerals, in ascending order of hardness:

1. talc
2. gypsum
3. calcite
4. fluorite
5. apatite
6. orthoclase
7. quartz
8. topaz
9. corundum
10. diamond

For example, a rock or mineral that can be scratched by topaz but not by quartz has a Mohs hardness of 7-8. Supplementary tests can be made with such objects as a fingernail (2+), a copper penny (approximately 3), a knife blade (5+), window glass (5½), or a steel file (6½).

COMPRESSIVE, SHEARING, AND BENDING STRENGTHS

Compressive strength, or bulk modulus, is measured on the basis of the highest pushing pressure or stress a rock can withstand per square unit, usually measured in pounds per square inch (psi). This strength is generally greater than transverse, tensional, or shearing strength. Limestones average about 15,000 psi, granites about 25,000 psi, quartzite about 30,000 psi, and basalts and traprock about 48,000 psi. This strength will vary considerably from rock to rock of the same composition because of variation in structural properties, as most rocks have fractures and voids and differ in grain size and shape.

Shearing strength, or modulus of rigidity, is measured on the basis of the highest lateral stress in pounds a rock can withstand per square inch. An example of shearing stress is the stress exerted by a car sideswiping another car. Values for limestones average about 2,000 psi and granites about 3,000 psi.

Bending, or transverse, strength is defined as the strength of a slab loaded at the center and supported only by adjustable knife edges. This strength is determined by the "modulus of rupture," which is a function of the rupture load in pounds, the length of a slab, the width or breadth of a slab, and the thickness of a slab.

ELASTICITY AND LINEAR EXPANSION

The so-called modulus of torsional rigidity, or elasticity (Young's modulus), is measured in pounds per square inch times 10^6 (millions) or dyne-centimeters^2. This property is a measure of how easily a material can return to its original shape when stressed. If stressed beyond the limits of this modulus, permanent changes in shape or deformation occur. This modulus is extremely variable, ranging typically from about 3×10^6 psi to 6×10^6 psi for limestone, to about 6×10^6 psi to 8×10^6 psi for granite. Elasticity refers to the ability of a material to spring back to its original shape without plastic deformation or rupture. Beyond a certain value of stress, rupture or flow will occur.

The coefficient of thermal expansion measures expansion along a line through the rock with increase in temperature. It is also termed linear expansion. The property is expressed as a ratio of change in length divided by unit length times change in temperature. Typical values for common rocks used in crushed stone or building stone range from about 4×10^{-6} (four millionths) to 12×10^{-6} (twelve millionths). Knowledge of the coefficient of thermal expansion is very important to engineers who design structures such as highways and bridges. On bridges, expansion-contraction joints are commonly used in consideration of this property.

RESISTIVITY AND FLUID ABSORPTION RATE

A measurement important to petroleum geologists, geophysicists, and engineers is electrical resistivity of rocks. The resistivity may be defined as the reciprocal of electrical conductivity. Resistivity is measured in ohm-centimeters; that is, electrical resistance along a centimeter's length. It will vary greatly depending on whether a rock is dry, contains fresh water in its pores or cracks, contains saline water, or contains organic compounds such as petroleum or natural gas. Igneous rocks such as basalts and traprock will vary from about 1×10^4 to 4×10^5 ohm-centimeters, and granites about 10^7 to 10^9 ohm-centimeters. Basalts have less resistivity and greater conductivity than granite because they generally contain more abundant amounts of dark, nonmetallic minerals than do granites. Limestones and sandstones commonly range from 10^3 to 10^5 ohm-centimeters, but those containing much saline water will have values much less, because water with dissolved salts is a good conductor of electricity. Metallic ore deposits will show very low resistivities because of the high conductivity of metals.

Absorption rate of fluids by rocks is another property often measured. A dry rock sample of known weight (or dry aggregate of rock chips) is soaked in water for twenty-four hours, dried under surface conditions, and weighed. The weight of water absorbed is given as a percentage of the dry weight of the rock. This property is extremely variable in rocks. For example, limestones may vary from 0.03 percent to 12 percent absorption rate. Very fine-grained limestones will commonly have a higher absorption rate than coarse-grained limestones.

SOUNDNESS, TEXTURE, SPACING BETWEEN FRACTURES

Soundness, or resistance to weathering, is important because rock materials are exposed to outside conditions, where temperature and moisture conditions may vary considerably. Chemicals may also attack the rock, and in Arctic or temperate climates, freezing and thawing may occur. In a climate where much freezing and thawing occur, the breakage and deterioration of cement and rock materials caused by this phenomenon will be noticed. Engine-driven vehicles and equipment would not last very long in winter if antifreeze were not added to the engine block and radiator. Examples of chemical weathering can be found in an old cemetery, where gravestone inscriptions of the same age on chemically resistant rocks such as quartzite are much clearer than those on chemically reactive rocks such as marble.

Texture relates to grain size and grain shape, as well as fabric, which is the relationship of the

SOME PHYSICAL PROPERTIES OF ROCKS

| Rock | Type | Strength (psi) | | Porosity (%) | Specific Gravity |
		Compressive	Tensile		
Basalt	Igneous	25,000-30,000	500-1,000	1	2.75
Granite	Igneous	30,000-50,000	500-1,000	1	2.67
Limestone	Sedimentary	2,000-20,000	400-850	2-20	2.2-2.5
Marble	Metamorphic	10,000-30,000	700-1,000	0.5-2	2.5-2.8
Quartzite	Metamorphic	15,000-40,000	700-1,000	1-2	2.5-2.6
Sandstone	Sedimentary	5,000-15,000	100-200	5-30	2.1-2.5
Shale	Sedimentary	5,000-10,000	100-200	7-25	1.9-2.4
Slate	Metamorphic	15,000-30,000	700-1,000	0.5-5	2.6-2.8

SOURCE: Data are from *McGraw-Hill Encyclopedia of Earth Sciences*, ed. Sybil P. Parker (2d ed., McGraw-Hill, Inc., 1987).

grains to one another. Appearance, resistance to weathering, absorption rate characteristics, and strength are all related to texture.

Spacing between fractures is a significant structural property of rocks. For building stone, it is usually desirable to have massive rock or rock separated at intervals of tens of meters by layering or bedding planes or by vertical cracks or joints. Preferably, the distance between planes will be fairly constant and the cracks flat and even. At least some fractures are necessary, because one face of the block must be blasted, and some incipient cracks are needed to drive in the charges. Other faces may be sawed out by means of wire saws and silica grit. For some uses of building stone, such as roofing or facing slates and paving stones, smooth, closely spaced fractures are desirable, provided the stone is relatively strong and free from fractures in between.

WAVE TRANSMISSION ABILITY

Ability to transmit waves is an important rock property to construction engineers designing buildings, roads, bridges, and dams in earthquake-prone areas and to petroleum geologists and engineers. This property depends on the density and the strength of the rock, especially its compressibility and its rigidity (shear) modulus. Basically, there are two types of waves that result from earthquakes or human-made explosions: longitudinal and transverse waves. Longitudinal waves, similar to sound waves, create pushing and pulling effects on molecules while traveling along a straight line. Transverse waves move sideways as they travel. Compressive strength (pushing strength) is important in determining the speed of longitudinal waves, whereas modulus of rigidity, or shear, strength is more important in determining the speed of transverse waves. In general, the greater the strength, the greater the wave speed through a material; the greater the density, the less the wave speed through a material or rock under a given temperature and pressure. The density of darker igneous rocks is generally greater than that of lighter-colored igneous and sedimentary rocks. Also, the density of rocks increases toward the Earth's center, but velocity is not always less, because strength tends also to increase. This factor may equal or even outweigh the density factor and produce a net increase in velocity or speed.

RADIOACTIVITY AND MELTING POINT

Radioactivity, another physical-chemical property, is demonstrated by those rocks that spontaneously emit radiant energy (which results from the disintegration of unstable atomic nuclei). Scientists use the heat generated by absorption of the radiation emitted by these rocks to establish the temperature of the Earth's interior and its thermal evolution. Radioactivity has been of interest since the early 1980's because of the health hazard posed by radon. Radon is an intermediate product of radioactive decay, and it can seep into basements and cause lung cancer. Proper designing or repair of basement interiors may be necessary to reduce rates of radon infiltration.

Rock melting point is a rock property that is generally of interest to petrologists, engineers on deep-drilling projects, and engineers designing furnaces or kilns that require rock or brick that has a melting point higher than that of the materials being processed in the kiln. Except in the case of dry, completely one-mineral rocks, melting will occur over a range of temperatures rather than at merely one temperature. This type of melting occurs for three reasons: Several types of minerals, each with a different melting point when pure, usually occur in rocks; each mineral mutually affects the melting points of the other minerals, usually lowering the melting points and extending them over a range of temperatures; and water and other fluids may mutually lower the melting point of a rock.

PHYSICAL PROPERTY TESTS

A great variety of tests are used to determine the degree or types of physical properties of rocks. Hardness may be determined by comparing a rock with known hardness points made from minerals of known hardness or from ceramic cones of various known hardnesses, either on the Mohs ten-point scale or on a more elaborate 1,000-point industrial scale. Another test is called the Dorry hardness test. Rock cones 2.5 centimeters in diameter are loaded with a total weight of about 1,000 or 1,250 grams, then subjected to abrasive action of fine quartz of known size fed upon a rotating cast-iron disk. The loss in weight of the core after 1,000 revolutions is used to compute value of hardness. One of the most widely used machines to test hardness is the Los Angeles Rattler (abrasion testing machine). It has a steel drum 70 centimeters in diameter and 50 centimeters long. The sample is inserted with a certain number of steel balls, and the drum is rotated five hundred revolutions at thirty revolutions per minute. The material is then sized into further grades, and the coarser sets of remaining fragments are subjected to one thousand further revolutions using various numbers of steel balls weighing about 5,000 grams. Another test, the Brinell test, determines hardness by pressing a small ball of hard material on the sample and measuring the size of the depression.

Toughness is tested on samples with a diameter of about 2.5 centimeters. The sample is held in a test cylinder on an anvil and subjected to the fall of a steel hammer or plunger weighing 2 kilograms. The first fall is from a 1-centimeter height and is progressively increased by 1 centimeter until breakage occurs. Height of fall is then expressed as toughness.

Specific gravity is measured by placing the sample in water. It is measured, dried to obtain dry weight, then immersed in water for twenty-four hours, surface dried, and weighed. Finally, it is weighed again in water. Loss of weight from dried weight initially is true specific gravity. The second step, involving the twenty-four-hour immersion, is necessary to ascertain that pore spaces in the sample are filled with water so that a true measure of buoyancy can be taken in the third step.

Compressive strength is measured on test rock cylinders 5 centimeters in diameter and 5.6 centimeters long. The sample is placed in a cylindrical casing in a press, and the press is lowered at a specified rate. Amount of compression in pounds per square inch at failure is the compressive strength.

The so-called soundness test measures resistance to chemical weathering or, alternatively, resistance to a major type of mechanical weathering, called ice or frost wedging. In the first case, the sample (usually about 1,000 grams) is covered with a saturated solution of sodium or magnesium sulfate for eighteen hours. Then it is oven-dried. The test is repeated usually five times. If the sample decomposes, it is called unsound; if it shows signs of decomposition but is still intact, it is called questionable. If it shows no or extremely minor signs of decomposition, it is called sound. In the

case of freeze-thaw tests, different-sized materials are tested by freezing in water to an air temperature of -22 degrees Celsius, then thawed at room temperature (about 20 degrees Celsius). A first cycle lasts twenty-four hours and may be repeated. Damage to the sample is then noted.

SEISMOGRAPHIC STUDY

Wave velocities, or speeds through rocks, are dependent on density and strength. These velocities can be determined theoretically or in the laboratory if density and strength can be measured. Conversely, compressive and shearing strength may be measured knowing rock density and velocity of wave transfer.

Geophysicists and engineering geologists may determine wave speeds through the Earth's crust by setting up a seismograph station in the field. Explosives are set off, and travel time and velocity of waves are monitored by a system of geophones that relay the wave energies to the seismograph, which then measures the waves on a recording drum with a stylus. Characteristic wave reflections occur at certain depths. By knowing the times of explosive detonation and the time of arrival of the two kinds of Earth body waves (primary and secondary waves) and by understanding that distance traveled for the two waves is the same, scientists can determined the wave speeds. This information is vital when drilling for oil or gas and digging deep foundations, especially in limestone sinkhole areas or in areas prone to earthquakes.

ENGINEERING AND CONSTRUCTION APPLICATIONS

Knowledge of particular rock physical properties is important to geologists, engineers, geophysicists, hydrologists, builders, architects, industrial city and regional planners, and the general public in a variety of ways. Rock properties are especially important in regard to the construction of buildings, dams, roads, airstrips, human-made lakes, and tunnels; drilling for petroleum and natural gas; mining and quarrying; monitoring earthquake hazards; measuring properties of aggregate and construction materials; determining rates of heat flow and mechanical strain in the Earth; monitoring radioactive hazards; studying chemical makeup and physical structure of the Earth's interior; and evaluating aesthetic properties of ornamental construction materials.

Within earthquake-prone areas (as in California) or areas affected by wind shear (as at the tops of high buildings), the nature and strength of material surrounding foundations, the foundations themselves, groundwater behavior and its effects on building strength, and the ability of building materials to withstand vibrations are of utmost importance. Engineering codes have been enacted within some of these areas so that earthquake-resistant or wind-shear-resistant buildings, dams, tunnels, and highways may be constructed.

David F. Hess

CROSS-REFERENCES

Andesitic Rocks, 1263; Anorthosites, 1269; Basaltic Rocks, 1274; Batholiths, 1280; Carbonatites, 1287; Continental Rift Zones, 579; Eruptions, 739; Granitic Rocks, 1292; Igneous Rock Bodies, 1298; Igneous Rock Classification, 1303; Island Arcs, 712; Kimberlites, 1310; Komatiites, 1316; Krakatau, 752; Lamproites, 1321; Magmas, 1326; Mount Pelée, 756; Mount St. Helens, 767; Orbicular Rocks, 1332; Plate Tectonics, 86; Plutonic Rocks, 1336; Pyroclastic Rocks, 1343; Recent Eruptions, 780; Silica Minerals, 1354; Stratovolcanoes, 792; Ultramafic Rocks, 1360; Volcanic Hazards, 798; Xenoliths, 1366.

BIBLIOGRAPHY

Birch, Francis, J. F. Schairer, and H. Cecil Spicer. *Handbook of Physical Constants*. Geological Society of America Special Paper 36. Baltimore: Waverly Press, 1942. An excellent, if older, compilation of physical and chemical properties of rocks and minerals. Some prefatory discussion is given with most of the tables. Especially important are discussions and tables concerning rock strength and wave propagation. Contains an extensive section on chemical properties of rocks and minerals. Suitable for college-level students.

Carmichael, Robert S., et al. *Practical Handbook of Physical Properties of Rocks and Minerals*. Boca Raton, Fla.: CRC Press, 1989. A more recent compilation of major physical and chemical properties of rocks and minerals. Of special interest are the sections on radioactive and electrical properties of rocks. Most of the tables are preceded by discussion of the particular property illustrated. Indispensable for geologists and engineers.

Dorn, Ronald I. *Rock Coatings*. New York: Elsevier, 1998. Dorn covers the physical properties of rocks and their distribution patterns on the surface of the Earth. The illustrations and maps are particularly useful, as is the extensive, detailed bibliography.

Gillson, Joseph L. *The Carbonate Rocks in Industrial Minerals and Rocks*. New York: American Institute of Mining, Metallurgical, and Petroleum Engineers, 1960. This source is a classic one, discussing physical properties, occurrence, uses, resource amounts, and quarrying methods for limestones, dolomites, and marbles. Includes an excellent discussion of tests for physical properties of these rocks and of other rock types as well. Also includes a few tables of physical properties. Suitable for high school readers but excellent at all levels.

Potts, P. J., A. G. Tindle, and P. C. Webb. *Geochemical Reference Material Compositions: Rocks, Minerals, Sediments, Soils, Carbonates, Refractories, and Ore Used in Research and Industry*. Boca Raton, Fla.: CRC Press, 1992. This advanced text focuses on the composition of rocks, determinative mineralogy, and analytical geochemistry as they apply scientific research and industrial use. Includes a useful bibliography.

Severinghaus, Nelson. *Crushed Stone in Industrial Minerals and Rocks*. New York: American Institute of Mining, Metallurgical, and Petroleum Engineers, 1960. Discussion in this source centers on the physical properties of good crushed stone, occurrence, economics of use, uses and methods of quarrying and processing, and testing. Some physical properties are listed, as well as tests for determining hardness, toughness, and strength of crushed stone and aggregate. Suitable for high school readers but excellent at all levels.

SILICA MINERALS

The silica minerals are a group of chemical substances composed largely of silicon dioxide and found mainly in rocks. Quartz is the most important and widely distributed mineral of this group. The opals are among the significant gem varieties of silica.

PRINCIPAL TERMS

CRYSTAL: a solid exhibiting flat surfaces or faces and consisting of atoms in a regularly repeating three-dimensional pattern

FELDSPAR: a family name for a group of common minerals found in such rocks as granite and composed of silicates of aluminum together with potassium, sodium, and calcium

MINERAL: a homogeneous inorganic substance with a definite chemical composition, distinctive physical properties, and characteristic crystal structure

POLYMORPHISM: the capacity of certain substances of the same chemical composition to occur as two more different crystal forms, as a result of different atomic arrangements

SILICA: a compound of silicon and oxygen whose chemical formula is SiO_2

SILICATE: a compound whose crystal structure contains tetrahedral units of silicon surrounded by four oxygen atoms (SiO_4)

TWINNING: a phenomenon in which two or more intergrown crystal parts, though they have the same atomic arrangement and properties, differ in orientation, for example, by rotation about a common axis

X RAY: a form of radiation that differs from light in its much shorter wavelength (and much higher energy)

OCCURRENCE OF SILICA MINERALS

Silica minerals are major constituents of the Earth's crust. In terms of mineral abundance, they are second to the feldspars. Silica minerals are found in such igneous rocks as granite, such sedimentary rocks as sandstone, and such metamorphic rocks as quartzite. Silica crystallizes in many forms, but quartz is by far the most common. This hard, glasslike mineral is so universally distributed that it occurs in rocks of almost all categories (with the exception of certain limestones and basalts). Geologists have discovered quartz in many colors and sizes, in microscopic crystals and in crystals weighing a ton or more. Quartz veins are widespread in many ore deposits, where they have often been transported by steam or hot waters. Silica minerals are even found in certain living things: Such primitive organisms as diatoms and dinoflagellates contain significant amounts of silica, as do such higher plants as the horsetails. Such animals as the glass sponges secrete spicules of silica, and silica has even been detected in human bones.

When pure, the silica minerals are colorless, transparent, and lustrous. They resist solution in most strong acids (except hydrofluoric acid); they are hard and strong; when subjected to mechanical stress, they shatter irregularly rather than cleave in definite crystal directions. Scientists explain this property of the silica minerals in terms of their crystal structure, which consists of silicate (SiO_4) tetrahedra yoked together through the corner oxygen atoms. Therefore, to break a quartz crystal, it is necessary to break some strong silicon-oxygen bonds, which accounts for the hardness of the mineral. The silica minerals are also nonconductors of electricity, and when placed in an inhomogeneous magnetic field, they tend to move away from the strong-field region (in other words, they are diamagnetic substances).

Silica also occurs in cryptocrystalline form. A cryptocrystalline mineral has crystal grains so small that they are difficult to see, even with a microscope. The silica minerals are often contaminated with impurities, and these impure forms are scattered throughout the world in many cryptocrystalline varieties, distinguished from one another mainly by their colors. These varieties are composed of microcrystals that were deposited long ago by precipitation from water solutions in the interstices and cavities of volcanic and sedimentary rocks. One of the important crypto-crystalline varieties is termed chalcedony after the historic Greek town of Chalcedon, where large amounts of this material were bought and sold. Chalcedony is composed of fibrous quartz that forms rounded crusts in the open spaces of igneous and sedimentary rocks. Among the gem varieties of chalcedony are agate and onyx. Agate, the most common variety, has conspicuous concentric color bands. Onyx is a type of agate in which alternate bands of color, often black and white, are parallel rather than curved.

POLYMORPHS

Silica minerals occur naturally in several different varieties or polymorphs, and others have been made synthetically. Despite the simplicity of their chemical formula (SiO_2), the structures of the polymorphs are rather complex. The principal differences among the various polymorphs derive from the detailed arrangement of the SiO_4 tetrahedra. For example, quartz has a relatively dense packing of silica tetrahedra compared to tridymite, which has fairly large open cavities. Therefore, the density of quartz is greater than that of tridymite. Quartz is the most abundant of the silica polymorphs, representing 12 percent by volume of the Earth's crust. Cristobalite and tridymite are uncommon, of interest to scientists for the puzzle of their origin, structures, and properties, and to collectors because of their rarity.

If any polymorph of silica is melted and the molten silica is then cooled, it generally fails to crystallize at the original melting point. Instead, as the temperature drops still further, the increasingly viscous liquid eventually turns into an amorphous glass. This material is called silica glass, or sometimes quartz glass or fused quartz (it is incorrect to call this vitreous silica quartz, as quartz is a crystal substance with properties vastly different from silica glass). Vitreous silica does not have the properties of a crystal—it neither cleaves nor forms crystal faces nor exhibits directional properties. The reason for this behavior is that its constituent atoms are not arranged in a rigid order, as in a crystal, but in a random way, as in a liquid. Indeed, glass is

a supercooled liquid so stiff that it cannot flow. Upon prolonged heating at temperatures exceeding 1,100 degrees Celsius, silica glass devitrifies—its glassy quality is destroyed and it becomes brittle and opaque—because the material has turned into minute crystals of chalky cristobalite.

QUARTZ AND TRIDYMITE

Each of the polymorphs of silica is stable within a range of temperatures and pressures. Quartz is the stable form of silica under atmospheric pressure and for temperatures up to 867 degrees Celsius. The transformation of quartz into a chalky white product as a result of prolonged heating at a very high temperature was a phenomenon recognized and made use of for centuries in ceramic and metallurgical processes, but it was not investigated scientifically until the end of the nineteenth century, when Henri Le Châtelier, a French chemist, discovered that quartz possessed an inversion temperature at which its properties changed abruptly. Specifically, he noticed that the thermal expansion of clay containing approximately 15 percent quartz changed distinctly in character at roughly 600 degrees Celsius. After further work, he attributed this peculiarity not to the clay but to the quartz. By measuring the thermal expansion of pure quartz, he discovered a sudden change in the size of his sample at 573 degrees Celsius. The form of quartz stable up to 573 degrees Celsius is called low-quartz (sometimes, alpha-quartz), and the high-quartz first observed by Le Châtelier (sometimes called beta-quartz) was later shown to be stable up to 867 degrees Celsius.

In 1926, the English physicist Reginald E. Gibbs found that the tetrahedral silica groups spiral around the long axis of the quartz crystal. When heated at 573 degrees Celsius, this structure opens, changing to the high-quartz form. The low-high (or high-low) transformations involve only slight displacements in bond directions and no bond breaking, so this change of the quartz polymorphs occurs rapidly and reversibly over a small temperature interval.

Because it involves the disruption of existing bonds and the subsequent formation of new ones, a sluggish transformation occurs at 867 degrees Celsius, when high-quartz changes into tridymite. In the late nineteenth and early twentieth centuries, scientists found that tridymite is the stable

Quartz is the most important and widely distributed mineral of the silica group. (© William E. Ferguson)

phase of silica above 867 degrees Celsius and below 1,470 degrees Celsius and that it exists in both a stable and an unstable variety. Because both varieties have low-high thermal inversion points, tridymite exists in a complex set of polymorphs. According to some scientists, tridymite must be regarded as a material of indefinite properties unless it has been carefully examined by X-ray diffraction or its thermal history is known. Some modern chemists believe that tridymite is not a silica mineral but most likely a solid solution in which silica and other compounds are dispersed through one another in a way analogous to liquid solutions.

CRISTOBALITE AND OPAL

In contrast to tridymite, the classification of cristobalite as a silica mineral is on much firmer ground. At atmospheric pressure, cristobalite is the stable form of silica above 1,470 degrees Celsius and below 1,723 degrees Celsius, its melting point. Cristobalite is more abundant in nature than is tridymite, and, like tridymite, cristobalite is a mineral often contained in cavities of volcanic rock. The German scientist G. vom Rath discovered, described, and named cristobalite in 1884. Its name derives from its discovery in rock specimens that vom Rath collected in Mexico's Cerro San Cristóbal. Cristobalite's polymorphs, called low- and high-cristobalite, were first observed in 1890. The interconversion of quartz and cristobalite on heating is sluggish, as it involves breaking and re-forming chemical bonds. In cristobalite, as in tridymite, the packing of the SiO_4 tetrahedra is much less constricted than in quartz; consequently, such foreign atoms as sodium and aluminum easily slip into their more open structures.

Traditionally, opal has been the most valued of the gem varieties of silica. Unlike quartz, opal is a form of silica that contains variable amounts of water. More specifically, opal is essentially an aggregate of microcrystals of cristobalite whose submicroscopic pores contain water. Chemical analysis of opal reveals amounts of water ranging from 4 to 20 percent, and these variable amounts cause many scientists to classify opal as a mineraloid (a mineral substance that is not truly crystalline because it does not have a definite chemical composition). Opal is one of the most beautiful gems formed by the silica minerals, and up to the nineteenth century, its monetary value was close to that of diamond. It owes its attractiveness to a spectrum of colors diffracted from the stone's interior.

COESITE, KEATITE, STISHOVITE

Since 1953, scientists have discovered three other polymorphs of silica. In the course of experiments on various mineral substances at very high pressures, Loring Coes, Jr., a research chemist in Worcester, Massachusetts, found a crystalline modification considerably denser than quartz, cristobalite, or tridymite, later named coesite for its discoverer. Although some mineralogists wanted to maintain the convention that names ending in "ite" should be reserved for natural minerals, chemists ignored the convention. With regard to coesite, the point became irrelevant when natural coesite was found at Meteor Crater in Arizona and at other places where large meteorites had once struck the Earth. Structural studies of coesite revealed that it contains chains of four-membered rings of silica tetrahedra.

Extremely high pressures also led to the discovery of two other polymorphs of silica. In 1954 while at Rutger University, Paul P. Keat produced the first new polymorph. The crystal structure of keatite, which was named for its discoverer, contains spirals of silica tetrahedra. Keatite has not yet been found in nature. In 1961, two Russian investigators, S. M. Stishov and S. V. Popova, working on the assumption that a denser phase of coesite was possible, heated a hydrated form of silica to temperatures of 1,200-1,400 degrees Celsius under pressure exceeding 160,000 atmospheres. They produced a new crystalline phase of silica that was almost twice as dense as quartz. Stishovite, named for its principal investigator, has the hardness of diamond and is the densest polymorph of silica. Its structure turned out to be unique among the silica minerals, for each silicon atom is octahedrally surrounded by six oxygen atoms. Stishovite has been found as a natural mineral in specimens from Meteor Crater and from other meteorite impact sites, and scientists have used this high-pressure mineral as an indicator of the shock wave associated with meteoritic impact or nuclear explosion. In sediments roughly 66 million years old, some scientists uncovered mineral deposits of stishovite resulting from a meteorite that crashed into the Earth at approximately the time that other scientists theorize an asteroid impact caused dinosaurs to become extinct.

DEVELOPMENT OF STUDY METHODS

Early mineralogists identified the silica minerals by their physical properties (hardness, density, cleavage, crystal form, and so on), and they determined the chemical compositions of these minerals by gravimetric methods, which involved dissolving samples and then selectively precipitating and weighing individual components. In the nineteenth century, research on optical methods of analysis had important implications for crystallography and, in particular, for the silica minerals. Polarized light proved to be very helpful in studying quartz crystals. Jean Baptiste Biot had discovered polarized light in 1816 when he passed a beam of ordinary light through a crystal of calcite and found that it split in two, each beam becoming plane-polarized (that is, the light's vibrating electric field lies in one plane). Because quartz is subject to chiral twinning, which consists of one twin portion of a crystal mirroring another, as in two gloves of a set, it became the object of investigations employing polarized light.

In the late nineteenth century, the development of thermal methods and optical microscopy both helped and hindered silica-mineral research. On the one hand, thermal methods of study were helpful in the discovery of high-quartz; on the other hand, the improved microscope brought about a multiplication in the number of new polymorphs of silica, advanced on the basis of optical characteristics but later discredited. In 1880, the brothers Jacques and Pierre Curie, while studying the electrical conductivity of certain crystals at the University of Paris, discovered the phenomenon now known as piezoelectricity. In measuring the electrical conductivity of quartz plates, they observed that pressure on the plates produced a deflection in an electrometer. The piezoelectrical property of quartz would later become practically beneficial when scientists found that quartz slices would vibrate at radio frequencies.

The application of X rays to the analysis of crystal structures in the twentieth century was the most important development in the history of crystallography and in the study of the silica minerals. In 1912, Max von Laue at the University of Munich discovered that X rays, in passing through crystals, are scattered from planes of atoms and on emerging yield information about the atomic structure of the crystal. The British father-and-son team of William Henry Bragg and William Lawrence Bragg amplified Laue's X-ray diffraction techniques in mapping the positions of atoms in many important crystals. Further application of X-ray diffraction by other crystallographers resulted in an understanding of the basic structural relations between the various polymorphs of the silica minerals. The identification of opal as a variety of cristobalite and the clarification of the relation of other cryptocrystalline varieties of silica to quartz and to cristobalite were also achieved primarily through X-ray methods.

GEOLOGIC AND INDUSTRIAL VALUE

The silica minerals are among the most important substances on Earth. They are important geologically because silica is a major component of most of the rock-forming minerals. They are valuable technologically for their essential role in construction, ceramics, and scientific instrumentation. Of all the silica minerals, quartz is clearly the most significant both in nature and in human civilizations. Its diverse forms and its abundance in igneous, sedimentary, and metamorphic rocks make it an extremely valuable material for scientists and for general use; millions of tons of it are used annually.

Construction industries consume the greatest amount of quartz. Because quartz is a major component of sand, and because sand is a major component of cement and mortar, much quartz ends up in buildings and roads. Sandstone and quartzite, which are used as building stones, also contain high percentages of quartz. Road and railway construction utilizes large quantities of both crushed sandstone and quartzite. Because of quartz's properties of low thermal expansion, high melting point, and low cost, the metallurgical industries, especially iron and steel, employ huge amounts of quartz in such refractory (heat-resistant) products as firebricks and foundry molds. Quartz is also extremely hard, and this property makes it very useful as an abrasive in sandpaper, sandblasting, and the polishing of glass, stone, and metal.

Although many industries can use the impure quartz in sand and common rocks, glass and porcelain manufacturers require pure quartz. These industries transform high-purity quartz into vitreous silica, which has low thermal expansion, high elasticity, and wide transparency to many wavelengths of light, especially ultraviolet light. These qualities of vitreous silica make it popular for lenses, optical fibers, precision-instrument components, and premium-grade chemical glassware. For example, silica glass is an essential part of optical instruments that employ ultraviolet light. Fibers of vitreous silica are also essential for such precision instruments as balances and galvanometers. Untwinned quartz crystals serve as oscillators to control radio frequencies. Such crystals are also commonly found in quartz clocks, where they help to drive electronic oscillators.

As in ancient times, beautifully colored varieties of quartz, such as the violet-hued amethyst, remain popular for their value as jewelry. Some common varieties find use as ornamental stones, whereas certain black opals and fire opals are treasured as precious gems.

Robert J. Paradowski

CROSS-REFERENCES

Andesitic Rocks, 1263; Anorthosites, 1269; Basaltic Rocks, 1274; Batholiths, 1280; Building Stone, 1545; Carbonatites, 1287; Cement, 1550; Dams and Flood Control, 2016; Earthquake Engineering, 284; Granitic Rocks, 1292; Igneous Rock Bodies, 1298; Igneous Rock Classification, 1303; Kimberlites, 1310; Komatiites, 1316; Lamproites, 1321; Land-Use Planning, 1490; Magmas, 1326; Minerals: Physical Properties, 1225; Orbicular Rocks, 1332; Plutonic Rocks, 1336; Pyroclastic Rocks, 1343; Radioactive Minerals, 1255; Radon Gas, 1886; Rock Magnetism, 177; Rocks: Physical Properties, 1348; Seismometers, 258; Stress and Strain, 264; Ultramafic Rocks, 1360; Xenoliths, 1366.

BIBLIOGRAPHY

Deer, W. A., R. A. Howie, and J. Zussman. *Rock-Forming Minerals.* Vol. 4, Framework Silicates. New York: John Wiley & Sons, 1988. The authors intend this book, which is part of a five-volume set, to be a reference tool for advanced students and research workers in the geological sciences. For each mineral, they summarize the best physical, chemical, and optical properties, and they also discuss the crystal structures as well as the rock types in which the mineral occurs. A fifty-page section in this volume treats the silica minerals in considerable detail. There is an extensive and excellent set of references at the end of this section.

Frondel, Clifford. *The System of Mineralogy of James Dwight Dana and Edward Salisbury Dana.* Vol. 3, Silica Minerals. 7th ed. New York: John Wiley & Sons, 1962. James Dwight Dana was a nineteenth century American geologist; his son Edward Salisbury Dana became a prominent American mineralogist of the early twentieth century. The father's System of Mineralogy, first published in 1837, set standards in the field, and the son took over the task of issuing later editions. Frondel's version is one of the more recent attempts to keep this classic work up to date and relevant. This volume describes extensively the mineralogy of the polymorphs of silica. The treatment is detailed and most suitable for students with some background in physics, chemistry, and geology, but the data collected by the author (who examined the periodical literature up to 1960) should be useful to a broader audience.

Frye, Keith. *Mineral Science: An Introductory Survey.* New York: Macmillan, 1993. This basic text, intended for the college-level reader, provides an easily understood overview of mineralogy, petrology, and geochemistry, including descriptions of specific minerals. Illustrations, bibliography, and index.

Heany, P. J., C. T. Prewitt, and G. V. Gibb, eds. *Silica: Physical Behavior, Geochemistry, and Materials Applications.* Washington, D.C.: Mineralogical Society of America, 1994. This lengthy, multiauthored volume is a good source of information on all aspects of silica minerals for those who have already read and understood introductory material on the subject. Illustrations and bibliography.

Hurlbut, Cornelius S., Jr. *Minerals and Man.* New York: Random House, 1969. Hurlbut, who was for years a professor of mineralogy at Harvard University, fulfilled a long-standing ambition in writing this book; its purpose is to capture the beauty and excitement of minerals for a wide readership. In this nontechnical work, Hurlbut emphasizes the stories of those minerals that helped to shape modern civilization, but he also discusses other minerals that enthrall him with their beauty or curiosity value. Quartz is the silica mineral that is given the most detailed treatment. A major attraction of this book is the many photographs, most of them in color.

Klein, Cornelis, and Cornelius S. Hurlbut, Jr. *Manual of Mineralogy.* 21st ed. New York: John Wiley & Sons, 1999. The authors designed this textbook to be an undergraduate's first course in mineralogy. They consequently stress the basic concepts that underlie the science.

They also devote considerable space to such aspects as the study of important rocks and minerals. Continuing in the tradition of the classic Dana handbook of the nineteenth century, Klein and Hurlbut use crystal chemistry as their unifying theme—the elucidation of the relationship between chemical composition, physical properties, and crystal structure.

Legrand, Andrae P., ed. *The Surface Properties of Silicas.* New York: John Wiley, 1998. This somewhat technical text includes essays on adsorption properties, fractal characteristics, and ionization of silica surfaces, as well as the health effects of silica. Illustration, bibliography, and index.

Sorrell, Charles A. *Minerals of the World.* New York: Golden Press, 1973. This field guide and introduction to the geology and chemistry of minerals is part of the popular and widely praised Golden Field Guide series. The author wrote this book to fill the gap between popular books and college textbooks. He discusses the chemical relationships and crystal structures of many minerals at an elementary level. The hundreds of full-color illustrations and diagrams provide the novice with pictorial aids for grasping complex concepts and structures, along with a reference catalog for mineral identification.

Sosman, Robert B. *The Phases of Silica.* New Brunswick, N.J.: Rutgers University Press, 1965. This book is an updated version of *The Properties of Silica* published by the American Chemical Society in 1927. The book's purpose was to bring together in one place, for the convenience of the ceramist, mineralogist, chemist, and petrologist, all the relevant information on the polymorphs of silica. This monograph went out of print in the 1930's. Sosman, a member of the School of Ceramics at Rutgers University, began his revision in 1947, but as his work progressed, he realized that, because of the immense amount of new material, he would be able to revise only the first fourteen chapters of *The Properties of Silica.* The book is intended for scientists, but the historical sections are accessible to general readers.

ULTRAMAFIC ROCKS

Ultramafic rocks are dense, dark-colored, iron- and magnesium-rich silicate rocks composed primarily of the minerals olivine and pyroxene. They are the dominant rocks in the Earth's mantle but also occur in some areas of the crust. Ultramafic rocks are important for what they contribute to the understanding of crust and mantle evolution. They also serve as an important source of economic commodities such as chromium, platinum, nickel, and diamonds, as well as talc and various decorative building stones.

PRINCIPAL TERMS

CRUST: the upper layer of the Earth, composed primarily of silicate rocks of relatively low density compared to those of the mantle

MANTLE: the middle layer of the Earth, composed of dense, ultramafic rocks, located from about 5-50 kilometers below the crust and extending to the metallic core

OLIVINE: a silicate mineral abundant in many ultramafic rocks; gem-quality, clear green olivine is called peridot

PERIDOTITE: a principal type of ultramafic rock, composed mostly of the minerals olivine and pyroxene

PYROXENE: a group of calcium-iron-magnesium silicate minerals that are important constituents of ultramafic and mafic rocks

DEFINITION

Ultramafic rocks are dense, dark-colored, iron- and magnesium-rich rocks that constitute a volumetrically and scientifically important class of Earth material. To qualify as ultramafic, a rock must contain 90 percent or more of the so-called mafic minerals, olivine and pyroxene. The term "mafic" is derived from "magnesium" and "ferric" (the latter meaning "derived from iron"); "ultramafic" is an appropriate term for these minerals as they are very rich in both magnesium and iron. Another class of common rocks in the Earth's crust, the mafic rocks (including the black lava called basalt), are also rich in magnesium and iron but not to the extent shown by ultramafic rocks. A typical basalt may contain an average of about 20 percent iron plus magnesium (as the oxides ferrous oxide and magnesium oxide), but typical ultramafic rocks may contain more than twice that amount. Thus, the prefix "ultra" is well deserved. In addition, ultramafic rocks are generally—but not always—lower in silica than are mafic rocks, which is particularly true for olivine-rich ultramafic rocks, which average less than 45 percent silica; basaltic rocks average about 50 percent silica.

ULTRAMAFIC ROCK CATEGORIES

Although petrologists (geologists who study rocks) recognize several types of ultramafic rocks, the type called peridotite is the most common and, thus, considered the most important of all ultramafic rocks. Peridotites consist of more than 40 percent of the light green mineral olivine, an iron-magnesium silicate. The gem-quality variety of olivine is called peridot, from which the term "peridotite" is derived. The remaining minerals consist of various pyroxene minerals or hornblende. Pyroxene is a group of silicate minerals that contain calcium, iron, and magnesium as principal constituents; hornblende is similar to high-calcium varieties of pyroxene, but it contains water as part of its chemical structure. Variable amounts of spinel-group minerals (oxides of chromium, magnesium, iron, and aluminum), garnet, and other minerals commonly occur in minor amounts in peridotites. Peridotite is the principal constituent of the Earth's upper mantle, which makes it one of the most abundant types of rock in the Earth. In the crust, however, peridotite is relatively rare, occurring only in specific geological environments. It is this property of peridotites, among others, that makes these rocks so important and interesting as subjects of scientific investigation.

Peridotite is not the only type of ultramafic rock. Another important category of ultramafic rocks is the pyroxenites, which, as the name implies, are rocks composed mostly of pyroxene (greater than 90 percent). Pyroxene is not one mineral but is rather a family of related minerals, all having single chains of silicon-oxygen tetrahedra as their central structural aspect. This mineral group consists of two principal structural-chemical types: the low-calcium orthopyroxenes (with orthorhombic crystal symmetry) and the high-calcium clinopyroxenes (with monoclinic crystal symmetry). Pyroxenites can have variable amounts of either of these two pyroxene types or may have one more or less exclusive of the other. Orthopyroxenites contain mostly orthopyroxene and clinopyroxenites contain mostly clinopyroxene; the term "websterite" is used to describe pyroxenites that contain an appreciable mixture of both minerals.

Peridotites can also be subdivided into different types based on their pyroxene components. Those that contain mostly orthopyroxene as their pyroxene type are called hartzburgites. If the major pyroxene is clinopyroxene, the rock is called wehrlite; if both pyroxenes occur along with olivine, the rock is called lherzolite. One special and fairly rare type of peridotite is dunite: a nearly pyroxene-free, olivine-rich peridotite (90 percent or greater olivine). Most peridotites from the mantle contain minor amounts of an aluminum-rich mineral, generally either plagioclase (calcium-sodium aluminum silicate), spinel (magnesium-aluminum oxide), or pyrope garnet (magnesium-iron aluminum silicate). Many of the peridotites formed in the crust by igneous processes commonly contain the chromiumiron spinel-group mineral called chromite, the principal ore of chromium metal.

ULTRAMAFIC ROCK ORIGINS

Geologists' knowledge of mantle rocks comes predominantly from either of two sources: chunks of rock torn from the walls of deep mantle conduits by molten magma that eventually carries these mantle "xenoliths" (Greek for "stranger rock") out of the Earth enclosed in lava flows, or thick slices of ultramafic mantle material that were thrust up to the surface during mountain-building processes. Scientific examination of these mantle samples shows that most of them are metamorphic rocks that crystallized under high pressures and temperatures in the solid state. Some mantle samples, however, show evidence of an earlier history of crystallization from molten magma.

Ultramafic rocks that formed in the crust mostly originated by purely igneous processes and thus show evidence of crystallization from magma. Most crustal ultramafic rocks are found in layered complexes: crystallized bodies of magma trapped deep within the crust that display a "layer-cake" structure, with different kinds of igneous rocks composing the various layers. The production of many kinds of rocks from a single parent magma is called differentiation; the differentiation involved in the formation of layered complexes largely results from the process of gravity differentiation. In this process, early-formed minerals sink to the lower reaches of the magma reservoir, with later minerals forming progressively higher layers until the magma is wholly crystallized. The parent magma from which all layered complexes form is basaltic in composition—the same material that forms the islands of Hawaii, for example, or the steep cliffs of the Columbia River gorge in Washington State. Because the iron-magnesium-rich (mafic) minerals, olivine and pyroxene, form relatively early in basalt magmas, these minerals collect in the first-formed, lower layers of layered complexes. Resulting rocks in these layers are typically peridotites and pyroxenites. The igneous rocks that form later, above the dark ultramafic rocks, consist of gabbros (plagioclase-pyroxene rocks) and other mafic rocks. Knowledge of the ultramafic and other rocks in layered complexes comes from the fact that some of these deep-seated, crystallized magma bodies have been uplifted in mountainous areas and uncovered by erosion.

MANTLE-TYPE ULTRAMAFIC ROCKS

Where one would travel to see examples of ultramafic rocks would depend upon which of the two major categories—mantle or crustal—is of interest. In North America, the best places to see mantle-type ultramafic rocks are the Coast Ranges and the Sierra Nevada of California and the Klamath Mountains of western Oregon. Belts of ultramafic rocks are also found in some areas of the Canadian Rocky Mountains and on Canada's east coast, particularly in Quebec and Newfoundland. Exposures of ultramafic rocks also occur at various locations within the Appalachian Mountains in the United States, particularly in Vermont and Virginia. Worldwide, the Mediterranean area lays claim to some of the most impressive areas of exposed mantle-derived ultramafic rocks, notably the Troodos complex in Cyprus and the Vourinos complex in Greece. Similar ultramafic belts extend in scattered exposures from France and Italy eastward to Turkey and Iran. Other prominent occurrences are known from India, Tibet, and China. Most if not all of these exposures are known as ophiolite complexes. Ophiolites are slices of ocean crust (basalt and sediments) and underlying mantle (mostly hartzburgite and lherzolite peridotite and some ultramafic gravity-stratified rocks) that become involved in mountain-building activities where two lithospheric plates (consisting of upper mantle and crust) collide. This collision area, called a subduction zone, is commonly located near continental margins, which is why ophiolites generally occur in mountain belts (like the Coast Ranges or Appalachians) that parallel a coast. The Mediterranean ophiolites are thrust up in an area that is being deformed by the collision of Africa (on a lithospheric plate moving slowly northward) with Eurasia. Unfortunately for the scientists wishing to study them, most ultramafic rocks in ophiolites, especially the peridotites, are altered to varying degrees to serpentine (a hydrated magnesium silicate similar structurally to clay), producing a metamorphic rock called serpentinite. It is produced before or during mountain building by the interaction of olivine with water, much of which probably consists of seawater trapped in peridotites and associated rocks as they lay below the ocean floor.

Another source of mantle-derived ultramafic rocks are the xenoliths brought to the surface by lava flows. These mantle samples are particularly valuable as objects for scientific study and as potential sources of gem-quality peridot because they commonly show few if any effects of alteration to hydrous minerals such as serpentine. Well known to scientists who study ultramafic rocks are the excellent xenolith localities at San Carlos, Arizona; Salt Lake crater, Hawaii; Sunset crater near Flag-staff, Arizona; and the garnet-bearing xenoliths incorporated in diamond-bearing kimberlite deposits in South Africa. Kimberlite deposits are chaotic masses of broken fragments of ultramafic rocks and their constituent minerals mixed with serpentine, other hydrated silicate minerals, and carbonate minerals (mostly calcite and calcium carbonate). They originate in the upper mantle, between about 150 and 200 kilometers depth. Propelled rapidly upward as carbon-dioxide- and water-charged mixtures of mantle and crustal rocks encountered during their ascent, they explode violently at the surface and form large craters. The eroded remnants of these kimberlite pipes (also called diatremes) are mined for diamonds, which form at very high pressures and temperatures in the mantle. Diamond-bearing kimberlites occur in the United States in Arkansas; diamond-free kimberlites occur in Missouri, Oklahoma, and Kansas.

CRUSTAL-TYPE ULTRAMAFIC ROCKS

Large layered complexes containing ultramafic rocks are exposed in only a few places on the world's continents. Probably the best example in the United States is the Stillwater complex, located in the Beartooth Mountains of western Montana. Estimated to have originally measured 8 kilometers thick, it is now exposed as a 5-kilometer-thick strip approximately 30 kilometers long. The Stillwater has been the object of intense scientific research over several years and is a source of economic chromium deposits. Platinum deposits also occur within the ultramafic rocks of the Stillwater. An even more intensely studied layered intrusive body is the Skaergaard intrusion of eastern Greenland. Discovered in 1930, it is exposed over an area of glaciated outcrops 3.2 kilometers thick. It is believed to represent basaltic magma trapped beneath a rift zone created as the North Atlantic ocean basin opened up about 50 million years ago. This magma differentiated into a complicated layered zone consisting of various kinds of peridotites, pyroxenites, and gabbros.

No discussion of layered ultramafic complexes would be complete without mentioning the famous Bushveld complex of South Africa. Measuring 270 by 450 kilometers in area, it is estimated to have been originally about 8 kilometers thick, like the Stillwater. Much of the world's supply of chromium comes from the chromite deposits associated with peridotites in this monstrous intrusive body. The Bushveld is also the world's largest source of platinum, contained in a 1- to 5-meter-thick assemblage of olivine, chromite, orthopyroxene, and sulfides known as the Merensky Reef. This single, thin unit extends for a total of 300 kilometers. Other notable layered complexes with well-studied ultramafic rocks are the funnel-shaped Muskox intrusion in northern Canada and the Great Dyke of central Zimbabwe. With an average width of 5.8 kilometers, the Great Dyke extends for 480 kilometers. Like the Bushveld, it is a source of chromium ore.

Another type of crustal ultramafic rock is the rare but very important komatiite ultramafic lava flow. Mostly restricted to very old Precambrian terrains (most komatiite ages lie between about 2.0 and 2.5 billion years), these rocks represent nearly completely melted mantle material, a feat that requires extremely high temperatures. Present temperatures in the upper mantle are not sufficient to produce komatiites; basalt magma is produced instead by much lower degrees of melting of mantle peridotite. The restriction of komatiites to Precambrian terrains suggests that the mantle was much hotter billions of years ago compared to more recent times. Excellent exposures of these ultramafic lava flows occur at Yilgarn Block, Australia, and the Barberton Mountains, South Africa.

ANALYSIS OF ULTRAMAFIC ROCK SAMPLES

Ultramafic rocks are studied using analytical techniques commonly applied to any kind of igneous or metamorphic rock type. These techniques include detailed microscopic analysis; analyses for major (1 percent or more by weight), minor (less than 1 percent), and trace (measured in terms of parts per million or parts per billion) elements; and analysis for various critical isotopes, such as those of rubidium and strontium, and the rare-earth elements neodymium and samarium. The latter isotope systems are the most commonly used to date ultramafic rocks by radiometric methods. What kinds of analyses are performed on any given ultramafic rock depends upon the type of ultramafic rock in question and the objectives of the specific research project. For example, ultramafic rocks from the mantle would probably not be approached from the same scientific aspect as

would ultramafic rocks formed in the crust. Analysis of mantle samples might reveal clues as to how the early Earth formed originally and then differentiated into a core, mantle, and crust. Because many igneous magmas now residing as igneous rocks in the crust (especially those of basaltic composition) were originally produced by partial melting of ultramafic mantle rocks, the study of mantle samples provides a glimpse into how the Earth's crust evolved. The detailed analysis of layered complexes, komatiites, kimberlites, and associated crustal rocks gives information about the complexities of igneous differentiation (making many kinds of rocks from a common parent) and also bears on the processes involved in crustal formation. These studies can also be applied to some ultramafic meteorites to show how other planets evolved in comparison to the Earth.

Because many crustal rocks ultimately originated in the mantle, mantle ultramafic rocks have been subjected to laboratory experiments that seek to simulate the production of various magmas in the mantle. In these experiments, mantle samples (or artificially concocted facsimiles) are heated to the point of melting at pressures calculated to occur at different depths in the mantle. These experiments have shown that different kinds of basaltic magma can be produced by simply changing the pressure (thus, the depth) at which mantle materials melt. Varying the proportions of constituents such as water and carbon dioxide can also produce different magma compositions for a given pressure. These experiments have been correlated with rocks collected in the field and their particular environments of emplacement. For example, experimental evidence shows that rocks like kimberlites and alkaline lavas (high in potassium and sodium, very low in silica) originate at great depth in areas of the mantle enriched in carbon dioxide, which helps to explain the explosive, gas-rich nature of kimberlite deposits and the high carbon dioxide contents in gases released upon extrusion by alkaline lavas.

Mantle samples also give scientists clues as to the very early history of the planet, because their geochemistry records the earliest differentiation events, including the formation of the nickel-iron core. The process of core formation was not perfect; some elements that could have gone to the core stayed in the mantle, later to be erupted to the crust by volcanoes and other igneous processes. At the same time, geochemists note that laboratory analyses of some other elements show abundances similar to certain very old, "primitive" meteorites called chondrites. These meteorites are believed to have crystallized from gas that made up the solar nebula, the gaseous cloud from which the solar system eventually evolved. Chondrites are considered primitive because they contain abundances and ratios of most elements that are similar to those measured for the Sun. Thus, chondrites are the rocky building blocks of most planets and asteroids. Analysis of mantle ultramafic rocks shows that their calcium-aluminum ratio, for example, is similar to that of chondrites, suggesting that mantle materials were originally composed of chondrites or chondrite-like material that later differentiated to make the core and crust. Interestingly, the mineralogical makeup of chondrites is similar to that of the mantle: mostly olivine and pyroxene. Much of the mantle, therefore, probably represents highly metamorphosed chondritic meteorites or similar precursor rocks.

COMMERCIAL USES

Although ultramafic rocks are not commonplace constituents of the Earth's crust, most people have seen examples of them, as they are used as building stone in large department stores, churches, banks, and various public buildings. The best known of these decorative stones is verde antique, which literally means "old green." Verde antique is considered to be a type of marble, but it mostly consists of dark green serpentine (a hydrous alteration of olivine) swirled together with white, gray, or pink calcite (calcium carbonate) that may also crosscut the serpentine as thin veins.

Ultramafic rocks may also serve as a source of gemstones. The semiprecious gemstone peridot originates as large, perfectly clear, green olivine crystals in peridotites, which can be incorporated into rings, bracelets, and necklaces, some of the finest examples being the silver with peridot jewelry made by members of the San Carlos (Arizona) Apache Tribe.

Perhaps the most important commerical use

for ultramafic rocks lies in their tendency to harbor rich deposits of certain metallic ores, namely chromium, vanadium, platinum, and nickel. Chromium and vanadium (in the mineral chromite) are mined from peridotite units, usually from chromite-rich seams. The Stillwater complex in western Montana and the Bushveld complex in South Africa are some of the richest sources of this ore. Platinum is mined as native platinum (metallic) and as various platinum sulfides from the Bushveld, as well as the Yilgarn Block area of Western Australia. Nickel is commonly associated with platinum deposits in ultramafic rocks, but large quantities of nickel are also contained in so-called lateritic nickel deposits. These deposits form by intensive weathering of ultramafic rocks in tropical areas, causing the nickel to be released from olivine as it reacts with water to form serpentine and clay. Important nickel laterite deposits occur in Cuba, the Dominican Republic, and Indonesia.

Finally, certain nonmetallic minerals, particularly talc, are also mined or quarried from ultramafic deposits. These include the talc deposits in Vermont and Virginia. In Virginia, the talc occurs in soapstone, a corrosion-resistant, durable stone commonly used for laboratory tables or for artistic carvings. Large talc deposits also occur in Italy and France.

John L. Berkley

CROSS-REFERENCES

BIBLIOGRAPHY

Ballard, Robert D. *Exploring Our Living Planet.* Washington, D.C.: National Geographic Society, 1983. This book is an excellent guide to the dynamic processes that shape the Earth. It does not have specific sections on ultramafic rocks, except for a brief item on the Bushveld complex. On the other hand, it is one of the best resources for the layperson on plate tectonics and volcanism, both of which involve ultramafic rocks. Extensively illustrated with color photographs and well-drafted, easily understood diagrams to show how the mantle is involved in processes of mountain building and volcanism. Includes an adequate glossary and extensive index.

Best, Myron G. *Igneous and Metamorphic Petrology.* 2d ed. Cambridge, Mass.: Blackwell, 1995. This is a college-level textbook, but it should be accessible to most general readers. Part 1 contains comprehensive treatments of all major plutonic rock bodies, complete with drawings, diagrams, and photographs detailing the essential features of each pluton type. Chapter 1 contains a section on how petrologists study rocks. The appendix contains chemical analyses of plutonic rocks and descriptions of important rock-forming minerals. One of the best books available for the serious student of igneous and metamorphic rocks.

Hall, Anthony. *Igneous Petrology.* 2d ed. New York: John Wiley & Sons, 1996. A standard text discussing the general aspects of the occurrences, composition, and evolution of igneous rocks and magmas. Each chapter discusses a different rock group. The role of water in magmas is emphasized in the sections on granites. Text is college-level, but the illustrations are useful to the general reader.

Press, Frank, and Raymond Siever. *Understanding Earth.* 2d ed. New York: W. H. Freeman, 1998. This is one of many introductory textbooks in geology for college students. It has a good chapter on igneous rocks.

Smith, David G., ed. *The Cambridge Encyclopedia of Earth Sciences.* New York: Cambridge University Press, 1981. This compendium of knowledge about the Earth is a useful reference for laypersons and specialists alike. The color and black-and-white photographs are excellent, as are the color diagrams and charts. The section on peridotites and other mantle rocks is sufficiently comprehensive to include stable isotopes and trace elements. Contains an extensive glossary and index.

Symes, R. F. *Rocks and Minerals.* Toronto: Stoddart, 1988. This book, written by Dr. Symes and the staff of the Natural History Museum of London primarily with the young reader in mind, is lavishly illustrated with colored photographs and diagrams on every page. The photographs of museum rock and mineral specimens are combined with descriptive text giving characteristics, occurrences, and uses of the illustrated specimens. Most of the rocks and minerals important to the subject of ultramafic rocks are described and illustrated, including olivine and pyroxene, peridotite and serpentine, and basalt/gabbro. A visual delight, this book is highly recommended.

Wyllie, Peter J., ed. *Ultramafic and Related Rocks.* Huntington, N.Y.: Robert E. Krieger, 1979. This compendium of several articles by a number of authors is one of the most comprehensive works on ultramafic rocks for more advanced readers. Every chapter is preceded by a foreword by Dr. Wyllie, summarizing the content and significance of the article. Has an extensive list of references (all pre-1967, the original publication year) and an author and subject index. This volume is an ideal source of seminal information on ultramafic rocks for the serious student, but there are a few topics not covered concerning ultramafic and associated rocks. Most college-educated persons should gain an appreciation for the significance and complexity of ultramafic rock associations by at least a cursory scanning of this book.

XENOLITHS

Xenoliths are blocks of pre-existing rocks within a magma. Consequently, xenoliths provide a sampling of materials through which the magma has traversed on its rise toward the surface. Some xenoliths originate within the mantle and provide the only means of obtaining samples of this elusive material.

PRINCIPAL TERMS

ASSIMILATION: the absorption of chemical components of wall rock or xenoliths into a magma

COUNTRY ROCK: rocks through which a magma is intruding; also known as "wall rock"

DIATREME: a pipelike conduit in the crust of the Earth filled with fragmented rock produced by gas-rich volcanic eruptions

ECLOGITE: rock composed principally of garnet and pyroxene that formed at high pressures associated with great depths

FELSIC ROCKS: igneous rocks rich in potassium, sodium, aluminum, and silica, including granites and related rocks

MAFIC ROCKS: igneous rocks rich in magnesium and iron, including gabbro, basalt, and related rocks

MAGMA: a naturally occurring silicate-rich melt beneath the surface of the Earth

MANTLE: the intermediate zone between the crust and the core of the Earth

PERIDOTITE: a class of ultramafic rocks made up principally of pyroxene and olivine, with subordinate amounts of other minerals

SEGREGATION: the concentration of early-formed minerals in a magma by crystal settling or crystal floating

ULTRAMAFIC: a term for any rock consisting of more than 90 percent ferromagnesium minerals, including olivine and pyroxene

OCCURRENCE OF XENOLITHS

Xenoliths are fragments of preexisting rocks that have been incorporated in a magma as it makes its way into higher levels of the crust. The term "xenolith" is derived from the Greek roots *xeno* and *lith,* meaning "strange or foreign" and "rock." These rock fragments are pieces of previously formed rocks that become incorporated into the magma and perhaps removed from their source as the magma moves. The xenoliths may retain their original identity with minor alteration, or they may be greatly altered by attendant heat and fluids present in the magma. Xenolithic inclusions may be preserved near to their original sources along the borders of an intrusive magma, or they may be carried for great vertical distances from where they originated. In this manner, fragments of deep crust and mantle material from as much as 200 to 300 kilometers below the Earth's surface have been brought to the surface by volcanic eruptions.

Xenoliths represent fragments of the rocks through which a magma has moved to its site of final emplacement and crystallization. They may be found in products of explosive volcanism such as volcanic tuff and breccia, within crystalline igneous rocks as in lava flows, and within shallow and deep-seated igneous rocks. Explosive volcanic materials are ejected by highly gas-charged eruptions that produce diatremes and maar-type volcanoes. These are volcanic craters in the form of inverted conelike or dishlike depressions in the surface surrounded by a rim of ejected deposits. Xenoliths are found as angular or rounded blocks embedded in ash tuffs or volcanic breccia in the rim and in the pipelike conduit underlying the crater. In certain types of basalt and related magmas, such as kimberlite, that originate deep in the mantle, rare fist-sized fragments of mantle material are transported upward from near the source of the magma origin. Fragments may also be collected from rocks traversed by the magma along its path of vertical ascent through the crust. Xenoliths in crystalline igneous rocks are embedded within the rock and are not exposed until erosion exposes the xenolith by removing overlying material.

Granite rocks typically contain large xenoliths of metamorphic or sedimentary rocks. Such xenoliths reflect the typical intrusive process that produces granites. In this process, subsurface magma chambers expand and move upward by physically plucking country rocks from the wall and roof.

WALL-ROCK XENOLITHS

Three basic varieties of xenoliths are recognized: wall-rock xenoliths, cognate xenoliths, and mantle xenoliths. Wall-rock xenoliths are represented by blocks and pieces of the adjacent country rock that have been incorporated into the magma. Cognate xenoliths are inclusion of chilled margins of the magma or comagmatic segregations in which early-formed crystals are segregated within the magma chamber and are later incorporated in more energetic magmatic motions. Mantle xenoliths are presumed to be pieces of the mantle that become incorporated in magmas that are formed by partial melting deep within the Earth.

Magmatic intrusions make room for themselves by three processes: forceful injection, stoping, and assimilation. Stoping occurs when the magmatic front advances by injection into fractures and surrounding blocks of country rock. These blocks may sink or float in the magma, and they may be slightly altered or totally assimilated within the magma depending on the characteristics of the magma and the wall rock. Most wall-rock xenoliths in felsic magmas do not move far from their source. Xenoliths are abundant near the margins of most intrusions. Because the margins are more likely to be losing heat and cooling at rates faster than the interior of the intrusion, the magma is more viscous, and xenoliths are less likely to move very far from the source. In contrast, fast-moving magmas in volcanic conduits often carry a wide variety of xenoliths from country rock traversed by the magma. In this way, a wide variety of crustal rocks cut by the volcanic vent may be brought to the Earth's surface.

Xenoliths usually show effects of the high temperatures to which they are exposed. Preexisting minerals within a xenolith react to form new minerals that are in equilibrium with the magma. Thus, most xenoliths are brought to a high-grade metamorphic state unless they are composed of high-temperature refractory minerals to begin with or unless they are exposed to high temperatures for short periods of time. The German term *Schlieren* is used to describe hazy, ill-defined streaks of nearly completely assimilated xenoliths. Materials caught up in low-temperature felsic magmas are more likely to be altered by reactive assimilation, in which there is an exchange of ions between the xenolith and the magma. Inclusion of large blocks of country rock may alter the composition of magma by enriching it with elements that were not originally abundant.

Buried salt beds are capable of plastic flowage in response to the weight of overlying rocks. Often the salt, which has a lower density than the enclosing sedimentary rocks, will rise many thousands of feet through overlying sediments to form salt domes or salt plugs. The rise of large masses of salt is very much like the rise of magma. Pieces of wall rocks and subsalt rocks may be incorporated into the salt as xenoliths. In

Amphibolite xenoliths in medium-grained pink granite, west of Whirlwind Lake, Northwest Territories. (Geological Survey of Canada)

this manner, salt domes in the Persian Gulf have brought up blocks of sedimentary, igneous, and metamorphic rocks from great depths in the crust. Ultramafic igneous xenoliths in the Weeks Island salt dome in southern Louisiana are thought to be fragments of mantle-derived ultramafic intrusions emplaced along fault zones prior to deposition of the salt.

COGNATE XENOLITHS

Cognate xenoliths, also called "autoliths," are xenoliths from parts of the magma that have previously crystallized. Magmas solidify over a wide range of temperatures. Large bodies of magma may require tens of thousands of years to crystallize fully. Material on the outer edge of the magma will cool and crystallize more rapidly than that of the interior, resulting in chilled margins. Elsewhere within the magma, early formed crystals in some magmas will either float or sink depending on the specific gravity differential with the magma. Feldspar crystals tend to float and collect near the top of the magma chamber, and mafic minerals such as olivine or pyroxene tend to sink to the bottom of the chamber. Some magmas undergo energetic degassing because of the reduction in confining pressure as they approach shallow levels in the crust. The rapid evolution of dissolved volatiles may disrupt previously crystallized portions of magma (chilled margins or crystal segregations) and mix solid cognate xenoliths with the mobile fluid phase.

MANTLE XENOLITHS

Perhaps the most exotic xenoliths are those that originate within the mantle. The mantle lies at depths of five to forty kilometers below the surface and extends down to the top of the outer core nearly three thousand kilometers beneath the surface. No drill has penetrated to the mantle; therefore, these materials are completely inaccessible for direct sampling. The probable composition and mineralogical makeup of the mantle is postulated from calculations of the density, pressure, and temperatures that exist at mantle depths and by comparisons with meteorites, which are pieces of asteroids that have been fragmented by collisions with other asteroids. The interior of asteroids are thought to reproduce conditions similar to those of the mantle. Fortunately, pieces of

the upper mantle are delivered to the surface of the Earth as xenoliths in some magmas that originate by partial melting deep within the Earth. For these rocks to make it to the surface without significant alteration by the host magma, they must be delivered to the surface in a fairly short period of time. Thus, it is not surprising that mantle xenoliths are found in volcanic rocks associated with rift zones and interplate magmatic zones that allow rapid rise of gas-rich magmas to the surface.

Typical mantle xenoliths are composed of peridotite (olivine and pyroxene-rich rocks) incorporated in mafic volcanic rocks such as basalt. Basalts are formed from magma that originates by partial melting of mantle materials. They often incorporate xenoliths from their place of origin as well as fragments of crustal rocks torn from walls of the conduit along which they are rising. Most notable of these magmas are varieties of peridotites known as "kimberlites" and "lamproites." Kimberlite (mica peridotite) is a potassic ultramafic rock that occurs in intrusive pipes and plugs and in explosively formed volcanic craters that overlay them. Lamproite is a porphyritic, ultrapotassic ultramafic rock that occurs in dikes and small intrusions. Kimberlite and lamproite magmas typically contain up to 75 percent xenoliths and xenocrysts. Kimberlites contain abundant xenoliths of lherzolite—a mantle peridotite with magnesian olivine, pyroxene, and minor calcium-plagioclase, spinel, or garnet.

The range of xenoliths in basalt is more restricted than that in kimberlites, because basalts form at shallower levels of the mantle. Thus, basalts incorporate less of the upper mantle on their way to the surface. Spinel lherzolites are common in alkali basalts such as those found in Hawaii, Arizona, the Rio Grande Rift, and Central Europe. Basalts also contain xenoliths of harzburgite, dunite, and eclogite. Harzburgite (a peridotite with magnetite and spinel) and dunite (an ultramafic rock that consists of mostly olivine with minor chrome-bearing spinel as an accessory mineral) probably represent residual melts following fractionation of lherztolite.

The occurrence of mantle xenoliths in volcanics is limited to intraplate magmatic environments such as oceanic islands (Hawaii and Tahiti) and continental volcanic provinces or rifts (the southern Colorado Plateau and the Eifel District,

Germany). Kimberlite intrusions favor old, stable, thick continental crust (South Africa and central North America). These environments are characterized by simple plumbing systems in which magmas rise rapidly to the surface; otherwise, xenoliths would sink in the host magma. Xenoliths are more rare in complex systems found in interplate environments such as collisional magmatic provinces or island arcs.

SCIENTIFIC VALUE

The chief scientific value of xenoliths rests in their being samples of materials collected as magma ascended through the Earth's mantle and crust. Wall-rock xenoliths provide samples of country rock that remain close to their original source. More important, xenoliths in rapidly ascending, volatile-rich volcanic magmas may provide samples of crustal rocks from the walls of the conduit throughout its entire path. These materials are brought to the surface in relatively unaltered states. In addition to mantle xenoliths, some localities, such as Kilbourne Hole, New Mexico, and Williams, Arizona, contain significant xenoliths of granite gneiss representing crustal basement rocks. Cognate xenoliths provide information on the earliest parts of the magma to crystallize. Basalt, kimberlite, and lamproites contain xenoliths from the mantle. Distribution of nodule occurrences is not random. Siliceous basalts rarely contain nodules, whereas alkali basalts commonly contain eclogite and spinel peridotites. Kimberlites contain abundant nodules, including garnet peridotite. These differences suggest different depths of origin for the different magmas. Mantle xenoliths are the only means of obtaining samples of the mantle. Study of these materials can lead not only to determining the composition of the mantle but also to developing an understanding of its physical state and some of its processes.

From a study of these rare rocks, various processes and conditions of the lower crust and upper mantle can be inferred. The mineralogical combinations serve as geobarometers and geothermometers. Controlled crystallization studies at a variety of temperatures and pressures are employed to characterize the stable mineral assemblage within differing crustal and mantle environments. A mixture of minerals or rocks exposed to elevated temperatures, as found in the lower crust and upper mantle, recrystallize into a mineral assemblage that is in equilibrium with the higher temperatures. Or, by examining altered rocks, such as some altered xenoliths, it is possible to determine their prior state before metamorphism by the magma.

ECONOMIC VALUE

The sole economic value associated with xenoliths and xenocrysts is as a source of diamonds that formed deep within the mantle. Diamonds occur as xenocrysts in kimberlite, lamproite, and in alluvial gravels derived from kimberlites. Diamonds form at very high pressures. Minimum conditions required are pressure greater than 40 kilobars, which is equivalent to depths greater than 120 kilometers, and temperatures of approximately 1,000 degrees Celsius. Diamonds are associated with magnesia garnet-bearing lherzolites and coesite.

Diamonds make up only one part in twenty million of a typical diamond-bearing kimberlite, and many kimberlites are devoid of diamonds. By far, most diamond production is from kimberlites in West Africa, South Africa, and Siberia, but diamonds are also mined in Australia, Brazil, and India. Small concentrations of diamonds have been found in kimberlite and lamproite bodies in Arkansas, Wyoming, Montana, Michigan, and Canada. After unsuccessful attempts to develop mining operations at the lamproite body at Murfreesboro, Arkansas, the area has been turned into the Crater of Diamonds State Park, where several small diamonds are found by tourists each year.

René De Hon

CROSS-REFERENCES

Andesitic Rocks, 1263; Anorthosites, 1269; Basaltic Rocks, 1274; Batholiths, 1280; Carbonatites, 1287; Diamonds, 1561; Evolution of Earth's Composition, 386; Experimental Petrology, 468; Granitic Rocks, 1292; Igneous Rock Bodies, 1298; Igneous Rock Classification, 1303; Kimberlites, 1310; Komatiites, 1316; Lamproites, 1321; Magmas, 1326; Metamorphic Mineral Deposits, 1614; Ophiolites, 639; Orbicular Rocks, 1332; Platinum Group Metals, 1626; Plutonic Rocks, 1336; Pyroclastic Rocks, 1343; Rocks: Physical Properties, 1348; Silica Minerals, 1354; Ultramafic Rocks, 1360.

BIBLIOGRAPHY

Best, Myron G. *Igneous and Metamorphic Petrology.* 2d ed. Cambridge, Mass.: Blackwell, 1995. This is a college-level textbook, but it should be accessible to most general readers. Part 1 contains comprehensive treatments of all major plutonic rock bodies, complete with drawings, diagrams, and photographs detailing the essential features of each pluton type. Chapter 1 contains a section on how petrologists study rocks. The appendix contains chemical analyses of plutonic rocks and descriptions of important rock-forming minerals. One of the best books available for the serious student of igneous and metamorphic rocks.

Dawson, J. B. *Kimberlites and Their Xenoliths.* New York: Springer-Verlag, 1980. A technical treatment of xenoliths and their scientific significance as samples of the mantle.

Legrand, Jacques. *Diamonds: Myth, Magic, and Reality.* New York: Crown, 1980. This amply illustrated book provides comprehensive coverage of diamonds, including their worldwide occurrence, geology, crystallography, mining, and cutting.

Mitchell, Roger H. *Kimberlites, Orangeites, and Related Rocks.* New York: Plenum Press, 1995. Mitchell provides a good introduction to the study of kimberlites and related rocks, including an extensive bibliography that will lead the reader to additional information.

_____. *Kimberlites, Orangeites, Lamproites, Melilitites, and Minettes: A Petrographic Atlas.* Thunder Bay, Ont.: Almaz, 1997. This book covers the worldwide distribution of kimberlites and related rock types. Color illustrations and bibliography.

Morris, E. M., and J. D. Pasteris. *Mantle Metasomatism and Alkaline Magmatism.* Boulder, Colo.: Geological Society of America, 1987. This collection of papers presented at the Symposium on Alkali Rocks and Kimberlites provides a technical discussion of the chemistry, mineralogy, and petrology of the mantle as determined from mantle xenoliths.

Nixon, Peter H., ed. *Mantle Xenoliths.* New York: John Wiley & Sons, 1987. This collection of highly technical articles covers a wide range of topics that are related to kimberlites. Considerable attention is paid to regional kimberlite occurrences and to the foreign rocks that are brought up by the kimberlite diatremes. Several of the articles are general enough to suit a beginning reader, but the work is best suited for the undergraduate and graduate student.

Raymond, L. A. *Petrology: The Study of Igneous, Sedimentary, and Metamorphic Rocks.* Dubuque, Iowa: William C. Brown, 1995. This book, written for undergraduate students, provides comprehensive coverage of the field of petrology. Xenoliths are discussed in the sections on igneous rocks.

Sinkanka, John. *Gemstones of North America.* Vols. 1 and 2. Princeton, N.J.: Van Nostrand, 1959 and 1975. These companion books, written for the amateur collector, describe gemhunting localites in North America, including several diamond localities in the United States.

3
METAMORPHIC ROCKS

ANATEXIS

As a crustal process, anatexis records the thermal culmination of prograde regional metamorphism and produces most of the granitic plutons and migmatites in mobile belts less than 600 million years old. Anatectic melting probably accounts for the major tracts of granulite facies rocks in Precambrian shield areas.

PRINCIPAL TERMS

FACIES: a part of a rock or group of rocks that differs from the whole formation in one or more properties, such as composition, age, or fossil content

LIQUIDUS TEMPERATURES: the temperature at which the last crystal of a rock disappears into the melt phase

MAGMA: molten rock material that crystallizes (solidifies) to form igneous rocks

METAMORPHIC CULMINATION: the point in time when the thermal maximum is reached during a prograde metamorphic event

MINIMUM MELT: the composition of the initial melt, formed at the solidus temperature, during progressive heating of a rock

MOBILE BELT: a linear belt of igneous and deformed metamorphic rocks produced by plate collision at a continental margin

PROGRADE METAMORPHISM: recrystallization of regional-scale solid rock masses induced by rising temperature

REFRACTORY MINERAL: a mineral with a sufficiently high melting temperature that is unaffected by anatexis and remains in the solid residue

REGIONAL METAMORPHIC FACIES: the particular pressure-temperature conditions under which metamorphism occurred

RETROGRADE METAMORPHISM: the reversal of prograde mineral reactions caused by the reintroduction of water and/or carbon dioxide during the period of declining temperature following metamorphic culmination

SOLIDUS TEMPERATURE: the temperature at which melting begins

PARTIAL MELTING

Anatexis refers to any process that leads to gradual melting of preexisting crustal or mantle rocks. In general, solid rocks near but below their solidus temperatures can be induced to melt in response to a temperature increase or a pressure decrease (decompression) or by a change in chemical composition (especially the addition of water)— or combinations of these processes. Since all common rocks are granular aggregates of several mineral species, each with its own melting behavior, melting of the total rock is a heterogeneous process. Therefore, a rock subject to slow progressive heating will melt in a stepwise fashion over a temperature interval determined by the melting points of the least and most stable minerals in the rock. The temperature at which melting begins is the solidus temperature; the temperature at which the last crystal disappears into the melt phase is the liquidus temperature. For most rocks, this melting interval extends over a range of about 200-250 degrees Celsius. Between the solidus and liquidus temperatures, a rock system consists of a melt, which is enriched in the more soluble and lowest-melting-point components, and a coexisting solid residue of the more refractory minerals. This condition of partial melting or partial fusion has become synonymous with anatexis and is responsible for the generation of magma in both the mantle and the crust.

Experiments have clearly shown that anatexis plays a major role in the production of basaltic magma from peridotite parent rock in the Earth's upper mantle. Most discussions of anatexis, however, concern processes operating in the crust in which partial melting is viewed as the culmination of regional metamorphism. In this context, the term "ultrametamorphism" is also used synonymously with "anatexis." As a crustal phenomenon, anatexis occurs under confining pressures that generally are less than 10 kilobars (roughly the pressure at a depth of 35 kilometers). This limit could extend as high as 15 kilobars if exceptional crustal thickening is produced by the governing plate-collision event. Experiments indicate that melting will begin in a wide variety of common metamorphic rocks under moderate pressures (4-7 kilobars) when the temperature reaches the vicinity of 700 degrees Celsius. These temperature-pressure conditions are attained only near the upper limit of the amphibolite facies of regional metamorphism. Upper amphibolite facies rocks, in general, mark the highest grades of metamorphism attained in Phanerozoic mobile belts (those less than 600 million years old). Thus, no crustal rock may be subjected to anatexis without first undergoing progressive metamorphism through greenschist facies and amphibolite facies conditions.

SOLID-MELT RATIOS

A large amount of experimental data show that the initial composition of an anatectic melt is that of granite. Granite is the "minimum melt" in any parent rock that contains both quartz and feldspar. In the region where melting occurs, the product will be a mixture of granitic melt and solid residual material. The melt-to-solid ratio depends upon the degree of melting (1 percent, 5 percent, 20 percent, and so on), which is a function of temperature, confining pressure, bulk composition of the parent rock, and the availability of free water. Among these controlling variables, water content is paramount because even small quantities of water (relative to rock saturation) significantly depress the solidus of a rock and therefore facilitate anatexis. Given enough free water, almost any rock will melt during high-grade metamorphism; under dry conditions, anatexis is virtually impossible in the crust.

The solid-melt mixture resulting from a significant degree of partial melting forms a low-density "mush" with a strong tendency to rise from the melt site, en masse, and form granitoid plutons or even volcanic rocks if it can successfully reach the surface before complete crystallization. The dynamics of rising magma bodies and their crystallization history in transit are major subjects of study in igneous petrology. The metamorphic petrologist, however, is concerned with the conditions that lead up to melting and the geological traits of the melt site itself.

AMPHIBOLITE FACIES ROCKS

In deeply eroded mobile belts of Phanerozoic age, conditions of peak metamorphic intensity are best recorded by pelitic schists of the upper amphibolite facies. These "pelites" are the metamorphic equivalents of common clay-rich shales and mudstones transformed into garnet-mica schists containing the distinctive zone mineral sillimanite. If this sillimanite zone is traversed in the direction of increasing metamorphic intensity, one normally encounters pods, veins, lenses, and small, discontinuous layers of granitic rock with increasing frequency. These rocks of "mixed" igneous-metamorphic appearance are migmatites, the origin of which has been debated for nearly a century. Migmatites are classified on the basis of the relationship between the leucosome (light-colored, granite-looking material) and the melanosome (dark mica-rich/amphibole-rich material with metamorphic texture).

There is broad agreement on the anatectic origin of migmatites and associated S-type granites (granites with a sedimentary parentage) in the sillimanite zone of regional metamorphic belts. For these migmatites, the leucosome is interpreted as the crystallization product of an anatectic melt, and the melanosome is interpreted as the refractory components of parent rock depleted in low-melting-point constituents (essentially quartz and alkali feldspar). The salient feature of migmatites as a rock class is the bewildering range of structural relationships existing between the leucosome and melanosome components. These very complexities are interpreted by some geologists as further evidence of origin by anatexis. They argue that a correlation exists between the degree of partial melting of the parent rock and the degree of complexity and intimacy manifested by the leucosome and melanosome. A common pattern is that of stromatic migmatites, in which the leucosome forms thin, discontinuous layers more or less parallel to the schistosity of the enclosing rock. Veinitic migmatites exhibit a melanosome densely traversed by a network of tiny leucosome veinlets. Nebulitic migmatites, common in some localities, are characterized by poorly defined "patches" of leucosome that pass by gradations, over a transition distance of a few centimeters, into vaguely defined melanosome patches. It is important to bear in mind that migmatites are exceptionally diverse in their features and mode of occurrence. The majority occur in vast Precambrian terranes where evidence for a simple origin by anatexis is lacking or equivocal. Such migmatites may be products of a number of subsolidus mechanisms; among these are metasomatism, metamorphic segregation coupled with polyphase folding, and injection by unrelated granitic magmas followed by intense deformation.

GRANULITE FACIES ROCKS

In contrast to Phanerozoic mobile belts, ancient Precambrian shield terranes are dominated by metamorphic rocks of basic composition that record temperature/pressure conditions of the granulite facies. Most of these terranes are bounded either by major faults or profound unconformities and do not exhibit rocks transitional to the lower-grade amphibolite facies. Minor lenses and zones of amphibolite facies rocks do occur within granulite terranes, but these are almost invariably the result of postmetamorphic rehydration of the granulites. Typical Archean (older than 2.5 billion years) and Proterozoic (older than 600 million years) granulites exhibit compositional layering deformed by large-scale recumbent folds. Migmatites, in the form of granitic lenses, are commonly present within layered granulite sequences, but there are no transitions from high-grade migmatized rocks to unmigmatized rocks of lower grade, as in Phanerozoic mobile belts. In fact, there is rarely an indication of the downgrade or upgrade directions. Pelitic and carbonate rocks are conspicuously rare or absent in granulite terranes.

The field relations and traits of Precambrian granulites have long puzzled geologists. Their study is seriously hindered by the lack of stratigraphic markers (destroyed by deformation) and the rarity of areas recording the transition to amphibolite facies conditions. Darjeeling, India, and Broken Hill, Australia, are two localities where this important transition has been recognized and studied. Granulite facies rocks are recognized by a definitive mineral assemblage which, for basic rocks, consists of clinopyroxene, orthopyroxene, and plagioclase. This assemblage may be accompanied by a variety of minor phases, including quartz, garnet, cordierite, sillimanite, kyanite, alkali feldspar, calcite, and olivine. Hornblende and biotite are absent or, at most, rare. This mineralogy reveals the outstanding characteristic of all granulite facies rocks: They are virtually anhydrous. They are, in fact, the driest rocks on Earth.

Laboratory melting experiments indicate that the refractory residue produced by a significant degree (perhaps 20-30 percent) of partial melting would possess properties similar to those of basic granulite facies rocks. These include anhydrous character, low radioactivity, relatively high density, and refractory mineralogy. The key link between anatexis and granulites is the behavior of water during melting. Prograde reactions under upper amphibolite-granulite facies conditions will certainly liberate small amounts of water by the successive destruction of muscovite, biotite, and hornblende (all hydrous phases). Once liberated, this water depresses the solidus temperature of

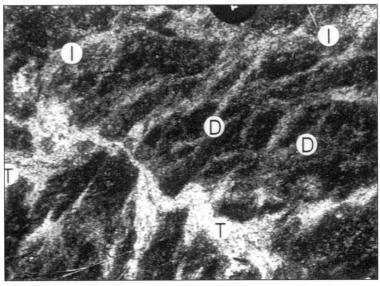

Anatectic melting probably accounts for the major tracts of granulite facies rocks in Precambrian shield areas, as in this outcrop showing different types of granulite veins (identified by letter markings). (Geological Survey of Canada)

the enclosed rock and permits melting. As temperature increases at any given pressure in the range 4-10+ kilobars, water is released in a stepwise fashion by the dehydration reactions, inducing further melting. This situation has the appearance of a runaway "crustal meltdown," and it would be, except for one important factor. Water, in the presence of an undersaturated silicate melt, is very efficiently absorbed by that melt, which effectively removes it from the presence of the restite. This stabilizes or even raises the solidus temperature of the rock enclosing the melt. At these pressures, granitic melts can absorb nearly 10 percent water (by weight); as a result, the rate of melting rapidly diminishes to zero. If the water-bearing granitic melt is capable of upward movement, it leaves behind a highly desiccated, refractory residue such as granulite facies rocks. The rarity of pelitic and carbonate lithologies in granulite terranes appears to be the result of metamorphism, dehydration reactions, anatexis, and the mobilization of granitic magma. Increasing evidence suggests that Precambrian granulite terranes represent the "root zones" of major batholith systems that have since been removed by erosion. Seismic data from the root zones of present-day batholiths such as the Sierra Nevada tend to support this hypothesis.

EVIDENCE OF ANATAXIS

Significant volumes of granitic migmatites and plutons are normally encountered in zones of highest metamorphic grade in Phanerozoic mobile belts. Vast amounts of experimental data indicate that, given the availability of water, prograde metamorphism to upper amphibolite facies conditions should culminate with anatexis. In many well-studied localities, granites and migmatites are restricted to upper amphibolite facies host rocks (including the northeastern United States, southern Greenland, Scotland, and the Black Forest of Germany). In the past, the mere spatial association of these rocks was accepted as evidence of crustal fusion during prograde metamorphism (that is, ultrametamorphism), but that is no longer the case. Modern studies have returned to the field in the classic localities to find evidence for availability of sufficient water to form the exposed granites.

The formation of sillimanite, a distinctive needlelike mineral, is a key factor in modern field studies. In most areas, sillimanite appears initially by the simple reaction: kyanite → sillimanite. Traversing up-grade from the rocks which are recording that reaction, researchers frequently discover that the rocks have undergone a second sillimanite-producing reaction that involves the breakdown of muscovite: muscovite + quartz → sillimanite + potassium feldspar + water. By careful study, the geologist draws lines on a geological map, delineating the appearance of sillimanite by both of these reactions. These lines are known as the first sillimanite isograd and the second sillimanite isograd, respectively. The second reaction is the key water-producing reaction that must precede melting if it occurs. It is demonstrated by mapping the first appearance of the pair sillimanite + potassium feldspar. Migmatites frequently appear very close to this isograd and are usually interpreted as evidence for the reaction: muscovite + quartz + water → sillimanite + melt. This third reaction constitutes anatexis, or ultrametamorphism, and could be expected to occur at depths of about 15 kilometers and temperatures in the vicinity of 700 degrees Celsius if water is available. This reaction must cease if the melt absorbs all the available water. Such is the paradoxical relationship between anatexis and regional metamorphism. The melt formed by the third reaction would be close in composition to the granite minimum but, because it would likely contain a refractory assemblage of biotite, hornblende, garnet, and plagioclase crystals, its bulk composition (melt + solid residue) could be closer to granodiorite. If the second and third reactions can produce sufficient melt volume, the melt may segregate, collect, and rise a small distance to form a granitic pluton. If it does not segregate and collect efficiently, migmatites are the likely result. Melts that remain immobile until they are water-saturated have a very limited capacity for upward movement (decompression induces crystallization). The potential for rising is also limited by the fact that as the second and third reactions are taking place, the entire rock mass is being intensely deformed by regional compression.

Beyond demonstrating a water-producing reaction such as the second reaction, the case for ultrametamorphism also requires proof that the appearance of granitic plutons is synchronous with, or slightly later than, metamorphic culmination and that the resulting granitic rocks have S-type traits (traits that indicate sedimentary parentage). Syntectonic plutons are recognized by their poor development or lack of thermal aureoles, semiconcordant contacts with enclosing rocks, and shared late-stage deformational features with their metamorphic parent rocks. The S-type characteristics will be reflected in the rock chemistry, in the mineralogy (especially the presence of garnet, cordierite, and muscovite), and by abundant unmelted inclusions of metamorphic rock.

ROLE IN CRUSTAL GROWTH

All rocks are subject to partial melting, or anatexis. In fact, the Earth's crust developed largely in response to partial melting of the upper mantle. Once a stable crust developed, its maximum temperature and thickness were limited by anatectic melting of the crustal rocks. The plate tectonics theory provides a satisfying basis for understanding large-scale mass transfer between the crust and mantle and the critical role played by anatexis. In all spreading-plate systems, anatectic magmas are generated at both leading and trailing plate margins. At present, new oceanic crust is forming at a rate of 3×10^{16} grams per year along oceanic-ridge axes. This new crust is basaltic in composition and derived by partial melting of underlying mantle peridotite. Additions to the continental crust occur on an even larger scale along continental margins in plate-collision settings. The most obvious manifestation of crustal growth in collisional settings is intensive, andesite-dominated volcanism, such as that currently taking place in the Andes of South America. Less obvious but more important contributions to crustal growth occur beneath the thick volcanic cover of youthful mobile belts such as the Andes. At depth, deformed sequences of metasedimentary rocks are intruded by gigantic granitic batholiths of anatectic origin.

As a crustal process, anatexis is logically viewed as the ultimate stage of progressive metamorphism and is often termed ultrametamorphism. In mobile belts, partial melting of metasedimentary rocks occurs at intermediate depths (am-

phibolite facies conditions) to form S-type granitic batholiths such as those that dominate the mountainous terrains of Western Europe and the Far East. These batholiths represent recycled crustal material rather than mantle-derived additions to the crust. They are, nevertheless, major components of many mobile belts and hosts to important metallic ores (tin and tungsten especially). In contrast, I-type granitic batholiths are produced by anatexis of metaigneous rocks at substantially greater depths (granulite facies conditions). The production of I-type granitoids is poorly understood but most likely involves melt contributions from the lowermost crust, the subducted slab associated with plate collisions, and the "mantle wedge" separating these two regimes. Most authorities agree that I-type plutonism involves significant volumes of mantle-derived melt and is, therefore, a major factor in crustal growth.

The production and destruction of oceanic crust appears to be balanced on a global basis (that is, subduction cancels ocean-ridge volcanism), but it is clear that the ratio of continental to oceanic crust has not remained constant throughout geological time. The past 2.5 billion years have witnessed major, episodic, lateral growth and thickening of continental crust around each of the world's ancient Precambrian shields. This record of growth clearly requires sustained production of I-type granitic magmas by anatexis.

Gary R. Lowell

CROSS-REFERENCES

BIBLIOGRAPHY

Best, Myron G. *Igneous and Metamorphic Petrology.* 2d ed. Cambridge, Mass.: Blackwell, 1995. A popular university text for undergraduate majors in geology. A well-illustrated and fairly detailed treatment of the origin, distribution, and characteristics of igneous and metamorphic rocks. Chapter 4 treats granite plutons and batholiths, and chapter 12 treats a broad spectrum of metamorphic topics.

Fyfe, W. S. "The Generation of Batholiths." *Tectonophysics* 17 (March, 1973): 273-283. An influential article by one of the major figures in petrological research. An overview of anatectic thinking that can be understood by college-level readers familiar with igneous and metamorphic processes. The periodical *Tectonophysics* will be found in major university libraries.

Hopgood, A. M. *Determination of Structural Successions in Migmatites and Gneisses.* Boston: Kluwer Academic, 1999. Hopgood provides a fairly technical book for geologists and students who are interested in the interpretation of complex structural relationships in highly deformed deep crustal rocks. Includes a bibliography and indexes.

Presnall, D. C. "Fractional Crystallization and Partial Fusion." In *The Evolution of the Igneous Rocks: Fiftieth Anniversary Perspectives*, edited by H. S. Yoder, Jr. Princeton, N.J.: Princeton University Press, 1979. A technical article dealing with theoretical aspects of rock melting and crystallization of silicate melts. Not for readers without a strong background in geology or chemistry.

Press, Frank, and Raymond Siever. *Understanding Earth.* 2d ed. New York: W. H. Freeman, 1998. This is one of many introductory geology textbooks for college students. Several sections provide readers with no background in geology an introduction to igneous and metamorphic processes, including the concept of partial melting and a discussion of migmatites.

White, A. J. R., and B. W. Chappel. "Ultrametamorphism and Granitoid Genesis." *Tectonophysics*, November 15, 1977: 7-22. An excellent article providing an overview of S-type and I-type granite genesis by the authors who first recognized these distinctions (now regarded as fundamental). Not overly technical, but requires some background in geology. This particular issue is entirely devoted to anatexis (ultrametamorphism), so those interested in the topic should scan its contents. This periodical is found in major university libraries.

Wiley, P. J. "Petrogenesis and the Physics of the Earth." In *The Evolution of the Igneous Rocks: Fiftieth Anniversary Perspectives*, edited by H. S. Yoder, Jr. Princeton, N.J.: Princeton University Press, 1979. Reviews fifty years of progress in experimental petrology. Production of basaltic magma (page 506) and granitic magma (page 511) by partial melting are concisely summarized from the experimental viewpoint. Technical—requires prior knowledge of melting phenomena.

Winkler, Helmut G. F. *Petrogenesis of Metamorphic Rocks.* 3d ed. New York: Springer-Verlag, 1974. A traditional text for undergraduate majors in geology. Chapter 18 deals with anatexis, migmatites, and granite magmas in some detail. This reference should be consulted after Press and Siever and before the more technical works.

Yardley, Bruce. *An Introduction to Metamorphic Petrology.* New York: Halsted Press, 1989. An excellent undergraduate text, with a good index and very extensive, current references. Chapter 3 deals with metamorphism of pelitic rocks, including anatexis and migmatization. The author is a prominent leader in migmatite research.

BLUESCHISTS

Blueschists are a class of metamorphic rocks that recrystallize at depths of 10 to 30 kilometers or more where ocean-floor-capped lithosphere rapidly subducts into the Earth's interior. Blueschists are important because they contain minerals indicating that metamorphism occurred under conditions of unusually high confining pressures and low temperatures. Their presence in mountain belts is the primary criterion for recognizing ancient subduction zones.

PRINCIPAL TERMS

ACCRETIONARY PRISM: the accumulation over time of variably deformed and metamorphosed sediments and ocean islands near a trench at a subduction zone

GEOTHERM: a curve on a temperature-depth graph that describes how temperature changes in the subsurface

LITHOSPHERE: the outer rigid shell of the Earth that forms the tectonic plates, whose movement causes earthquakes, volcanoes, and mountain building

METAMORPHISM: the alteration of the mineralogy and texture of rocks because of changes in pressure and temperature conditions or chemically active fluids

PROGRADE and RETROGRADE METAMORPHISM: metamorphic changes that occur primarily because of increasing and then decreasing temperature conditions

RECRYSTALLIZATION: the formation of new crystalline grains in a rock

SUBDUCTION: the process of sinking of a tectonic plate into the interior of the Earth

TECTONISM: the formation of mountains because of the deformation of the crust of the Earth on a large scale

TRENCH: a long and narrow deep trough on the sea floor that forms where the ocean floor is pulled downward because of plate subduction

VOLCANIC ARC: a linear or arcuate belt of volcanoes that forms at a subduction zone because of rock melting near the top of the descending plate

MINERAL CONTENT

Blueschists are a distinctive class of metamorphic rock containing one or more of the minerals lawsonite, aragonite, sodic amphibole (glaucophane), and sodic pyroxene (omphacite and jadeite plus quartz). These minerals indicate that recrystallization occurred in the temperature range of 150-450 degrees Celsius and pressures of 3-10 kilobars or more. Blueschists of basaltic composition typically contain abundant glaucophane, a mineral that can give a rock a striking blue color. Other minerals commonly found in blueschists include quartz, mica, chlorite, garnet, pumpellyite, epidote, stilpnomelene, sphene, and rutile. The abundance of these and rarer minerals depends, as it does in all metamorphic rocks, upon rock composition, the exact pressures and temperatures of recrystallization, and the nature of chemically active fluids that have affected the rocks.

STRUCTURAL FEATURES

Blueschists in one place or another display a remarkably wide variety of structural features. In the field, many are complexly deformed, with intricate folding and refolding of compositional layering at scales of millimeters to tens of meters. Commonly, folded rocks also display a thickening and thinning of the layering, forming an interesting structure known as boudinage. Many blueschists are faulted, some so intensely that they are fragmented rocks known as breccias. When flaky minerals such as mica are lined up at the microscopic scale, a rock has a scaly foliation known as a schistosity. A parallel alignment of rod-shaped amphiboles gives the rock a lineation. Blueschists typically have schistosities, and many also have lineations. The development of these features depends upon the magnitude of penetrative rock flowage by microscopic crystal deformation con-

Blueschists display a wide variety of structural features, and many are complexly deformed, with intricate folding and refolding of compositional layering at scales of millimeters to tens of meters. Seen here are examples of of boudinage (a thickening and thinning of the layering) and chevron folds in Queretaro, Mexico's El Doctor limestone. (U.S. Geological Survey)

current with metamorphic recrystallization (a geologic process known as dynamic metamorphism). Although most blueschists were so highly deformed during metamorphic recrystallization that all original features in the rocks were destroyed, some retain features from the rock's premetamorphic history, such as ripple marks, delicate fossils, or volcanic flow layering. Blueschists bearing these features were recrystallized but not highly deformed. Hence, the diversity of minerals and deformational features of the class of rocks known as blueschists is great. Many are truly blueschists, but some are neither blue in color where glaucophane is lacking nor schistose in texture when deformation was minor.

METAMORPHIC CONDITIONS OF FORMATION

Metamorphic recrystallization near 200 degrees Celsius causes anorthite (calcium-rich plagioclase) in combination with water to recrystallize as lawsonite at approximately 3 kilobars, calcite to transform to aragonite at 5 kilobars, and albite (sodium-rich plagioclase) to recrystallize as jadeite plus quartz near 7 kilobars. At higher temperatures, these changes occur at higher pressures. Experiments combined with other measures of metamor-

phic temperature conditions indicate that most blueschists were metamorphosed at temperatures of 150-450 degrees Celsius and minimum confining pressures of 3-10 kilobars, respectively. Metamorphic pressures of 3-10 kilobars correspond to burial depths for recrystallization of 10-30 kilometers. Thirty kilometers is near the base of typical continental crust. Ultra-high-pressure blueschists containing relics of the mineral coesite, a dense mineral having the same composition as quartz forming at extremely high confining pressures, have been found in small areas of the Alps, Norway, and China. Laboratory experiments indicate that confining pressures of 25-30 kilobars are required to transform quartz to coesite. Although coesite-bearing blueschists are rare, their occurrence is very important because they indicate that some blueschists recrystallized at depths of 75-90 kilometers, depths in the Earth which are very near the base of the lithosphere.

The mineralogy of blueschists indicates that the metamorphic conditions for their formation within the Earth would be equivalent to geothermal gradients of 10-15 degrees Celsius per kilometer depth or less. Such ratios of temperature to depth do not exist in the interior of normal lithospheric plates, because geothermal gradients are typically 25-35 degrees Celsius per kilometer depth. The plate tectonic setting for the generation of blueschists is thus very unusual.

SUBDUCTION

Regional terranes of blueschist extending for hundreds of kilometers in length and tens of kilometers in width are found in California, Alaska, Japan, the Alps, and New Caledonia. Smaller bodies of blueschist that are probably remnants of once-extensive terranes are found at numerous other sites around the world. Blueschists are found as fault-bounded terranes juxtaposed against deposits of unmetamorphosed sediments, igneous batholiths, or sequences of basalt, gabbro, and peridotite thought to be fragments of ocean crust (ophiolites) or other metamorphic terranes. The common feature of all occurrences is that they are regions that were probably the sites of ancient lithospheric plate convergence, a tectonic process commonly known as subduction. Subduction carries surficial rocks into the depths of the Earth, where the increase in pressure and temperature causes metamorphism. Plate convergence involves localized shearing action between the descending plate and the overriding plate. As a result, blueschists and associated rocks typically undergo a complex deformational history, sometimes forming chaotic mixtures known as mélanges where deformation was particularly intense. It is of special interest that nearly all extensive blueschist terranes are of Mesozoic age or younger (less than about 250 million years).

Sites of plate subduction in the modern world are marked by ocean trenches, great earthquakes, and arcs of andesitic volcanoes. The region between the trench and volcanic arc is known as the arc-trench gap. Typically a fore-arc basin is on the arc side of the gap and an accretionary prism is on the trench side. The fore-arc basin sits atop the overriding plate and becomes filled largely with basaltic to andesitic volcanic debris generated in the nearby arc. Fore-arc basin deposits are essentially undeformed and unmetamorphosed. In striking contrast, the accretionary prism is directly above the descending plate and consists largely of variously deformed and metamorphosed sediments that were bulldozed off the descending plate during plate convergence. Most typically, blueschists are found in the arcward parts of an accretionary prism, locally faulted directly against fore-arc basin deposits.

The unusual conditions of very low temperatures for a given depth of burial can develop within subduction shear zones because plate convergence at speeds of tens of kilometers per million years (centimeters per year) transports cold lithosphere downward faster than the Earth's interior heat is conducted upward through it. As a result, after a few tens of millions of years of subduction, the front of the overriding plate cools, and the local geothermal gradients become greatly depressed. After fast plate convergence has occurred for a few tens of millions of years, temperatures less than 200 degrees Celsius at depths of 30 kilo-

meters or more can be attained. The subduction zone metamorphism that creates blueschists is also known as high-pressure/low-temperature metamorphism.

OFFSCRAPING AND UNDERPLATING

Because blueschists are found within accretionary prisms, it is important to understand how prisms grow and deform. Subduction accretion occurs by both offscraping and underplating. Offscraping is the process of trenchward growth or widening of the prism by addition at its toe of incoming sediments and seamounts (which are ocean islands such as Hawaii). It occurs by bulldozer-like action that causes the incoming pile of oceanic and trench-axis sediments to be folded and thrust-faulted. Offscraped rocks are weakly metamorphosed with the development of zeolite-group minerals and, at somewhat greater depths, the minerals prehnite and pumpellyite. Underplating is the process of addition of material to the bottom of a prism and, at greater depths, the bottom of the overlying crystalline plate. Underplating thickens and uplifts the overriding block and occurs concurrent with the shearing motions driven by the movement of the descending plate. Blueschists form in the region of underplating.

Underplating appears to be the basic process that drives both the thickening of accretionary prism and the uplift of included masses of blueschist. Underplating by itself, however, does not bring blueschists nearer the surface. The presence of a steep trench slope (5-10 degrees) causes a prism to thin by gravity-driven downslope spreading, much like a glacier thins as it flows down a mountain. Prism thinning seems to occur by a combination of normal faulting and rock flowage. Over a period of tens of millions of years, underplating-driven thickening at the base of the prism and gravity-driven thinning near the surface of the prism would slowly uplift a large terrane of blueschist near the edge of the overriding plate. The actual exposure of blueschist bedrock over a substantial area typically occurs only after subduction ceases and the top of the prism has become exposed to erosion.

The type of high-pressure/low-temperature metamorphism varies with depth. At the shallower depths of offscraping, temperatures and pressures are low, and only zeolites, prehnite, and pumpellyite develop. Surficial rocks subducted to depths of 10-30 kilometers and temperatures of 150-350 degrees Celsius are continuously metamorphosed into blueschists. At depths of 30-40 kilometers and more, and at higher temperatures, the blueschists turn into the class of rock known as eclogite.

PROGRADE AND RETROGRADE METAMORPHISM

The sequence of change from the zeolite to prehnite-pumpellyite to blueschist and finally to eclogite mineral assemblages is known as prograde metamorphism. Rocks in accretionary prisms commonly show all gradations of the progressive sequence. Overall, prograde metamorphism causes a general decrease in rock water content, destruction of the original minerals by recrystallization, increase in rock density, and increase in size of recrystallized crystals. At depths where the basalts and gabbros in the ocean crust (or ophiolite) at the top of the descending plate change from blueschist into eclogite, there is a large increase in the bulk density of the descending plate. This transformation decreases the buoyancy of the descending plate to such an extent that it may be the primary driving force of plate subduction and mantle convection.

When the descending plate reaches depths of 100-125 kilometers, magmas are generated near its upper surface. They rise to the surface to form a volcanic arc of basaltic to andesitic composition. The presence of ultra-high-pressure blueschists directly confirms that some sediments are actually dragged down to (and returned from) very near the typical depths of arc magma origin. The intrusion of hot arc magmas near the surface and the eruption of volcanoes causes heating of the wall rocks, creating metamorphic rocks known as greenschists and amphibolites. This near-surface prograde metamorphism is of a low-pressure/high-temperature type. As a result, many ancient subduction zones are delineated on a regional scale by parallel belts of high-pressure/low-temperature and low-pressure/high-temperature metamorphic belts, a distinctive association known as paired metamorphic belts.

Plate convergence stops either when the relative motions between the descending and overriding plates become such that the margin becomes a transform plate margin or when a buoyant continent or island arc is conveyor-belted into a trench and "plugs up" the subduction zone. Transform plate motion occurs largely by horizontal movement along steep faults, a type of movement known as strike-slip faulting. In the process, some fault blocks rise and blueschists are eroded while others subside and blueschists become buried, reheated, and remetamorphosed as more normal geothermal conditions are reattained (a process known as retrograde metamorphism). Postsubduction destruction of blueschists by either erosion or retrograde metamorphism is the probable explanation for why most extensive terranes of blueschist are of Mesozoic age or younger.

GEOLOGICAL MAPPING OF BLUESCHISTS

Geologists study blueschists in the field, in the confines of the laboratory, and with computer modeling. Fieldwork involves going to the sites where blueschists are exposed in rock outcrops. Geological maps are made to show the field relations between blueschists and associated rocks. The first stage of geological mapping is recording on a topographic map the distribution of the major types of rocks, the orientation of bedding, and the locations and orientations of major faults and folds. Representative rock samples are collected for later laboratory study. The second stage of mapping is typically of much smaller areas. These detailed maps delineate additional variations in the types of rocks, the orientation of minor faults and folds, and associated schistosities and lineations. This stage of analysis usually provides the basis for determining the detailed movement patterns of the blueschists during subduction-zone deformation.

STUDY OF COMPONENT MINERALS

Laboratory studies of blueschists include the analysis of thin sections of the rock samples collected in the field, the geochemical studies of mineral compositions, and experiments to determine the stability limits of minerals under different conditions of pressure, temperature, and fluid composition. Thin sections of the rocks are examined under polarized light with a petrographic microscope. Different minerals display different colors and other optical properties that enable their identification.

Minerals, particularly finely crystalline ones, are also identified by X-ray diffraction. The analysis of the scattering pattern of a beam of X rays focused upon the sample enables the researcher to identify minerals and, for minerals such as feldspar, pyroxene, or chlorite, to estimate their elemental composition. The elemental composition of powdered rock samples is commonly determined using X-ray fluorescence. A focused X-ray beam causes atoms in a powder to emit other X rays whose type and intensity depend upon the types and amounts of atoms in the powder. The elemental composition of individual mineral grains is determined by analysis with the electron microprobe. A beam of high-energy electrons is focused on a 100-square-micron portion of a crystal in a highly polished thin section. As for X-ray fluorescence, the type and intensity of emitted X rays depend upon the types and amounts of atoms in a small spot in the crystal. Measurement of the composition of many spots in a traverse across a crystal enables determination of the variation in mineral composition from its core to rim, a variation known as compositional zoning. Zoning is a sensitive measure of the pressure and temperature history of the growing minerals and, hence, both their prograde and retrograde metamorphic history.

Mass spectrometers are used to determine the isotopic ratios of the component minerals of blueschists for the calculation of the age of metamorphism. Isotopic ratios of neodymium 143 to neodymium 144 and strontium 87 to strontium 86 are indicators of the geologic setting in which igneous rocks were erupted. Measurements of the ratios of oxygen 18 to oxygen 16 in coexisting minerals are indicators of the temperature of metamorphic recrystallization. The age of metamorphism for blueschists is determined from the analysis of radioactive isotopes and their daughter decay products in certain crystals. Examples are potassium 40, which decays into argon 40; rubidium 87, which decays into strontium 87; and uranium 238, which decays into lead 206. The measurement of the ratio of parent to daughter elements in either the whole rock or component minerals can be used to calculate the metamorphic age or ages of the rocks.

Experiments are conducted under controlled conditions in the laboratory to determine the stability limits and compositional relations for minerals at different pressures, temperatures, and fluid compositions. The goal is to simulate physical conditions deep in the Earth. Experimental studies are also performed to determine how the ratios of oxygen isotopes in quartz and other minerals vary with different temperatures and oxygen pressure conditions. Laboratory calibration of elemental and isotopic compositions of minerals under controlled laboratory conditions is the basis for estimating the pressures and temperatures of metamorphism.

Computer simulations of the temperature conditions within subduction zones give an understanding of how temperatures change with time. Computer models that employ the principles of continuum mechanics are used to simulate the long-term tectonic deformation of an accretionary prism and the uplift of blueschist terranes. Geochemical computer models employ the principles of thermodynamics and are used to calculate what types of minerals should develop during prograde and retrograde subduction-zone metamorphism.

INDICATORS OF GEOLOGIC PROCESSES

The study of blueschists is important because they are direct indicators of the geologic processes that occur deep within subduction zones that become mountain belts. Their creation indicates that abnormally cold geothermal conditions develop arcward of the ocean trenches where rapid plate convergence occurs for tens of millions of years. Their preservation indicates that tectonic movements by faulting, folding, and rock flowage can be such that they become uplifted to near the surface while geothermal conditions remain very cold. Understanding the deformational history of blueschists is important because many of the world's largest and most destructive earthquakes occur at subduction zones at the very depths where blueschists are forming today. An understanding of how they deform and recrystallize during their downward and upward paths in ancient subduction zones will eventually provide new understanding of how destructive subduction-zone earthquakes are nucleated and, hence, better earthquake prediction.

Subduction zones are the sites where ocean-floor-capped lithosphere plunges back into the Earth to be recycled. Blueschists are direct indicators that some of the sediment on top of the descending plate is also dragged to near the base of the lithosphere. Their presence in paired metamorphic belts is the primary way that geologists recognize ancient subduction zones. Blueschists are a key part of the geologic story of how continents grow by the addition of accretionary prisms along their edges.

Mark Cloos

CROSS-REFERENCES

Anatexis, 1373; Batholiths, 1280; Contact Metamorphism, 1385; Continental Crust, 561; Continental Growth, 573; Foliation and Lineation, 1389; Granitic Rocks, 1292; Magmas, 1326; Metamorphic Rock Classification, 1394; Metamorphic Textures, 1400; Metasomatism, 1406; Pelitic Schists, 1412; Plate Tectonics, 86; Regional Metamorphism, 1418; Sub-Seafloor Metamorphism, 1423; Subduction and Orogeny, 92.

BIBLIOGRAPHY

Bebout, Gray E., et al., eds. *Subduction Top to Bottom*. Washington, D.C.: American Geophysical Union, 1996. This collection of essays provides a fine introduction to subduction and plate tectonics. Maps, Illustrations, bibliography.

Best, Myron G. *Igneous and Metamorphic Petrology*. 2d ed. Cambridge, Mass.: Blackwell, 1995. A popular university text for undergraduate majors in geology. A well-illustrated and fairly detailed treatment of the origin, distribution, and characteristics of igneous and metamorphic rocks.

Cox, Allan, ed. *Plate Tectonics and Geomagnetic Reversals*. San Francisco: W. H. Freeman, 1973. This book is a collection of papers discussing the basic principles of the theory of plate tectonics and how it was developed. Chapters are introduced by short articles that discuss the importance of the following group of papers. The text is suitable for college-level students.

Davis, George H. *Structural Geology of Rocks and Regions*. New York: John Wiley & Sons, 1984. A structural geology textbook that discusses how folds, faults, and rock flowage occurs. Chapter 6 covers the theory of plate tectonics and contains a short section on blueschists and their occurrence. The text is suitable for college-level students.

Ernst, W. G. *Earth Materials*. Englewood Cliffs, N.J.: Prentice-Hall, 1969. An elementary Earth science book that discusses rocks and minerals. Sections on regional metamorphism, chemistry of metamorphic rocks, physical conditions of metamorphism, and metamorphism and the rock cycle include discussions specifically referring to blueschists. The text is suitable for high school students.

Ernst, W. G., ed. *Subduction Zone Metamorphism*. Stroudsburg, Pa.: Dowden, Hutchinson and Ross, 1975. A collection of technical papers that cover the topic of subduction-zone metamorphism around the world. Most of the papers focus on blueschists. The editor has written a series of summaries that introduce groups of related papers and explain their relative importance. This text is for advanced college-level students.

Evans, Bernard W., and Edwin H. Brown, eds. *Blueschists and Eclogites*. Boulder, Colo.: Geological Society of America, 1986. This book reports the nature of blueschists and eclogites at many sites around the world. Contains numerous pictures and diagrams of structures and mineral textures found in blueschists. The text is suitable for advanced college-level students.

Hallam, A. *A Revolution in the Earth Sciences*. New York: Oxford University Press, 1973. An elementary Earth science book that discusses continental drift and the theory of plate tectonics. Blueschists are discussed in the chapter on the origin of mountain belts. The text is suitable for high school-level readers.

Lima-de-Faria, Josae. *Structural Minerology: An Introduction*. Dordrecht, the Netherlands: Kluwer, 1994. This book provides a good college-level introduction to the basic concepts of crystal structure and the classification of minerals. Illustrations, extensive bibliography, index, and a table of minerals on a folded leaf.

Miyashiro, Akiho. *Metamorphism and Metamorphic Belts.* New York: John Wiley & Sons, 1973. An advanced textbook on metamorphic petrology that contains much discussion of blueschists, their occurrence, and their plate tectonic setting. The text is for advanced college-level students who want to understand the principles that control the formation of minerals in metamorphic rocks.

Perch, L. L., ed. *Progress in Metamorphic and Magmatic Petrology.* New York: Cambridge University Press, 1991. Although intended for the advanced reader, several of the essays in this multiauthored volume will serve to familiarize the new student with the study of metamorphic rocks. In addition, the bibliography will lead the reader to other useful material.

Press, Frank, and Raymond Siever. *Understanding Earth.* 2d ed. New York: W. H. Freeman, 1998. This is one of many introductory textbooks in geology for college students. The section on plutonism and metamorphism contains a discussion of blueschists. An extensive glossary of geological terms is included. The text is aimed at students at the advanced high school and freshman college levels.

Windley, B. F. *The Evolving Continents.* 2d ed. New York: John Wiley & Sons, 1984. This book focuses on the origin and evolution of the continents. Several sections on blueschists are listed in a comprehensive index. The text is for college-level students.

CONTACT METAMORPHISM

Contact metamorphism is caused by the temperature rise in rocks adjacent to magmatic intrusions of local extent that penetrate relatively shallow, cold regions of the Earth's crust. Many economically important metallic mineral deposits occur in contact metamorphic zones.

PRINCIPAL TERMS

AUREOLE: a ring-shaped zone of metamorphic rock surrounding a magmatic intrusion

CONTACT METAMORPHIC FACIES: zones of contact metamorphic effects, each of which is characterized by a small number of indicator minerals

FACIES: a part of a rock or group of rocks that differs from the whole formation in one or more properties, such as composition, age, or fossil content

HORNFELS: the hard, splintery rocks formed by contact metamorphism of sediments and other rocks

LITHOLOGY: the general physical type of rocks or rock formations

CONTACT METAMORPHIC FACIES

Contact metamorphic rocks are moderately widespread; they occur at or near the Earth's surface, where magmas of all kinds intrude low-temperature rocks. Minerals in contact metamorphic rocks are similar to those in regional metamorphic rocks of comparable metamorphic grade. Contact metamorphic effects are divided into facies (zones), each of which is characterized by a small number of concentric indicator mineral rings surrounding an intrusive rock. Nonsymmetrical zones imply special conditions, such as less (or more) chemically or thermally reactive rocks or a nonvertical intrusive body.

Development of contact metamorphic facies reflects both the history of pressure and temperature changes and the bulk-rock chemistry. Thus, by stating that a rock belongs to a particular facies, scientists convey much about the rock's history. This information is vital to exploration for metallic and industrial minerals that commonly occur in contact metamorphic aureoles and for general understanding of regional geology.

Contact metamorphic mineral facies have counterparts with regional metamorphic zones. In addition to the bulk chemical composition of a rock, temperature and pressure are two variable factors. These two factors can be independent. Thus, for example, one can find low-pressure, high-temperature facies or, alternatively, high-pressure, low-temperature facies. With contact metamorphism, these facies occur very close to the intrusive rock. With regional metamorphism, however, the effect is widespread and may not be related to an intrusive rock.

AUREOLES

Contact metamorphic rocks are recognized by their location adjacent to igneous bodies and by evidence indicating a genetic temporal relationship. Contact metamorphic rocks are commonly massive. Granitic rocks are the most common intrusive material. The most frequent depth of the solidification of a granitic magma is 3 kilometers, corresponding to a load pressure of 800-2,100 bars. There are intrusions that solidify at a greater or shallower depth; a depth of 1 kilometer corresponds to a load pressure of 250 bars. Consequently, the load pressures effective during contact metamorphism range from 200 to 2,000 bars in most cases. In contrast, load pressures prevailing during regional metamorphism are generally greater.

When a magma intrudes into colder regions, the adjacent rocks are heated. If the heat content of the intruded magma is high and the volume of the magma is not too small, there will be a temperature rise in the bordering rock that lasts long enough to cause mineral reactions to occur. The rocks adjacent to small intrusions of dikes and sills are not metamorphosed (only baked), whereas larger plutonic rocks give rise to a distinct contact aureole of metamorphic rocks. Several zones of increasing temperature are recognized in contact aureoles.

The contact metamorphic zones surround the intrusion in generally concentric rings that approximate the shape of the intrusion. Those zones which correspond to the highest-temperature minerals are closest to the intrusion; the zone corresponding to the lowest temperature is located farthest from the intrusion. The lowest-temperature contact metamorphic zone gradually grades into unmetamorphosed country rock.

MINERAL DEVELOPMENT

Contact metamorphic rocks are characteristically massive because of lack of deformation; most are fine-grained except for a special variant called a skarn. Skarns may contain metallic mineralization in sufficient concentration that they can be worked for a profit. Such mineral concentrations are called ores.

Contact metamorphic rocks lack schistosity. The very fine-grained, splintery varieties are called hornfels. The large metamorphic gradient, decreasing from the hot intrusive contact to the unaltered country rock, gives rise to zones of metamorphic rocks differing markedly in mineral constituents. The intensity and mineral assemblage in contact metamorphic zones are dependent on several factors: the chemical composition of the intrusive rock; its temperature and volatile content; the composition and permeability of the host units; the structural or spatial relationship of the reactive units to intense contact effects and solutions conduits, or traps; and the pressure or depth of burial.

Argillaceous limestone is usually receptive to contact metamorphism, as its diverse rock chemistry is amenable to mineral development over a broad set of physical and chemical conditions. A typical contact metamorphic mineral assemblage formed from rocks originally of this composition is a plagioclase-garnet-epidote rock. This kind of rock frequently hosts important ore deposits. Shale usually converts to hornfels that is characterized by the minerals cordierite, biotite, and chlorite and that is essentially nonreactive and nonpermeable. Member lithology usually does not host ore deposits. Porous limestone is often exceptionally susceptible to solution and to replacement by ore. Rocks formed from this mate-

rial contain magnetite, garnet, and pyroxene.

Clean, massive limestone has only thin, contact metamorphic zones developed in it. Clean limestones merely recrystallize, forming coarse-grained marble in the highest-temperature contact metamorphic zones adjacent to an intrusion. Dolomite is usually poorly mineralized and nonreactive. Siliceous dolomite, however, may act as a good host, with assemblages characterized by tremolite, diopside, serpentine, and talc. Mafic rocks, such as andesites, diabases, and diorites, can be reactive hosts capable of producing some types of ore deposits. Secondary biotite is the key alteration mineral in mafic rocks. It is associated with ore minerals and is present in lieu of sericite as the alteration product. The biotite zone may be very broad.

DIAGNOSTIC MINERALS

Contact metamorphic facies are identified by the mineral assemblages that are developed in the metamorphosed rocks. Albite and chlorite are the contact metamorphic minerals restricted to the albite-epidote hornfels facies. Calcite, epidote, and talc also occur in this facies. Andalusite may occur in the highest-temperature part of the albite-epidote hornfels.

Anthophyllite-cummingtonite is restricted to the hornblende hornfels facies. Muscovite is present in this facies as well as in the albite-epidote hornfels facies and sometimes in the lowermost pyroxene hornfels facies. Grossular-andradite garnet and idocrase, sometimes with vesuvianite, biotite, and almadine garnet, are present here as well as in the pyroxene hornfels facies. Sillimanite may occur in the highest-temperature part of the facies at higher pressure; staurolite may occur in high-pressure, iron-rich rocks. Calcite is present, but not with tremolite-actinolite or epidote or plagioclase. Under certain chemical compositions of the original rocks, other minerals that may be present include anthophyllite, cummingtonite, phylogopite, biotite, diaspore, and scapolite.

Orthoclase with andalusite or sillimanite is restricted to the pyroxene hornfels facies. Sillimanite is present here and also in the upper hornblende hornfels facies. Hyperstine and glass present in the pyroxene hornfels may also be present in the sanidinite facies. Muscovite is present only in the lowest-temperature part of the facies. For silica-deficient rocks in the pyroxene hornfels facies, dolomite, magnesite, and talc may occur only in the lowest-temperature part of the facies or at high carbon dioxide pressures.

Diagnostic minerals of the sanidinite facies are sanidine, mullite, tridymyte, and pigeonite. In silica-deficient rocks at this facies, wollastonite, grossularite, and plagioclase are present. Other minerals that may occur under special conditions are perovskite, spinel, diopside, and pseudobrookite.

STUDY AND EXPLORATION

Contact metamorphic zones are studied by standard geologic techniques, including the preparation of geologic maps through field study. Aerial photographs and satellite images are frequently used to identify contact metamorphic zones through detecting rock alteration in the contact metamorphic zone. Satellite sensors measure, analyze, and interpret electromagnetic energy reflected from the Earth's surface for subsequent computer analysis. Geophysical techniques (gravity and electrical methods) are used to locate mineralized zones containing relatively heavy metallic ore minerals. The gravity contrast of these zones can be measured with surface instruments and then mapped. Some of these same minerals, particularly the ore minerals, transmit electric current in an anomalous manner. These anomalies plotted on base maps may be an additional clue to the presence of ore deposits in contact metamorphic zones.

Subsequent laboratory work involves the determination of mineral relationships by studying thin slices of rock through which light passes under a microscope (petrography). Frequently, the chemistry of entire rock samples is determined to help scientists understand the presence or absence of minerals as a guide to mineral exploration and to the composition of the original rock. Because of significant advances in laboratory instruments, detailed mineral chemistry analyses are routinely performed that determine chemical makeup of microvolumes of minerals. These determinations permit the Earth scientist to understand the conditions under which the contact metamorphic zones formed and to interpret the history of the contact metamorphic zone.

This information provides critical insight into the potential for metallic ore deposits within the contact metamorphic zone. Such deposits are explored by means of surveys that detect and record variations in geochemistry, gravity, and plant types and abundances. Anomalously high element concentrations in rocks, soils, and plant tissues may indicate mineral deposits that are not exposed. Many ore deposits have an anomalous gravity signature because of heavy associated silicate minerals and metallic minerals. Maps of localized variations may lead to subsurface exploration, which is conducted by drilling techniques. Continuous cylindrical samples, to a depth of as much as several thousand feet, are taken, and the study of such subsurface samples leads to the evaluation of possible minable concentrations of certain metals and other elements used by modern civilizations.

ECONOMIC AND GEOLOGIC VALUE

Contact metamorphic zones frequently contain metallic mineral deposits without which modern civilization could not exist. Many significant mineral deposits worldwide occur in these zones. The metals extracted from deposits of this type include tin, tungsten, copper, molybdenum, uranium, gold, silver, and, in some cases, refractory industrial metals. Specialized surveys assess the potential of metallic ore deposits before expensive subsurface sampling by drilling.

Recognition and understanding of the contact metamorphic environment can lead to a significantly improved understanding of regional geology and the regional geologic history. In some areas, contact metamorphic zones are associated with igneous intrusives of only one geologic age. Specialized laboratory studies provide supporting data to determine the geologic history and potential for economic mineralization.

Jeffrey C. Reid

CROSS-REFERENCES

BIBLIOGRAPHY

Best, Myron G. *Igneous and Metamorphic Petrology.* 2d ed. Cambridge, Mass.: Blackwell, 1995. A popular university text for undergraduate majors in geology. A well-illustrated and fairly detailed treatment of the origin, distribution, and characteristics of igneous and metamorphic rocks.

Deer, W. A., R. A. Howie, and J. Zussman. *An Introduction to Rock-Forming Minerals.* 2d ed. New York: Longman, 1992. Standard references on mineralogy for advanced college students and above. Each chapter contains detailed descriptions of chemistry and crystal structure, usually with chemical analyses. Discussions of chemical variations in minerals are extensive. *An Introduction to Rock-Forming Minerals* is a condensation of a five-volume set originally published in the 1960's.

Hyndman, D. E. *Petrology of Igneous and Metamorphic Rocks.* New York: McGraw-Hill, 1972. Suitable for introductory-level college education in the Earth sciences. Key concepts are summarized at the ends of chapters. Appropriate for advanced students with teacher guidance.

Oldershaw, Cally. *Rocks and Minerals.* New York: DK, 1999. This small, 53-page volume is filled with color illustrations and is therefore of great use to new students who may be unfamiliar with the rock and mineral types discussed in classes or textbooks.

Perch, L. L., ed. *Progress in Metamorphic and Magmatic Petrology.* New York: Cambridge University Press, 1991. Although intended for the advanced reader, several of the essays in this multiauthored volume will serve to familiarize the new student with the study of metamorphic rocks. In addition, the bibliography will lead the reader to other useful material.

Pough, F. H. *A Field Guide to Rocks and Minerals.* 4th ed. Boston: Houghton Mifflin, 1976. Provides an excellent overview of the environments in which minerals are formed and found, and simple laboratory tests that can be conducted in the field or laboratory. Offers the user a basic grasp of the subject of mineralogy. Profusely illustrated.

Sorrella, C. A. *Minerals of the World.* Racine, Wis.: Western Publishing, 1973. This well-illustrated softcover book presents an excellent overview of rocks and minerals. It provides information about the minerals expected in a variety of environments, including contact metamorphic zones.

Voll, Gerhard, ed. *Equilibrium and Kinetics in Contact Metamorphism: The Ballachulish Igneous Complex and Its Aureole.* New York: Springer-Verlag, 1991. The editor brings together a number of essays that deal with contact metamorphism in the Ballachulish Igneous Complex in the Scottish Highlands, thereby providing a useful and detailed case study of a specific region. Includes illustrations, as well as two foldout maps.

Winkler, H. G. F. *Petrogenesis of Metamorphic Rocks.* 4th ed. New York: Springer-Verlag, 1976. An introductory text at the college level. Portions may be of interest to advanced students.

FOLIATION AND LINEATION

Planar and linear penetrative rock fabrics are called foliation and lineation, respectively. Such rock fabrics are the most common evidence that rocks flow in the solid state at depth in the Earth. Foliation and lineation can form by a wide variety of processes depending on pressure, temperature, and the rate of rock deformation.

PRINCIPAL TERMS

CLEAVAGE: a foliation that imparts a preferred direction of fracturing to a rock

KINEMATICS: the pattern of movements in a mass undergoing strain

PENETRATIVE: a feature of a rock mass that occurs not only on surfaces but also throughout the rock volume

PREFERRED ORIENTATION: a systematic bias or regularity in the orientation of mineral grains in a rock

ROCK FABRIC: a penetrative geometrical property of a rock, such as compositional layering or preferred orientation of platy grains

SCHISTOSITY: a foliation defined by preferred orientation of elongate or platy mineral grains visible to the naked eye

STRAIN: distortion of a rock or other material in response to stress

STRESS: forces applied to a rock or other material that tend to cause it to change its shape

FORMATION PROCESSES

Foliation and lineation are the common record of strain of a rock mass. Such fabrics are common, though not ubiquitous, in igneous bodies in which flow of the molten rock before it solidifies completely can cause mineral grains to become aligned to define a foliation or lineation. Foliation and lineation are mainly formed, however, during solid-state flow of rocks in response to tectonic stress, the forces that cause crustal movements. Such solid-state flow, or ductility, of rocks is closely related to the observation that metals are malleable, or able to be molded into different shapes while in the solid state without fracturing. Many of the solid-state flow processes in rocks are the same as those found in metals. In rocks, however, these processes operate very slowly and only at high temperatures.

Three aspects of foliation and lineation are important to consider: the processes by which these fabrics form, descriptive types of rock fabrics, and kinematic significance of rock fabrics. Fracture cleavage, cataclastic foliation, and pressure-solution cleavage are types of foliations that form at low temperatures near the Earth's surface. Fracture cleavage forms when a rock is cut into millimeter-scale sheets by closely spaced cracks, which generally are tiny faults. This is not a common phenomenon, however, and most of what has previously been called fracture cleavage has since proven to be pressure-solution cleavage.

CATACLASTIC FOLIATION

Cataclasis is a process in which a rock "flows" by the creation and sliding of microscopic cracks (microcracks). Cataclasis is common near faults. The resulting rocks (cataclasites) may be foliated if they contain significant amounts of platy minerals such as mica or clay minerals. Such minerals need not be original constituents of the rock and commonly form during cataclasis by chemical breakdown of original feldspar minerals. The platy mineral grains are physically rotated into alignment with the plane of the fault to define a foliation. This foliation is usually somewhat irregular, such that the resulting rock appears scaly. A lineation can form within the foliation by streaking out of mineral grains in the direction of fault slip.

Cataclastic foliation thus defines the orientation of the plane along which masses slip past each other. This type of deformation is called a shear, and the plane along which it occurs, delineated by the cataclastic foliation, is called the shear plane. The lineation defines the direction of slip within the shear plane. A shearing deformation need not be restricted to a distinct plane in space but can be distributed through a volume of rock. Therefore, cataclastic foliation along a fault is a particular case of a shear foliation in which the shear is localized along a particular plane in space, the plane of the fractures.

PRESSURE-SOLUTION FOLIATION

In pressure solution, stress applied to a rock causes the solubility in water of mineral grains to be higher on the sides of the grains that support higher stress than on the sides of those that support lower stress. This increased solubility results in dissolution of material from the sides of grains oriented at a high angle to the direction in which the rock is most strongly compressed. This causes flattening of the grains and formation of a foliation. Pressure-solution foliations tend to be composed of distinct surfaces, spaced a few millimeters to a fraction of a millimeter apart. These surfaces presumably were conduits for movement of the water in which the solid material dissolved. Cleavage in slate forms is predominantly by pressure solution. Pressure-solution cleavages do not define a shear plane. Instead, they form in an orientation perpendicular to the direction of greatest shortening of the rock body. Slates deformed in shear sometimes contain both a cataclastic and a pressure-solution foliation, at an angle to each other.

When a pressure-solution cleavage (or any other foliation) cuts across bedding, the intersection of the two planar fabrics defines an intersection lineation, which in an outcrop can be recognized as the trace of bedding on a cleavage surface or the trace of cleavage on a bedding surface. More generally, intersection lineations occur in any rock body that contains two or more foliations. Further, when a pressure-solution cleavage forms in a rock that already contained a strong foliation such as a schistosity, the intersection lineation with older foliation appears as wrinkles, or crenulations, in the older foliation. The cleavage that forms the crenulations is called a crenulation cleavage.

DISLOCATION CREEP

At the high temperatures existing deep in the crust and in the mantle, and at the slow deformation rates characteristic of geologic processes, the mineral grains that comprise a rock themselves begin to flow slowly. The principal process of crystalline flow is called dislocation creep. Dislocation creep occurs by formation and movement of crystal defects called dislocations and proceeds at slow but steady rates at small stresses. Dislocation creep is believed to be the main process by which the flow in the mantle occurs that causes the lithospheric plates to move and is the main process by which metamorphic rocks in the crust flow.

In dislocation creep, virtually every grain in a rock may be flattened into alignment with a plane perpendicular to the direction of greatest shortening of the rock mass, defining a foliation, and may be elongated along the direction of greatest stretching, defining a lineation. The resulting rocks are usually sufficiently coarse-grained to be called schists, or gneisses, if they are compositionally layered. Rocks deformed by dislocation creep not only have preferred orientations of grain shapes to define a schistosity but also typically acquire strong preferred orientations of the crystallographic axes of constituent mineral grains. Crystallographic preferred orientation is uniquely the result of dislocation creep, and its widespread presence in metamorphic rocks is one of the strongest lines of evidence that dislocation creep is a very important process in the Earth.

MYLONITES

The result of very high degrees of deformation by dislocation creep is a fine-grained rock called a mylonite. A mylonite is usually intensely foliated and/or lineated. Mylonites are commonly found in ductile shear zones, which are planar zones of intense shear deformation. Ductile shear zones are geometrically equivalent to faults but form by ductile flow instead of by fracturing. The origin of mylonites was for a long time controversial because of confusion between fine-grained rocks formed by crushing of mineral grains, such as cataclasites, and mylonites, in which the small grain size results from the growth of new crystals at high temperature (recrystallization). A wide variety of observations, including strong crystallographic preferred orientation and a lack of evidence for microfracturing, show that natural mylonites form by dislocation creep at high temperatures and are not closely related in origin to cataclasites. This conclusion has been confirmed by creation of artificial mylonites in high-temperature laboratory experiments.

Rock fabrics in mylonites have recently been the subject of intense research interest. Mylonites have been found to contain multiple foliations that appear to have formed more or less contemporaneously. One foliation, termed the S foliation by some researchers, is defined by grain flattening and stretching. This foliation forms obliquely in the shear zone in which the mylonite formed, in an orientation that depends upon the direction and amount of shearing. A second foliation, called C foliation, is a spaced foliation that forms essentially parallel to the shear plane. The C foliation contains a lineation in the shear direction. It is composed of a multitude of closely spaced, microscopic shear zones that are miniature analogues of the larger one in which the mylonite formed. Recognition and understanding of the relationships of rock fabrics in mylonites have recently led to great strides in deducing the kinematics of structures formed by ductile flow in the middle and deep crust.

STUDY METHODS

Foliation and lineation are studied in the field as well as in the laboratory. Field studies focus primarily on measuring the orientations of rock fabrics, determining the geometrical and age relationships between different fabrics, and deducing the relationships of foliations to other structures, such as folds or shear zones. Commonly, rock samples are collected in the field for later laboratory studies to complement the field studies.

The laboratory tool most important to the study of rock fabrics is the polarizing petrographic microscope. Thin sections (30-micron slices of rock) are viewed with the petrographic microscope, using either transmitted light (the sample is lit from behind) or reflected light (light is reflected off of the polished top surface of the thin section). Observations using the petrographic microscope give the identities, sizes, shapes, and geometrical relationships of the mineral grains that make up a rock. This provides crucial information about processes by which a fabric formed. Such microscopic study may also include the use of a universal stage, an attachment that allows the thin section to be viewed under the microscope from any direction. The universal stage is especially im-

portant for recognizing and measuring crystallographic preferred orientation.

In very fine-grained rocks such as some slates, the crystals that make up the rock are so small that little can be seen with a petrographic microscope. In this case, X-ray diffraction (XRD) has been used to determine the nature and intensity of the preferred orientation that defines the foliation. In XRD, a beam of X rays is directed at a small sample of the rock which deflects (diffracts) the X rays as they pass through it. The orientations and intensities of diffracted X rays are measured with a device called an X-ray goniometer. The measured pattern of diffracted X rays can then be related to the strength of the preferred orientation of crystals in the sample.

Transmission electron microscopy (TEM) has been used more recently to study atomic-scale processes that operate when rocks flow by dislocation creep. In TEM, a beam of electrons is directed at a slice of a deformed crystal only tens to hundreds of atoms thick. As in XRD, the electrons in the beam are deflected as they pass through the crystal. The part of the crystal immediately surrounding a defect deflects the electrons differently from the way the rest of the crystal deflects electrons, so that by using TEM, one can actually take pictures of individual defects. This allows the scientist to determine if dislocations, a particular kind of crystal defect, are present in a sample. Dislocations are formed by deformation of crystals; TEM provides conclusive evidence for dislocation creep having operated in the deformation of a sample.

The laboratory techniques above are applied not only to naturally deformed rocks but also to experimentally deformed rocks. Experimenters have reproduced in the laboratory many of the features of rocks deformed at high temperatures in nature, including foliations of various kinds. The experiments provide quantitative data regarding solid-state flow processes in rocks.

PRACTICAL APPLICATIONS

Rock fabrics have many practical applications. Slate blackboards and roofing and paving slates are possible because deformation has aligned the platy mineral grains in shale to define a very planar foliation along which the slate splits into thin sheets. Similarly, although some flagstone is quarried from thin-bedded sandstone that is not foliated, the platiness of flagstone in many cases reflects a strain-induced foliation.

Rock fabrics can also contribute to geologic hazards. Because foliations are commonly favored directions of rock fracture, strongly foliated bedrock may cause problems with slope stability. Preventing landsliding onto a road from a road cut through slate can be a major engineering problem. The foliation may be not only a direction of weakness but also a weakness itself, because where it occurs, water is allowed to flow through the rock. Therefore, bridges and dams must be sited very carefully in areas of foliated rocks, because percolation of water along the foliation may lead to undermining and ultimate failure.

John M. Bartley

CROSS-REFERENCES

Aerial Photography, 2739; Anatexis, 1373; Blueschists, 1379; Contact Metamorphism, 1385; Earth Resources, 1741; Experimental Petrology, 468; Metamorphic Rock Classification, 1394; Metamorphic Textures, 1400; Metasomatism, 1406; Pelitic Schists, 1412; Regional Metamorphism, 1418; Sub-Seafloor Metamorphism, 1423.

BIBLIOGRAPHY

Borrodaile, G. J., M. B. Bayly, and C. M. Powell. *Atlas of Deformational Textures and Metamorphic Rock Fabrics.* New York: Springer-Verlag, 1982. This is a compendium of images of deformed rocks, ranging in scale from outcrop photographs to submicroscopic electron micrographs. The images are contributed, described, and interpreted by more than ninety rock-deformation experts from around the world. The text is intended for professional geologists, but the pictures, which are beautifully reproduced, are of general interest.

Davis, George H. *Structural Geology of Rocks and Regions.* New York: John Wiley & Sons, 1984. Chapter 12, "Cleavage, Foliation, and Lineation," describes rock fabrics and relationships to strain and to other geologic structures. The illustrative photographs are excellent. Both text and figure captions are written in a casual style that makes them relatively unintimidating and accessible to college-level readers.

Fry, Norman. *The Field Description of Metamorphic Rocks.* New York: Halsted Press, 1984. This is a compact handbook of practical techniques for the description of metamorphic rocks as they occur in outcrop. Chapters 4, 6, and 9 focus on aspects related to rock fabrics. Suitable for college students.

Hobbs, Bruce E., Winthrop D. Means, and Paul F. Williams. *An Outline of Structural Geology.* New York: John Wiley & Sons, 1976. This college-level text provides a complete and detailed treatment of rock fabrics and the processes by which they form. More advanced, detailed, and complete than the book by Davis and somewhat more difficult. Chapter 2 covers deformation processes and their products at a microscopic level. Chapters 5 and 6 give detailed descriptions and interpretations of foliation and lineation, respectively.

Holness, Marian B., ed. *Deformation-Enhanced Fluid Transport in the Earth's Crust and Mantle.* London: Chapman and Hall, 1997. This book deals with the mechanics and evolution of rock deformation. Many essays focus on the role of fluid dynamics in the Earth's mantle and crust, as well as the resultant geophysical conditions. Illustrations, bibliography, and index.

Oldershaw, Cally. *Rocks and Minerals.* New York: DK, 1999. This small, 53-page volume is filled with color illustrations and is therefore of great use to new students who may be unfamiliar with the rock and mineral types discussed in classes or textbooks.

Press, Frank, and Raymond A. Siever. *Understanding Earth.* 2d ed. New York: W. H. Freeman, 1998. This introductory-level college text examines the basic descriptive properties of metamorphic rocks and theories of their origin at an elementary level. It includes a brief discussion of foliation and various types of foliated rocks.

Ranalli, Giorgio. *Rheology of the Earth.* 2d ed. London: Chapman and Hall, 1995. Ranalli offers a look into the geodynamics of rocks and rock deformation, including geophysical processes associated with foliation and lineation. Intended for the reader who does not have a strong background in the Earth sciences.

Spry, Alan. *Metamorphic Textures*. Oxford, England: Pergamon Press, 1969. This is the standard college-level text and reference work covering the sizes, shapes, and geometrical arrangement of the mineral grains that comprise metamorphic rocks. Chapters 8 through 11 discuss textures of foliated and lineated rocks.

METAMORPHIC ROCK CLASSIFICATION

Metamorphic rocks bear witness to the instability of the Earth's surface. They reveal the long history of interaction among the plates that comprise the surface and of deep-seated motions within the plates. Among the metamorphic rocks are found many ores and stones of value to human civilization.

PRINCIPAL TERMS

CONTACT METAMORPHISM: metamorphism characterized by high temperature but relatively low pressure, usually affecting rock in the vicinity of igneous intrusions

FACIES: a part of a rock or group of rocks that differs from the whole formation in one or more properties such as composition, age, or fossil content

FOLIATION: a texture or structure in which mineral grains are arranged in parallel planes

METAMORPHIC FACIES: an assemblage of minerals characteristic of a given range of pressure and temperature; the members of the assemblage depend on the composition of the protolith

METAMORPHIC GRADE: the degree of metamorphic intensity as indicated by characteristic minerals in a rock or zone

METAMORPHISM: changes in the structure, texture, and mineral content of solid rock as it adjusts to altered conditions of pressure, temperature, and chemical environment

PELITIC ROCK: a rock whose protolith contained abundant clay or similar minerals

PRESSURE-TEMPERATURE REGIME: a sequence of metamorphic facies distinguished by the ratio of pressure to temperature, generally characteristic of a given geologic environment

PROTOLITH: the original igneous or sedimentary rock later affected by metamorphism

REGIONAL METAMORPHISM: metamorphism characterized by strong compression along one direction, usually affecting rocks over an extensive region or belt

TEXTURE: the size, shape, and relationship of grains in a rock

METAMORPHISM

A significant part of the Earth's surface is made up of rocks quite different from sedimentary or igneous rocks. Many of them have distinctive textures and structures, such as the wavy, colored bands of gneiss or the layered mica flakes of schist. They often contain certain minerals not found or not common in igneous or sedimentary rocks, such as garnet and staurolite. Studies of their overall chemical composition and their relations to other rocks in the field show that they were once igneous or sedimentary rocks, but, being subjected to high pressure and temperature, they have been distorted and altered or recrystallized through a process called metamorphism. Metamorphism involves both mechanical distortion and recrystallization of minerals present in the original rock, the protolith. It can cause changes in the size, orientation, and distribution of grains already present, or it can cause the growth of new and distinctive minerals built mostly from materials provided by the destruction of minerals that have become unstable under the changed conditions. The chemical components in the rock are simply reorganized into minerals that are more stable under higher pressure and temperature.

DESCRIPTIVE CLASSIFICATION OF METAMORPHIC ROCKS

Metamorphic rocks can be classified in a purely descriptive fashion according to their textures and dominant minerals. Because the growing understanding of metamorphic processes can be applied to interpret the origin and history of the rocks, they are also classified according to features related to these processes. The most common classification schemes, in addition to the purely descriptive, categorize the rocks by general metamorphic processes, or metamorphic environments; by the original rocks, or protoliths; by metamorphic intensities, or grades; by the general pressure and temperature conditions, called facies; and by the ratios of pressure to temperature, called pressure-temperature regimes.

The oldest classification is purely descriptive, based on the rock texture (especially foliation) and mineral content. Foliation is an arrangement of mineral grains in parallel planes. The most

SYSTEMS OF OF METAMORPHIC ROCK CLASSIFICATION

System	Categories	Comments/Examples
Descriptive	Foliated	Mineral grains arranged in parallel planes. Includes slate, schist, gneiss.
	Nonfoliated	Lack parallel structure. Includes quartzite, marble, amphibolite, serpentinite, hornfels.
Metamorphic Process	Regional	Compression causing foliation, typified by the foliated rocks slate, schist, and gneiss.
	Contact	Heat from intrusive magma causes recrystallization, including metasomatism. Includes quartzite, marble, skarn, hornfels.
	Cataclastic	Occurs along fault zones, where grains are intensely sheared and smeared out by stress. In deep parts of the fault, the sheared grains recrystallize to a fine-grained, finely foliated rock called mylonite.
	Burial	Occurs where pressure and temperature are high enough to form fine grains of zeolite minerals.
	Hydrothermal	Hot water infiltrates rock; the water and dissolved substances may be incorporated into the crystals of certain minerals. Includes serpentine.
Protolithic Origin	Pelitic	Metamorphism yields metapelites.
	Basaltic	Metamorphism yields metabasalts.
	Carbonates	Metamorphism yields metacarbonates.
Grade (intensity)	—	Presence of certain "index" minerals indicates intensity of pressure and temperature; most commonly used grades are named for index minerals in pelitic rocks.
Facies	—	Based on characteristic mineral assemblage within a given range of pressure and temperature. Includes greenschist facies, named for low-grade metamorphosed basalt; amphibolite facies, the range of pressures and temperatures that would give the staurolite and kyanite grades in pelitic rocks; and the granulite facies, corresponding to extreme conditions bordering on anatexis.
Pressure-temperature regimes	Barrovian	Greenschist, amphibolite, and granulite facies include mineral assemblages of increasing metamorphic intensity whose pressure and temperature rise together approximately as they would with increasing depth under most areas of the Earth's surface.
	Abukuma	Temperature is much cooler for any given pressure. Blueschist facies is typical of this series.
	Hornfels	Temperature rises much faster than pressure, corresponding to contact metamorphism.

common foliated rocks are slate, schist, and gneiss. In slate, microscopic flakes of mica or chlorite are aligned so that the rock breaks into thin slabs following the easy cleavage of the flakes. Schist contains abundant, easily visible flakes of mica, chlorite, or talc arranged in parallel; it breaks easily along the flakes and has a highly reflective surface. Gneiss contains little mica, but its minerals (commonly quartz, feldspar, and amphibole) are separated into different-colored, parallel bands, which are often contorted or wavy. The foliated rocks can be described further by naming any significant minerals present, such as "garnet schist." Nonfoliated metamorphic rocks lack parallel structure and are usually named after their dominant minerals. Common types are quartzite (mostly quartz), marble (mostly calcite), amphibolite (with dominant amphibole), serpentinite (mostly serpentine), and hornfels (a mixture of quartz, feldspar, garnet, mica, and other minerals). Quartzite breaks through its quartz grains, whose fracture surfaces give the break a glassy sheen. Marble breaks mostly along the cleavage of its calcite crystals, so that each flat cleavage surface has its own glint. Hornfels often exhibits a smooth fracture with a luster reminiscent of a horn.

CLASSIFICATION BY PROCESSES OR ENVIRONMENTS

The second classification of metamorphic rocks is based on the general processes that formed them or the corresponding environments in which they are found. The recognized categories are usually named regional, contact, cataclastic, burial, and hydrothermal metamorphism.

Regional metamorphism is characterized by compression along one direction that is stronger than the pressure resulting from burial. The compression causes foliation, typified by the foliated rocks slate, schist, and gneiss. These rocks are found in extensive regions, often in long, relatively narrow belts parallel to folded mountain ranges. According to the theory of plate tectonics, folded mountain ranges like the Appalachians and the Alps began as thick beds of sediments deposited in deep troughs offshore from continents. The sediments were later caught up between colliding continents, strongly compressed, and finally buckled up into long, parallel folds. The more deeply the original sediment is buried, the more

intense is the metamorphism. Clay in the sediment recrystallizes to mica, oriented with the flat cleavage facing the direction of compression. Thus, pelitic or clay-bearing sediments become slates and schists with foliation parallel to the folds of the mountains. At higher temperatures, the mica recrystallizes into feldspar, and the feldspar and quartz migrate into light-colored bands between bands of darker minerals so that the rock becomes gneiss. At yet higher temperatures, some of the minerals melt (a process called anatexis), and the rock, called a migmatite, becomes more like the igneous rock granite.

Contact metamorphic rocks are commonly found near igneous intrusions. Heat from the intrusive magma causes the surrounding rock, called country rock, to recrystallize. Though the rock is under pressure because of burial, there is usually no tendency toward foliation because the pressure is equal from all directions. Some water may be driven into the rock through fine cracks or, conversely, water may be driven from the rock by the heat. Especially mobile atoms such as potassium can migrate into the rock and combine with its minerals to form new crystals (a process called metasomatism). Usually, however, the chemical content is not greatly altered, and recrystallization chiefly involves atoms from smaller crystals or the cement migrating into larger crystals or forming new minerals. In quartz sandstone, for example, the quartz crystals grow to fill all the pore space in a tight, polygonal network called crystalloblastic texture, and the rock becomes quartzite. Similarly, the tiny crystals in limestone or dolomite grow into space-filling calcite crystals, forming marble or, if other minerals are present, the mixed rock called skarn. Pelitic rocks recrystallize to hornfels, containing a variety of minerals, such as quartz, feldspar, garnet, and mica.

Cataclastic (or dynamic) metamorphism occurs along fault zones, where both the rock and individual grains are intensely sheared and smeared out by stress. In deep parts of the fault, the sheared grains recrystallize to a fine-grained, finely foliated rock called mylonite.

Burial metamorphism occurs in very deep sedimentary basins, where the pressure and temperature, along with high water content, are sufficient to form fine grains of zeolite minerals among the sedimentary grains. The process is intermediate

METAMORPHIC ROCK CLASSIFICATION BY TEXTURE AND COMPOSITION

	Non-foliated			Foliated			
	Non-layered			Layered	Non-layered		
Texture	Fine to coarse grained	Fine to coarse grained	Fine grained	Coarse grained	Coarse grained	Fine grained	Very fine grained
Composition	Calcite				Chlorite		
			Mica		Mica		
			Quartz				
			Feldspar				
			Amphibole				
			Pyroxene				
Rock Name	Marble	Quartzite	Hornfels	Gneiss	Schist	Phyllite	Slate

between diagenesis, which makes a sediment into a solid rock, and regional metamorphism, in which the texture of the rock is modified.

Hydrothermal metamorphism (which many prefer to call alteration) is caused by hot water infiltrating the rock through cracks and pores. It is most common near volcanic and intrusive activity. The water itself, or substances dissolved in the water, may be incorporated into the crystals of certain minerals. One important product is serpentine, formed by the addition of water to olivine and pyroxene, which is significant in sub-seafloor metamorphism.

CLASSIFICATION BY ORIGINAL ROCKS

A third classification is based on the original rocks, or protoliths. This classification is possible because relatively little material is added to or lost from the rock during metamorphism, except for water and carbon dioxide, so the assemblage of minerals present depends on the overall chemical composition of the original rock. The most abundant protoliths are pelitic rocks (from clay-rich sediments, usually with other sedimentary minerals), basaltic igneous rocks, and limestone or other carbonate rocks. Each kind of protolith recrystallizes into a different characteristic assemblage of minerals. The categories can be named, for example, metapelites, metabasalts, and metacarbonates.

CLASSIFICATION BY GRADE

Metamorphic intensity, or grade, is the basis for a fourth classification scheme. As the pressure and temperature increase, certain minerals become unstable, and their chemical components reorganize into new minerals more stable in the surrounding conditions. The presence of certain minerals, called index minerals, therefore indicates the intensity of pressure and temperature. The grades most commonly used are named for index minerals in pelitic rocks; in the late nineteenth century, they were described by George Barrow in zones of metamorphic rocks in central Scotland. These Barrovian grades, in order of increasing intensity, are marked by the first appearances of chlorite, biotite, garnet, staurolite, kyanite, and sillimanite.

Metamorphism is a slow process, however, especially at low temperatures, and conditions some-

times change too rapidly for the mineral assemblage to come to equilibrium. It is not uncommon to find crystals only partially converted into new minerals or to find lower-grade minerals coexisting with those of higher grade. At low grade, some structures of the original rock, such as bedding, may be preserved. Even the outlines of earlier crystals may be seen, filled in with one or more new minerals. High-grade metamorphism usually destroys earlier structures.

All metamorphic rocks available for study are at surface conditions, so the pressures and temperatures that caused them to recrystallize have been relieved. If conditions were relieved slowly enough, and especially if water was available, the rock may have undergone retrograde metamorphism, reverting to a lower grade and thus adjusting to the less intense pressure and temperature. Retrograde metamorphism is usually not very complete, and some evidence of the most intense conditions almost always remains. For example, the distinct outline of a staurolite crystal might be filled with crystals of quartz, biotite mica, and iron oxides.

CLASSIFICATION BY FACIES

A fifth classification scheme categorizes the rocks according to the intensity of pressure and temperature, or facies, without reference to protoliths. The concept of facies was developed by Penti Eskola, working in Finland about 1915, who enlarged on the work of Barrow in Scotland and V. M. Goldschmidt in Norway. Eskola realized that each protolith has a characteristic mineral assemblage within a given facies, or range of pressure and temperature. The facies are named for one of the assemblages within a specified range of conditions; for example, the greenschist facies, named for low-grade metamorphosed basalt, refers to equivalent low-grade assemblages from other protoliths as well. Other examples are the amphibolite facies, the range of pressures and temperatures that would give the staurolite and kyanite grades in pelitic rocks; and the granulite facies, which corresponds to extreme conditions bordering on anatexis.

CLASSIFICATION BY PRESSURE-TEMPERATURE REGIMES

Finally, the facies themselves, or the metamorphic assemblages in them, can be classified according to ratios of pressure to temperature, called pressure-temperature regimes. The greenschist, amphibolite, and granulite facies include mineral assemblages of increasing metamorphic intensity whose pressure and temperature rise together approximately as they would with increasing depth under most areas of the Earth's surface. This sequence is sometimes referred to as the Barrovian pressure-temperature regime because Barrow's metamorphic grades in pelitic rocks fall in these facies. In another regime, called the Abukuma series after an area in Japan, the temperature is much cooler for any given pressure. The blueschist facies, characterized by blue and green sodium-rich amphiboles and pyroxenes, is typical of this series. The converse situation, a regime in which temperature rises much faster than pressure, corresponds to contact metamorphism, and is called the hornfels facies.

INVESTIGATING ROCK FEATURES AND HISTORIES

Initial studies of metamorphic rocks are almost always done in the field. The tectonic or structural nature of the region suggests the processes to which the rock has been subjected. For example, an area of folded mountains can be expected to exhibit regional metamorphism; a volcanic area, some contact metamorphism; and a fault zone, some cataclastic metamorphism. Some features are obvious at the scale of an outcrop, such as banding and foliation, or the halo of recrystallized country rock abutting an igneous intrusion. Some textural features, such as foliation, large crystals such as the garnet in garnet schist, or the luster of a fracture surface as in quartzite or marble, are easily seen in a hand specimen. Similarly, a preliminary estimate of mineral content can be made from a hand specimen.

Many features, however, are best seen in thin section under a petrographic microscope. Usually all but the finest grains can be identified. From the relative abundance of the various minerals and their known chemical compositions the overall chemical composition of the rock can be calculated. The protolith can then be identified by comparing the calculated composition to the known compositional ranges of igneous and sedimentary rocks.

Textures seen under the microscope reveal much about the history of the rock. Foliation, for example, usually indicates regional metamorphism. A space-filling, polygonal texture can show contact or hydrothermal recrystallization, and a crumbled, smeared-out cataclastic texture indicates faulting.

More recent methods of investigation sometimes applied to metamorphic rocks are X-ray diffraction and electron microprobe analysis. The pattern of X rays scattered from crystals depends on the exact arrangement and spacing of atoms in the crystal structure, which is useful for identifying minerals. X-ray diffraction can be used to identify crystals that are too small or too poorly formed to be identified with a microscope. The microprobe can analyze the chemical composition of crystals even of microscopic size. Determination of exact composition or of variation in composition within growth zones of a single crystal can be especially useful for identifying variations in conditions during crystal formation.

RECOGNIZING METAMORPHIC ROCKS

Most people encounter metamorphic rocks while traveling through mountains and other scenic regions. Recognizing these rocks is easier if one has a general idea of where the various types occur and how they appear in outcrops.

Regional metamorphic rocks are best exposed in two kinds of localities: the continental shields and the eroded cores of mountain ranges. Ancient basement rocks of the continental platform, composed of regionally metamorphosed and igneous rocks, are exposed in shields without a cover of sedimentary rock. The Canadian Shield, extending from northern Minnesota through Ontario and Quebec to New England, is the major shield of North America. Similar shields are exposed in western Australia and on every other continent. The old, eroded mountains of Scotland and Wales contain abundant outcrops in which some of the pioneering studies of metamorphic rocks were conducted. The Appalachians have even larger exposures, extending from Georgia into New England. Somewhat smaller outcrops occur in many parts of the Rocky Mountains and the Coast Ranges. The foliated rocks schist, slate, and gneiss make up the bulk of these exposures. Rock cleavage parallel to the foliation is an important clue to recognizing outcrops of these rocks; slopes parallel to foliation tend to be fairly smooth and straight, while slopes eroded across the foliation are ragged and steplike. Bare slopes of schist can reflect light strongly from the many parallel flakes of mica. Slate is usually dull and dark-colored but characteristically splits into ragged slabs. The colorful, contorted bands of gneiss are easily recognized.

Contact metamorphic rocks are much less widespread and are generally confined to areas of active or extinct volcanism. The Cascades and the Sierra Nevada show many examples, but some of the best exposures are found near ancient intrusives in the Appalachians and in New England. Contact metamorphic rocks are more of a challenge to recognize because they generally lack foliation. The best clue is physical contact with a body of igneous rock. They often form a shell or halo around an igneous body, most intensely recrystallized at the contact and extending outward a few centimeters to a few kilometers (depending mostly on the size of the igneous body), until they eventually merge into the surrounding unaltered country rock. The halo is generally similar to the country rock but, because it is recrystallized, it is usually harder, more compact, and more resistant to erosion. Broken surfaces can be distinctive. Depending on the nature of the country rock, one might look for the glassy sheen of fractured quartzite, the glinting cleavage planes of marble, or the smooth, hornlike fracture of hornfels.

Hydrothermally altered rocks from sub-seafloor metamorphism, when finally exposed on land, are found among regional metamorphic rocks and appear much like them except for the greenish-gray colors of chlorite, serpentine, and talc. Terrestrial hydrothermal alteration is most easily rec-ognized in areas of recent volcanic activity. Good examples are exposed in the southwestern United States, such as the so-called porphyry copper deposits. Many such areas contain valuable deposits, and so may have been mined or prospected. The outcrops are often much fractured and veined near the intrusion. Where alteration is most intense, the rock may appear bleached; farther from the intrusion, it may have a greenish hue because of low-grade alteration. Quartz and sulfide minerals such as pyrite are common in the veins, but weathering often leaves a rusty-looking, resistant cap called an iron hat or gossan over the deposit.

Rocks formed by cataclastic and burial metamorphism require specialized equipment for their recognition. Zones many miles wide containing mylonite, the product of cataclastic metamorphism, are exposed along the Moine fault in northwestern Scotland (where mylonite was first studied) and along the Brevard fault, extending along the Appalachians from Georgia into North Carolina. Examples of burial metamorphic rocks are found under the Salton Sea area of California and the Rotorua area of New Zealand.

James A. Burbank, Jr.

CROSS-REFERENCES

BIBLIOGRAPHY

Bates, Robert L. *Geology of the Industrial Rocks and Minerals.* Mineola, N.Y.: Dover, 1969. Somewhat technical but readable and practical descriptions of the geological occurrence and production of metamorphic rocks and minerals (among others), listing their principal uses. Representative rather than comprehensive. Topical bibliography and good index.

Best, Myron G. *Igneous and Metamorphic Petrology.* 2d ed. Cambridge, Mass.: Blackwell, 1995. A popular university text for undergraduate majors in geology. A well-illustrated and fairly detailed treatment of the origin, distribution, and characteristics of igneous and metamorphic rocks. Chapter 4 treats granite plutons and batholiths, and chapter 12 treats a broad spectrum of metamorphic topics.

Blatt, Harvey, and Robert J. Tracy. *Petrology: Igneous, Sedimentary, and Metamorphic.* New York: W. H. Freeman, 1996. Undergraduate text in elementary petrology for readers with some familiarity with minerals and chemistry. Thorough, readable discussion of most aspects of metamorphic rocks. Abundant illustrations and diagrams, good bibliography, and thorough indices.

Compton, Robert. *Geology in the Field.* New York: John Wiley & Sons, 1985. A standard undergraduate text with a chapter devoted to the interpretation of metamorphic rocks and structures in the field. Some knowledge of mineralogy is assumed. The illustrations are helpful.

Ehlers, Ernest G., and Harvey Blatt. *Petrology: Igneous, Sedimentary, and Metamorphic.* New York: W. H. Freeman, 1982. Undergraduate text in elementary petrology for readers with some familiarity with minerals and chemistry. Thorough, readable discussion of most aspects of metamorphic rocks. Abundant illustrations and diagrams, good bibliography, and thorough indices.

Oldershaw, Cally. *Rocks and Minerals.* New York: DK, 1999. This small, 53-page volume is filled with color illustrations and is therefore of great use to new students who may be unfamiliar with the rock and mineral types discussed in classes or textbooks.

Perch, L. L., ed. *Progress in Metamorphic and Magmatic Petrology.* New York: Cambridge University Press, 1991. Although intended for the advanced reader, several of the essays in this multiauthored volume will serve to familiarize the new student with the study of metamorphic rocks. In addition, the bibliography will lead the reader to other useful material.

Pough, Frederick H. *A Field Guide to Rocks and Minerals.* 4th ed. Boston: Houghton Mifflin, 1976. The best of the most widely available field guides, authoritative but easy to read. Color plates of representative mineral specimens, and sufficient data to be useful for distinguishing minerals. Very brief description of rocks, including metamorphic rocks. Elementary crystallography and chemistry are presented in the introduction.

Strahler, Arthur N. *Physical Geology.* New York: Harper & Row, 1981. The chapter on metamorphic rocks is a good intermediate-level approach to classification and metamorphic processes. Related chapters on geological environments may interest the reader. Excellent bibliography for the beginning student, with thorough glossary and index.

Tarbuck, Edward J. *Earth: An Introduction to Physical Geology.* Upper Saddle River, N.J.: Prentice Hall, 1999. 6th ed. One of the better Earth science texts for beginning college or advanced high school readers. Good elementary treatment of metamorphic rocks and, in other chapters, of related environments. Color pictures throughout are excellent. Bibliography, glossary, and a short index.

METAMORPHIC TEXTURES

Metamorphic textures are important criteria in the description, classification, and understanding of the conditions under which metamorphic rocks form. Textures vary widely and develop as a result of the interaction between deformation, recrystallization, new mineral growth, and time.

PRINCIPAL TERMS

COHERENT TEXTURE: an arrangement allowing the minerals or particles in a rock to stick together

FOLIATION: a planar feature in metamorphic rocks

MICA: a silicate mineral (one silicon atom surrounded by four oxygen atoms) that splits readily in only one direction

STRAIN: deformation resulting from stress

STRESS: force per unit of area

TECTONICS: the study of the processes and products of large-scale movement and deformation within the Earth

METAMORPHISM TYPES

In the classification and description of sedimentary, igneous, or metamorphic rocks, texture is a very important factor. Texture has a relatively straightforward definition—it is the size, shape, and arrangement of particles or minerals in a rock. How texture is related to the rock fabric and rock structure may be confusing. Rock structure may refer only to features produced by movement that occurs after the rock is formed. Rock fabric sometimes refers to both the structure and the texture of crystalline (most igneous and metamorphic) rocks. There is so much disagreement that the three terms are used interchangeably. Yet, although there is no agreement on the categories of metamorphism, some are more generally accepted than others. Regional, contact, and cataclastic are the three principal types. Shock, burial, sub-seafloor, and hydrothermal metamorphism tend to be more controversial.

Metamorphic textures develop in response to several factors, the most influential being elevated temperatures and pressures and chemically active fluids. In each type of metamorphism, the relative influence of each of these three factors varies greatly. Generally, the effect of temperature on metamorphism is easy to understand. Converting dough to bread by baking provides a common analogy. The roles of pressure and fluids, however,

are more difficult to explain. Two types of pressure, confining and directed, play a role in metamorphism. Confining pressure is the force exerted on a rock at depth by the overlying material. Directed pressure is a larger force exerted by some external factor, usually tectonic in origin. Of these two types, directed pressure produces more apparent textural changes in metamorphic rocks. The primary role of chemically active fluids is to facilitate the metamorphic reactions and textural changes that occur.

REGIONAL METAMORPHISM

Of all the types of metamorphism, the most pervasive is regional metamorphism. Regional metamorphism occurs in the roots of mountain belts as they are forming. Thousands of square kilometers can be involved. Temperature and directed pressure are the critical agents in producing the new mineral assemblages and textures distinctive of this type of metamorphism. The most diagnostic feature in regionally metamorphosed rocks is a planar fabric, which forms as platy and elongate minerals develop a preferred orientation. This planar fabric, or foliation, generally forms only in the presence of (and in an orientation that is perpendicular to) the directed pressure. Directed pressure may also impart a linear element on some rocks. This lineation may develop because elongate minerals or groups of minerals are recrystallized, deformed into a preferred direction, or both. Lineations may also develop where two planar fabrics intersect. Of all the different rock types, shales and other fine-grained rocks are the most sensitive to increasing conditions of metamorphism and show marked textural changes as metamorphic conditions increase. Not all regionally metamorphosed rocks develop foliations or lineations; many rock types show very little textural change from the lowest to highest conditions of metamorphism.

As metamorphism begins, shale is converted to slate, which diagnostically splits along smooth surfaces called cleavage. Cleavage takes two forms: flow cleavage and fracture cleavage. Flow cleavage is a pervasive or penetrative fabric; directed pressure influences every portion of the rock, with most of the micaceous minerals undergoing recrystallization. A variety of penetrative fabrics characterize regionally metamorphosed rocks formed at different metamorphic conditions. In fracture cleavage, the distinct planar fractures are separated by discrete, relatively undeformed segments of rock.

At the lowest grades of regional metamorphism, as the penetrative fabric develops in slate, very fine-grained micaceous minerals start to grow in a preferred orientation, thus producing planes of weaknesses along which the slate readily splits. At slightly higher metamorphic conditions, the micaceous minerals continue to form. The rock typically takes on a pronounced sheen, although the individual mineral grains remain mostly too small to see with the unaided eye. This texture is referred to as a phyllitic texture, and the rock is called a phyllite. As metamorphic conditions continue to increase at ever greater depths, the micaceous minerals eventually grow large enough to be seen with the naked eye, and the preferred orientation of these platy minerals becomes obvious. This texture is referred to as a schistosity, and the rock is called a schist. As the conditions of metamorphism approach very high levels, the micaceous minerals start to break down to form other minerals that are

not platy, and the rock begins to lose its property to split along foliation surfaces. A new type of foliation develops, however, because light and dark minerals tend to separate into alternating bands. This banded texture is characteristic of rocks called gneisses, and the texture is referred to as a gneissosity. At even higher conditions, rocks begin to melt in a process called anatexis. Anatexis occurs at conditions that are considered to be at the interface between metamorphism and igneous processes.

CONTACT METAMORPHISM

Although not as pervasive as regional metamorphism, contact metamorphism is of interest because many of its deposits are economically important. Contact metamorphism occurs in areas of relatively shallow depths where a hot, molten igneous body comes in contact with the cooler rock that it has invaded. The metamorphism takes place outside the margins of the igneous mass, with temperature as the key agent of metamorphism and recrystallization as the dominant process modifying the original rock. The area metamorphosed around the intrusion is called the aureole. The size of this aureole is a function of the size, composition, and temperature of the igneous body, as well as the composition of the host

Light and dark minerals tend to separate into alternating bands. This banded texture is characteristic of metamorphic rocks called gneisses, and the texture is referred to as gneissosity. (© William E. Ferguson)

rock and depth of the intrusion. The intensity of the metamorphism is greatest at the contact between the two rock bodies and decreases away from the source of heat. Aureoles may be as thin as a meter or as wide as 2-3 kilometers.

A number of rock types and textures are characteristic of contact metamorphism. Most rocks produced by contact metamorphism are fine-grained because the processes by which they were formed are relatively short-lived. Typically, the aureole is made of a hard, massive, and fine-grained rock called a hornfels. Characteristically, the hornfels is more fine-grained than is the rock that was metamorphosed. The uniform grain size gives the rock a granular appearance, particularly when magnified; this texture is termed granoblastic or hornfelsic. Since hornfels undergo no deformation, textures in the original rock may be preserved, even when recrystallization is complete.

When shales are metamorphosed, very large crystals of minerals such as andalusite and cordierite may develop in an otherwise fine-grained granoblastic rock. These large crystals constitute a porphyroblastic texture and are themselves called porphyroblasts. When schists or slates are subjected to contact metamorphism, remnant porphyroblasts or ovoid masses of mica may produce a spotted appearance, and the rock is named a spotted slate or schist. If the aureole contains silicon-rich carbonate rocks (impure limestones), much coarser textures are likely to form at or near the contact of the igneous body. These coarser-grained rocks are termed tactites or skarns.

CATACLASTIC METAMORPHISM

The third type of metamorphism is cataclastic or dynamic metamorphism, which occurs in fault or deep-shear zones. Directed pressure is the key agent, with temperature and confining pressure playing variable roles. The textures in cataclastic rocks are divided into four groups: incoherent, nonfoliated, mylonitic, and foliated-recrystallized. Incoherent texture develops at very shallow depths and low confining pressures. Based on the degree of cataclasis (from least to most), fault breccia and fault gouge are the rock types formed. Nonfoliated (or mortar) texture forms coherent rocks typified by the presence of porphyroclastic minerals (large crystals that have survived cataclasis) surrounded by finely ground material. Rocks with this texture

lack obvious foliations. Also included in this group are cataclastic rocks with glassy textures. Unusually intense grinding produces enough frictional heating to melt portions of the rock partially. Mylonitic texture occurs in coherent rocks that show a distinct foliation (flow structure). The foliation typically forms because alternating layers show differing intensities of grinding. Textural subdivisions and rock names are based on the degree of grinding, such as protomylonite (least amount of grinding), orthomylonite, and ultramylonite (most amount of grinding). The final group is the foliated-recrystallized textures. Cataclastic rocks that illustrate considerable recrystallization are also typically foliated. The degree of grinding determines the rock name. Mylonite gneiss shows the least cataclasis, while the blastomylonite is more crushed.

SHOCK METAMORPHISM

Shock metamorphism is the rarest type of metamorphism. It is characteristically associated with meteorite craters and astroblemes. Astroblemes are circular topographic features that are inferred to represent the impact sites of ancient meteorites or comets. Some geologists consider shock metamorphism as a type of dynamic metamorphism because in both, directed pressure plays the essential role, whereas temperature may vary from low to extremely high.

A number of textural features are associated with shock metamorphism. Brecciation (breaking into angular fragments), fracturing, and warping of crystals are common. On a microscopic scale, the presence of two or more sets of deformation lamellae in quartz crystals is considered conclusive evidence of a meteorite impact. (Deformation lamellae are closely spaced microscopic parallel layers that are partially or totally changed to glass or are sets of closely spaced dislocations within mineral grains.) Shock metamorphism is also detected when minerals are partially or completely turned to glass (presumably with and without melting) and when shatter cones are present. Shatter cones are cone-shaped rock bodies or fractures typified by striations that radiate from the apex.

STRUCTURAL ANALYSIS

Although much of the work of describing metamorphic textures comes from the field of metamorphic petrology, the means of studying how these textures, fabrics, or structures are generated is more within the area of structural geology called structural analysis. Structural analysis relies heavily on the fields of metallurgy, mechanics, and rheology to describe and interpret the process by which rocks are deformed. Structural analysis considers metamorphic textures and larger features, such as faults and folds, from three different perspectives: descriptive analysis, kinematic analysis, and dynamic analysis. Synthesis of the information obtained from these three analyses typically leads to comprehensive models or hypotheses that explain problems concerning metamorphic textures.

Descriptive analysis is the foundation of all studies of metamorphic textures. In descriptive analysis, geologists consider the physical and geometric aspects of deformed rocks, which include the recognition, description, and measurement of the orientations of textural elements. Typically, descriptive analysis involves geologic mapping, in which geologists measure the orientations of metamorphic structures with an instrument that is a combination compass-clinometer. (A clinometer is an instrument or scale used to measure the angle of an inclined line or surface in the vertical plane.) These structural measurements are plotted in a variety of ways to determine if the data show any statistical significance. Field studies may also include analysis of aerial photographs or satellite imagery to determine if textures or structures that are evident on the microscopic scale, or those of a rock outcrop, are related to much larger patterns.

In the laboratory, descriptive analysis of metamorphic textures can involve several techniques. One of the most important involves the petrographic microscope, in which magnified images of textures result from the transmission of polarized light through thin sections of properly oriented rock specimens. The thin section is a paper-thin slice of rock that is produced by gluing a small cut and polished block of the rock specimen to a glass slide. This block, commonly a little less than 2.7 millimeters by 4.5 millimeters, is then cut and ground to the proper thickness.

Kinematic analysis is the study of the displacements or deformational movements that produce features such as metamorphic textures. Several types of movement are recognized, including distortion and dilation. The study of distortion (change of shape) and dilation (change of volume) in a rock is called strain analysis. Strain analysis involves the quantitative evaluation of how original sizes and shapes of geological features are changed. Although strain commonly expresses itself as movement along preexisting surfaces, distortion also creates new surfaces such as cleavages and foliations in metamorphic rocks.

Dynamic analysis attempts to express observed strains in terms of probable patterns of stress. Stress is the force per unit of area acting upon a body, typically measured in kilograms per square meter. One important area of study in dynamic analysis involves experimental deformation of rocks under various levels of temperature, confining pressure, and time. Such experiments are typically conducted on short cylindrical specimens on a triaxial testing machine, which attempts to simulate natural conditions with the variables controlled. By jacketing specimens in impermeable coverings, pressures can be created hydraulically to exceed 10 kilobars, which are comparable to those near the base of the Earth's crust, or outermost layer. Dynamic analysis also involves scale-model experiments, using clays and other soft substances to attempt to replicate naturally occurring structural features.

Applications in Building and Construction

Although a number of metamorphic minerals and rocks are used by society, in most cases, the metamorphic textures of these materials play no role in their utility. One notable exception is slate. Because metamorphic processes impart a strong cleavage to this very fine-grained rock, it readily splits apart along these thin, smooth surfaces. The combination of grain size, cleavage, and strength makes slate useful in a number of products, including roofing shingles, flagstones, electrical panels, mantels, blackboards, grave vaults, and billiard tables. Most other metamorphic rocks, however, have foliations or other textural features that make the rocks structurally weak, and limit their usefulness.

In other cases, however, metamorphic rocks are useful because they lack pronounced foliations or do not readily break along planes of weakness. Many widely used marbles, particularly those quarried in Italy, Georgia, and Vermont, are products of regional metamorphism. They are prized in part because they are generally massive (lacking texturally induced planes of weakness) and can be cut into very large blocks. Several quartzites and gneisses also tend to be massive enough to be used as dimension stone or as aggregate because of high internal strength and relative chemical inertness.

Aside from direct usage, problems can develop in mountainous or hilly regions when construction occurs where weak, intensely foliated metamorphic rocks exist. When road or railway cuts are excavated through foliations inclined toward these cuts, rock slides are possible. In some cases, the roads can be relocated, or, if not, slide-prone slopes may be modified or removed. In other places, slide-prone exposures or mine walls and ceilings can be pinned and anchored.

Construction of buildings in metamorphic ar-

eas also requires careful evaluation. Where weak foliations run parallel to slopes, it must be determined whether the additional weight of structures and water for lawns and from runoff are likely to induce rock slides or other forms of slope instability. If so, land-use plans and zoning restrictions need to be adopted to indicate that these areas are potentially hazardous.

Ronald D. Tyler

CROSS-REFERENCES

Anatexis, 1373; Blueschists, 1379; Contact Metamorphism, 1385; Foliation and Lineation, 1389; Hydrothermal Mineralization, 1205; Metamorphic Mineral Deposits, 1614; Metamorphic Rock Classification, 1394; Metasomatism, 1406; Pelitic Schists, 1412; Petrographic Microscopes, 493; Plate Tectonics, 86; Regional Metamorphism, 1418; Sub-Seafloor Metamorphism, 1423.

BIBLIOGRAPHY

Best, Myron G. *Igneous and Metamorphic Petrology.* 2d ed. San Francisco: W. H. Freeman, 1995. A readable advanced college-level text that generally should not be beyond the general reader who wants to learn more about metamorphic textures. This illustrated treatment of metamorphism includes the topics of mineralogy, chemistry, and the structure of metamorphic rocks. Both field relationships and global-scale tectonic associations are discussed.

Blatt, Harvey, and Robert J. Tracy. *Petrology: Igneous, Sedimentary, and Metamorphic.* New York: W. H. Freeman, 1996. An upper-level college text that deals with the descriptions, origins, and distribution of igneous, sedimentary, and metamorphic rocks. Includes an index and bibliographies for each chapter.

Chernikoff, Stanley. *Geology: An Introduction to Physical Geology.* Boston: Houghton-Mifflin, 1999. This is a good overview of the scientific understanding of the geology of the Earth and surface processes. Includes the address of a Web site that provides regular updates on geological events around the globe.

Davis, George H. *Structural Geology of Rocks and Regions.* New York: John Wiley & Sons, 1984.

An upper-level geology text that provides an excellent treatment of the deformational features in metamorphic rocks. Includes chapters on descriptive, kinematic, and dynamic analyses. Provides a well-illustrated treatment of cleavage, foliation, and lineation. Good bibliography and author and subject index.

Dietrich, Richard V., and B. J. Skinner. *Rocks and Rock Minerals.* New York: John Wiley & Sons, 1979. This short, readable college-level text provides a relatively brief but excellent treatment of regional, contact, and cataclastic metamorphic rocks, with good coverage of metamorphic textures. Very well illustrated. Includes a subject index and a modest bibliography.

Dolgoff, Anatole. *Physical Geology.* Lexington, Mass.: D. C. Heath, 1996. This is a comprehensive guide to the study of the Earth. Although this is an introductory text for college students, it is written in a style that makes it understandable to the interested layperson. Extremely well illustrated, with a glossary and index.

Ernst, W. G. *Earth Materials.* Englewood Cliffs, N.J.: Prentice-Hall, 1969. Chapter 7 in this very fine, compact, introductory-level book provides a succinct treatment of the topic of metamorphism. Metamorphic structures, cataclastic rocks, contact metamorphism, and regional metamorphism are all briefly discussed. Includes a subject index and a short bibliography.

Hyndman, Donald W. *Petrology of Igneous and Metamorphic Rocks*. 2d ed. New York: McGraw-Hill, 1985. A college-level text intended for the undergraduate geology student. Not overly technical. Hyndman clearly covers the traditional themes (categories, processes, and conditions) necessary to understand metamorphic rocks. Text provides information on the locations and descriptions of metamorphic rock associations. Includes an index and an exhaustive bibliography.

Spry, A. *Metamorphic Textures*. Oxford, England: Pergamon Press, 1969. Advanced college-level text that treats metamorphism in terms of textural changes and largely disregards chemical interactions. Although parts of the book may be beyond the introductory reader, several chapters deal with very basic principles. Provides excellent illustrations and photographs. Includes an extensive bibliography and author and subject indexes.

Suppe, John. *Principles of Structural Geology*. Englewood Cliffs, N.J.: Prentice-Hall, 1985. A well-illustrated college text intended for the geology major. Chapters 10 and 11 provide clear discussions of "Fabrics" and "Impact Structures," respectively, that are very readable and that do not generally require extensive background to understand. Includes a bibliography and subject index.

Williams, Howel, F. J. Turner, and C. M. Gilbert. *Petrology: An Introduction to the Study of Rocks in Thin Sections*. 2d ed. San Francisco: W. H. Freeman, 1982. Although the bulk of this upper-level college text deals with the description of textures and mineral associations in igneous, sedimentary, and metamorphic rocks beneath the petrographic microscope, the authors provide excellent, understandable descriptions of the major types of metamorphism and associated textures.

METASOMATISM

Metasomatism is produced by circulation of aqueous solutions through rock undergoing metamorphic recrystallization. The solutions cause chemical losses and additions by dissolving and precipitating minerals along their flow paths. Metasomatic processes have produced the world's major ore bodies of tin, tungsten, copper, and molybdenum, as well as smaller deposits of many other metals.

PRINCIPAL TERMS

AQUEOUS SOLUTION, HYDROTHERMAL FLUID, and INTERGRANULAR FLUID: synonymous terms for fluid mixtures that are hot and have a high solvent capacity, permitting them to dissolve and transport chemical constituents; they become saturated upon cooling and may precipitate metasomatic minerals

CONTACT METASOMATISM: metasomatism in proximity to a large body of intrusive igneous rock, or pluton

DENSITY: the ratio of rock mass to total rock volume; usually measured in grams per cubic centimeter

METEORIC WATER: water that originally came from the atmosphere, perhaps in the form of rain or snow, as contrasted with water that has escaped from magma

PERMEABILITY: the capacity to transmit fluid through pore spaces or along fractures; high-fracture permeability is generally requisite for metasomatism, as porosity is greatly reduced by metamorphic recrystallization

POROSITY: the ratio of pore volume to total rock volume; usually reported as a percentage

PROGRADE METAMORPHISM: recrystallization of solid rock masses induced by rising temperature; differs from metasomatism in that bulk rock composition is unchanged except for expelled fluids

RECRYSTALLIZATION: a solid-state chemical reaction that eliminates unstable minerals in a rock and forms new stable minerals; the major process contributing to rock metamorphism

REGIONAL METASOMATISM: large-scale metasomatism related to regional metamorphism

PROCESSES OF CHEMICAL CHANGE

Metasomatism is an inclusive term for processes that cause a change in the overall chemical composition of a rock during metamorphism. Such processes may be described as positive or negative depending upon whether a net gain or a loss is produced in the affected rock body. Where chemical changes are slight, the minerals in the original rock remain unchanged or register only very subtle changes. On the other hand, intense metasomatism may result in total destruction of the original mineral assemblage and its replacement by new, and different, minerals. In these extreme cases, metasomatism is usually difficult to detect, particularly if large rock volumes are affected. Metasomatic effects are most obvious when the original rock texture and mineral assemblage is partially destroyed. In such cases, the resulting rock will exhibit an unusually large variety of minerals as well as microscopic evidence of incomplete chemical reactions. In normal metamorphic reactions, chemical migration occurs on the scale of a single mineral grain (a few millimeters at most) during recrystallization. In contrast, metasomatism involves chemical transport on the scale of a few centimeters or more. In areas where metasomatism has been intense, it can often be demonstrated that the scale of chemical transport ranged from about 100 meters to as much as several kilometers. It is the movement of chemical components through rocks on this larger scale which distinguishes metasomatism from metamorphism. Dehydration (water-releasing reaction) and decarbonation (carbon dioxide-releasing reaction) are the commonest types of chemical reactions during metamorphism at the higher grades. These reactions certainly produce significant changes in rock composition and involve large-distance chemical transport, but because they typify normal prograde metamorphism, they do not constitute metasomatism. If, on the other hand, water and/or carbon dioxide were reintroduced into a rock which had previously experienced prograde metamorphism, this would constitute a fairly common type of metasomatism.

DIFFUSION AND INFILTRATION METASOMATISM

The reshuffling of the chemical components into new mineral assemblages during metamorphism, particularly when gaseous phases are lost, necessitates major changes in rock volume (usually volume reductions) and a corresponding change in rock porosity and density. Metasomatic replacement of a metamorphic rock will induce additional changes in volume (either reductions or increases), porosity, and density. Although metasomatism is defined as a process of chemical change, physical changes in rock properties also occur, and these are an essential aspect of metasomatism. For convenience, introductory textbooks treat metamorphic and metasomatic reactions as "constant volume" processes. This assumption may hold true in specific instances, but it has no general validity, particularly when volatile constituents are involved in the reactions. It has been demonstrated on theoretical grounds that chemical transport on the scale of a few centimeters or more requires the presence of an intergranular pore fluid that can effectively dissolve existing minerals and deposit others while the rock as a whole remains solid. The relative mobility of this fluid phase leads to two theoretical types of metasomatism: diffusion metasomatism and infiltration metasomatism. In diffusion metasomatism, the chemical components move through a stationary aqueous pore fluid permeating the rock by the process of diffusion. The effects are limited to a distance of a few centimeters from the surface of contact between rocks of sharply contrasting composition. This process cannot produce large-scale metasomatic effects. Infiltration metasomatism involves a mobile aqueous fluid that circulates through pores and fractures of the enclosing rock and carries in it dissolved chemical components. This fluid actively dissolves some existing minerals and deposits new ones along its flow path. The scale of infiltration metasomatism is thus determined by the circulation pattern of the fluid, and

rock compositions over distances of several kilometers can be easily altered. For convenience, the term "metasomatism" will be used in place of "infiltration metasomatism" and the effects of diffusion metasomatism will be ignored.

CONDITIONS FAVORABLE TO METASOMATISM

The degree of chemical change that accompanies recrystallization is closely related to the fluid/rock ratio prevailing during the process. Since regional metamorphism is a deep-seated process, a low fluid/rock ratio generally prevails, and the process is "rock-dominated" in the sense that minerals dominate the composition of the fluid circulating through the rocks. A significant degree of metasomatism under such conditions is unlikely unless the fluid is very corrosive. Small quantities of fluorine or chlorine can produce corrosive aqueous fluids, but these elements are rarely important in regional metamorphism. Contact metamorphism, however, occurs close to the Earth's surface, where large volumes of groundwater circulate in response to gravity. Groundwater will mix with any water given off by a high-level crystallizing pluton; it follows that contact metamorphism takes place under conditions of high fluid/rock ratio. Such a process will be "fluid-dominated" in the sense that the circulating fluid will exert control over the compositions of the minerals formed by recrystallization. These conditions are highly favorable for metasomatism.

Intense, pervasive metasomatism will develop under the following conditions: when energy is available to provide temperature and pressure gradients to sustain fluid movement; when a generally high fluid/rock ratio prevails; when the aqueous fluid has the solvent capacity to dissolve minerals in its flow path; and when the enclosed rocks possess, or develop, sufficient permeability to permit fluid circulation. These conditions must be sustained for a sufficient time period, or recur with sufficient frequency, to produce metasomatism.

GREISENIZATION AND FENITIZATION

Such conditions are commonly attained in rocks adjacent to large bodies of intrusive igneous rock, or plutons, particularly those which expel large quantities of water. As an example of contact metasomatism, consider the effects produced in the fractured roof zone of a peraluminous, or

S–type granite pluton. Hot water vapor, concentrated below the roof by crystallization, is often enriched in corrosive fluorine accompanied by boron, lithium, arsenic, silicon, tungsten, and tin. This reactive fluid migrates up fractures, enlarging pathways by dissolving minerals and seeping into the adjacent rock. The result is a network of quartz-muscovite-topaz-fluorite replacement veins that may contain exploitable quantities of tin, tungsten, and base metal ore. Fluorine-dominated metasomatism is known as greisenization. Greisen effects may extend 5 to 10 meters into the wall rocks from vein margins. Within this zone, the original rock textures, minerals, and chemical composition will be profoundly modified by metasomatism. In many cases, the entire roof of the granite pluton is destroyed and replaced by greisen minerals. Some of the world's major tin and tungsten deposits—such as those in Nigeria, Portugal, southwest England, Brazil, Malaysia, and Thailand—were formed by just such a process. On a larger scale, spectacular sodium and potassium metasomatism develops adjacent to intrusions of ijolite and carbonatite plutons. This intense alkali metasomatism is called fenitization and is developed on a regional scale in the vicinity of Lake Victoria in East Africa. In this region, ijolite-carbonatite complexes are plentiful, and aureoles, or ring-shaped zones, of fenite extend outward 1 to 3 kilometers from the individual plutons. The width of the fenite zone around a source pluton is largely determined by the fracture intensity in the surrounding rocks; where they are highly fractured, large-scale, and even regional, fenitization is present. Massive, unfractured rocks resist fenitization, and the resulting aureoles are narrow. Metasomatism around ijolite intrusions is dominated by the outflow of sodium dissolved in an aqueous fluid expelled by the magma and an apparent back-flow migration of silicon to the source intrusion. A typical result would be an inner zone, 20 to 30 meters wide, of coarse-grained "syenite fenite" composed of aegirine, sodium feldspar, and sometimes nepheline. Beyond the outer limit of this zone, there would be a major aureole of shattered host rock veined by aegirine, sodium amphibole, albite, and orthoclase, which might extend well beyond 1 kilometer from the parent intrusion. Fenitization around carbonatite intrusions is dominated by potassium metasoma-

tism, which converts the country rocks into orthoclasite (a coarse-grained metasomatic rock composed almost exclusively of potassium feldspar). The resulting fenite aureole is typically less than 300 meters wide and is roughly proportional to the diameter of the parent carbonatite. The chemical composition of the country rocks appears to exert little influence on the progression of fenitization. The rocks adjacent to ijolite intrusions are driven toward a bulk chemical composition approaching that of ijolite, while those adjacent to carbonatite intrusions are driven toward orthoclasite regardless of their initial composition.

METASOMATIC MINERAL ZONES

Intensely metasomatized rocks usually exhibit mineral zoning, which is more or less symmetrical around the passageways that controlled fluid migration. These passages may be networks of vein-filled fissures, major fault zones, shattered roof zones of igneous plutons, or the fractured country rocks adjacent to such plutons. Emphasis is placed on fracture permeability as a control of fluid migration, as most metamorphosed rocks have negligible porosity. The metasomatic mineral zones are often dominated by single, coarse-grained mineral species, which is obviously "exotic" with respect to the original mineral assemblage. Inner zones often cut sharply across outer zones, and the resulting pattern reflects a systematic increase in metasomatic intensity as the controlling structure is approached. By means of foot traverses across the metasomatized terrain, geologists carefully map the mineral zoning pattern and its controlling fractures. Such maps provide insight into the fracture history of the area, show the distribution and volume of the metasomatic products, and indicate the relative susceptibility of the various rock types present to the metasomatic process. If ore bodies are present, geologic maps provide essential information for exploring the subsurface extent of the ores through drilling. They also provide a basis for systematically sampling the metasomatic zones as well as the country rock beyond the limit of metasomatism; the unaltered country rock is called the protolith.

The sample collection provides material for study in the laboratory after field studies are complete. A paper-thin slice is cut from the center of each rock sample and is mounted on a glass slide for microscopic examination. From such examinations, the scientist can identify both fine- and coarse-grained minerals and can determine the order of metasomatic replacement of protolith minerals.

STUDY AND SAMPLING OF PROTOLITHS

The objectives in a study of metasomatism are to determine the chemical changes that have taken place in the altered rocks and to reconstruct the history of fluid-rock interaction. To this end, it is essential to determine the chemical compositions of the various protoliths affected by metasomatism so that additions and losses in the metasomatized rocks may be calculated. The ideal protolith is a rock unit that is both chemically uniform over distances comparable to the scale of zoning and highly susceptible to metasomatism. Considerable attention must, therefore, be devoted to the study and sampling of protoliths during the field stage of a project. Ideally, the samples should be collected just beyond the outer limit of metasomatism in order to avoid metasomatic contamination and to minimize the effect of a lack of protolith uniformity. Unfortunately, it is only possible to approximate the position of this outer limit in the field, because the decreasing metasomatic effects merge imperceptibly with the properties of the unaltered protolith. Sampling problems are further compounded in study areas where rock exposures are poor or where exposed rocks are deeply weathered. Rock weathering promotes chemical changes indistinguishable from metasomatism, so weathered samples cannot be accepted for chemical study. The sample requirements for metasomatic research are more stringent than for any other type of geological study, and meeting them always taxes the ingenuity of a geologist.

The samples, having been cut in half for the microscope slides, are then prepared for chemical analysis and for density measurement. One half is stored "as is" in a reference collection for future use. The other half is cleaned of all traces of surface weathering and plant material. The samples are then oven-dried, and bulk densities are obtained by weighing and coating them with molten paraffin (to seal pore spaces), followed by immersion in water to determine their displacement volumes. Next, the samples are crushed and ground to fine powder. The average mineral grain density is determined by weighing a small amount of rock powder and measuring its displacement volume in water. The porosity of each sample is determined, and the powdered samples (20-30 grams each) are then sent to a laboratory specializing in quantitative chemical analysis. Because it is desirable to study as many as sixty-five different chemical elements, the laboratory will use several modern instrumental techniques as well as the traditional "wet method" to determine their concentrations.

Modern studies use the mass balance approach, which relates the physical, volumetric, and chemical properties between the altered rocks and their protoliths. As an example, consider a particular element in a given volume of protolith. The mass of this element is the product of volume × bulk density × element concentration. If the element is totally insoluble in the circulating fluid, then its mass must remain constant during metasomatism of the enclosing rock. It follows that for this immobile element, the product of volume × bulk density × element concentration in the altered rock must equal that of the parent protolith. This provides an objective test for determining which elements were mobile and which immobile during the metasomatic event. From this point, it is a simple matter to calculate gains and losses of each element for the altered rocks.

SKARN DEPOSITS

Metasomatic deposits of a wide range of metals and industrial minerals are commonly found at or near the contact between igneous plutons and preexisting sedimentary rocks. Ore deposits concentrated by contact metasomatic processes are collectively known as skarn deposits. Skarn deposits are major sources of tungsten, tin, copper, and molybdenum. Important quantities of iron, zinc, cobalt, gold, silver, lead, bismuth, beryllium, and boron are also mined from skarn deposits. Additionally, such deposits are a source of the industrial minerals fluorite, graphite, magnetite, asbestos, and talc. For the most part, skarn deposits are found in relatively young mobile belts that are not yet deeply eroded. The most productive skarns are generally those in which granitic magma has invaded sedimentary sequences dominated by layers of carbonate rocks.

The physical and chemical principles of metasomatism have been deduced by research geologists over many decades from thousands of individual field and laboratory studies. The knowledge derived from the studies is put to practical use by economic geologists who explore remote areas in search of new skarn deposits or who exploit known skarn ores at producing mines. These economic geologists test, on a daily working basis, the theoretical hypotheses and generalizations formulated by research geologists regarding metasomatism.

Gary R. Lowell

CROSS-REFERENCES

Anatexis, 1373; Astroblemes, 2648; Blueschists, 1379; Contact Metamorphism, 1385; Experimental Rock Deformation, 208; Folds, 624; Foliation and Lineation, 1389; Metamorphic Rock Classification, 1394; Metamorphic Textures, 1400; Mountain Belts, 841; Pelitic Schists, 1412; Regional Metamorphism, 1418; Stress and Strain, 264; Sub-Seafloor Metamorphism, 1423.

BIBLIOGRAPHY

Augustithis, S. S. *Atlas of Metamorphic-Metasomatic Textures and Processes.* Amsterdam: Elsevier, 1990. A thoughtful treatment of metamorphic rocks, minerology, and metasomatism. Excellent illustrations add to understanding concepts and processes. Appropriate for the college student or layperson with an interest in the Earth sciences.

Blatt, Harvey, and Robert J. Tracy. *Petrology: Igneous, Sedimentary, and Metamorphic.* New York: W. H. Freeman, 1996. An upper-level college text that deals with the descriptions, origins, and distribution of igneous, sedimentary, and metamorphic rocks. Includes an index and bibliographies for each chapter.

Burnham, C. W. "Contact Metamorphism of Magnesian Limestones at Crestmore, California." *Geological Society of America Bulletin* 70 (1959): 879-920. A classic account of metasomatism at one of the best-known mineral-collecting localities in the world, having more than one hundred species of rare minerals. This periodical can be found in university libraries. Written for professional geologists but can be understood by those with some knowledge of geology.

Chernikoff, Stanley. *Geology: An Introduction to Physical Geology.* Boston: Houghton-Mifflin, 1999. This is a good overview of the scientific understanding of the geology of the Earth and surface processes. Includes the address of a Web site that provides regular updates on geological events around the globe.

Einaudi, M. T., L. D. Meinert, and R. J. Newberry. "Skarn Deposits." In *Economic Geology: Seventy-fifth Anniversary Volume.* El Paso, Tex.: Economic Geology Publishing, 1981. A comprehensive review of skarn deposits and skarn theory written for professional geologists. The text is technical, but the extensive bibliography is useful to those interested in learning more about specific types of skarn and skarn localities.

Fyfe, W. S., N. J. Price, and A. B. Thompson. *Fluids in the Earth's Crust.* New York: Elsevier, 1978. The best overview of the theoretical side of fluid-rock interaction. The authors argue that subduction and seafloor spreading may be viewed as large-scale metasomatic processes, the end product of which is the Earth's crust. Written for advanced students of geology or chemistry. Extensive bibliography.

LeBas, Michael John. *Carbonatite-Nephelinite Volcanism: An African Case History.* New York: John Wiley & Sons, 1977. A detailed account of the geology in the Lake Victoria region of East Africa, containing excellent descriptions of volcanoes and calderas formed by recent alkalic and carbonatitic magmas. Fenites are discussed throughout in great detail. Can be understood by those with a modest background in geology, provided that they first familiarize themselves with the nomenclature of alkalic rocks.

Mason, Roger. *Petrology of the Metamorphic Rocks.* Winchester, Mass.: Allen & Unwin, 1978. A very good college-level text dealing with major aspects of metamorphism. Chapter 5 is devoted to contact metamorphism and metasomatism. Excellent glossary and index.

Oldershaw, Cally. *Rocks and Minerals*. New York: DK, 1999. This small, 53-page volume is filled with color illustrations and is therefore of great use to new students who may be unfamiliar with the rock and mineral types discussed in classes or textbooks.

Press, Frank, and Raymond A. Siever. *Understanding Earth*. 2d ed. New York: W. H. Freeman, 1998. This book introduces the high school and college reader without a geology background to the subject of metamorphism and metasomatism. Metasomatism is treated in very general terms. Good index and illustrations.

Sigurdsson, Haraldur, ed. *Encyclopedia of Volcanoes*. San Diego, Calif.: Academic Press, 2000. This book contains a complete summery of the scientific knowledge of volcanoes. It contains eighty-two well-illustrated overview articles, each of which is accompanied by a glossary of key terms. Although this is a college-level text, it is written in a clear and comprehensive style that makes it generally accessible. Cross-references and index.

Taylor, Roger G. *Geology of Tin Deposits*. New York: Elsevier, 1979. An overview of tin deposits. Chapter 6 deals authoritatively with metasomatism and its application to exploration for tin. Requires some background in geology and chemistry.

PELITIC SCHISTS

Pelitic schists are formed from fine-grained sedimentary rocks. They are an important metamorphic rock type because they undergo distinct textural and mineralogical changes, used by geologists to gauge the temperatures and pressures under which the rocks are progressively modified.

PRINCIPAL TERMS

EQUILIBRIUM: a situation in which a mineral is stable at a given set of temperature-pressure conditions

INDEX MINERAL: an individual mineral that has formed under a limited or very distinct range of temperature and pressure conditions

ISOGRAD: a line on a geologic map that marks the first appearance of a single mineral or mineral assemblage in metamorphic rocks

METAMORPHIC ZONE: areas of rock affected by the same limited range of temperature and pressure conditions, commonly identified by the presence of a key individual mineral or group of minerals

MICA: a platy silicate mineral (one silicon atom surrounded by four oxygen atoms) that readily splits in one plane

MINERAL: a naturally occurring chemical compound that has an orderly internal arrangement of atoms and a definite chemical formula

PROGRESSIVE METAMORPHISM: mineralogical and textural changes that take place as temperature and pressures increase

SEDIMENTARY ROCK: a rock formed from the physical breakdown of preexisting rock material or from the precipitation—chemically or biologically—of minerals

SHALE/MUDSTONE: sedimentary rock composed of fine-grained products derived from the physical breakdown of preexisting rock; shales break along distinct planes and mudstones do not

TEXTURE: the size, shape, and arrangement of crystals or particles in a rock

AGENTS OF METAMORPHISM

Metamorphism literally means "the change in form" that a rock undergoes. More precisely, metamorphism is a process by which igneous and sedimentary rocks are mineralogically, texturally, and, occasionally, chemically modified by the effects of one or more of the following agents or variables: increased temperature, pressure, or chemically active fluids. Pelitic schists have long been recognized as one of the best rock types to gauge and preserve the wide range of mineralogical and textural changes that take place as metamorphic conditions progressively increase. It is perhaps easiest to understand the conditions under which metamorphism occurs by excluding those conditions that are not generally considered metamorphic. At or near the Earth's surface, sedimentary rocks form from the physical or chemical breakdown of preexisting rocks at relatively low pressures and temperatures (less than 200 degrees Celsius). Igneous rocks form from molten material at high temperatures (650-1,100 degrees Celsius) and at low pressures for volcanic rocks to high pressures for those formed at great depth. The conditions that exist between these two are those that are considered metamorphic.

Although all the agents of metamorphism work together to produce the distinctive textures and minerals of regionally metamorphosed rocks, each has its own special or distinctive role. Temperature is considered the most important factor in metamorphism. It is the primary reason for recrystallization and new mineral growth. Two types of pressure affect metamorphic rocks. The pressure caused by the overlying material and uniformly affecting the rock on all sides is referred to as the confining pressure. An additional pressure, referred to as directed pressure, is normally caused by strong horizontal forces. While confining pressure has little apparent influence on metamorphic textures, directed pressures are considered the principal reason that several types of metamorphosed rocks develop distinctive textures. Chemically active fluids are important because they act both as a transporting medium for chemical constituents and as a facilitator of chemical reactions during metamorphism. As metamorphic conditions increase, fluids play an important role in aiding metamorphic reactions, even though they are concurrently driven out of the rocks. Most metamorphic rocks preserve mineral assemblages that represent the highest conditions achieved. As metamorphic conditions start to fall, reactions are very slow to occur because adequate fluids are normally not available. If the metamorphism involves significant additions and losses of constituents other than water, carbon dioxide, and other fluids, this process is referred to as metasomatism. For example, if the magnesium content of a rock markedly increases during metamorphism, this is referred to as magnesium metasomatism.

TYPES OF METAMORPHISM

Geologists recognize several types of metamorphism, but one, regional metamorphism, is predominant. Regional metamorphism takes place at depth within the vast areas where new mountains are forming. The products of regional metamorphism are exposed in a variety of places. One of the best is on shields—broad, relatively flat regions within continents. There, the thin veneer of sedimentary rocks has been stripped away, exposing wide areas of deeply eroded ancient mountain belts and their regionally metamorphosed rocks, complexly contorted and intermingled with igneous rocks. The cores of older eroded mountain belts, such as the Appalachians and the Rockies, also commonly expose smaller areas of very old regionally metamorphosed rocks. Pelitic schists mainly occur in regionally metamorphosed rocks. They may also occur in contact metamorphic deposits, which are small, baked zones that formed immediately adjacent to igneous bodies that invaded pelitic rocks as hot molten material. In contrast to the thousands of square kilometers typically occupied by regionally metamorphosed rocks, individual contact metamorphic deposits are rarely more than several kilometers wide.

FOLIATION

Fine-grained sedimentary rocks (shales and mudstones) that formed from the physical breakdown of other rock types were once commonly referred to as pelites. Although the terms "pelite" and "pelitic" are largely absent in modern treatments of sedimentary rocks, they are widely used in the literature on metamorphism. Pelitic rocks undergo very marked textural and mineralogical changes that geologists use to gauge the conditions of formation in a particular area. "Schist," as used in the term "pelitic schist," has several definitions. The name may be applied broadly to any rock that is foliated or contains minerals that have a distinct preferred planar orientation. "Schist" also has a much narrower definition, referring only to foliated, coarse-grained rocks in which most mineral grains are visible to the unaided eye. This foliation, or preferred orientation, is im-

Kyanite (Al$_2$SiO$_5$), one of three key minerals found in pelitic schists. (© William E. Ferguson)

TEMPERATURE-PRESSURE SCHEMATIC SHOWING MINERAL ZONES

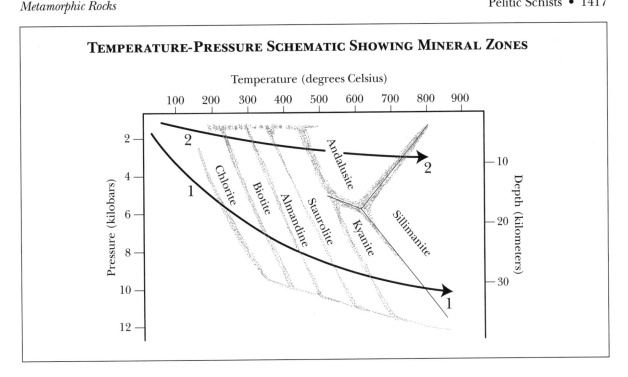

parted upon the rock because of deformation and recrystallization in response to a pronounced directed pressure at elevated temperatures.

The development of foliation is the most diagnostic feature of rocks that have undergone regional metamorphism. Directed pressure produces a variety of foliations that change as a function of the conditions of formation. An unmetamorphosed pelitic rock is typically a very fine-grained, soft rock that may or may not be finely layered (containing closely spaced planes). At low temperatures and pressures (low grades of metamorphism), pelitic rocks remain fine-grained, but microscopic micaceous minerals form or recrystallize and align themselves perpendicular to the directed pressure. That produces a dense, hard rock called a slate, which readily splits parallel to the preferred orientation. Any layering may be partially or totally obliterated. The accompanying texture is called a flow or slaty cleavage. At slightly higher conditions, micaceous minerals become better developed but remain fine-grained. The foliated rock produced takes on a pronounced sheen and is referred to as a phyllite. As conditions continue to increase, grain size increases until it is visible. This change produces a texture referred to as a schistosity, and the rock is called a

schist. At very high conditions, the micas that were diagnostic of lower conditions start to become unstable. The rock, referred to as a gneiss, remains foliated but does not readily split as the slates, phyllites, and schists do. The foliation, or gneissosity, takes the form of an alternating light and dark banding. At higher conditions, gneissic rocks gradually grade into the realm where igneous rocks form.

MINERALOGIC ZONES

The first study to show that a single rock type can undergo progressive change on a regional scale was that of British geologist George Barrow toward the end of the nineteenth century in the Highlands of Scotland, southwest of Aberdeen. His work indicated that pelitic schists and gneisses contained three discrete mineralogic zones, each represented by one of three key minerals: staurolite, kyanite, and sillimanite. Barrow suggested that increasing temperature was the controlling agent for the zonation observed. Later work by Barrow and other geologists confirmed this hypothesis and broadened the mapping over the entire Highlands, revealing additional mineral zones that formed at lower metamorphic conditions. From lowest to highest temperatures, several mineralogical zones were recognized. In the chlorite zone, there are slates, phyllites, and mica schists generally containing quartz, chlorite, and muscovite. In the biotite zone, mica schists are marked by the appearance of biotite in association with chlorite, muscovite, quartz, and albite. Biotite occurs in each of the higher zones. A line marking the boundary of the chlorite and biotite zones is referred to as the biotite isograd. (An isograd is a line that marks the first appearance of the key index mineral that is distinctive of the metamorphic zone.) In the almandine (garnet) zone, mica schists are characterized by the presence of almandine associated with quartz, muscovite, biotite, and sodium-rich plagioclase. The garnet first appears along the almandine isograd and occurs through all the higher zones. In the staurolite zone, mica schists typically contain quartz, muscovite, biotite, almandine, staurolite, and plagioclase. In the kyanite zone, mica schists and gneisses contain quartz, muscovite, biotite, almandine, kyanite, and plagioclase. Staurolite is no longer stable in this zone. In the sillimanite zone, mica schists and gneisses are characterized by quartz, muscovite, biotite, almandine, sillimanite, and plagioclase. Sillimanite forms at the expense of kyanite. These zones—also called Barrovian zones—are recognized throughout the world.

STUDY OF PELITIC ROCKS

The study of pelitic schists can be conducted from a number of perspectives. The foundation of all studies, however, is the kind of fieldwork that George Barrow conducted in Scotland, which includes careful mapping, sample collection, rock description, and structural measurement.

Once the data and samples are returned to the laboratory, other methods of investigation may be employed. Rocks are commonly studied under the binocular microscope or powdered and/or made into thin sections to be analyzed in polarized light transmitted through the sample beneath the petrographic microscope. The preparation of the standard thin section involves several steps: cutting a small block of the sample so that it is 2.7 millimeters by 4.5 millimeters on one side; polishing one side and glueing it to a glass slide; and cutting and grinding the glued sample to a uniform thickness of 0.03 millimeter. Observations and descriptions of these thin sections are essential to the study of pelitic rocks because they commonly provide information about the interrelationships among the minerals present and clues to the metamorphic history of the sample.

Whole-rock chemical analyses of pelitic rocks are also important. Standard analytical procedures, atomic absorption, and X-ray fluorescence spectroscopy are all methods used to determine the amounts of the elements within these rocks. Chemical analyses of individual minerals also provide important information about pelitic rocks.

MINERAL ANALYSES

Although mineral analyses have traditionally been conducted by the same analytical techniques as whole-rock analyses, problems obtaining material pure enough for analysis have plagued scientists. Since the 1960's, almost all published mineral analyses have been produced on the electron microprobe, which greatly improves the accuracy of the analyses. These whole-rock analyses and individual mineral analyses are commonly plotted together in several ways on triangular graph paper in order to illustrate the kinds of mineral associations that can exist at different metamorphic conditions.

In experimental petrology, metamorphic minerals are grown under a variety of controlled equilibrium conditions to provide geologists with a better understanding of the actual conditions for their formation. Although such studies have greatly enhanced knowledge of naturally occurring metamorphic reactions, the field is not without controversy. The behavior of the reactions of kyanite to sillimanite, andalusite to sillimanite, and kyanite to sillimanite as temperatures increase shows the great complexity of this endeavor. The various studies conducted on these minerals that occur in pelitic schists are not in agreement, so many geologists remain deeply divided about the exact conditions under which these three minerals are stable.

Physical chemistry and thermodynamics provide geologists with tools to predict how minerals in idealized, simplified chemical reactions theoretically behave. Such comparisons are extremely important in making judgments about more complex natural environments.

ECONOMIC USES

The only important pelitic rock type that is useful as such is very low-grade slate that is quarried for flagstone, roofing material, countertops, blackboards, billiard tables, and switchboard panels. Yet, pelitic rocks do contain a number of minerals that are extracted for their usefulness.

Staurolite, almandine-garnet, corundum, and kyanite are all minerals from pelitic rocks that have some use as gemstones. In addition, staurolite crystals grow together and may form crosses, which are sold as amulets called fairy stones, although most objects sold as "fairy stones" are not genuine staurolite. (The name "staurolite" is derived from the Greek word meaning "cross" because of the mineral's diagnostic cruciform twin.) The gem forms of corundum occur as two important varieties: the ruby and the sapphire. The ruby is characteristically red; sapphires, which are traditionally considered blue, occur in a wide variety of colors. Corundum also has limited use as an abrasive, particularly for the production of optical lenses. Garnet, from pelitic schists, also has some use as an abrasive for sandblasting and spark plug cleaning. In addition, sillimanite and kyanite are used as refractory (high-temperature) materials in porcelain and spark plugs.

Graphite is another mineral that may occur in or is associated with pelitic schists. Deposits near Turin, Italy, occur in micaceous phyllites, schists, and gneisses. Graphite has a variety of uses, including the production of refractory crucibles for the making of bronze, brass, and steel. It is used with petroleum products as a lubricant and blended with fine clay in the "leads" of pencils. It is also used in steel, batteries, generator brushes, and electrodes, and for electrotype.

Another relatively uncommon mineral, pyrophyllite, is found in very low-grade pelitic rocks and has properties and uses similar to those of talc. Pyrophyllite is used in paints, paper, ceramics, and insecticides, and as an absorbent powder. A special variety of pyrophyllite, called agalmatolite, is prized by the Chinese for the carving of small objects.

Ronald D. Tyler

CROSS-REFERENCES

Anatexis, 1373; Blueschists, 1379; Contact Metamorphism, 1385; Foliation and Lineation, 1389; Groundwater Movement, 2030; Hydrothermal Mineralization, 1205; Metamorphic Mineral Deposits, 1614; Metamorphic Rock Classification, 1394; Metamorphic Textures, 1400; Metasomatism, 1406; Regional Metamorphism, 1418; Sub-Seafloor Metamorphism, 1423.

BIBLIOGRAPHY

Bates, Robert L. *Geology of the Industrial Rocks and Minerals.* Mineola, N.Y.: Dover, 1969. A reference book that deals with rocks and minerals that are extracted because of their economic importance. Slate, graphite, pyrophyllite, kyanite, and corundum, all products of metamorphosed pelitic rocks, are considered in terms of their properties and uses, production, and occurrences. Includes an index and an excellent bibliography.

Blatt, Harvey, and Robert J. Tracy. *Petrology: Igneous, Sedimentary, and Metamorphic.* New York: W. H. Freeman, 1996. An upper-level college text that deals with the descriptions, origins, and distribution of igneous, sedimentary, and metamorphic rocks. Pelitic rocks figure prominently in the treatment of the occurrences, graphic representations, the processes of metamorphism, and the mineral changes during metamorphism. Includes an index and bibliographies for each chapter.

Dietrich, Richard V., and B. J. Skinner. *Rocks and Rock Minerals.* New York: John Wiley & Sons, 1979. A college-level book that focuses on the description and identification of rocks and minerals through simple methods intended for use in the field. Chapter 6 provides a good overview of metamorphism, emphasizing the important role that pelitic rocks play in the study of regional metamorphism.

Ehlers, Ernest, and H. Blatt. *Petrology: Igneous, Sedimentary, and Metamorphic.* San Francisco: W. H. Freeman, 1982. An upper-level college text that deals with the descriptions, origins, and distribution of igneous, sedimentary, and metamorphic rocks. Pelitic rocks figure prominently in the treatment of the occurrences, graphic representations, the processes of metamorphism, and the mineral changes during metamorphism. Includes an index and bibliographies for each chapter.

Ernst, W. G. *Earth Materials.* Englewood, N.J.: Prentice-Hall, 1969. A compact but excellent introduction to the study of rocks and minerals. Chapter 7, which treats metamorphism, emphasizes the importance of textural and mineralogical changes that pelitic rocks undergo during regional metamorphism. Chapter 2 also briefly addresses the mineral reactions among the index minerals kyanite, sillimanite, and andalusite.

Gillen, Cornelius. *Metamorphic Geology.* Winchester, Mass.: Allen & Unwin, 1982. A readable, well-illustrated introduction to metamorphism that focuses on textures and field relationships. Provides a good discussion of pelitic rocks and Barrovian zones. Also emphasizes the strong tie between metamorphic processes and mountain-building events. Includes a glossary and a short bibliography.

Howie, Frank M., ed. *The Care and Conservation of Geological Material: Mineral Rocks, Meteorites, and Lunar Finds.* Oxford: Butterworth-Heinemann, 1992. This collection of essays written by leading geologists examines mineralogy and the processes associated with the field. It provides a good treatment of the fine minerals that make up schists and pelitic schists.

Lindsley, Donald H., ed. *Oxide Minerals: Petrologic and Magnetic Significance.* Washington, D.C.: Mineralogical Society of America, 1991. An excellent overview of petrology and petrogenesis studies. Many of the essays focus on oxide minerals and their magnetic properties. There is a detailed section on the geochemical makeup of schists.

Mason, R. *Petrology of the Metamorphic Rocks.* Winchester, Mass.: Allen & Unwin, 1978. A college-level text intended for the second-year geology student that deals with metamorphic rocks from the perspective of descriptions in the field and beneath the petrographic microscope. Discussion of pelitic rocks covers both regional and contact metamorphism. A glossary, bibliography, and index are included.

Miyashiro, Akiho. *Metamorphism and Metamorphic Belts.* New York: John Wiley & Sons, 1973. An excellent college-level text that treats several types of metamorphism but focuses on regional metamorphism. Early in the text, emphasis is on the mineralogical reactions that occur in pelitic rocks (and other rock types) that best illustrate the concepts of progressive metamorphism. Includes an index and a fine bibliography.

Oldershaw, Cally. *Rocks and Minerals.* New York: DK, 1999. This small, 53-page volume is filled with color illustrations and is therefore of great use to new students who may be unfamiliar with the rock and mineral types discussed in classes or textbooks.

Press, Frank, and Raymond A. Siever. *Understanding Earth.* 2d ed. New York: W. H. Freeman, 1998. This book introduces the high school and college reader without a geology background to the subject of metamorphism and metasomatism. Good index and illustrations.

Tennisen, A. C. *Nature of Earth Materials.* 2d ed. Englewood Cliffs, N.J.: Prentice-Hall, 1983. A text written for the nonscientist that covers the nature of atoms and minerals and igneous, sedimentary, and metamorphic rocks. Also includes a section on the uses of these materials. Well illustrated. Bibliography and subject index are included.

Winkler, H. G. F. *Petrogenesis of Metamorphic Rocks.* 5th ed. New York: Springer-Verlag, 1979. An upper-level college text that addresses the chemical and mineralogical aspects of metamorphism. Emphasizes the principle that mineral reactions in common rock types can be used to determine metamorphic conditions. Separate chapters deal with pelitic and other important rock types. Includes index and short bibliographies after each chapter.

REGIONAL METAMORPHISM

Regional metamorphism, which takes place in the roots of actively forming mountain belts, is a process by which increased temperatures and pressures cause a rock to undergo recrystallization, new mineral growth, and deformation. These changes diagnostically cause minerals to develop a preferred orientation, or foliation, in the rock.

PRINCIPAL TERMS

EQUILIBRIUM: a situation in which a mineral is stable at a given set of temperature-pressure conditions

FACIES: a part of a rock or group of rocks that differs from the whole formation in one or more properties, such as composition, age, or fossil content

IGNEOUS ROCK: a rock formed from the cooling of molten material

MICA: a platy silicate mineral (one silicon atom surrounded by four oxygen atoms) that readily splits in one plane

MINERAL: a naturally occurring chemical compound that has an orderly internal arrangement of atoms and a definite formula

SEDIMENTARY ROCK: a rock formed from the physical breakdown of preexisting rock material or from the precipitation, chemically or biologically, of minerals

SHALE: a sedimentary rock composed of fine-grained products derived from the physical breakdown of preexisting rock material

STRAIN: change in volume or size in response to stress

STRESS: force per unit of area

TEXTURE: the size, shape, and arrangement of crystals or particles in a rock

ZEOLITES: members of a mineral group with very complex compositions: aluminosilicates with variable amounts of calcium, sodium, and water

ROCK TEXTURE AND COMPOSITION

Metamorphism is a process whereby igneous or sedimentary rocks undergo change in response to some combination of increased temperature, pressure, and chemically active fluids. Several types of metamorphism are recognized, but regional metamorphism is the most widespread. Regionally metamorphosed rocks in association with igneous rocks make up about 85 percent of the continents. In order to describe regionally metamorphosed rocks and to understand the processes by which they form, one must understand two properties of these rocks: texture and composition. Texture refers to the size, shape, and arrangement of crystals or particles in a rock. Composition refers to the minerals present in a rock.

The effect of temperature and pressure on texture and composition is aided by fluids. Fluids transport chemical constituents and facilitate chemical reactions that take place in the rocks. As metamorphic conditions increase, metamorphic reactions readily occur if fluids are present; however, fluids are gradually driven off as recrystallization reduces open spaces. Once fluids are eliminated and temperatures and pressures decrease, the reactions generally occur very slowly. Therefore, in most metamorphic rocks, the minerals preserved are those that represent the highest temperatures and pressures attained, assuming equilibrium when adequate fluids were still available.

Regionally metamorphosed rocks characteristically develop distinctive textures, largely because of the role of two types of pressure in concert with elevated temperatures: confining pressure, or the force caused by the weight of the material overlying a rock, and a stronger directed pressure, or a horizontal force created by mountain-building processes.

FOLIATION, SCHISTOSITY, GNEISSIC TEXTURE

As metamorphism commences, rocks become deformed, and original features are strongly distorted or obliterated. Original minerals either re-

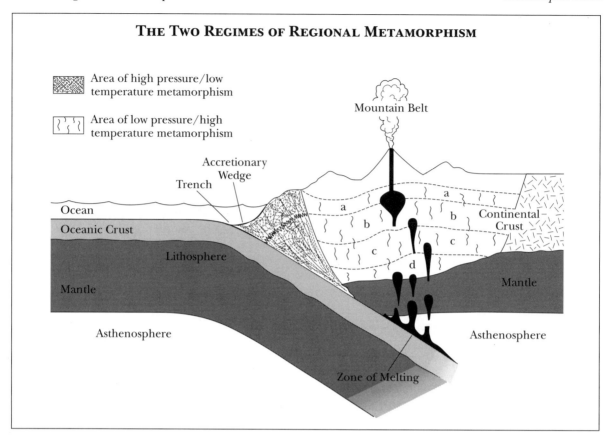

THE TWO REGIMES OF REGIONAL METAMORPHISM

Area of high pressure/low temperature metamorphism

Area of low pressure/high temperature metamorphism

Mountain Belt

Accretionary Wedge

Trench

Ocean

Oceanic Crust

Lithosphere

Mantle

Asthenosphere

Continental Crust

Mantle

Asthenosphere

Zone of Melting

crystallize or react to form new minerals and develop a pronounced preferred orientation, or foliation, in response to the directed pressure. Preferentially aligned platy minerals (such as the micas), formed during the early stages of metamorphism, enhance the development of foliation. Foliation assumes different forms depending on the degree of metamorphism.

Although the minerals in most rock types characteristically develop preferred orientations, shales and other fine-grained rock types display more distinct changes of texture. Shales undergo textural changes as they pass from conditions of low-grade metamorphism (low temperatures and pressures) at shallow depths to high grades (higher temperatures and pressures) deeper in the Earth. At low grades of regional metamorphism, microscopic platy minerals grow and align themselves perpendicular to the orientation of the directed pressure. This realignment allows this fine-grained rock, called a slate, to split or cleave easily along this preferred direction. Under some-

what higher metamorphic grades deeper in the Earth, the platy minerals continue to develop parallel arrangements in response to the directed pressures, but the resultant mineral grains, typically the micas, are now large enough to be identified with the unaided eye. This texture, a schistosity, forms in a rock called a schist. At higher grades, the platy minerals react to form new minerals, and a gneissic texture develops. This texture is distinguished by alternating layers of light and dark minerals that give the rock, called a gneiss, a banded appearance.

INDEX MINERALS

Just as textural changes in metamorphic rocks are generally predictable as grade increases, minerals also appear or disappear in a systematic sequence and become even better gages of the conditions of metamorphism. A progression of key minerals in shales, called index minerals, form as the grade of metamorphism increases. These index minerals, therefore, occur in rocks metamorphosed at a given range of temperature-pressure conditions. When mapped in the field, areas containing index minerals are called metamorphic zones. As the metamorphic grades increase, the typical sequence of the index minerals is chlorite, biotite, garnet, staurolite, kyanite, and sillimanite.

Since metamorphic zonation, based on index minerals, relies on the presence of shales, if shales are not abundant, zonation can be limited. Fortunately, another type of metamorphic zonation is recognized that is not based on single minerals in one rock type but on sequences of mineral assemblages in groups of associated rocks within ranges of temperature-pressure conditions. These ranges of temperatures and pressures in which diagnostic mineral assemblages in different rock types may exist are called metamorphic facies. The zeolite facies represents the lowest conditions of regional metamorphism. At the lowest extremes of this facies, conditions considered sedimentary merge with those that are metamorphic. This facies is named for the zeolite group of minerals that form within this facies. As metamorphic conditions increase, the assemblages of minerals distinctive of the zeolite facies break down, and new assemblages form that are distinctive of either the blue-schist or the greenschist facies. As the names of these facies imply, the textures formed under these conditions are commonly schistose, and a number of the minerals impart either blue or green tints to the rocks in their respective facies. In a given area of regional metamorphism, it is typical to find rocks that represent a sequence of metamorphic facies that follow either of two pathways. One path has temperature increasing relatively more rapidly than pressure, so the combination of facies that it crosses is referred to as a low pressure/high temperature sequence. Assemblages that represent an alternate pathway, in which temperature increases more slowly relative to pressure, are termed a high pressure/low temperature sequence.

UNDERSTANDING REGIONAL METAMORPHISM PROCESSES

By understanding how mountains are formed, scientists can learn how regionally metamorphosed rocks are generated. Regional metamorphism occurs in the roots of actively forming mountain belts. Linear zones thousands of kilo-meters long and hundreds of kilometers wide can be involved. It will take millions of years before the products of regional metamorphism are exposed in the Andes or Cascades. On all the continents, however, broad areas of exposed rocks, called shields, now reveal the products of numerous ancient mountain-building events.

In the 1960's, the concept of plate tectonics revolutionized geology and scientists' understanding of the processes in regional metamorphism. For decades, geologists have known that the Earth is divided into several concentric layers. The outermost zone, the crust, consists of relatively low-density rocks and averages 35 kilometers thick on the continents and 10 kilometers thick under the oceans. Beneath the crust is the mantle (nearly 3,000 kilometers thick), and then the core.

With the advent of plate tectonics, another important subdivision was recognized. The upper 60-100 kilometers, which includes the crust and a portion of the upper mantle, behaves in a relatively rigid fashion and is composed of about twenty pieces that move independently of one another. This rigid zone is called the lithosphere, and the pieces are called lithospheric plates. Because each plate is moving relative to its neighbors, three types of boundaries with other plates exist: those in collision, those pulling apart, and those sliding by one another. The rigid lithosphere apparently floats on a soft plastic layer, called the asthenosphere, that extends from 60-100 kilometers to at least 250 kilometers into the mantle.

CONVERGENT BOUNDARIES

Regional metamorphism is restricted to margins in collision, called convergent boundaries. The collision of two lithospheric plates produces several profound effects. One plate subducts, or sinks, under the other but manages to crumple the edge of the continent into a mountain belt. An oceanic trench forms at the point of contact. As one plate plunges downward, melting occurs where temperatures sufficiently rise. The molten material rises through the lithosphere to produce volcanoes at the surface and masses of igneous rock at depth. It is in this realm—extending from the zone of collision to well beyond the igneous bodies—that regional metamorphism occurs. Within this area, two distinct sequences of regional metamorphism develop. The high pressure/low temperature sequence forms in the region nearest the trench. As sediments and volcanic rocks are carried rapidly (in geological terms) downward into the subduction zone, they attain relatively high pressures but remain relatively cold because they have not had the time to heat up. The minerals that form in these high-pressure conditions are those of the zeolite and blueschist facies. If the downward movement continued uninterrupted, the rocks would eventually heat up. Before that can happen, however, thin, cold slabs with high-pressure mineral assemblages within them are chaotically pushed or thrust back toward the surface. Why this thrusting occurs is not entirely clear, but it is not uncommon to find a region 100 kilometers wide, adjacent to a trench, composed entirely of highly deformed high pressure/low temperature facies rocks. This region is called an accretionary wedge because of its inferred shape in a vertical slice through the Earth. It is this material that allows the continents to grow larger through time.

Study of Regionally Metamorphosed Rocks

The study of regionally metamorphosed rocks is conducted from several different perspectives. At the core of these studies are the field observations. Descriptions of rock units and structures provide information on how these rocks have formed. Much of this fieldwork is synthesized into a geological map.

In the laboratory, samples from the field are studied by a variety of physical and chemical methods. Studies with the petrographic microscope yield information on mineral compositions and textural relationships that provide clues for classifying and determining modes of origin for these rocks. X-ray powder diffraction techniques are commonly used to identify metamorphic minerals not readily identified by visual inspection. Since metamorphic rocks are markedly changed texturally and mineralogically from their original state, chemical analysis provides data that can be compared to probable premetamorphic rock types. Standard classical chemical methods of analysis may be used, but a number of more sophisticated, faster, and simpler spectrographic methods are popular. Atomic absorption and X-ray fluores-

cence spectroscopy are two of the more widely applied techniques to determine elemental abundances, although emission spectrographic and neutron activation analyses are also used. Another important tool is the electron microprobe. Most published mineral analyses are generated by electron microprobe analyses. Mass spectroscopic analyses also provide information on the distribution of isotopes in metamorphic rocks and minerals. These data are useful in determining age and conditions of formation.

Rock deformation experiments provide information about the mechanical properties of metamorphic rocks. Rock samples subjected to stress and strain tests yield data on properties such as plasticity, strength, and viscosity. These data are particularly important in understanding how directed pressures influence textures. Another phase of the study of metamorphic rocks is experimental petrology. Here, metamorphic minerals are synthesized under controlled-equilibrium conditions. From these studies, geologists gain knowledge about the actual ranges of stability of these minerals in naturally occurring environments. The study of the crystal chemistry and thermodynamics provides information regarding actual conditions of formation and potential reactions. The roles of variables—such as entropy, volume change, and heats of reaction—provides clues to the behavior of mineral assemblages at varying temperature and pressure conditions. Theoretical petrology, the treatment of data on metamorphic rocks by mathematical models or the principles of theoretical physics, also provides important information in the study of regional metamorphism.

In addition to analyzing the samples collected, scientists analyze field data in the laboratory. Statistical analysis of orientations of foliations and other structural features is greatly facilitated by the computer. Laboratory analyses are recast to generate a variety of graphical treatments to check for chemical trends or metamorphic facies relationships.

Applications for Industry and Engineering

The products of regional metamorphism touch people's lives in many ways. Marbles, gneisses, and slates are used as building and cut ornamental stones, and some quartzites, marbles, and gneisses are used as aggregate. The minerals graphite, talc, vermiculite, and asbestos are four of the more commonly used mineral products of regional metamorphism, whereas wollastonite, garnet, kyanite, emery, and pyrophyllite are mineral products of limited use. Graphite is most commonly used in the metallurgical industry, with lesser amounts used as lubricants, in paints, in batteries, as pencil "leads," and in electrodes. Talc is also a product with diverse uses. The ceramics industry is the largest consumer, followed by paint and paper manufacturers. Vermiculite is a common product in thermal and acoustic insulation. Asbestos, despite concerns about its health risks, remains an important fire-retardant material for appropriate uses.

Regionally metamorphosed rocks provide certain advantages in construction but also present unique engineering problems. A scan of the skyline of the island of Manhattan in New York City reveals that the largest skyscrapers are restricted to several areas of the island, surrounded by expanses of much shorter buildings. The reason that buildings such as the Empire State Building occupy only certain areas is that they sit directly on structurally strong regionally metamorphosed rocks that occur at or near the surface. In other areas of the island, these rocks are too deeply buried, and the overlying sediments are too weak, to support the larger structures. In other cases, inclined foliations paralleling slopes in hilly or mountainous regions represent potential planes of slippage that can give way and produce massive rock slides, particularly when road, mine, or dam construction modifies the landscape. In areas where foliations are not adequately taken into account, the cost can be millions of dollars in repairs and/or great loss of life. Special techniques and restrictive measures need to be applied when construction occurs in areas where foliations present potential problems.

Ronald D. Tyler

CROSS-REFERENCES

Anatexis, 1373; Blueschists, 1379; Contact Metamorphism, 1385; Foliation and Lineation, 1389; Metamorphic Rock Classification, 1394; Metamorphic Textures, 1400; Metasomatism, 1406; Mountain Belts, 841; Pelitic Schists, 1412; Sub-Seafloor Metamorphism, 1423.

BIBLIOGRAPHY

Augustithis, S. S. *Atlas of Metamorphic-Metasomatic Textures and Processes.* Amsterdam: Elsevier, 1990. A thoughtful treatment of metamorphic rocks, minerology, and metasomatism. Excellent illustrations add to understanding concepts and processes. Appropriate for the college student or layperson with an interest in the Earth sciences.

Bates, Robert L. *Geology of the Industrial Rocks and Minerals.* Mineola, N.Y.: Dover, 1969. Readable text that concentrates on the occurrences and uses of rocks and minerals. Key chapters 4, 9, and 11 deal with metamorphic rocks, metamorphic minerals, and minor industrial minerals, respectively. All provide excellent summaries of how the products of regional metamorphism are utilized. Includes an index and good bibliography.

Best, Myron G. *Igneous and Metamorphic Petrology.* 2d ed. San Francisco: W. H. Freeman, 1995. A readable advanced college-level text that generally should not be beyond the general reader who wants to learn more about metamorphic textures. This illustrated treatment of metamorphism includes the topics of mineralogy, chemistry, and the structure of metamorphic rocks. Both field relationships and global-scale tectonic associations are discussed.

Blatt, Harvey, and Robert J. Tracy. *Petrology: Igneous, Sedimentary, and Metamorphic.* New York: W. H. Freeman, 1996. An upper-level college text that deals with the descriptions, origins, and distribution of igneous, sedimentary, and metamorphic rocks. Includes an index and bibliographies for each chapter.

Chernikoff, Stanley. *Geology: An Introduction to Physical Geology.* Boston: Houghton-Mifflin, 1999. This is a good overview of the scientific understanding of the geology of the Earth and surface processes. Includes the address of a Web site that provides regular updates on geological events around the globe.

Dietrich, R. V., and B. J. Skinner. *Rocks and Rock Minerals.* New York: John Wiley & Sons, 1979. A college-level book that describes rocks and minerals in very clear, concise terms, emphasizing simple identification techniques. Gives an excellent but succinct overview of metamorphic rocks and their conditions of formation, notable occurrences, and uses.

Ernst, W. G. *Earth Materials.* Englewood Cliffs, N.J.: Prentice-Hall, 1969. A short, highly regarded introductory-level book that uses crystal structures, the concepts of chemical equilibria, physical chemistry, and thermodynamic principles to explain the formation of minerals in the three major rock types. This textbook is concerned with the stability of metamorphic mineral assemblages and textures in terms of temperature-pressure conditions.

Gillen, Cornelius. *Metamorphic Geology: An Introduction to Tectonic and Metamorphic Processes.* London: Allen & Unwin, 1982. This short, well-illustrated book provides an introduction to metamorphism and indicates the strong relationships among mountain building, plate tectonics, and metamorphic processes. Regional examples are largely European, but this is one of the few introductory-level texts that is fully devoted to metamorphism. Includes glossary and short bibliography.

Mason, R. *Petrology of the Metamorphic Rocks.* London: Allen & Unwin, 1978. In addition to short sections on experimental and theoretical petrology, isotope and electron-microprobe analyses, the focus of this second-year college-level text is the description of metamorphic rocks in the field and under the microscope. This illustrated book of modest length is written from the perspective of a European geologist.

Miyashiro, A. *Metamorphism and Metamorphic Belts.* New York: John Wiley & Sons, 1973. Readable college-level text that deals with the basic concepts, characteristics, and problems of metamorphic geology. Focuses on the use of shales and specific igneous rocks to help explain metamorphism. Summarizes classic metamorphic belts throughout the world.

Tennisen, A. C. *Nature of Earth Materials.* 2d ed. Englewood Cliffs, N.J.: Prentice-Hall, 1983. A text written for the nonscientist that covers the nature of atoms and minerals as well as igneous, sedimentary, and metamorphic rocks. Also includes a section on the uses of these materials. Well illustrated, with a bibliography and subject index.

SUB-SEAFLOOR METAMORPHISM

Sub-seafloor metamorphism of oceanic ridge basalts by magma-driven, convecting seawater induces significant changes in the chemical composition of both rock and circulating fluid. These changes are a major factor in the exchange of elements between the lithosphere and the hydrosphere and play a critical role in the origin of exhalative ore deposits.

PRINCIPAL TERMS

BLACK SMOKERS: active hydrothermal vents along seafloor ridges, which discharge acidic solutions of high temperature, volume, and velocity, charged with tiny black particles of metallic sulfide minerals

CONVECTION: fluid circulation produced by gravity acting on density differences arising from unequal temperatures within a fluid; the principal means of heat transfer involving fluids of low thermal conductivity

FACIES: a part of a rock or a rock group that differs from the whole formation in one or more properties, for example, composition

METASOMATISM: chemical changes in rock composition that accompany metamorphism

OPHIOLITE COMPLEX: an assemblage of metamorphosed basaltic and ultramafic igneous rocks that originate at marine ridges and are subsequently emplaced in mobile belts by plate-collision tectonics

PILLOW BASALT: a submarine basaltic lava flow in which small cylindrical tongues of lava break through the surface, separate into pods, and accumulate downslope in a formation resembling a pile of sandbags

REGIONAL METAMORPHIC FACIES: the particular pressure and temperature conditions prevailing during metamorphism as recorded by the appearance of a new mineral assemblage

THERMAL GRADIENT/GEOTHERMAL GRADIENT: the rate of temperature increase with depth below the Earth's surface

WATER-TO-ROCK RATIO: the mass of free water in a given volume of rock divided by rock mass of the same volume; processes occurring at water-to-rock ratios less than 50 are "rock-dominated," while those greater than 50 are "fluid-dominated"

IDENTIFICATION OF THE PROCESS

The seafloor spreading process generates an estimated 3×10^{16} grams of new basaltic crust per year along the world's ocean ridge system. Magma rising from the mantle produces this juvenile ocean-floor crust, which consists of upper layers of permeable lavas and lower layers of dike networks cutting gabbroic intrusive bodies. The thermal gradients and permeabilities of rocks in the vicinity of ocean ridges are both high, resulting in an environment favorable for convective cooling of the rock sequence by circulating seawater. Intensive studies of heat-flow patterns across ocean ridge systems in the early 1970's confirmed that large-scale seawater circulation through hot basaltic crust is actually occurring. During the same period, advances in structural and chemical studies of ophiolite complexes on land led geologists to conclude that these peculiar rock sequences are fragments of oceanic crust and upper mantle that were tectonically emplaced by plate collisions.

A principal line of evidence used to support this conclusion is the fact that ophiolitic basaltic rocks had been metamorphosed and intensely veined prior to emplacement. The ophiolitic metabasalts were, in fact, identical in all respects to samples obtained in dredge hauls along the axial valleys of ocean ridges. As a result, a new form of rock metamorphism, regional in extent but restricted to the marine ridge environment, was recognized. The first scientific paper describing this phenomenon in a comprehensive and unified fashion appeared in 1973; this landmark paper referred to the newly recognized metamorphic process as sub-seafloor metamorphism. Subsequent researchers have introduced alternative terms, such as "submarine hydrothermal alteration" and "ocean-ridge metasomatism." For historical reasons, the original terminology is the most convenient; for brevity, the acronym SFM will be used.

It is important to recognize that SFM is a process involving both heat and mass transfer. Geochemists were quick to realize that large-scale, continuous alteration of ridge basalts by convecting seawater could have profound effects on the exchange of chemical elements between the Earth's lithosphere and hydrosphere. This hypothesis was tested by direct sampling of discharge fluids from submarine hot springs in the Galápagos Rift and the East Pacific in the late 1970's. Compared with normal seawater, these hot springs vent a fluid that is distinctly acidic; strongly depleted in magnesium (Mg) and sulfate (SO_4); enriched in silica (SiO_2), calcium (Ca), and hydrogen sulfide (H_2S); and enriched in a wide range of metallic elements. These compositional differences are the result of rock-seawater interaction and prove that SFM involves large-scale metasomatism. This is in marked contrast to regional metamorphism in continental mobile belts, which is a quasi-isochemical process.

REGIONAL METAMORPHIC FACIES

Samples recovered by dredging the steep escarpments of submarine ridge valleys represent the common regional metamorphic facies (zeolite, prehnite-pumpellyite, greenschist, and amphibolite), but the vast majority record the conditions of the greenschist facies. Most samples are metabasalts with mineralogy indicative of a temperature range of 100-450 degrees Celsius and very low pressures (400-500 bars) relative to continental metamorphism. The original mineralogy of ridge basalts, which consists of tiny crystals of calcic plagioclase, clinopyroxene, and olivine set in a glassy matrix, is converted by SFM to complex mixtures of actinolite, tremolite, hornblende, albite, chlorite, epidote, talc, clay, quartz, sphene, and pyrite that, however, tend to preserve the original igneous texture of the rock. Such assemblages are well known in basaltic rocks of continental mobile belts and ophiolite complexes, and they record greenschist facies conditions. For the majority of samples, greenschist facies metamorphism has produced an alteration assemblage which consists of albite-actinolite-chlorite-epidote and is accompanied by a small amount of quartz and pyrite. The

major mineralogical transformations recorded in these rocks appear to be plagioclase to albite to chlorite or to albite plus epidote; plagioclase plus clinopyroxene to chlorite plus epidote; olivine to chlorite plus pyrite; clinopyroxene to actinolite; and matrix glass to chlorite plus actinolite.

The intense hydrothermal veining observed in some dredge samples and especially in ophiolite complexes appears to record former high fracture-permeability, which is prerequisite to extensive basalt-seawater interaction. The major vein minerals are chlorite, actinolite, epidote, quartz, pyrite, and, less commonly, sulfides of iron, copper, and zinc. On the basis of mineralogy, greenschist facies alteration is described as chlorite-rich (chlorite greater than 15 percent of rock, epidote less than 15 percent) or epidote-rich (epidote greater than 15 percent, chlorite less than 15 percent). Chlorite-rich alteration is dominant in the dredge-haul samples recovered so far, and this preponderance is taken to mean that the chlorite-forming reactions occur at lower temperatures and closer to the seawater-basalt interface, where seawater influx of the basaltic crust begins. In other words, the transition from chlorite-rich to epidote-rich alteration records prograde metamorphism that, in the case of ridge basalts, correlates with increasing depth and temperature.

MAGNESIUM-METASOMATISM

The mechanics of ridge-basalt emplacement ensure that all such rocks have an opportunity to react with seawater and undergo metasomatism. Although the chemical composition of average mid-ocean-ridge basalt (MORB) is known, it is unlikely that it reflects exactly the pristine composition of the initial magma. The metasomatic nature of SFM has mainly been determined on the basis of dredge samples of pillow basalt. Individual pillows typically exhibit intensely altered rims and cores of relatively fresh basalt. On this basis, it is known that chlorite-rich alteration shows greater departures from parent-rock composition than does the higher-grade epidote-rich alteration. Chlorite-rich samples exhibit significant gains of magnesium and corresponding losses of calcium, but their state of oxidation is, for the most part, unaffected by the alteration process. Epidote-rich samples generally undergo minor losses of magnesium and gains of calcium but become somewhat oxidized by the alteration reactions.

Evidently, magnesium is removed from seawater and held in magnesium-rich secondary phases (especially chlorite) at the water-rock interface. The extent of this reaction depends on both temperature and water-to-rock mass ratio. At water-to-rock ratios less than 50 (in other words, rock-dominated conditions), magnesium removal from seawater is complete, the reaction rate being proportional to temperature. The form of this reaction is suggested by

$$3Mg^{++} + 4SiO_2 + 4H_2O = Mg_3Si_4O_{10}(OH)_2 = 6H^+$$
$$\text{(talc)}$$

where talc is used as a simple chemical analogue for the more complex formulas of the chlorites that actually form. Note that H^+ (positively charged hydrogen) appears on the product side of this reaction; this outcome means that the seawater solution will be acidic as long as magnesium remains in solution to drive the reaction. For rock-dominated conditions, however, magnesium is soon depleted and the acidic conditions associated with the reaction are short-lived. However brief, the acidic stage is important because it permits the circulating seawater to leach metallic elements from the enclosing basalt. At water-to-rock ratios in excess of 50 (that is, fluid-dominated conditions), magnesium cannot be completely removed from solution, so that the acidic, metal-leaching conditions of the reaction are maintained indefinitely. Electrical neutrality of the solution is balanced by the loss of calcium, sodium, and potassium from basalt because these elements are not utilized in the formation of chlorite. In contrast, iron and aluminum are retained in the rock because they participate in the formation of chlorite. The rate and magnitude of the reaction both increase with temperature and, because thermal gradients near ridges are very high, it would be expected that magnesium-metasomatism would be restricted to relatively shallow crustal depths. This limitation exists because fluids downwelling from the crust-seawater interface would become quickly heated.

SODIUM-METASOMATISM

For temperatures in excess of 350 degrees Celsius, experiments predict that magnesium-metasomatism will be replaced by sodium-metasomatism of a slightly more complex nature. In this case, seawater sodium (Na) replaces calcium in plagioclase crystals of basalt, which in turn permits the liberated calcium to form epidote and maintain acidic conditions independent of the fluid-to-rock ratio. This reaction takes the form:

$$Na^+ + 2CaAl_2Si_2O_8 + 2SiO_2 = H_2O$$
$$\text{(calcium-plagioclase)}$$
$$= NaAlSi_3O_8 = Ca_2Al_3Si_3O_{12}(OH) = H^+$$
$$\text{(albite)} \qquad \text{(clinozoisite)}$$

where clinozoisite is a simple analogue for epidote. This reaction shows that the more albite forms, the more calcium is recycled from plagioclase to epidote, and the more acidic the aqueous solution becomes. This reaction depends upon the availability of silica, which must be supplied to the solution by the rock. Because basalts with glassy matrices are more susceptible to silica leaching than comparable rocks with crystalline matrices, it is expected that sodium-metasomatism (albitization or spilitization) is most pronounced in basalts that were formerly glassy.

The complex fluids currently venting from submarine ridge hot springs are viewed as the end product of a sequence of chemical reactions that begin with seawater infiltration of hot, fractured basalt some distance from the ridge axis. Convective circulation, supported by magmatic heat, drives the downwelling fluids to depths of perhaps 1 or 2 kilometers, during which time they are heated to temperatures in excess of 350 degrees Celsius and react extensively with the surrounding rocks. The fluids ultimately are returned to the seafloor by upwelling through narrow, focused zones along ridge axes and discharged into the sea as highly evolved hot-spring solutions. Many chemical and physical details of this complex process are yet to be explained, but it seems clear that a calcium-fixing reaction, such as the second reaction described, must play a major role in producing the high-temperature, acidic, metal-charged solutions that vent from ridge-crest hot springs.

OBSERVATION AND EXPERIMENTATION

Prior to the late 1970's, scientific data were largely gathered along ocean ridges from specially designed research ships operating on the surface. The methods utilized included bathymetry, heat-flow and magnetic measurements, dredge sampling, and core drilling of bottom sediments and ridge basalts. It is impossible to overemphasize the success of these techniques, which provided the confirming evidence for seafloor spreading and the foundation for modern plate tectonic theory. A new era began, however, in 1977, when the manned submersible *Alvin* was utilized to make direct observations of submarine hydrothermal activity along the Galápagos Rift. That was soon followed by the immensely successful 1979 RISE expedition, which used *Alvin* to photograph and sample the "black smokers" on the East Pacific Rise. Manned submersibles, supported by conventional research vessels, offer many advantages. Chief among them are that small-scale geological phenomena may be directly observed and revisited to study temporal effects; a wide range of geophysical data may be measured directly on the seafloor; samples of sediment, fluid, and rock may be collected and re-collected directly from precisely known sample sites; the physical characteristics of each data-measurement site can be observed and described; and interactions between biological and geological systems can be observed.

Direct observations, coupled with detailed laboratory experimentation on basalt-seawater reactions, are valuable as a means of investigating processes operating at the rock-seawater interface along ocean ridges where crustal growth occurs. Additional data derive from studies of hot brines in deep boreholes of active geothermal areas such as Salton Sea (California), Iceland, and New Zealand. Useful as these approaches are, their value is limited because they cannot tell how metamorphic conditions vary with depth beneath the rock-seawater interface; neither do they predict the effects of prolonged (for example, over millions of years) rock-seawater interaction. Full characterization of SFM as a petrologic process requires traditional geological study of rocks on land that record the entire range of SFM effects. Fortunately, such rocks, known as ophiolite complexes, are relatively common in major orogenic belts such as the Alps, Appalachian-Caledonian system, and numerous circum-Pacific mountain ranges.

VOLCANIC-EXHALATIVE ORE BODIES

From the economic point of view, studies of sub-seafloor metamorphism have greatly advanced knowledge of the important "volcanic-exhalative"

class of ore deposits. Volcanic-exhalative ore bodies are major stratabound metallic deposits that precipitate directly onto the seafloor from metal-laden solutions discharged by submarine hot springs. The resulting ores are mineralogically variable but are generally sulfide-rich and, for this reason, are often called massive sulfide. Of the several types of exhalative ore deposits now recognized, the Cyprus-type is most closely identified with igneous and metamorphic processes operating along submarine ridges. On the island of Cyprus, massive sulfide copper-zinc ores are hosted by ancient, metamorphosed, seafloor rocks called ophiolites and blanketed by a thin layer of iron-manganese-rich marine chert. The Cyprus ores are lensoidal to podlike bodies, 200-300 meters across and up to 250 meters thick. The largest of these massive sulfide lenses contain 15-20 million tons of ore that typically assays about 4 percent copper, 0.5 percent zinc, 8 parts per million each of silver and gold, 48 percent sulfur, 43 percent iron, and 5 percent silica.

Cyprus-type massive sulfides are of worldwide occurrence, and hundreds of these deposits are now known. As a class, they constitute a major mineral resource produced as a by-product of sub-seafloor metamorphism. They are considered, by most geologists, to form directly on the seafloor from the discharge of black smokers such as those discovered along the East Pacific Rise. The smokers are 2- to 10-meter-high chimneys that pump dense, black plumes of hot, acidic, metal-laden brine onto the seafloor. The turbulent black plumes, for which this special class of hot spring is named, are composed of fine particles of pyrite, pyrrhotite, and sphalerite. Calculations indicate that, at their present high flow rates, ten black smokers such as those operating on the East Pacific Rise could produce the largest known Cyprus-type massive sulfide body in only two thousand years.

Gary R. Lowell

CROSS-REFERENCES

BIBLIOGRAPHY

Best, M. G. *Igneous and Metamorphic Petrology.* 2d ed. New York: W. H. Freeman, 1995. A popular university text for undergraduate majors in geology. Well-illustrated and fairly detailed treatment of the origin, distribution, and characteristics of igneous and metamorphic rocks. Sub-seafloor metamorphism is covered on pages 427-432, but the treatment is somewhat dated in this rapidly developing field.

Cann, J. R., H. Elderfield, and A. Laughton, et al., eds. *Mid-Ocean Ridges: Dynamics of Processes Associated With Creation of New Ocean Crust.* Cambridge, England: Cambridge University Press, 1997. This collection of essays written by leading experts in their respective fields describes indepth the processes of ocean ridges and ocean crusts. There is a clear explanation of the Earth's crust and its metamorphism. Illustrations, bibliography, and index.

Condie, Kent C. *Plate Tectonics and Crustal Evolution.* 4th ed. Oxford: Butterworth-Heinemann, 1997. This textbook provides a detailed description of plate tectonics and the ocean crust. The reader will find a variety of useful illustrations and maps. Includes bibliographical references.

Fyfe, W. S., N. J. Price, and A. B. Thompson. *Fluids in the Earth's Crust.* New York: Elsevier, 1978. The best overview available of the theoretical side of fluid-rock interaction. The authors argue that subduction and seafloor spreading may be viewed as large-scale metasomatic processes, the end product of which is the crust. Written for advanced students of Earth sciences. Extensive bibliography.

Guilbert, J. M., and C. F. Park, Jr. *The Geology of Ore Deposits.* New York: W. H. Freeman, 1986. A splendid edition of a traditional text for undergraduate majors in geology. Chapter 13 treats ore deposits formed by submarine volcanism in comprehensive fashion. Cyprus-type ore bodies are specifically described on pages 598-603. Excellent index and illustrations.

Haymon, R. M. "Growth History of Hydrothermal Black Smoker Chimneys." *Nature* 301 (1983): 695-698. A brief technical article describing the chemistry and mineralogy of the black smokers discovered by the 1979 RISE expedition. This prestigious British science magazine can be found in most university libraries.

Humphris, S. E., and G. Thompson. "Hydrothermal Alteration of Oceanic Basalts by Seawater." *Geochimica et Cosmothimica Acta* 42 (January, 1978): 107-125. This paper is a major contribution to the growing field of sub-seafloor metamorphism, although it requires some background in geochemistry. The authors examine the long-term implications of chemical exchange between seawater and the lithosphere.

MacLeod, P. A. Tyler, and C. L. Walker, eds. *Tectonic, Magmatic, Hydrothermal, and Biological Segmentation of Mid-Ocean Ridges.* London: Geological Society, 1996. This compilation of essays written by leading geologists looks at the processes of mid-ocean ridges and sub-seafloor spreading. Includes an examination of tectonics and hydrothermal vents, as well as a look at submarine geology and magmatism.

Mottl, M. J. "Metabasalts, Axial Hot Springs, and the Structure of Hydrothermal Systems at Mid-Ocean Ridges." *Geological Society of America Bulletin* 94 (1983): 161-180. An excellent article that summarizes the status of sub-seafloor metamorphism. Although technical, much can be gained from this article by college-level readers who have some background in this subject area. This periodical will be found in any university library.

Seyfried, W. E., Jr. "Experimental and Theoretical Constraints on Hydrothermal Alteration Processes at Mid-Ocean Ridges." *Annual Review of Earth and Planetary Sciences* 15 (1987): 317-335. This review paper is on the status of sub-seafloor metamorphism. Emphasis is on experimental aspects of basalt-seawater interaction. Technical, with some background in geochemistry needed. This journal is an annual periodical, published in hardcover book form, and should be found in most major libraries.

Spiess, F. N., and RISE Project Group. "East Pacific Rise: Hot Springs and Geophysical Experiments." *Science* 207 (March 28, 1980): 1421-1432. This article is the report on the 1979 RISE expedition that discovered the now-famous black smokers and explored the rift system in the vicinity of 21 degrees north latitude by manned submersibles. Recommended for general readers.

Spooner, E. T. C., and W. S. Fyfe. "Sub-Sea-Floor Metamorphism, Heat, and Mass Transfer." *Contributions to Mineralogy and Petrology* 42, no. 4 (1973): 287-304. This paper is the classical work in the ophiolite terrane of the Ligurian Apennines that established a unified model for sub-seafloor metamorphism. Essential reading at a modest technical level.

Yardley, B. W. D. *An Introduction to Metamorphic Petrology.* New York: John Wiley & Sons, 1989. An excellent undergraduate text with a good index and extensive current references. For college-level readers, a good place to begin studies of any type of metamorphism.

4
SEDIMENTARY ROCKS

BIOGENIC SEDIMENTARY ROCKS

Biogenic sedimentary rocks represent the accumulation of skeletal material, produced by the biochemical action of organisms, to form limestone, chert, and phosphorites or the accumulation of plant material to form coal. Biogenic sedimentary rocks are economically important as sources of energy, fertilizers, and certain chemicals, and they contain evidence of former life on Earth.

PRINCIPAL TERMS

CARBONATE ROCKS: the general terms for rocks containing calcite, aragonite, or dolomite

CHERT: multicolored sedimentary rock composed of silica and formed in the ocean

COAL: dark brown to black sedimentary rock formed from the accumulation of plant material in swampy environments

DIAGENESIS: the chemical, physical, and biological changes undergone by a sediment after its initial deposition

LIMESTONE: light gray to black sedimentary rock composed of calcite and formed primarily in the ocean

LITHIFICATION: the process whereby loose material is transformed into solid rock by compaction or cementation

OPAL: a form of silica containing a varying proportion of water within the crystal structure

PHOSPHORITE: sedimentary rock composed principally of phosphate minerals

POLYMORPH: minerals having the same chemical composition but a different crystal structure

LITHIFICATION AND COALIFICATION

Numerous organisms, particularly those living in the oceans, produce a shell or some type of skeletal material composed of the minerals calcite or aragonite (polymorphs of calcium carbonate), silica (silicon dioxide), or phosphate minerals, principally apatite. After the organism dies and the soft tissue decays, the skeletal material accumulates as sediment. This sediment then undergoes lithification, a process involving compaction and chemical cementation. The time necessary for lithification is highly variable and ranges from a few decades to millions of years. Biogenic sedimentary rocks composed of calcite or aragonite minerals are commonly referred to as limestones; those composed of silica are cherts, and those composed of phosphate minerals are known as phosphorites or sedimentary phosphate deposits. Limestones and cherts may form as the result of direct chemical precipitation of calcium carbonate or silica from a saturated solution, but the primary mode of formation is biochemical. Limestones are the most abundant of the biogenic sedimentary rocks and, economically, the most important.

Coal also is a biogenic sedimentary rock. Most coal forms from the accumulation of woody plant material in anoxic (oxygen-lacking) environments, such as stagnant lakes, swamps, and bogs. The formation of coal is referred to as coalification and represents microbiological, physical, and chemical processes whereby the percentage of carbon is increased and the volatile content decreased. Limestones, cherts, phosphorites, and coal are often closely associated and may occur as alternating beds in a sequence of sedimentary rocks.

PHOSPHORITES

Because of their biological affinities, biogenic sedimentary rocks contain important evidence of former life on Earth, a record of organic evolution (changes in organisms through geologic time), and clues to the past environments in which the rocks were formed. Limestones may serve as reservoir rocks for oil and gas, which form from the chemical breakdown of microscopic organisms that accumulate with the sediment at the bottom of the ocean and in some lakes. Phosphates, which are contained in phosphorites, are one of the chief constituents of fertilizers and are widely used in the chemical industry. Important evidence of past swamp communities, including both plants and animals, has been preserved in coal beds.

Phosphorus is one of the essential elements of life, and when chemically linked with other elements to form phosphate minerals, especially apatite, it becomes one of the major constituents of all vertebrate skeletons and some invertebrate hard parts. Phosphate is a primary nutrient in marine waters and, therefore, controls organic productivity. Phosphorites often occur as nodules, which are highly variable in size and shape; they may be several centimeters in diameter and up to a few meters in length. These nodules are usually rich in vertebrate skeletal debris, especially that of fish, and fecal material. Bedded phosphorites are commonly interbedded with limestones. Thin beds of phosphorite are rich in bones, fish scales, and fecal material. Modern phosphorites are forming in areas where cold, nutrient-rich waters rise from the ocean depths toward the surface, such as off the west coasts of North and South America and Africa.

CHERTS

Cherts, which are occasionally referred to as flint or navaculite, are not nearly as abundant as limestones; however, chert has two properties that have made it one of the most important rocks to humans. Like quartz, chert is very hard, and it tends to break along smooth, curved surfaces. With a little practice, one can produce a sharp edge by chipping a piece of chert with another hard object. Although early humans used numerous rock types for tools and weapons, chert was the most important for the production of arrowheads, knives, and scrapers. Consequently, chert might be considered to have contributed as much to the rise of civilization as did the development of the steam engine. In addition to its utility, Precambrian cherts (1.8-3.4 billion years old) are interbedded with important iron deposits known as banded iron formations. These cherts contain evidence of some of the earliest life to evolve on Earth.

Chert—which may be any color, depending on the presence of impurities—occurs, principally, as spherical to oblong nodules 2-25 centimeters in diameter in limestones and dolostones or as thin beds 2-25 centimeters deep. Nodular cherts are commonly parallel to bedding or stratification. Bedded cherts may extend laterally for great distances and are commonly interbedded with limestones and dolostones. Most geologists believe that nodular chert forms from the replacement of limestone and that bedded chert forms either by the complete replacement of carbonate-rich beds or by the diagenetic alteration of siliceous ooze.

The source of the silica may be the chemical precipitation of silicon dioxide, volcanic ash, or skeletal material. Biogenic silica is opaline, which means that it may contain up to 10 percent water. Only skeletal silica will be considered in this article. The principal biogenic sources of silica for chert are sponges, diatoms, and radiolarians. Skeletal materials from these organisms are common constituents of oceanic sediments. As this siliceous skeletal material accumulates, the opaline silica is diagenetically converted to crystalline opal and reprecipitated as bladed crystals. Continued diagenesis results in the formation of quartz chert, a mosaic of microscopic quartz crystals. The final diagenetic change often obliterates the shape or structure of the original skeletal material.

Chert strata in the Franciscan Formation near San Francisco. Chert is a common form of biogenic sedimentary rock. (© William E. Ferguson)

Sponges, which are abundant in most marine environments, contain as part of their supportive structure microscopic rods of opaline silica called spicules. When the sponge dies, these spicules accumulate and become part of the sediment. During burial, the opaline silica undergoes diagenesis and precipitation of crystalline opal occurs within pores in the sediment. As crystalline opal is converted into quartz chert, chemical replacement of the surrounding sediment, usually calcite, occurs, resulting in the formation of chert nodules.

Radiolarians and diatoms are microscopic organisms with disk-shaped, elongate, or spherical tests (shells) with spines and surface ornamentation composed of opaline silica. Radiolarians occur as part of the marine zooplankton, and diatoms are part of the marine and nonmarine phytoplankton. Radiolarian and diatom oozes accumulate on the ocean floor. In time, these beds of silica-rich ooze are diagenetically converted into thin, bedded cherts. Diatoms may also occur in great numbers in lakes and accumulate to form diatomaceous earths, or diatomites. Diatomaceous earths have a wide variety of uses, such as in filtering agents, absorbents, and abrasives.

LIMESTONES

Limestones are the most abundant of the chemical sedimentary rocks. A minor amount of limestone forms from the inorganic precipitation of calcite from seawater or the deposition of calcite in caves and around hot springs. The majority of limestones form as the result of biological and biochemical processes that produce aragonite or calcite. Later, this material becomes part of the carbonate sediment. Once deposited, the carbonate sediment is often modified by the chemical and physical processes of diagenesis.

Numerous animal phyla, such as mollusks, brachiopods, echinoderms, bryozoans, coelenterates, and certain protozoans, produce aragonite or calcite as part of their skeletal structure. The skeletal remains of these organisms are important constituents of carbonate sediments and, eventually, of most limestones. Some marine, bottom-dwelling algae are partially or wholly composed of aragonite or calcite. These calcareous algae represent a significant contribution to carbonate sediment. Calcareous algae called coccoliths and calcareous foraminifera are found in great numbers

in the plankton of the open ocean. Accumulation of this calcareous material at the bottom of the ocean contributes greatly to the formation of chalk, a type of biogenic limestone. Modern marine organisms principally secrete aragonite; consequently, modern calcareous sediments are composed predominantly of aragonite. Aragonite, however, is unstable and quickly undergoes diagenesis to calcite.

It is not uncommon to find that ancient limestones have been partially or completely transformed into dolostones; this process is known as dolomitization. Dolomitization occurs when calcium carbonate minerals are diagenetically converted into dolomite. Diagenesis may take place soon after the calcite or aragonite has been deposited or a long time after the deposition. The diagenesis is the result of the magnesium-bearing waters (seawater or percolating meteoric water) moving through the carbonate sediment or limestone. One important aspect of dolomitization is that it often leads to the formation of pores, cavities, and fissures, which enable the rock to serve as a reservoir for oil, gas, and water.

Under certain conditions, limestones are relatively easy to dissolve. In areas such as Florida, where limestones are abundant and there is adequate rainfall, cave systems may develop. Regions with extensive cave systems and related features are referred to as karst.

STUDY OF ROCK SAMPLES

Initially, the study of biogenic sedimentary rocks involves field investigations. Where outcrops of these rocks occur, geologists plot and map their distribution and thicknesses; note changes in the rocks' character, both vertically and horizontally; observe their association with other types of rocks; and collect samples of both the rock and any fossils that may be present. When drilling for oil, gas, or water, engineers and geologists study the same phenomena.

Samples, which are brought back to the laboratory, are analyzed with a vast array of techniques, including X-ray analysis, electron microprobe analysis, scanning electron microscopy, cathodoluminescence, and thin-section examination with a polarized microscope. Samples containing large fossils are observed with a magnifying glass or dissecting microscope. Some samples, particularly limestones, are dissolved in acid to recover both megafossils and microfossils. When samples have been ground thin enough to allow light to pass through the rock, microfossils can be observed with the help of microscopes. Scanning electron microscopy is particularly useful for observing extremely small fossils, particularly radiolarians, diatoms, and coccoliths.

AREAS OF RESEARCH

In general, the chemistry of biogenic sedimentary rocks is simple; however, diagenesis can alter the structure, texture, and mineralogy of a sediment during its deposition, lithification, and burial. The analysis of diagenetic changes, particularly in limestones, is one of the most important avenues of investigation. Limestones are greatly influenced by diagenetic modifications and undergo changes in sediment size, porosity (the spaces between sediment grains), and mineralogy. Early diagenetic changes include the conversion of unstable aragonite, which is produced by most calcium-carbonate-producing organisms, into calcite and changes in the calcite's magnesium content. (Since the atoms of calcium and magnesium have similar properties, they can often substitute for each other in the crystal structure.) Dolomitization may also occur. The replacement of carbonate minerals by silica, so that chert is formed, is another important aspect of diagenesis. Stages of diagenesis can best be observed by using cathodoluminescence, which bombards a thin slice of rock with an electron beam. This process causes minerals within the rock to luminesce (emit light energy for a short interval after the energy source has been removed). The luminescences indicate various stages of diagenesis and are particularly important in studying the details of cements and crystal growth. X-ray analysis and electron microprobe techniques allow scientists to detect subtle differences in chemistries and the presences of trace elements or rare-earth elements.

Another major area of research is the classification of biogenic sedimentary rocks. The classification of limestones is of primary importance. Limestones have been classified using several criteria, but in general, classifications have focused on chemical and mineralogical composition; fabric features, such as fossils and cements; and special

physical parameters, such as porosity. When examining chert, the geoscientist must ask: What was the source of the silica (volcanic ash, skeletal grains, chemical precipitate)? What was the time and rate of conversion of siliceous ooze to chert? What was the environment of deposition?

ECONOMIC AND EVOLUTIONARY SIGNIFICANCE

Biogenic sedimentary rocks are among the most important of all sedimentary rocks. Because of their biological affinities, they have preserved an important record of past life on Earth. In addition, some of these rocks were used for tool-making by early humans; others are important resources for construction materials, fertilizers, and chemicals. Oil and natural gas, which also have biological affinities, are inseparably linked to biogenic sedimentary rocks.

Although not widespread, biogenic phosphorites are economically important, particularly to the fertilizer and chemical industries. Biogenic chert was of major importance in the past. Early humans used chert, or flint, to make tools and weapons. Biogenic limestone is the most common and widespread biogenic sedimentary rock. Economically, limestones are very important. Hydrocarbons (oil and gas) are commonly recovered from porous limestones and dolostones, and in some areas, limestones are reservoirs for groundwater supplies. Limestones often serve as host rocks for important mineral deposits, such as lead. Limestones are quarried as building stone; crushed to form construction materials, such as gravel; or processed into lime and cement. Biogenic limestones are the most important sedimentary rock containing fossils and, therefore, are the most important record of the evolution of life.

Larry E. Davis

CROSS-REFERENCES

Chemical Precipitates, 1440; Diagenesis, 1445; Earth's Lithosphere, 26; Flood Basalts, 689; Heat Sources and Heat Flow, 49; Hydrothermal Mineralization, 1205; Limestone, 1451; Magmas, 1326; Metasomatism, 1406; Ocean Ridge System, 670; Oceanic Crust, 675; Ophiolites, 639; Regional Metamorphism, 1418; Sedimentary Rock Classification, 1457; Siliciclastic Rocks, 1463; Spreading Centers, 727.

BIBLIOGRAPHY

Birch, G. F. "Phosphatic Rocks on the Western Margin of South Africa." *Journal of Sedimentary Petrology* 49 (1979): 93-100. This paper provides a good starting point for a review of phosphorites. Suitable for college-level readers. Available in most university libraries.

Blatt, Harvey, and Robert J. Tracy. *Petrology: Igneous, Sedimentary, and Metamorphic.* New York: W. H. Freeman, 1996. Although written for advanced undergraduates and beginning graduate students, this text provides easily understood background material and has an extensive bibliography for each category of sedimentary rocks.

Chernikoff, Stanley. *Geology: An Introduction to Physical Geology.* Boston: Houghton-Mifflin, 1999. This is a good overview of the scientific understanding of the geology of the Earth and surface processes. Includes the address of a Web site that provides regular updates on geological events around the globe.

Ham, W. E., ed. *Classification of Carbonate Rocks: A Symposium.* Tulsa, Okla.: American Association of Petroleum Geologists, 1962. This collection of ten papers is the standard reference for limestone classification. Suitable for college students. Available in most university libraries.

McBride, E. F. *Silica in Sediments: Nodular and Bedded Chert.* Tulsa, Okla.: Society of Economic Paleontologists and Mineralogists, 1979. This collection of sixteen papers provides the most complete discussion available of chert. Each paper contains an extensive bibliography. Available in most college and university libraries. Suitable for college-level readers.

Oldershaw, Cally. *Rocks and Minerals.* New York: DK, 1999. This small, 53-page volume is filled with color illustrations and is therefore of great use to new students who may be unfamiliar with the rock and mineral types discussed in classes or textbooks.

Prothero, Donald R., and Fred Schwab. *Sedimentary Geology: An Introduction to Sedimentary Rocks and Stratigraphy.* New York: W. H. Freeman, 1996. A thorough treatment of most aspects of sediments and sedimentary rocks. Well illustrated with line drawings and black-and-white photographs, it also contains a comprehensive bibliography. Chapters 11 and 12 focus on carbonate rocks and limestone depositional processes and environments. Suitable for college-level readers.

Reading, H. G., ed. *Sedimentary Environments: Processes, Facies, and Stratigraphy.* Oxford: Blackwell Science, 1996. A good treatment of the study of sedimentary rocks and biogenic sedimentary environments. Suitable for the high school or college student. Well illustrated, with an index and bibliography.

Tucker, Maurice E. *Sedimentary Rocks in the Field.* New York: John Wiley & Sons, 1996. Presents a concise account of biogenic sedimentary rocks and other sedimentary rocks. Classification of sedimentary rocks is well covered. Depositional environments are only briefly discussed. References are well selected. Suitable for undergraduates.

CHEMICAL PRECIPITATES

Chemically precipitated sediments precipitate from water that has become super-saturated with certain dissolved ions. Upon burial, these sediments harden to become chemically precipitated rock. The most common examples are limestone and salt, though a wide variety of other types include some of the most valuable ore deposits on Earth. Industrial society depends upon these chemical sedimentary ores.

PRINCIPAL TERMS

CARBONATE SEDIMENT: a chemically precipitated rock, largely composed of minerals containing carbon and oxygen

DIAGENESIS: the conversion of unconsolidated sediment into consolidated rock after burial by the processes of compaction, cementation, recrystallization, and replacement

DIATOM: an alga (single-celled photosynthetic plant) that precipitates silica within itself

EVAPORITE: a rock largely composed of minerals that have precipitated upon evaporation of seawater or lake water

HYDROCARBON: an organic compound composed only of carbon and hydrogen; the number of hydrogen atoms always is two more than twice the number of carbon atoms

IRONSTONE: a chemically precipitated sedimentary rock that contains more than 15 percent iron

LIMESTONE: a chemically precipitated sedimentary rock that contains more than 50 percent calcite, aragonite, or both

REEF: a wave-resistant structure composed of organisms that precipitate calcium carbonate

SILICA: a glassy combination of silicon and oxygen

SUPERSATURATION: a condition of a fluid that has been concentrated, usually by evaporation, to the point at which it should start to precipitate minerals

TEXTURE: the size, shape, and arrangement of individual grains of a rock

EVAPORITES

When seawater dries on skin, the skin becomes covered with a thin layer of white salt. This salt is chemical sediment that precipitated because evaporation of the seawater caused supersaturation of the ions that are dissolved in all seawater. If seawater is weighed and allowed to evaporate completely, the weight of the remaining salt crystals would be about 3.5 percent of the original weight of the seawater. About the same percentage by weight of salt can be found in seawater anywhere in the world and at any depth in the ocean. Moreover, the proportions of the most abundant chemical elements in seawater are always the same.

Evaporation of seawater has produced salt within lagoons and along ancient desert coasts, and the accumulated salt generally exhibits a chemical composition similar to modern seawater salt. Burial of salt causes it to harden into a rock that is called evaporite. The resulting chemically precipitated rock is fairly abundant within rock

layers that have accumulated during the past 5 percent of Earth history, but the proportion of salt to other types of chemically precipitated rock progressively decreases within progressively older rock layers. The generally increasing scarcity of salt in progressively older sedimentary rock sequences is most likely a result of the continual slow dissolution of salt by slow-moving fluids deep in the Earth.

Although there is a general decrease in the proportion of evaporites within older rock sequences, at certain times in the Earth's history, for example, 200-280 million years ago, the accumulation of evaporites was so great that rocks of this age range still contain a particularly high proportion of evaporites. In some parts of the world, these old evaporites are mined to supply table salt, a mineral composed of sodium and chlorine. In southern New Mexico, evaporites of this age are being used for quite a different purpose: as a repository for nuclear waste.

CARBONATES

The most abundant chemically precipitated rocks are not evaporites but carbonates. Carbonate rocks are largely composed of one or more of the following carbonate minerals: calcite, aragonite, and dolomite. The chemical formula for calcite is identical to that of aragonite, calcium-carbon-oxygen. Dolomite differs in that it also contains magnesium. The crystal structure of aragonite differs from that of calcite in that aragonite is denser. Calcite therefore can be converted to aragonite by subjecting it to extreme pressure. Aragonite can be converted to calcite with less effort, because only calcite is chemically stable under earth-surface conditions. The only reason that aragonite occurs abundantly on the modern ocean floor is that specific animals selectively precipitate it for unknown reasons. Precipitation of calcium carbonate by inorganic processes, such as the evaporation of seawater, generally produces calcite.

An easy place to find chemically precipitated carbonate minerals is on a beach, in the form of clamshells. Many clams selectively precipitate aragonite from the calcium and carbon-oxygen ions dissolved in seawater. Once formed, this aragonite may last for millions of years before it eventually converts to the more stable crystal form of calcite. It is common to find limestone that contains both calcite and aragonite but only a tiny proportion of other minerals. Limestone is one of the most abundant sedimentary rock types and is surpassed only by the predominant clastic sedimentary rocks, shale and sandstone.

DOLOSTONE

Dolomite is the most abundant magnesium-bearing carbonate mineral. A rock that is largely composed of dolomite is called either dolostone or, although the term is potentially confusing, dolomite. For unknown reasons, the proportion of dolostone to limestone generally increases with the age of sedimentary rocks. Virtually no dolostone is now being produced on the seafloor. The lack of modern examples impedes the interpretation of the processes that have produced ancient dolostone. For the same reason, there is no consensus about the origin of another type of chemically precipitated rock, ironstone.

Some dolostone exhibits fossils of animals that always precipitate their shells from either calcite or aragonite. These shells clearly have been partially or completely replaced by dolomite through the precipitation of magnesium from water passing through the buried sediment. Any change in sediment under conditions of shallow burial is called diagenesis, and so this process is called diagenetic dolomitization. Other ancient dolostone appears to have accumulated on the seafloor directly rather than by diagenetic replacement of calcite or aragonite. This apparent precipitation of dolostone from seawater remains unexplained, mainly because dolostone is not presently precipitating anywhere from seawater of normal salinity.

CORAL REEFS

Carbonate sediments (limestone and dolostone) exhibit a wide range of beautiful textures because of the great variety of carbonate fossils and diagenetic alterations. Given that calcite and aragonite are quite soluble, these carbonate minerals may dissolve within groundwater and then reprecipitate as crystals, which radiate away from some surface, such as the surface of a fossil shell. One of the most common carbonate fossils is coral. Most carbonate rocks largely composed of fossil corals are remnants of buried reefs.

Modern examples of coral reefs occur along the southern edge of the Florida Keys and along the northeastern coast of Australia (the Great Barrier Reef). The most spectacular reefs in the Americas occur along the coast of Belize and the northeastern coast of Andros, the largest island in the Bahamas. Study of these reefs has greatly enhanced scientists' understanding of the accumulation and diagenesis of this type of chemically precipitated rock. Myriad colorful life-forms induce the carbonate sedimentation.

Commonly, old coral reefs partially dissolve diagenetically, leaving large holes that become filled with fluids or valuable minerals. Buried reefs are prime exploration targets for both petroleum geologists and lead-zinc miners. Petroleum-bearing ancient reefs abound in northwestern Texas, and lead-zinc-bearing reefs occur at Pine Point in the Northwest Territories of Canada.

DIATOMS

Although calcite and aragonite constitute the bulk of the minerals that are chemically precipitated by marine organisms, some plants and animals precipitate silica. The most prolific precipitator of silica in the modern oceans is a group of algae called diatoms. Although a microscope is needed to see an individual diatom, these tiny plants occur in such extraordinary numbers that the silica that they secrete locally composes most of the sediment on the seafloor. The greatest concentration of diatoms is near Antarctica, partly because silica inherently tends to precipitate in cold water, whereas carbonate minerals tend to precipitate in warm water. A chemically precipitated rock that is largely composed of silica is called chert. Only a small proportion of all chert exhibits fossils of diatoms or tiny silica-precipitating animals, because all these fossils commonly are dissolved during diagenesis and reprecipitate as featureless silica.

STUDY OF CHEMICALLY PRECIPITATED ROCKS

Chemically precipitated rocks are usually studied by making a thin slice of the rock and studying that slice with a high-magnification microscope. Identification of any observed fossils helps to determine the age of the rock and the environment in which the sediment accumulated. If the rock is being studied by a petroleum geologist, it may be chemically analyzed for traces of hydrocarbons. If it is being studied by a lead-zinc miner, a polished slab of the rock would be studied with a microscope under reflecting light, because the lead-zinc minerals exhibit characteristic colors in reflected light. If the miner were to find any valuable minerals, the rock would be analyzed chemically for such elements as lead, zinc, gold, and silver.

INDUSTRIAL USES

About 20 percent of all sedimentary rocks are chemically precipitated. Of these, the great bulk are carbonate rocks. Carbonate rocks are most widely used as crushed stone and in the production of cement. The widespread use of concrete depends on the quarrying of carbonate rocks. Given the great abundance of carbonate rocks, this resource is essentially inexhaustible. Exploration continues for new quarries, however, because cement is best produced from limestone of a particular chemical composition and because transportation costs are so great that geologists try to identify the carbonate resources that are closest to potential markets.

In addition to the common calcium-bearing carbonate minerals, there is a host of rarer carbonates that are mined for other metals: iron carbonate, manganese carbonate, magnesium carbonate, and sodium carbonate. These minerals also occur in chemically precipitated sedimentary rocks. Evaporites are mined for table salt, or sodium chloride, and for plasterboard, which is made of a calcium-sulfur-oxygen mineral known as gypsum. Plasterboard (also known by the trade name Sheetrock) is produced by pressing a powder of white gypsum between cardboard, and it may be nailed onto studs to make walls. Previously, walls were made from dehydrated gypsum that was hydrated into plaster just before its application. Agriculture has come to depend on phosphorus that is mined from chemically precipitated rocks to produce fertilizer.

Carbonate rocks, particularly fossil reefs, may host petroleum or a wide variety of valuable metals. Chert is also commonly associated with valuable metals and may even be a host for petroleum if holes for the petroleum have formed by breakage of the chert during an earthquake. Unlike carbonate rocks, chert is usually too insoluble to develop large holes when groundwater pervades it during diagenesis.

Prior to the commercial exploitation of petroleum, wood and fossil wood (peat and coal) were virtually the only sources of fuel for industry. Peat is recently buried sediment that consists almost entirely of plant matter. Peat and coal are chemically precipitated deposits of carbon. Like carbonate sediments, peat characteristically undergoes diagenesis. Peat contains abundant oxygen and hydrogen, but the diagenetic transformation of peat to coal during millions of years causes the sediment to lose most of that oxygen and hydrogen until it consists of little other than carbon. Unfortunately, most coal also contains an iron-sulfur mineral that decomposes to produce sulfur gas upon combustion of the coal. Release of this sulfur into the atmosphere is a major cause of acid rain.

Michael M. Kimberley

CROSS-REFERENCES

BIBLIOGRAPHY

Alexander, Clark R., Richard A. Davis, and Vernon J. Henry, et al., eds. *Tidalites: Processes and Products*. Tulsa, Okla.: Society for Sedimentary Geology, 1998. An examination of chemical sedimentology and tidalites with a strong emphasis placed on geochemistry and geochemical processes. This book is intended for the college-level reader with some background in chemistry or Earth sciences.

Bathurst, Robin G. C. *Carbonate Sediments and Their Diagenesis*. 2d ed. New York: Elsevier, 1975. Modern shallow-water carbonate sediments in the Bahamas and the Persian Gulf are particularly well reviewed in this thorough discussion of carbonate sediments.

Baturin, G. N. *Phosphorites on the Seafloor: Origin, Composition, and Distribution*. Translated by Dorothy B. Vitaliano. New York: Elsevier, 1982. Chemical sedimentation of phosphorus onto the seafloor is rare but produces economically valuable deposits called phosphorite. Most phosphorite deposits are tens to hundreds of millions of years old, but some phosphorite occurs on the modern seafloor, as reviewed in this book. Phosphorite is mined to produce fertilizer.

Berner, Robert A. *Principles of Chemical Sedimentology*. New York: McGraw-Hill, 1971. The origin of chemical sedimentary rocks is explained with the aid of diagrams that incorporate the most basic principles of thermodynamics.

Carozzi, Albert V. *Carbonate Rock Depositional Models: A Microfacies Approach*. Englewood Cliffs, N.J.: Prentice-Hall, 1989. A large number of black-and-white drawings and photographs are used to illustrate the relationship between the textures of carbonate sediments and the corresponding sedimentary environments in which these textures form.

Leeder, Mike R. *Sedimentology and Sedimentary Basins: From Turbulence to Tectonics*. Oxford: Blackwell Science, 1999. A good look into the study of sedimentology and chemical sedimentology, and their relationship with tectonics. This is a good introduction to sedimentology for the layperson without much background in the Earth sciences.

Oldershaw, Cally. *Rocks and Minerals*. New York: DK, 1999. This small, 53-page volume is filled with color illustrations and is therefore of great use to new students who may be unfamiliar with the rock and mineral types discussed in classes or textbooks.

Peryt, T., ed. *Coated Grains*. Berlin: Springer-Verlag, 1983. Many chemically precipitated sedimentary rocks are agglomerations of concentrically layered, or coated, grains; one variety of these grains that is particularly common is the ooid. Ooids may be composed of carbonate minerals, silica, iron minerals, manganese minerals, or calcium phosphate. This book consists of fifty-two papers that describe such ooids and other coated grains.

Rahmani, Ray A., and Romeo M. Flores, eds. *Sedimentology of Coal and Coal-Bearing Sequences*. Oxford, England: Blackwell Scientific, 1985. This special publication of the International Association of Sedimentologists is one of the few modern collections to discuss the origins of coal, a chemically precipitated rock that is most often viewed only in an economic context. Coal and coal-forming processes from a number of localities around the world are described.

Scholle, Peter A. *A Color Illustrated Guide to Carbonate Constituents, Rock Textures, Cements, and Porosities*. Tulsa, Okla.: American Association of Petroleum Geologists, 1978. The study of carbonate sediments with a microscope reveals spectacularly beautiful shapes that record ancient life on Earth. This book consists of color photographs that record the diversity and grandeur of life within environments that were precipitating chemical sediment.

Scoffin, T. P. *An Introduction to Carbonate Sediments and Rocks*. New York: Chapman and Hall, 1986. This book is an update on the book by Robin Bathurst. Scoffin goes further in relating modern carbonate sediments to ancient carbonate rocks. This relationship is difficult to deduce, because extensive diagenesis generally makes it difficult to understand what ancient carbonate rocks may have looked like when they were soft sediment on the seafloor.

Sengupta, Supiya. *Introduction to Sedimentology.* Rotterdam, Vt.: A. A. Balkema, 1994. An excellent introduction to the practices, policies, and theories used within the field of sedimentology. The book is filled with illustrations and maps that help to clarify concepts and processes. Index and bibliographical references included.

Sonnenfeld, Peter. *Brine and Evaporites: Depositional Environments of Precipitates in Hypersaline Brines.* Orlando, Fla.: Academic Press, 1984. Evaporation of seawater eventually causes it to precipitate salt. This book describes the environments of evaporation, the sequence of precipitated salts, and the nature of the chemically precipitated rocks that consist of salt.

DIAGENESIS

Diagenesis refers to the physical, chemical, and biological changes that sediment undergoes after it is deposited. These processes change loose sediment into sedimentary rock and occur in the upper several hundred meters of the Earth's crust.

PRINCIPAL TERMS

BAR: a unit of pressure equal to 100 kilopascals and very nearly equal to 1 standard atmosphere

CARBONATE: a mineral with CO_3 in its chemical formula, such as calcite ($CaCO_3$)

LITHIFICATION: the hardening of sediment into a rock through compaction, cementation, recrystallization, or other processes

PORE FLUIDS: fluids, such as water (usually carrying dissolved minerals, gases, and hydrocarbons), in pore spaces in a rock

POROSITY: the amount of space between the sedimentary grains in a rock or sediment

SEDIMENT: loose grains of solid, particulate matter resulting from the weathering and breakdown of rocks, chemical precipitation, or secretion by organisms

SEDIMENTARY ROCK: a rock resulting from the consolidation of loose sediment that has accumulated in flat-lying layers on the Earth's surface

DISTINGUISHING FROM METAMORPHISM

Diagenesis refers to the physical, chemical, and biological processes that occur in sediment after deposition as it is buried and transformed into sedimentary rock. These processes alter the texture, porosity, fabric, structure, and mineralogy of the sediment. Through diagenesis, sand is changed into sandstone, mud is changed into shale, and carbonate sediments are changed into limestone and dolomite. The processes and degree of alteration depend in part on the initial sediment composition and the depth of burial.

As sediment is buried to increasing depths, the temperatures and pressures increase, and diagenesis becomes metamorphism. The increase in temperature with depth is referred to as the geothermal gradient. Although the exact limits separating diagenesis and metamorphism are not strictly defined, diagenesis can be considered to occur under temperatures ranging from those at the Earth's surface up to nearly 300 degrees Celsius and under pressures ranging from atmospheric pressure to at least 1 kilobar (1,000 bars). These conditions occur at depths approximating 10 kilometers. Some classifications restrict the zone of diagenesis to about 0 to 1 kilometer, grading down into the zones of catagenesis (several kilometers deep, with temperatures of 50 to 150 degrees and

pressures of 300 to 1,000 or 1,500 bars), metagenesis (up to about 10 kilometers deep), and metamorphism. Temperature is an important control on many diagenetic processes, because it influences chemical reactions such as the dissolution and precipitation of minerals, recrystallization, and authigenesis.

COMPACTION

The primary physical diagenetic process is compaction. Compaction presses sedimentary grains closer together under the load of overlying sediment, causing pore space to be decreased or eliminated and squeezing out pore fluids. In sandstones, compaction occurs by the rotation and slippage of sand grains, the breakage of brittle grains, and the bending and mashing of ductile (soft, easily deformed) grains. Brittle grains include thin shells, skeletal fragments, and feldspar grains. Ductile grains include clay or shale chips, fecal pellets, and some metamorphic rock fragments, such as slate. Compaction also causes some mineral grains to interpenetrate, producing irregular stylolitic contacts. In general, sands compact much less than muds. That is true since the average sandstone has a high percentage of hard grains, such as quartz; muds typically have a high initial water content, and water is squeezed out during compaction. The compaction of sands, however, is influenced by the nature of the sand grains present; sands with a large percentage of ductile grains are more susceptible to compaction.

CEMENTATION, AUTHIGENESIS, REPLACEMENT

Chemical diagenetic processes include cementation, the growth of new minerals (authigenesis), replacement, neomorphism (recrystallization and inversion), and dissolution. Cementation is the precipitation of minerals from pore fluids. These minerals glue the grains of sediment together, forming a rock. The most common cements are quartz (silica), calcite, and hematite, but other cements occur, including clays, aragonite, dolomite, siderite, limonite, pyrite, feldspar, gypsum, anhydrite, barite, and zeolite minerals. The type of cement is controlled by the composition of the pore fluids.

Authigenesis refers to the growth of new minerals in the sediment and the transformation of one mineral into another. Some of the most common authigenic minerals in sandstones are calcite, quartz, and clay cements. Other than cements, authigenic minerals include glauconite, micas, and clay minerals. Glauconite is a green mineral which forms on the seafloor when sedimentation rates are low. Authigenic micas and clays typically form in the subsurface at higher temperatures and pressures. Often, one clay mineral is transformed into another as a result of dehydration (water loss) or chemical alteration by migrating fluids. Clays may also be formed from the alteration of feldspars or volcanic ash and rock fragments. Authigenesis also includes the alteration of iron-bearing minerals (such as biotite, amphibole, or pyroxene) to pyrite, under reducing conditions, or to iron oxide (limonite, goethite, or hematite), under oxidizing conditions.

Replacement is the molecule-by-molecule or volume-for-volume substitution of one mineral for another. Replacement generally involves the simultaneous dissolution of an original mineral and precipitation of a new mineral in its place. Fossils which were originally calcium carbonate may be replaced by different minerals, such as quartz, pyrite, or hematite. Many minerals are known as replacement minerals, including calcite, chert, dolomite, hematite, limonite, siderite, anhydrite, and glauconite. Factors controlling replacement include pH, temperature, pressure, and the chemistry of the pore fluids.

NEOMORPHISM AND DISSOLUTION

Neomorphism is a term meaning "new form"; it refers to minerals changing in size, shape, or crystal structure during diagenesis. The chemical composition of the minerals, however, remains the same. Neomorphism includes the processes of recrystallization and inversion. Recrystallization alters the size or shape of mineral grains without changing their chemical composition or crystal structure. Recrystallization can occur in any type of sedimentary rock, but it is most common among the carbonates. Limestones are commonly recrystallized during diagenesis, producing a coarsely crystalline rock in which original sedimentary textures and structures may be fully or partially obliterated. The reason that minerals recrystallize is not well understood, but it may be related to energy stored in strained crystals or to a force arising from the surface tension of curved crystal boundaries. Inversion is a process in which one mineral is changed into another with the same chemical composition but a different crystal structure. The two minerals involved are called polymorphs, meaning "multiple forms." Aragonite and calcite are polymorphs. Both have the same chemical composition, but each has a different crystal structure: Aragonite is orthorhombic and calcite is rhombohedral. Aragonite, with time, will become calcite by inversion. Inversion may occur along a migrating film of liquid, causing the simultaneous dissolution of one mineral and precipitation of its polymorph, or by solid-state transformation (switching of the positions of ions in the crystal lattice).

Dissolution refers to the dissolving and total removal of a mineral, leaving an open cavity or pore space in the rock. This pore space may persist, or it may become filled by another mineral at a later time. Some of the more soluble minerals are the carbonates and the evaporites, such as halite and gypsum. Large-scale dissolution of limestone leads to the formation of caves and caverns. Pressure solution is the dissolution of minerals under the pressure of overlying sediment. Stylolites, a common result of pressure solution, commonly occur in carbonate rocks. Stylolites are thin, dark, irregular seams with a zigzag pattern that separate mutually interpenetrating rocks. The dark material along the seam is a concentration of insoluble material such as clay, carbon, or iron oxides. Pressure solution can result in a 35 to 40 percent reduction in the thickness of carbonate rocks. The carbonate removed by pressure solution is frequently a source of carbonate cements.

BIOLOGICAL DIAGENETIC PROCESSES

Biological diagenetic processes occur soon after sediment is deposited and consist of the activities of organisms in and on the sediment. Bacteria are particularly important to the chemical diagenetic processes. Bacteria living in the sediment control many chemical reactions involving mineral precipitation or dissolution; they are involved in the breakdown or decomposition of organic matter (one of the steps in the formation of oil and gas), and can cause the pH of the pore fluids to increase or to decrease, depending on the kinds of microorganism, type of organic matter, decomposition products, and availability of oxygen. For example, in aerobic environments (those where oxygen is present), decay of organic matter generally causes decreasing pH—increasing acidity—which may lead to the dissolution of carbonate minerals such as calcite. Under anaerobic conditions, organic decay generally raises the pH and may lead to the precipitation of calcite cement. The formation of pyrite is also influenced by the activity of bacteria. Sulfate-reducing bacteria in anoxic environments change sulfate into hydrogen sulfide. If iron is present, it reacts with the hydrogen sulfide to form iron sulfides, such as pyrite.

Bioturbation is the disturbance of the sediment by burrowing (excavation into soft sediment), boring (drilling into hard sediment), the ingestion of sediment and production of fecal pellets, root penetration, and other activities of organisms. Bioturbation generally occurs shortly after deposition and causes mixing of sediment that was originally deposited as separate layers, destruction of primary sedimentary structures and fabrics, and breakdown or clumping of grains. In some cases, chemical alteration of the sediment accompanies bioturbation. For example, light-colored halos may form around burrows or roots, particularly in red or brown sediments, because of the reduction of iron.

Diagenesis may decrease or increase the porosity and permeability of the sediment. Porosity is decreased by compaction, the precipitation of cements in pore spaces, and bioturbation. Porosity is increased by dissolution. Zones of increased porosity are particularly favorable for oil and gas accumulations.

STUDY TECHNIQUES

Diagenesis is primarily studied using sedimentary petrography, which is the microscopic examination of thin sections of sedimentary rocks. Thin sections are slices of rock, typically 30 micrometers thick, bonded to glass slides, which are examined with a petrographic microscope. In this way, minerals can be identified based on their optical properties, and textural relationships can be studied, such as the size, shape, and arrangement of grains; the geometry of cements and pore spaces; the character of contacts between grains; the presence of dissolution features; and mashing or fracturing of grains. Thin sections may be enhanced, to allow easier identification of minerals, with various staining and acid etching techniques. In addition, acetate peels may be prepared of etched and stained rock surfaces for examination with the microscope.

There are a number of other techniques which can be used in conjunction with petrography to obtain more specific diagenetic data. Cathodoluminescence microscopy can provide information about the spatial distribution of trace elements in rocks. Luminescence is the emission of light from a material which has been activated or excited by some form of energy. Cathodoluminescence works by activating various parts of a polished thin section with a beam of electrons. The electron beam excites certain ions, producing luminescence. This technique can reveal small-scale textures and inhomogeneities of particles and cements through differences in their luminescence, which is related to differing concentrations of trace element ions.

Scanning electron microscopy can magnify seventy thousand times or more, permitting detailed study of extremely small particles that cannot be adequately examined using a petrographic microscope. Scanning electron microscopy involves reflections of an electron beam from a rock or mineral surface. Fine details of cements and grains may be readily observed and photographed.

X-ray diffraction is used to determine the mineralogy of sedimentary rocks, particularly fine-grained rocks such as shales. The technique is based on reflections of X rays from planes in the crystal structure of minerals. Each mineral has a characteristic crystal structure and produces a distinctive X-ray diffraction pattern consisting of peaks of different position and intensity, which are plotted on chart paper by the X-ray diffractometer.

Fluid inclusions are extremely small droplets of fluid encased within crystals or mineral grains. The fluids are a small sample of the original pore fluids from which the mineral was precipitated. By examining fluid inclusions using heating and freezing devices attached to a microscope, the geologist can determine the composition of the original pore fluids and the temperature at which the mineral was precipitated. This technique reveals that many minerals and cements were precipitated from hot, saline pore fluids.

Stable isotopes of oxygen and carbon are commonly used to determine the chemistry of the pore fluids and the temperatures under which precipitation of cements or authigenic minerals occurred. Studies of the stable isotopes of microfossils have also provided information on past climatic changes. Isotopes are different forms of elements which vary in the number of neutrons present in the nucleus; hence, the various isotopes of an element have different atomic weights. By comparing the ratios of oxygen 16 and oxygen 18 or carbon 12 and carbon 13 in minerals such as calcite, it is possible to determine whether the minerals precipitated from fresh water or marine water, or to determine the temperature of the fluid from which the mineral precipitated.

Pamela J. W. Gore

CROSS-REFERENCES

Acid Rain and Acid Deposition, 1803; Biogenic Sedimentary Rocks, 1435; Biostratigraphy, 1091; Building Stone, 1545; Carbonates, 1182; Chemical Precipitates, 1440; Evaporites, 2330; Karst Topography, 929; Limestone, 1451; Microfossils, 1048; Reefs, 2347; Sedimentary Mineral Deposits, 1629; Sedimentary Rock Classification, 1457; Silica Minerals, 1354; Siliciclastic Rocks, 1463.

BIBLIOGRAPHY

Boggs, Sam, Jr. *Principles of Sedimentology and Stratigraphy.* Columbus, Ohio: Merrill, 1987. A textbook designed for undergraduate geology majors. It is clearly written, comprehensive, and well illustrated, with a long chapter on the topic of diagenesis. The chapter discusses the major diagenetic processes, the diagenetic environment (temperatures and pressures, as well as the chemical composition of subsurface waters), major controls on diagenesis, and the major effects of diagenesis (physical, mineralogic, and chemical changes). Contains background information on other aspects of sedimentary rocks.

Jonas, E. C., and E. McBride. *Diagenesis of Sandstone and Shale: Application to Exploration for Hydrocarbons.* Austin: Department of Geological Sciences, University of Texas, Continuing Education Program, 1977. This book provides clear coverage of the diagenesis of sandstones and shales. It is well illustrated, with photomicrographs of thin sections showing the evidence for various diagenetic processes. It also has graphs and line drawings, which make the processes easier to envision. It is written for persons with a basic background in sedimentology, but it can be understood by the nonspecialist.

Larsen, Gunnar, and George V. Chilingar, eds. *Diagenesis in Sediments and Sedimentary Rocks.* New York: Elsevier, 1979. A two-volume book which is number 25 in the Developments in Sedimentology series. The series covers a number of aspects of sedimentary geology in depth, and each volume contains papers written by and for specialists in the field. The first chapter of volume 1 is an introduction to the diagenesis of sediments and rocks; subsequent chapters deal with more specialized subjects, such as the diagenesis of sandstones, coal, and carbonate rocks. In volume 2, there are specialized chapters on the various phases of diagenesis, low-grade metamorphism, and the diagenesis of shales, deep-sea carbonates, and iron-rich rocks. The book is suitable for advanced college students. The first chapter is probably the most useful for the nonspecialist.

Leeder, Mike R. *Sedimentology and Sedimentary Basins: From Turbulence to Tectonics.* Oxford: Blackwell Science, 1999. A good look into the study of sedimentology and chemical sedimentology, and their relationship with tectonics. This is a good introduction to sedimentology for the layperson without much background in the Earth sciences.

McDonald, D. A., and R. C. Surdam, eds. *Clastic Diagenesis.* Memoir 37. Tulsa, Okla.: American Association of Petroleum Geologists, 1984. This book is divided into three parts: basic concepts and principles of diagenesis, changes in porosity, and applications of diagenesis in the exploration and production of hydrocarbons. Most of the articles are case histories of the diagenesis of particular rock units. Most are technical in content, but there are some which provide general overviews of specialized aspects of diagenesis. Suitable for advanced college students.

Oldershaw, Cally. *Rocks and Minerals.* New York: DK, 1999. This small, 53-page volume is filled with color illustrations and is therefore of great use to new students who may be unfamiliar with the rock and mineral types discussed in classes or textbooks.

Pettijohn, F. J. *Sedimentary Rocks.* 3d ed. New York: Harper & Row, 1975. This textbook provides an introduction to the basic types of sedimentary rock and touches on various aspects of diagenesis. Of particular interest is a chapter on concretions, nodules, and other diagenetic segregations. The chapter explains more about secondary sedimentary structures than most other textbooks on sedimentology, and it is a useful guide for persons curious about the origin of geologic oddities and those who would like a better background on the formation of sedimentary rocks. Suitable for college and advanced high school students.

Scholle, Peter A. *A Color Illustrated Guide to Carbonate Constituents, Rock Textures, Cements, and Porosities.* Memoir 27. Tulsa, Okla.: American Association of Petroleum Geologists, 1978. A superbly illustrated book on various types of carbonate rock as seen in thin section. It illustrates the major carbonate grains, along with dolomite, evaporite, silica, iron, phosphate, and glauconite minerals. Cements and carbonate rock textures are also covered. It provides background information on porosity and techniques for studying carbonate rocks. There is a brief explanatory caption and geologic locality data for each photograph.

_____. *A Color Illustrated Guide to Constituents, Textures, Cements, and Porosities of Sandstones and Associated Rocks.* Memoir 28. Tulsa, Okla.: American Association of Petroleum Geologists, 1979. A color picturebook of various types of sandstone as seen in thin section, this source illustrates all the major detrital sand grains, textures, cements, replacement or displacement fabrics, compaction and deformation fabrics, and porosity. Clays and shales, chert, and other types of sediment are also included. There is a brief explanatory caption and geologic locality data for each photograph. Basic information on sandstone classification and various techniques for studying sedimentary rocks is offered.

Scholle, Peter A., and P. R. Schluger, eds. *Aspects of Diagenesis.* Special Publication 26. Tulsa, Okla.: Society of Economic Paleontologists and Mineralogists, 1979. This book is divided into two major sections. The first covers the determination of diagenetic paleotemperatures; the second, the diagenesis of sandstones (in particular, hydrocarbon reservoirs). The papers are the result of symposia that were held on these topics. They include both general review articles and specific examples. Suitable for advanced college-level readers.

Sengupta, Supiya. *Introduction to Sedimentology.* Rotterdam, Vt.: A. A. Balkema, 1994. An excellent introduction to the practices, policies, and theories used within the field of sedimentology. The book is filled with illustrations and maps that help to clarify concepts and processes. Index and bibliographical references included.

Tucker, M., ed. *Techniques in Sedimentology.* Boston: Blackwell Scientific Publications, 1988. This volume covers techniques used by sedimentologists to study diagenesis and other areas of sedimentology. Chapters are included on the collection and analysis of field data, grain-size data and interpretation, microscopical techniques, cathodoluminescence microscopy, X-ray powder diffraction, scanning electron microscopy, and chemical analysis. The book explains the techniques used to study diagenesis and also discusses diagenetic fabrics produced by compaction, cementation, dissolution, alteration, and replacement. It is well indexed and profusely illustrated. Includes a number of excellent photomicrographs of rock thin sections illustrating diagenetic textures. For college-level students.

LIMESTONE

Limestone, the third most common sedimentary rock, is composed mostly of calcium carbonate, typically of organic origin. Limestone is usually fossiliferous and thus contains abundant evidence of organic evolution; it is also important as a construction material, groundwater aquifer, and oil reservoir.

PRINCIPAL TERMS

CALCITE: the main constituent of limestone, a carbonate mineral consisting of calcium carbonate

CARBONATES: a large group of minerals consisting of a carbonate anion (three oxygen atoms bonded to one carbon atom, with a residual charge of two) and a variety of cations, including calcium, magnesium, and iron

CEMENTATION: the joining of sediment grains, which results from mineral crystals forming in void spaces between the sediment

DEPOSITION: the settling and accumulation of sediment grains after transport

DEPOSITIONAL ENVIRONMENT: the environmental setting in which a rock forms; for example, a beach, coral reef, or lake

DIAGENESIS: the physical and chemical changes that occur to sedimentary grains after their accumulation

GRAINS: the individual particles that make up a rock or sediment deposit

LITHIFICATION: compaction and cementation of sediment grains to form a sedimentary rock

TEXTURE: the size, shape, and arrangement of grains in a rock

WEATHERING: the disintegration and decomposition of rock at the Earth's surface as the result of the exertion of mechanical and chemical forces

LIMESTONE IDENTIFICATION AND IMPORTANCE

Limestones are a diverse group of sedimentary rocks, all of which share a common trait: They contain 50 percent or more calcium carbonate, either as the mineral calcite or as aragonite. Both are composed of calcium carbonate; however, they have different atomic arrangements. Other carbonate minerals may also be present; siderite (iron carbonate) and dolomite (calcium-magnesium carbonate) are especially common. Although carbonate minerals can form other rocks, limestone is easily the most common and important carbonate rock. Many geologists use the terms "carbonate rock" and "limestone" almost interchangeably, because most carbonate rocks are limestones.

Limestone may be of chemical or biochemical (organic) origin and can form in a wide variety of depositional environments. A limestone's texture and grain content are often useful clues for determining how and where it formed; however, diagenesis can easily obscure or destroy this evidence.

Texture and grain content remain the basis for naming numerous varieties of limestone. These include dolomitic limestone, fossiliferous limestone, and crystalline limestone. Other common varieties include chalk, a very soft, fine-grained limestone; travertine, a type of crystalline limestone that forms in caves; and calcareous tufa, which forms by precipitation of calcium carbonate at springs.

Most limestones contain fossils, and many are highly fossiliferous. Limestones are perhaps our best record of ancient life and its evolutionary sequence. They are important sources for building and crushed stone and often contain large supplies of groundwater, oil, and natural gas. Weathering of limestone helps to develop distinctive landscapes as well.

LIMESTONE FORMATION

Limestones form in one of three ways: chemical precipitation of crystalline grains, biochemical precipitation and accumulation of skeletal and nonskeletal grains, or accumulation of fragments of preexisting limestone rock. Chemical precipitation occurs when the concentration of dissolved calcium carbonate in water becomes so high that the calcium carbonate begins to come out of solution and form a solid, crystalline deposit. The concentration of calcium carbonate in the water may change for a number of reasons. For example, evaporation, an increase in water temperature, or decreasing acidity can all cause precipitation. Crystalline limestone forms in the ocean, in alkaline lakes, and in caves.

One of the famous chalk (limestone) cliffs of Dover, England. (© William E. Ferguson)

Certain marine organisms are responsible for the formation of many kinds of limestone. Their calcareous (calcium carbonate) skeletons accumulate after death, forming carbonate sediment. Many limestones are nothing more than thousands of skeletal grains joined to form a rock. The organisms that contribute their skeletons to carbonate sediments are a diverse group and include both plants and animals. Among these are algae, clams, snails, corals, starfish, sea urchins, and sponges. Some marine animals also produce nonskeletal carbonate sediments. An animal's solid wastes, or fecal pellets, may accumulate to form limestones if they contain abundant skeletal fragments or compacted lime mud. Limestones composed of skeletal grains, or of nonskeletal grains produced by living organisms, are called "organic limestones."

Recycling of preexisting limestones is a third source for carbonate grains. Weathering and erosion produce limestone fragments, or clasts, that may later be incorporated into new limestone deposits. Limestones consisting of clasts are clastic, or detrital, limestones; they are probably the least common of the three types of limestone.

LITHIFICATION

The processes that turn loose sediment grains into sedimentary rock are known as "lithification." These may include either compaction, cementation, or both. The grains (crystals) in a chemically formed limestone are usually joined together into an interlocking, solid matrix when they precipitate; thus, they do not undergo further lithification. The grains in organic and clastic limestones, however, are usually loose, or unconsolidated, when they first accumulate and so must be lithified to form rock.

Limestones, unlike most other sedimentary rocks, are believed to undergo lithification during shallow burial rather than when deep below the Earth's surface. Some may be lithified within a meter or two of, or even at, the surface. Therefore, lithification in most limestones consists of cementation without significant compaction of grains. In most cases, the cement is calcium carbonate. If the spaces, or pores, between the grains become cement-filled without much compaction, cement can be as much as 50 percent or more of the volume of a limestone. The cement forms by precipitation, much like the formation of a crystalline limestone.

FORMATION AND PRESERVATION FACTORS

A number of factors control the formation and preservation of carbonate sediments. These include the water temperature and pressure, the amount of agitation, concentrations of dissolved carbon dioxide, noncarbonate sedimentation, and light penetration. Cold, deep water with high levels of dissolved carbon dioxide tends to discourage the formation and accumulation of carbonate sediment. Warm, clear, well-lit, shallow water tends to promote formation and accumulation.

Certain periods of geologic history also favored limestone formation. Generally, the greatest volumes of ancient limestones formed when the global sea level was higher than today, so that seas covered large areas of the continents, and when global temperatures were also higher than at present. This combination of factors was ideal for producing thick, extensive deposits of carbonate rocks. Such limestones are exposed throughout the world today and provide a glimpse into Earth's distant past. Their abundant marine fossils are especially useful to paleontologists and biologists, as they allow them to piece together the sequence of biological evolution for a variety of plants and animals.

Modern carbonate sediments accumulate in ocean waters ranging in depth from less than a meter to more than five thousand meters and at nearly all latitudes. However, most ancient limestones now exposed at the Earth's surface formed in low-latitude, tropical, shallow marine environments; for example, in reefs or lagoons. This is probably because elsewhere noncarbonate sediments, such as quartz sand or silt, diluted the carbonates so that limestone did not form. Carbonates from many other environments are also less likely to be preserved, so they are not a major part of the ancient rock record. This lack of preservation may result from weathering and erosion or from destruction by plate tectonic activities. Other factors may also play a role but are probably less significant.

One of the world's largest modern accumulations of carbonate sediment and rock is located in the Great Barrier Reef off the northeast coast of Australia. This reef tract, the largest in the world, contains thick sequences of carbonate sediment deposited during the last few thousand years draped over even older carbonate rocks formed by coral reef organisms in the more distant past. As long as these reefs continue to thrive, carbonate sediment production will also continue. As a result, this mass of limestone will grow even thicker, and the older rock will continue to subside, sinking deeper into the subsurface.

DIAGENESIS

No matter where they form or what their origin, carbonate sediments are all subject to diagenesis. Diagenesis consists of those processes that alter the composition or texture of sediments after their formation and burial and before their eventual re-exposure at the Earth's surface. Therefore, lithification is a part of diagenesis, and weathering is not.

One of the great mysteries of geology concerns the origin of dolomite, the calcium-magnesium carbonate mineral, and dolostone, the dolomitic equivalent of limestone. Many geochemists believe dolomite and dolostone owe their origin to the diagenesis of limestone. Dolostones are relatively common in the ancient rock record, and yet the formation of dolomite by direct crystallization is rare. This creates a dilemma: Where did all this ancient dolomite come from? Many geochemists believe that the answer lies in alteration (diagenesis) of relatively pure limestone to form dolostone, which contains at least 15 percent dolomite. However, there is no general agreement as to how this process, known as "dolomitization," actually occurred.

Limestone at Minerva's Terraces at Yellowstone National Park. (© William E. Ferguson)

JOINTS

Limestone weathers rapidly in humid environments. Rainwater, because of its content of dissolved carbon dioxide, is slightly acidic; it attacks and chemically weathers limestone quite readily. Since limestone is dense but of low strength, it also breaks, or fractures, easily. This characteristic allows even more rapid chemical weathering to occur, and the fractures quickly widen, becoming what are known as "joints."

In the subsurface, horizontal and vertical joints widen, as downward-flowing surface water and laterally flowing groundwater dissolve away limestone, creating increasingly large void spaces, or caves, in the rock. The largest and most extensive cave systems in the world, such as Mammoth Cave in Kentucky, form in limestones. The largest caves usually form where multiple joints intersect in the subsurface.

Where joints intersect at the Earth's surface, conical depressions, or sinkholes, begin to form. Sinkholes may be exposed at the surface or covered by a layer of soil. Some sinkholes grow and subside only very slowly, while others may collapse in one rapid, catastrophic event. They range in size from a few meters wide and deep to sinkholes that are large enough to swallow several large buildings should they collapse. The resulting irregular, pockmarked landscape is called "karst topography." Karst topography is easily recognized by the presence of sinkholes, disappearing streams (which flow into sinkholes), caves, and springs.

CARBONATE PETROGRAPHY

Geologists study limestones for a variety of reasons and at a variety of scales. Most early studies of limestones focused on their fossils. Many limestones contain abundant, well-preserved fossils; some are famous for the exceptional quality of the specimens they contain. Visible, or macroscopic, fossils provide evidence for the sequence of evolution of many invertebrate organisms. Fossils are clues to a limestone's depositional environment as well; however, more detailed information concerning depositional environments can often be gathered by examining limestones at either smaller or larger scales.

Carbonate petrography involves the study of limestones for the purpose of description and classification. This usually involves using a microscope to determine a limestone's grain content; that is, the types of carbonate and noncarbonate grains present and their mineral composition (mineralogy). Carbonate petrology deals with the origin, occurrence, structure, and history of limestones. This involves petrographic studies of limestone as well as field studies of one or more outcrops. Carbonate stratigraphy applies the concepts of petrology at even larger scales and attempts to determine the physical and age relationships between rock bodies that may be separated by great distances.

Carbonate petrography is commonly performed by observing a thin slice of a rock through a light microscope. A small block of the rock is cemented onto a microscope slide, then ground down and polished until the slice is about thirty microns (0.03 millimeter) thick. The slice is then thin enough for light to pass easily

In the Yankee Belle formation near Clear Mountain in Canada, lighter beds of limestone (center) are overlain by thin-bedded shale, grading upward into limestone that can be seen at top right. (Geological Survey of Canada)

through it. Microscopic examination of a thin section can reveal a limestone's mineralogy and its microfossil or other grain content. Other observable traits include cement types, the presence or absence of lime mud, the purity of the limestone, and the types and degree of diagenesis. Use of special stains along with microscopy can reveal even more details of mineralogy. Stains allow easy identification of particular minerals—for example, Alzarin Red S colors calcite red and dolomite purple—so that their percentages can be determined.

STUDY OF DEPOSITIONAL ENVIRONMENTS

Field study of exposures, or outcrops, of limestone also provides useful information. Along with the macroscopic fossil studies mentioned above, geologists can study sedimentary structures to learn about a limestone's depositional environment. Sedimentary structures are mechanically or chemically produced features that record environmental conditions during or after deposition and before lithification. For example, ripple marks indicate water movements by either currents or waves, and their shape and spacing suggests the depth of water and velocity of water movement. Careful study of sedimentary structures provides detailed information to methodical observers.

Outcrops also contain evidence of the lateral and vertical sequence of environments responsible for limestone deposition. By studying the lateral changes in a series of limestone outcrops, it is possible to interpret the distribution of environments, or paleogeography, in an area. For example, a researcher might determine in what direction water depths increased, or where a coastline might have been located. The vertical sequence of limestones at an outcrop indicates the paleogeography through time. By interpreting changes in sedimentary structures and other characteristics, a vertical sequence of limestones may indicate that, during deposition, a lagoon existed initially but gave way first to a coral reef and finally to an open ocean environment.

Many researchers conduct even larger-scale studies of limestone sequences. Using advanced technology developed for locating and studying petroleum reservoirs, geophysicists can produce cross sections showing limestone distribution in the subsurface. This research technique, sequence stratigraphy, allows geologists to see very large-scale features located thousands of meters below the surface and so provides even better insights into regional paleogeography.

STUDY BY GEOCHEMISTS, MINERALOGISTS, ENGINEERS

Geochemists and mineralogists study limestones to determine their mineral composition. Simple techniques might involve dissolving a sample of limestone in stages, using a series of different acids. At each stage, the scientist weighs the remaining solid material. From this, approximate percentages of the limestone's mineral components can be determined. More advanced techniques can involve the use of X rays and high-energy particle beams to determine a mineral's atomic structure and precise composition. Such analysis might, for example, allow a chemist to suggest new industrial uses for a particular limestone deposit.

Engineers also study limestones, usually to determine their suitability as a construction or foundation material. Numerous tests are available; engineering tests generally involve determination of physical and chemical properties such as composition, strength, durability, porosity, permeability, solubility, and density. Results from such tests help to predict the behavior of limestones under certain conditions. For example, testing may indicate how a particular limestone would perform as a building foundation in an area with a humid climate and highly acidic soils.

Clay D. Harris

CROSS-REFERENCES

Biogenic Sedimentary Rocks, 1435; Carbonates, 1182; Chemical Precipitates, 1440; Clays and Clay Minerals, 1187; Contact Metamorphism, 1385; Diagenesis, 1445; Fluid Inclusions, 394; Geothermometry and Geobarometry, 419; Metamorphic Rock Classification, 1394; Microfossils, 1048; Minerals: Structure, 1232; Oil and Gas Origins, 1704; Regional Metamorphism, 1418; Rocks: Physical Properties, 1348; Sedimentary Mineral Deposits, 1629; Sedimentary Rock Classification, 1457; Siliciclastic Rocks, 1463; Stress and Strain, 264; Weathering and Erosion, 2380.

BIBLIOGRAPHY

Brown, G. C., C. J. Hawkesworth, and R. C. L. Wilson, eds. *Understanding the Earth: A New Synthesis.* Cambridge, England: Cambridge University Press, 1992. Chapter 17, "Limestones Through Time," summarizes how limestone deposition, mineralogy, and diagenesis have changed through geologic time. Contains numerous small diagrams and photographs that help illustrate ideas presented in the text. A short bibliography of more advanced sources is included. Suitable for college-level readers.

Chesterman, C. W. *The Audubon Society Field Guide to North American Rocks and Minerals.* New York: Alfred A. Knopf, 1978. Well written and organized, beautifully illustrated with color photographs, this is an excellent introduction to both rocks and minerals. Contains systematic descriptions of occurrences, chemical formulas, physical traits, and more. Also contains a mineral identification key, a glossary of terms, a short bibliography, and a list of localities from the text. Suitable for high school readers.

Dougal, D. *The Practical Geologist.* New York: Simon & Schuster, 1992. An easy-to-read, well-illustrated book that covers all areas of geology at an introductory level. Related topics are presented in a methodical manner and are integrated with other concepts through illustrations and examples. Discusses the tools geologists use to investigate minerals, rocks, and fossils and how they use them. Contains a short glossary of terms. Suitable for high school readers.

Grotzinger, J. P. "New Views of Old Carbonate Sediments." *Geotimes* 38 (September, 1993): 12-15. This short article discusses how ancient limestones, unlike modern ones that are produced primarily by organic activity, may contain evidence of inorganic carbonate production and a record of the development of carbonate-producing organisms. Suitable for college-level readers.

Oates, Joseph A. H. *Lime and Limestone: Chemistry and Technology, Production and Uses.* New York: Wiley-VCH, 1998. An excellent look into the geochemical makeup, evolution, and uses of limestones. The emphasis on chemistry, geochemistry, and geophysics makes this book useful for the reader with some scientific background. However, the book is filled with useful illustrations, charts, and maps that clarify difficult concepts.

Oldershaw, Cally. *Rocks and Minerals.* New York: DK, 1999. This small, 53-page volume is filled with color illustrations and is therefore of great use to new students who may be unfamiliar with the rock and mineral types discussed in classes or textbooks.

Prothero, Donald R., and Fred Schwab. *Sedimentary Geology: An Introduction to Sedimentary Rocks and Stratigraphy.* New York: W. H. Freeman, 1996. A thorough treatment of most aspects of sediments and sedimentary rocks. Well illustrated with line drawings and black-and-white photographs, it also contains a comprehensive bibliography. Chapters 11 and 12 focus on carbonate rocks and limestone depositional processes and environments. Suitable for college-level readers.

Scholle, P. A., D. G. Bebout, and C. H. Moore. *Carbonate Depositional Environments*, Memoir 33. Tulsa: American Association of Petroleum Geologists, 1983. Well illustrated with color photographs, figures, and tables, this is an excellent treatment of limestone deposition. Each chapter covers a specific example of a different depositional environment and contains a general description of the physical traits of the environment and the limestones that form there, as well as a thorough bibliography for that environment. Suitable for college-level readers.

Wilson, J. L. *Carbonate Facies in Geologic History.* New York: Springer-Verlag, 1975. A comprehensive treatment of most aspects of limestone deposition throughout geologic time. Includes examples from a variety of environments and geographic locations. Contains many line drawings and black-and-white photographs, as well as a detailed bibliography. Suitable for college-level readers.

SEDIMENTARY ROCK CLASSIFICATION

Because sedimentary rocks are formed by several different processes—for example, precipitation, crystallization, and compaction—no single classification scheme is applicable to all of them. Used in various combinations, the main elements of classification are mode of origin, mineralogy, the size of the individual mineral grains that make up the rock, and the origin of these grains.

PRINCIPAL TERMS

ARKOSE: a sandstone in which more than 10 percent of the grains are feldspar or feldspathic rock fragments; also called feldspathic arenite

CARBONATE ROCK: a sedimentary rock composed of grains of calcite (calcium carbonate) or dolomite (calcium magnesium carbonate)

CLASTIC ROCK: a sedimentary rock composed of broken fragments of minerals and rocks; typically a sandstone

CLAYSTONE: a clastic sedimentary rock composed of clay-sized mineral fragments

GRAYWACKE: a sandstone in which more than 10 percent of the grains are mica or micaceous rock fragments; also called lithic arenite

LIMESTONE: a carbonate sedimentary rock composed of calcite, commonly in the form of shell fragments or other aggregates of small calcite grains

ORTHOQUARTZITE: a sandstone in which more than 90 percent of the grains are quartz

SANDSTONE: a clastic sedimentary rock composed of sand-sized mineral or rock fragments

EVAPORITES

Because there are several very different processes that lead to the formation of sedimentary rocks, no single classification scheme is suitable to all sedimentary rocks. The main elements of classification, however, are mode of origin, mineralogy, the size of the individual mineral grains making up the rock, and the origin of the individual grains. These elements are used in various combinations to categorize several major groups of sedimentary rocks.

One of the major groups is the evaporites. All natural waters contain some dissolved solids that will precipitate when the water evaporates. The crust that forms in a teakettle that has been used for a long time is an example. Seawater contains about 33 parts per thousand dissolved solids and is the major source of the sedimentary rocks classified as evaporites. When a body of seawater in an area with low rainfall is cut off from the sea, as when a sandbar builds up across the mouth of a bay, the trapped seawater tends to evaporate. During this evaporation, several minerals are precipitated in a predictable order. The first to precipitate is the mineral gypsum, hydrated calcium

sulfate. If the water is very hot, the precipitating calcium sulfate will not be hydrated, and the mineral anhydrite will form. Further evaporation will cause the precipitation of halite (sodium chloride, or ordinary table salt), and still further evaporation will lead to the precipitation of a complex series of potassium and magnesium salts.

In some environments, most commonly closed depressions in desert areas (the Dead Sea for example), fresh water evaporates to produce evaporite minerals that are quite different from the minerals produced by the evaporation of seawater. The natron (hydrated sodium carbonate) that was used by the Egyptians in embalming mummies and the borax that originally made Death Valley, California, famous are freshwater evaporite minerals.

From the standpoint of classification, the rocks of evaporative origin—that is, masses of individual crystals of minerals produced by evaporation of seawater or fresh water—are not usually given distinctive names. A fist-sized piece of evaporative sedimentary rock composed of gypsum is normally called gypsum. When it is necessary to indicate clearly that a rock, rather than a mineral, is being mentioned, the term "rock gypsum" is used. An exception is a rock composed entirely of crystals of halite, which is almost invariably referred to as rock salt.

CLASTIC ROCKS

A second major group of sedimentary rocks is the clastic rocks. "Clastic" comes from the Latin for "broken"; the individual grains of clastic rocks are the product of the mechanical and chemical breakdown, or weathering, of older rocks. Beach sands are composed of such grains. They are composed of residue from weathering of granites and many other kinds of igneous, sedimentary, and metamorphic rocks. Common soil, or mud, also is a residue of the weathering of rocks. Mud differs from beach sand primarily in its content of fine-grained clay minerals. Clay minerals are similar to the mineral mica, but the individual grains are very small, by definition less than 0.004 millimeter in size. When subject to prolonged attack by water and the atmosphere, many minerals that are common in igneous and metamorphic rocks—principally feldspars—are changed into micalike clay minerals. Between beach sands and the clay-sized component of muds are an intermediate size of mineral grains called silts. Silts range in size, again

by definition, from 0.0625 millimeter to 0.004 millimeter. Muds are mixtures of silt-sized and clay-sized mineral grains. Grains larger than 2 millimeters in diameter are classified as granules, pebbles, and boulders with increasing size.

The primary basis for classification of clastic sedimentary rocks is grain size. Coarse-grained rocks composed of pebbles and granules are called conglomerates, rocks composed of sand-sized grains are called sandstones, rocks composed of mud-sized materials are mudstones, and rocks composed only of clay-sized grains are called claystones. Mudstones and claystones that split readily along flat planes, the bedding planes, are referred to as shales.

Sandstones are further classified by mineral composition. In most sandstones, more than 90 percent of the sand grains are quartz, but some sandstones contain appreciable amounts of feldspar grains, volcanic rock fragments, mica, and micaceous rock fragments. These grains are the basis for further classification. Sandstones that are nearly pure quartz sand (more than 90 percent quartz grains, by one common definition) are called orthoquartzites, or quartz arenites ("arenite" is from the Latin for "sand"). Sandstones containing more than 10 percent feldspar grains and

SEDIMENTARY ROCK TYPES

Type	Definition	Subcategories	Explanation	Examples
Clastic	Rocks that consist of fragments of other rocks	Conglomerates	Grains are boulder-, cobble-, or gravel-sized	Breccia
		Sandstone	Grains are sand-sized	Quartz sandstone, arkose, graywacke
		Mudstone	Grains are silt- or clay-sized	Mudstone, shale
Precipitates	Chemically precipitated or replaced; inorganic in origin	Evaporites	Solids have precipitated after evaporation of water in which they were dissolved	Salt, gypsum, anhydrite, borax, potash
		Carbonates	Compounds of calcium or magnesium	Calcite (inorganic limestone), dolomite
		Siliceous	Chemically precipitated silicas	Chert: flint, jasper
Organic (biochemical)	Remains of plant or animal organisms	—	—	Coal, organic limestone

A red sandstone formation of Permian age in Monument Valley, northeastern Arizona. Sandstones are among the most common sedimentary rocks. (© William E. Ferguson)

volcanic rock fragments are called arkoses or feldspathic arenites, and sandstones with more than 10 percent mica flakes and micaceous rock fragments are called graywackes or lithic arenites.

LIMESTONES

Limestones are another major group of sedimentary rocks. Limestones are composed predominantly of the mineral calcite (calcium carbonate), although some limestones may include some clastic material, typically quartz sand or clay. A similar group of rocks is the dolomites. Dolomites consist predominantly of the mineral dolomite (calcium-magnesium carbonate) and appear invariably, or nearly invariably, to form by chemical alteration of preexisting limestone. Therefore, many workers prefer to lump the limestones and dolomites together in a rock group named carbonates.

The greater part of the calcite in limestones is secreted by marine organisms that make their shells from the mineral; clams and oysters are good examples of such organisms. Once these organisms die, their shells are washed about by waves and currents and are broken and abraded into fragments. The fragments may range in size from pebbles to mud, and most limestones are composed of this biogenic detritus.

Two common components of limestones appear to be of inorganic origin. Oölites are round grains, composed of very small crystals of calcite, that have a superficial resemblance to fish eggs. A shell fragment or quartz grain, the nucleus, at the center of the oölite, is surrounded by a coating of fine-grained calcite crystals layered like tree rings. It appears that oölites grow by inorganic deposition of calcite, directly from seawater, on the surface of the nucleus. Field observation of modern oölites suggests that they form only on sea bottoms that are shallow and are periodically agitated by strong waves or currents. It also appears that mud-sized calcite crystals precipitate directly from seawater in some circumstances.

Finally, small organisms, principally marine worms, ingest calcite mud, to extract whatever useful organic matter it may contain, and excrete it as fecal pellets. The fecal pellets, held together by mucus from the gut of the organism, survive if the bottom currents are not too strong.

Most limestones consist of aggregates of the materials described above. One classification, originally introduced by Robert L. Folk in 1959 and the most widely used, is based on the nature of the aggregates and the material that occurs between the grains and cements them together. Mud-sized calcium carbonate tends to accumulate in quiet waters, and a limestone consisting of only mud-sized material is called micrite. Sand-sized calcite grains are deposited in areas with stronger currents, generated by waves, winds, or tides. After final deposition, open spaces between sand-sized grains are often filled by inorganic calcite cement, called spar. In Folk's classification, the abbreviated name of the sand-sized material followed by the name of the material between the sand-sized grains is the rock name, with the traditional rock-name ending "-ite." Typical examples are oösparite, pelmicrite, and biosparite (where "bio-" refers to shell fragments).

CHEMICAL ROCKS

A fourth major group of sedimentary rocks is the chemical rocks. These rocks are divided into two subgroups: chemical precipitates and chemical replacements. Chemical precipitates are sediments that accumulate directly on the sea bottom as a result of chemical reactions that do not involve evaporation. Deposits of iron minerals (most typically the iron oxide mineral hematite) and phosphate minerals (most typically calcium phosphate, or the mineral apatite) are the most common, and economically important, examples. The process of formation of these rocks is not well understood, but research suggests that in most cases bacteria are involved in producing the proper chemical environment for formation of these important, although rather rare, deposits.

The second subgroup is the chemical replacements. In some cases, the original sediment is dissolved and a new mineral takes its place. Typical examples are the solution of calcite and its replacement by dolomite and the solution of calcite and its replacement by fine-grained quartz. The replacement of calcite by fine-grained quartz in limestones is especially common. The replacement product, the very fine-grained quartz replacement, is called chert, but most people are more familiar with the popular term "flint."

IDENTIFYING SEDIMENTARY ROCKS AND COMPONENT MINERALS

Of the more than twenty thousand minerals identified, only twenty-two of them are common in rocks at the Earth's surface; these are easily identified in the field by observation and simple tests of physical properties. Even most fine-grained rocks can be identified in the field with the aid of a twelve-power hand lens. Gypsum, for example, is easily identified by its satiny sheen and the fact that it can be scratched by a fingernail. Calcite and dolomite have very similar appearances and physical properties, but a drop of dilute hydrochloric acid will cause calcite, but not dolomite, to effervesce, or fizz. A pocket-sized dropper bottle of dilute hydrochloric acid is standard equipment for the sedimentary geologist.

More detailed studies are done on samples brought to the laboratory. A very common method is the study of thin sections of rocks. To prepare a thin section, a thin slice, about a centimeter thick, is cut from the sample. The slice is glued to a glass slide and further thinned. When the thickness has been reduced to about 0.03 millimeter, a thin cover glass is glued to the top of the slice, and the thin section is complete.

Thin sections are studied with a microscope, usually a petrographic microscope that has a polarized light source. The effect of passage of the polarized light through individual mineral crystals can be analyzed and much information about the arrangement of the atoms in the crystals obtained. For example, passage of the light through halite does not affect the planar vibration of the polarized light, but when it passes through quartz, it is forced to vibrate in two planes that are perpendicular to each other and parallel to the two crystallographic axis directions of quartz. Therefore, even microscopic-sized grains of halite and quartz are easily distinguished.

Clay minerals, which are too small to be studied effectively by optical methods, are commonly studied by X-ray diffraction. X rays have very short wavelengths and can be diffracted by the regularly arranged planes of atoms in a crystal in the same way that light is diffracted by a diffraction grating. The sample to be analyzed is irradiated with X rays of a single wavelength at steadily varying angles of incidence. The angles at which diffraction occurs represent planes of atoms at different spacings. If the clay mineral kaolin, for example, is present in the sample, a strong diffraction will occur at an angle that corresponds to an atomic plane spacing of 7.2×10^{-10} meters (usually expressed as 7.2 angstroms), and weaker diffractions will occur at angles corresponding to several other spacings that are characteristic of the mineral.

Scanning electron microscopy is a powerful tool for the study of all types of sedimentary rocks but is especially useful for the very fine-grained varieties. Photographic images at 50,000 magnifications are routinely obtained, revealing remarkable details of individual grains and the openings between grains. In addition, semiquantitative chemical analyses of individual grains for sodium and elements heavier than sodium can be made.

INDUSTRIAL APPLICATIONS

Classification of sedimentary rocks has led to the recognition of predictable associations of sedimentary rock types with particular geologic conditions. Sandstones that are derived from, and deposited near to, mountains with granitic cores—the Front Ranges of the Rocky Mountains, for example—most commonly contain more than 10 percent feldspar, mostly of the potassium-rich variety, and are classified as arkoses. Sandstones derived from mountains with metamorphic rock cores, such as the Appalachian Mountains, contain more than 10 percent mica and are classified as graywackes, or lithic arenites. Sandstones deposited far from any mountain chain normally are nearly pure quartz, in some cases more than 99.9 percent quartz. For quality-control purposes, glass manufacturers prefer sand that is very nearly pure quartz, and the sedimentary geologist who specializes in industrial minerals will begin the search for glass sands far from any large mountain chain.

The organic-matter content of muds that lie on the sea bottom for long periods of time tends to be destroyed by scavenging organisms. Sedimentary geologists who specialize in petroleum exploration know that one of the requirements for the generation of petroleum is a mudstone with a high organic-matter content. They look for mudstones that were buried by new mud shortly after deposition—that is, mudstones that had a high rate of sedimentation. As a general rule, such mudstones will be associated with arkoses or graywackes rather than with nearly pure quartz sandstones.

Robert E. Carver

CROSS-REFERENCES

Biogenic Sedimentary Rocks, 1435; Building Stone, 1545; Carbonates, 1182; Caves and Caverns, 870; Chemical Precipitates, 1440; Diagenesis, 1445; Dolomite, 1567; Karst Topography, 929; Limestone, 1451; Petroleum Reservoirs, 1728; Siliciclastic Rocks, 1463.

BIBLIOGRAPHY

Blatt, Harvey. *Sedimentary Petrology.* 2d ed. San Francisco: W. H. Freeman, 1992. A very well illustrated work offering complete coverage of the subject of sedimentary rocks. Intended to be a college-level textbook but perfectly accessible to the interested high school student or layperson. The final two chapters cover the design and conduct of research projects.

Blatt, Harvey, and Robert J. Tracy. *Petrology: Igneous, Sedimentary, and Metamorphic.* New York: W. H. Freeman, 1996. Stated to be intended for the college sophomore or junior but perfectly suited to the interested high school student or general reader. The section on sedimentary rocks is clear and very well illustrated. An excellent introduction to the subject of sedimentary rocks in general and classification in particular.

Carver, Robert E., ed. *Procedures in Sedimentary Petrology.* New York: John Wiley & Sons, 1971. Covers the methods commonly used to study sedimentary rocks, including the methods of grain-size analysis and mineralogical analysis that are the basis of their classification. Where appropriate, a chapter on obtaining the data is followed by a chapter on mathematical or statistical analysis of the data. The mathematics involved is not difficult.

Ehlers, E. G., and Harvey Blatt. *Petrology: Igneous, Sedimentary, and Metamorphic.* San Francisco: W. H. Freeman, 1982. Stated to be intended for the college sophomore or junior but perfectly suited to the interested high school student or general reader. The section on sedimentary rocks is clear and very well illustrated. An excellent introduction to the subject of sedimentary rocks in general and classification in particular.

Fairbridge, Rhodes W., and Joanne Burgeois, eds. *The Encyclopedia of Sedimentology.* New York: Van Nostrand Reinhold, 1982. An excellent reference for the reader who needs more information on any specific aspect of the classification of sedimentary rocks. For example, the information on clay-pebble conglomerates would not be easily found in any other reference. Extensively and usefully cross-referenced.

Ham, W. E., ed. *Classification of Carbonate Rocks: A Symposium.* Tulsa, Okla.: American Association of Petroleum Geologists, 1962. A classic work on the title subject by the originators of various schemes of classification of limestones and dolomites, written at a time when there was still much discussion of what was the most appropriate classification. Required reading for all students of carbonate rock classification.

Hatch, F. H., R. H. Rastall, and J. T. Greensmith. *Petrology of the Sedimentary Rocks.* Rev. 4th ed. London: Thomas Murby, 1965. An older reference from the time of widespread fascination with classification. Therefore, it contains details of composition and classification of some rock groups—for example, the carbonaceous group (coal and related rocks)—that cannot be found elsewhere. The illustrations are mostly line drawings of thin sections (a lost art), but these are very informative.

Pettijohn, F. J., P. E. Potter, and R. Siever. *Sand and Sandstone.* 2d ed. New York: Springer-Verlag, 1987. A thorough, well-illustrated, and clearly written treatment of the subject. Some preliminary study of the general subject is advised, because a general familiarity with the geology and mineralogy of sandstones is required.

Tucker, M., ed. *Sedimentary Rocks in the Field.* 2d ed. New York: Wiley & Sons, 1996. A clearly written and well-illustrated introductory text aimed at British college undergraduates, who learn principally through independent study. The greater part of the text is devoted to classification in terms understandable to the general reader.

Williams, Howel, F. J. Turner, and C. M. Gilbert. *Petrography: An Introduction to the Study of Rocks in Thin Sections.* 2d ed. San Francisco: W. H. Freeman, 1982. An introductory college text that is yet accessible to younger students and to general readers who have some familiarity with the literature. As in the work by Hatch, Rastall, and Greensmith (above), the sedimentary rock illustrations are primarily line drawings of thin sections, but they make the point very well.

SILICICLASTIC ROCKS

Siliciclastic rocks, which include siltstone, sandstone, and conglomerate, are second only to shales in abundance among the world's sedimentary rock types. They form major reservoirs for water, oil, and natural gas and are the repositories of much of the world's diamonds, gold, and other precious minerals. In their composition and sedimentary structures, siliciclastic rocks reveal much about paleogeography and, consequently, Earth history.

PRINCIPAL TERMS

CLASTIC ROCK: a sedimentary rock composed of particles, without regard to their composition; this term is sometimes used, incorrectly, as a synonym for "detrital"

DETRITAL ROCK: a sedimentary rock composed mainly of grains of silicate minerals as opposed to grains of calcite or clays

DIAGENESIS: chemical and mineralogical changes that occur in a sediment after deposition and before metamorphism

LITHIC FRAGMENT: a grain composed of a particle of another rock; in other words, a rock fragment

SHALE: a rock composed of abundant clay minerals and extremely fine siliciclastic material

WEATHERING AND EROSION

Weathering and erosion constantly strip the Earth's surface of its rocky exterior. Rocks experience fluctuating temperatures, humidity, and freeze-thaw cycles that physically disaggregate (break up) mineral grains. Simultaneously, these mineral grains are exposed to water, oxygen, carbon dioxide, and dissolved acids or bases that chemically attack them. As a result, a hard, rocky surface is eventually transformed to a collection of mineral grains, clay minerals, and ions dissolved in water.

Weathering and erosion are slow, but continuous, agents of destruction. Because mountains and other high regions are constantly being uplifted, weathering and erosion provide a more or less steady supply of sediment to streams and rivers and, from there, into the ocean for deposition. Deposited sediment is eventually compacted and cemented into rocks. The rocks that are formed by this process are known collectively as siliciclastic rocks because they are made mostly of particles of silicate minerals.

SILICICLASTIC ROCK COMPOSITION

Siliciclastic rocks differ in the sizes of their grains and their composition. Three main size classes are recognized: silt, sand, and gravel. Silt includes particles between 0.004 and 0.0625 milli-meter in diameter; sand is composed of particles between 0.0625 and 2 millimeters in diameter; and gravel generally includes particles larger than 2 millimeters. Most sedimentologists refer to sand and sandstone when discussing siliciclastic rocks.

Siliciclastic rocks are composed of three broad classes of material: framework grains, matrix, and cement. Framework grains are particles of minerals or small fragments of other rocks that usually make up the bulk of a siliciclastic rock. Matrix is extremely fine-grained material such as clay that is deposited at the same time as framework grains. Cement is any material precipitated within the spaces, or pores, between grains in a sediment or rock. Framework grains and matrix are primary deposits, whereas cement is a secondary deposit because it is precipitated in pores after the primary material.

QUARTZ AND FELDSPAR MINERALS

The main minerals that compose siliciclastic rocks are quartz, various types of feldspar, and micas. The ferromagnesian minerals, such as olivine, pyroxene, and amphibole, are not as common in sedimentary rocks as they are in igneous and metamorphic rocks. Ferromagnesian minerals are more easily dissolved during weathering, and they fracture more readily during erosion, than do the more durable quartz and feldspar.

Consequently, quartz and feldspar (and, to a lesser extent, the micas) are concentrated in sediments because the other minerals are selectively removed beforehand. Carbonate rocks dissolve very easily and seldom form much residue to contribute to sand.

Quartz is the most chemically and physically stable mineral that forms siliciclastic rocks. Although quartz is uncommon in some igneous and metamorphic rocks, it is sufficiently abundant in granite and gneiss to contribute a large amount of bulk to almost all sands. Subtle differences may provide useful clues as to the origin of some quartz grains. For example, quartz from plastically deformed metamorphic rocks (schist, gneiss) usually occurs as polycrystalline aggregates or displays a distorted crystal structure, revealed by the petrographic microscope. Quartz from volcanic rocks often possesses planar crystal faces or embayments (deep, rounded indentations) that were formed as the crystal grew in magma. Quartz grains eroded from older sedimentary rocks sometimes retain the previous rock's quartz cement as a rind, which is called an inherited overgrowth.

Feldspar minerals are also common in most siliciclastic rocks. Two main groups of feldspar are recognized: plagioclase feldspar and the potassium feldspars (microcline, orthoclase, and sanidine). The plagioclase group is actually a collection of many similar minerals with different chemical compositions. Anorthite is a plagioclase feldspar composed of calcium, silicon, and oxygen, whereas albite is a plagioclase composed of sodium, silicon, and oxygen. Calcium and sodium substitute rather easily for each other, so most plagioclase feldspars have some calcium and some sodium. In general, plagioclase with more calcium is more susceptible to chemical weathering than is sodium-rich plagioclase. Consequently, sodium-rich albite is more common than anorthite in most siliciclastic rocks.

The potassium feldspars all have the same chemical composition of potassium, silicon, and oxygen. They differ from one another in their chemical structure. Microcline is the most highly organized, crystallographically, followed by orthoclase and then by sanidine. Microcline is formed in igneous and metamorphic rocks that crystallize very slowly, permitting the greatest amount of crystallographic ordering. At the other extreme, sanidine is formed in volcanic rocks where little time is available for ions to get into their "proper places." During chemical weathering of detrital potassium feldspars, sanidine dissolves much more readily than does microcline or orthoclase. Microcline is usually about equal in abundance to orthoclase in siliciclastic rocks, whereas sanidine is usually rare.

In this Cambrian sequence at the head of Nordenskiold Fjord, Greenland, a layer of carbonate grainstones (RG2) is overlain by siliciclastic rocks at the top. (Geological Survey of Canada)

MICAS AND ACCESSORY MINERALS

The mica minerals, biotite and muscovite, are common in some siliciclastic rocks and rare in others. In general, muscovite (white mica) is more durable and, consequently, more abundant than is biotite (brown mica). Chlorite is a mineral with a sheetlike crystal structure similar to the micas. It is common in a few sandstones but generally less abundant than the micas. A large variety of grains is often present in siliciclastic rocks as "accessory minerals." Altogether, they seldom constitute more than 1 percent of any siliciclastic rock. Sometimes these accessory minerals are called heavy minerals because most of them have a much greater density than does quartz or feldspar. This group includes zircon, tourmaline, rutile, garnet, the ferromagnesian minerals pyroxene, amphibole, and olivine, and iron oxides such as hematite, magnetite, and limonite. If one examines a handful of sand, the accessory minerals are usually the dark grains.

Accessory minerals are important in several ways. First, they may reveal information about the source rock from which the sand was eroded. For example, the accessory mineral chromite is generally formed in basalt, so its presence in a sand indicates that basalt (usually from the seafloor) had previously been uplifted and eroded nearby. Similarly, garnets of particular composition may be representative of particular metamorphic source rocks. Second, accessory minerals may tell geologists how "mature" a sand is: in other words, how great has been its exposure to chemical and physical weathering. Zircon, tourmaline, and rutile are far more durable than are the other accessory minerals. If they are the only accessory minerals in a sand, then the sand probably experienced severe weathering in the period before its final deposition. Third, accessory minerals may be economically valuable. Diamonds and gold are accessory minerals in a few cases. Much of the United States' titanium comes from unusual concentrations of rutile and ilmenite in beach sands along the coast of South Carolina. Zircon in sands provides an important source of zircomium, used in high-temperature ceramics.

LITHIC FRAGMENTS AND MATRIX

Pieces of rocks (lithic fragments) may form an important part of coarse-grained sandstone and conglomerate. Lithic fragments may be composed of pieces of almost any igneous, metamorphic, or sedimentary rock. Their presence is a powerful indicator of the source rock from which the sand or gravel was weathered and eroded. As one might expect, not all rock fragments are equally durable in the sedimentary environment. Limestone and shale fragments disintegrate very rapidly, whereas granite and gneiss rock fragments are quite robust. Consequently, both the presence and the absence of certain rock fragments may reveal information about the origin of a siliciclastic sediment.

Matrix in siliciclastic rocks is usually clay or fine silt. This substance is the "mud" that often accompanies sand or gravel during deposition. It is commonly deformed between the more rigid framework grains as the sediment is compacted. Some types of matrix were not originally deposited as fine-grained matrix. This "pseudomatrix" is actually squashed fragments of soft grains of, for example, shale or schist. During compaction of the sediment, these fragments are more easily deformed than are durable quartz and feldspar, and the fragments superficially resemble original matrix. Considerable skill is required to distinguish between true matrix and pseudomatrix.

SANDSTONES

Sand-sized siliciclastic rocks are often classified based on their mineral composition. Sandstones may be named with the term "arenite" (after the Latin *harena*, sand). Relative abundances of quartz, feldspar, and rock fragments determine the name. One of the major types of sandstone is quartz arenite, which is composed almost entirely of quartz, with less than 5 percent of other framework minerals and less than 15 percent clay matrix (usually much less). This rock is usually white when fresh. Another is feldspathic arenite, which is sometimes called arkose. This rock has abundant feldspar and up to 50 percent lithic fragments. Typical composition is perhaps 30 percent feldspar, 45 percent quartz, and 25 percent lithic fragments. Up to 15 percent detrital clay matrix may also be present. This rock is often pinkish-gray in color. A third type is lithic arenite, a sand composed largely of lithic fragments and up to 50 percent feldspar. A typical composition might be 60 percent lithic fragments, 30 percent feldspar, and 10 percent quartz. Up to 15 percent matrix may be present as well. Almost all conglomerates are lithic arenites. Lithic arenites are often dark gray, sometimes with a salt-and-pepper appearance from the lithic fragments. A fourth type is wacke, sometimes called graywacke; this rock has between 15 percent and 75 percent clay matrix at the time of deposition. The term "wacke" may be modified by prefixing with the name of the dominant nonmatrix component, such as "quartz wacke," "feldspathic wacke," or "lithic wacke." Some wackes may be "formed" diagenetically from lithic arenites by alteration of shale or schist fragments so that they appear similar to originally detrital clay matrix. This rock is usually gray in color (thus, the name "graywacke").

The composition of sandstones holds important clues as to the kinds of rocks from which they were derived. These clues may help petrologists to reconstruct the plate tectonic setting of the sand's source region even though the source region has long since disappeared because of erosion or tectonic destruction. For example, quartz arenite often reflects sedimentation on or near a stable craton (piece of the Earth's crust), where physical and chemical weathering are permitted to eliminate unstable minerals for an extended period of time. At the other extreme, a lithic arenite with abundant basaltic rock fragments, calcic plagioclase, and rare quartz may represent detritus shed from an island arc during plate collision. Careful observation of the mineral composition of sandstones has become an important part of reconstructing the Earth's history.

DIAGENESIS

Diagenesis of siliciclastic rocks is a complex collection of processes that include cementation, replacement, recrystallization, and dissolution. Diagenesis includes chemical reactions that begin at the time of deposition and may continue until metamorphism takes place. For many sedimentary petrologists, low-temperature cementation is part of diagenesis, whereas for others, diagenesis includes only those reactions that occur at temperatures exceeding 100 degrees Celsius. The upper range of diagenesis grades into metamorphism; sedimentary petrologists and metamorphic petrologists share some common territory in diagenesis.

Siliciclastic sediments are transformed to rocks by compaction and cementation. Common cements in siliciclastic rocks are quartz, calcite, clays, and iron oxides such as limonite or hematite. It is common for a rock to have several cements, each deposited at different times or under different conditions of pore-water chemistry. Cement is introduced into sediment in the form of elements or compounds dissolved in water. For example, silica may be dissolved from quartz grains or given off during low-grade metamorphism and may saturate water in the pores of a rock. One means of dissolving silica from quartz grains is called pressure solution. Quartz grains exert very high pressures upon one another at their points of contact. Quartz is more soluble under high pressure, so it is more easily dissolved where grains touch each other. Well-compacted sands often display flattened grain-to-grain contacts as a consequence of pressure solution. The quartz lost from these grains is carried away, dissolved in pore water, to be precipitated elsewhere.

As the water moves slowly through the pores of the rock, it may encounter different temperatures, pressures, acidity, or gas concentrations. Changes in these conditions may cause the water to lose some of its ability to dissolve silica, and the silica may therefore precipitate as crystals. Similar factors influence the dissolution and precipitation of other cements in siliciclastic rocks. Cement not only holds sediment grains together, forming a rock, but it also fills the pores between grains, which may inhibit further flow of fluids. As a consequence, the porosity and permeability of most rocks are reduced by cementation. Rocks with abundant cement generally make poor reservoirs for petroleum or water.

In some cases, the pore-water chemistry conducive to precipitation of one cement is capable of dissolving another cement. A common example is the generally inverse relationship between quartz and calcite cement. Pore water that precipitates quartz often dissolves calcite at the same time, or the other way around. Sometimes, dissolution occurs without immediate precipitation of another mineral type in the void that is formed. Pores formed by dissolution are called secondary porosity; such porosity may be the main cause of porosity in many sandstones. Minerals commonly dissolved include carbonates such as calcite and aragonite, silicates such as feldspars and ferromagnesian minerals, and, though rarely, accessory minerals such as garnet. In some cases, earlier cements may be dissolved to form secondary porosity.

RECRYSTALLIZATION AND REPLACEMENT

Recrystallization is a common feature in diagenesis. In recrystallization, one mineral is transformed into another mineral with the same (or nearly the same) chemical composition or into different sized (usually larger) crystals of the original mineral. The most common example is the transformation of aragonite to calcite in limestones. In siliciclastic rocks, microcrystalline quartz may recrystallize into more coarsely crystalline quartz, or rutile may recrystallize into anatase.

A more prevalent form of diagenesis among ancient sandstones is replacement. In this process, one mineral is replaced volume-for-volume by another mineral. In some cases, it appears that the replacing mineral has combined material from its host grain with dissolved ions from pore water. The most common replacement examples involve clay minerals. For example, kaolinite may replace potassium feldspar, forming a grain of kaolinite that looks superficially like a potassium feldspar grain.

EXAMINING SILICLASTIC ROCKS

The principal means of studying siliciclastic rocks is with the petrographic microscope. This microscope is similar to other microscopes except that light passes through the specimen rather than illuminates its surface. Rocks are normally not transparent, however, and they must be cut into very thin slices in order for light to pass through them. Magnifications of up to four hundred times are routinely used. In thin section, it is possible to view closely individual grains and their relations to other grains, to identify the mineral grains in the rock, and to view matrices and cements. Thin sections are routinely used to estimate the amount of pore space in rocks, revealing their potential as reservoirs for fluids such as water, oil, and natural gas.

When illuminated by polarized light, different minerals display characteristic interference colors that reveal the identity of a mineral to the experienced observer. Other clues are the existence and pattern of cleavages in the minerals, alteration products resulting from diagenesis, and overall grain shape. In addition, it is possible to view cements that surround grains and partially or completely fill pores. The sequence of cementation can be interpreted by the relative positions of cements filling pores. Knowing the sequence of cements can help geologists understand how deeply buried the rock was when it was cemented and how the chemical composition of the pore water changed through time.

Another means of examining siliciclastic rocks is with the scanning electron microscope (SEM). The SEM permits examination of clastic grains with much greater magnification than with the conventional petrographic microscope. Magnifications of up to ten thousand times are easily accomplished, with details as small as 0.001 micrometer being easily resolved. The SEM is very useful in examining the surface texture of minerals. The presence of pits, grooves, and cracks can reveal clues as to the environment under which the sand grains were deposited. Were the sand grains blown by wind, carried by water, or transported by glaciers? In some cases, this question may be answered by examination under the extreme magnification of the SEM.

ELECTRON MICROPROBE AND X-RAY STUDY

When accurate data concerning chemical composition of minerals are needed, the electron microprobe is used. The microprobe is actually an elaborate version of an SEM. It differs in that its beam does not normally scan but instead remains fixed in one spot, and its detectors are carefully calibrated to give accurate readings of the elemental composition of the material being examined. Microprobes collect X rays and can determine from them the amounts of different elements in almost any mineral. In some instances, the abundance of trace elements in minerals such as zircon, tourmaline, garnet, or rutile may provide clues as to the source region from which the grains were derived, giving an indication of the source for a sedimentary formation. The electron microprobe is a very useful tool in these studies.

There are limitations to the microprobe. It cannot analyze for elements lighter than sodium, so the amount of oxygen or carbon cannot be determined. Furthermore, rock material must be very carefully prepared for readings to be accurate. Also, electron microprobes are rather expensive and require considerable maintenance; however, X rays are also employed in the study of rocks. X-ray diffraction of powdered or whole specimens can identify minerals and their degree of crystal ordering. X-ray fluorescence identifies the chemical composition of minerals. Both techniques are commonly used in the study of clays in the matrix of siliciclastic rocks. Sample preparation is relatively easy, and beginners can operate most modern X-ray machines. X-ray analysis usually requires destruction of the sample by grinding, however, so the sizes, shapes, and relationships between grains are lost.

ECONOMIC SIGNIFICANCE

Sandstone and its cousins siltstone and conglomerate are common sedimentary deposits. These rocks, collectively known as siliciclastic rocks, are important for a number of reasons. For example, they are porous—that is, they contain small spaces between grains that may be filled with valuable fluids. Siliciclastic rocks are the major type of aquifer in most of the world. Also, they form reservoirs for oil and natural gas. Sand and gravel are essential parts of modern construction because they form the "bulk" in concrete. In this role, they represent the most economically significant mineral resource in the United States, ahead of petroleum, coal, and precious metals.

Michael R. Owen

CROSS-REFERENCES

Aluminum Deposits, 1539; Biogenic Sedimentary Rocks, 1435; Building Stone, 1545; Cement, 1550; Chemical Precipitates, 1440; Clays and Clay Minerals, 1187; Coal, 1555; Diagenesis, 1445; Earth Resources, 1741; Evaporites, 2330; Fertilizers, 1573; Glacial Deposits, 880; Limestone, 1451; Oil and Gas Exploration, 1699; Oil Shale and Tar Sands, 1717; Petroleum Reservoirs, 1728; Sand, 2363; Sedimentary Mineral Deposits, 1629; Sedimentary Rock Classification, 1457; Stratigraphic Correlation, 1153; Weathering and Erosion, 2380.

BIBLIOGRAPHY

Blatt, Harvey. *Sedimentary Petrology.* 2d ed. San Francisco: W. H. Freeman, 1992. A useful introduction to sedimentary petrology. Suitable for general readers.

Blatt, Harvey, and Robert J. Tracy. *Petrology: Igneous, Sedimentary, and M-etamorphic.* New York: W. H. Freeman, 1996. A standard textbook on sedimentary rocks, covering more than siliciclastics. Includes a complete discussion of the sources of grains, their transportation, deposition, and deformation.

Houseknecht, David W., and Edward D. Pittman, eds. *Origin, Diagenesis, and Petrophysics of Clay Minerals in Sandstones.* Tulsa, Okla.: Society for Sedimentary Geology, 1992. A collection of essays written by leading experts in their respective fields, this book examines the geochemical and geophysical properties of sandstone, clay minerals, and other sedimentary rocks. Filled with illustrations and includes an index and bibliographical references.

McDonald, D. A., and R. C. Surdam, eds. *Clastic Diagenesis.* Memoir 37. Tulsa, Okla.: American Association of Petroleum Geologists, 1984. A thorough discussion of sandstone diagenesis, complemented by well-documented case studies. Superbly illustrated.

Pettijohn, F. J., P. E. Potter, and R. Siever. *Sand and Sandstone.* 2d ed. New York: Springer-Verlag, 1987. Perhaps the most comprehensive, authoritative, and readable book on sandstone in English. Very well illustrated, with copious references. The basic reference for all sandstone studies.

Prothero, Donald R., and Fred Schwab. *Sedimentary Geology: An Introduction to Sedimentary Rocks and Stratigraphy.* New York: W. H. Freeman, 1996. A thorough treatment of most aspects of sediments and sedimentary rocks. Well illustrated with line drawings and black-and-white photographs, it also contains a comprehensive bibliography. Chapters 11 and 12 focus on carbonate rocks and limestone depositional processes and environments. Suitable for college-level readers.

Scholle, Peter A. *Color Illustrated Guide to Constituents, Textures, Cements, and Porosities of Sand-*

stones and Associated Rocks. Memoir 28. Tulsa, Okla.: American Association of Petroleum Geologists, 1979. An excellent guide to the major types of grains, matrices, and cements common in sandstones. Includes microphotographs, complete with short description of the features they show.

Scholle, Peter A., and D. R. Spearing, eds. *Sandstone Depositional Environments.* Memoir 31. Tulsa, Okla.: American Association of Petroleum Geologists, 1982. Well-illustrated compendium of the environments in which siliciclastic rocks may be found.

Tucker, M., ed. *Sedimentary Rocks in the Field.* 2d ed. New York: Wiley & Sons, 1996. A very useful collection covering the major techniques, practices, and policies of sedimentary petrology. Well referenced.

5
LAND AND SOIL

DESERTIFICATION

Desertification comprises a variety of natural and human processes that cause the impoverishment of ecosystems, manifested in reduced biological productivity, rapid deterioration of soil quality, and associated declines in the condition of regional human economic and social systems.

PRINCIPAL TERMS

ALBEDO: the fraction of visible light of electromagnetic radiation that is reflected by the properties of a given type of surface

AQUIFER: a water-bearing bed of rock, sand, or gravel capable of yielding substantial quantities of water to wells or springs

DESALINIZATION: the process of removing salt and minerals from seawater or from saline water occurring in aquifers beneath the land

surface to render it fit for agriculture or other human use

MOVING DUNES: collections of coarse soil materials that result from wind erosion and threaten marginal vegetation and settlements as they move across deserts

SALINIZATION: the accumulation of salts in the soil

DEFINITION OF DESERT CONDITIONS

Desertification—or, as some authorities prefer, desertization—is a complex set of interactions between natural and cultural forces and most often occurs in the borderlands of large, natural desert environments. The broadest conventional definition of desert conditions refers to precipitation levels: Regions that receive less than 100 millimeters of precipitation per year, for example, usually exhibit commonly recognized surface features such as flat, gravelly, or pebbly tables and accumulations of sand dunes. The long-term absence of concentrated vegetation cycles and water erosion features results in only thin, marginally fertile soils, although in places along internal drainage systems these soils may be quite productive in the short run.

The term "desert" inevitably evokes references to land and soil, but in terms of precipitation levels and their consequences, the term can also be applied to surface regimes such as glacier fields and ice caps. Annual precipitation in Antarctica, for example, is well below 100 milimeters. Both ocean surfaces, in the sense of annual precipitation input, and various ocean levels and seabed, in terms of the presence of nutrients analogous to soil on land, may be classified as desert or as undergoing forms of the desertification process, resulting from both natural and human-induced causes. Land deserts typically feature dispersed, perennial vegetation, some of which may be in a dormant state much of the time. Vegetation may be somewhat denser in regions where groundwater aquifers are near the surface, or may be absent entirely, as in major dune fields.

Human activities along the desert fringe, at least until recent times, were also remarkably accurate indicators of the boundaries of desert conditions. The 100-millimeter line of annual precipitation marks the effective limit of economical agriculture without the use of systematic irrigation. The presence of irrigation, therefore, is frequently a danger signal that human exploitation has ignored the limiting factors of arid environments.

WIND AND WATER EROSION

Desertification is primarily the result of wind and water erosion. The first step is usually the loss of diffuse, arid zone vegetation that normally is sufficient to protect the soil surface from dramatic erosion or to impede the wind transport of a sufficient amount of airborne sand and dust from other areas to balance the loss of local soil matter. This vegetation loss may occur from natural or human causes. The former may include long-term climatic shifts resulting in reductions in annual precipitation or in shorter-term climatic or meteorological phenomena such as abnormally long drought or erratic cloudbursts.

Once protective vegetation is lost, wind erosion can remove even the coarser materials in a soil layer in a matter of a few years. Much of this material will accumulate in moving dunes that become a threat to surviving marginal vegetation. Finer materials may be carried thousands of miles, even to other continents. Dust from the Sahara, for example, may be carried deep into Europe or even across the Atlantic Ocean to Florida and the Caribbean littoral.

Erosion of soil matter exposes a surface of pebbles and gravels extremely sensitive to further damage, particularly that resulting from human actions. In parts of the North African Sahara, vast areas that were scenes of armored engagements in World War II still exhibited ruts and surface disturbance from heavy-tracked vehicles nearly half a century later. Mass motorcycle races in the Mojave Desert of California have caused extensive, and most likely permanent, surface damage.

Assessment of the desert environment in terms of average conditions often obscures the potential of water erosion to accelerate desertification. Hard stone or clay desert surfaces can become impermeable to water absorption by the action of raindrops on the finer soil materials, which seal the surface to the point where nearly all precipitation is lost by runoff. This condition not only destroys remnant vegetation beneath the surface by depriving it of water but also vastly increases the volume of water passing through internal drainage systems in periodic flash floods, which wreak havoc on surviving vegetation clinging to the banks of usually dry riverbeds.

Still another cause of desertification is accumulation of salts and alkaline substances in the soil through a process called salinization. Although this process may develop naturally along interior drainage systems, the most extreme cases result from prolonged irrigation. Virtually all water sources contain some level of dissolved minerals. In regions of poor drainage, an endemic characteristic of many arid and semiarid zones, these minerals accumulate over the years and eventually render the soil useless for agriculture, encouraging its abandonment to the forces of wind erosion. Thousands of square kilometers have been lost to productive agriculture through salinization in an-

cient agricultural centers such as the Near East and Pakistan as well as in more recently exploited regions in California and Australia.

HUMAN EXPLOITATION OF DESERT ENVIRONMENTS

In traditional or premodern societies, human exploitation of desert environments relied upon pastoral economies, wherein grazing livestock converted plant matter for human consumption in the form of dairy products. Herders moved their animals from region to region in a form of transhumance (the seasonal movement of herds from pasture to pasture). Social mechanisms that discourage intensive environmental exploitation and continually redistribute wealth were common in these societies.

The introduction of alien social values and exotic animal species into arid environments has greatly accelerated the desertification process. Human population growth has resulted in unprecedented numbers of people making demands on the land. Conversion of a pastoral economy from dependence on a species such as the dromedary, which is a solitary and desultory grazer, to social herders and voracious grass-eaters such as sheep or goats, together with expanding human numbers, can destroy the equilibrium in a marginal zone in a few seasons. Sheep, for example, have been known to stimulate erosion simply because entire herds repeatedly follow the same path to a water source and destroy the vegetation and soil cover along the way.

As fossil fuel prices have soared, peasant populations have opened a new assault on marginal lands by cutting down the few available woody plants for firewood. Ecologists estimate that an average-sized family living in the arid borderlands of the African-Asian desert belt consumes the wood production of 1-3 hectares of land each year and that more than 25 million hectares per year may be denuded of trees and woody growth to provide fuel for burgeoning peasant societies.

FINDINGS FROM CLIMATOLOGY AND METEOROLOGY

The central problem in desertification research is to determine to what degree expansion of deserts results from long-term, natural fluctuations in the global environment and to what extent the phenomenon is a result of human intervention, and to distinguish the two. All the techniques that may be applied to reconstruction of climatic history and tracking of meteorological patterns therefore pertain to desertification studies.

Because of the generally brief period covered by modern meteorological records and an imperfect knowledge of the broader patterns of global climatic change, many theories have emerged that accord varying importance to natural and human-induced factors in desertification. Some climatologists propose that during the twentieth century a global swing toward greater aridity became apparent. Northern Hemisphere records suggest an expansion of polar air masses into lower latitudes with a consequent depression of moist equatorial air masses closer to the equator, so that the convergence between these systems—which roughly marks the northernmost penetration of tropical monsoons into North Africa and the Middle East—does not come as far north as it once did.

Conversely, the input of solar energy into tropical ecosystems, which drives the monsoons, could be decreasing. The GARP Atlantic Tropical Experiment (GATE), a massive, international research effort conducted in the mid-1970's to learn more about tropical environments and weather patterns, provided a knowledge base from which alarming conclusions may be drawn concerning the rapid deforestation of the tropics because of human exploitation, the consequent decrease in photosynthesis activity and increase in the albedo of the tropics, and the overall decrease in solar energy entering tropical ecosystems that results from these changes. Appreciation of the enormous importance of the tropics, as opposed to high latitudes, in understanding middle-latitude desertification was a major outcome of GATE.

Seeking Solutions

Research into desertification contains a strong applied component in that its major objective is the discovery of procedures that may slow, or even reverse, what specialists in all fields agree is a relentless advance of desert frontiers. Conditions in sandy deserts with moving dunes have received particular attention in applied research because of the danger they present to settlements and installations. Careful analysis of the mineral and chemical composition as well as the physical sizes of particles, all of which may vary widely, is crucial in determining the most appropriate measures for possible fixation. Some dunes may be anchored by carefully chosen plant cover. Liquid binders and emulsions have also been tried with some success. Ironically, since intensive human exploitation is responsible for much contemporary desertification, applied research is frequently directed toward new dimensions of human adaptation to desert conditions. Results have included attempts to exploit desert plant species, inexpensive means of desalinization to utilize often plentiful sources of water in desert aquifers, and more efficient solar energy technologies.

Remote sensing techniques, using special films in aerial photography and various optical sensors in satellite observations, now provide a season-by-season record of conditions in deserts and borderlands. Infrared-sensitive films are capable of recording early germination of plant life, concentrations of various substances in the soil composition, and the presence of certain contaminants. Landforms and topographic configurations not readily apparent on the ground, and which may have a bearing on the presence of water or the beginnings of vegetation deterioration, or which may make certain areas particularly vulnerable to erosion, frequently reveal themselves through remote sensing. Small-scale mapping of vast areas and careful monitoring of conditions are keys to identifying desertification in its early stage, when reversal may still be possible.

The Sahel

Research on desertification in the past few decades has concentrated on the Sahel, the fragile zone on the southern margins of the Sahara where true desert conditions become transitional to the grassland environments of tropical Africa. Beginning in the 1970's, a series of calamitous droughts struck the Sahel, bringing death and misery to huge numbers of people and animals and destroying entire pastoral societies. The impact of drought in the Sahel is magnified tremendously by the fact that the greatest breadth of the African continent is precisely in this zone, so that changes in northernmost penetration of the tropical monsoons of only a few minutes of latitude can spell catastrophe over thousands of square kilometers.

Research in the Sahel demonstrates the difficulty of separating natural from human causes of desertification. In the Democratic Republic of Sudan, for example, there is well-documented evidence, even in the recent colonial period, for retreat of the boundary zone between grassland and desert in some areas several hundred miles to the south and concomitant decreases in rainfall. Yet there is also a much broader picture, derived from archaeological studies, suggesting that this pattern may extend many centuries into the past and that modern socioeconomic and demographic change, therefore, cannot be entirely responsible.

Despite evidence that desertification processes have been at work for many centuries, the pressure of recent population increase on the Sahel and other marginal areas has created an atmosphere of crisis around applied research programs, so that most are directed toward fundamental changes in local practices in agriculture and animal husbandry or toward climate modification and desert reclamation. Both strategies often have the characteristic of uncontrolled experiments. Socioeconomic intervention tends to stress traditional or small-scale economic practices, often on the partly subjective conviction that they are better adapted to local conditions.

Schemes for climate and weather modification frequently require drastic environmental change. Among the more moderate proposals for the Sahel are construction of tree shelter belts, local vegetation modification, and cloud seeding designed to force the monsoon effect northward. Other schemes demand fundamental intervention in the patterns of atmospheric circulation that broadly determine the location of major desert systems. Most of these require gross changes in surface features. The creation of huge lakes in the natural drainage basins of the Sahara, through dams, evaporation control, and exploitation of groundwater aquifers, might increase precipitation levels. One research team has proposed paving large strips or "islands" of the Sahara with asphalt. The surface albedo of these areas would be much lower than sand or even vegetation, and their higher temperatures presumably would heat the air above them, promoting cloud formation

and precipitation. Others have proposed periodic releases of carbon dust into the atmosphere over the Sahara to increase heat absorption and cloud formation. None of these schemes has been attempted, and all of them woudl be likely to generate scientific controversy because of their impact on environment and their unknown potential.

Vulnerability to Desertification

Desertification is not confined merely to marginal lands bordering on true deserts. Careless agriculture or animal husbandry can create near-desert conditions in vast areas of grassland, as in the case of the Dust Bowl phenomenon of the 1930's on the central plains of the United States. Enormous amounts of top-quality soil are lost each year to water and wind erosion encouraged by unwise farming practices.

Much of the land most vulnerable to desertification, however, lies in lower or tropical latitudes, where soil regimes and microenvironments are not nearly as resilient as they are in Europe or North America. In many of these areas, moreover, the human issues arising from desertification involve not so much a lowering of living standards as the very survival of local populations. They call for massive intervention schemes with multiple economic and social dimensions, usually on the assumption that the indigenous social fabric or capacity to respond to emergencies of this magnitude is nonexistent or has broken down.

Ronald W. Davis

Cross-References

BIBLIOGRAPHY

Allan, J. A., ed. *The Sahara: Ecological Change and Early Economic History.* Outwell, England: Middle East and North American Studies Press, 1981. A review of ecological changes in this great desert region and human utilization of the desert margins in preclassical times. This careful reconstruction of historical environmental conditions is useful as a comparison to modern desert limits and the impact of long-term settlement.

Brooks, George E. "A Provisional Historical Schema for Western Africa Based on Seven Climate Periods, ca. 9000 B.C. to the Nineteenth Century." *Cahiers d'études africaines* 16 (1986): 46-62. This is a summary of the historical evidence for climatic change in the Sahel, including maps projecting fluctuations of desert limits in various periods.

Bryson, Reid A., and Thomas J. Murray. *Climates of Hunger: Mankind and the World's Changing Weather.* Madison: University of Wisconsin Press, 1977. An excellent general work on climatic change and the possible extent of human-induced factors, with emphasis on deforestation and desertification around the world.

Chouha, T. S. *Desertification in the World and Its Control.* Jodhpur, India: Scientific Publishers, 1992. Chouha deals with desertification on a global scale but focuses much attention on India. Several sections deal with tactics that have been used and proposed to control the problem. Maps, index.

Climate, Drought, and Desertification. Geneva, Switzerland: World Meteorological Organization, 1997. This brief booklet, prepared by the World Meteorological Organization, deals with the climatic factors, such as drought, that lead to desertification worldwide. Includes color illustrations.

Crawford, Clifford S., and James R. Gosz. "Desert Ecosystems: Their Resources in Space and Time." *Environmental Conservation* 9 (1982): 181-195. This article discusses the tensions between continuous exploitation of desert environments and the intermittent character of vital resources such as water and vege-

tation in a given location.

Eckholm, Erik, and Lester R. Brown. *Spreading Deserts: The Hand of Man.* New York: Worldwatch Institute, 1971. A brief, useful summary of world desertification patterns, emphasizing careless human exploitation and the interruption of natural processes that limit the spread of deserts.

Evenari, Michael, Leslie Shanan, and Nephtali Tadmor. *The Negev: The Challenge of a Desert.* Cambridge, Mass.: Harvard University Press, 1982. The best comprehensive study of attempts to develop settlement styles suitable to desert resources, based on a combination of archaeological research on ancient settlement patterns and the most recent findings in the Earth sciences.

Glantz, Michael H., ed. *Desertification: Environmental Degradation in and Around Arid Lands.* Boulder, Colo.: Westview Press, 1977. An excellent discussion of a broad selection of related phenomena, including nature and causes of desertification, theories of natural versus human-induced environmental change, case studies of desertification, experiments in weather modification, and efforts of international agencies to treat desertification as a unified, global problem. Includes an extensive bibliography.

Mairota, Poala, John B. Thornes, and Nichola Geeson, ed. *Atlas of Mediterranean Environments in Europe: The Desertification Context.* New York: John Wiley & Sons, 1996. This multi-authored text focuses on desertification in the Mediterranean region. Includes chapters on seminatural environments and processes, socioeconomic processes and change, and field studies. Bibliography, index, maps, and plant species glossary.

Oliver, F. W. "Dust-Storms in Egypt and Their Relation to the War Period, as Noted in Maryut, 1939-1945." *Geographical Journal* 106-108 (1945-1946): 26-49, 221-226. An early study of the extent to which intensive and extensive human movement and occupation of the desert may result in short-term deterioration of arid environments.

United Nations Conference on Desertification, Nairobi, Kenya. *Desertification: Its Causes and Consequences.* Elmsford, N.Y.: Pergamon Press, 1977. Includes four in-depth studies on the relationship between desertification and climate, ecological change, population, and technology, each with an extensive bibliography.

Walls, James. *Land, Man, and Sand: Desertification and Its Solution.* New York: Macmillan, 1980. An extensive collection of case studies from around the world, each illustrating facets of desertification and pragmatic, localized solutions. Contains numerous examples of how physical, biological, and social science approaches may be combined in comprehending and controlling deterioration.

EXPANSIVE SOILS

Expansive soils, soils that expand and contract with the gain and loss of water, cause billions of dollars in damage to houses, other lightweight structures, and pavements, exceeding the costs incurred by earthquakes and flooding.

PRINCIPAL TERMS

DIFFERENTIAL MOVEMENT: the unequal movement of various parts of a building or pavement in response to swelling or shrinkage of the underlying soil

EVAPOTRANSPIRATION: the movement of water from the soil to the atmosphere in response to heat, combining transpiration in plants and evaporation

LOADING: an engineering term used to describe the weight placed on the underlying soil or rock by a structure or traffic

MOISTURE CONTENT: the weight of water in the soil divided by the dry weight of the soil, expressed as a percentage

MONTMORILLONITES: a group of clay minerals characterized by swelling in water; the primary agent in expansive soils

SOIL STABILIZATION: engineering measures designed to minimize the opportunity and/or ability of expansive soils to shrink and swell

SOILS AND FOUNDATION ENGINEERING: the branch of civil engineering specializing in foundation and soil subgrade design and construction

HYDRATION: a process whereby, when soils become wet, water is sucked into the spaces between the particles, causing them to grow several times their original size

SHRINKAGE: an effect opposite to hydration, caused by evapotranspiration

SOIL SWELLING AND SHRINKAGE

Certain types of soil expand and contract with the gain and loss of water, primarily through seasonal changes. One meter of expansive soil has sufficient power to lift a 35-ton truck 5 centimeters. Many cracks in walls and concrete surfaces are assumed to be caused by settlement, when in fact they are produced by the lifting force of expansion.

Soils vary enormously in their ability to expand, ranging from zero in very sandy soils to highly expansive in montmorillonite clays. By contrast, kaolinite clays do not expand. Some key indicators of expansive soils are a bricklike hardness when dry, including great resistance to crushing, and stickiness and weakness when wet; also, they have a glazed surface when cut with a knife. Soil cracks and popcornlike surface appearance are other key signs. Where soils from volcanic ash produce primarily montmorillonite clays, bentonite is formed. This extremely expansive soil prevents even vegetation from growing and is easily visible from a road, with its barren, popcornlike surface and painted-desert coloration.

The ability of montmorillonite clays to expand comes from their physical and chemical structure. Clay minerals are like leaves, or sheets, of paper. Some bond very strongly from layer to layer and have little swell potential. Expansive clay minerals, however, have very thin and broad shapes, with weak bonds between the layers. These layers are negatively charged and repel one another but are bonded by positively charged ions (cations) such as sodium and calcium dissolved in the soil water. When soils become wet, water is sucked into the spaces between the clay particles, a process called hydration, causing them to grow up to seven times their original size. The opposite effect, shrinkage, is caused by evapotranspiration.

In climates with great variation in seasonal moisture, especially marginally humid and semiarid lands, large changes occur in the moisture content of the soil, with subsequent swelling and shrinkage. In the midlatitudes and subtropics, winter and spring are the wet seasons, as vegetation is dormant and rain and melting snow are abundant. Summer and fall are the dry seasons, soilwise; evapotranspiration demand is high, and much water is transferred from the soil to the atmosphere. Human activity can also produce localized effects through excessive watering, leaking pipes, poor drainage around buildings, and use of trees and bushes for landscaping. It is these volume changes in the swelling soils beneath structures, caused by the gain and loss of water, that make expansive soils so damaging.

DIFFERENTIAL MOVEMENT AND VARIABLE LOADING

Expansive soils do their damage primarily through differential movement and variable loading. If the entire structure is lifted and sinks uniformly, there is little problem. Unfortunately, swelling and shrinkage are seldom uniform. Many buildings are located on sloping land, and, as a result, groundwater is more likely to be encountered on the uphill side during wet seasons. Consequently, more expansive force is exerted on that side, cracking walls and slabs. The construction season can also be a factor on sloping land, as the portion of the building pad cut into the hillside and the portion built up with fill tend to respond differently to changes in moisture. Even on level land, deep foundations may be stable while grade beams and slabs resting upon the surface rise and fall with the seasons.

While much attention is given to damaged buildings, damage to roads may actually have a greater impact on the public. The life of a road is measured by the total weight it carries over a period of time, divided by the strength of the roadbed. Many older roads are built directly upon the natural grade, including highly expansive soils. Because these soils are weak when wet, the heavier loading at concentrated points under truck tires causes the pavement to break down more rapidly. Once a cracking pattern is visible, a chuckhole is but a few wetting cycles away, as water can now easily reach and further weaken the soil below. Major roadbed problems may also occur near the boundaries between highly expansive and less expansive soils as a result of differential swelling. Water seepage at concentrated points beneath the roadbed also undermines the strength of the roadbed and contributes to differential movement.

Damage to structures from expansive soils can be summarized as reduced performance, economic loss, and massive failure. Massive failure is rare, but the threat of it is usually enough to demand removal of the danger. High on this list would be leaning chimneys and massively cracked concrete channels. Also, raised portions of patios, sidewalks, and driveways are hazardous to pedestrians. Reduced performance occurs when doors and windows cannot be closed because of misalignment, and unsightly cracks appear on walls and concrete slabs. Economic losses occur from such unsightly cracks when potential buyers are discouraged from purchasing property. Also, higher heating and cooling bills result from energy losses through cracks, especially around windows and doors. For taxpayers, a little-recognized cost is incurred through more frequent road repairs.

A slump of expansive soil after heavy rains. (© William E. Ferguson)

ON-SITE GEOTECHNICAL INVESTIGATION

Fortunately, the problems caused by expansive soils are reasonably well understood by soils and foundation engineers. Typical procedures include an on-site geotechnical investigation, laboratory testing of samples obtained during the on-site investigation, and design recommendations specific to the anticipated usage of a site. The importance of sound engineering here cannot be overemphasized. The cost of such work is small compared to the cost of the project and is far cheaper than potential remedial work. Inasmuch as the soil ends up underneath the building, roadway, or other structure, it is obviously inaccessible when problems arise; it is far easier and cheaper, therefore, to compact the soil properly and install piers and drains before rather than after.

An experienced soils engineer will generally know what to expect within a given part of a city or region. Soils are notoriously diverse, however, and the only sure way to secure the accurate information needed for a sound, cost-effective design is an on-site investigation. Key information is provided by borehole samples, extracted by a drilling rig similar to that used for wells. The selection and number of borehole locations are determined from architectural layouts and the size of the project. One hole may be sufficient for a house, whereas numerous holes may be ordered for a high-rise building. Three kinds of geotechnical information are obtained directly from these holes: a description of the type and thickness of soil and rock; resistance to penetration (influenced by moisture content), which gives a crude measure of soil strength; and, if the hole is deep enough, depth to the water table.

With distressed structures, one of the most important judgments a soils engineer can make is whether remedial work is needed or will be cost-effective. Cracks often appear in new buildings and pavement; though frustrating, they may be of little consequence. By contrast, serious problems may be avoided if timely repairs are made before incipient failures, such as leaning chimneys, reach the critical point.

TESTS FOR EXPANSIVE POTENTIAL

The most common measure of expansive potential is the Atterberg limits test, run in the laboratory. Though crude, it has proven its value over many years as a predictor of expansive and nonexpansive soil characteristics. The test essentially measures the difference between the resistance of a soil sample to flow as it becomes wetter and wetter, called the liquid limit, and disintegration under rolling as it becomes drier and drier, called

the plastic limit. The numbers used are moisture content percentages, obtained by computing the weight of water as a percentage of the dry weight of the soil. The difference between these two percentages is called the plasticity index, referred to as the PI. Variances among plastic limit numbers are generally small, but liquid limit numbers vary enormously. The stickier the clay, the more moisture will be required to make it flow, thus the higher the liquid limit. One characteristic of expansive soils worth remembering is their stickiness when wet. Consequently, the higher the liquid limit, the larger the PI will be and the greater the swell potential.

The Atterberg limits test is relatively easy and inexpensive to run. It is the simplest way to test large numbers of samples, and such tests are conducted by the thousands in soils laboratories. Where precise and detailed information is needed from key samples, however, a consolidation-swell test is run. This test is much more complex, time-consuming, and expensive, but the information gained is vital for the design of more expensive structures. The test mimics the behavior of soils under the loading applied by a building and their response to swell pressure as water is absorbed. Additional tests, such as confined and unconfined compression, are also commonly run on the more expensive projects. These test results, combined with experience and input from colleagues, form the basis for the recommended design. Good designs anticipate and avoid problems, minimize construction costs, and recommend specifications for quality-assurance testing during construction.

DESIGN ALTERNATIVES

Design alternatives include strengthening the structure, stabilizing the soil, some combination of the first two, or anticipating movements to isolate their impact. If a slab can be strengthened so that it moves as a unit, the effects of differential movement will be mitigated. In larger buildings, reinforced grade beams are commonly suspended on piers, with space deliberately left under the grade beam to allow room for soil expansion. Major attention is given to stabilizing the soil. The goal here is to compact the soil in its expanded condition, balancing the need for strength with the need to allow some room for additional expansion. A key problem is to keep the soil from drying out before it is sealed, usually by a concrete slab or asphalt. Drains are commonly installed around the perimeter and covered by concrete, asphalt, or plastic sheeting. Three types of drains are used: peripheral drains around the perimeter; interceptor drains to control subsurface water from uphill sources; and sump drains, which require a pump and are usually located under a basement floor slab.

A combination of increased strength and soil stabilization is often used. Much attention is being given to the water content and density of the soil beneath the structure. When feasible, select fill (stable soil having low expansion and good binding qualities) is imported for the top layer. Lime or fly ash may be mixed with the soil to increase both its strength and its stability. This treatment is especially effective under pavement (parking areas and roadbeds). Special equipment has been invented to mix these additives into the soil efficiently. Laboratory tests are commonly run to measure the effectiveness of treatment in terms of PI reduction and strength; a common rule for lime treatment is to reduce the PI to 10 or less. Lime is the easiest to handle and its effectiveness is easiest to test, though fly ash, a by-product of coal burning in electrical power plants, is growing in use because of its low cost. The problem with fly ash is that its use requires meticulous application, beyond the abilities of many inexperienced operators. Engineering supervision of the mixing and compaction, with strict attention to moisture and time specifications, is critical to the successful use of fly ash.

Where construction cost is a key concern, it is sometimes feasible to design the structure so that slabs and interior walls can freely move without affecting the foundation or other parts of the building. Trim work can mask such movement so that aesthetic concerns are not a factor. Once structures are completed, correction of problems is far more difficult. Correction may range from simple drains and moisture barriers around perimeters to tearing out the pavement and redoing the soil subgrade. An accurate diagnosis is even more essential to success in this case. Consequently, skilled engineers study the problem in detail, and heroic efforts are sometimes made to obtain soil samples. The engineer's judgment of the seriousness of the problem and the consequent remedial action are especially vital. A skilled engineer can tell much from cracks and other distress patterns.

GROWING IMPORTANCE OF FOUNDATION ENGINEERING

Only since the 1970's has the severe threat represented by swelling soils been recognized by the construction industry. Several changes are responsible for this growth. First, buildings with slab-on-grade floors (concrete resting directly upon soil) have replaced basements and crawl spaces in many areas. Inasmuch as seasonal moisture changes are concentrated in the top 1 meter and essentially disappear at a depth of 3 meters, this change created conditions far more susceptible to expansion and contraction. Second, heavier trucks and increased usage have put increasing stress on roadbeds and parking areas. Once the pavement is broken, it is much easier for water to reach the soil below, and failure soon follows. Third, population growth and rising costs of land have encouraged development of nonfarmland, which increases the odds of encountering expansive soils. Still, this extended use of land is fine, as long as the dangers are understood and taken into account in designing and planning projects.

A major issue when one is confronted with any hazard is some idea of when a problem is serious enough to require expert help. Foundation engineering is a little-recognized but invaluable part of any construction program involving expansive soils. Knowledge of soil conditions prior to new construction is well worth the relatively small costs. Well-informed individuals provide an invaluable service to themselves personally, to employers, and to the organizations they serve as knowledgeable citizens. As part of the joint federal-state National Cooperative Soil Survey, the U.S. Department of Agriculture and the Soil Conservation Service provide booklets on regional surveys. In map and table format accompanied by explanatory text, these publications present detailed information about agricultural and engineering soil conditions.

Nathan H. Meleen

CROSS-REFERENCES

Aerial Photography, 2739; Alluvial Systems, 2297; Aquifers, 2005; Atmosphere's Global Circulation, 1823; Climate, 1902; Desert Pavement, 2319; De-

BIBLIOGRAPHY

Fenner, Janis L., Debora J. Hamberg, and John D. Nelson. *Building on Expansive Soils.* Fort Collins: Colorado State University, 1983. This thorough and well-illustrated booklet explains the swelling mechanism, includes a map of the United States showing the occurrence and distribution of potentially expansive materials, and gives identification techniques, design alternatives, and remedial measures. Especially recommended for builders.

Harpstead, Milo I., and Francis D. Hole. *Soil Science Simplified.* Ames: Iowa State University Press, 1980. Excellent primer for the nonspecialist. Well illustrated and suitable for high school Earth science curricula. Thorough coverage of soil topics relevant to agriculture, engineering, and geology.

Holtz, Wesley G., and Stephen S. Hart. *Home Construction on Shrinking and Swelling Soils.* Denver: Colorado Geological Survey, 1978. A short primer on the problems and possible solutions related to buildings situated on expansive soils. Includes good illustrations of typical problems, drain design, and recommended interior construction techniques.

Jochim, Candace L. *Home Landscaping and Maintenance on Swelling Soil.* Special Publication 14. Denver: Colorado Geological Survey, 1981. Prepared for the homeowner, this beautifully illustrated publication has large type and clear, simple illustrations, including a map of swell potential in Colorado, especially the Denver area. Also provides information about groundwater and how to maintain proper drainage around a house.

Katti, R. K. *Behaviour of Saturated Expansive Soil and Control Methods.* Rotterdam, Netherlands: A. A. Balkema, 1994. This book, published under the auspices of the Central Board of Irrigation and Power in India, focuses on soil stabilization, soil dynamics, and black cotton soil in that country. Includes an extensive bibliography and an index.

Koerner, Robert M. *Construction and Geotechnical Methods in Foundation Engineering.* New York: McGraw-Hill, 1984. Typical of introductory textbooks in the field, this book is part of a series in construction engineering and geotechnical engineering and attempts to address this mixed audience. Heavy mathematical treatment.

Legget, Robert F. *Cities and Geology.* New York: McGraw-Hill, 1973. Designed as a textbook for courses in urban or environmental geology, this book is highly relevant to those interested in urban planning. Uses case histories worldwide to demonstrate the effects of geology on urban growth and problem solution. An important chapter on geologic hazards is included.

Nelson, John D. *Expansive Soils: Problems and Practice in Foundation and Pavement Engineering.* New York: John Wiley and Sons, 1992. Nelson deals primarily with the interaction between soil and human-made structures such as building foundations and roadways.

Olson, Gerald W. *Soils and the Environment: A Guide to Soil Surveys and Their Application.* New York: Chapman & Hall, 1981. Designed for a worldwide audience and general readership, this book explains how soil surveys are prepared, what the goals are, and how to understand the most important applications.

Sobolevsky, Dmitry Yu. *Strength of Dilating Soil and Load-Holding Capacity of Deep Foundations: Introduction to Theory and Practical Application.* Rotterdam, Netherlands: A. A. Balkema, 1995. Sobolevsky has written a detailed survey of soil mechanics, soil testing, swelling soils, and foundations suitable for the student who already has some background in the field.

LAND MANAGEMENT

Land management, the control of land use and the preservation of land, is essential to the proper maintenance of this limited natural resource.

PRINCIPAL TERMS

LAND USE: the application to which a tract of land is subjected

LAND: any part of the Earth's surface that may be owned as goods and everything annexed to it, such as water, forests, and buildings

MULTIPLE USE: the use of land for more than one purpose or activity at the same time

SUBSURFACE FEATURES: land features or characteristics that are not visible or apparent, lying beneath the land surface, such as minerals, oil and gas, and structural features

TAXATION: a land-management tool that usually reflects the perceived best use of land

TOPOGRAPHY: the collective physical features of a region or area, such as hills, valleys, streams, cliffs, and plains

ZONING: a land-management tool used to limit uses and define conditions and extent of use

TYPES OF LAND USE

Land is not reproducible; it is present in only a fixed amount. In addition, its location—important to its use and value—cannot be changed. Aspects of land include its topography, which controls many uses; its soil, which is vital to agricultural and other applications; its subsurface structure and composition, which can prove to be either beneficial or problematic; and the availability of minerals, oil and gas, and other natural resources.

Land management is the science that has developed in response to the need for control of land use and preservation. Land-management programs attempt to organize, plan, and manage land-use activities. Well-developed programs are concerned with land and water and address the issues of water-use rights, subsurface rights, surface rights, and above-surface rights. Land-management activities can take place at local, state, and federal levels. They can be geared to single- or multipurpose land uses and can be oriented toward rural or urban settings.

Land uses can be either reversible or irreversible. Reversible uses—such as agricultural activities, grazing of livestock, forestry, recreation, and mining—can allow for reversion to former or alternate uses. These uses are frequently applied in multiple-use programs of land management and are generally compatible with one another. The purpose of programs such as these is to maximize use while allowing for the greatest good for the most people. Irreversible land uses result in permanent changes in the character of the land such that it cannot revert to a former condition or use. The filling-in of swamps or other water bodies, the building of cities, and the development of non-reclaimable surface mines result in irreversible, permanent changes. Activities such as these generally result in single-use situations and preclude the development of alternative-use plans.

LAND-USE ISSUES

Major land-use issues requiring the application of land-management policies can be identified as relating to new growth, declining growth, reclamation, resource exploitation and/or utilization, preservation of natural or cultural resources, plans for maintenance of stable populations, or environmental, economic, or social concerns. Each land-use issue and attendant land-management policy may require specialized knowledge and specific approaches. Science and technology are vital in developing land-use plans; science provides the knowledge, while technology provides the means to implement that knowledge.

All land-use issues can present or generate one or more uncertainties with respect to future applications of the land, represent a problem or an opportunity, be subject to the effects of supply and demand, and be dealt with systematically or conceptually. The degree of uncertainty in any land-use issue is a direct result of the availability of data regarding the use and prior experience with the issue. One of the focal points of well-developed land-management policies is to reduce or remove the uncertainties related to land use. Opportunities in land use are those activities that benefit a large segment of the population, either directly or indirectly. Examples range from the establishment of a national park to the development of a new airport. Problems in land use might include the subsurface disposal of radioactive waste or the threat to wildlife habitat from construction of a reservoir system. Supply will dictate how much land should be subjected to a specific use in response to the perceived demand.

ECOLOGICAL CONSIDERATIONS

Ecological diversity is an important aspect of land-use planning and also of land-management policies. The ultimate goal of all land-management programs should be to put land to its multiple best uses. In the process, the ecological "carrying capacity" of a regional environment should not be exceeded. Natural resources should be maintained in a state of availability, and development should be encouraged only in areas best suited for it; development should be discouraged in areas of significant resource value. Development should also be discouraged in areas of natural or human-made hazards.

Land-management programs and policies are largely a result of the location of the land to be managed and the anticipated impact the programs might have on a given population. Economics will frequently dictate the preferred use of land, sometimes at the expense of wildlife, aesthetic beauty, and other ecological factors. Land-use issues can be addressed at different levels. Factors that help to determine the level at which any particular issue might be addressed include the number of people and locality or localities that might be affected by the issue; the magnitude of the potential cumulative effects that may result from the issue; and the threshold at which an issue becomes significant. The availability (or lack) of water and the effect of water pollution are examples of factors that can have a cumulative effect, while air pollution is an example of an issue that has reached a threshold, elevating it from a local to an international concern.

REGULATORY LAND MANAGEMENT

Land management is an ongoing activity, and policies may require change and/or modification with time. Land has an intrinsic (cash or exchange) value and an extrinsic (inherent or judgmental) value, both of which must be considered when dealing with or formulating a management plan. Effective planning and subsequent management, public or private, require that land-use controls be regulated and supported by sufficient authority.

Lands are generally managed and their use controlled by taxation, police and regulatory powers, and strategic considerations. Taxation serves as a management tool because taxes levied are generally a reflection of the perceived best use. Changes in tax status frequently result in changes in land use, as in the case of agricultural land that is converted to urban use as a direct result of an increase in taxes and subsequent cessation of agricultural activities. Police and regulatory powers dictate what can and cannot be done on a specific piece of land. Subdivision regulations, environmental laws, and zoning ordinances are the most common form of regulatory land management. Master plans also control land use and are required by most local governments. They assist cities and counties in coordinating the regional implementation of statutes and/or regulations and include a statement of goals, an outline of societal needs, and a list of specific objectives. They are generally collective plans backed by extensive information and by many independent studies. Master plans also outline mechanisms by which the objectives are to be reached. Strategically located lands can affect the use of adjoining parcels. The presence of industrial areas could, for example, preclude adjoining residential development, while the existence of parks and golf courses could discourage adjacent industrial development. Airports, ski areas, forests, and rangelands can also have strategic value if situated properly.

A drive through cities and suburban developments shows that some areas have been relegated to industrial, commercial, or residential uses. These use areas are the result of zoning, taxation, and other management tools that attempt to encourage certain types of development in relation to the carrying capacity or suitability of the land

and its annexed improvements. The availability of deep-water ports and rail transportation is, for example, more important to commercial and industrial development than to residential land use. By the same token, certain soils and other natural factors might favor residential development. Land reclamation, soil-erosion prevention measures, and imposed land-use limitations are all part of land management. The state of Georgia, for example, requires that all mine sites provide reclamation of as many acres as were actually mined during a given year, although the reclaimed acreage need not be that which was mined. Soil-erosion prevention programs are incorporated as part of nearly every development or activity plan that will result in disturbance or modification of a soil profile, including plans for wilderness or forest roads, residential subdivisions, construction along waterways or coastlines, and agricultural activities.

MANAGEMENT OF FEDERAL LANDS

Federal and state land-management programs are widely recognized forms of land-use planning. These programs have a direct impact on the use of public land such as parks, forests, seashores, and inland waterways. The U.S. federal government has owned lands that were not required for its own activities since October 29, 1782. Approximately one-fifth of the public domain (lands owned by the federal government) was eventually granted to individual states. These lands were set aside for schools, hospitals, mental institutions, and transportation or were swamps and flooded lands—all part of an overall land-management plan.

The management of federal lands is largely custodial. It is carried out under the provisions of numerous statutes and regulations, including the Multiple Use-Sustained Yield Act of 1960, which legalized the multiple use of federal lands; the Wilderness Act of 1964, which set aside wilderness areas; the Classification and Multiple Use Act of 1964, which allowed for the classification of land for determining the best use and determining which lands should be retained or discarded; and the National Environmental Policy Act of 1969, which required the filing of an Environmental Impact Statement (EIS) for major actions that would significantly affect the quality of the human environment.

Grazing is the oldest use of federal lands, but oil and gas activities generate the largest revenues. Mining of nonfuel minerals is governed by the Mining Laws of 1866 and 1872, while the Mineral Leasing Act of 1920 provides for competitive and noncompetitive leasing of land containing oil and gas, oil shale, coal, phosphate, sodium, potash, and sulfur. The United States Forest Service administers all federal forest lands, while the Bureau of Land Management (BLM) administers all other lands.

Issues that must be addressed in all federal land-management programs include fraud and trespass (relating to illegal harvesting of timber or other valuable materials), resource depletion, reserved rights on lands that have been discarded, multiple use of lands, equity for future generations, the ability to maintain lands and retain their value, and the ideal of private land ownership. Policy issues that are closely related include how such land should be acquired by the federal government, how much land should be discarded, to whom the lands should be granted and what rights (such as access to minerals) should be retained and for how long, what should be the terms of land disposal, how much should be spent to maintain lands that are retained, to what use should retained lands be put, who should share in the benefits which accrue from lands retained, and who should develop the land-management plan and execute it. Policy issues change with time, as do approaches to land management.

FOREST MANAGEMENT

Forests are an important target of land-management activities because they occupy approximately one-third of the total land area of the United States. Of that area, nearly two-thirds is occupied by commercial forests, while the remainder is reserved from harvest. Forests are used as watershed areas, renewable consumable resources, recreational areas, and wilderness preserves. Policy issues that affect the management of forests include questions regarding how much forest to maintain, how much to restore, how much to withdraw from use, and how they should be harvested. Several criteria must be met to establish practical forest policy. They are physical and biological feasibility of an action, economic efficiency, economic equity, social acceptability, and operational practicality. Not all uses of forests are compatible in a potential multiuse scenario; some uses will necessarily exclude others, which must be considered in a forest-management program. For example, interactive effects must be considered in the harvesting of timber, as it affects the watershed, soil, regenerative growth, and wildlife. Policy areas directly related to the maintenance and management of forest resources include taxation, often a large cost of forest ownership; housing programs, which affect the demand for forest products; foreign trade with attendant import duties, quotas, or tariffs affecting the merchantability of forest products; transportation, which affects the marketability of forest products; direct aids to forest development programs, such as research, education, and production subsidies; and the administration of public forests.

COASTAL LAND MANAGEMENT

Coastal land management is as complex as the management of inland areas, if not more so. In Florida, for example, the value of shore properties frequently dictates the reclaiming of lost lands or the creation of new lands for urban use. Swamps and intertidal areas along coastlines may be filled in at the expense of what is frequently a fragile environment. At issue is whether development can take place in such a manner that people can live in an area without destroying the natural features and the beauty that attracted them in the first place. One approach has been to set aside areas of land to be used as parks or conservation areas. This approach is increasingly popular in the formulation of land-management policy.

On a stroll along the beach, one may observe a person fishing from a jetty or surfcasting from the base of a seawall. Sailboats cruise the inner harbor, protected from the sea by the distant breakwater. All these physical structures—the jetty, seawall, and breakwater—are part of the coastal land-management program. The jetty attempts to prevent beach erosion by the longshore current that runs nearly parallel to the shoreline, while the seawall aids in the maintenance of a stable coastline that might otherwise erode under the constant battering of winter storms. The breakwater helps to maintain quiet waters in the shallow inner harbor area, otherwise subjected to high, frequently damaging waves.

MINING AND AGRICULTURE

Mining activities generally require extensive land-use planning and must be carried out under well-defined land-management policies. These policies control mining activities from the earliest stages of exploration through the actual mining and production of mineral materials and finally through reclamation. Most of the management policies are in place to help preserve the character of the land to the greatest possible degree. The routing and design of access roads, when on federal lands, must generally be approved by either the Forest Service or the Bureau of Land Management. Some restrictions also apply on private lands, requiring special permits.

Mineral-exploration activities frequently must be limited in size on federal land so that they do not interfere with natural wildlife habitats or other approved uses of the land. Once a valuable deposit has been identified, mining permits must be applied for, EIS's may have to be prepared, and reclamation procedures must be outlined prior to the extraction of the deposit. Once mining has been completed, the land must be reclaimed in accordance with an approved plan. All these activities take place under the land-management plan affecting the mine area.

Crop rotation and strip farming are land-management mechanisms employed in agriculture. Different crops require different kinds and levels of nutrients for proper development. Crop rotation, or changing the type of crop grown on a particular tract of land with each growing season, allows for the greatest yield of nutrients from the soil. Strip farming regenerates nutrients and also serves as a soil-erosion prevention measure. Early farming techniques put all lands under cultivation and, therefore, all were subject to wind erosion. Strip farming leaves alternating strips vegetated or cultivated, helping to prevent erosion.

Kyle L. Kayler

CROSS-REFERENCES

BIBLIOGRAPHY

Burby, Raymond J., ed. *Cooperating With Nature: Confronting Natural Hazards With Land-Use Planning for Sustainable Communities.* Washington D.C.: Joseph Henry Press, 1998. Burby's book pays close attention to the role that natural disasters and hazards play in the development of new communities. Even though this is intended for the reader with some background in the field, the layperson with an interest in land-management policies will also find it useful. Illustrations, bibliography, and index.

Clawson, Marion. *The Federal Lands Revisited.* Washington, D.C.: Resources for the Future, 1983. A discussion of federal land policies and management. The author explores all aspects of land management and introduces many new concepts as solutions to the current problems in the field. Major policy issues and present usages are discussed, including wildlife, grazing, minerals extraction, oil and gas production, watershed protection, and recreation. Numerous data tables and figures are included in an easily understood format, and an index is provided. Written for the nonspecialist.

_____. *Forests for Whom and for What?* Baltimore: Johns Hopkins University Press, 1975. A well-developed discussion regarding forest management from both a public and a private viewpoint. Issues addressed include timber production, recreational usage, wildlife protection, and watershed management, as well as other economic, social, and environmental concerns. Public forest policy is discussed in detail, and impacts of land conversion, restoration, and clearcutting are considered. Includes several data tables, an index, and a bibliography. Geared toward an intellectual but nonspecialist audience.

_____. *Man, Land, and the Forest Environment.* Seattle: University of Washington Press, 1977. A compilation of three essays that were originally delivered as public lectures in 1976. The topics discussed include land-use planning and control, as well as the private and federal ownership of forested land. Sugges-

tions for future land management are presented, as are discussions of past "mismanagement." A short bibliography is included along with several graphs and data tables. The essays are targeted toward an intellectual but nonspecialist audience.

The Conservation Foundation. *State of the Environment: An Assessment at Mid-Decade.* Washington, D.C.: Author, 1984. An insightful report on the status of the environment in 1984, this book deals with underlying trends in conditions and policy and addresses environmental contaminants, natural resources, future problems, and the assessment of environmental risks. An extensive bibliography and an index are included. Geared toward a diverse audience with interests in politics, statistics, and the environment.

Dasmann, Raymond F. *No Further Retreat.* New York: Macmillan, 1971. A discussion of the development of Florida from the late 1960's to 1971. Concentrates on coastal and inland waterway land-management efforts. Conservation efforts that are discussed include Everglade protection, wildlife preservation, and the development of the Florida Keys. Offers reasonable approaches for controlling land use and instituting proper planning. Includes several photographs and maps and is well indexed. Suitable for any interested layperson, it uses no technical language.

Davis, Kenneth P. *Land Use.* New York: McGraw-Hill, 1976. This book discusses concepts of land, land ownership, land use and classification, and land-use controls. Examines the planning process with respect to land management and valuation, as well as the attendant decision-making processes. Several case histories are presented as examples. Appropriate for many levels of reader interest, from the layperson to the technical specialist.

Economic Commission for Europe. *Land Administration Guidelines: With Special Reference to Countries in Transition.* New York: United Nations, 1996. This United Nations publication describes the land-use and land-tenure policies that have been enacted in Central and

Eastern European countries during their transition from communist to democratic governments.

Fabos, Julius Gy. *Land-Use Planning.* New York: Chapman and Hall, 1985. This book examines land-use planning from many perspectives. Planning issues are discussed in detail, as are the roles of science and technology in land-use planning. The evolution of land-use planning is addressed from the standpoint of interaction between disciplines, public versus private planning, and types of planning. Regional and local considerations are discussed, as are future prospects in land-use planning. Contains a good bibliography. Appropriate for all levels of interested readership.

Healy, Robert G. *Competition for Land in the American South.* Washington, D.C.: The Conservation Foundation, 1985. A carefully organized discussion of land use and development in the southern United States. Divided into chapters that address the issues of competition for land, the economic uses of agriculture, wood protection, animal agriculture, and human settlement. Each land use is analyzed with respect to future demands. The effects on soil, water, wildlife, and aesthetics are discussed. Contains a summary which is well thought out. Includes a reference list as well as an index. Very readable, this book is targeted toward a nonspecialist audience.

Knight, Richard L., and Peter B. Landres, eds. *Stewardship Across Boundaries.* Washington D.C.: Island Press, 1998. A good analysis of the conservation of natural resources and management of public lands in the United States. Maps and illustrations.

Paddock, Joe, Nancy Paddock, and Carol Bly. *Soil and Survival.* San Francisco: Sierra Club Books, 1986. This book analyzes what the authors view as a lack of human commitment to the land. Examines the threats to American agriculture, the drive for greater land efficiencies at the expense of natural beauty, and the technical loss of land through erosion, chemical usage, and development. Attitudes about ethics are stressed, as is land stewardship and environmental concerns. Well indexed and contains footnotes. Geared toward an environmentalist audience.

Steel, Brent S., ed. *Public Lands Management in the West: Citizens, Interest Groups, and Values.* Westport, Conn.: Praeger, 1997. This collection of essays explores the volitile conflict between environmentalists and "wise use" groups in the western United States. Suitable for the reader without any background in environmental studies. Index and bibliography.

LAND-USE PLANNING

Land-use planning is that part of the broader process of comprehensive planning that deals with the types and locations of existing and future land uses and their impacts on the environment.

LAND USE: OPPORTUNITIES AND LIMITATIONS

Land-use planning is a process that attempts to ensure the organized and wise use of land areas. Two things are certain: First, land has great value in modern society; second, there is only a limited amount of it. Land is indeed a valuable resource, deserving of careful management. It is also true that the physical environment influences the location and types of human settlements, transportation routes, and economic endeavors. The hills, ridges, valleys, and depressions of a landscape create potential opportunities and/or limitations for human use. It is important to realize that preexisting land uses likely will affect an area's future possibilities. Land-use planning is part of the master-planning process: It deals with the types and the distribution of existing and future land uses, their relationship to other planning elements, such as transportation networks, and the interactions between land use and the environment.

This last point deserves elaboration. While it is certain that existing physical and cultural aspects of a landscape affect land use, humans have an enormous capacity to alter their surroundings. The behavior and wants of people greatly influence emerging land uses. Additionally, each land use affects not only the users but the nonusers as well. An industrial park, for example, may result in increased traffic, slower travel times for commuters passing through the area, and the eventual construction of a multilane highway requiring expenditure of public funds. Furthermore, the effects of land use tend to be cumulative: Even small changes can, over time, combine to produce large and long-lasting impacts.

Almost without exception, everyone is affected by growth, development, and changing land-use patterns. Plans may determine how far residents must drive to shop, the location of a new park or school, or where houses may be built. Land-use plans may be as simple as efforts to protect citizens from hazardous locations. Such plans might call for setback zones along eroding sea cliffs that prohibit construction, thus preventing structures from being destroyed as the cliff retreats. Conversely, plans may be very complex attempts to guide and direct the types, rates, and locations of change in an area. Such plans have the potential to influence the area's economic and cultural characteristics, its environmental quality, and the way its residents will live in the near and distant future.

GOAL DEFINITION AND DATA COLLECTION

William Spangle and others in a report issued by the U.S. Department of the Interior, *Earth-Science Information in Land-Use Planning*, recognize five separate phases of the land-use planning process: the identification of problems and definition of goals and objectives; data collection and interpretation; plan formulation; review and adoption of plans; and plan implementation. At each phase, feedback occurs so that modifications can be made as the process progresses. Even implemented plans are subject to review and redefinition as information accumulates.

Once the goals and objectives have been defined, the problems of the acquisition and interpretation of the data on which the plan is based must be addressed. For example, Earth science information, in the form of an Environmental Impact Statement (EIS), is needed throughout the planning process. At least a basic understanding of the climate, hydrology, geology, and soils of the area is essential. A considerable amount of data may already be available and may need to be consolidated from existing sources, such as published reports. Most often, however, many of the needed data must be collected specifically for the proposed plan. All data—whether already available or newly developed—must be evaluated and analyzed. Not all information is of equal quality or compatible with the needs of the plan. Some data may be of poor quality and must be eliminated. Some is not useful although it may be of high quality. For example, in an assessment of an area's capability to support structures of various kinds, a list and discussion of the fossils found in the bedrock are not useful, whereas the engineering properties of the same rocks are important. Still other types of data may be incompatible with the needs of the plan because they were collected for different purposes, by different groups, or with different systems. EIS's also must be in sufficient detail and at the appropriate map scale (the relationship between the same distances on the map and on the land surface).

PLAN FORMULATION, REVIEW, AND IMPLEMENTATION

Once high-quality and appropriate data are accumulated, they can be used to produce maps that show the capability of the area to support each po-

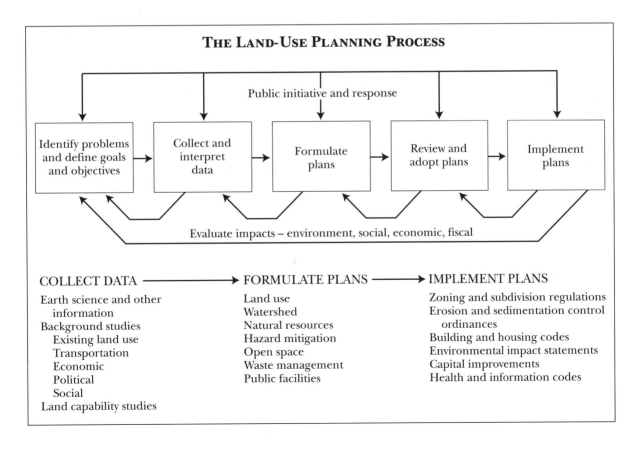

THE LAND-USE PLANNING PROCESS

Public initiative and response

| Identify problems and define goals and objectives | Collect and interpret data | Formulate plans | Review and adopt plans | Implement plans |

Evaluate impacts – environment, social, economic, fiscal

COLLECT DATA ⟶ FORMULATE PLANS ⟶ IMPLEMENT PLANS

COLLECT DATA
Earth science and other
    information
Background studies
    Existing land use
    Transportation
    Economic
    Political
    Social
Land capability studies

FORMULATE PLANS
Land use
Watershed
Natural resources
Hazard mitigation
Open space
Waste management
Public facilities

IMPLEMENT PLANS
Zoning and subdivision regulations
Erosion and sedimentation control
    ordinances
Building and housing codes
Environmental impact statements
Capital improvements
Health and information codes

tential use. These land-capability maps are analyzed together with projections and with economic, social, and political factors to evaluate alternative land-use patterns. Maps are prime aids for land-use planning because they present the location, size, shape, and distribution patterns of landscape features. Specialized or derivative maps can be prepared by combining several environmental factors, such as geology, soil types, and slope, to illustrate the best possible use or specific problems of an area. Furthermore, maps can be used to guide further development by outlining areas reserved for particular land uses.

The most desirable or feasible alternatives are adopted to prepare a suitable plan on which future decisions can be based. The formulated plan must then be reviewed and adopted by the commissioning agency or governmental body. At this phase, the technical personnel responsible for the development of the plan must be available to answer questions, respond to criticism, and make any further changes considered reasonable.

The final, and perhaps most critical, phase is that of plan implementation. Plans that are adopted, but not implemented, serve little purpose. There must be ways through which the plan influences the formal and informal processes of decision making. Zoning ordinances, construction regulations, and building codes based on the plan should be in place and enforced. Responsibilities and guidelines for preparation and evaluation of required reports, impact assessments, and proposals must be established. This requires a staff that not only reviews proposals but also requests additional information or modification of project proposals.

DATA COLLECTION METHODS

Data are collected for land-use planning purposes in many ways. Some information can be obtained through door-to-door or mailed surveys. Often, technical personnel, such as geologists or hydrologists, will conduct field investigations in the area. Observations and measurements are made and recorded, and samples are collected for laboratory analysis. Water samples might be analyzed to determine the quality of the water resources. Rock and soil samples may be tested to yield information about engineering properties, such as strength.

Remote-sensing methods are among the most useful sources of collecting data for land-use studies. Remote sensing is a method of imaging the Earth's surface with instruments operated from distant points, such as airplanes or satellites. Many different types of remote-sensing instruments, such as aerial photography equipment and radar, are applicable to land-use purposes. Basically, remote-sensing systems collect and record reflected or emitted energy from the land surface. Aerial photography, for example, collects visible light reflected from the Earth's surface to produce images on film. Line-scanning systems can record a wide range of energy types. For example, infrared devices can be used to detect small differences in temperature, which reflect differences of soil, water, or vegetation on the Earth's surface. Another technique, side-looking airborne radar (SLAR), uses pulses of microwave energy to locate surface objects by recording the time necessary for the energy transmitted to the object to be returned to the radar antenna.

Remote-sensing techniques are applicable to many data-collection efforts. Inventories of existing land uses, crop patterns, or vegetational types can be accomplished relatively quickly. Locations and extents of environmental hazards, such as floods or forest fires, can be delineated and landscape changes—crop rotations, shoreline shifts, and forest clearings—can be documented over a period of time.

DATA ANALYSIS

Because they can store and manipulate large data sets, computers are ideally suited to comparing and combining many types of data into a final integrated or interpretive map. The application of a computer system to land-use planning is accomplished by changing EIS's and other data into a form that can be entered into the machine. The data can be entered manually or by more sophisticated optical scanning methods. One straightforward method uses a grid system to enter data. Basically, maps of the area of interest portraying different types of information can be subdivided into a grid formed by equally spaced east-west and north-south lines. The size of the grid squares depends on the purpose of the study and the nature of the data. The intersection points of the grid lines or the centers of the grid squares can be used as data entry locations. An illustration of this operation is the overlaying of a grid on a map of soil types and the entry of the soil type at each line intersection or the dominant soil type in each square.

When all the available types of data are in the computer, they can be analyzed, combined, and applied to indicate areas best or least suited for a particular use. The information can also be used to make predictions. A planner may need to suggest the location of future solid-waste disposal sites. Although such a task involves knowledge of waste sources and waste volumes as well as of transportation distances, the performance capability of the various sites is a primary concern. The planner can select those data sets considered to be important to a site's capability to contain waste effectively and have the computer evaluate the area in terms of those factors. Bedrock geology, soil type and depth, distance to groundwater, distance from surface water bodies, and other factors may be selected. Each factor is assigned a weighted value, a measure of its relative importance to the specific purpose. Depth to the water table, for example, might be considered more important than the distance to surface water bodies. The computer incorporates the relative ranking of the variables into the data analysis to produce a composite map of favorable and unfavorable areas for solid-waste disposal. Such comprehensive systems, known as geographic information systems (GIS's), are being developed widely for many purposes. To summarize, a GIS consists of computer storage files of data, a program to analyze the data, and another program to map the stored data and produce several forms of output from the analysis program.

EFFECTIVE LAND-USE PLANNING

Land-use planning attempts to ensure organized and reasonable development of areas as the pressures of potential multiple uses mount and the value of land increases. Planning must be based on a thorough understanding of the environment and of economic, political, and cultural factors to produce a prediction of future land-use needs and ways to satisfy those needs.

An effective land-use plan consists of four basic parts. First, it includes a discussion of land-use issues and a statement of goals and objectives. This section provides the background and establishes what the plan seeks to accomplish. Second, there

is a discussion of the methods of data collection, evaluation, and analysis. This will include a description not only of the collection techniques, methodology, and the sources of the information, but also a consideration of the limitations of the data used in the study. Third, a land classification map is produced. The classification used to complete the map will depend on the goals of the plan and on the needs of each area. This map will serve as the basis for many kinds of decisions—from the location of new facilities to regulatory policies and tax structure. Finally, a report is included that provides a framework to the plan and discusses sensitive issues or areas. Any environmentally significant sites that may be adversely affected by change are considered, and appropriate policies are recommended.

POLITICAL CONSIDERATIONS

Land-use decisions are made by individuals, groups, industry, and governmental bodies. Individuals want the freedom to choose where they live and what to do with their property, but most also do not want an incompatible land use located next door. Land-use plans based on high-quality data and sound interpretations of that data, with abundant public input during formulation, adoption, and implementation of the land-use plan, can help to avoid costly and time-consuming confrontations. There will seldom be unanimous agreement, but planning can provide a framework in which to make decisions that will have far-ranging effects.

Land-use planning or management is a controversial subject. Landowners, developers, environmentalists, and government agencies have very different views on what is the best use of a parcel of land and who should make the decision. These are issues of individual versus group rights and benefits and they are not easily resolved. It is clear that as land resources dwindle and use pressures increase, some form of planning or management is necessary. It seems equally clear that while decisions among multiple uses will often be difficult, the best decisions can only be based on sufficient high-quality data.

Ronald D. Stieglitz

CROSS-REFERENCES

BIBLIOGRAPHY

Burby, Raymond J., ed. *Cooperating With Nature: Confronting Natural Hazards With Land-Use Planning for Sustainable Communities*. Washington, D.C.: Joseph Henry Press, 1998. Burby's book pays close attention to the role that natural disasters and hazards play in the development of new communities. Even though this is intended for the reader with some background in the field, the layperson with an interest in land-management policies would also find it useful. Illustrations, bibliography, and index.

Davidson, Donald A. *Soils and Land Use Planning*. New York: Longman, 1980. A short text for middle-level students that shows how soils information can be used in land-use planning.

Dluhy, Milan J., and Kan Chen, eds. *Interdisciplinary Planning: A Perspective for the Future*. New Brunswick, N.J.: Center for Urban Policy Research, 1986. Land-use planning is covered, along with other types of planning. The complex nature of the planning process is illustrated.

Economic Commission for Europe. *Land Administration Guidelines: With Special Reference to Countries in Transition*. New York: United Nations, 1996. This United Nations publication describes the land-use and land-tenure policies that have been enacted in Central and Eastern European countries during their transition from communist to democratic governments.

McHarg, I. L. *Design with Nature*. Garden City, N.Y.: Doubleday, 1969. An older but still useful text. Provides the basis for environmentally based planning technologies.

Marsh, William M. *Landscape Planning: Environmental Applications*. New York: John Wiley & Sons, 1986. A useful introduction to landscape planning. A wide range of topics are discussed and numerous case histories are provided.

Rhind, D., and R. Hudson. *Land Use*. London: Methuen, 1980. This is a well-illustrated treatment of land-use issues and the planning process. Good discussions of data needs and planning models are included.

Seabrooke, W., and Charles William Noel Miles. *Recreational Land Management*. 2d ed. New York: E. and F. N. Spon, 1993. This college-level text examines the management of land intended for recreational use in urban environments. Includes illustrations, bibliographies, and index.

So, Frank S., ed. *The Practice of Local Government Planning*. 5th ed. Chicago: American Planning Association, 1979. A thorough treatment of all aspects of planning at the level of cities and counties. Discussion is clear but advanced.

So, Frank S., Irving Hand, and Bruce McDowell. *The Practice of State and Regional Planning*. Chicago: Planners Press, 1986. Designed to be a companion volume to the work above. Thorough treatment of the practices of planning at higher government levels that are not as well defined.

Spangle, William, and Associates, F. Beach Leighton and Associates, and Baxter, McDonald and Company. *Earth-Science Information in Land-Use Planning: Guidelines for Earth Scientists and Planners*. Geological Survey Circular 721. Arlington, Va.: U.S. Department of the Interior, Geological Survey, 1976. An excellent discussion of the sources, accuracy, and applications of Earth science information in the land-use planning process.

LAND-USE PLANNING IN COASTAL ZONES

In order to designate appropriate coastal land uses, policymakers must take into consideration geological processes, recognizing that the coastal zone is one of rapidly evolving landforms, sedimentary deposits, and environments. Much of the American coastal zone has undergone rapid development without the benefit of landscape evaluation. Future development, as well as corrective measures, must be grounded in the knowledge of the underlying geology and its climatic and oceanographic interactions.

PRINCIPAL TERMS

COASTAL WETLANDS: shallow, wet, or flooded shelves that extend back from the freshwater-saltwater interface and may consist of marshes, bays, lagoons, tidal flats, or mangrove swamps

COASTAL ZONE: coastal waters and lands that exert a measurable influence on the uses of the sea and its ecology

ESTUARINE ZONE: an area near the coastline that consists of estuaries and coastal saltwater wetlands

ESTUARY: a thin zone along a coastline where freshwater system(s) and river(s) meet and mix with an ocean

GROUNDWATER: water that sinks into the soil, where it may be stored in slowly flowing underground reservoirs

LAND-USE PLANNING: a process for determining the best use of each parcel of land in an area

MASS WASTING: the downslope movement of Earth materials under the direct influence of gravity

SALTWATER INTRUSION: aquifer contamination by salty waters that have migrated from deeper aquifers or from the sea

WATER TABLE: the level below the Earth's surface at which the ground becomes saturated with water

EVALUATION OF COASTAL ZONES

The dependency of the United States on its coastal zone cannot be exaggerated, a lesson taught to national leaders in 1969 when the Stratton Report, "Our Nation and the Sea," revealed that coastal areas contained more than 50 percent of the nation's population (projected to reach 80 percent by the year 2000), seven of the nation's largest cities, 60 percent of the petroleum refineries, 40 percent of industry, and two out of three nuclear or coal-fired electrical generating plants. Not surprisingly, the Stratton Report became the primary impetus for the passage of the National Coastal Zone Management Act (NCZMA) of 1972 and 1980. NCZMA provides federal aid to thirty coastal states and five territories for development and implementation of voluntary, comprehensive programs for the management and protection of coastlines. In 1982, Congress passed the Coastal Barrier Resource Act, which declared 195 acres of beachfront on various barrier islands ineligible for federal infrastructure funding.

The first step of all land-use planning, the provision of geologic information and analysis, is complicated in the coastal zone by the fact that the major value of the area is water. Indeed, water is the overwhelming determinant of intrinsic suitabilities of the various microenvironments under consideration. Landscape evaluation must also recognize that the zone is not static but is the locus of rapid geologic change. Such change involves a complex sediment dispersal system that responds rapidly to human modifications. The coastal zone also includes fundamental legal boundaries between private and public ownership that are also high-energy geological boundaries. The presence of barrier islands demands a highly specialized set of land-use practices for these dynamic systems. Natural hazards in the coastal zone include periodic hurricane attacks on the Atlantic and Gulf coasts and landslides on the cliffed coasts of the Pacific. Superimposed upon the dynamic processes of the coastal zone is a global sea-level rise, initiated some eighteen thousand years

Cape Hatteras, North Carolina. A groin traps sand that normally moves along the shoreline. (U.S. Geological Survey)

ago with the termination of the Late Wisconsin glaciation. Sea-level rise is currently accelerating, possibly in response to the "greenhouse effect," human-induced atmospheric change.

A major resource of the area is the estuarine zone, encompassing less than 10 percent of the total ocean area but containing 90 percent of all sea life. Estuaries trap the nutrients that rivers wash down from the land and use them to produce an extraordinary quantity of biomass that sustains a variety of marine life. It is estimated that 60-80 percent of all edible seafood is dependent on the estuary for survival. Estuarine zones and coastal wetlands also serve as natural flood control devices by absorbing the energy of damaging storm waves and storing floodwaters. If not overloaded, these wetlands have the ability to remove large quantities of pollutants from coastal waters. This contribution alone was conservatively valued at $75,000 an acre in 1989; with the addition of monies generated by sport and commercial fishing, an acre was worth $83,000.

Beach Erosion and Shoreline Retreat

The great financial potential of beachfront properties makes the consequences of beach erosion, as well as efforts to halt shoreline retreat, of vital concern in coastal zone land-use issues. Although some beach erosion is rightly attributed to

sea-level rise, much is the direct result of human activities. The damming of major rivers has essentially cut off the sediment supply to the nation's beaches. Loss of dredged sediment through maintenance of shipping channels results in rapid erosion on the downdrift beaches to which wave action would normally deliver sands. Both jetties and groin fields act as dams to sediment transport and create erosion on adjacent beaches. Dune destruction removes the natural barriers to storm surges and reservoirs for sand storage. In addition to degradation of the aesthetic and recreational character of the shoreline, seawalls accelerate erosion by increasing wave energy. Emplacement of sand on the beach from offshore or inland sources (beach renourishment) and pumping sediment across navigation channels (inlet bypassing) represent more acceptable "soft engineering" alternatives to the armoring of the shoreline.

On the Atlantic and Gulf coasts, hurricanes account for many of the sediment distribution patterns. These tropical cyclones modify barrier islands by removal of beach sand offshore as well as over the dunes. Overwash, in which large quantities of sediment are moved inland, is the major means by which barrier islands retreat before the rising sea. The inevitability of hurricane attack mandates land use compatible with predicted flooding patterns, building codes based on hurricane-force winds and tidal storm surges, and up-to-date evacuation plans.

Retreat of the shoreline along the West Coast is more frequently related to seacliff erosion. Erosion along cliffed coasts is caused by a combination of wave scour at the base and landsliding higher on the bluffs. In winter, the large tides and waves force the rain-weakened cliffs to retreat. The degree of failure depends on wave energy, hardness of the cliff rock, and internal fractures and faults. Weathering processes further weaken rocks and aid erosion. Human activities that accelerate cliff retreat include septic tank leaching, landscape irrigation, alteration of drainage patterns, and introduction of nonnative vegetation.

GULF AND ATLANTIC COASTAL DEVELOPMENT

The low, sandy Gulf and Atlantic coasts, characterized by estuaries, marshes, and inlets, are separated from the open ocean by barrier islands. Major values of the lower coastal plain include abundant forests, a diversity of land and water wildlife habitats, and a unique potential for water-based recreation. The area is largely unsuited for urbanization, because of the high water table, but remnant ancient barrier island chains, or "terraces," represent elevated sites suitable for building. The largely sandy soils are poorly suited for widespread agriculture, though some fertile alluvial, or river valley, soils are present. The high water table makes aquifer pollution an ever-present threat; the abundant forests, which support a major pulp and paper production region, are fire-prone.

Because the broad coastal plains of the depositional coasts were formed by barrier island migration during the sea-level fluctuations of the Pleistocene "ice ages," the resulting sediments are largely barrier island sands or marsh muds and peats. Rapid, unplanned growth in the coastal zone has resulted in widespread use of septic tanks in these unsuitable sandy soils, allowing waste wa-

ter to percolate too rapidly and releasing improperly treated effluent. The degradation of coastal waters by septic tanks can be avoided by tertiary treatment of waste water, which may in turn be used for wetlands recharge or for irrigation. A sanitary landfill properly sited in mud or peat has low permeability, which limits the flow of leachate, contaminated landfill runoff. The most heavily developed resort areas, with the greatest need for landfill space, also have the highest land prices, making suitable land acquisition difficult.

MARSHLAND DEVELOPMENT

The vast marshes of the coastal zone grade from fresh to brackish to salt water. The river-swamp system, interrelated with the marsh-estuary system through flows of water, sediments, and nutrients, plays a major role in determining the impact of inland development on the estuary. In their natural state, swamps buffer the coast from the many impacts of land-use activities. If drained, filled, or altered, this contribution is lost. An increase in stormwater runoff related to urbanization reduces their ability to hold and absorb floodwaters. This problem could be avoided by placing limits to development and requiring that nonhighway "paving" be of permeable material, such as crushed limestone, instead of asphalt. The river swamp's ability to filter and absorb pollutants also improves the quality and productivity of downstream coastal waters; if overloaded with pollutants from industrial effluents, agricultural practices, or improperly sited septic tanks and landfills, the closing of shellfish beds is inevitable.

The "low marsh," inundated daily by the tides, is well recognized as an area so valuable and productive that conservation is mandatory. Although protected by state laws, estuarine marshland continues to experience some destruction as a result of marina construction and the disposal of dredged sediments from navigation channels. Marshland is also lost

Jetties are built to keep harbors from filling with sand and to protect them from storm damage. (© William E. Ferguson)

to the boat-wake erosion that undercuts the banks, causing slumping of the rooted clays.

At a slightly higher elevation is the "high marsh," sandy marshland occasionally wetted by storms or extreme tides. This marsh, unprotected by law, is known to be an important wildlife habitat but has not been widely studied. Indeed, the role of the high marsh and its relationship to the low marsh is unknown. The widespread exploitation of this area for residential and commercial development will make it impossible for the low marsh to reestablish itself landward in response to sea-level rise. The resultant loss in areal extent of the salt marsh will cause a great decline in estuarine-dependent marine life and a corresponding rise in pollution.

BARRIER ISLAND DEVELOPMENT

The barrier islands that front low-lying coastal plains are such dynamic geologic features that they are clearly unsuited for widespread development. These islands are composed of unconsolidated sediments that continuously seek to establish equilibrium with the waves, winds, currents, and tides that shape them. Early settlements were wisely constructed on the landward-facing, or back-barrier, portion of the islands. The beachfront was considered too dangerous because of hurricanes.

Ideally, high-density barrier island development should be confined to the back-barrier part of the island, there protected from storms and hurricanes. Low-density development should thin out in both directions, and open island should be preserved along the high-energy beachfront. That would ensure the scenic and recreational value of the island, as well as the sand supply and the natural maintenance of the system. If stabilization of inlets for shipping is necessary, the sand built up on the updrift side should be bypassed downdrift to reenter the system. Many islands should be preserved totally free of development for sand storage, as well as for educational, recreational, and aesthetic benefits.

The barrier island uplands, or high ground, consist of maritime forest on ancient oceanfront dune ridges; these dunes were left behind as the island built seaward, caused by a fortuitous combination of sediment supply and fluctuating sea level. This forested region is appropriate for environmentally sensitive development, if care is taken

to conserve ample wildlife preserves. Freshwater sloughs and ponds are located in the low swale areas between dune ridges, where rainfall floats on salt water within the porous material of the island subsurface. These wetlands represent important wildlife habitat and, on developed islands, serve as natural treatment plants for storm runoff and as natural storage areas for floodwaters. They may also be recharged with tertiary treated waste water, if not overloaded. Unfortunately, there has been a great loss of freshwater wetlands, on the mainland as well as the islands, to developers who have drained and filled them to produce construction sites.

BEACHFRONT DEVELOPMENT

Any beachfront construction should first be considered within the context of readily available historical shoreline change data. If the area under consideration has a suitable history of stability or accretion, development should be limited to a setback line landward of the heavily vegetated, stable dunes. Construction should be compatible with historical hurricane storm surge data, and, at elevations below recorded storm surges, dwellings should be elevated on a foundation of pilings.

Access to the beach should be via dune walkovers to avoid damage to the plants that stabilize the fragile dune environment. The dunes that are closest to the ocean and tied to the beach for their windblown sand supply are constantly changing landforms. Storm waves scour away the front of the dunes, and fair-weather waves return the lost sand to the beach to rebuild the dunes. These ephemeral features, vital as buffers to storm attack, should be protected not only from construction but also from human activity of any sort.

The updrift and downdrift inlet beachfronts are best utilized as nature preserves or public beaches, free of construction, because of the natural instability of both zones. The updrift inlet is constantly reestablishing its channel in response to tide-deposited mounds of sand, causing alternating and far-flung advances and retreats of the north-end shoreline. The downdrift, or south end, represents the geologically young part of the island. Because the area is barely above sea level, hurricane storm surges can readily cut through an elongated south-end spit, creating a new inlet and freeing the sand to migrate downdrift.

PACIFIC COASTAL DEVELOPMENT

The geologic instability of the rocky Pacific coast demands land-use planning strategies that will protect its rugged beauty and minimize the threats of natural hazards. Development in the available coastal zone regions of cliff tops, dune fields, and beachfront are all potentially hazardous.

On the beachfront, a wide beach between cliff and shoreline should not be mistaken for a permanent feature. Developed beachfront areas at the bases of cliffs are subject to periodic floodings by large waves combined with high tides; after much of the beach sand is removed by the storm, the wave energy is expended against the buildings. Houses in this zone have collapsed when their foundations were undermined or have smashed through their pilings after being uplifted by waves. If beachfront development is allowed, it should be based on a comparison of the site elevation and expected tidal ranges, storm surges, and storm-wave heights. A secure piling foundation below any potential wave scour should elevate the structure above any maximum inundation.

Although the presence of dunefields is restricted on young mountain range coasts, existing dunes have not escaped urbanization, and the resultant problems are identical to those of the East Coast. Where homes and condominiums have been built on conventional foundations, periodic dune erosion has undermined and threatened the structures, forcing emplacement of sea walls constructed of boulders. The dynamic nature of sand dunes negates any notion of wise land-use planning other than prohibition.

The zone consisting of bluff or cliff tops represents the West Coast environment with the greatest potential for development. Construction should not proceed here without a large enough setback behind the cliff edge so that the structure should endure for at least one hundred years, based on long-term erosion rates. The increased runoff associated with urbanization, if not collected and diverted away from the seacliff, results in serious slope failure. Homes, patios, swimming pools, and other construction also decrease the stability of the seacliff by increasing the driving forces, forces that tend to make Earth materials slide. Although seacliff erosion is a natural process that cannot be completely controlled, regardless of financial investment, it can be minimized by sound land-use practices.

Martha M. Griffin

CROSS-REFERENCES

Aerial Photography, 2739; Dams and Flood Control, 2016; Desertification, 1473; Environmental Health, 1759; Expansive Soils, 1479; Land Management, 1484; Land-Use Planning, 1490; Landfills, 1774; Landslides and Slope Stability, 1501; Remote Sensing and the Electromagnetic Spectrum, 2802; Soil Chemistry, 1509; Soil Erosion, 1513; Soil Formation, 1519; Soil Profiles, 1525; Soil Types, 1531; Volcanic Hazards, 798.

BIBLIOGRAPHY

Burby, Raymond J., ed. *Cooperating with Nature: Confronting Natural Hazards with Land-Use Planning for Sustainable Communities.* Washington D.C.: Joseph Henry Press, 1998. Burby's book pays close attention to the role that natural disasters and hazards play in the development of new communities. Even though this is intended for the reader with some background in the field, the layperson with an interest in land-management policies would also find it useful. Illustrations, bibliography, and index.

Dolan, Robert. "Barrier Islands: Natural and Controlled." In *Coastal Geomorphology,* edited by Donald R. Coates. Binghamton: State University of New York, 1972. A comparison of the stabilized islands (Cape Hatteras National Seashore) and the natural islands (Core Banks) of the Outer Banks of North Carolina. The responses of these two systems to coastal processes, particularly major storms, are of great interest for both preservation and management reasons. Suitable for college-level readers.

Griggs, Gary, and Lauret Savoy, eds. *Living with the California Coast.* Durham, N.C.: Duke University Press, 1985. This book, part of the Living with the Shore series (sponsored by the

National Audubon Society), addresses the problems resulting from coastal development during a relatively storm-free period without analysis of historical storms or long-term erosion rates. It includes conclusions and recommendations of coastal geologists and other specialists. Suitable for those with an interest and concern for the California coast.

Hoyle, Brian, ed. *Cityports, Coastal Zones, and Regional Change: International Perspectives on Planning and Management.* New York: Wiley, 1996. This collection of essays covers the planning and management of coastal zones, with particular emphasis on harbors and commercial ports. Illustrations and maps.

Kaufman, Wallace, and Orrin H. Pilkey, Jr. *The Beaches Are Moving: The Drowning of America's Shorelines.* Durham, N.C.: Duke University Press, 1983. An introduction to environmental coastal geology, nearshore processes, and the effects of human modifications on the shorelines of the United States, this book serves as background to the Living with the Shore series, covering the basic issues that are applied to specific shorelines. Suitable for those interested in the shorelines of the United States.

Keller, Edward A. *Environmental Geology.* 5th ed. Columbus, Ohio: Merrill, 1988. This work introduces physical principles basic to applied geology and reviews major natural processes and geologic hazards, including how society deals with them. Previous exposure to the geological sciences is not necessary for comprehension. Suitable for college-level readers.

Knight, Richard L., and Peter B. Landres, eds. *Stewardship Across Boundaries.* Washington D.C.: Island Press, 1998. A good analysis of the conservation of natural resources and management of public lands in the United States, including coastal regions. Maps and illustrations.

Kraus, Nicholas C. *Shoreline Changed and Storm-Induced Beach Erosion Modeling.* Springfield, Va.: National Technical Information Service, 1990. This text details the U.S. Army Engineer Waterways Experiment Station's work in developing mathematical models to study coastal changes, beach erosion, and storm surges, and their effects on coastal development.

McHarg, Ian L. *Design with Nature.* Garden City, N.Y.: Doubleday, 1971. An explanation of the ecological land-use planning method, proving that necessary human structures can be accommodated within the existing natural order. (National Book Award Nominee, 1971.) Suitable for high-school-level readers.

Neuman, A. Conrad. "Scenery for Sale." In his *Coastal Development and Areas of Environmental Concern.* Raleigh: North Carolina State University Press, 1975. A statement on the science, scenery, and selling of a barrier island system like the Outer Banks of North Carolina, cleverly illustrated by the author. Free of technical terminology and readable for all.

LANDSLIDES AND SLOPE STABILITY

Landslides occur under specific geological conditions that are usually detectable. Site assessments done by qualified geologists are important to land-use planning and engineering design; much of the tragedy and expense of landslides is preventable.

PRINCIPAL TERMS

ANGLE OF REPOSE: the maximum angle of steepness that a pile of loose materials such as sand or rock can assume and remain stable; the angle varies with the size, shape, moisture, and angularity of the material

AVALANCHE: any large mass of snow, ice, rock, soil, or mixture of these materials that falls, slides, or flows rapidly downslope

COHESION: the strength of a rock or soil imparted by the degree to which the particles or crystals of the material are bound to one another

CREEP: the slow, more or less continuous downslope movement of Earth material

EARTHFLOW: a term applied to both the process and the landform characterized by fluid downslope movement of soil and rock over a discrete plane of failure; the landform has a hummocky surface and usually terminates in discrete lobes

HUMMOCKY: a topography characterized by a slope composed of many irregular mounds (hummocks) that are produced during sliding or flowage movements of Earth and rock

LANDSLIDE: a general term that applies to any downslope movement of materials; landslides include avalanches, earthflows, mudflows, rockfalls, and slumps

MUDFLOW: both the process and the landform characterized by very fluid movement of fine-grained material with a high water content

SLUMP: a term that applies to the rotational slippage of material and the mass of material actually moved

FALLS, SLIDES, AND FLOWS

Slope failure, or landsliding, is the gravity-induced downward and outward movement of Earth materials. Landslides involve the failure of Earth materials under shear stress and/or flowage. When slope failures are rapid, they become serious hazards. Areas of the United States that are particularly susceptible to landslides include the West Coast, the Rocky Mountains of Colorado and Wyoming, the Mississippi Valley bluffs, the Appalachian Mountains, and the shorelines and bluffs around the Great Lakes. Downslope movement of soil and rock is a natural result of conditions on the planet's surface. The constant stress of gravity and the gradual weakening of Earth materials through long-term chemical and physical weathering processes ensure that, through geologic time, downslope movement is inevitable.

Slope failures involve the soil, the underlying bedrock, or both. Several types of movements (falling, sliding, or flowing) can take place during the failures. Simple rockfalls, or topples, may occur when rock overhangs a vertical road cut or cliff face. Other failures are massive and include flows and slides. Slides involve failure along a discrete plane. The failure planes in soils are usually curved, as in the illustration of a rotational slide or slump. The failure planes in bedrock can be curved or straight. Failures often follow planes of weakness, such as thin clay seams, joints, or alignment of fabric in the rock. Slides may be slow or rapid but usually involve coherent blocks of dry material. Flows, on the other hand, behave more like a fluid and move downslope much like running water. Earthflows, mudflows, sand flows, debris flows, and avalanches occur when soils or other unconsolidated materials move rapidly downslope in a fluidlike manner. The movement destroys the vegetative cover and leaves a scar of hummocky deposits where the flow occurred. Although flows usually involve wet materials, rare exceptions, such as the destructive flows in Kansu, China, occur in certain types of dry materials.

Slides and flows are terms applied to failures that produce rapid movement. Rock slides are those slides that involve mostly fresh bedrock; debris slides include those movements that are mostly rock particles larger than sand grains but with significant amounts of finer materials; mudslides involve even finer material and water, but the failure plane is straight. Earthflows involve mostly the soil overburden and move over a slope or into a valley rather than failing along a rock bedding plane; mudflows involve more water than earthflows and have a downslope movement much like flowing water.

CREEP

"Creep" is a term given to very slow movement of rock debris and soils. Creep in itself does not usually pose a life-threatening danger. When creep occurs beneath man-made structures, however, it leads to economic damage that requires repair or reconstruction in a new location. Examples include the gradual cracking and destruction of buildings, disalignment and breaking of power lines and fences, the filling of drains along highways, and the movement of topsoils into streams and reservoirs. Sometimes creep precedes a very rapid failure, and therefore new evidence of creep requires careful monitor-

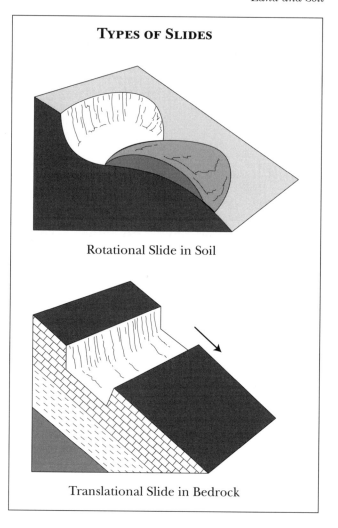

TYPES OF SLIDES

Rotational Slide in Soil

Translational Slide in Bedrock

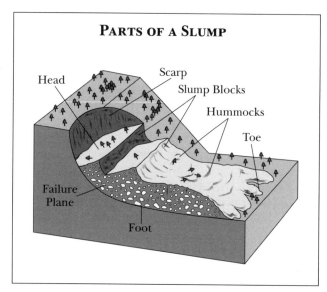

PARTS OF A SLUMP

Head
Scarp
Slump Blocks
Hummocks
Toe
Failure Plane
Foot

ing and an evaluation of the conditions that produce it.

Solifluction is a special type of creep that occurs in cold climates where the soil is frozen most of the year, permafrost regions. In summer, the ice in the upper layer of soil melts, and the soil becomes waterlogged and susceptible to downslope movement. Solifluction is an important consideration for the design of structures in cold climates: For example, the Alaska pipeline, used for conveyance of petroleum, could not be buried but had to be set above ground on supports that were anchored at a depth below the seasonal depth of thaw to escape solifluction movement. Houses and other buildings in such areas must be set on supports and insulated

Sherman Glacier overlain by a rockslide avalanche that resulted from the collapse of Shattered Peak (middle distance), following Alaska's Good Friday earthquake in 1964. (U.S. Geological Survey)

exert pressure on either side of the joint and decrease the cohesion there, thus causing the block above the discontinuity to move downslope.

Loose volcanic materials often can absorb so much water that they flow quickly down even gentle slopes. A mudflow destroyed the Roman city of Herculaneum at the base of Mount Vesuvius in A.D. 79; mudflows generated during the Mount St. Helens eruption of May 18, 1980, destroyed many properties.

ROLE OF HUMAN DEVELOPMENT

Designers may unwittingly assemble, in a man-made structure, conditions that produce slope failures. Dry materials such as mine wastes have sometimes been stacked into piles that are steeper than their angle of repose after saturation. Much later, a rainstorm or Earth tremor can send the piles into motion, destroying all structures around them. When the supporting toe is removed from the base of a slope, during excavations such as occur for a highway or building foundation, this action pro-

in order to keep heat from the structures from melting the underlying soils.

ROLE OF WATER

Water is an important agent in promoting instability in slopes. Where soils are saturated, water in large pores will flow naturally in a downward direction. The resulting pressure of the water pushing against the soil grains is called pore pressure. As pore pressures increase, the grains are forced apart and cohesion decreases. The saturated soils become easier to move downslope. Water flowing along a bedding plane or joint can also

Solifluction lobes on the side of a kame in the Nunatarssuaq region of Greenland. Solifluction is a special type of creep that occurs in cold climates where the soil is frozen most of the year. In summer, the ice in the upper layer of soil melts, and the soil becomes waterlogged and susceptible to downslope movement. (U.S. Geological Survey)

duces many landslides, evidence of which can be found on most highways constructed through hilly terrain. An unstable slope can be set into motion by loading the slope from above, which occurs when a structure such as a building, a storage tank, or a highway is built on materials that cannot remain stable under the load. Catastrophic failures have occurred where mudflows were produced when dams built from mine tailings burst as a result of slope failure. The mine wastes used for these dams were susceptible to swelling, absorption of water, and weakening over time.

Human development alters the natural drainage of an area and increases runoff. Occasionally, water from storm drains, roof gutters, septic tanks, or leaking water mains reaches a sensitive slope and generates movement. This instability is particularly likely to occur where intensive housing development takes place in several levels on a long slope.

REGIONAL STUDIES

The stability of slopes is evaluated over large regions from aerial photographs, satellite photographs, and images made by remote sensing techniques. Investigators look for telltale signs, such as hummocky topography and old scars left by slides, that may not be evident when viewed from the ground.

The regional study involves an evaluation of the history of past landslides within the region. That history often reveals particular geological formations that have an association with landslides. For example, the shales of the Pierre formation are well known by engineering geologists in the area of Denver, Colorado, as materials in which many slope failures occur. A geologic map that shows where this formation is exposed at the ground surface reveals potentially dangerous areas. Ignoring evidence of past landslides invites disaster.

The regional study also defines loose surficial

A TIME LINE OF DEADLY LANDSLIDES, MUDSLIDES, AND ROCKSLIDES

Nov. 5, 1926	PEREIRA, COLOMBIA: 100 dead, 60 injured when a landslide crashes down on houses, burying the inhabitants alive.
Jan. 10, 1931	HUIGRA, ECUADOR: 165 railroad workers dead when, after two days of rain, a small landslide covers about 200 yards of the Quito-Guayaquil railroad. The next morning, workers, clearing debris from the tracks in the narrow Cleanchan River Valley, are killed when the almost-vertical 2,000-foot walls collapse in a second slide.
Dec. 13, 1941	HUARAZ, PERU: 2,000 dead when, at 7:30 A.M., the swollen Santa River results in mudflow that kills more than a quarter of the town's population, followed by an earthquake. The 150-foot-deep and 0.5-mile-wide liquid mass leaves no survivors.
Oct. 22, 1954	BERLY, HAITI: 262 dead after Hurricane Hazel blows into southwestern Haiti, flattening the town of Jérémie and rendering roads impassable. Rain-drenched soils in mountains 20 miles south of Port-au-Prince give way and bury the town of Berly.
Oct. 18, 1955	ATENGUIGUE, MEXICO: 100 dead after torrential rains and strong winds cause many landslides, one of which buries much of this town.
Oct. 23, 1958	NEAR SPRING HILL, NOVA SCOTIA: 84 dead as a result not of rockfall but rock upheaval. In higher levels of a mine, an upward pressure pushes the floor into the roof, causing coal and sandstone to tumble down to the deeper expanses.
Jan. 22, 1960	NEAR JOHANNESBURG, SOUTH AFRICA: 550 dead after a mile-long shaft collapses in Coalbrook, South Africa, causing a ventilating fan to explode and the roofs of two tunnels to drop. At 7:30 P.M. two gas explosions cause an audible cave-in, shutting off a mile of tunnel and causing parts of the Earth's surface to drop 3 feet.
Mar. 13, 1961	KIEV, UKRAINE: 145 dead when winds force water over an earthen dam in the Babi Yar Ravine, washing thousands of tons of mud into a Kiev suburb. A tram station, houses, shops, and factories are destroyed.

materials that are likely to fail. Soils rich in clay minerals that swell and expand when wet are notorious for slumping and flowing. Usually, movement occurs in these soils in the spring, when the soil is very saturated from soil thaw and snowmelt. Other soils fail simply because they have low cohesion and large amounts of open space (pore space) between the tiny soil grains. Collapse of these soils requires no wetting; strong vibrations can trigger the movement. The most tragic example of this type took place in 1920, when thick deposits of fine loess (a type of soil deposited by the wind) settled rapidly during an earthquake in Kansu, China, and the resulting flows toppled and buried the many homes built upon them. More than 100,000 deaths resulted from the flows that occurred in the loess.

Regional studies also look at the earthquake history of an area, because a tremor, even a fairly mild one, can provide the coup de grâce to a slope that has been resting for decades in a state of marginal instability. Huge blocks of the shoreline slid beneath the ocean at Valdez, Alaska, during the 1964 earthquake when the rotational slumping of materials occurred below sea level. Landslides triggered by that same earthquake destroyed much of Anchorage, Alaska.

Finally, the regional study includes a history of weather events. When the right geological conditions exist, periods of intense rainfall can trigger the movement of unstable slopes. In the southern and central Appalachian Mountains, periods of increased frequency of slides often coincide with severe local summer cloudbursts and thunderstorms. Intense rainfall events associated with hurricanes which have moved inland also trigger landslides over larger areas. Studies in the Canadian Rockies reveal a definite link between rainstorms and rockfalls, and landslides are particularly abundant during the rainy season along the West Coast of North America.

ON-SITE INVESTIGATION

Once knowledge is collected on the region, more specific questions about the local site itself

June 27-July 1, 1961	CENTRAL JAPAN: Poor city planning is a major factor in 244 dead, 972 injured, $70 million in damages that result from seven days of heavy rain.
Oct. 9, 1963	BELLUNO, ITALY: A slab 2 kilometers wide by 1.6 kilometers long and 250 meters thick slides into the Vaiont Reservoir in Switzerland in less than one minute. The rockslide pushes water over the dam, producing a downstream flood that kills almost 3,000 people.
Jan. 11-13, 1966	RIO DE JANEIRO, BRAZIL: 239 dead when the heaviest rains since 1883 wreak havoc on the city, with slum dwellers bearing the brunt. By sundown, 114 corpses have been recovered from the rubble in shantytowns across the city.
Oct. 21, 1966	ABERFAN, SOUTH WALES, UNITED KINGDOM: 147 dead, 32 injured, and a school and 8 houses destroyed after a 700-foot "tip" (mound of mining wastes) slides over town.
Feb. 17-20, 1967	RIO DE JANEIRO, BRAZIL: 224 dead, 4,000 homeless after thirty hours of rain wash the earth from beneath a mammoth boulder. The rock rolls down a gully, crashes through a house, then tears out the foundation of an eight-story apartment building, which in in turn falls onto a four-story building.
Mar. 8, 1968	LUHONGA, CONGO: 150 die when a mudslide sweeps away this small African village. The nearly mile-long mudflow is 15 feet deep when it engulfs the village and washes over its fields.
July 30, 1971	KHINJAN PASS, AFGHANISTAN: 100 dead after inundations in northern Afghanistan cause a lake in the Hindu Kush Mountains to overflow its shores, sending floodwater, then mud and rocks, onto a village at Khinjan Pass.
June 18, 1972	HONG KONG: 100 dead after three days of heavy rain cause landslides that bury squatters' huts in a "resettlement estate" north of the Kai Tak international airport and topple two houses and an apartment building above Victoria City Center.

are considered. The investigator first looks at the steepness of the slopes and the Earth materials present. In the case of loose materials and soils, the angle of repose is very important. Dry sand poured carefully onto a table to form an unsupported conical pile cannot achieve a cone with sides steeper than approximately 40 degrees, because the cohesion between loose dry sand grains is not strong enough to allow the material to support a steeper face. The 40-degree angle is the maximum angle of repose for dry sand. The angle of repose changes with water content, mineral content, compaction, grain shape, and sorting. Soils that contain clay may be tough and cohesive when dry and have natural repose angles greater than 40 degrees. When wet, their angle of repose may be only 10 degrees. This is particularly true if the soils contain clay minerals such as montmorillonite that absorb large amounts of water. Those soils, sometimes called "quick clays," can fail instantaneously and flow downslope almost as rapidly as pure water.

The orientation of discontinuities in rocks is as important in determining the stability of a slope as is the type of rock involved. Bedding planes that dip downslope serve as directions of weakness along which failure may occur. Other planar weaknesses may develop along joints and faults and along parallel fabrics produced by the alignment of platy and rodshaped minerals that are oriented downslope.

The investigator will check to see if natural processes are removing the supporting material at the bases of slopes. Landslides are particularly common along stream banks, reservoir shorelines, and large lake and seacoasts. The removal of supporting material by currents and waves at the base of a slope produces countless small slides each year. Particularly good examples are found in the soft glacial sediments along the shores of the Great Lakes of the United States and Canada.

Finally, the investigator will look for evidence of actual creep at the site. Damage to structures already on the site, curved tree trunks (where tilting occurring as a result of soil movement is compensated for by the tree's tendency to resume vertical growth), the offset of fences and power lines, or the presence of hummocky topography on slopes can demonstrate the presence of recent motion at the site.

COSTS AND REMEDIATION OF STABILITY PROBLEMS

An annual economic loss of between $1 billion and $1.5 billion is a reasonable estimate for costs of landslides within the United States. Expenses include the loss of real estate around large lakes, rivers, and oceans; loss of productivity in agricultural and forest lands; depreciated real estate in areas of slide development; public aid for victims of large landslides; and the contribution of sediment to streams that decreases water quality, injures aquatic life, and results in the loss of reservoir storage space. In the United States, approximately twenty-five lives are lost each year from landslides. Elsewhere, in densely populated areas, single landslide events cause death tolls in the thousands.

The remediation of slope stability problems involves contributions from both geologists in the investigation of the site and civil engineers in the design of the project. Geologists employed by state geological surveys and the U.S. Geological Survey provide a tremendous service by constructing geological and slope stability maps based on knowledge of the soils and rock formations, use of remote-sensing methods such as satellite and high-altitude photography, and by field study of suspect areas. These maps are made readily available to engineers, contractors, developers, and home owners. These maps show color-coded areas of active and potentially active landslides. Such maps have been produced for many areas with a high population density. Residents in the United States may contact their local state's geological survey, which serves as a distributor for such maps.

Edward B. Nuhfer

CROSS-REFERENCES

BIBLIOGRAPHY

Casale, Riccardo, and Claudio Margottini, eds. *Floods and Landslides: Integrated Risk Assessment.* New York: Springer, 1999. A strong examination based on case studies of the risks associated with floods and landslides in regions, primarily European, prone to these natural disasters. Illustrations and maps.

Close, Upton, and Elsie McCormick. "Where the Mountains Walked." *National Geographic* 41 (May, 1922): 445-464. Kansu, China, 1920 earthquakes and landslides—a graphic account of the most devastating landslide in history.

Costa, J. E., and G. F. Wieczorek, eds. *Debris Flows/Avalanches: Process, Recognition, and Mitigation.* Reviews in Engineering Geology 7. Boulder, Colo.: Geological Society of America, 1987. A number of case studies from various parts of the United States, Canada, and Japan. The content of the text is intended for professionals, but there are many photographs and illustrations that will interest students and laymen.

Costa, J. E., and V. R. Baker. *Surficial Geology: Building with the Earth.* New York: John Wiley & Sons, 1981. A very well-written text for undergraduates in environmental geology and engineering geology courses. Its particular strength is the use of simple numerical problems to illustrate concepts quantitatively. It may be understood by students with a background in algebra, trigonometry, and introductory geology or Earth science. It is well illustrated and contains a good bibliography.

Cummans, J. *Mudflows Resulting from the May 18, 1980, Eruption of Mount St. Helens.* U.S. Geological Survey Circular 850-B. Washington, D.C.: Government Printing Office, 1981. This is a good illustration of the devastation caused by mudslides associated with volcanism.

Hays, W. W. *Facing Geologic and Hydrologic Hazards.* U.S. Geological Survey Professional Paper 1240-B. Washington, D.C.: Government Printing Office, 1981. A well-written and beautifully illustrated booklet. It can be appreciated by readers from grade school through professionals.

Hoek, E., and J. W. Bray. *Rock Slope Engineering.* 3d ed. Brookfield, Vt.: IMM/North American Publications Center, 1981. This is an engineering reference often used by professionals. Although much of it involves a solid quantitative approach, the descriptive sections are graphic and well written. Some parts may be understood by secondary students, and most of the book may be understood by junior and senior undergraduates in geology and civil engineering.

Kalvoda, Jan, and Charles L. Rosenfeld, eds. *Geomorphological Hazards in High Mountain Areas.* Boston: Kluwer Academic, 1998. This collection of essays examines the processes used in determining which mountain areas are at risk of landslides. It provides a clear understanding of the hazards involved and what to look for in order to predict them.

Keefer, D. K. "Landslides Caused by Earthquakes." *Geological Society of America Bulletin* 95 (April, 1984): 406-421. A good review of the relationship between earthquakes and major landslides.

Keller, E. A. *Environmental Geology.* 5th ed. Columbus, Ohio: Merrill, 1988. Chapter 7, on landslides and related phenomena, is highly recommended for beginners. It is well illustrated and written in a simple and descriptive manner. A good set of references follows each chapter.

Kennedy, Nathaniel T. "California's Trial by Mud and Water." *National Geographic* 136 (October, 1969): 552-573. A graphic account of the interaction between landslides, earthquakes, and heavy seasonal rainfall.

Kiersch, G. A. "Vaiont Reservoir Disaster." *Civil Engineering* 34 (1964): 32-39. An enthralling account of one of the world's most tragic landslides. Excerpts of Kiersch's original article have been reprinted in many engineering geology texts.

McDowell, Bart. "Avalanche!" *National Geographic* 121 (June, 1962): 855-880. A graphic account of avalanches in Peru.

Radbruch-Hall, Dorothy H. *Landslide Overview Map of the Conterminous United States.* U.S. Geological Survey Professional Paper 1183. Washington, D.C.: Government Printing Office, 1981. A map, with accompanying text, that illustrates the major landslide areas within the United States.

Rahn, P. H. *Engineering Geology: An Environmental Approach.* New York: Elsevier, 1986. This text is for the serious undergraduate or graduate student interested in a solid quantitative approach. The text is well illustrated and well referenced.

Schuster, Robert L., ed. *Landslide Dams: Processes, Risk, and Mitigation.* New York: American Society of Civil Engineers, 1986. A particularly interesting compilation of papers that includes case studies of slides at Thistle Creek, Utah, and the control of the new Spirit Lake, which was produced by a landslide at Mount St. Helens.

Schuster, Robert L., and Keith Turne, eds. *Landslide: Investigation and Mitigation.* Washington, D.C.: National Academy Press, 1996. This National Research Council and Transportation Research Board special report offers a thorough analysis of landslide hazards and slope stability. There is a good description of the processes involved in pinpointing unstable areas. Illustrations, bibliographical references, and index.

SOIL CHEMISTRY

Soils are complex chemical factories. Regardless of the type of soil, chemical processes such as plant growth, organic decay, mineral weathering, and water purification are ongoing processes.

PRINCIPAL TERMS

CATION: an element that has lost one or more electrons such that the atom carries a positive charge

CLAY MINERAL: a group of minerals, commonly the result of weathering reactions, composed of sheets of silicon and aluminum atoms

LYSIMETER: a simple pan or porous cup that is inserted into the soil to collect soil water for analysis

STABLE ISOTOPES: atoms of the same element that differ by the number of neutrons in their nuclei yet are not radioactive

WEATHERING: reactions between water and rock minerals, at or near the Earth surface, that result in the rock minerals being altered to a new form that is more stable

WEATHERING PROCESS

Soil chemistry has been studied as long as there has been sustainable agriculture. Although they did not recognize it as such, those first successful farmers who plowed under plant stalks, cover crops, or animal wastes were actively managing the soil chemistry of their fields. These early farmers knew that to have productive farms in one location season after season, they had to return something to the soil. It is now understood that soil chemistry is a complex of chemical and biochemical reactions. The most obvious result of this complex of reactions is that some soils are very fertile whereas other soils are not. Soil itself is a unique environment because all the "spheres"—the atmosphere, hydrosphere, geosphere, and biosphere—are intimately mixed there. For this reason, soil and soil chemistry are extremely important.

Soil chemistry begins with rock weathering. The minerals comprising a rock exposed at the Earth's surface are continually bathed in a shower of acid rain—not necessarily polluted rainwater but naturally occurring acid rain. Each rain droplet forming in the atmosphere absorbs a small amount of carbon dioxide gas. Some of the dissolved carbon dioxide reacts with the water to form a dilute solution of carbonic acid. A more concentrated solution of carbonic acid is found in any bottle of sparkling water.

Most of the common rock-forming minerals, such as feldspar, will react slowly with rainwater. Some of the chemical elements of the mineral, such as sodium, potassium, calcium, and magnesium, are very soluble in rainwater and are carried away with the water as it moves over the rock surface. Other chemical elements of the mineral, such as aluminum, silicon, and iron, are much less soluble. Some of these elements are dissolved in the water and carried away; most, however, remain near the original weathering, where they recombine into new, more resistant minerals. Many of the new minerals are of a type called clays.

Clay minerals tend to be very small crystals composed of layers of aluminum and silicon. Between the layers of aluminum and silicon atoms are positively charged ions (cations) of sodium, potassium, calcium, and magnesium. The cations hold the layers of some clays together by electrostatic attraction. In most cases, the interlayer cations are not held very tightly. They can migrate out of the clay and into the water surrounding the clay mineral, to be replaced by another cation from the soil solution. This phenomenon is called cation exchange.

The weathering reactions between rainwater and rock minerals produce a thin mantle of clay mineral soil. The depth to fresh, unweathered rock is not great at first, but rainwater continues to fall, percolating through the thin soil and reacting with fresh rock minerals. In this way, the weathering front (the line between weathered minerals and fresh rock) penetrates farther into the rock, and the overlying soil gets thicker.

BIOLOGICAL PROCESSES

Throughout the weathering process, biological processes contribute to the pace of soil formation. In the very early stages, lichens and fungi are attached to what appear to be bare rock surfaces. In reality, they are using their own acids to "digest" the rock minerals. They absorb the elements of the mineral they need, and the remainder is left to form soil minerals. As the soil gets thicker, larger plants and animals begin to colonize. Large plants send roots down into the soil looking for water and nutrients. Some of the necessary nutrients, such as potassium, are available as exchangeable cations on soil clays or in the form of deeper, unweathered minerals. In either case, the plant obtains the nutrients by using its own weathering reaction carried on through its roots. The nutrient elements are removed from minerals and become part of the growing plant's tissue.

Without a way to replenish the nutrients in the soil, the uptake of nutrients by plants will eventually deplete the fertility of the soil. Nutrients are returned to the soil through the death and decay of plants. Microorganisms in the soil, such as bacteria and fungi, speed up the decay. Since the bulk of the decaying plant material is found at the surface (the dead plant's roots also decay), most of the nutrients are released to the surface layer of the soil. Some of the nutrients are carried down to roots deep in the soil by infiltrating rainwater. Most of the nutrients, however, are removed from the water by the shallow root systems of smaller plants. The deeper roots of typically large plants can mine the untapped nutrients at the deep, relatively unweathered soil-rock boundary.

The soil and its soil chemistry are now well established, with plants growing on the surface and their roots reaching toward mineral nutrients at depth. Water is flowing through the soil, carrying dissolved nutrients and the soluble by-products of weathering reactions.

STUDY OF SOIL WATER

The study of soil chemistry is concerned with the composition of soil water and how that composition changes as the water interacts with soil at-

mosphere, minerals, plants, and animals. Soil and its chemistry can be studied in its natural environment, or samples can be brought into the laboratory for testing. Some tests have been standardized and are best conducted in the laboratory so that they can be compared with the results of other researchers. Most of the standardized tests, such as measures of the soil's acidity and cation-exchange capacity, are related to measures of the soil's fertility and its overall suitability for plant growth. These tests measure average values for a soil sample because large original samples are dried and thoroughly mixed before smaller samples are taken for the specific test.

Increasingly, soil chemists are looking for ways to study the fine details of soil chemical processes. They know, for example, that soil water chemistry changes as the water percolates through succeeding layers of the soil. The water flowing through the soil during a rainstorm has a different chemical composition from that of water clinging to soil particles, at the same depth, several days later. Finally, during a rainstorm, the water flowing through large cracks in the soil has a chemical composition different from that of the same rainwater flowing through the tiny spaces between soil particles.

SAMPLING TECHNIQUES

Soil chemists use several sampling techniques to collect the different types of soil water. During a rainstorm, water flows under the influence of gravity. After digging a trench in the area of interest, researchers push several sheets of metal or plastic, called pan lysimeters, into the wall of the trench at specified depths below the surface. The pans have a very shallow V shape. Soil water flowing through the soil collects in the pan, flows toward the bottom of the V, and flows out of the pan into a collection bottle. Comparing the chemical compositions of rainwater that has passed through different thicknesses of soils (marked by the depth of each pan) allows the soil chemist to identify specific soil reactions with specific depths.

After the soil water stops flowing, water is still trapped in the soil. The soil water clings to soil particles and is said to be held by tension. Tension water can spend a long time in the soil between rainstorms. During that time, it reacts with soil mineral grains and soil microorganisms. Tension water is sampled by placing another type of lysimeter, a tension lysimeter, into the soil at a known depth. A tension lysimeter is like the nozzle of a vaccum cleaner with a filter over the opening. Soil chemists actually vacuum the tension water out of the soil and to the surface for analysis.

DETERMINATION OF ISOTOPIC COMPOSITION

Nonradioactive, stable isotopes of common elements are being used more often by soil chemists to trace both the movement of water through the soil and the chemical reactions that change the composition of the water. Trace stable isotopes behave chemically just the way their more common counterparts do. For example, deuterium, an isotope of hydrogen, substitutes for hydrogen in the water molecule and allows the soil chemist to follow the water's movements. Similarly, carbon 13 and nitrogen 15 are relatively rare isotopes of common elements that happen to be biologically important. Using these isotopes, soil chemists can study the influences of soil organisms on the composition of soil water. Depending on what the soil chemist is studying, the isotope may be added, or spiked, to the soil in the laboratory or in the field. Alternatively, naturally occurring concentrations of the isotope in rain or snowmelt may be used. Regardless, soil water samples are collected by one or more of the lysimeter methods, and their isotopic composition is determined.

THE SOIL CHEMICAL FACTORY

The wonderful interactions of complex chemical and biochemical reactions that are soil chemistry are one indication of the uniqueness of planet Earth. Without the interaction of liquid water and the gases in the atmosphere, many of the nutrients necessary for life would remain locked up in rock minerals. Thanks to weathering reaction, the soil chemical factory started to produce nutrients, which resulted in the exploitation of the soil environment by millions of organisms. The processes involved in soil chemistry—from weathering reactions that turn rock into new soil to the recycling of plant nutrients through microbial decay—are vital to every human being. Without fertile soil, plants will not grow. Without plants as a source of oxygen and food, there would be no animal life.

Because of the complex chemical interrelationships that have developed in the soil environment, it may seem that nothing can disrupt the "factory" operation. As more is understood about soil chemistry and the ways in which humans stress the soil chemistry by their activities, it is apparent that the factory is fragile. Not only do humans rely on soil fertility for their very existence, but they also are taking advantage of soil chemical processes to help them survive their own past mistakes. Soil has been and continues to be used as a garbage filter. Garbage, whether solid or liquid, has been dumped on or buried in soil for ages. Natural chemical processes broke down the garbage into simpler forms and recycled the nutrients. When garbage began to contain toxic chemicals, those chemicals, when in small quantities, were either destroyed by soil bacteria or firmly attached to soil particles. The result is that water—percolating through garbage, on its way to the local groundwater, stream, or lake—does not carry with it as much contamination as one might expect. Soil chemistry has, so far, kept contaminated garbage from ruining drinking water. There are well-known cases, however, where the volume and composition of waste buried or spilled were such that the local soil chemistry was overwhelmed. In cases of large industrial spills, or when artificial chemicals are spilled or buried, the soil needs help to recover. The recovery efforts are usually very expensive but, faced with the possible permanent loss of large parts of the soil chemical factory, humankind cannot afford to neglect this aspect of the environment.

Richard W. Arnseth

CROSS-REFERENCES

BIBLIOGRAPHY

Berner, Elizabeth K., and Robert A. Berner. *The Global Water Cycle: Geochemistry and Environment.* Englewood Cliffs, N.J.: Prentice-Hall, 1987. Despite the fact that this book is designed to teach environmental geochemistry to college students, the role of water in soil chemistry makes this a great resource. The authors emphasize the hydrologic cycle. There are good introductory chapters on the uniqueness of water and the water cycle. Of most interest are the chapters on rainwater and soil water. Each chapter ends with an extensive bibliography to more technical references.

Bohn, Heinrich, B. L. McNeal, and G. A. O'Connor. *Soil Chemistry.* 2d ed. New York: Wiley, 1985. A fairly widely used college text on soil chemistry. The introductory chapter and the chapter on weathering are accessible to any reader with a college background. The authors provide an annotated bibliography at the end of each chapter for further reading.

Brill, Winston. "Agricultural Microbiology." *Scientific American* 245 (September, 1981): 198. The indisputable link between soil chemistry and microbiological processes is emphasized in this article. Emphasis is on nitrogen biochemistry and genetic engineering. Several figures nicely illustrate the complexity and cyclic nature of soil nitrogen chemistry. Designed for those with a background in the terminology of microbiology.

Evangelou, V. P. *Environmental Soil and Water Chemistry: Principles and Applications.* New York: Wiley, 1998. An excellent examination of soil and water chemistry. The author looks at the pollution levels of water and soils and the chemical processes used to determine those levels.

Lloyd, G. B. *Don't Call It Dirt.* Ontario, Calif.: Bookworm Publishing, 1976. The author provides a backyard gardener's point of view on soil fertility, microorganisms, and the interrelationships between soil moisture and soil processes. Soil chemistry is never mentioned, but all the information is pertinent and accessible to the general reader.

McBride, Murray B. *Environmental Chemistry of Soils.* New York: Oxford University Press, 1994. This college text focuses on the chemical make-up and properties of soils in different regions. Written to be understood by the college-level reader.

Millot, Georges. "Clay." *Scientific American* 240 (April, 1979): 108. This article emphasizes the chemistry and industrial uses of clay minerals. A brief discussion of weathering as the source of clays and as part of the geologic cycling of clays, and an interesting discussion of how weathering reactions differ in different climates and yield different clay minerals in the local soils. Provides several spectacular scanning electron microphotographs of representative clay types. Accessible to the general reader.

Sparks, Donald S., ed. *Soil Physical Chemistry.* 2d ed. Boca Raton, Fla.: CRC Press, 1999. This widely used college textbook is an excellent introduction to soil chemistry. It offers a clear look into the practices, procedures, and applications of the field and is easily understood by the careful reader.

Tan, Kim Howard. *Principles of Soil Chemistry.* 3d ed. New York: M. Dekker, 1998. Tan's text has been expanded in this third edition. Written as an introduction to soil chemistry, the book explains the processes and application of the science in an easy-to-understand manner accompanied by plenty of illustrations. Includes a twenty-five-page bibliography, as well as an index.

SOIL EROSION

Soil erosion, which plays an important role in shaping the landscape, is a process that takes extended periods, although people have greatly accelerated it by removing vegetation. Controlling soil erosion is essential to maintaining the world's food supply.

PRINCIPAL TERMS

GRANULES: small grains or pellets

POROSITY: the ability to admit the passage of gas or liquid through pores

RILLS: small rivulets in channels

SUSPENSION: a condition in which particles are dispersed through a supporting medium

UNDERSTRATA: material lying beneath the surface of the soil

SHAPING THE LANDSCAPE

Soil erosion is a natural process whereby rock and soil are broken loose from the Earth's surface at one location and moved to another. Erosion creates and transforms land by filling in valleys, wearing down mountains, and making rivers appear and disappear. Generally, this process takes thousands or even millions of years.

The active forces of erosion are water and wind. Water erosion involves the movement of soil by the action of rainwater and melted snow moving rapidly over exposed land surfaces. The principal types of water erosion are splash erosion and gully erosion.

SPLASH EROSION

Splash erosion is the removal of thin layers, one at a time, from land surfaces. Fine-grained soil such as silt loams, fragile sandy soil, and all soils deficient in organic matter are especially vulnerable. Most troublesome are lands that tend to slope, lands subjected to heavy rainfall, and lands composed of shallow surface soils overlying dense clay subsoil.

When rainwater is absorbed into the ground, small sievelike openings are made in the soil. Eventually, fine particles carried by the water plug the openings, thereby causing the rainwater to flow off the land. As the soil is softened by the rainfall, clods, lumps, and granules break down and form a pasty mass, which resists penetration by rainwater. As a result, the runoff increases and forms a relatively impervious skinlike film sometimes referred to as puddled soil. As rain pro-

ceeds, the abrasive force results in cutting, which penetrates the skinlike layer and starts trenching.

When splash erosion reaches an advanced stage, it is referred to as rill erosion. Rill erosion can be defined as localized small washes in channels that can usually be eliminated by ordinary plowing. Rilling is the commonest form of erosion on soft, freshly plowed soils that are high in silt content where the slopes are deeper than 4 or 5 percent.

There are several specialized forms of splash erosion. Pedestal erosion occurs when a stone or tree root protects easily eroded soil from splash erosion. Consequently, isolated pedestals containing resistant material are left standing. This type of erosion occurs after several years primarily on bare patches of grazing land. The erosion patterns formed in highly erodible soils are called pinnacle erosion. This erosion is usually found in deep vertical rills in the sides of gullies. When these gullies cut back and join, they form pinnacles. Pinnacle erosion is usually found where there is some sort of imbalance, such as excessive sodium. When pinnacle erosion is present, reclamation is difficult. Another specialized form of splash erosion is piping, which is associated with the formation of continuous pipes or channels underground. Piping usually occurs in soil types that are subject to pinnacle erosion. It involves water penetrating the soil surfaces and moving downward until it reaches a less permeable layer. The fine particles of the more porous soil may be washed out if the water flows over the less permeable layer through an outlet. The more rapid lateral flow increases the sideways erosion, causing the entire surface flow to disappear down a vertical pipe. The water then flows underground until it reappears in the side of a gully. Once pipe erosion starts, it cannot be controlled. Fortunately, pipe erosion is restricted to the "bad lands." The last type of splash erosion is slump erosion. It is prominent in areas of high rainfall with deep soils. Slumping can become the chief agent in the development of gullies, probably as a result of flood flow in channels. Riverbank collapse and coastal erosion are the other main cases of slumping.

GULLY EROSION

The second principal type of water erosion is gully erosion. This type of erosion occurs in places where runoff from a slope is sufficient in volume and velocity to cut deep trenches. It also takes place in areas where concentrated water continues cutting in the same groove long enough to form deep rills. Gullies usually begin in slight depressions in or below fields where water concentrates, in ruts left by farm machinery, in livestock trails, or along furrows between crop rows. Gullies ordinarily carry water only during or following rains or melting snows. Most gullies cannot be removed by normal plowing because of their size; in fact, some of them take the form of huge chasms 15-30 meters deep. A field gully is a channel at least 45 centimeters wide and between 25 and 30 centimeters deep; a woodland channel, on the other hand, is considered to be a channel deep enough to expose the main lateral roots of trees. Woodland gullies develop in wooded areas that received water from cultivated slopes immediately above or on closely cut, severely burned-over woodland.

There are two main types of gullies. V-shaped gullies, which have sloping sides with narrow, V-shaped bottoms, are the normal type of gully. Gullies that have straight, more or less vertical sides with broad bottoms are U-shaped gullies. They are less winding than the V-shaped gullies because the soft materials at the base of the sides tend to give way to the impact of currents. Usually, the presence of gullies means that the land has been overused or abused. U-shaped gullies are usually the more serious because the soft, unstable materials commonly found in their lower depths, such as sand, loose gravel, or soft rocks, are easily cut out by floodwaters. This cutting near the bot-

Gully erosion on farmland. (© William E. Ferguson)

tom causes the banks to split off from above in great vertical blocks. When the caved-in material is washed out of the gully, the trench is left box-shaped. Most gullies tend to branch out as they grow, but the tendency is more serious in the U-shaped gullies. They are the most difficult to control because of the instability of the understrata.

Waterfall erosion is an important form of gully erosion because it does so much damage. It is caused by water cascading over the heads and sides of gullies, over dams, and over terraces whose channels have been filled with the debris of erosion. This type of erosion is most commonly associated with flooding.

Gullies in the United States are known by different names. In the east-central, southeastern, and south-central states, they are called gullies; in the northeastern and north-central states, they are called ditches. "Arroyo" and "barranca" are common names for gullies in the southwestern states, although the term "wash" is frequently used for gullies having channelways of shallow depth and exceptional width. In the northwestern states, gullies are called coulees.

WIND EROSION

The second active force of soil erosion is wind erosion. Wind erosion can occur in places where water erosion is also active, but it reaches its most serious proportions on both level and sloping areas during dry times. Like water erosion, wind erosion usually proceeds very slowly; however, wind erosion is likely to increase rapidly in relatively flat and gently undulating treeless regions, like the Great Plains. When grass is plowed up, the cultivated soil becomes much less cohesive. Organic material that normally collects under a cover of grass and that serves as a bonding agent when grass is present disappears by decay and oxidation when the grass is gone. After periods of drought, the soil turns into a dry powdery mass, which is easily swept up by the wind and lifted into the pathways of high air currents, which carry it hundreds, and, at times, thousands of kilometers. Coarser, heavier particles, known as ground drift, are blown along near the surface of the ground and pile up in drifts about houses, fences, farm implements, and clumps of vegetation. The fine materials that are blown away are dust; the coarser materials that are left behind are sand. The susceptibility of soils to wind erosion depends on the size of the particles and on the content of the organic matter. Coarse sands are more likely to blow away immediately after plowing than are heavy clays. Ironically, the finer textured soils, especially those of granular structure, show the greatest resistance. In fact, they sometimes remain undisturbed through years of cultivation until their organic material is disrupted.

The massive removal of soil particles through the action of the wind takes the form of dust storms. Early in 1934, a dust storm originating in the Texas-Oklahoma Panhandle covered a vast territory extending eastward from the Rocky Mountains to several hundred kilometers over the Atlantic. This one storm deprived the Great Plains states of 200-300 million tons of soil.

There are essentially five different types of wind erosion, although there is some overlapping, and several of the processes occur simultaneously. De-trusion is the wearing away of rocks and soil formations by fine particles carried away in suspension. This process often carves large rocks in deserts into grotesque shapes. Abrasion occurs close to the ground where the moving particles are larger and bound over the surface. The removal of very fine particles, carried off in suspension, is efflation, and the rolling away of large particles is extrusion. The removal of particles of intermediate size bouncing downwind is known as effluxion.

SEVERITY OF EROSION

The severity of erosion is determined by a great many factors: One of the least controllable is climate. The United States has more erosion problems than does England because it lacks England's gentle rains and mists. Many areas in the United States have rainstorms of sufficient strength and intensity to erode many centimeters of soil in minutes on a field not protected by vegetation. For example, rain-induced erosion may be severe in the Corn Belt (roughly, the states of Indiana, Illinois, Missouri, Nebraska, Kansas) because rains frequently occur in these states in June when the plants have not matured enough to protect the soil; in arid areas of the western United States, serious water erosion may exist because there is not enough rain to establish a protective ground cover for the infrequent rains.

Soil type and topography also influence the severity of erosion. Soils in the United States range from poorly drained to very dry, from sandy to clayish, from acid to alkaline, and from shallow to deep. They also vary in slope characteristics, porosity, organic content, temperature, and the capacity to supply nutrients. Land is better for crops if it is nearly level, with just enough slope for good drainage. About 45 percent of the cropland in the United States falls into this category.

A farmer's cropping practices can also greatly influence the severity of erosion. Erosion is likely to occur when fields are plowed in the conventional way: in straight rows regardless of the topography, with all plant cover and crop residue removed. Erosion can be retarded if the farmer uses the best land for crops and puts any other land to different uses. After that, any number of conservation practices, such as strip cropping, contour planting, crop rotation, terracing, and various conservation tillage practices, can be introduced.

MEASURING EROSION

Soil losses per acre are measured by either the Universal Soil Loss Equation (USLE) or the Wind Erosion Equation (WEE). These formulas have been developed from field experiments in various parts of the country. Both equations measure the tons of each soil type that are lost annually through the action of climate, cropping systems, management practices, and topography.

The use of these equations involves a number of limitations. Although both the USLE and the WEE measure the movement of the soil, they do not reveal the distance the soil has traveled or where the soil was deposited. Thus, the USLE could overestimate the severity of the erosion. Another drawback to the USLE is its failure to measure losses resulting from snowmelt. A revised version of the USLE, the RUSLE, has been developed to overcome these limitations.

The soil losses measured by the RUSLE or WEE are average figures taken from measurements over an extended period. They are usually reported in tons per acre. The computed losses are often connected with soil-loss tolerances, or T-values. These figures are the maximum soil losses that can be sustained without adversely affecting productivity. Excess soil erosion is frequently defined as amounts greater than T-values. The U.S. Department of Agriculture has assigned T-values that usually range from 1 to 5 tons per acre, depending on the properties of the soil. Unfortunately, the validity of these numbers in representing maximum sustainable soil losses is doubtful. A soil loss of 5 tons per acre per year translates into a net loss of 2.54 centimeters of soil every thirty years. T-values have been set too high on some soils to assure long-term maintenance of the soil. T-values are also limited in their value because they do not reflect the impact of technology on crop yields.

Erosion can also be studied according to how it will be affected by different types of rain and how it will vary for different types of soil. Therefore, the amount of erosion that occurs depends on a combination of the ability of the soil to withstand rain and the power of the rain to cause erosion. This relationship between factors can be expressed in mathematical terms. Erosion is a function of the erosivity (of the rain) and the erodibility (of the soil); that is, erosion = erosivity × erodibility.

For given soil conditions, one rainstorm can be compared quantitatively with another, and a numerical scale of values of erosivity can be created. For given rainfall conditions, one soil condition can be compared quantitatively with another, and a numerical scale of values of erodibility can be created. Erodibility of the soil can be subdivided into two parts: the inherent characteristics of the soil (that is, mechanical, chemical, and physical composition) and the way the soil is managed. Management may, in turn, be subdivided into land management and crop management.

COSTS AND BENEFITS OF EROSION

Soil erosion can be both beneficial and harmful. Erosion benefits people by contributing to the formation of soil by breaking up rocks. Erosion also creates rich, fertile areas as it deposits soil at the mouths of rivers and on the floors of valleys. Erosion is also important from the standpoint of aesthetics. The Grand Canyon, for example, was created over millions of years through the eroding action of the Colorado River.

Yet, soil erosion is one of the leading threats to the food supply because it robs farmland of productive topsoil. Soil scientists estimate that it takes nature between three hundred and one thousand years to produce about 2.5 centimeters of topsoil, although humans can replace topsoil at a much faster rate. Still, considering that about 70 percent of the United States is subject to erosion, the prospect of building productive soil becomes overwhelming. Once the topsoil has been removed, a heavy layer of clay often remains, which may not contain enough porosity to support a good crop.

Crops can also be damaged by the loss of nutrients to erosion. Most serious are losses of nitrogen, sulfur, and phosphorus. In addition to the natural nutrients taken from the soil, synthetic fertilizers and chemicals are often washed from fields and into lakes and rivers, thereby contributing to pollution.

Finally, serious erosion forming deep gullies presents problems to farmers. Gullies caused by flowing water are often too deep for farm machinery to cross. Unable to accommodate tractors and other farm equipment, fields riddled with gullies face ruin.

Alan Brown

CROSS-REFERENCES

Acid Rain and Acid Deposition, 1803; Clays and Clay Minerals, 1187; Desertification, 1473; Expansive Soils, 1479; Freshwater Chemistry, 405; Geochemical Cycle, 412; Groundwater Movement, 2030; Hydrologic Cycle, 2045; Land Management, 1484; Land-Use Planning, 1490; Land-Use Planning in Coastal Zones, 1495; Landslides and Slope Stability, 1501; Precipitation, 2050; Soil Chemistry, 1509; Soil Formation, 1519; Soil Profiles, 1525; Soil Types, 1531; Weathering and Erosion, 2380.

BIBLIOGRAPHY

Agassi, Menachem, ed. *Soil Erosion, Conservation, and Rehabilitation.* New York: Marcel Dekker, 1996. This collection of essays assesses soil erosion processes through an examination of rainfall and water runoff. Many of the essays deal with methods used in soil conservation and rehabilitation programs. Illustrations, index, and bibliography.

Batie, Sandra S. *Soil Erosion: Crisis in America's Croplands?* Washington, D.C.: The Conservation Foundation, 1983. This short government publication attempts to structure an improved public policy for soil conservation. The assessment of the extent and effects of erosion, as well as the explanation of the techniques for reducing soil erosion, is written in language that can be easily understood by the nonscientist. Also contains photographs and diagrams.

Bennett, Hugh Hammond. *Elements of Soil Conservation.* New York: McGraw-Hill, 1955. Although some of the information is dated, the chapter describing the processes and effects of erosion are up-to-date. Contains many photographs, but a few more diagrams would have been helpful. Also includes a bibliography of books, pamphlets, and films concerning soil conservation.

Boardman, John, and David Favis-Mortlock, eds. *Modelling Soil Erosion by Water.* New York: Springer-Verlag, 1998. This collection of essays covers the effects of climatic factors and climatic changes on soil erosion. Several experts discuss possible climatic causes and effects and use computer-simulated erosion models to illustrate complex processes. Technical at times.

Hudson, Norman. *Soil Conservation.* 3d ed. Ames: Iowa State University Press, 1995. This fully revised and updated college textbook covers the history of soil conservation and demonstrates how the methods of soil conservation that are practiced in the United States can be applied in developing countries. The author adopts an engineering approach to explain soil conservation. Includes many charts, illustrations, and photographs.

Lal, Rattan, ed. *Soil Quality and Soil Erosion.* Boca Raton, Fla.: CRC Press, 1999. Based on papers presented at a 1996 symposium on soil and water conservation, this book offers an examination of the quality, management, and erosion of soils and water. There are also sections dealing with methods of determining those levels.

Morgan, Royston Philip Charles. *Soil Erosion and Conservation.* 2d ed. New York: John Wiley, 1995. Morgan's book explains the procedures and protocol used in determining the rates and effects of soil erosion. He also describes how the application of these techiniques is used in soil conservation programs. Illustrations, maps, thirty-page bibliography, and index.

Simms, O. Harper. *The Soil Conservation Service.* New York: Praeger, 1970. This book is primarily a record of the activities of the U.S. Soil Conservation Service. The chapter entitled "Soil Conservation in Practice" discusses the methods that the government recommends for preventing soil erosion. For the general reader.

Stallings, J. H. *Soil: Use and Improvement.* Englewood Cliffs, N.J.: Prentice-Hall, 1957. This book is concerned with proper land management. Explains both past and present methods of soil conservation in terms that the general reader can understand.

Thompson, Louis M. *Soils and Soil Fertility.* New York: McGraw-Hill, 1957. A textbook for an introductory course in soils for students of agriculture. Covers every aspect of soil, including the physical properties of soil and soil conservation. Highly readable and well supplemented with charts, illustrations, and diagrams, although the reference sections are dated.

U.S. Soil Conservation Series. *Our American Land: Use the Land, Save the Soil.* Washington, D.C.: Government Printing Office, 1967. This publication promotes the need for soil conservation by providing statistical evidence of damage that has been done in the United States by erosion. Also recommends proper land management techniques. For the general reader.

SOIL FORMATION

A soil is formed when decomposed organic material is encompassed into weathered mineral material at the Earth's surface. The climate, the organisms living in the soil, the type of parent material, the local topography, and the amount of time the soil has been developing all influence the resulting soil characteristics. Changes in the soil arise from the main processes of soil formation: addition, removal, transfer, and transformation of parts of the soil.

PRINCIPAL TERMS

HORIZON: a layer of soil material approximately parallel to the surface of the land that differs from adjacent related layers in physical, chemical, and biological properties

LEACHING: the dissolving out or removal of soluble materials from a soil horizon by percolating water

SEDIMENT: rock fragments such as clay, silt, sand, gravel, and cobbles

SOIL PROFILE: a vertical section of a soil, extending through its horizons into the unweathered parent material

WEATHERING: the mechanical disintegration and chemical decomposition of rocks and sediments

STAGES OF FORMATION

Soil formation and development are continuing processes. A soil is a layer of unconsolidated weathered mineral matter on the Earth's surface that can support life. It is initially formed when the remains of plants are mixed into weathered mineral fragments. The soil characteristics can change, however, for the soil develops through time as the environment dictates.

Many processes contribute to the two stages of soil formation, which may grade indistinctly into each other. The first stage is the accumulation of rock fragments that are known as the parent material. These sediments may have formed in place from the physical and chemical weathering of the hard rocks below or by the accumulation of sediments moved to that place by wind, water, gravity, or glaciers. The second stage is the formation of soil horizons, which may either follow or occur simultaneously with the first stage.

The characteristics of the horizons and their degree of differentiation depend upon the relative strengths of four major processes in the soil: weathering, transfers, gains, and losses of the soil constituents. These processes have been observed directly in laboratory experiments but also are implied by relating physical, chemical, and morphological properties in different parts of the soil profile.

TRANSFORMATIONS AND TRANSFERS

Weathering is the fundamental process of soil formation, as it transforms soil constituents into new chemicals. The primary minerals found in the initial sediments and rocks are decomposed by physical and chemical agents into salts that are soluble in water and new minerals. Clay is a very important new mineral formed from the breakdown of primary minerals by the agents of water and air. Calcite, or calcium carbonate, is another mineral formed through weathering, especially of rocks such as basalt. Iron and aluminum oxides form a third major group of weathering-produced minerals. Organic matter is added to the soil when roots die and vegetation residues fall onto the surface. These are transformed by microorganisms into a mixture of chemically stable, organic substances called humus.

The transfer of soil constituents is generally in a downward direction. Rainwater is the major transporting agent. As it moves through the soil, it picks up salts and humus in solution and clays and humus in suspension. These materials are usually deposited in another horizon when the water is withdrawn by roots or evaporation or when the materials are precipitated as a result of differences in pH (degree of acidity) or salt concentration. The clays normally are oriented on the exteriors of sediments and line the interiors of pores. This transfer of materials through solution is commonly called leaching and needs a well-drained situation. If there is poor drainage in a soil, the soil profile becomes stratified according to the solubility of the dissolved substances. In forest soils, leaching generally moves iron, aluminum, and humus and is called podzolization.

Gains and Losses

Gains consist of additions of new materials to the soil. Wind can deposit fine sands and silts if there is a source nearby. Chemical reactions during weathering add oxygen and water to the soil minerals. Groundwater brings new minerals into the soil pores. Floods can add new sediment to the soils in a floodplain. The largest addition to soils,

Soil Horizons

O = organic debris

A = topsoil (humus, minerals)

B = subsoil (clay, iron oxide, carbonate calcium)

C = bedrock

however, comes from organic matter from the decomposition of plant material. Annual additions of new organic material between 255 and 22,900 kilograms per hectare have been estimated for soils in different environments ranging from deserts to tropical rain forests.

Losses from the soil profile generally occur when there is so much water moving through the system that the dissolved and suspended materials exit at the bottom of the profile and move downslope and eventually enter a stream. Soil erosion also occurs on the surface by water runoff, especially where there is minimal vegetation to control the sediment. Soil erosion is one of the major environmental problems facing the world today.

CLIMATE

After describing many soil profiles around the world and classifying them, scientists concluded that this weathered material at the Earth's surface was quite variable and differed from environment to environment. Hans Jenny summarized the work of early researchers and considered the five main soil-forming factors that created this variability to be climate, organisms, parent material, topography, and time. The rates and degrees of these processes in different soils depend upon these five factors. These factors are interdependent, and, therefore, very different soils may be formed from the same parent material.

Of the five factors, climate has probably the greatest effect on soil development. This dominance is best shown on a large-scale geographic basis, in which the distribution of soils on soil maps is related to climatic zones. Precipitation provides the water for the transfer processes in the soils. In dry climates, where little water is moving through the system, calcium carbonate and salts build up in the soils. In humid climates, abundant water moving through the systems allows leaching and clay movement processes to dominate in the soils. Temperature is the second important element of climate, as it controls the rates of chemical weathering reactions in the soils. In colder environments, soil weathering and soil development are slower because the weathering reactions are not active all year round. Some extreme climatic conditions can also control soil development, such as permafrost, which limits soil depth and

In Royal Natal National Park, South Africa, soil horizons A (topsoil) and B (subsoil) are visible, with bedrock below of shale and sandstone. (© William E. Ferguson)

drainage. Overall, climate controls vegetation, which also controls soil development.

ORGANISMS AND PARENT MATERIALS

Living organisms, primarily vegetation, control the distribution and types of humus that are formed in the soils. In turn, humus controls some of the important soil-forming processes. Grassland vegetation produces humus that is highest in mineral matter and not very acid, so resulting soils have thick A horizons that are high in organics. Conifer forests produce humus that is low in mineral constituents and highly acid. These forest soils are dominated by leaching and production of O, E, and Bs horizons. Deciduous forest soils are located somewhere between the conifer forest and the grassland soils in characteristics. Therefore, it is rare to find white, bleached E horizons under grassland vegetation and vice versa—it is rare to find thick A horizons under conifer forest vegetation. Through its leaves and root systems, vegetation also protects the soil from erosion and, therefore, promotes soil development.

The parent material is the rock or sediment deposit from which the soil has developed. The composition of the original minerals controls the complexity of the soil processes by controlling the weathering by-products and controls the rate of these processes by porosity. A parent material of pure quartz sand would undergo minimal weathering, producing a simple soil profile. A parent material of wind-deposited silt, or loess, produces a more complex soil, as it has many minerals initially that can weather to by-products such as clays, salts, carbonates, and oxides. A soil produced on granite would yield a fairly deep, sandy soil, whereas a soil produced on basalt would yield a fairly thin, clay-rich soil. A soil developed in shale would be clay-rich, whereas one developed on sandstone would be sandy. Soils formed on rocks rich in iron tend to produce very red soils. Soils formed on limestone are difficult to acidify and podzolize because the parent material neutralizes those chemical processes. Porosity, or the amount of pore space, of the parent material gives ready access of gases and liquids to weathering. In porous, or permeable, soils, weathering and transfers of soil constituents are more rapid than in soils with low permeability.

TOPOGRAPHY AND TIME

Topography is the geometric configuration of the soil surface and is defined by the slope gradient, orientation, and elevation. This factor essentially controls the microclimate. A slope modifies rainfall by allowing rapid runoff on a steep slope or ponding in low places. Slope orientation modifies soil temperatures by the differences at which the Sun's rays strike the soil. In the northern latitudes, southfacing slopes are warmer and drier than those on north-facing slopes. In mountain ranges, rainfall increases and temperature decreases with increases in elevation.

Time is the factor of the duration of the processes working in the soil. All the characteristics in the soil take time to form. The A horizon forms most rapidly, taking as little as 10 years with the aid of humankind to more than 1,000 years in a

cold and dry Arctic or alpine environment. B horizons take longer to form because minerals weather slowly. In the dry western parts of the United States, discoloration from the weathering of iron-rich minerals appears at about 1,000 years, but the strongest red colors require 100,000 years of development in the same environment. In the same part of the United States, initial detection of clay movement into the B horizon in a sandy parent material occurs after about 10,000 years of development, with maximum development taking more than 100,000 years. Calcium carbonate deposition in arid regions of the world takes between 100,000 and 500,000 years to produce a strongly cemented horizon. It is estimated that it takes more than 1 million years to produce a highly weathered lateritic (decayed iron-rich rock that is red in color) soil in the tropics.

Soils change with time just as humans change as they grow. Young or poorly developed soils in a grassland environment have only an A and C horizon. With time and the movement of iron oxides and clay into the lower parts of the profile, a Bw horizon develops between the A and C horizons. An old or well-developed soil in this environment would have a thick A horizon and a thick Bt horizon that had abundant clay that had moved into it, all overlying a C horizon.

LAND USE PLANNING

Scientists are finding that soils can be useful tools in the interpretation of past environments of an area and also in the formulation of land-use plans for that region. First, the soil scientists must determine which soils are in the area through a mapping program. During this project, they determine the regional developmental sequence of soils, that is, which are the poorly developed soils and the well-developed soils of the different places on the landscape. They assign ages to each soil in the sequence using radiocarbon-dated samples from the soils. As a result, when a particular soil is located in that region and compared with the sequence, an age can be assigned to the land surface upon which the soil had developed. Geologists commonly use soils to estimate the ages of deposits, especially on glacial landscapes. The idea of multiple glaciations has been developed through the use of soils. A poorly developed soil on glacial sediments represents a recent glaciation compared with a site with a well-developed soil that lies on sediment deposited by an older advance of the glaciers.

The frequency of geological hazard events such as landslides, rockfalls, and flooding can be estimated by the study of buried soils in sediment deposits. An A horizon within a soil profile means that some geological event covered an old land surface with sediment. If the organic matter of that buried A horizon is dated using radiocarbon dating, an approximate time of burial can be determined. If a valley bottom has many buried soils, it probably means that the nearby stream floods quite frequently. These decisions are important for land-use planning, for one does not want to build in an area where hazards occur frequently. For example, nuclear power plants must not be located in areas where there are active faults (ground breaks). Soils are commonly used to determine if faults are active in an area and, if so, how frequently the movement on the fault is.

The U.S. Department of Agriculture produces maps of soils across the country. The productivity of crop growth and manageability related to land use of each soil are described with the maps. These data help farmers grow crops that are best suited for their soils; they tell the farmers which soils are most susceptible to erosion so that extra precautions can be taken with the fragile soils. The data also help land-use planners keep the most productive soils in agriculture instead of paving them. In addition, the maps help to predict how much fertilizer to put on the fields. Natural soil fertility is related to geological activity and weathering, as the oldest soils are leached of most of their nutrients and are least fertile. Young soils have had little weathering and leaching and are most fertile, especially those on floodplains, young glacial deposits, and volcanic deposits. Soil maps are also used to recognize where shrink-swell clays occur in the soils. The presence of shrink-swell clays in the soil is important for home owners. If present, these clays can crack the foundation of a house. To prevent these damages, special foundations must be constructed and the clays stabilized with the use of lime.

Scott F. Burns

Cross-References

Bibliography

Birkeland, P. W. *Soils and Geomorphology.* 3d ed. New York: Oxford University Press, 1999. This book on soils is one of the best written from the perspective of a geologist. Many examples from around the world are given. More than half the book is devoted to soil formation and development. The final chapter on applications of soil formation to geology is excellent. Suitable for advanced high school and university-level readers.

Buol, S. W., F. D. Hole, and R. J. McCracken. *Soil Genesis and Classification.* 4th ed. Ames: Iowa State University Press, 1997. This volume is an update of a classic soil science text. New to the edition is an update on the new soil horizon symbol nomenclature. Includes an in-depth discussion on soil development. Suitable for university-level readers.

To ease understanding, however, they may want to read the first few chapters of the Birkeland or Singer books before reading this text.

Gerard, A. J. *Soils and Landforms.* London: Allen & Unwin, 1981. Even though this book goes over all factors of soil formation, it emphasizes the topographic factor. Excellent section on catenas (distribution of soils on a slope). Best suited to university-level readers.

Jenny, H. *Factors of Soil Formation: A System of Quantitative Pedology.* New York: Dover, 1994. This text is hard to follow, but the author has many important ideas. The whole book expands on the five factors of soil formation and development. Aimed at university-level readers. Illustrations, bibliography, and index.

_____. *The Soil Resource: Origin and Behav-*

ior. New York: Springer-Verlag, 1980. This edition is a rewrite of the classic book written by Jenny in 1941 that described variations in soils. At times, the text is hard to follow, but the author has many important ideas. The whole book expands on the five factors of soil formation and development. Aimed at university-level readers.

Paton, T. R. *Soils: A New Global View.* New Haven, Conn.: Yale University Press, 1995. Even though this book is aimed at readers at the university level, it is fairly easy to understand and has many good examples of soils and soil formation from around the world. Color illustrations, maps, bibliography, and index.

Simonson, R. W. "Outline of a Generalized Theory of Soil Genesis." *Soil Science Society of America Proceedings* 23 (1959): 152-156. This work is a classic paper in which the author outlines the four types of processes that take part in soil formation.

Singer, M. J., and D. N. Munns. *Soils.* New York: Macmillan, 1986. This book is one of the most readable elementary soil science texts available. Includes a good basic section on soil formation, and is suitable for high-school-level readers.

Soil Science Society of America. *Glossary of Soil Science Terms.* Madison, Wis.: Soil Science Society of America, 1978. This work is basically a dictionary of soil science terms written by people in the soil science area.

United States Department of Agriculture. Soil Survey Staff. *Soil Taxonomy.* Handbook 436. Washington, D.C.: Government Printing Office, 1975. This publication contains the most used soil classification system in the world, which is the official one used in the United States. Difficult to follow at times; read the books listed above before trying to understand this one. A source of abundant information.

Van Breemen, Nico, and Peter Buurman. *Soil Formation.* Boston: Kluwer Academic Publishers, 1998. This college-level text examines all stages and processes of soil formation. There are illustrations and diagrams to help clarify difficult concepts. Although intended for the college student, the layperson will also find much of this text useful.

SOIL PROFILES

A soil profile is the vertical section of a soil extending through its horizons into the unweathered parent material. Scientists use the different arrangements of horizons to give classification names to the soil profiles, which can be used to interpret landscape history and determine future land use.

PRINCIPAL TERMS

HORIZON: a layer of soil material approximately parallel to the surface of the land that differs from adjacent related layers in physical, chemical, and biological properties

LEACHING: the dissolving out or removal of soluble materials from a soil horizon by percolating water

SEDIMENT: rock fragments such as clay, silt, sand, gravel, and cobbles

STRUCTURE: the arrangement of primary soil particles into secondary units called peds

TEXTURE: the relative proportions of varying sediment sizes in a soil

WEATHERING: the mechanical disintegration and chemical decomposition of rocks and sediments

HORIZON NOMENCLATURE

The term "soil" has many definitions. To the soil scientist, it is anything that will support plant growth. To the civil engineer, it is an unconsolidated surficial material that can be penetrated by a shovel. To the geologist, it is an unconsolidated layer of weathered mineral matter arranged in layers at the Earth's surface. A soil profile is a vertical arrangement of soil horizons down to and including the parent material in which the soil has developed. Two sets of horizon nomenclature are used to describe soil profiles: symbols for field descriptions and diagnostic horizon names for classification. The field description symbols have been in use since the nineteenth century, whereas the diagnostic horizon names only came into use in the 1960's.

The field description symbols used to identify various kinds of horizons were developed in Russia in the 1880's. There, soil scientists such as V. V. Dokuchayev applied the letters A, B, and C to the main horizons of the black soils of the steppes of Russia. The A horizon was designated the zone of maximum organic material accumulation, the C horizon was the unaltered parent material, and the B horizon was the layer in between the A and the C. These concepts spread to the rest of the world, with the B horizon concept being modified to be a zone of accumulation of iron, aluminum,

and clays that had moved down the profile. Horizons may be prominent or so weak that they can only be detected in the laboratory; they can be thick or thin. The use of the A, B, and C master horizons has remained the central backbone of the horizon symbols. Three additional master horizons, however, have also been included: O, E, and R.

Since the 1930's, many sets of horizon symbols have been developed around the world. With the United States and many other countries adopting a horizon nomenclature similar to that produced by the Food and Agricultural Organization (FAO) of the United Nations in the early 1980's, the world has come closer to a universally accepted set of horizon symbols.

REFINING HORIZON NOMENCLATURE

There are three kinds of symbols used in soil horizon descriptions: capital letters, lowercase letters, and arabic numerals. The capital letters describe the master horizons, the lowercase letters depict some specific characteristic of a master horizon, and the numeral characterizes a further subdivision of a horizon or parent material layering. An example of a horizon symbol might be Bg1.

The "g" describes a "gleyed" horizon, or one where the iron has been removed during soil-forming processes or saturation with stagnant water in a high water table has preserved the reduced state of the iron, therefore a neutral or gray color. This letter also can describe a soil with red and gray specks from the oxidized and reduced forms of iron. A horizon that is more than 90 percent cemented is given the "m" designation. Horizons with abundant silica are described with a "q" and abundant sodium with an "n." If the horizon is very densely packed, which commonly occurs in silt-rich soils, the "x" designation is used. A "b" denotes a buried soil horizon. A "c" signifies that weathering has formed concretions, or nodules, in the soil.

Transitions between master horizons are commonly found in nature. Where the properties of both horizons are mixed in the same layer, both capital letters are used, with the first letter denoting the horizon whose properties dominate. The term BA horizon would be a transition between the A and B horizons, where the B horizon characteristics prevail. Where the horizon has distinct parts of both horizons, the two capital letters are separated by a slash mark (A/B, E/B, B/C).

ORGANIC SOIL HORIZONS

The O horizon is a master horizon in which there is an accumulation of mainly organic matter that overlies a mineral soil. It is dominated by fresh or partly decomposed organic litter such as twigs and needles, and many times the original form of most of the vegetative matter is visible to the naked eye. This horizon contains between 20 and 30 percent organic material, sometimes more, depending upon the clay content of the underlying mineral horizons. Three subdivisions of this master horizon are based on the amount of decomposition of the organic material and range from the Oi (least decomposition) to the Oe (intermediate decomposition) to the Oa (most decomposition).

The A horizon is a master horizon in which decomposed organic matter, called humus, is mixed with mineral sediments. The organic matter content is not great enough to be classified as an O horizon. This horizon is generally a surface horizon, except when located below an O horizon. Because of the presence of organic material in this horizon, this layer is as dark or darker than underlying horizons. The organic material is assumed to

be derived from the decomposition of plant and animal remains deposited on the surface. This layer is the zone of maximum biological activity. A subdivision is the Ap, or "plow horizon," where the A horizon has been disturbed by cultivation.

MINERAL SOIL HORIZONS

The E horizon is the master horizon commonly located below the O or A horizons in forest environments where leaching is dominant in the soil profile. Abundant water passing through the O horizon may become very acidic, and it leaches iron, aluminum, and organics from the A horizon as it passes to the lower part of the profile. The remaining mineral matrix is light-colored because of the color of the primary mineral grains of sand that remain. Sometimes the horizon may be almost pure white. This horizon for many years was called an A2 horizon but has been removed from the A horizon category because it is not a zone of organic material accumulation, which the A horizon concept implies.

The B horizon is the master zone of accumulation of materials that have been moved down by water from the O, A, and/or E horizons. These suspended materials, which have descended down the profile, include clay, iron, aluminum, and organic matter. The soil horizon shows little or no evidence of the original sediment or rock structure. Several kinds of B horizons are recognized depending upon the materials moved into them. The Bh horizon is an accumulation of abundant organic material and, therefore, is very black in color. In the Bs horizon, the deep red color depicts the accumulation of abundant iron and aluminum (the latter of which is colorless). The Bk horizon has an abundant amount of calcium carbonate and has a white color throughout. The Bt horizon has a large amount of clay that has moved down from the horizons above. Clay films are noted on the sediments, and the soil samples are very sticky. The By horizon is an accumulation of gypsum. The Bz horizon contains salts more soluble than gypsum. The Bw horizon is a weakly developed B horizon in which a reddish color has developed through weathering but few or no apparent materials have been moved into the horizon from above.

The C master horizon is the subsurface layer of partially weathered parent material. Soil-forming processes of the downward movement of particles and chemicals from the O, A, E, and B horizons have not affected this horizon, yet weathering of the sediments has slightly changed the color of the parent material. If the horizon material has the structure of the parent rock, yet is weathered enough to get a shovel through it, it is called a Cr horizon. This Cr horizon is also called a saprolite by geologists. The R master horizon is the unweathered, consolidated bedrock underlying the soil. This rock might be granite, basalt, or sandstone.

SOIL TAXONOMY

A second set of diagnostic horizon nomenclature developed in the latter half of the twentieth century in the United States to aid in soil classification using a new system called Soil Taxonomy. Information from diagnostic surface horizons—called epipedons—subsurface horizons, soil-moisture, and temperature regimes together are used to determine the soil classification. Soil Taxonomy is based mainly on measurable soil properties based in seventeen epipedons and subsurface horizons rather than on theories of soil formation, which form the basis for older classification systems. The classification system is very exact and results in exotic combinations of Latin and Greek names, such as Pachic Cryumbrepts and Natraqualfic Mazaquerts. Field observations must be supplemented by laboratory analyses of properties plus climatic data. Similar approaches using this type of horizon nomenclature have also been developed in Canada and by the FAO.

Four epipedons are described in Soil Taxonomy. The Mollic epipedon is a thick, dark-colored surface horizon rich in mineral nutrients. The Umbric epipedon is similar to the Mollic visually, but it is low in mineral nutrients. The Ochric epipedon is a thin surface horizon that does not meet the requirements for the Mollic or Umbric. The Histic epipedon is basically an O horizon.

The additional Soil Taxonomy diagnostic horizons are subsurface horizons. The Albic horizon is basically an E horizon, as it has been highly leached. An Argillic horizon is a B horizon that has been enriched significantly with clay from higher in the profile. The Natric horizon has a high abundance of salt in it. The Spodic horizon is a B horizon with abundant iron, aluminum, and/ or organics that have been moved from above. The Cambic horizon is a B horizon that is slightly enriched in red color but does not have an accumulation of products moved from above like the Spodic and Argillic horizons. The Oxic horizon is a B horizon with an abundance of highly weathered minerals. A Calcic horizon has an abundance of calcium carbonate in it. A Petrocalcic horizon is a cemented Calcic horizon. The Gypsic horizon has an abundance of gypsum in it. A Petrogypsic horizon is a cemented Gypsic horizon. A Duripan is cemented with silica. A Fragipan horizon is very dense and is many times formed in silt particles.

STUDYING A SOIL

The first step in the study of a soil is the production of a description of the soil profile. A scientist studying a particular soil exposes a soil profile either by digging a soil pit or by finding a road cut. First, the soil horizons are delineated (A, B, C, and so on). Next, each horizon is characterized as to its thickness, color, presence of films, and boundaries. Also recorded is the horizon texture (the relative amounts of sediment sizes). For example, a sandy loam soil has an abundance of sand with some silt in it. The structure of that horizon, or how the sediments are put together, is described also. Third, samples from each horizon are gathered to be analyzed in the laboratory. Last, an approximate classification is made based on the minimal information available in the soil pit.

The soil samples are returned to the laboratory for a chemical and physical characterization. The pH (degree of acidity) and amounts of the chemical nutrients are generally determined using pH meters and titrations. The exact texture is determined using sieves and water-settling columns. The percentage of organic matter is calculated using titration. Additional information can be obtained using an X-ray machine to determine the types of clays in the soil.

Once the field and laboratory data have been assembled, diagnostic horizons can be selected and the exact classification is determined using Soil Taxonomy. Eleven major classification groupings, or orders, are possible in this system. The Mollisols are soils with a Mollic epipedon. An Aridisol is a very dry soil. An Alfisol has an Argillic horizon and abundant nutrients, whereas the

Histosols are peaty soils that have thick histic epipedons. The peat here has been cut into blocks for use as fuel. (© William E. Ferguson)

of the soils that do not fit into any of the above groupings. Using the proper classification, the scientist can interpret the soil's age, future land use, the particular crops that can grow best on it, and its past history.

A FUNDAMENTAL TOOL

The soil profile is the basic descriptive tool that scientists who study soils have been using for years. This cross section constitutes the format from which scientists make their soil classifications. Without a proper soil profile description, one cannot do the soil classification that is needed before environmental interpretations can be made and soil maps can be produced. The symbols that are used to describe the horizons of the soil profile may have changed, but the basic A-B-C concept that was developed during the nineteenth century by the Russians has not changed much. Different systems of description of the soil profile and the resulting soil classification systems are present around the world, but, with time, they are becoming more similar.

In the United States, the Soil Taxonomy system has been officially used since 1975. In 1982, some modifications were made in describing the soil horizons of the profiles. This new Soil Taxonomy classification is a very exact hierarchical system similar to what biologists use to classify plants and animals. The eleven major orders are divided into suborders, great groups, subgroups, and finally

Ultisols are similar but have few nutrients. Histosols are peaty soils that have thick Histic epipedons. Spodosols are highly leached soils that have Spodic horizons. Oxisols are extremely weathered soils of the tropics that have Oxic horizons. Entisols are young soils that have no B horizon. Vertisols are high in clays that shrink and swell with seasonal moisture variation and exhibit extensive cracking in dry seasons. The Andisols are a newly formed order made of volcanically produced soils. The Inceptisol order is the catchall group for all the rest

An aridisol with alkaline crust left by evaporation of soil water. (© William E. Ferguson)

soil series. The subgroup name, such as "Typic Paleudult," is similar to the genus/species designation of a plant or animal. Soil Taxonomy is the most comprehensive soil classification system in the world and therefore is used more than any other approach. It is such an exact system that one can give a scientist who knows the taxonomy a subgroup name, and he or she can construct a complete soil profile description very close to the actual one in the field without even seeing the soil.

Scott F. Burns

BIBLIOGRAPHY

Birkeland, P. W. *Soils and Geomorphology.* 3d ed. New York: Oxford University Press, 1999. One of the best books written on soils from the perspective of a geologist, it gives many examples from around the world. The newly adopted soil profile description symbols are covered in the early chapters. A very elementary discussion of Soil Taxonomy in the early chapters. Suitable for university-level readers, but the early chapters are easily understood by high-school-level readers.

Buol, S. W., F. D. Hole, and R. J. McCracken. *Soil Genesis and Classification.* 4th ed. Ames: Iowa State University Press, 1997. This volume is an update of a classic soil science text. New to the edition is an update on the new soil horizon symbol nomenclature. Includes an in-depth discussion on soil development. Suitable for university-level readers. To ease understanding, however, they may want to read the first few chapters of the Birkeland or Singer books before reading this text.

Jenny, Hans. *Factors of Soil Formation: A System of Quantitative Pedology.* New York: Dover, 1994. This book describes the factors involved with variations in soils. Can be difficult to follow, but the author presents many important ideas. Suitable for university-level readers. Illustrations, index, and bibliography.

_____. *The Soil Resource: Origin and Behavior.* New York: Springer-Verlag, 1980. This rewrite of the classic book written by Jenny in 1941 describes variations in soils. Can be difficult to follow, but the author presents many important ideas. Suitable for university-level readers.

Paton, T. R. *Soils: A New Global View.* New Haven, Conn.: Yale University Press, 1995. Even though this book is aimed at readers at the university level, it is fairly easy to understand and has many good examples of soils and soil formation from around the world. Color illustrations, maps, bibliography, and index.

Singer, M. J., and D. N. Munns. *Soils.* New York: Macmillan, 1986. One of the most readable elementary soil science texts around, it includes well-explained soil profile description and classification. Suitable for high-school-level readers.

Soil Science Society of America. *Glossary of Soil Science Terms.* Madison, Wis.: Author, 1978. In what is basically a dictionary format, this volume of soil science terms was written by people in the soil science field.

Sparks, Donald S., ed. *Soil Physical Chemistry.* 2d ed. Boca Raton, Fla.: CRC Press, 1999. This widely used college text book is an excellent introduction to soil chemistry. It offers a clear look into the practices, procedures, and applications of the field and is easily understood by the careful reader.

Tan, Kim Howard. *Principles of Soil Chemistry.* 3d ed. New York: M. Dekker, 1998. Tan's text has been expanded in this third edition. Written as an introduction to soil chemistry, the book explains the processes and application of the science in an easy-to-understand manner accompanied by plenty of illustrations. Includes a twenty-five-page bibliography, as well as an index.

United States Department of Agriculture. Soil Survey Staff. *Soil Taxonomy*. Handbook 436. Washington, D.C.: Government Printing Office, 1975. Soil Taxonomy is the most used soil classification system in the world and the official one used in the United States. The text is difficult to understand at times, so it is recommended that the books listed above be read first. An abundant source of information.

SOIL TYPES

Classification of soils is so important that taking soil surveys is one of the primary tasks of the U.S. Department of Agriculture. Knowledge of soil types indicates which crops are best suited for a particular soil. Soil classification also provides farmers with the most appropriate methods for preventing soil erosion.

PRINCIPAL TERMS

A HORIZON: the surface soil layer; also known as topsoil

B HORIZON: the soil layer just beneath the topsoil

BELT: a geographic region that is distinctive in some way

LEACH: to remove, or be removed from, by the action of a percolating liquid

MONOLITH: a large soil profile created by dovetailing monolith containers together and combining small, individual profiles

PEDOLOGIST: a soil scientist

POLYPEDONS: bodies of individual kinds of soil in a geographic area

PROFILE: a representative soil sample containing different layers of soil

ZONE: an individual soil group within a horizon

THREE BASIC TYPES

Because soils consist of particles varying greatly in size and shape, specific terms are required to give some indication of their physical properties. Soils can be classified according to different characteristics. The simplest way to classify soils is by their texture. Three broad yet fundamental groups of soils are now recognized by pedologists: sands, clays, and loams. All of the additional class names that have been devised through years of soil study and classification are based on these three groups. The sand group includes all soils of which sand makes up more than 70 percent of the material by weight. In contrast to the heavier groups of soils, which are stickier and more clayey by nature, this group is characteristically sandy in texture. Two specific classes within the sand group are recognized: sand and loamy soil. Soils included in the clay group must consist of at least 35 percent clay and in most cases not less than 40 percent. The names are "sandy clay," "silty clay," or simply "clay," which is the commonest of all. Sandy clays often contain more sand than clay; similarly, the silt content of silty clays usually exceeds that of the clay fraction itself. The loam group, which is the most important for agriculture, is more difficult to explain. An ideal loam is a combination of sand, silt, and clay particles. It also exhibits light and heavy properties in about equal proportions.

In most cases, the quantities of sand, silt, or clay present require a modified class name. For example, a loam in which sand is present is classified as a sandy loam.

Qualifying attributes such as stone, gravel, and various grades of sand must be considered when placing some soils in classes. For example, one refers to stony-clay loams, gravelly sands, or fine-sandy loams. Pedologists also classify soils according to their composition. A great soil group consists of many soil types whose profiles have major features in common. Every soil type has the same numbers and kinds of definitive horizons, but they need not be expressed in the same profile to the same degree. For example, the Fayette, Dubuque, Downs, and Quandahl soils are all members of a single great soil group.

Geographic belts marked by certain combinations of great soil groups can be shown on maps of small scale. A schematic soil map of the world consists of six broad belts. One belt consists of rough landscapes, such as mountains, in which many of the soils are stony, shallow, or both. The patterns of soil types in these areas are especially complex. The other five broad belts have simpler patterns, but each includes a number of great soil groups. The soil types within a single farm, for example, commonly represent two or more great soil groups. The thousands of soil types in the United

States can be classified in about forty great soil groups. The soil groups in the world as a whole, however, number as many as sixty or more. Collectively, the groups have wide ranges in their characteristics. They also vary greatly in such qualities as fertility, ability to hold available moisture, and susceptibility to erosion.

PODZOLIC AND LATOSOLIC SOILS

Podzolic soils dominate a broad belt in the higher latitudes of the Northern Hemisphere and some smaller areas in the southern half of the world. They include the brown podzolic soils, gray-brown podzolic soils, and gray wooded soils. These groups were formed under forest vegetation in humid, temperate climates. The B horizons of these soils vary. Some podzolic soils have B horizons that are composed primarily of humus, sesquioxides, or both; others have B horizons that are mainly accumulations of clay. Podzolic soils are more strongly weathered and leached than many of the other soil groups. Because they are usually acidic, low in bases such as calcium, and low in organic matter, levels of fertility are moderate to low. As a group, however, these soils are responsive to scientific management.

The equatorial belts of Africa and South America are dominated by latosolic soils. These soils are also dominant in the southeastern parts of Asia and North America, in northeastern Australia, and in the larger islands of the western Pacific Ocean. Latosolic soils include the great soil groups known as laterites, reddish-brown lateritic soils, yellowish-brown lateritic soils, red-yellow podzolic soils, and several kinds of latosols. Red-yellow podzolic soils are so named because they share features with both the latosolic and podzolic groups, but they are more closely related to the latosols. Latosolic soils have been formed under forest and savanna vegetation in tropical and subtropical and humid to fairly dry climates. Although these soils do not extend into arid regions, they may be found in alternately wet and dry areas with low rainfall. The latosols are the most strongly weathered soils in the world. As a result of the large amounts of iron oxides formed through intense weathering, the profiles commonly display red and yellow colors. Except for a darkened surface layer, most of these soils lack distinct horizons; therefore, the profile may remain unchanged for many feet. Although supplies of plant nutrients are usually low, the capacity to fix phosphorus in unavailable forms is high. Most of these soils are resistant to erosion and are easily penetrated by plant roots. In fact, some tree roots in southeastern Brazil have extended more than 60 feet below the surface layer. The moisture capacities are mostly moderate to high in latosolic soils, although they are low in some. Without the benefit of modern science and industry, productivity is normally low.

CHERNOZEMIC AND DESERTIC SOILS

Chernozemic soils have been formed under prairie or grass vegetation in humid to semiarid and temperate to tropical climates. Although these soils are most extensive in temperate zones, some large areas exist in the Tropics. Chernozemic soils include the great soil groups known as chernozems, prairie soils (or brunizems), reddish prairie soils, chestnut soils, and reddish chestnut soils in temperate regions. In tropical and sub-

Podzolic soil is a highly acidic soil. (© William E. Ferguson)

tropical regions, these soils are often known as black cotton soils, grumusols, regurs, and dark clays. The A horizons of chernozemic soils are normally very thick, fertile, and slightly weathered. The B horizons are usually much less distinct. The A horizons of chernozemic soils are high in organic matter and nitrogen in temperate zones but not in tropical and subtropical zones. These soils are less acidic and higher in bases and in plant nutrients than are podzolic or latosolic soils. Therefore, they are among the most fertile soils in the world, producing about 90 percent of the grain. Within the United States, the chernozemic soils form the heart of the Corn Belt and wheat-producing regions. Because the soils extend from the margins of humid into semiarid zones, production varies with seasonal weather. Chernozemic soils in tropical and subtropical zones are not suitable for cultivation because they are high in clay, are plastic, and are subject to great shrinking and swelling. Handling these soils is difficult.

Desertic soils have been formed under mixed shrub and grass vegetation or under shrubs in arid climates ranging from hot to cold. These soils are commonly found in the great deserts of Africa, Asia, and Australia and in the smaller ones of North America and South America. They include the great soil groups known as desert soils, red desert soils, sierozems, brown soils, and reddish-brown soils. Desertic soils have been slightly weathered and leached because of the shortage of moisture. The lack of adequate rainfall also limits plant growth; consequently, these soils are low in organic matter and nitrogen. Limited rainfall is also reflected in the shallow profiles. Most of the horizons of these soils are faint. The A horizon is lighter in color than the B horizon because it has commonly lost carbohydrates and perhaps some bases and clay. The slightly darker B horizon has some accumulation of clay but is very low in organic matter. Levels of nutrients other than nitrogen are usually moderate to high in these soils. Available moisture capacities vary, depending on the thickness of the profiles and textures of horizons. Land management can have a tremendous impact on these soils.

SOIL SAMPLES

The simplest and commonest method of determining the class name of a soil is by its feel. In fact, rubbing the soil between the thumb and fingers can probably tell a scientist as much about a soil as other superficial means can. To estimate plasticity more accurately, it is helpful to wet the sample. This technique causes the soil to "slick out," thereby giving the observer a good idea of the amount of clay present. Much can be learned simply by the way the soil particles feel: Sand particles are gritty; silt has a floury or talcum-powder feel when dry and is only moderately plastic or sticky when wet; silt and clay often take the form of clods.

A surveyor can learn more about the nature of soils by using simple tools. The only tool on which the fieldworker can absolutely rely is the spade. Although many spades tend to compact the soil layers somewhat, the gouge or cheese-sampler types of spade obtain samples of near perfection. Good samples that reveal the differentiations of horizons can also be taken with a butcher's knife. A horizontal cut is made between the horizons, and then two vertical cuts are made in the face. Individual elements can then be picked out and used as representative samples of the zones.

A permanent record of the soil character can be taken by means of a monolith. A monolith container is made of steel and is 66 centimeters long. Since these containers are made to dovetail into each other, soil profiles can be taken at any depth. After the sample has been taken, a fixing solution is applied to preserve it. As soon as the sample has been "set," it can be erected in a vertical position so that the soil layers can be examined in their natural order. Individual soil fragments taken by hand can also be preserved. The specimen is first air-dried and then oven-dried at 50 degrees Celsius until it is at the same temperature throughout. While it is still warm, it is immersed in a bath containing a solution of 3 percent cellulose acetate in acetone until bubbles cease to appear. It is then allowed to drain and dry on a sieve. Specimens fixed in this manner may be handled without breaking down.

A field map illustrating the soil types that are present in a large area can also be made by taking samples. A surveyor bores a hole with an auger until sure that a representative sample of the site has been acquired. After describing the profile in a notebook and then marking the site with a triangle and number, the surveyor moves to a neighboring site and continues taking samples at regular intervals.

Aerial Photographs and Diagrams

Much can also be learned about soil characteristics from the "layout" of fields in aerial photographs. These are especially useful for examining soils in more inaccessible areas, such as vast forests. Aerial photographs directly show only the external or environmental characteristics of soil. With experience, however, enough may be deduced from the study of these features so that a very shrewd idea may be obtained about the nature of what lies underneath as well. Aerial photographs can generally be studied for the recognition of soils without the aid of sophisticated photographic apparatus. Quality prints at a scale of about 15 centimeters to 1.5 kilometers usually give good enough definition for general purposes. Later, simple stereoscopic study of "stereo pairs" of contact prints provide three-dimensional detail.

Probably the most accurate method of determining the class of soils has been devised by the U.S. Department of Agriculture. This method involves the creation of a diagram that can be obtained through mechanical analysis. The diagram reveals that a soil is a mixture of different sizes of particles and that a close correlation exists between particle size distribution and the properties of soils. A surveyor can readily check the accuracy of class designations once mechanical analyses of the field soils under scrutiny are available.

The U.S. Conservation Service groups soils into eight land-capability classes. Soils in classes I-III are generally suitable for cultivation. Soils in classes IV-VIII are severely limited in their usefulness for farming.

Farming Applications

Knowledge of soil types is essential for effective farming. Farming methods have to be tailored according to the type of soil that is being cultivated. As people from the eastern United States moved to the prairies and the plains, for example, the farming practices that they brought with them had to change according to the different types of soils that they encountered: chernozem, prairie, and chestnut soils. Even today, farming practices are still undergoing change because no one knows just what type of farming is best suited to the chestnut and brown soils.

Failure to adapt farming practices to the soil type can have a devastating impact on a community. A decline of productivity often occurs because of the inability of people in some regions to establish a satisfactory type of farming. For example, people attempt to force a pattern of use on soils that are not suited to it, as on the reddish-chestnut soils of the southern plains in the United States. Soil erosion and soil depletion, which often occur because of inappropriate farming practices, are symptoms of bad relationships between people and soils. It can be said, then, that not only do people live differently on different soils, but they also must live differently to live at all. This fact may be illustrated by civilizations that largely developed on one particular type of soil: the Egyptian on the alluvial soils along the Nile; the Greek-Roman on the terra rossa soils around the Mediterranean Sea; the Arabian on the soils of the deserts; and the Western on the gray-brown podzolic of western Europe, the northeastern United States, eastern Canada, and parts of Australia, New Zealand, and South Africa. Not only were the types of crops upon which these cultures relied determined in great part by the soil type but so were important artifacts, such as the kinds of pottery that they used.

Alan Brown

Cross-References

BIBLIOGRAPHY

Bear, Firman E. *Earth: The Stuff of Life*. Norman: University of Oklahoma Press, 1962. This introduction to soil science explains how soil came into being, but it is primarily a treatise on soils in relation to the growth of plants, animals, and humans as well as a plea for the sensible conservation of soil. Recommended for the general reader.

Birkeland, P. W. *Soils and Geomorphology*. 3d ed. New York: Oxford University Press, 1999. One of the best books written on soils from the perspective of a geologist, it gives many examples from around the world. The newly adopted soil profile description symbols are covered in the early chapters. Includes a very elementary discussion of soil taxonomy in the early chapters. Suitable for university-level readers, but the early chapters are easily understood by high-school-level readers.

Buckman, Harry O., and Nyle C. Brady. *The Natures and Properties of Soils*. 7th ed. New York: Macmillan, 1969. The most complete book on soil science available. Covers such facets of soil as soil composition, formation, soil reactions, and soil's effect on the world's food supply. Contains ample maps, diagrams, and photographs as well as a complete glossary. For the college-level student.

Buol, S. W., F. D. Hole, and R. J. McCracken. *Soil Genesis and Classification*. 4th ed. Ames: Iowa State University Press, 1997. This volume is an update of a classic soil science text. New to the edition is an update on the new soil horizon symbol nomenclature. Includes an in-depth discussion on soil development. Suitable for university-level readers.

Clarke, G. R. *The Study of Soil in the Field*. 5th ed. Oxford, England: Clarendon Press, 1971. A short (144-page) but comprehensive introduction to soil science, which includes such areas as soil site characteristics, soil sampling, soil surveys, and soil evaluation. The charts and diagrams help to clarify some of the esoteric terminology. Also includes a short section of references. For college-level students.

Kellogg, Charles E. *Our Garden Soils*. New York: Macmillan, 1952. Although the book is intended as a manual for gardeners, the first chapter, entitled "Natural Soils," explains the composition and characteristics of various types of soil. Also introduces the general reader to the terminology that scientists use to describe soils.

_____. *The Soils That Support Us*. New York: Macmillan, 1961. A comprehensive and easy-to-read introduction to soil science, with an emphasis on the relationship between people and soil. Includes photographs, charts, maps, and several useful appendices. For high school and college students.

Paton, T. R. *Soils: A New Global View*. New Haven, Conn.: Yale University Press, 1995. Even though this book is aimed at readers at the university level, it is fairly easy to understand and has many good examples of soils and soil formation from around the world. Color illustrations, maps, bibliography, and index.

Simonson, Roy W. "What Soils Are." In *Soil*, edited by the U.S. Department of Agriculture. Washington, D.C.: Government Printing Office, 1957. This fourteen-page article is a somewhat technical but very comprehensive explanation of soil formation and composition. Includes a chart and diagram. For the college-level student.

Sparks, Donald S., ed. *Soil Physical Chemistry*. 2d ed. Boca Raton, Fla.: CRC Press, 1999. This widely used college text book is an excellent introduction to soil chemistry. It offers a clear look into the practices, procedures, and applications of the field and is easily understood by the careful reader.

Thompson, Louis M. *Soils and Soil Fertility*. 2d ed. New York: McGraw-Hill, 1957. A textbook for an introductory college course in soils for students of agriculture. Covers every aspect of soil, including the physical properties of soil and soil conservation. Highly readable and well supplemented with charts, illustrations, and diagrams, although the reference sections are probably too dated to be of much use.

6
ECONOMIC RESOURCES

ALUMINUM DEPOSITS

Aluminum deposits provide humans with a lightweight, electrically conductive metal having such diverse uses as food wrap, engine blocks, and wire. The first major study of aluminum deposits attributed them to precipitation from warm acidic fluids, but most subsequent researchers have preferred ordinary soil-forming processes as the mechanism for aluminum concentration to ore grade.

PRINCIPAL TERMS

BAUXITE ORE: bauxite that can be mined for a profit; generally, it must contain at least 25 percent aluminum and not more than 15 percent iron or more than 3 percent silicon

BAUXITE: a deposit which is rich in aluminum and oxygen but poor in most of the other elements that are abundant in the Earth's crust, such as silicon, sodium, potassium, magnesium, and calcium

CHEMICAL SEDIMENTARY ROCK: rock that accumulated as sediment that chemically precipitated onto the bottom of a water body, typically onto the ocean floor

HYDROXIDE: a mineral that contains hydroxyl (oxygen plus hydrogen), for example, gibbsite (aluminum plus hydroxyl); an oxyhydroxide mineral contains oxygen in addition to hydroxide, for example, boehmite (aluminum-oxygen-hydroxide)

IGNEOUS ROCK: rock (for example, granite) that forms by cooling of molten matter

KARST: a type of topography developed above beds of limestone as a result of partial dissolution of the limestone by through-flowing water

SHALE: sedimentary rock with sediment grains less than one-sixteenth of a millimeter in diameter; it exhibits thin layers like pages in a book and usually is dark in color

SILICATE MINERAL: a mineral that contains both silicon and oxygen, for example, quartz

SYENITE: a coarse-grained igneous rock that resembles granite and is rich in aluminum, sodium, and potassium

MODELS EXPLAINING DEPOSITS

Aluminum is the third most abundant element in the crust of the Earth, constituting about 8 percent. The two more abundant elements are oxygen (48 percent) and silicon (27 percent). More than 95 percent of the aluminum in the crust occurs within silicate minerals, that is, minerals that contain both silicon and oxygen. In contrast, minable aluminum deposits (bauxite ores) are composed mostly of oxides and hydroxides of aluminum that lack silicon. Typical minerals in bauxite include gibbsite (an aluminum hydroxide) and boehmite and diaspore (both aluminum oxyhydroxides). Often, these minerals are collectively called aluminum hydroxides.

The first comprehensive study of aluminum deposits was published in 1871. Aluminum concentration was attributed to precipitation from fluids that had dissolved aluminum because they had acquired corrosive volatiles such as carbon dioxide.

Hot volatiles were envisioned to have risen from deep in the Earth. The mixing of hot volatiles with water produced a hydrothermal solution that dissolved aluminum from ordinary rocks. The solution then precipitated the aluminum upon removal (neutralization) of the acidity, which could have occurred by reaction of the solution with limestone.

Most twentieth century researchers have rejected the original hydrothermal model for aluminum deposits. As an alternative, aluminum deposits have been attributed to intensive weathering of aluminum-rich sediment or rock, generally in the humid tropics. If this mechanism were viable, aluminum ore deposits would be forming today in the humid tropics, but no such ore deposits have been found.

CHALLENGES TO WEATHERING MODEL

The popularity of the weathering hypothesis for aluminum-rich rock has been based on the observation that silicon is slightly more soluble than aluminum when silicate minerals are attacked by the organic acids that drain from decaying plant matter. This difference could become accentuated under conditions of intensive weathering. Such intensive weathering, however, would require particularly abundant plant life; the plant life, in turn, would require abundant nutrients, including potassium. Potassium is one of the three ingredients listed on any fertilizer bag. (For example, a common "garden-variety" fertilizer is labeled 8-8-8 because it contains 8 percent each of nitrogen, phosphorus, and po-

Bauxite ore, a principal source of aluminum. (© William E. Ferguson)

tassium.) Given the high solubility of potassium, this element tends to be completely removed during intensive weathering. Potassium may be retained within leaf litter on flat land, but several bauxite deposits have accumulated on hillsides.

Although modern soil-forming processes are not producing voluminous bauxite deposits, they do concentrate aluminum. The concentration of aluminum in tropical soils mostly results in the crystallization of kaolinite. Kaolinite is a clay mineral that consists of silicon, aluminum, oxygen, and hydrogen. Globally, kaolinite is one of the most abundant soil minerals within forty latitudinal degrees of the Earth's equator. In tropical soils, kaolinite is commonly accompanied by minor amounts of aluminum hydroxides; there is no significant portion of the Earth's present (weathered) surface in which aluminum hydroxides predominate over silicates. In contrast, aluminum hydroxides always predominate over silicates in bauxite ore. Kaolinite predominates only where bauxite ore grades vertically or laterally to other rock types.

The youngest major aluminum deposit overlies aluminum-poor sand above a major fault that forms the northern boundary of the Amazon River Basin. There is no obvious aluminum-rich material that could have weathered to form this deposit, but the underlying fault could have channeled rising acidic fluids. The lack of an obvious aluminum-rich parent similarly plagues the at-

tempted application of the tropical-weathering model to most other aluminum deposits. Ancient aluminum deposits commonly overlie aluminum-poor limestone.

SUPPORT FOR HYDROTHERMAL MODEL

The original hydrothermal model is preferred herein for the origin of aluminum deposits (bauxite ore deposits). Geologists generally agree that most ores of other metals similarly have precipitated from warm acidic (hydrothermal) solutions. Aluminum-rich hydrothermal fluids presumably have reached the Earth's surface where they locally have coated the land with aluminum minerals in a blanket resembling soil. The resulting deposits, however, do not display the horizons that characterize soil: the A, B, and C soil horizons. Moreover, blanket-type aluminum deposits either have a thin reaction zone at the bottom or are separated from underlying rock by aluminum-poor sediments. Layering does take place in some blanket-type deposits, but successively deeper layers are not consistently richer in aluminum, as would be expected if the layers represent soil horizons. Layers in aluminum deposits are interpreted to record episodes of outpouring of aluminum-rich fluids.

The hydrothermal hypothesis does not exclude the influence of weathering on blanket-type aluminum deposits. Even if originally deposited by the outpouring groundwater, blanket-type deposits would become subjected to weathering like everything else at the Earth's surface. This weathering, moreover, could enhance ore grade, given the insolubility of aluminum during weathering. The hypothetical hydrothermal fluid would have contained many soluble elements, including uranium, that would be removed by the infiltration of rainwater.

Aluminum deposits that have not accumulated as blankets on the land mostly have accumulated within caverns in limestone. An area with abundant limestone caverns is called karst. Karst bauxite is attributable to the same genetic process as advocated for blanket-type bauxite: precipitation from acidic fluids. Acidic fluids rapidly dissolve any limestone. Detailed studies have shown that the cavern-dissolution process in karst limestone commonly accompanied precipitation of aluminum minerals in bauxite. Any acidic solution may become neutralized (non-acidic) by dissolution of limestone. Neutralization would induce precipitation of aluminum because aluminum cannot remain in solution once an acidic solution becomes neutralized.

CONCENTRATIONS OF DEPOSITS

Despite the disagreement about hydrothermal versus weathering processes, all bauxite researchers agree that the aluminum occurred within aluminous silicate minerals prior to separation of the aluminum from silicon. Both silicon and aluminum are ubiquitously distributed throughout the Earth's crust. Among the abundant chemical elements in the crust, aluminum is one of the most evenly distributed in all types of igneous rock, averaging about 8.5 percent aluminum (15.6 percent aluminum-oxygen component). In contrast, iron in igneous rocks may range from negligible amounts to more than 15 percent. One of the most aluminum-rich igneous rock types is syenite, but even in syenite the aluminum content rarely exceeds 10 percent.

Among ordinary sedimentary rocks, aluminum is most concentrated in shale. Shale constitutes about half of all sedimentary rocks. The percentage of aluminum in the two other common types of sedimentary rock, sandstone and limestone, generally is less than half as much as that of typical shale. Average shale contains a slightly higher percentage of aluminum than does average igneous

rock, which contains an average of 8.5 percent aluminum. Unfractured shale generally is impermeable to groundwater, however, so it is an unlikely source of aluminum for the hypothesized acidic groundwater. Sandstone is much more permeable and contains scattered grains of the most common of all minerals in the Earth's crust, the aluminum-bearing mineral feldspar.

Of all the common rock types within the crust, limestone is one of the poorest in aluminum. Nevertheless, bauxite is as closely associated with limestone as with any other rock type. Limestone-hosted bauxite is abundant around the Mediterranean Sea and in Jamaica and the western Pacific islands. These karst-bauxite deposits do share one attribute: They generally formed at times when there was enough subsidence or melting of the adjacent crust to account for production of volatiles at depth.

AGE OF DEPOSITS

Unlike most types of ore deposits, aluminum deposits accumulate on exposed land or within open caverns that are close to the Earth's surface. Most other types of ore accumulate on the sea floor or deep within the continental crust. As a result, aluminum deposits become eroded more readily than other types of ore. The average age of aluminum deposits, therefore, is less than that of most other ore deposits. Nevertheless, some of the largest aluminum deposits are tens of millions of years old, having formed in the first half of the Cenozoic era. Other ore types are typically even older. The average age of iron deposits is about 2 billion years. By comparison, the age of planet Earth is about 4.6 billion years. Being younger than most other metallic ore deposits, aluminum ores characteristically are more porous and crumbly than other types of ore. This quality facilitates mining because less effort is required to loosen the rock. In many deposits, bulldozers or huge shovels may extract the bauxite ore without any loosening by explosives.

There appears to be a continuum among ore deposits of aluminum, phosphorus, and iron. Portions of some bauxite deposits contain more than 10 percent iron; portions of some phosphorus deposits (phosphorites) contain more aluminum than phosphorus. Aluminum concentrations in phosphorites commonly have been attributed to postdepositional weathering of the phosphorite. The alternative hypothesis is the initial precipitation of aluminum and phosphorus from a common warm fluid. The vast majority of iron deposits that have formed during the past 600 million years contain a significant percentage of chemically precipitated aluminum. Collectively, iron deposits younger than 600 million years probably contain more chemically precipitated aluminum than do bauxite ore deposits. None of this iron-bound aluminum, however, ever has been extracted commercially.

ALUMINUM MINING

Aluminum deposits are studied mostly by companies that assess the potential profit from mining. The prime factors regarding profitability are the chemical composition of the bauxite, nature of the aluminum minerals, volume of ore, depth of cover (if any) by aluminum-poor material, distance from the nearest port, labor cost, and political stability of the host country. Aluminum must exceed about 25 percent by weight and other impurities must be limited, particularly silicon. The minimum aluminum content for mining depends upon the dominant type of aluminum mineral because the amount of energy required to separate pure aluminum varies among the aluminum minerals. Mining of aluminum deposits generally occurs by digging downward with enormous shovels that can move several cubic meters in each shovelful. An aluminum ore deposit, therefore, must be at or near the Earth's surface. Although aluminum deposits generally are formed at or near the Earth's surface, some have become deeply buried under other sediment.

The volume of an aluminum ore deposit is determined by drilling small holes on a fixed grid over the deposit. At least several million tons of bauxite ore are needed to justify mining because each enormous shovel costs several million dollars. The distance of the ore deposit to the nearest port is crucial because a large volume of bauxite ore must be shipped, and the cost of transportation may exceed the cost of mining if the distance is greater than a few tens of kilometers.

The cost of living generally is low in humid tropical areas; mining, therefore, is particularly concentrated in these areas. The average cost of labor in aluminum mining is as low as the average cost of labor for any other type of mining. Unfortunately, political stability commonly is inversely proportional to the cost of labor. Thus, the political stability of the host country may be a major concern to potential investors.

ALUMINUM PRICING

Aluminum is the most widely used metal among those metals that essentially were unused prior to the twentieth century. Aluminum could not be processed readily into pure metal by melting, so refining of aluminum was delayed until electric energy became available.

Modern industrial society takes for granted that aluminum will be supplied continuously. A smaller proportion of aluminum, however, is mined in industrial countries than any other metal that is widely used by industry. This leaves the industrial countries potentially vulnerable to price-fixing by a cartel of Third World producers. Price-fixing of aluminum is unlikely in the foreseeable future because Third World producers are scattered around the globe, and there has been little movement toward collective action. Another factor that would counteract any sharp increase in price is that special plastics have been found to exhibit many of the desirable properties of aluminum. As long as petroleum is available for the plastics industry, substitution for aluminum potentially regulates prices.

Aluminum prices partly depend on the cost of electric power. Immediately following World War II, aluminum companies established hydroelectric power plants in isolated parts of the globe because aluminum refining requires substantial electric power, and these remote localities were too far from major cities to use their hydroelectric power. This situation has changed with the expansion of cities in previously sparsely populated areas and with the development of techniques for long-distance transmission of electric power. Hydroelectric power plants in remote areas now have the option of selling electricity to distant cities; thus, the aluminum-refining companies are facing competition that previously did not exist.

Michael M. Kimberley

CROSS-REFERENCES

BIBLIOGRAPHY

Aleva, G. J. J. "Essential Differences Between the Bauxite Deposits Along the Southern and Northern Edges of the Guiana Shield, South America." *Economic Geology* 76 (1981): 1142-1152. The world's youngest major aluminum deposit, along the Amazon River, is described and compared to older deposits in northeastern South America.

Augustithis, S. S., ed. *Leaching and Diffusion in Rocks and Their Weathering Products*. Athens: Theophrastus Publications, 1983. This volume was produced with the cooperation of the International Committee for the Study of Bauxites, Alumina, Aluminium. Many of the twenty-six papers therefore pertain to the concentration of aluminum by weathering. Aluminum deposits are described from Europe, India, China, and Africa.

Baardossy, Gyeorgy, and G. J. J. Aleva. *Lateritic Bauxites*. New York: Elsevier, 1990. An examination of the aluminum mining industry, the economics associated with aluminum, and the geochemical make-up and properties of lateritic bauxites. Color illustrations.

Brown, Martin, and Bruce McKern. *Aluminum, Copper, and Steel in Developing Countries*. Washington, D.C.: OECD Publications and Information Center, 1987. The profitability of the mining of aluminum deposits changes with the world economy. This book describes the economics of aluminum and recommends how aluminum producers (mostly Third World countries) should interact with aluminum refiners in industrial countries.

Evans, Anthony M. *An Introduction to Economic Geology and Its Environmental Impact*. Malden, Mass.: Blackwell Science, 1997. This college textbook provides a wonderful introduction to mines and mining practices, minerals, ores, and the economic opportunities and policies surrounding these resources. Illustrations, maps, index, and bibliography.

Frakes, L. A. *Climates Throughout Geologic Time*. New York: Elsevier, 1979. This book assumes that aluminum deposits form by intensive weathering under humid tropical conditions. The distribution of aluminum deposits is correspondingly interpreted (along with other indicators) to record the history of climatic change on the Earth.

LeLong, F., Y. Tardy, G. Grandin, J. J. Trescases, and B. Boulange. "Pedogenesis: Chemical Weathering and Processes of Formation of Some Supergene Ore Deposits." In *Handbook of Strata-Bound and Stratiform Ore Deposits*, edited by Karl H. Wolf. New York: Elsevier, 1976. This paper provides a thorough comparison of aluminum deposits to other earth-surface ore deposits. The chemical reactions involved in the weathering of aluminum-rich silicate minerals are reviewed, and aluminum deposits are compared to earth-surface (residual weathering) deposits of nickel and manganese.

Ogura, Y., ed. "Proceedings of an International Seminar on Laterite." *Chemical Geology* 60 (1987): 1-396. This collection of forty papers provides an up-to-date overview of the weathering processes that affect aluminum-bearing silicate minerals. A paper by M. P. Tole (pp. 95-100) describes the "Thermodynamic and Kinetic Aspects of Formation of Bauxites." This paper lists chemical reactions which characterize the weathering of aluminum-bearing silicate minerals.

Sehnke, Errol D., ed. *Bauxite Mines Worldwide*. Washington, D.C.: U.S. Department of the Interior, U.S. Bureau of Mines, 1995. This brief government publication takes a look at the state of aluminum mines and the aluminum mining industry in the United States. Includes bibliographical references.

Sposito, Garrison, ed. *The Environmental Chemistry of Aluminum.* Boca Raton, Fla.: CRC Press, 1989. This college-level text looks at the chemical make-up of aluminum and the changes it undergoes due to environmental factors such as water and soil erosion. Illustrations, index, and bibliography.

Stuckey, John A. *Vertical Integration and Joint Ventures in the Aluminum Industry.* Cambridge, Mass.: Harvard University Press, 1983. This book reveals that the aluminum industry in the non-Communist world is dominated by only a few companies. Prior to World War II, a virtual monopoly was maintained by the Aluminum Company of America (Alcoa). Two firms were created after the war: Reynolds Metals Company and Kaiser Aluminum and Chemical Corporation. Joint ventures among these three and smaller companies have helped to stabilize the cost of aluminum production, hence consumer prices.

Valeton, Ida. *Bauxites.* New York: Elsevier, 1972. This book remains a prime reference for the geology of aluminum deposits. Few other books subsequently have been published on the topic, so it remains the most widely quoted reference.

Wilson, R. C., ed. *Residual Deposits: Surface Related Weathering Processes and Materials.* London: Blackwell Scientific Publications, 1983. This volume describes various types of residual deposits; that is, deposits of insoluble elements that remain when soluble elements are preferentially removed by weathering. The debate about the origin of aluminum deposits hinges on the issue of a residual versus hydrothermal origin. A paper by Ida Valeton in this book (pp. 77-90) describes the "Paleoenvironment of Lateritic Bauxites with Vertical and Lateral Differentiation," in which some blanket-type aluminum deposits grade downward and sideways into other rock types.

BUILDING STONE

Building stone is any naturally occurring stone that is used for building construction. The three rock types (igneous, metamorphic, and sedimentary) are all utilized in building stone. The physical characteristics of each type of rock, such as hardness, color, and texture, determine how the stone is used.

PRINCIPAL TERMS

GRANITE: an igneous rock that is known for its hardness and durability; in modern times, it has been used on the exterior of buildings, as it is able to resist the corrosive atmospheres of urban areas

IGNEOUS: rock that was formed from a molten material originating near the base of the Earth's crust

LIMESTONE: a sedimentary rock that can be easily shaped and carved; currently it has gained acceptance as a thin veneer

MARBLE: a metamorphic rock that has been used since Grecian times as a preferred building stone; it is known for its ability to be carved, sculptured, and polished

METAMORPHIC: rock that was formed by heat and pressure; tectonic, or mountain-building, forces of the Earth's crust create and alter the mineral composition and texture of the original rock material

SANDSTONE: a sedimentary rock that is known for its durability to resist abrasive wear; it is likely to be used for paving stone

SEDIMENTARY: most commonly, rock that was formed by marine sediments in an ocean basin; it usually shows depositional features and may include fossils

SLATE: a metamorphic rock that has a unique ability to be split into thin sheets; some slates are resistant to weathering and are thus good for exterior use

VARIOUS FORMS

Building stone is any kind of rock that has supplied humankind with the material to erect monuments and edifices throughout history. The pyramids of Egypt, the temples of Greece, the skyscrapers of the modern world have all utilized various forms of building stone. The ancients used blocks of stone stacked like building blocks, whereas the modern builder uses a thin veneer of stone anchored to the exterior of a building frame. Special physical properties of the various stone materials lend themselves to these various usages. Because of these special physical properties, only a small percentage of the Earth's rock material can be classified and utilized as building stone. Granite, an igneous rock that forms the core of the continents, is a good example. Granite is abundant, but because of mineralogic variations or structural weakness, only a small percentage of the granite areas will yield quarry blocks of suitable dimensions for building stone.

Building stone is quarried, or excavated, in most countries and, from its point of origin, may be shipped around the world. Quite commonly, granite blocks from Brazil are shipped to Canada for sawing and finishing and then shipped to a construction site in one of the cities in the United States. Similarly, building stone from the United States has been shipped to Italy for sawing and finishing and then shipped to England for installation on buildings. The average quarry block weighs 15 to 20 tons. These quarry blocks are sawed into slabs that vary in thickness from a half inch to four inches. Whether the building stone is a limestone, marble, granite, slate, or a sandstone, the quarry block must be solid enough to yield slabs of competent rock. These slabs are further cut on diamond saws to the desired shape and dimensions and then finished to the requirements of the architect or owner.

DESIGN CRITERIA

Buildings constructed before or around the turn of the century were built with cubic pieces

6 inches thick or greater. Cubic stone forms both the supporting structure for the walls and the exterior protection to the building. Most buildings built since the 1950's, however, require that the building stone be cut into thin panels less than 3 inches thick and fastened to the exterior of the steel supporting structures. This later use of stone demands that more engineering and design criteria be used to establish acceptable versus unacceptable building stone. The processes by which the architect, engineer, and geologist select and qualify a particular stone for usage help to determine the aesthetic and physical characteristics of the stone. The architect usually establishes the design criteria for the size and shape of the stone. He or she looks at the stone from the point of view of aesthetics—for example, focusing on its color and texture. The engineer and geologist then determine the suitability of the selected stone for realizing the design plan.

Geologists and engineers who are involved with quarrying and fabricating building stone look for deposits of stone that are uniform in texture, color, and structural integrity. The exploration of new areas is accomplished by researching and reviewing existing geological maps, aerial photographs, and geological libraries. Ground sleuthing or field mapping and reconnaissance are the next steps in delineating a potential quarry locality. Rock sample collection, mineralogic identifica-

tion, and physical property testing are conducted to determine whether the stone would be suitable for the specific use. The end use of the building stone varies from a dimension stone block size of 100 cubic feet to a rubble stone size of 1 cubic foot. Some blocks smaller than 100 cubic feet can be used to produce tiles and novelties but, for the most part, the major emphasis is on quarrying the larger blocks of stone.

Quarrying large blocks of stone requires geological uniformity of the rock type over an ample enough area. This area needs to be adequate in size to open a quarry that will yield uniform material for an extensive period of time (ten to twenty years). To find an area of this size, the geologist must understand the various tectonic forces that have acted over geological time. Most metamorphic and sedimentary rocks have been folded and faulted during geological time. Most igneous rocks have internal stresses that were created during the molten or plastic stage of injection into the Earth's crust. Such forces will cause cracks and weakness in the rock masses that diminish the potential use for dimension stone.

COLOR AND TEXTURE

The color characteristic of the stone, so critical to the architect's design criteria, is studied by polished hand samples or test block extraction from the potential quarry area. The mineralogic content of the various granites will determine the color consistency and stability. Petrologic microscope work and thin-section studies are an important part of the search for suitable granite. Color stability in a building stone is as important a factor as color uniformity; there are numerous dark limestones that fade to a pale gray that are unsuitable for exterior use but are beautiful, for example, in an interior lobby.

Texture and mineral fabric are also important considerations in studying the suitability of a particular stone for building stone. The use of the stone will depend on whether

A slate quarry. (© William E. Ferguson)

Dolomitic marble at the Jazida do Urubu, Minas Gerais, Brazil. A steep reverse fault is accompanied by drag folds. (U.S. Geological Survey)

it has a fine or coarse texture. Interlocking mineral grains, a well-cemented matrix, and nonsoluble minerals are also important characteristics to identify. Many fine-textured marbles are "sugary" and soft because of poorly interlocked calcite grains, whereas a similar fine-grained marble may be tough and durable to atmospheric corrosion because of a well-interlocked mineral fabric and subsequent low porosity to surface water.

Soundness and Strength

The major criterion in determining whether a building stone will be suitable for a project is the soundness of the deposit. The soundness is a measure of the combination of natural fractures and the strength of the stone. Geological conditions during the formation of the various rock types will determine the soundness. In a sedimentary rock, the bedding (foundation or stratification) and jointing (fracturing without displacement) will determine the block sizes capable of being produced. In an igneous and metamorphic rock, the schistosity (tendency to split along parallel planes), cleavage (tendency to split along closely spaced planes), and jointing will be contributing factors in determining the degree of soundness. Also, in an igneous and metamorphic rock mass, the internal stresses need to be measured and understood before determining block sizes that may be capable of being produced. In the years prior to diamond-cutting tools and hydraulic handling

equipment, cleavage, jointing, and bedding were used to help pry out blocks of stone. With the development of such tools, the quarry workers can cut out more solid masses of rock for building stone.

Rock strength is another criterion of a building stone that needs definition. Specific laboratory equipment is used by a test engineer to determine the flexural and the compressive strength for each building stone. Depending on the planned usage, water absorption and abrasive wear test data are also important for knowing how a particular rock will react in certain applications. The test program may also include producing test specimens of the finished product and installing them in mock-up panels.

The Building Stone Industry

Building stones have been used by humans to produce comfortable habitations ever since cave dwellers blocked up cave entrances for protection and warmth. In the modern world, architects utilize building stone to clad the exterior of modern buildings as a protective shield for the structural steel supports. Designers take polished building stone and clad interior lobbies for durability and beauty in the entrances of these same buildings. Slate, marble, granite, limestone, and sandstone are all utilized as building stone. These materials occur throughout the world in easily exploitable deposits. From the mid-1800's to the mid-1900's, the building stone industry in North America was both the impetus and nucleus for towns, railroads, and machine tool industries across the continent. From the 1900's through the 1930's, many technological advances in quarry drilling and finishing equipment were fostered by the building stone industry. The 1950's through the 1970's saw many adaptations to the demands of architects and designers. Innovative technology was developed to gain the competitive advantage over other producers as well as to satisfy the need to exploit deeper and more solid deposits of stone. This technology went from the muscle power of workers in the mid-1800's to the use of steam, electricity, and compressed air by the early 1900's. The equipment and tools developed from the early 1900's into the 1930's remained relatively unchanged into the 1950's. Since the 1950's, the more sophisticated use of carbides and diamonds for cutting and a high grade of steel for drilling has allowed for the more efficient excavation of stone. The use of hydraulics and electronics has also allowed for a higher degree of automation and cost savings for the manufacturer.

In North America, the building stone industry has shrunk from more than a thousand active quarries in the 1920's, with more than a dozen localities where quarries, mills, and shops were an integral part of the local economy, to a situation in which active quarries number in the hundreds and there are few areas where integrated quarries, mills, and shops are an important part of the local economy. Yet, the trend toward the increased use

of stone in the late 1980's has suggested to many that the building stone industry would once again become an important factor in local economies.

Lance P. Meade

CROSS-REFERENCES

Aluminum Deposits, 1539; Cement, 1550; Chemical Precipitates, 1440; Coal, 1555; Diamonds, 1561; Dolomite, 1567; Earth Resources, 1741; Fertilizers, 1573; Gold and Silver, 1578; Groundwater Movement, 2030; Hydroelectric Power, 1657; Hydrothermal Mineralization, 1205; Industrial Metals, 1589; Industrial Nonmetals, 1596; Iron Deposits, 1602; Karst Topography, 929; Manganese Nodules, 1608; Metamorphic Mineral Deposits, 1614; Mining Processes, 1780; Oxides, 1249; Pegmatites, 1620; Platinum Group Metals, 1626; Salt Domes and Salt Diapirs, 1632; Sedimentary Mineral Deposits, 1629; Soil Chemistry, 1509; Uranium Deposits, 1643; Weathering and Erosion, 2380.

BIBLIOGRAPHY

Barton, William R. *Dimension Stone.* Information Circular 8391. Washington, D.C.: Government Printing Office, 1968. A very good description of the dimension (building) stone industry in North America before major technological advances. Appropriate for high school students.

Bates, Robert L. *Stone, Clay, Glass: How Building Materials Are Found and Used.* Hillside, N.J.: Enslow, 1987. A good, concise (64-page) reference for junior high school students and above.

Bates, Robert L., and Julia A. Jackson. *Our Modern Stone Age.* Los Altos, Calif.: William Kaufmann, 1982. A good reference on industrial minerals. Building stone is included in the discussion.

Chacon, Mark A. *Architectural Stone: Fabrication, Installation, and Selection.* New York: Wiley, 1999. Chacon offers a clear look at stones used in the construction of buildings. He provides information on the tests to select the stones, as well as the steps involved in installation. Although this book is a hands-on manual, there are nice treatments of the selection and use of building stones.

Evans, Anthony M. *An Introduction to Economic Geology and Its Environmental Impact.* Malden, Mass.: Blackwell Science, 1997. This college textbook provides a wonderful introduction to mines and mining practices, minerals, ores, and the economic opportunities and policies surrounding these resources. Illustrations, maps, index, and bibliography.

Meade, Lance P. "Defining a Commercial Dimension Stone Marble Property." In *Twelfth Forum on the Geology of Industrial Minerals.* Atlanta: Georgia Department of Natural Resources, 1976. An objective discussion of the important criteria for developing a dimension stone property. For the college-level reader.

Newman, Cathy, and Pierre Boulat. "Carrara Marble: Touchstone of Eternity." *National Geographic* 162 (July, 1982): 42-58. An excellent photo essay on building stone quarrying. The Italian quarries described provided Michelangelo with the marble for his sculptures.

Shadmon, Asher. *Stone: An Introduction.* 2d ed. London: Intermediate Technology, 1996. This college-level text examines the properties and features of stones, including those used as building materials. In addition to the chapters that deal with the chemical and geophysical aspects of building stones, there are chapters on restoration, conservation, and deterioration.

Smith, Mike R., ed. *Stone: Building Stone, Rock Fill, and Armourstone in Construction.* London: Geological Society, 1999. This book deals with the use of stone in architecture and construction. Includes chapters on geological engineering processes. Color illustrations, maps, index, and bibliography.

CEMENT

Cement is a common construction material used to bond mineral fragments in order to produce a compact whole. The most common types of cement result from the reaction of lime and silica. These are called hydraulic cements because of their ability to set and harden under water.

PRINCIPAL TERMS

AGGREGATE: a mineral filler such as sand or gravel that, when mixed with cement paste, forms concrete

ALUMINA: sometimes called aluminum sesquioxide, alumina is found in clay minerals along with silica; tricalcium aluminate acts as a flux in cement manufacturing

CLINKER: irregular lumps of fused raw materials to which gypsum is added before grinding into finely powdered cement

CONCRETE: a composite construction material consisting of particles of an aggregate bound by a cement

GYPSUM: a natural mineral, hydrated calcium sulfate; it helps control the setting time of cement

HYDRAULIC CEMENT: any cement that sets and hardens under water; the most common type is known as portland cement

LIME: a common name for calcium oxide; it appears in cement both in an uncombined form and combined with silica and alumina

SILICA: silicon dioxide; it reacts with lime and alkali oxides and is a key component in cement

TYPES OF CEMENT

By far the most common type of cement is that which is called portland cement. Portland cement consists primarily of lime, silica, and alumina. These materials are carefully ground, mixed, and heated to produce a finely powdered gray substance. In the presence of water, these ingredients react to form hydrated calcium silicates that, after setting, form a hardened product. Such a product is classified as a hydraulic cement because of its ability to set and harden under water.

While portland cement is by far the most common type of cement, there are others. Most of these, such as high-early-strength cements, slag cements, portland-pozzolan cements, and expansive cements, are variations upon the basic portland cement. Manufacturers can produce these specialized cements through slight variations in the basic chemical composition of portland cement and through the use of various additives. Each of these cements is designed for specific uses, and their advantages include lower cost, higher strength, and faster setting times. Another type, high-alumina cement, is not based on portland cement. Formed by the fusion of limestone and bauxite, high-alu-mina cement hardens rapidly and withstands the corrosive effects of sulfate waters (unlike basic portland cement). Its early promise as a structural material has diminished because of a number of failures, but its ability to withstand high temperatures makes it quite useful in constructing furnaces.

HYDRATION

Adding water to dry cement creates a paste that eventually hardens. The reaction of water with cement is known as the hydration process. This reaction involves much more than water molecules attaching themselves to the constituent elements of cement. Rather, the constituents are reorganized to form new, hydrated compounds. One of the first reactions involves the aluminates, particularly tricalcium aluminum ($3CaAl_2O_3$, hereafter abbreviated as C_3A). Although C_3A is undesirable in cement, it is necessary as a flux during the manufacturing process. If allowed to react unchecked with water, it would lead to flash setting, not allowing time for working the cement or concrete product. To preclude this difficulty, a carefully controlled amount of gypsum is added to the cement at the time of manufacture. The resultant sulfo-aluminates slow the hydration of C_3A, giving the more important calcium silicates time to react with water.

Hydration of the calcium silicates occurs more slowly than that of the aluminates but forms the basic, strength-giving structure of hardened portland cement. Two different calcium silicates $3CaO \cdot O_2$ (abbreviated as C_3S) and $2CaO \cdot SiO_2$ (abbreviated as C_2S) are present in the cement. Both forms of calcium silicates react with water to produce the hydrated calcium silicate ($C_3S_2H_3$). This product is sometimes called tobermorite because of the resemblance its molecular arrangement bears to that of the natural but rare mineral of the same name (taken from Tobermorey, Scotland). Another product of the reaction is calcium hydroxide, or $Ca(OH)_2$, which is integral to the microstructure of the hardened cement. It is important to note that as the exact ratios of the different constituents vary, so does the composition of the product. What is actually being produced, at different places and by different manufacturers, is a family of hydrated calcium silicates rather than one precise formula.

STAGES OF PRODUCTION

From the time water is added to the time the cement paste sets and fully hardens, cement goes through four general phases. Four main compounds are present through all these stages: a gel of the above-mentioned hydrated compounds, crystals of calcium hydroxide, unhydrated cement, and water. The proportions of these compounds change with time; the cement gains rigidity as the percentage of hydrated calcium silicate increases, with the attendant drop in the amount of free water.

These compounds eventually arrange themselves in loose, crumpled layers. For the first few minutes after adding water to cement, the two form a paste in which the cement is in suspension within the water. At this stage, the cement dissolves in the water. The next stage, sometimes called the dormant period, lasts for one to four hours. During this time, the cement forms a gel and begins to set, thus losing its pliability. The individual cement grains build a coating of hydration products, and loose, crumpled layers of hydrated calcium silicate begin to form. In the microscopic spaces between the cement grains, water is held by the surface forces of the cement particles. Larger spaces, called capillary pores, hold free water, which the cement slowly absorbs for use in the hydration process. During the third stage, which peaks about six hours after water is added to the cement, the coating of hydration products around the cement grains ruptures, exposing unhydrated cement to water, thus further building the calcium silicate layers. As these layers grow, they entrap water, which continues to react with the cement particles. They also contain calcium hydroxide (a by-product of the hydration of calcium silicates), which fills the larger pores and thus apparently contributes to the overall strength of the cement. The fourth stage produces the final setting and the hardening of the cement. The hydration process may continue for years, and there will in all probability be a small percentage of the cement that never hydrates. Hydration of C_3S and C_2S occurs at different rates. In the first four weeks, the hydration of the C_3S contributes most to the strength of the hardening cement,

Portland cement. (© William E. Ferguson)

while after that the hydration of the C₂S contributes more to the cement's strength. After about a year, the hydration rates of the two compounds are roughly equal.

Quality Measurement

The quality of portland cement can be measured by four principal physical characteristics: fineness, soundness, time of set, and compressive strength. The desired specifications for each may vary from one country to another and will certainly vary for the different types of cement, but generalizations about them may safely be made. The fineness of the cement plays a large role in determining the rate of hydration. The finer the cement, the larger the surface area of the cement particles, and the faster and more complete the

Materials for cement include lime, silica, and alumina, mined at quarries such as this one in the Bois Blanc Formation at Point Colborne, Canada. (Geological Survey of Canada)

hydration. A method has recently been developed that measures fineness in terms of the specific surface (the surface area of cement particles measured in square centimeters per gram of cement). The two most common ways of measuring specific surface are the Wagner turbidimeter test and the Blaine air permeability test. To use the turbidimeter, a sample of cement is dispersed in kerosene inside a tall glass container. A beam of light is then passed through the kerosene at given elevations at a specified time, and the concentration of cement is measured by a photoelectric cell. The specific surface can then be calculated from the photoelectric cell readings.

The air permeability test relies on the fact that the number and size of pores are functions of the size of the particles and their distribution. A given volume of air is drawn through a bed of cement, and the time it takes for the air to pass through the cement is used to calculate the specific surface of that sample.

Soundness is another important characteristic of cement. A sound cement is one that will not crack or disintegrate with time. Unsoundness is often caused by the delayed hydration and subsequent expansion of lime. The usual method of testing cement for soundness is in an autoclave. A small sample is placed in the autoclave after curing for twenty-four hours and is subjected to extremely high pressure for three hours. After the sample has cooled, it is measured and compared to its original length. If it has expanded less than 0.8 percent, the cement is usually considered sound.

Time of setting is tested on fresh cement paste as it hardens. The ability of the paste to sustain a given weight on a needle of a given diameter (usually a 300-gram load on a 10-millimeter diameter needle) can easily be correlated to its setting time.

Concrete, like other masonry products, is much stronger in compression (for example, a vertical pillar in which gravity pushes the pillar in on itself) than in tension (for example, a horizontal beam that is bowed such that the bottom half wants to pull apart). Thus it is most useful to study the compressive strength of concrete. The usual test is to make a two-inch cube of cement and sand (in a 1:2.75 mixture) and compress it until it breaks. Much can be learned by measuring the breaking load, type of fracturing, and other results of this test.

USES IN CONSTRUCTION

Cement has been used for construction purposes since at least 4000 B.C.E. The Romans used a hydraulic cement based on slaked lime and volcanic ash in many of their construction projects, some of which are still standing. Hydraulic cement disappeared with the Roman Empire and did not reappear until the middle of the eighteenth century. The famed British engineer John Smeaton rediscovered the use of hydraulic cement in 1756 as he rebuilt the Eddystone Lighthouse. Since that time, cement has become a crucial building material.

One important use of cement is as mortar, or cement mixed with sand. Mortar is the substance that binds bricks, stone, and other masonry products. Cement is also the primary ingredient in grout, such as that used between tiles. The most important and common use of cement, however, is in making concrete.

Concrete is the product of mixing cement paste with a mineral aggregate. The aggregate, which acts as a filler, can be a wide variety of materials but is usually a sand or gravel. As a construction material, concrete has many advantages. First, it is inexpensive and readily available. The energy costs alone are a fraction of what they would be for a substance such as steel, and the raw materials for concrete are often available near the construction site, thus saving considerable transportation costs. Another important advantage of concrete is the ability to form it in a wide variety of shapes and sizes, quite often on the job site. Concrete is also known for its long life and low maintenance, as a result of the strong binding characteristics of cement and its resistance to water. Finally, concrete's high strength in compression and proven long-term performance make it a good choice for many structural components.

Concrete for structural uses comes in four major forms. Ready-mixed concrete is transported to a construction site as a cement paste and is then poured into forms to make roadways, driveways, floor slabs, foundation footings, and many other types of structural foundations. Precast concrete can be used for anything from a birdbath to wall slabs for a building, which are cast at a concrete work and then transported to their intended site. A common example of a precast member might be the beams of a highway overpass. Reinforced concrete is any concrete to which reinforcement (usually steel rods) has been added in order to increase its strength. Prestressed concrete is a relatively new form of concrete. Developed in the 1920's, prestressed concrete is put under compression through the use of jacks or steel cables, such that a beam is always in compression, whereas an unstressed beam in the same place would be experiencing tension.

These forms of concrete are used, often in combination, in a variety of ways. Slabs, walls, pipes, dams, spillways, and even elegant vaulted roofs (known as thinshell vaulting) are all made of concrete. Cement, especially as it is used in concrete, has played a crucial role in shaping the physical environment. Concrete is the most widely used manufactured construction material in the world. In most modern countries, the ratio of concrete consumption to steel consumption is at least ten to one. Although concrete is often taken for granted, the world is literally built upon it.

Brian J. Nichelson

CROSS-REFERENCES

BIBLIOGRAPHY

Blanks, Robert F., and Henry L. Kennedy. *The Technology of Cement and Concrete.* New York: John Wiley & Sons, 1955. A somewhat dated but still useful work on cement and concrete. The introductory chapter has some helpful information on the history, economy, and general background on the subject. These same themes are woven into the remainder of the text.

Gani, Mary S. J. *Cement and Concrete.* London: Chapman and Hall, 1997. A look at the processes used to make cement and concrete, as well as the uses and applications of the materials. Suitable for the layperson. Illustrations, index, and bibliography.

Gartner, E. M., and H. Uchikawa, eds. *Cement Technology.* Westerville, Ohio: American Ceramic Society, 1994. A thorough examination of the processes and protocol involved in the manufacture and use of cement.

Mehta, P. Kumar. *Concrete: Structure, Properties, and Materials.* Englewood Cliffs, N.J.: Prentice-Hall, 1986. This book is well illustrated with tables, charts, photographs, and drawings. Definitions are abundant and clear. Also contains numerous examples, all illustrated with photographs, of modern-day projects, ranging from sculpture to dams.

Mindess, Sidney, and J. Francis Young. *Concrete.* Englewood Cliffs, N.J.: Prentice-Hall, 1981. A carefully written and detailed explanation of concrete covering the different cements, chemical reactions, aggregates, and all aspects of making and using concrete. Chapter 2, "Historical Development of Concrete," includes a brief but useful historical overview of cement. Each chapter has a separate bibliography.

Neville, A. M. *Properties of Concrete.* 2d ed. New York: John Wiley & Sons, 1973. Another work on concrete that includes a lengthy discussion of cement. This is a fairly technical work, but it is clearly written and contains good definitions. Chapter 1 is a detailed discussion of portland cement, and chapter 2 discusses other types of cement. References are included at the end of each chapter.

Orchard, Dennis Frank. *Concrete Technology.* Vol. 1, Properties of Materials. 4th ed. London: Applied Science Publishers, 1979. Despite its title, almost one-third of this book discusses cement. Harder reading than some other books on the subject and with less detail but gives a suitable overview for the general reader. Refers to specifications and practices in both Great Britain and the United States.

Popovics, Sandor. *Strength and Related Properties of Concrete: A Quantitative Approach.* New York: Wiley, 1998. A good examination of the properties and uses of cement. This book is accompanied by a computer disk that helps to illustrate concepts.

Portland Cement Association. *Principles of Quality Concrete.* New York: John Wiley & Sons, 1975. Designed to educate persons for employment in the concrete industry, this work explains cement and concrete in the simplest terms. Provides a brief historical overview of the development of cement and concrete as well as of innovations such as reinforcing and prestressing. Also contains useful discussions of applications of cement and concrete.

Taylor, Harry F. W. *Cement Chemistry.* 2d ed. London: T. Telford, 1997. This college textbook looks at the chemical make-up and properties of cements and concretes, as well as chemical changes that may affect their usefulness. The emphasis on advanced chemistry and mathematics may make this book difficult for the reader without some background in these areas.

Troxell, George Earl, Harmer E. Davis, and Joe W. Kelly. *Composition and Properties of Concrete.* 2d ed. New York: McGraw-Hill, 1968. This book contains a section on cement. It is written clearly enough and with sufficient definitions of terms and other aids to allow the general reader to explore the subject.

COAL

Coal is a sedimentary rock composed of altered plant debris. Its principal uses are for fueling steam power plants, as a source of coke for smelting metals, and for space heating and industrial process heat. Synthetic gas and oil are manufactured from coal on a large scale.

PRINCIPAL TERMS

BRITISH THERMAL UNIT (BTU): the amount of heat required to raise the temperature of one pound of water by 1 degree Fahrenheit at the temperature of maximum density for water

CELLULOSE: the substance forming the bulk of plant cell walls

FIXED CARBON: the solid, burnable material remaining after water, ash, and volatiles have been removed from coal

HUMIC ACID: organic matter extracted by alkalis from peat, coal, or decayed plant debris; it is black and acidic but unaffected by other ac-ids or organic solvents

LIGNIN: a family of compounds in plant cell walls, composed of an aromatic nucleus, a side chain with three carbon atoms, and hydroxyl and methoxyl groups

MOLECULE: the smallest entity of an element or compound retaining chemical identity with the substance in mass

ORGANIC MOLECULES: molecules of carbon compounds produced in plants or animals, plus similar artificial compounds

VOLATILES: substances in coal that are capable of being gasified

ORGANIC MATTER

Coal is a heterogeneous mixture of large, complex, organic molecules. It is mostly carbon but contains significant amounts of hydrogen, nitrogen, sulfur, and water. Coal is derived from plant debris that accumulated as peat (plant remains in which decay and oxidation have ceased). When covered by sediments, peat begins to lose its water and more volatile organic compounds. It also compacts and progressively becomes more chemically stable. Thereafter, peat may successively alter to lignite, bituminous coal, anthracite, or graphite as deeper burial, deformation of the Earth's crust, or igneous intrusion increases temperature and pressure.

Enzymes, insects, oxygen, fungi, and bacteria convert plant debris to peat. If unchecked, they can quickly destroy the deposit. Thus, permanent accumulation is limited to situations where oxygen is excluded and accumulated organic waste products prevent further decay. Rapid plant growth, deposition in stagnant water, and cold temperatures promote peat accumulation. Bacteria, the principal agents of decay, operate under a wide range of acidity and aeration; eventually, they remove oxygen and raise acidity so that decay stops.

Peat is a mixture of degraded plant tissue in humic acid jelly. All protoplasm, chlorophyll, and oil have decayed. Carbohydrates have been seriously attacked: First starch, then cellulose, and finally lignin are destroyed. Epidermal tissue, seed coats, pigments, cuticles, spore and pollen coats, waxes, and resins are most durable, but they occur in relatively small amounts. Thus, peat is dominated by lignin, the most resistant carbohydrate, with an enhanced proportion of durable tissues.

TYPES OF COAL

Lignites range from brown coal, which closely resembles peat but has been buried, to black or dark brown lignite, which is similar to higher-ranking coal. Lignite is partially soluble in ammonia. Its resins and waxes dissolve in organic solvents. Water content is high, and there generally is less than 78 percent carbon and more than 15 percent oxygen on an ash-free basis. Woody structure may be obvious and well-preserved. Lignite yields less than 8,300 Btu per pound.

Further compression and heating progressively

Coal-bearing Upper Cretaceous clastic rocks in the lower Babbage River area of Canada's Yukon Territory. (Geological Survey of Canada)

material; and fusain, a dull black, powdery material. Bright coal is dominated by vitrain. Banded coal, which is the most abundant, is dominated by clairain. Dull coals are mostly durain, and fusain is referred to as mineral charcoal. Microscopic study reveals materials derived from woody or cortical tissues called vitrinite or fusinite. Vitrinite is dominant in vitrain and in the "bright" laminae of clairain. Fusinite characterizes durain and the "dull" laminae in clairain. Other microscopic entities, or macerals, include exinite, coalified spores and plant cuticles; resinite, fossil resin and wax; sclerotinite, fungal sclerotia; and alginite, fossil algal remains. Micrinite is unidentified vegetal material. The chemical composition of coal is poorly known because the large, complex, organic molecules in coal break down under attempts to separate them as well as in the process of analysis. The molecular composition of many derivative molecules, however, is known.

convert lignite to subbituminous coal. Fibrous, woody structure gradually disappears, color darkens, the coal becomes denser and harder, water content goes down, and carbon content increases. There is a pronounced decrease in alkali solubility and susceptibility to oxidation. Subbituminous coal, ranging from 8,300 to 13,000 Btu per pound, still weathers significantly and is subject to spontaneous combustion. Like lignite, subbituminous coal burns to powdery ash.

Bituminous coals range from 46 to 86 percent fixed carbon and from 11,000 to about 15,000 Btu per pound. They burn to fused or "agglomerating" ash, resist weathering, and do not spontaneously ignite. Anthracite ranges from 86 to 92 percent fixed carbon, having lost almost all water and volatiles. In addition, it is nonagglomerating, and heating values are about 12,500-15,000 Btu per pound.

Coal is composed of vitrain, a shiny, black material with a glassy luster; durain, a dull, black, granular material; clairain, a laminated, glossy black

MINERAL MATTER

Mineral matter in coal includes all admixed minerals as well as inorganic elements in the coal itself. The organic elements—carbon, hydrogen, oxygen, nitrogen, and sulfur, which form the organic matter in the coal—also occur in compounds, such as iron sulfide, which are part of the mineral matter. Ash is altered mineral matter remaining after the coal is burned and is not synonymous with "mineral matter." Carbonate minerals such as calcite (calcium carbonate) lose their carbon dioxide. Sulfides such as pyrite (iron sulfide) break down to yield sulfur dioxide. Clay minerals lose their water and are drastically altered in molecular structure. Furthermore, the minerals and inorganic elements in the coal react with one another to produce an ash of mixed oxides, silicates, and glass.

Clays are the most abundant minerals in coal. Some clay is washed into the coal swamp, but much arises from chemical reactions occurring in the peat and coal during and after coalification. Sulfides generally are half as abundant as is clay, with the iron sulfides—pyrite and marcasite—most widespread. Sulfides of zinc and lead also may be abundant. Pyrite and marcasite may originate during plant decay and coalification as hydrogen sulfide generated from organic sulfur combines with iron. Hydrogen sulfide also may result from decay of marine organisms as the swamp is invaded by the sea, thus producing more pyrite. Coals associated with marine rocks generally have higher sulfur content than do those coals from wholly alluvial deposits. Carbonates of calcium (calcite), iron (siderite), and magnesium (dolomite) generally are half as abundant as are sulfides. Quartz is ubiquitous, ranging from small amounts to as much as one-fifth of the mineral matter. More than thirty additional minerals have been noted as abundant or common in coal.

Trace elements such as zinc, cadmium, mercury, copper, lead, arsenic, antimony, and selenium are associated with sulfides. Others, such as aluminum, titanium, potassium, sodium, zirconium, beryllium, and yttrium, are associated with mineral grains washed into the swamp. Still others find their way into the peat within plant tissues or later are concentrated from waters circulating through either the peat swamp or the coal seam. These elements include germanium, beryllium, gallium, titanium, boron, vanadium, nickel, chromium, cobalt, yttrium, copper, tin, lanthanum, and zinc. Coal ash has been a source of germanium and vanadium, and both uranium and barium have been mined from some coal seams.

Iron, selenium, gallium, zinc, and lead occurrences in coal have been investigated as possible sources of these metals, and, inasmuch as release of metals from coal mining or combustion is generally deleterious, their recovery in pollution control may be feasible.

CURRENT FORMATION

Peats form today in two very different environments. Poor drainage in areas of recent glaciation, coupled with low temperature, facilitates peat formation in high latitudes. In warm temperate and tropical regions that are poorly drained, vigorous forests may produce peat. Coastal plains and shoreline deposits, such as the Dismal Swamp and Everglades; deltas and alluvial plains, such as the

Hills of coal at the Ohio Valley Coal Company. (AP/Wide World Photos)

Mississippi Delta; and tropical alluvial plains, as in the upper Amazon Basin, are all good examples. Most of the high-latitude peat is unlikely to be incorporated in major sedimentary accumulations and, thus, is unlikely to become coal. Other modern peats, however, are an extension of the sort of peat accumulation that occurred in the geologic past and, therefore, provide a guide to understanding coal formation.

Individual coal beds or seams may be very widespread or of limited extent. The Illinois #2 Coal, for example, is recognizable from western Kentucky to northeastern Oklahoma. Other coal beds cover only a few square kilometers. Coals range from a few millimeters to more than 100 meters thick. They generally are tabular but may be interrupted by filled stream channels or rolls, which are protuberances of overlying rock apparently forced into the coal while in a plastic state. Thin layers of clay (splits) or cracks filled with clay (clay veins) interfere with mining and dilute mined coal with extraneous rock. Coal beds may be subhorizontal or may be steeply inclined, depending on deformation in the area. They also may be continuous or may be offset by faults—fractures of the crust along which movement has occurred. Inclination or interruption of the bed interferes with mining.

Testing

The distribution and character of coal beds are determined by standard field geologic methods. Surface exposures are plotted on maps, and their geometric orientation is recorded. Thereafter, the coal beds are projected geometrically into the subsurface. Wherever possible, their position is verified in wells and mine shafts so that the full regional extent, depth, and attitude of the coal are illustrated. If detailed information is required for mining, specially drilled coal test holes will be utilized to locate channel fillings and faults interrupting the coal as well as to determine changes in thickness and quality. These test holes make it possible to plan mining efficiently. In addition, the relationship of the coal to rocks above and below will be investigated so that mining methods may be adjusted to potential geologic hazards, such as caving roofs and incursion of underground water.

Standardized, practical tests define quality and/

or suitability for specific uses. Burning coal samples under controlled conditions at 750 degrees Celsius produces a standard American Society for Testing Materials (ASTM) ash, which defines the total ash content and its nature. Coals with low-ash content are preferred. The character of the ash—agglomerating produces a glassy clinker, and nonagglomerating produces a powder—determines the type of grate on which the coal can be burned. The heating value of the coal, figured on an ash-free basis, is determined by controlled combustion in a calorimeter: a closed "bomb" fitted with temperature sensors. The amount and quality of volatile materials are analyzed by carbonizing the coal at a standard temperature in a closed vessel and measuring the amount and kinds of substances driven out of the coal. These data are required to classify the coal according to rank and grade and to fix its value in the market.

Analysis

More precise analytical techniques and tools are employed in research, as opposed to routine coal testing. The physical composition of the coal may be determined by microscopic study of thin sections (slices of coal mounted on glass and reduced to a thickness allowing transmission of light) or by examining polished surfaces with a reflecting microscope. In this way, the components of the coal may be distinguished and examined separately. More detailed study may utilize either transmission or scanning electron microscopy. Coal rank also may be determined very precisely by means of measuring reflectance—in this case, the amount of light reflected from polished vitrinite.

Mineral matter in the coal, as distinguished from ash, may be directly examined by these techniques but also may be recovered from the coal employing a low-temperature asher. The asher is an electronic device that vaporizes combustible materials without significantly raising the temperature of the sample. In this way, clays, carbonates, sulfides, and other minerals that are significantly altered by heat are delivered for study in their original state.

Elemental chemical analysis of coal provides information as to its composition at a level of limited value to coal investigators. The organic compounds in coal, however, suffer substantial alteration under almost any analytical technique that can be applied to them. The volatiles recovered by coal carbonization in the absence of oxygen may be separated by distillation, but they are not the compounds originally present in the coal. Solid organic materials may be selectively dissolved and the resultant materials subjected to organic analysis, but, again, these materials are not the ones that were present in the original coal. In spite of these limitations, however, some conception of the original chemistry of the coal is obtained, and very good information applicable to coal utilization is developed. X-ray diffraction and other spectroscopic techniques have begun to uncover the structure of coal molecules without altering them. Concepts of coal molecular structure, however, remain rudimentary.

FUEL SOURCE

Coal is a major source of heat energy and a significant source of organic compounds of practical use—from drugs to plastics. It fueled the Industrial Revolution and, thus, is responsible for the appearance of modern industrial society. Coal was the principal source of energy until World War I. Although its use declined thereafter under competition from oil and gas, coal is again increasing in relative importance. By the end of the twentieth century, coal provided approximately 39 percent of the world's electricity. In contrast to oil and gas, whose reserves are limited and for which production is expected to peak early in the twenty-first century, coal reserves appear to be adequate at least until well into the twenty-second century.

In spite of its large reserves, coal presents several significant problems. Coal is a solid fuel and is inappropriate for use in domestic heating and vehicular transport. Therefore, substantial research on converting coal to liquid and gaseous fuel is needed and is underway. Furthermore, coal combustion produces gaseous and solid wastes that must be managed. Sulfides are converted to metallic oxides and sulfur dioxide, which combine with water in the atmosphere to produce acid rain. This precipitation kills plants, causes respiratory difficulties, and destroys aquatic life. The carbon in coal, as well as in wood or hydrocarbons, combines with oxygen to form carbon dioxide when burned. Enough carbon dioxide has already been produced by burning wood, coal, and hydrocarbons so that a change in global atmospheric composition has been detected. Resultant climatic changes are feared, even though ultimate results are not yet predictable. Other deleterious elements are released, either into the atmosphere or in ash, such as lead, cadmium, arsenic, and mercury. Although these elements are emitted in small amounts, the environmental effects must be understood so that appropriate action may be taken. Mine hazards, gas and cave-ins especially, are geologically controlled. Subsidence of underground mines—its extent, timing, and ultimate cost to surface values—also is geologically controlled. Strip-mining reclamation and the problem of mine waters entering both surface and subsurface water supplies are additional geologic concerns.

Ralph L. Langenheim, Jr.

CROSS-REFERENCES

Aluminum Deposits, 1539; Building Stone, 1545; Cement, 1550; Diamonds, 1561; Dolomite, 1567; Earth Resources, 1741; Fertilizers, 1573; Future Resources, 1764; Gold and Silver, 1578; Hydrothermal Mineral Deposits, 1584; Industrial Metals, 1589; Industrial Nonmetals, 1596; Iron Deposits, 1602; Manganese Nodules, 1608; Metamorphic Mineral Deposits, 1614; Pegmatites, 1620; Platinum Group Metals, 1626; Salt Domes and Salt Diapirs, 1632; Sand, 2363; Sedimentary Mineral Deposits, 1629; Silica Minerals, 1354; Uranium Deposits, 1643.

BIBLIOGRAPHY

Averitt, P. "Coal." In *United States Mineral Resources,* U.S. Geological Survey Professional Paper 820. Washington, D.C.: Government Printing Office, 1973. A concise account of coal formation, its rank, sulfur content, and minor elements occurring in coal. United States and world coal resources also are reviewed. Written for the informed general public.

Dalverny, Louis E. *Pyrite Leaching From Coal and Coal Waste.* Pittsburgh, Penn.: U. S. Department of Energy, 1996. This brief government publication looks at coal-mining policies and protocol, as well as the state of disposal procedures used by coal-mine operators. Illustrations and bibliography.

Francis, Wilfrid. *Coal: Its Formation and Composition*. London: Edward Arnold, 1961. Describes the tissues of coal-forming plants, how they accumulate to form peat, and formation and composition of peat, lignites, and higher-ranking coals. Reviews the composition of coal, including inorganic constituents, and discusses coal-forming processes. Moderately technical. Outdated but usable.

Galloway, W. E., and D. K. Hobday. *Terrigenous Clastic Depositional Systems*. New York: Springer-Verlag, 1983. Discusses the character and processes of nonmarine sedimentary environments, excepting glacial environments, as related to coal deposition. Reviews the character and composition of coal as affected by depositional origin. Written for college students but intelligible to the general reader.

Gayer, Rodney A., and Jierai Peesek, eds. *European Coal Geology and Technology*. London: Geological Society, 1997. A complete examination of the technologies used in European coal mines and the coal mining industry. Illustrations, maps, index, and bibliographical references.

James, P. *The Future of Coal*. London: Macmillan Press, 1982. Summary account of the origin of coal, mining methods, its use and markets, and environmental and health problems associated with coal use. Also a discussion of coal occurrence and use on a worldwide basis. Written for the nontechnical reader.

Keefer, Robert F., and Kenneth S. Sajwan. *Trace Elements in Coal and Coal Combustion Residues*. Boca Raton, Fla.: Lewis Publishers, 1993. An examination of the chemical make-up and trace elements found in coal and coal residues. Suitable for the beginner. Includes illustrations, maps, index, and bibliographical references.

Stach, E. *Stach's Textbook of Coal Petrography*. 3d ed. Berlin-Stuttgart: Gebruder Borntraeger, 1983. Includes discussion of origin and formation of peat and coal, a detailed account of coal's physical constituents and their origin, a discussion of trace elements in coal, the methods of coal petrography, and practical applications of coal petrography. The most comprehensive account in English. Written for professional coal geologists.

Stewart, W. N. *Paleobotany and the Evolution of Plants*. New York: Cambridge University Press, 1983. A well-illustrated account of the evolution of plants, including those responsible for coal. Written for beginning college-level students.

Swaine, Dalway J. *Trace Elements in Coal*. Boston: Butterworth, 1990. A chemical analysis of the trace elements and properties of coal. Although much attention is focused on chemistry, this book is suitable for the non-chemist. Illustrations help to clarify concepts. Bibliographic references.

U. S. Department of Labor, Mine Safety and Health Adminstration. *Coal Mining*. U. S. Department of Labor, Mine Safety and Health Adminstration, 1997. This brief government publication examines the state of mines and coal mines in the United States. There is a strong focus on the health risks associated with coal mining. Illustrations and maps.

Van Krevelen, D. W. *Coal: Typology-Chemistry-Physics-Constitution*. Amsterdam: Elsevier, 1961. A comprehensive treatise on the physical and chemical properties of coal, its constitution and classification, and its geology and petrology. Somewhat dated, but still the best single reference. Written at the technical level.

Williamson, I. A. *Coal Mining Geology*. London: Oxford University Press, 1967. General geology as it pertains to coal, along with details of mine geology, coal composition, and Carboniferous fossils and stratigraphy. Written for elementary geology and mining students.

DIAMONDS

Diamond is an important industrial mineral as well as the most valued of gemstones. Natural diamonds crystallize only at very high pressures and are brought to the Earth's surface in kimberlite, an unusual type of igneous rock that forms in the upper mantle. Kimberlite also contains "gems" of another sort: rare pieces of the Earth's deep crust and mantle.

PRINCIPAL TERMS

BORT: a general term for diamonds that are suitable only for industrial purposes; these diamonds are black, dark gray, brown, or green in color and usually contain many inclusions of other minerals

COESITE: a mineral with the same composition as quartz (silicon dioxide), but with a dense crystal structure that forms only under very high pressures

CRYSTAL: a solid that possesses a definite orderly arrangement of its atoms; it differs from an amorphous solid such as glass, and all true minerals are crystalline solids

GRAPHITE: a crystalline variety of the element carbon, characterized by its softness and ability to cleave into flakes; the carbon atoms are arranged in sheets that are weakly bonded together

KIMBERLITE: an unusual, fine-grained variety of peridotite that contains trace amounts of diamond

METASTABLE: crystalline solids are said to be metastable if they exist outside the temperature and pressure conditions under which they formed; thus, diamond forms at very high pressures within the Earth but is metastable at the Earth's surface

PERIDOTITE: a dense, dark-green rock that is composed mainly of magnesium- and iron-rich silicates, particularly olivine; the Earth's mantle and the ultramafic nodules derived from it are composed of peridotite

ULTRAMAFIC: rocks such as peridotite that contain abundant magnesium- and iron-rich minerals

AVAILABILITY OF DIAMONDS

In the eighteenth century, virtually all the world's diamonds came from India and were hoarded by royalty; few people had ever seen a diamond, let alone possessed one. Brazil became an important producer in the late 1700's, but diamonds were still unavailable to most people. In 1866, a farm boy discovered a bright pebble on the banks of the Orange River in South Africa and unknowingly started a chain of economic, social, and political upheavals that continue to this day; the stone was later determined to be a 21-carat diamond (one carat equals 200 milligrams). Within a decade, South Africa's mines would be producing 3 million carats a year to a world market; by the turn of the century, diamonds would become an important industrial commodity. Today, several South American countries also export diamonds, but South Africa remains the world's foremost diamond producer. North America has no major diamond deposits.

Early diggings were concentrated along the Orange and Vaal rivers of South Africa, where prospectors staked small claims and shoveled into the diamondiferous gravel deposits. Fines were washed away in a rudimentary method known as wet digging, and the remaining gravel was spread out to be examined, pebble by pebble. Around 1870, a diamond was found approximately 100 kilometers from the nearest river, and prospectors began to speculate that the stones along the riverbanks did not originate there but were washed in from elsewhere. Increasing numbers of miners went into the bush to pursue diamonds, finding them in local patches of weathered rock known as yellow ground. As geologists would soon understand, the yellow ground was merely the uppermost layer of deep, funnel-shaped pipes of diamond-bearing igneous rock that had been injected into the Earth's crust by ancient volcanoes. The bluish-gray rock was named kimberlite for a nearby South African town.

INCLUSION IN KIMBERLITE

At present, more than 90 percent of the world's diamonds are found in river gravels, beach sands, and glacial deposits of many geological ages. Only in kimberlite pipes are diamonds found in the original rock in which they were formed. Kimberlite pipes are rather insignificant features, seldom having diameters greater than 1 kilometer. Mining has shown them to be carrotlike bodies whose vertical dimensions far exceed the sizes of their surface outcrops. Mine shafts have penetrated about 1.5 kilometers into kimberlite pipes, but minerals contained in the kimberlite, including the coveted diamonds, suggest that the pipes extend all the way through the crust and into the Earth's upper mantle to a total depth of about 250 kilometers. This is deeper than any other variety of igneous rock.

Beneath the weathered yellow ground, fresh kimberlite is a hard, dark bluish-gray rock that miners call blue ground. Its texture gives strong evidence of an igneous origin, indicating that kimberlite was injected into the Earth's crust as a molten liquid and then quickly solidified against cooler rocks surrounding the pipe. The major constituents of kimberlite are silicate minerals: compounds of silicon and oxygen with other metal ions. Kimberlite is a variety of peridotite ("peridot" is an ancient word for olivine), and hence its major constituent is the mineral olivine, a magnesium-iron silicate. The olivine is usually altered to the mineral serpentine, giving the rock its characteristic blue-gray color.

Exotic rocks contained within the kimberlite matrix are perhaps more interesting than the kimberlite itself. Diamonds are one such inclusion, although they comprise a minuscule proportion of the total rock. Typical diamond contents in minable kimberlite range from about 0.1 to 0.35 carat per ton; even the famous Premier Mine in South Africa has produced only about 5.5 tons of diamonds from 100 million tons of rock, which is about 0.000005 percent.

FORMATION OF DIAMONDS

Three independent lines of evidence indicate that kimberlite is formed in the upper mantle, at depths and pressures far greater than for any other type of igneous rock. To calculate pressures of formation, it is usually necessary to know or assume the temperature of formation. The continental geotherm supplies the needed temperature information. At temperatures along the continental geotherm, diamond is stable only at depths greater than about 100 kilometers; at lower pressures, graphite forms instead. Diamond exists at low pressures only because it is metastable (otherwise there could be no diamonds on the Earth's surface), but diamond does not form naturally at low pressures.

A second line of evidence for great pressure involves the minerals that have been trapped inside the diamonds during their growth. Diamonds sometimes contain inclusions of coesite, a mineral with the same chemical composition as quartz (silicon dioxide) but with a more compact structure that forms only at very high pressures. At temperatures along the continental geo-

Raw diamonds. (Geological Survey of Canada)

therm, laboratory studies have shown that coesite, in turn, gives way to the silica mineral stishovite at depths greater than about 300 kilometers. Because stishovite has never been found in diamonds, the diamonds must have formed at depths less than about 300 kilometers. Hence, diamonds seem to have formed at depths between about 100 and 300 kilometers in the Earth's upper mantle. The pressures and temperatures that have been calculated from minerals in the ultramafic nodules are in agreement with this depth range: Most of the nodules seem to have formed at depths of 100-250 kilometers in the upper mantle and at temperatures of about 1,100-1,500 degrees Celsius.

PHYSICAL PROPERTIES

The unusual physical properties of diamond are a reflection of its crystalline structure. Diamond is a three-dimensional network of elemental carbon, with each carbon atom linked to four equidistant neighbors by strong covalent bonds. The dense, strongly bonded crystal structure gives diamond its extreme hardness. Another mineral made of pure carbon is graphite, the writing material in lead pencils. In graphite, sheets of carbon atoms are weakly bonded and are separated by relatively large distances. Thus, graphite has a lamellar structure and is very soft.

The physical properties of diamond are remarkable in comparison to virtually all other materials. It is the hardest substance known. With a hardness of 10, it tops the Mohs scale of relative hardness and is actually about forty-two times as hard as corundum, its nearest neighbor, with a hardness of 9. Its luster is adamantine to greasy, and it cannot be wetted by water—a property that is of great practical benefit in separating diamonds from waste rock. Diamonds vary in color from water-clear (most valuable) to pale blue to yellow to deep yellow or brown; industrial varieties are brown or grayish-black. Raw diamonds occur most often as octahedra (eight-sided polygons) or as cubes but are also found as tetrahedra (four-sided polygons) and dodecahedra (twelve-sided polygons) as well as in slender, irregular shapes. Diamond can be cleaved in four directions, parallel to its octahedral faces. Its refractive index of 2.42 is the highest of all gems, producing strong reflections in cut stones, and its very high dispersion (ability to separate white light into the colors

of the spectrum) gives cut diamonds their "fire." Diamond is also triboelectric (becomes electrically charged when rubbed) and fluorescent (emits visible light when struck by ultraviolet rays).

Diamonds are separated from waste rock by first crushing the kimberlite, wetting it, then passing it over a series of greased bronze tables. Diamond cannot be wetted by water and is the only mineral that sticks to the grease, which is later scraped off as the diamonds are extracted. Another way of separating diamonds takes advantage of their fluorescent property. As the crushed rock is passed beneath ultraviolet lamps, the diamonds are spotted by photosensors, which trigger jets of compressed air that eject the diamonds into bins. Gem-quality stones of less than 1 carat are called melee. Less than 5 percent of all diamonds are suitable for cut stones of 1 carat or larger.

DIAMOND CUTTING AND GRADING

Diamonds have been fashioned into precious jewels for several millennia, but diamond cutting has become a major industry only during the past century, in response to worldwide demand and to the abundant supply of South African diamonds. Five basic steps are involved in diamond cutting: marking, cleaving, sawing, girdling, and faceting. The diamond is first carefully studied, sometimes for months, in order to identify its cleavage planes and to map out any inclusion-rich areas that will affect how it is to be cut. Large diamonds are usually irregular in shape and are seldom left whole. A central, master stone is commonly envisioned within the mass, and the "satellite" offcuts become fine gems in their own right. Lines for cleaving or sawing are marked in black ink, and the diamond is then sent to the cutter. If the diamond is to be cleaved, a thin groove is first established using a saw charged with diamond dust. The diamond is mounted in a dop or clamp, and a steel wedge is inserted into the groove and is struck sharply with a mallet. A misdirected blow can shatter the stone: It is said that Joseph Asscher, after successfully cleaving the 3,100-carat Cullinan diamond in 1908, swooned into the arms of an attending physician. Sawing is a slower, if not safer, alternative to cleaving; it uses a thin, circular bronze blade that is charged with diamond dust. Next, the diamond is girdled by placing it in a lathe and grinding it against another diamond to make it round; the "girdle" is where the upper and lower sets of facets meet. Finally, the diamond is faceted: The facets are cut and polished by clamping the stone in a holder and placing it against rotating laps that are charged with diamond dust. The most popular cut is the "brilliant"—a round stone with fifty-eight facets. Other cuts include the marquise (oval), emerald (rectangular), and pear.

Cut stones are graded under strict rules, using an elaborate system of four criteria: cut, color, clarity, and carat. Cut refers to the skill with which the gem has been shaped—its symmetry and reflective brilliance. Color refers to the tint of the stone: The most valuable gems are water-clear, but many fine stones are pale yellow, blue, or pink; colored diamonds are called "fancies." Clarity refers to the size, number, and locations of any inclusions that may be present. Inclusions do not necessarily degrade a stone if they are small or are located in inconspicuous places. Carat refers to the weight of the stone. The largest known diamond was the Cullinan, which weighed 0.6 kilogram before it was cleaved and fashioned into an assemblage of stones that are now part of the British crown jewels.

STUDY OF DIAMONDS

Diamonds have been studied using the same analytical methods that are applied to other crystalline solids, notably X-ray diffraction, to identify their crystalline structure. Mineral inclusions in diamond (mostly coesite and garnet) have been examined with the electron microprobe, a device that employs a tiny electron beam to measure the percentages of elements that are present in a mineral. The compositions of coexisting minerals in kimberlite and in ultramafic nodules have similarly been analyzed.

Igneous petrology is a subdiscipline of geology concerned with the description and origin of igneous rocks. Petrologists use all the methods listed above for individual minerals, plus larger-scale observations of entire rock masses. Minerals are assemblages of atoms, and rocks are assemblages of minerals; hence, understanding the origins of minerals, including diamond, involves an understanding of how the enclosing rock was formed. Individual rock masses such as kimberlite, ultramafic nodules, and xenoliths of the crust are glued onto glass slides, ground into thin slices, and studied under the microscope. In addition to the minerals they contain, the textures of rocks also reveal much about their origins. The results of mineralogical and textural observations are then interpreted in the light of even larger-scale observations, involving geological mapping on the surface and underground. Finally, all mineralogical, petrological, and geological observations must be interpreted within the constraints of geophysical data on the internal constitution of the Earth.

The quest for diamonds has fostered much inquiry of the rock in which diamonds are found. Kimberlite is a very unusual type of igneous rock that forms deep in the Earth's mantle, at depths up to 250 kilometers. The ultimate source of diamonds, kimberlite also contains "gems" of a different sort: pieces of the Earth's deep continental crust and upper mantle that would be inaccessible by any other means. Volcanic pipes of kimberlite are therefore windows into the Earth's interior, and the rock and its inclusions are avidly studied in order to learn more about the internal constitution of the Earth. Diamonds contain trapped inclusions of liquids, mineral solids, and gases, mostly carbon dioxide. Analyses of these inclusions have led to a better understanding of the conditions under which diamond is formed and of the volatiles that are present in the Earth's mantle.

INDUSTRIAL USES

The hardness, brilliance, and fire of diamonds have made them unsurpassed as gems; the use of diamonds as industrial materials is perhaps less well known. All stones not suitable for use as gems are destined for industrial use. "Bort" refers to dense, hard, industrial diamonds, and carbonado is diamond that has a lower specific gravity than does normal diamond. Such diamonds vary in color from off-white to black. Most industrial diamonds are used as abrasives. Crushed into various sizes, they are used for grinding wheels, grinding powders, polishing disks, drill bits, and saws. Diamond is indispensable for grinding the tungsten carbide cutting tools that have been in use since the 1930's. Industrial diamonds are also sorted for shape: Blocky stones are suited for more severe grinding operations such as rock-drill bits, and more splintery ones are reserved for grinding tungsten carbide.

In 1955, the General Electric Company succeeded in manufacturing synthetic diamonds at low pressure. Initially more expensive than natural stones, synthetic diamonds are now widely used in grinding wheels to sharpen tungsten carbide. Synthetic diamonds are smaller but have rougher surfaces than natural stones and are now manufactured in a variety of shapes and sizes for specialized abrasive applications.

Diamonds are also used as dies for drawing out the fine tungsten filaments of incandescent light bulbs, as scalpels for eye surgery, and as stereo phonograph needles. Unrivaled as a heat conductor, diamond is an important component in the miniature diodes that are used in telecommunications. Diamond has even served as a tiny instrument window on the Pioneer space probe to Venus, as it tolerates the extremes of heat and cold in outer space.

William R. Hackett

CROSS-REFERENCES

BIBLIOGRAPHY

Cox, K. G. "Kimberlite Pipes." In *Volcanoes and the Earth's Interior*, edited by Robert Decker and Barbara Decker. San Francisco: W. H. Freeman, 1982. This article is reprinted from *Scientific American* (April, 1978). Its main points can probably be understood by the high school student with a strong science background. An excellent summary of the origin of Kimberlite and its inclusions, including diamonds. Mostly covers the geological aspects of kimberlite and diamond.

Decker, Robert, and Barbara Decker. *Volcanoes.* New York: W. H. Freeman, 1997. This 240-page paperback is geared to the general public and contains much useful and interesting information. A few pages focus on diamonds and kimberlite pipes, described within the context of volcanoes. Includes a glossary, selected references for each chapter, and an index.

Harlow, George E., ed. *The Nature of Diamonds.* Cambridge, England: Cambridge University Press, 1998. A good examination of the evolution of diamonds and diamond mines. This is a thorough introduction to diamonds and their properties suitable for the layperson.

Color illustrations, maps, index, and bibliography.

Hurlbut, C. S., Jr., and G. S. Switzer. *Gemology.* New York: John Wiley & Sons, 1979. A well-illustrated and complete introduction to gemstones, describing the methods of gem study. Includes a section with descriptions of minerals and other materials prized as gems. A good introduction to mineralogy for the nonscientist.

Kluge, P. F. "The Man Who Is Diamond's Best Friend." *Smithsonian* 19 (May, 1988): 72. This very interesting article follows the strategy and execution of cutting a 900-carat diamond into one large stone and ten smaller ones. Color photographs document the procedure from start to finish. Readable for all.

Mitchell, Roger H. *Kimberlites, Orangeites, and Related Rocks.* New York: Plenum Press, 1995. Mitchell provides a good introduction to the study of kimberlites and related rocks, including an extensive bibliography which will lead the reader to additional information.

_____. *Kimberlites, Orangeites, Lamproites, Melilitites, and Minettes: A Petrographic Atlas.* Thunder Bay, Ontario: Almaz, 1997. This book covers the worldwide distribution of kimberlites and other related rock types. Color illustrations and bibliography.

O'Neil, Paul. *Gemstones.* Alexandria, Va.: Time-Life Books, 1983. Readable by everyone, this superbly written and thoroughly entertaining book is a treasure trove of color photographs. Contains much historical and geological information about many important gemstones; the chapters on diamonds are outstanding. Bibliography and index.

Smith, D. G., ed. *The Cambridge Encyclopedia of Earth Sciences.* New York: Crown, 1981. Although there is little mention of diamonds, the section on physics and chemistry of the Earth provides a good foundation for understanding the Earth's interior and hence the origin of diamonds and kimberlite. This well-illustrated and carefully indexed volume is suitable for college-level readers.

DOLOMITE

Dolomite is a common rock-forming carbonate mineral of uncertain origin. Its chemically basic nature makes it a useful raw material for industrial applications; its refractory properties render it invaluable for metallurgical uses.

PRINCIPAL TERMS

CARBONATE MINERAL.: a mineral compound with a fundamental structure that includes the CO_3^{-2} anion

CONNATE WATER: water trapped in the pore spaces of a sedimentary rock at the time of its deposition

DEAD-BURNED: a term used to describe a carbonate material that has been heated until it contains less than 1 percent carbon dioxide

DIAGENESIS: the conversion of unconsolidated sediments into rock after burial; the process includes compaction, cementation, and replacement

DOLOMITIZATION: the process by which a deposit of calcite or aragonite reacts with magnesium-rich waters to become partially or wholly replaced with dolomite

DOLOSTONE: a sedimentary rock composed chiefly of the mineral dolomite; this rock is often associated with or interbedded with limestone

METEORIC WATER: water derived from atmospheric origins; rainwater and snowmelt

PENECONTEMPORANEOUS: formed shortly after sediment deposition and before sediments consolidate into rock

PROTODOLOMITE: a form of dolomite in which magnesium and calcium cations share the same crystallographic planes; an order-disorder polymorph of dolomite

REFRACTORY: a material able to withstand contact with corrosive substances at high temperatures

SABKHA: an arid or semiarid coastal environment just above high-tide level, characterized by evaporite-salt, tidal-flood, and wind-borne deposits

DOLOMITE STRUCTURE

Dolomite, or calcium magnesium carbonate $(CaMg(CO_3)_2)$, is a common rock-forming mineral. It derives its name from Déodat de Dolomieu, the eighteenth century French geologist who first studied the rocks of the Dolomite Alps of northern Italy. Dolomite belongs to a family of minerals called "carbonates," compounds with a fundamental structure that includes the carbonate anion (CO_3^{-2}). Dolomite is one of the most common varieties of carbonate mineral, along with calcite and aragonite (both calcium carbonate, $CaCO_3$); together, these minerals make up approximately 2 percent of the Earth's crust. Pure dolomite contains almost equal amounts of calcium and magnesium, making it intermediate in composition between calcite and magnesite (magnesium carbonate, $MgCO_3$).

Dolomite's crystal structure resembles that of calcite. Both have a rhombohedral crystal structure; however, in calcite, layers of calcium cations (atoms with a positive charge) separate parallel layers of CO_3 anions (groups of atoms with a negative charge), whereas in dolomite, the cation sites are alternately occupied by calcium and magnesium. Some forms of dolomite, called "protodolomites," exhibit a disordered structure, with calcium and magnesium sharing the same crystallographic planes. This phenomenon, in which two crystalline substances have the same composition but differing states of order within their atomic arrangements, is known as "order-disorder polymorphism." Protodolomites occur in younger deposits (those formed within the past ten thousand years), while older dolomites have the more orderly structure.

A number of cations are known to substitute within the dolomite structure. A portion of the magnesium in dolomite may be replaced by ferrous iron or, to a lesser extent, by manganese.

Zinc and cobalt may also substitute for magnesium in minor quantities. Small amounts of barium and lead may substitute for the calcium. Boron, rubidium, strontium, and uranium are other elements that have been known to occur within the dolomite structure.

Dolomite may be colorless, white, pale pink, gray, green, brown, or black, with a glassy, pearly, or dull luster. It occurs as crystals or as granular masses. Its crystal faces are often convexly curved; aggregates of small, curved crystals sometimes form a distinctive saddle shape. Dolomite has a perfect rhombohedral cleavage, a hardness ranging from 3.5 to 4 on the Mohs scale, and a specific gravity of 2.86. The mineral may be identified by its weak effervescence (fizzing) in dilute hydrochloric acid. Some dolomites exhibit a property called "triboluminescence"; that is, they glow if crushed, scratched, or rubbed.

DOLOSTONE

Rock made up principally of the mineral dolomite is also sometimes called "dolostone" or "dolomite rock." Dolostone is a sedimentary rock similar in composition and characteristics to limestone, and it is commonly found in association with or interbedded with limestone. A rock is classified as dolostone if dolomite accounts for more than 90 percent of its weight and calcite accounts for under 10 percent; by contrast, a rock is considered to be limestone if it consists chiefly of calcite, with or without magnesium carbonate. Intermediate rock compositions are often found in nature. A limestone containing at least 90 percent calcite and 5 to 10 percent dolomite, for example, would be considered a magnesian limestone, owing to its appreciable magnesium content. If a limestone includes a significant amount (10 to 50 percent) of dolomite in its composition, it is known as "dolomitic limestone." (The Dolomite Alps are actually composed of dolomitic limestone.) A dolostone in which calcite is conspicuous (10 to 50 percent) but dolomite predominates (50 to 90 percent) is a calcitic dolostone.

DOLOMITIZATION

The intermediate compositions are believed to be the result of incomplete dolomitization. Dolomitization, the process by which dolomite replaces the aragonite and calcite in calcium carbonate sediment or limestone, is a topic of ongoing debate in the geologic literature. It is known that calcium carbonate precipitates from seawater as calcite or aragonite, either inorganically or through marine organisms, to be deposited as sea-bottom sediment. This material undergoes diagenesis, a series of changes including compaction, cementation, and replacement. Calcite replaces most or all of the aragonite, and ultimately the sediments harden to become limestone. While the conversion of this sediment or rock to dolomite is generally accepted in the scientific community, the exact mechanism by which this secondary replacement occurs has yet to be determined. However, because direct precipitation of dolomite from seawater under natural conditions has never been observed or chemically demonstrated and is believed to contribute relatively minor amounts of the mineral at best, most researchers focus on secondary replacement of calcium carbonate as the

Dolomite. (© William E. Ferguson)

most likely mechanism for the formation of extensive dolostone deposits.

Modern protodolomites are known to form in sabkhas, arid or semiarid coastal environments just above high-tide level. Sabkhas are characterized by evaporite-salt, tidal-flood, and wind-borne deposits. Dolomite formation has also been observed in association with hot springs and with the precipitation of calcite, aragonite, magnesium calcite, gypsum, and anhydrite in highly saline lagoons (that is, in bodies of water that are rich in mineral salts). These localities where modern dolomitization is taking place are geographically restricted, relatively rare depositional environments; yet somehow, thick, extensive deposits of dolostone and other dolomitic carbonate rocks have developed throughout the geologic record, particularly in major sequences of marine strata, and no indication has been found to suggest they developed under unusual geologic conditions. It is worth noting that the relative proportion of dolostone to limestone increases with age in carbonate rocks. It is unknown whether this trend indicates that past environments were more conducive to dolomitization or that limestones are more likely to become dolomitized the older they grow. The exact relationship between ancient dolostones and today's protodolomites is one geologists and geochemists are still trying to define.

MODELS EXPLAINING DOLOMITIZATION

Several models explaining dolomitization have been proposed; most involve seawater, an obvious, plentiful source of magnesium for the replacement reaction. Some models require this seawater to be concentrated by evaporation into a highly saline brine. Magnesium-charged meteoric waters (those derived from the atmosphere as rain or snow) have also been suggested as a possible dolomitizing fluid, as have connate waters (the waters trapped in the interstices of a sedimentary rock at the time of its deposition). Some theories of dolomitization suggest that a combination of waters (for example, meteoric and sea waters, or fresh and salt waters) in certain ratios is an important part of the process. The fluids that bring about dolomitization must be able to carry magnesium cations to the calcium carbonate deposit and remove excess calcium cations from it.

Fluid flow is another key component of these models; there must be a way for the magnesium-rich solutions to reach the parent materials and to move on, carrying with them the excess calcium cations. Proposed fluid-flow mechanisms include sea-level fluctuations, gravity-driven percolation through permeable material, periodic tidal influx, and density differences between fluids. An example of the latter is evaporative reflux, a process by which the evaporation of highly saline waters eventually produces magnesium-rich brines dense enough to sink, displacing connate waters and seeping through the underlying sediment or rock.

INVESTIGATIONS INTO DOLOMITIZATION

Laboratory investigations into the dolomitization process have identified catalysts and inhibitors that are consistent with observations of naturally formed dolomites. These catalysts include an increase in temperature, in the ratio of magnesium cations to calcium cations in the parent solution, and in the surface area of the material being replaced. The precipitation of evaporite minerals may encourage dolomitization by positively affecting the cation ratios. Experiments have also found that dolomitization occurs more readily with aragonite than with calcite.

Fluid inclusions, minuscule cavities within a mineral that contain liquid or gas, have yielded additional insights into dolomitization. Fluid inclusions within a dolomite crystal provide geochemists with a sample of the spent solution from which the dolomite crystallized. Through laboratory analysis of fluid inclusions, researchers gain information on the origin and timing of the dolomitization and learn what temperatures and pressures are linked to the process.

DOLOMITE IDENTIFICATION

In the field, geologists often use dilute hydrochloric acid in identifying dolomite. Like other carbonates, dolomite effervesces in dilute hydrochloric acid: The rock fizzes as the acid's hydrogen reacts with the carbonate ions to produce water and carbon dioxide. In the case of dolomite, however, the reaction proceeds relatively slowly, whereas calcite effervesces rapidly and strongly. In some cases, the dolomite must be powdered or the acid heated before a reaction is evident. This characteristic difference in effervescence is often used as a quick and easy method for distinguishing dolomite from calcite.

In the laboratory, staining techniques are a common method used in dolomite identification. In fact, one of the earliest rock-staining methods, developed by J. Lemberg in the late 1800's, was designed for the study of dolomite and other carbonates. Staining techniques employ a dye or reagent, or a series thereof, to stain a rock sample selectively, coloring only those components having a certain composition or chemical property. Many staining methods are best suited for use on a hand specimen with a polished surface, or on rock sliced and ground down to a thin section; however, some staining methods will accommodate rock chips or individual mineral grains. Staining used in conjunction with microscopy is particularly helpful where dolomite and calcite are intimately interlayered within a carbonate rock.

X-ray diffraction is another laboratory method commonly employed in identifying and studying dolomite. Different crystal structures will diffract a beam of X rays in different, characteristic ways, depending on the spacing of the planes of atoms within the crystal. An older dolomite with a well-ordered crystal lattice will yield an X-ray diffraction pattern distinctly different from that of a structurally similar calcite crystal.

INDUSTRIAL APPLICATIONS

Dolomite and dolomite-containing rocks are important natural resources that are used for a variety of purposes. Like limestone, dolostone and rocks of intermediate composition are quarried, cut, and shaped into blocks and slabs that are widely used for construction. Coarsely crushed, they become aggregate, used as a component of concrete and other construction materials, as railroad ballast, as fill material, and as sewage filter beds.

Dolostone and rocks of similar composition are a source of dolomitic quicklime (calcium and magnesium oxides) and related compounds, such as type N and type S dolomitic hydrates. Dolomitic quicklime is similar to lime (calcium oxide), a well-known basic industrial compound obtained from limestone, and shares many of its uses. Dolomitic quicklime is produced by heating crushed dolostone or other dolomitic carbonate rocks to above 1,000 degrees Celsius. This process,

called "calcining," causes the rock to dissociate; it loses carbon dioxide and yields dolomitic quicklime. Adding water to stabilize dolomitic quicklime yields the slaked, or hydrated, forms. The glass, paper, chemical, construction, and waste-treatment industries are among the major consumers of these compounds. Dolomitic carbonates are also a key component in the ferrosilicon (Pidgeon) process for obtaining magnesium metal.

Another industry in which dolomite and dolomite-containing rocks are extensively used is metallurgy. Dolostone and similar rocks are commonly employed as flux stone in smelting operations such as the blast-furnace process that produces pig iron from iron ore. Flux is a material that lowers the melting temperature of the metal and purifies it. In iron smelting, dolomitic flux is introduced into the blast furnace along with the iron ore and coke. It combines with incombustible impurities from the ore and ash from the coke to form a liquid waste material called "slag." The slag floats on top of the molten iron and is drawn off separately. Limestone was previously the more widely used flux stone, because dolostone was believed by some to be less effective as a flux and purifier, to impede the fluidity of the slag, and to necessitate higher furnace temperatures. However, environmental considerations have helped to increase dolostone's popularity as a flux. Unlike limestone, which produces a hardened slag that disintegrates in water, dolostone forms a solidified slag that can be employed as a lightweight aggregate or for other uses.

Additional metallurgical applications for dolomitic materials make use of their refractory properties. A refractory is a substance that can withstand a corrosive and high-temperature environment, such as that found within a metallurgical furnace. Heating dolostone or rocks of similar composition in a blast furnace or special kiln at about 1,650 degrees Celsius drives away the little carbon dioxide that might remain in ordinary dolomitic quicklime, yielding what is called "dead-burned" dolomite. This material is used for manufacturing refractory brick to line the basic open-hearth furnaces and converters used in producing steel.

ROLE IN CARBON CYCLE

Most important, dolomitic carbonates play a role in the Earth's carbon cycle. Some of the vast amounts of carbon dioxide that pass from the atmosphere into the oceans ultimately become bound up in carbonate rocks and minerals. Dolostones and the other carbonate rocks serve as the largest repository of carbon in the cycle. The carbon remains locked in these rocks until they are subducted, in which case the carbon may be vented to the atmosphere in volcanic gases, or until they are uplifted to the Earth's surface, in which case the carbon is released during erosion. The slow turnover in the carbonate-rock reservoir helps to regulate the composition of the Earth's atmosphere and oceans.

Karen N. Kähler

CROSS-REFERENCES

BIBLIOGRAPHY

Bates, Robert L. *Geology of the Industrial Rocks and Minerals.* New York: Dover, 1969. The subchapter on limestone and dolomite compares and contrasts dolostone with limestone and describes dolomite's uses and physical and chemical properties. Although primary deposition is given more emphasis than is found in many recent works, this remains a useful, clearly written overview. Includes extensive references.

Brownlow, Arthur H. *Geochemistry.* 2d ed. Upper Saddle River, N.J.: Prentice-Hall, 1996. The chapter on sedimentary rocks includes a straightforward, comprehensible discussion of dolomitization and dolostones within the context of diagenesis. Includes references.

Evans, Anthony M. *An Introduction to Economic Geology and Its Environmental Impact.* Malden, Mass.: Blackwell Science, 1997. This college textbook provides a wonderful introduction to mines and mining practices, minerals, ores, and the economic opportunities and policies surrounding these resources. Illustrations, maps, index, and bibliography.

Hubbard, Harold A., and George E. Ericksen. "Limestone and Dolomite." In Donald A. Brobst and Walden P. Pratt, eds. *United States Mineral Resources (U.S. Geological Survey Professional Paper 820).* Washington, D.C.: U.S. Government Printing Office, 1973. Discusses dolomite and dolostone as natural resources. Summarizes the occurrence, geologic environments, desirable properties, and uses of dolomitic carbonates and limestone. Includes references.

Klein, Cornelius, and Cornelius S. Hurlbut, Jr. *Manual of Mineralogy.* 21st ed. New York: John Wiley & Sons, 1999. The subchapter on carbonate minerals includes a description of dolomite's crystallography, physical properties, composition, structure, occurrence, and uses. A later chapter on mineral assemblages discusses dolostones and other carbonate rocks. This lucid, classic college text includes references and suggested readings.

Krauskopf, Konrad B., and Dennis K. Bird. *Introduction to Geochemistry.* 3d ed. New York: McGraw-Hill, 1995. The chapter on carbonate sediments includes a brief but helpful discussion of "the dolomite problem." Includes references and suggestions for further reading. Suitable for the college-level reader.

Morse, John W., and Fred T. Mackenzie. *Geochemistry of Sedimentary Carbonates.* Amsterdam: Elsevier, 1990. Dolomite, dolostone, and dolomitization are discussed throughout the text; chapter 7 in particular provides a good overview of what is known regarding dolomite formation. While the material is technical in nature, it is clearly presented.

Purser, Bruce, Maurice Tucker, and Donald Zenger, eds. *Dolomites: A Volume in Honor of Dolomieu*. Oxford, England: Blackwell Scientific Publications, 1994. Although many of the articles in this volume are highly technical, the general college-level reader should find the introductory articles accessible, particularly the opening discussion of the problems, progress, and future research regarding dolomites and dolomitization.

Selley, Richard C. *Applied Sedimentology*. London: Academic Press, 1988. The chapter on autochthonous sediments includes a substantial section on carbonate rocks and minerals. Primary and secondary dolomites and the chemical constraints on dolomite formation are discussed. Suitable for the college-level reader.

Tucker, M. E. *Sedimentary Petrology: An Introduc-tion to the Origin of Sedimentary Rocks*. 2d ed. Oxford, England: Blackwell Scientific Publications, 1991. The chapter on limestone includes a section that discusses dolomite and the various theories of dolomitization. A technical text appropriate for the college-level reader.

Zenger, Donald H., and John B. Dunham. "Concepts and Models of Dolomitization—An Introduction." In *Concepts and Models of Dolomitization*, edited by Donald H. Zenger, John B. Dunham, and Raymond L. Ethington. Tulsa, Okla.: Society of Economic Paleontologists and Mineralogists, 1980. A good overview of various proposed models for dolomitization, suitable for the college-level reader. Includes copious references.

FERTILIZERS

Fertilizers increase food, feed, and fiber production by providing minerals needed by plants. Organic refuse, gases, rock phosphate, and salts from ancient seas are the chief sources of materials for fertilizers.

PRINCIPAL TERMS

COMPOUND: a chemical combination of elements with distinct properties that may differ from the elements from which it formed

ELEMENT: a pure substance that cannot be broken down into anything simpler by ordinary chemical means

GRADE: the percent by weight of a nutrient present in a fertilizer, expressed as a standard element or compound

LEACH: to dissolve from the soil

MACRONUTRIENT: a substance that is needed by plants or animals in large quantities; nitrogen, phosphorus, and potassium are macronutrients

MICRONUTRIENT: a substance that is needed by plants or animals in very small quantities

MINERAL: a naturally occurring substance with definite chemical and physical properties

NUTRIENT: a substance required for optimal functioning of a plant or animal; foods, vitamins, and minerals essential for life processes are nutrients

pH: a term used to describe the hydrogen ion activity of a system; a solution of pH 0-7 is acid, pH of 7 is neutral, pH 7-14 is alkaline

SEDIMENTARY: formed from material that has been deposited from solution or eroded from previously existing rocks

---

TYPES AND GRADES

The major nutrients needed by plants are carbon, oxygen, hydrogen, and nitrogen. They form 95 percent of a plant's dry weight. Carbon and oxygen can be taken in directly from the air. Hydrogen comes from water absorbed by plant roots. Nitrogen, although present in the atmosphere, is not taken in by plants in the gaseous form. It must come from soluble nitrogen-containing compounds in the soil or be produced by bacteria that live in the roots of some plants. The remaining 5 percent of plant dry weight is composed of phosphorus, potassium, calcium, magnesium, silicon, sodium, sulfur, and chloride. Selenium, aluminum, silicon, iron, boron, molybdenum, zinc, and manganese are necessary to most plants in very small amounts. Because plants cannot take in nitrogen directly from the air, and nitrates are easily leached from soils, nitrogen fertilizers are the most commonly used fertilizers. Phosphorus and potassium are also commonly added to soil by fertilizers.

Two types of fertilizers may be used. Organic fertilizers, which are produced by plants and animals, include bone meal, blood meal, humus, peat, domestic sewage sludge, and manure. Guano, the accumulated wastes of birds or bats, is also an organic fertilizer. Inorganic fertilizers begin with atmospheric gases, natural gas, or deposits of sedimentary rock. The rocks precipitate from mineral-rich coastal waters or form when shallow lakes and seas evaporate. Inorganic materials are chemically processed to yield commonly used fertilizers such as liquefied ammonia, urea, and superphosphates.

The price and recommended use of commercial fertilizers depend on their grade. Both organic and inorganic fertilizers are labeled with a grade. Fertilizer grade is usually written as three numbers, such as 10-16-10 or 10-0-10. Each number tells the percentage by weight of one of the three macronutrients. They are always given alphabetically: nitrogen, phosphorus, potassium. The first number represents total nitrogen (in the form of elemental nitrogen) in percent by weight. The second macronutrient, phosphorus, is usable by plants when it is soluble in ammonium citrate. Usable phosphate is calculated as phosphorus pentoxide. Potassium, the third major nutrient, is calculated as water-soluble potassium oxide. A fertilizer with a grade of 10-5-10 would contain 10 percent by weight of nitrogen, 5 percent by weight of available phosphorus, and 10 percent by weight of water-soluble potassium. A fertilizer may contain more of a nutrient than is stated on the label. If chemical tests show that the nutrient is in a form not usable by plants, the amount is not used in calculating the grade.

NITROGENOUS FERTILIZERS

Ammonia is the principal nitrogenous fertilizer. Pure ammonia is a colorless, pungent, irritating gas. An inorganic fertilizer, it becomes usable to plants when liquefied and injected into the soil. Most fertilizers sold are ammonia or a compound derived from ammonia. Anhydrous (without water) ammonia is made by combining atmospheric nitrogen and natural gas.

Urea contains 45-46 percent nitrogen. The nitrogen in urea must be converted to the ammonium ion by soil enzymes and held in humus or clay to be useful to plants. Urea can be applied as pellets or in a water solution. Chemical manufacturing plants synthesize urea from ammonia and carbon dioxide. Another solid source of nitrogen is ammonium nitrate. It contains 33.5 percent nitrogen and includes both ammonium and nitrate forms of nitrogen. The nitrate can be used directly by plants, but it is easily leached. Ammonium fixed onto clays becomes available to plant roots. Ammonium nitrate is synthesized from ammonia and nitric acid. Natural deposits occur in Chile. Ammonium sulfate is made from coke-oven gases. It is 21 percent nitrogen as the ammonium ion, NH_3^+. Ammonium sulfate supplies sulfur as well as nitrogen and is often used on wet soils.

Often, urea or other solid fertilizers are coated with sulfur or compounded from materials of different solubilities. The sulfur coating slows the solution of the fertilizer granule. When materials of different solubilities are blended, some will dissolve immediately, while others take longer to dissolve. These fertilizers release nitrogen continuously and save on the costs of application when supplements are needed over a long period of time. Nurseries and turf farms prefer continuous-release fertilizers.

PHOSPHATE FERTILIZERS

A second macronutrient supplied by fertilizers is phosphorus. Phosphate fertilizers are available in both inorganic and organic forms. Rocks rich in phosphorus are mined in the southeastern and western United States. Rock phosphate fertilizers contain about 4-5 percent equivalent of phosphorus pentoxide. The rock is merely ground into particles small enough to be applied to the soil. Bone meal, the phosphorus fertilizer that has been in use the longest time, contains about the same amount of available phosphorus as rock phosphate. Bones from slaughterhouses provide the raw materials for bone meal.

Rock phosphate can be enriched in several ways. When mixed with sulfuric acid, it becomes superphosphate with 16-20 percent phosphorus pentoxide. Mixtures with phosphoric acid yield triple superphosphate, with a 0-45-0 grade. Other phosphate fertilizers are diammonium phosphate (46-53 percent phosphorus pentoxide and 18-21 percent nitrogen), monomonium phosphate (48 percent phosphorus pentoxide and 11 percent nitrogen), and nitric phosphate (20 percent phosphorus pentoxide and 20 percent nitrogen). Nitric phosphates are made by treating phosphate rock with nitric acid and phosphoric acid. The addition of muriate of potash (potassium chloride) to nitric phosphates gives a 10-10-10 grade fertilizer containing all three basic nutrients. Polyphosphate fertilizers, another inorganic source of phosphorus, contain high amounts of phosphorus pentoxide. They form soluble combinations with iron and aluminum. These chemicals, called complexes, are usable by plants. Other phosphate fertilizers may form insoluble compounds in soils that have high iron and aluminum content. Thus, the phosphate is not available to plants.

Phosphate rock is minable in Florida, Tennessee, North Carolina, Idaho, Wyoming, and Montana. North Africa and the region west of the Ural Mountains also have large deposits. In the United States, geoclinal (West Coast) deposits form in deep areas at the margins of continents. The geoclinal deposits of the northern Rocky Mountains extend over hundreds of square kilometers. Platform (East Coast) deposits were formed closer to shore where phosphate-rich waters warmed on approaching land. Weathered or residual deposits form when calcium-rich rocks are leached and less soluble phosphate remains. The "brown-rock" deposits of Tennessee are of this type. Sometimes, the phosphate goes into solution and is deposited lower in the soil profile. The Pliocene Bone Valley formation of Florida contains phosphatic pebbles and nodules reworked from the underlying Miocene Hawthorn formation. Phosphatic guano occurs on some Pacific islands where old deposits of bird excreta have lost their nitrogen through decomposition. Guano that has decomposed and then has been leached by percolating waters may reach 32 percent phosphorus pentoxide.

POTASSIUM FERTILIZERS

The third nutrient commonly supplied in fertilizers is potassium. The most widely used potassium supplements contain muriate of potash, sulfate of potash, or sulfates of potassium and magnesium. These are all inorganic salts. Muriate of potash, chemically potassium chloride, is 95 percent muriate of potash, equivalent to 60 percent potassium oxide. Although it is the most commonly used potassium source, muriate of potash can be toxic to some plants. Potatoes, tobacco, and avocados are sensitive to excess chlorine. Sulfate of potash contains 48-50 percent available potassium oxide. Sulfate of potash-magnesia analyzes at about 18 percent magnesium oxide and 22 percent potassium oxide. Other potassium fertilizers include potassium nitrate and potassium polyphosphate.

Potassium salts are mined in Canada, New Mexico, Utah, and elsewhere from deposits left behind by the withdrawal or evaporation of ancient shallow seas. Brines from the Great Salt Lake are evaporated to yield magnesium and potassium chlorides. Potash salts are the last salts to come out of solution when seawater evaporates. This fact, combined with their solubility in water, makes potash salts more rare than gypsum and rock salt. The largest mineral deposits of potash salts occur in the Devonian muskeg or Prairie formation of western Canada, North Dakota, and Montana. The Saskatchewan deposits occur nearest the surface and so are the leading producer. The Permian Zechstein evaporites of England, Germany, and Poland are equally famous. Guano derived from the wastes of birds or bats is an organic source of potassium nitrate fertilizer.

SPECIALTY FERTILIZERS

Specialty fertilizers supply nutrients that may be depleted by fertilizer use, soil management prac-

tices, irrigation, or heavy crop production. Sulfur deficiencies can be alleviated by using ammonium or potassium sulfate fertilizers. The use of ammonia fertilizers, leaching, or excessive moisture may lead to acidic soils. Dolomitic limestone adds calcium and magnesium and corrects soil pH. Some ammonia fertilizers include recommendations for sweetening soils that may become acidic because of the production of hydrogen ions. Boron is supplied by borax. Borates, from which borax is made, occur in bedded deposits beneath old playas, brines of saline lakes, and hot springs or fumaroles. Glass particles added to fertilizers also provide boron.

Micronutrient metals such as molybdenum, iron, zinc, manganese, and copper can be added to inorganic fertilizer as needed. Soil tests, chemical analysis of crop samples, and crop performance help indicate need. Molybdenum provided by sodium or ammonium molybdates can be added in small quantities to other fertilizers or mixed with water and used to soak seed. Iron, zinc, and copper sulfates supply soluble metals.

The formation of water-insoluble chelates may lead to shortages of micronutrient metals. Chelates are organic chemicals that bond to a metal to keep it from forming a precipitated salt. Helpful chelates are soluble in water. Manure contains water-insoluble chelates that may tie up metals. Excessive use of manure as a fertilizer, therefore, can cause deficiency of metals.

Commercial solid fertilizers combine several types of ingredients. The "carrier" contains the nutrient element. Nitrates, ammonium sulfates, and ammonium carbonate carry nitrogen. Phosphates carry phosphorus. Potassium can be carried with nitrogen in potassium nitrate. The second ingredient of many fertilizers is a conditioner. Conditioners prevent caking and assure good flow. Vermiculite and organic wastes condition by absorbing water. Diatomaceous Earth, oils, plastic, and waxes maintain flow. Often, inert substances are selected so that reactions will not affect the solubility of the carrier. Neutralizers, such as ground dolomitic limestone, may be added to counteract acidity. Fillers, such as sand, bring the product up to a standard weight. Special additives—including micronutrients, fungicides, herbicides, and insecticides—help save on labor costs or prevent unwanted plants from absorbing nutrients intended for the crop. Fertilizers are often blended for the needs of specific soils growing specific crops.

AGRICULTURAL AND ENVIRONMENTAL CONCERNS

Most current research on fertilizers is related to more efficient agricultural use and to assessing and reducing the effects of fertilizer on the environment. Coated fertilizers and combination fertilizers with timed release of nutrients are being developed. Studies show that for some crops, applying fertilizer as a liquid and at certain stages of plant growth is preferable to solid application. Micronutrients have been found to be of importance in the budding and setting of fruit.

Areas of environmental concern include assessment of fertilizer's effect on the groundwater and the health of developing nations. Using locally available materials such as greensand, manure, chitin from crustacean shells, or phosphate rock may boost Third World agricultural production without incurring foreign debt or environmental damage. In the western United States, excessive use of fertilizers combined with irrigation from salty river waters has left some soils too salty to grow certain crops. Research continues on the use of low-salt fertilizers and salt-tolerant crop varieties. Finally, research on the use of sewage sludge as a fertilizer and conditioner is important. Accumulation and uptake of heavy metals may be prevented by the addition of fungi, which concentrate metals and prevent them from entering crop plants. Fertilizer use becomes more efficient as researchers learn more about plant metabolism and the role of specific nutrients at different stages of the plant life cycle.

A soil test can reveal the need for fertilizers. Agricultural specialists suggest the best type of fertilizer for a particular soil and crop. Some nutrient carriers, such as muriate of potash, may harm salt-sensitive crops such as tobacco and potatoes. Crops such as trees or turf grass may require a continuous-release form of fertilizer. Carefully planned fertilization avoids waste and minimizes hazards to the applicator and to the environment.

Dorothy Fay Simms

CROSS-REFERENCES

Aluminum Deposits, 1539; Building Stone, 1545; Carbonates, 1182; Carbonatites, 1287; Cement, 1550; Chemical Precipitates, 1440; Coal, 1555; Diamonds, 1561; Dolomite, 1567; Geochemical Cycle, 412; Gold and Silver, 1578; Hydrothermal Mineral Deposits, 1584; Industrial Metals, 1589; Industrial Nonmetals, 1596; Iron Deposits, 1602; Limestone, 1451; Manganese Nodules, 1608; Metamorphic Mineral Deposits, 1614; Pegmatites, 1620; Platinum Group Metals, 1626; Reefs, 2347; Salt Domes and Salt Diapirs, 1632; Seawater Composition, 2166; Sedimentary Mineral Deposits, 1629; Sedimentary Rock Classification, 1457; Uranium Deposits, 1643.

BIBLIOGRAPHY

Bates, Robert L. *Geology of the Industrial Rocks and Minerals.* Mineola, N.Y.: Dover, 1960. Provides information on the occurrence and use of nonmetallic minerals, including mineral fertilizers.

Birkeland, P. W. *Soils and Geomorphology.* 3d ed. New York: Oxford University Press, 1999. One of the best books written on soils from the perspective of a geologist, it gives many examples from around the world. The newly adopted soil profile description symbols are covered in the early chapters. A very elementary discussion of Soil Taxonomy in the early chapters. Suitable for university-level readers, but the early chapters are easily understood by high-school-level readers.

Buol, S. W., F. D. Hole, and R. J. McCracken. *Soil Genesis and Classification.* 4th ed. Ames: Iowa State University Press, 1997. This book is the latest edition of a classic soil science text. The new edition contains an update on the new soil horizon symbol nomenclature. Provides an in-depth discussion of Soil Taxonomy classification that is fairly easy to read. Suitable for university-level readers, but it would be helpful to read the first few chapters of the Birkeland or Singer and Munns book before reading this text.

Colwell, J. D. *Estimating Fertilizer Requirements: A Quantitative Approach.* Wallingford, England: CAB International, 1994. This book provides mathematical and statistical models to test and estimate the properties of fertilizers and soils. This book is quite technical at times and would be most useful to students with some familiarity with the subject.

Donahue, Roy L., Raymond W. Miller, and John C. Shickluna. *Soils: An Introduction to Soils and Plant Growth.* 4th ed. Englewood Cliffs, N.J.: Prentice-Hall, 1977. Includes chapters on soil chemistry, fertilizers, and plant nutrition.

Foth, Henry D., and Boyd G. Ellis. *Soil Fertility.* 2d ed. Boca Raton, Fla.: CRC Lewis, 1997. A good introduction to soil and fertilizers, this book describes soil chemistry, fertilizers, and plant nutrition, as well as ways to test for fertility.

Govett, G. J. S., and M. H. Govett, eds. *World Mineral Supplies: Assessment and Perspective.* New York: Elsevier, 1976. A compilation of the location and availability of industrial and agricultural minerals.

Jensen, Mead L., and Alan M. Bateman. *Economic Mineral Deposits.* 3d ed. New York: John Wiley & Sons, 1979. Discusses the world's best-known mineral deposits.

Pettijohn, F. J. *Sedimentary Rocks.* 3d ed. New York: Harper & Row, 1975. A thorough discussion of the origin, classification, and uses of sedimentary rocks.

Sides, Susan. "Grow Powder." *The Mother Earth News* 114 (November/December, 1988). Discusses organic and unprocessed or minimally processed fertilizers.

Sopher, Charles D., and Jack V. Baird. *Soils and Soil Management.* Reston, Va.: Reston Publishing, 1978. A concise description of the major fertilizers with some crop recommendations.

GOLD AND SILVER

Silver and gold have played important roles in human economies, industries, and finances for thousands of years. They have served widely in applications as diverse as medicine and jewelry and remain two of the most highly sought and useful of metals.

PRINCIPAL TERMS

AMALGAM: an alloy of mercury and another metal; gold and silver amalgams occur naturally and have been synthesized for a variety of uses

CARAT: a unit of measure (abbreviated "k") of the purity of gold; it is equal to $\frac{1}{24}$ part gold in an alloy

ELECTRUM: a term commonly used to designate any alloy of gold and silver containing 50-80 weight percent gold

FINENESS: a measure of the purity of gold or silver expressed as the weight proportion of these metals in an alloy; gold fineness considers only the relative proportions of gold and silver present, whereas silver fineness considers the proportion of silver to all other metals present

LODE DEPOSIT: a primary deposit, generally a vein, formed by the filling of a fissure with minerals precipitated from a hydrothermal solution

PLACER DEPOSIT: a mass of sand, gravel, or soil resulting from the weathering of mineralized rocks that contains grains of gold, tin, platinum, or other valuable minerals derived from the original rock

NATURAL FORMS

Gold, a deep yellow metal, is chemical element number 79, with an atomic weight of 196.967. Natural gold consists of a single isotope, gold 107, and in a pure form exhibits a density of 19.3 grams per cubic centimeter. Silver, a brilliant white metal, is chemical element number 47, with an atomic weight of 107.87. Natural silver consists of two isotopes, silver 107 (51.4 percent) and silver 109 (48.6 percent), and in a pure form exhibits a density of 10.5 grams per cubic centimeter. Both metals are very malleable, and this property has led to the common use of gold in gilding, a process by which extremely thin films are applied to surfaces of metal, ceramic, wood, or other materials for decorative purposes. Gold is so malleable that it can be pounded into translucent films so thin that it would take more than 300,000 of them to form a pile 1 inch (2.5 centimeters) high.

Gold and silver occur in a variety of minerals. Native gold and electrum, the gold-silver alloy, are the most commonly occurring natural forms of gold in ore deposits. Although laboratory studies indicate that all compositions of gold-silver alloys may be synthesized, naturally occurring gold and electrum compositions nearly always lie between 50 and 95 weight percent gold. The most commonly encountered compositions contain about 92 percent gold.

PRINCIPAL GOLD AND SILVER MINERALS

Gold Minerals		Silver Minerals	
Native gold	Au	Native silver	Ag
Electrum	(Au,Ag)	Acanthite	Ag_2S
Calaverite	$AuTe_2$	Tetrahedrite	$(Cu,Ag)_{12}(Sb,As)_4S_{13}$
Montbrayite	Au_2Te_3	Dyscrasite	Ag_3Sb
Krennerite	$(Au,Ag)Te_2$	Polybasite	$(Au,Cu)_{16}Sb_2S_{11}$
Petzite	Ag_3AuTe_2	Proustite	Ag_3AsS_3
Sylvanite	$AuAgTe_4$	Pyrargyrite	Ag_3SbS_3
Aurostibite	$AuSb_2$	Smithite	$AgAsS_2$
Maldonite	Au_2Bi	Stephanite	Ag_5SbS_4

ECONOMIC VALUE

Although gold and silver are commonly thought of as precious metals, they differ greatly in their chemical properties and economic value. Gold is a true noble metal; that is, it is almost totally resistant to chemical attack and dissolution. Gold forms few chemical compounds and minerals. On the other hand, silver is less noble, meaning that it is considerably easier to dissolve and that it forms several compounds and minerals. Gold is about sixty times costlier than silver; this difference is partly because of gold's superior chemical properties but mostly because of its rareness relative to silver. These differences affect the geochemical and geological behavior of these metals as well as the intensity of the mining and mineral exploration activities directed at their recovery.

The average abundance of gold in the Earth's crust is about 0.004 part per million, and the average abundance of silver is 0.08 part per million, about twenty times that of gold. These amounts of gold and silver are far too small to pay for the extraction of these metals from common rock. In some places in the Earth's crust, however, hydrothermal solutions have circulated, collected some of the gold and silver from the rocks, and transported them to sites of deposition. The rocks at these sites contain gold and silver values many times greater than those of average rocks and are sought out as ore deposits. In order for gold deposits to be mined economically, the concentrations of the gold generally must be on the order of 1-3 parts per million (about 0.1 troy ounce per ton), but some of the richest deposits contain ores with 8-10 parts per million (about 0.2 troy ounce per ton). Because silver is worth much less, deposits must generally contain 150-300 parts per million (5-10 troy ounces per ton) if they are mined solely for silver. Much silver, however, is extracted as a by-product from copper, lead, and zinc ores in which the silver is present in concentrations of 1-50 parts per million.

Because economic concentrations of gold are so low, very sensitive chemical, analytical techniques must be used to determine the amount of gold present. One of the oldest but still widely used is fire assaying. In this method, lead oxide is mixed with a powdered ore sample and melted until the lead oxide is converted to metallic lead that picks up the gold and silver and settles to the bottom of the crucible. The lead, gold, and silver bottom is melted in a porous cup where the lead is again converted to lead oxide that is absorbed by the cup, leaving a gold and silver mixture known as doré metal. The doré metal is weighed and the silver is dissolved by nitric acid. The remaining pure gold is reweighed to determine the amount of gold.

MINING AND EXTRACTION

When economic gold-bearing deposits are found, the method of mining the ore and extracting the metals (some silver is always present) varies, depending on the location of the deposit and its mineralogy. If the gold occurs as large enough grains, it is readily concentrated by gravity techniques, and the high density of the gold (15-19 grams per cubic centimeter) allows it to be physically separated from the other minerals that have much lower densities. The gold pans and sluices used by early miners and some mines today are small-scale examples of gravity methods. If, on the other hand, the gold is present in very small, sometimes submicroscopic grains, chemical solvent techniques are utilized. The most commonly employed method applies a sodium cyanide solution that dissolves the gold. The gold is removed from the solution by activated charcoal, and the solution is reused. The gold is redissolved from the charcoal and then electrochemically precipitated on steel wool. The steel wool is mixed with silica, borax, and niter and melted to produce a doré metal of the gold and silver and a slag with all the impurities. The doré metal is sent to a refinery to separate the gold and silver.

Silver-rich ores usually contain much higher concentrations of silver-bearing minerals than do gold mines. After grinding the ores to 0.01-0.05 inch (0.2-1.0 millimeter), the silver minerals are separated by flotation techniques in which small bubbles generated in soaplike solutions are used to pick up and concentrate the small grains. The concentrated materials are shipped to refineries for chemical separation of the metals and sulfur.

The principal gold- and silver-producing areas in the United States are the Carlin district, Nevada, and the Coeur d'Alene district, Idaho, respectively. South Africa has the largest gold deposit in the world, called the Witwatersrand district. This

Gold in its native state. (U.S. Geological Survey)

district is thought to contain at least 760 million troy ounces (24,000 metric tons) of gold and produces about 22 million troy ounces (700 metric tons) of gold annually. The Real de Angeles Mine in Mexico is the world's largest producer of silver. This district produces about 13 million troy ounces of silver annually. Gold and silver are produced in many other areas of the world as well. In addition, significant amounts of gold and silver are recovered as by-products during the processing of the ores of other metals such as copper.

HYDROTHERMAL SYSTEMS

There are many types of hydrothermal systems that produce deposits of gold, silver, or both. Many of these hydrothermal systems are ancient analogues of modern geothermal systems such as the one at Yellowstone National Park, Wyoming. Geologists deduce the nature of the hydrothermal system that produced a particular deposit by studying its size, shape, mineralogy, ore grade, and tectonic setting. The size and shape of the ore body are determined from maps of the surface outcrop and from geological cross sections of the subsurface extent of the ore body, prepared by consulting drill core logs, and maps of outcrops in underground workings. The mineralogy of the ore (mineralized rock from which metals can be economically extracted), the gangue (minerals that have no economic value and that were depos-

ited by the hydrothermal solutions), the country rock (the unmineralized rock surrounding the deposit), and the alteration zone (country rock that was changed chemically or mineralogically by reactions with the hydrothermal solutions) are determined by hand sample observations, microscopic (transmitted and reflected light) observation, X-ray diffraction analysis, and electron microprobe analysis. The types, compositions, and associations of these minerals are used to infer the chemical conditions at the time of the formation of the ore deposit. Fluid inclusions, or small bubbles of the hydrothermal solution trapped in growing mineral grains, are commonly observed features that are particularly useful in determining the nature of the fluids from which the ore minerals precipitated. The gold and/or silver content of the rocks (ore grade) is determined by chemical analysis of systematically selected samples from various parts of the ore body. Some ore bodies show sharp cutoffs of grade between the mineralized rock and the country rock. Others show a gradual decrease from the mineralized zone to the background levels of the country rock. The tectonic (deformational crust) and structural setting of the ore body give evidence of the general geologic framework for the ore-forming process. Nearly all hydrothermal ore deposits are associated with tectonically active areas, where the rate of heat flow from the Earth's interior to the surface is quite high. It is this heat flow that increases the temperatures of the hydrothermal solutions and drives them through the rocks. Mid-ocean ridges and rises and island arcs associated with subduction zones (where the edge of one crustal plate descends below the edge of another) are recognized as tectonically active areas that have high rates of heat flow and are thus settings for the potential formation of hydrothermal ore deposits.

Weathering breaks down rocks both chemically and physically. The products of weathering are ions that are carried away by solutions and solids that are eroded from the surface by rainwater. When hydrothermal ore deposits weather, most of the ore minerals are destroyed by oxidation, and their components are carried away in solution. If these solutions penetrate downward into the outcrop, they sometimes encounter more reducing conditions at depth, and the ore elements reprecipitate, or separate out again, to form an enriched zone. This process is called secondary enrichment. The grade both of gold and especially of silver deposits can be improved by this process. Most of the time, however, gold is simply too unreactive to be destroyed by chemical weathering and instead concentrates in the soil horizon to form an eluvial placer. As the soil erodes, the gold grains are carried into streams and rivers where they are transported until they encounter areas of slower water velocity. There, they are dropped by the stream and accumulate, because these grains have a much greater density than do common silicates. Gold and electrum grains have densities ranging from 15.5 to 19.3 grams per square centimenter, whereas common silicate minerals have densities ranging from 2.6 to 3.0 grams per square centimeter. Thus, the common silicates are simply washed away, leaving the gold and electrum grains behind. The Witwatersrand deposits are paleoplacer deposits; that is, they are placer deposits that formed about 2 billion years ago when rivers and streams carrying gold grains flowed into a large basin. As the waters of these rivers and streams entered this basin, their velocity slowed and they deposited their load of gold-bearing gravels and sands. These deposits were eventually covered by other rocks and became lithified, or changed to stone. The very large extent and relatively high grade (8-10 parts per million) of these deposits make the Witwatersrand district the largest gold district in the world.

One of the more interesting problems surrounding the formation of silver and lode gold deposits is determining how these elements were transported by aqueous solutions. Gold is a noble metal, meaning that it is very resistant to the natural chemical attacks that would dissolve it. Silver is not so noble as gold and therefore tends to form compounds (minerals) with a wide variety of elements. These silver minerals as well as native silver have low solubilities. The solubility of gold and silver and their minerals does seem to increase with temperature; however, even at very high temperatures, their solubility in pure water is much too low to account for the amounts of the metals transported into ore deposits. The most likely explanation for the ability of hydrothermal solutions to transport these metals is that they react with other ions in the solution, such as chloride, bisulfide, and perhaps others, to form very stable complexes. The formation of complexes increases the stability of the gold or silver species in the aqueous solution and therefore makes gold and silver soluble enough to account for their transport to form ore deposits. When solutions containing these complexes are cooled, mixed with solutions containing lower concentrations of complexing ions, oxidized by reactions with other minerals, or boiled to remove chloride or bisulfide as vapors, the complexes tend to break down, lowering the solubility of gold and/or silver and causing these elements to precipitate in metallic form or as constituents of minerals.

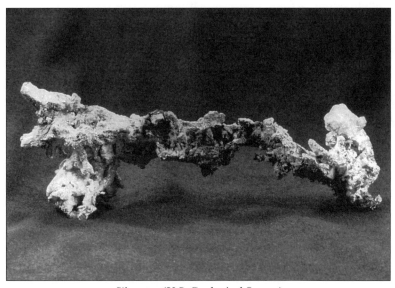

Silver ore. (U.S. Geological Survey)

CONTINUING DEMAND

Gold and silver were among the earliest of metals used by humans. They were employed as amulets and jewelry as early as 5000 B.C.E., and their use as money and in coinage had begun in Asia Minor and Greece by about 600 B.C.E. Although these practices continue today, gold and silver have found many additional uses and have diverged in their principal applications. In the United States, gold today is used primarily in jewelry and in the arts (about 55 percent of the total usage); other major uses include solid-state electronic devices (34 percent), dental supplies (12 percent), and investment products (0.1 percent). In contrast, silver is primarily an industrial metal, with 43 percent being used in photography and 35 percent being used in electrical products. Other important silver uses include jewelry (6 percent), sterling ware (6 percent), and coinage (4 percent). Gold and silver coins were widely minted in many countries, especially in the 1700's and 1800's, but were generally dropped from use in the early 1900's. The United States minted gold coins from 1849 until 1933 and silver coins from 1794 until 1964. Since 1986, the United States has minted gold and silver bullion coins that do not have assigned denominations but rather contain specified amounts of gold or silver and vary in value as metal market prices fluctuate.

Throughout history, the quest for gold and silver has played a very important role in the development and expansion of human civilization, culture, and enterprise. By the time of the Egyptian empire, humankind had already developed sophisticated mining and metallurgical processes for producing precious metals. The wealth and prosperity of Athens was based in large part upon the silver production from the mines at Laurion in southeastern Attica (Greece), and the Roman empire flourished as a result of gold and silver obtained from the Iberian Peninsula. Christopher Columbus' encounter with gold-bearing native Americans on his first voyage provided a powerful stimulus for the Spanish to explore and exploit the New World. The new gold and silver stocks (181 metric tons of gold and 18,000 metric tons of silver) shipped from the New World to Europe from 1500 to 1650 produced a revitalization of the economic system that helped fuel the later stages of the Renaissance. The search for gold and silver also played a major role in the development of the United States, especially the Gold Rush of 1849 that lured thousands to California. Subsequent gold discoveries in the Black Hills of South Dakota in 1874 and in the Klondike and Yukon in 1896 drew additional treasure hunters. The quest for gold also led to the opening of Australia (1851) and South Africa (1886). Currently, gold exploration and gold rushes in Brazil have produced an invasion of much of the tropical rain forests.

Gold and silver are expensive metals, and thus there is a constant search to find less expensive substitutes. Many jewelry, electrical, and electronic products can be made by alloying other metals with gold and silver or by merely coating base metals. Other products can be redesigned to maintain their utility while using smaller amounts of gold. In many cases, less expensive palladium and silver can be substituted for gold. Although it is even more expensive, platinum is sometimes used instead of gold for coins, jewelry, and electrical products. Aluminum and rhodium are less expensive substitutes for silver in mirrors. Surgical plates, pins, and sutures can be made of tantalum instead of silver. Silver tableware can be replaced by stainless steel products. Video cameras, silverless black and white film, and xerography reduce the silver demand of copying and photography. Even with these substitutions, it is unlikely that the demand for these metals will diminish.

J. Donald Rimstidt and James R. Craig

CROSS-REFERENCES

Aluminum Deposits, 1539; Biogenic Sedimentary Rocks, 1435; Building Stone, 1545; Carbonates, 1182; Cement, 1550; Chemical Precipitates, 1440; Coal, 1555; Diagenesis, 1445; Diamonds, 1561; Dolomite, 1567; Earth Resources, 1741; Environmental Health, 1759; Fertilizers, 1573; Groundwater Pollution and Remediation, 2037; Hydrothermal Mineral Deposits, 1584; Industrial Metals, 1589; Industrial Nonmetals, 1596; Iron Deposits, 1602; Land Management, 1484; Manganese Nodules, 1608; Metamorphic Mineral Deposits, 1614; Minerals: Physical Properties, 1225; Pegmatites, 1620; Platinum Group Metals, 1626; Salt Domes and Salt Diapirs, 1632; Seawater Composition, 2166; Sedimentary Mineral Deposits, 1629; Soil Chemistry, 1509; Soil Profiles, 1525; Uranium Deposits, 1643.

BIBLIOGRAPHY

Boyle, Robert W. *The Geochemistry of Silver and Its Deposits.* Geological Survey of Canada Bulletin 160. Ottawa: Queen's Printer, 1968. A general semitechnical presentation of the chemical behavior of silver in various geological environments and of the minerals that contain silver. Describes many of the world's major silver deposits and the techniques employed in prospecting for silver.

_____. *Gold: History and Genesis of Deposits.* New York: Van Nostrand Reinhold, 1987. This volume provides a fine overview of the history of gold and its mode of formation from primitive to modern times. Also presents annotations on the scientific articles covering the general geochemistry of gold and the discussion of various theories of the mechanisms by which deposits form. Intended for the scientist or the serious layperson.

Brooks, Robert R., ed. *Noble Metals and Biological Systems: Their Role in Medicine, Mineral Exploration, and The Environment.* Boca Raton, Fla.: CRC Press, 1992. This collection of essays explores not only the economic importance of precious metals but also the importance of the roles they play in biogeochemical and physiological research and applications. Slightly technical at times and therefore recommended for someone with some background in the field.

Cotton, Simon. *Chemistry of Precious Metals.* London: Blackie Academic and Professional, 1997. This college-level textbook clearly illustrates the procedures and protocols used to determine the chemical make-up and properties of precious metals. Illustrations, bibliography, and index.

Evans, Anthony M. *An Introduction to Economic Geology and Its Environmental Impact.* Malden, Mass.: Blackwell Science, 1997. This college textbook provides a wonderful introduction to mines and mining practices, minerals, ores, and the economic opportunities and policies surrounding these resources. Illustrations, maps, index, and bibliography.

Gasparrini, Claudia. *Gold and Other Precious Metals: From Ore to Market.* New York: Springer-Verlag, 1993. Gasparrini offers a step-by-step account of the procedures used in mining precious metals. All aspects of the field are covered, from mining the ores to the metals' economic value in the marketplace. A good book for someone without a background in the field.

St. John, Jeffrey. *Noble Metals.* Alexandria, Va.: Time-Life Books, 1984. An excellent general treatment of gold, silver, platinum, and other precious metals. Discusses the geological origin of these metals, methods of exploration, means of extraction, and uses. Also develops a useful historical perspective on the role of these metals in society. Spectacular illustrations of natural grains, nuggets, and crystals as well as manufactured precious metal objects, jewelry, and coins.

U.S. Bureau of Mines. *Mineral Facts and Problems.* Bulletin 675. Washington, D.C.: Government Printing Office, 1985.

_____. *Yearbook, 1986.* Washington, D.C.: Government Printing Office, 1986. Chapters on gold and silver provide concise discussions and many data tables on uses, resources, technology, and production relationships for gold and silver.

Watkins, Tom H. *Gold and Silver in the West: The Illustrated History of an American Dream.* Palo Alto, Calif.: American West, 1971. An abundantly illustrated historical account of the search for and discovery of gold and silver in North America. Traces the paths westward and provides a wonderfully readable and accurate history of those who sought gold and silver and the boom towns they built. Written for the layperson.

White, Benjamin. *Silver: Its History and Romance.* Detroit, Mich.: Tower Books, 1971. This book gives the history of silver mining and use from prehistoric times to the present. Contains a fair amount of production and monetary statistics as well as moderately detailed historical accounts of most of the important silver-producing regions of the world. Probably the most interesting chapters deal with the various uses of silver in jewelry and in coins by various cultures.

HYDROTHERMAL MINERAL DEPOSITS

Hydrothermal mineral deposits are formed through processes related to the flow of hot water or fluid. They are often, but not always, formed in association with magmatic activity. Deposits of this nature are important sources of numerous elements, including copper, iron, lead, zinc, gold, silver, mercury, tin, uranium, and nickel.

PRINCIPAL TERMS

COUNTRY ROCK: nonmineralized rock surrounding or penetrated by mineral veins

GANGUE: mineral material associated with ore but having no economic value

MINERAL DEPOSIT: the natural occurrence of mineral material of sufficient extent and concentration to warrant exploitation

ORE: an economically minable concentration of minerals

PLUTON: a body of igneous rock that has formed beneath the Earth's surface by consolidation from magma

STRATABOUND: a tabular or lenticular ore body within the layers of stratified or sedimentary rocks

CLASSIFICATION OF ORE DEPOSITS

Ore deposits can be classified on the basis of mineralization and associated country rock and structural elements. Various types of deposits can be found grading into one another where genetic boundaries cannot be precisely defined. Zoned mineral suites are likely to reflect differences in pressure and temperature of formation in different crustal zones, and some ore deposits may have a complex history suggesting multiple origins. Ore is generally classed, therefore, based upon the characteristics of the majority of the ore body. Several classification schemes have been developed to categorize ore deposits, and no two are alike. Swiss-born American geologist Waldemar Lindgren introduced a system in 1913 that is widely used in the United States. Modifications to Lindgren's original system have added terms that allow one to address issues not fully covered in the 1913 classification.

Hydrothermal ore deposits can be classed based upon their presence in bodies of rock or in bodies of water. They can be further grouped on the basis of temperature and pressure of formation. Hypothermal deposits are those where ore deposition and concentration have taken place at great depths or in the presence of high temperatures and pressures. Mesothermal deposits are formed at intermediate depths under high-pressure conditions but only moderately high temperatures. Epithermal deposits reflect a further decrease in temperature and pressure, while telethermal deposits are generally considered to have been formed in near-surface conditions with low temperatures and pressures. The telethermal deposits are considered to represent the upper terminus of the hydrothermal range. Xenothermal deposits are formed at shallow depths in association with plutonic intrusions, and volcanogenic deposits are formed under low-pressure conditions in association with moderately high-temperature underwater springs related to volcanism.

IDENTIFICATION AND EXPLORATION

Hydrothermal mineral deposits are variable in their mineralogy, their physical structures, and their settings. No single criterion is sufficient to characterize and classify a deposit because of the complexity and variability of factors involved. Certain associations of ore and gangue minerals, as well as a consideration of certain wall rock alteration products, are used, therefore, to identify deposits and characterize depositional zones.

Exploration for hydrothermal mineral deposits is keyed toward identifying physical settings that are typical of the wide range of known deposits. Base- and precious-metal sulfide ores associated with modern and ancient volcanism are known from a variety of tectonic settings, including continental margins, mid-ocean ridges, back-arcs, fore-arcs, within arcs, and in the vicinity of intraplate seamounts. Prospecting for deposits involves identification of the appropriate tectonic setting, regional mapping to identify likely zones of concentration (faults or other localizing structures or features must be present), and material sampling and analysis. Magnetic surveys, geochemical surveys, and petrologic investigations may also be required to identify likely ore material.

HYPOTHERMAL AND MESOTHERMAL MINERAL DEPOSITS

Hypothermal mineral deposits are generally coarse grained, having been formed at high temperatures (300 to 500 degrees Celsius) and having been subjected to an extended period of cooling under high-pressure conditions. Mineral veins in these deposits grade into the country rock without sharp boundaries. The ore may be present in other geometries, but veins and tabular shapes are most common. The ore is frequently concentrated along the crests of folds or within shear zones in the country rock. Hypothermal deposits are usually found in older rocks that have been subjected to metamorphism associated with the emplacement of plutons.

A variety of minerals may be found in hypothermal deposits, comprising ores of gold, tin, tungsten, iron, lead, arsenic, uranium, cobalt, and nickel. Quartz is the most common gangue mineral, and pyrite is normally abundant. Examples of hypothermal mineral deposits can be found throughout the world. Significant deposits include the galena and sphalerite ores at the Sullivan mine in British Columbia, the gold ores of the Homestake mine in South Dakota, and the Kolesar gold fields in Mysore, India.

Mesothermal mineral deposits are formed under lower pressures than hypothermal deposits; they can, in fact, be formed under near-surface conditions. They represent an intermediate zone of ore deposition, with temperatures of formation ranging from 200 to 300 degrees Celsius. Although these deposits may have an association with igneous rocks, host rocks may also be metamorphic and are commonly sedimentary. Mesothermal mineral deposits may be in the shape of veins or pipes, or they may be present as disseminated veinlets or spots of ore within the host rock. Many deposits are formed as a replacement of a favorable bed within the host rock stratigraphy.

Copper, lead, zinc, molybdenum, silver, and gold are commonly found in mesothermal deposits. Gangue associated with the ore minerals may include quartz, pyrite, or a variety of carbonate minerals. Alteration zones around the deposits may be extensive. The copper deposits at Matahambre, Cuba, are a good example of mesothermal deposition. The silver deposits of the Coeur d'Alene area in Idaho and the porphyry copper deposits at Chuquicamata, Chile, are also mesothermal in origin.

EPITHERMAL AND TELETHERMAL MINERAL DEPOSITS

Epithermal mineral deposits are formed at shallow depths (within 900 meters of the surface) and under low temperature conditions (50 to 200 degrees Celsius). Most are found in younger rocks in areas of tertiary volcanism. The volcanic environment has, in some instances, yielded hot waters in working mines.

An open-pit mine; the ore body is a hydrothermal deposit of copper minerals. (© William E. Ferguson)

Both primary and replacement ores can be formed under epithermal conditions. Deposits may be found as fillings in veins or irregular fissures, and as stockworks (highly fractured rock with many veins and veinlets) or breccia pipes. Rounded and layered colloform textures resulting from precipitation of fine-grained materials from a fluid are common in these deposits as a result of ore-bearing fluids flowing through the host rock with relative ease. The relatively high porosity of the country rock in these deposits frequently contributes to widespread, conspicuous alteration of the rock.

Silver, gold, mercury, and native copper are important products of epithermal mineral deposits. Notable occurrences include the gold and silver ores of the Comstock Lode in Nevada, mercury mines in California, antimony ores in China, and the extensive copper deposits of the Keweenaw Peninsula in Michigan.

Telethermal mineral deposits are formed from nearly spent hydrothermal solutions. Fluids forming these deposits have migrated so far from their source that they have lost much of their heat and their ability to react chemically with surrounding rocks. The deposits are formed under shallow conditions at low temperatures, and there may be little or no evidence of ore mineral deposition from ascending fluids. Many telethermal deposits are composed of suites of replacement minerals in carbonate bodies. Most are stratiform or lenslike with simple structures. Wall rock alteration is generally lacking, and the textures and structures of the ore materials may be nondiagnostic of the origin of the ore body.

Ore mineral assemblages in telethermal deposits may vary greatly. Sphalerite, galena, chalcopyrite, pyrite, marcasite, chalcocite, native copper, barite, and fluorite are some of the more common minerals making up ores of zinc, lead, copper, iron, barium, and fluorine. Exceptional lead-zinc and fluorite deposits of telethermal origin have been mined in the Mississippi Valley and the tri-state area of Tennessee, Kentucky, and Illinois. Other telethermal ores include the uranium-vanadium deposits of the Colorado Plateau and Wyoming, and the Kennecott copper deposits in Alaska.

XENOTHERMAL AND VOLCANOGENIC MINERAL DEPOSITS

Xenothermal mineral deposits are considered to have formed at shallow depths. They are associated with shallow plutonic intrusions that expelled high-temperature fluids into low-pressure environments. Ore fluids cooled rapidly during the formation of these deposits; it is possible, therefore, to find both low-temperature and high-temperature ore minerals in close proximity to each other.

Xenothermal deposits are generally composed of composite veins of mineralization associated with relatively young volcanic rocks. The host rocks are typically fractured or sheared, and ore minerals are fine-grained as a result of the rapid cooling of ore-bearing fluids. The mineral deposit may be zoned if the cooling takes place slowly enough, but it is common to find both high-temperature

MINERALS FOUND IN HYDROTHERMAL ORES AND ASSOCIATED ELEMENTS

Mineral	Element
Argentite	Silver
Arsenopyrite	Iron
Bismuthinite	Bismuth
Cassiterite	Tin
Chalcopyrite	Copper
Cinnabar	Mercury
Cobaltite	Cobalt
Enargite	Copper
Galena	Lead
Hematite	Iron
Magnetite	Iron
Molybdenite	Molybdenum
Native bismuth	Bismuth
Native gold	Gold
Native silver	Silver
Pentlandite	Nickel
Pyrite	Iron
Pyrrhotite	Iron
Rhodochrosite	Manganese
Rhodonite	Manganese
Scheelite	Tungsten
Siderite	Iron
Smaltite	Cobalt
Sphalerite	Zinc
Sulfo salts	Silver
Tellurides of gold and silver	Gold
Tetrahedrite	Copper
Uraninite	Uranium
Wolframite	Tungsten

and low-temperature minerals "dumped" together within the deposit. Tin, tungsten, molybdenum, and silver ores are common to xenothermal mineral deposits. Deposits of this type also yield ores of gold, copper, lead, zinc, bismuth, and arsenic. Important occurrences of these types of ores can be found in Ikuno and Akenobe in southwestern Honshu, Japan.

Volcanogenic mineral deposits are formed by processes and activities associated with thermal springs beneath bodies of water. These deposits are formed in near-surface, low-pressure environments under temperatures that may be high to moderate. Elements necessary for hydrothermal volcanogenic ore development are a source of heat (magma chamber or intrusive igneous body), a source of fluids (seawater or water derived from the atmosphere), a source of ore elements (seawater, country rock, or magma), and faults or other localizing structures.

Volcanogenic mineral deposits are found along ancient or recent crustal spreading centers (mid-ocean ridges) or other oceanic tectonic settings. Mid-ocean ridges are divergent crustal plate margins with rifts that provide conduits for the flow of seawater into hot rocks surrounding basaltic magma chambers below the seafloor. The seawater is heated as it penetrates downward and reacts with the country rock. Rock-water reactivity increases with increasing temperatures and pressure until the elementally rich fluids reach relatively impermeable rock. The fluids then rise back to the seafloor, where they mix with seawater and precipitate ore-forming metals. Vent water temperatures may be as high as 350 degrees Celsius.

Volcanogenic deposits vary in size and grade (degree of concentration of the ore-forming elements or minerals), partly as a function of the persistence of hydrothermal activity. Fast-spreading ridges provide more persistent activity than slow-spreading ridges and are more capable of creating larger accumulations of ore material. Deposits are normally massive exhalative, stratabound bodies and are not usually found in the form of veins.

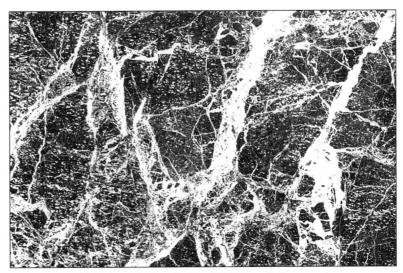

Hydrothermal mineral veins. (© William E. Ferguson)

Volcanogenic deposits are generally made up of massive sulfide associations of iron, manganese, zinc, and copper-bearing minerals. Silver, lead, and barite may also be recovered in economic quantities from these deposits. The Kuroko ores of Japan are a massive mix of iron, zinc, lead, and barite minerals. Other deposits of similar genetic origin can be found in the Philippines and Europe. New deposits are currently being created along the Mid-Atlantic Ridge, the East Pacific Rise, and the Gorda Ridge off the coast of California and Oregon.

SIGNIFICANCE

Hydrothermal mineral deposits are important because they represent some of the richest concentrations of ore-forming materials. These materials are present throughout the Earth's crust, but economically minable deposits are limited in their distribution and accessibility.

Hydrothermal processes serve to concentrate many ore-forming elements such that their grade and distribution make mining and processing profitable. Ores obtained from hydrothermal mineral deposits yield elements and compounds that are used to manufacture jewelry, electronic circuitry, batteries, wire, pipes, gasoline additives, paints, television screens, computer parts, rubber, building materials, cutting tools, lamp filaments, chemicals, pharmaceuticals, and a number of other consumer products.

Kyle L. Kayler

CROSS-REFERENCES

BIBLIOGRAPHY

Humphris, Susan E., et al., eds. *Seafloor Hydrothermal Systems: Physical, Chemical, Biological, and Geological Interactions.* Geophysical Monograph 91. Washington, D.C.: American Geophysical Union, 1995. This publication contains a broad collection of studies on hydrothermal vents, hydrothermal deposits, and hydrothermal vent ecology. It contains illustrations and some color maps.

Kesler, Stephen E. *Mineral Resources, Economics, and the Environment.* New York: Maxwell Macmillan International, 1994. This book contains information relative to mineral resources, their origins and availability, economics, and related environmental issues. Ore-forming systems and geochemistry are discussed, as is mineral availability as a function of economic conditions.

Krasnov, S. G., I. M. Poroshina, and G. A. Cherkashev. "Geologic Setting of High-temperature Hydrothermal Activity and Massive Sulphide Formation on Fast- and Slow-Spreading Ridges." In *Hydrothermal Vents and Processes*, edited by L. M. Parson, C. L. Walker, and D. R. Dixon. Geological Society Special Publication 87. London: Geological Society, 1995. This paper includes a discussion of the occurrence of massive sulfide elements associated with hydrothermal vent activity at oceanic ridge systems. Deposit genesis is addressed, and methods of prospecting are introduced.

Lindgren, Waldemar. *Mineral Deposits.* New York: McGraw Hill, 1913. Lindgren discusses the genesis of a variety of mineral deposits and introduces a system of classification that can be applied to the majority of deposits.

McMurray, Gregory R., ed. *Gorda Ridge: A Seafloor Spreading Center in the United States' Exclusive Economic Zone: Proceedings of the Gorda Ridge Symposium, May 11-13, 1987, Portland, Oregon.* New York: Springer-Verlag, 1990. A compendium of papers on the subjects of geology and geophysics of a submarine volcanic ridge. Hydrography, mineral deposits, economic development potential, prospecting, exploration, and mining technology are discussed.

Mills, Rachel A. "Hydrothermal Deposits and Metalliferous Sediments from TAG, 26 degrees north Mid-Atlantic Ridge." In *Hydrothermal Vents and Processes*, edited by L. M. Parson, C. L. Walker, and D. R. Dixon. Geological Society Special Publication 87. London: Geological Society, 1995. The Trans-Atlantic Geotraverse hydrothermal site on the Mid-Atlantic Ridge is investigated as a site of volcanogenic mineralization in a tectonically active environment. Deposit genesis is discussed in detail.

Park, Charles F., and Roy A. MacDiarmid. *Ore Deposits.* 3d ed. San Francisco: W. H. Freeman, 1975. The authors provide a comprehensive discussion of ore deposits and their origins. The book contains numerous examples of deposits from around the world, including many illustrations, diagrams, and tables.

Scott, S. D., and R. A. Burns. "Hydrothermal Processes and Contrasting Styles of Mineralization in the Western Woodland and Eastern Manus Basins of the Western Pacific." In *Hydrothermal Vents and Processes*, edited by L. M. Parson, C. L. Walker, and D. R. Dixon. Geological Society Special Publication 87. London: Geological Society, 1995. The authors discuss precious metal and sulfide ore associations, both modern and ancient, known from a variety of tectonic settings. Examples from Japan, Canada, Tasmania, and Sweden are considered.

INDUSTRIAL METALS

Metals have played a major role not only in human survival but also in the high standard of living that most cultures enjoy today. Humans have developed an understanding of the geologic conditions under which minerals form in nature and have learned to prospect for and produce those minerals from which metals are derived. A use for virtually every metallic element has been found.

PRINCIPAL TERMS

BASIC: a term to describe dark-colored, iron- and magnesium-rich igneous rocks that crystallize at high temperatures, such as basalt

BY-PRODUCT: a mineral or metal that is mined or produced in addition to the major metal of interest

GRANITE: a light-colored, coarse-grained igneous rock that crystallizes at relatively low temperatures; it is rich in quartz, feldspar, and mica

HYDROTHERMAL SOLUTION: a watery fluid, rich in dissolved ions, that is the last stage in the crystallization of a magma

LATERITE: a deep red soil, rich in iron and aluminum oxides and formed by intense chemical weathering in a humid tropical climate

NODULE: a spherical to irregularly shaped, chemically precipitated mass of rock

ORE MINERAL: any mineral that can be mined and refined for its metal content at a profit

PEGMATITE: a very coarse-grained igneous rock that forms late in the crystallization of a magma; its overall composition is usually granitic, but it is also enriched in many rare elements and gem minerals

PLACER: a surface mineral deposit formed by the settling from a water current of heavy mineral particles, usually along a stream channel or a beach; gold, tin, and diamonds often occur in this manner

PRIMARY: a term to describe minerals that crystallize at the time that the enclosing rock is formed; hydrothermal vein minerals are examples

REFRACTORY: a term to describe minerals or manufactured materials that resist breakdown; most silicate minerals and furnace brick are examples

RESERVE: that part of the mineral resource base that can be extracted profitably with existing technology and under current economic conditions

RESOURCE: a naturally occurring substance in such form that it can be currently or potentially extracted economically

SECONDARY: a mineral formed later than the enclosing rock, either by metamorphism or by weathering and transport; placers are examples

VEIN: a mineral-filled fault or fracture in rock; veins represent late crystallization, most commonly in association with granite

MAJOR METALS

The industrial metals may, for convenience, be divided into three groups: the major nonferrous metals, the ferroalloys, and the minor metals. The major metals are copper, tin, lead, and zinc. Copper has been used longer than any metal except gold. It occurs in at least 160 different minerals, of which chalcopyrite is the most abundant ore. The world's most important source of copper is in large masses of granite rock known as porphyry copper bodies, which are found throughout the western United States but are especially numerous in southern Arizona. Hydrothermal solutions deposited copper-bearing minerals in openings and cracks throughout these masses. Copper also occurs in sedimentary rocks. Such deposits account for more than one-fourth of the world's copper reserves, with the best examples in the Zambian-Zairean copper belt of Africa. A future resource is the copper in manganese nodules that cover large portions of the ocean floor, especially in the North Pacific. Chile, the United States, and Can-

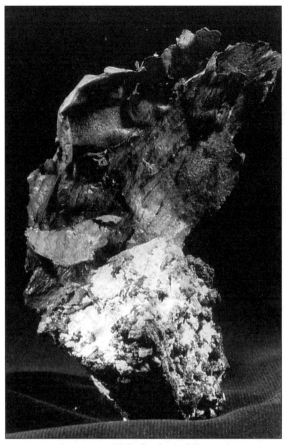

Copper ore. (U.S. Geological Survey)

ada, in that order, are the world's leading producers of copper.

Like copper, tin has long been used by humans. The principal ore mineral of tin is the oxide cassiterite, "tinstone." Primary tin-bearing deposits include granite pegmatites and hydrothermal veins. Much more important from a commercial standpoint, however, are secondary stream placer deposits. Although domestic mine production is negligible, the United States is the world's leading producer of recycled tin. World leaders in the mining of tin are Malaysia, Brazil, Indonesia, Thailand, China, and Bolivia.

Zinc and lead now rank just behind copper and aluminum as essential nonferrous (not containing iron) metals in modern industry. The most important geologic occurrences of lead and zinc are within stratified layers of metamorphic or carbonate rocks. The deposits at Ducktown, Tennessee, and Franklin, New Jersey, are in metamorphic rocks, while those in the upper Mississippi Valley, southeastern Missouri (the Virburnum Trend), and the Tri-State mineral district (Missouri-Oklahoma-Kansas) are found in carbonate rocks. This latter type of occurrence is now referred to as Mississippi Valley-type deposits. The principal ore minerals for lead and zinc are the sulfides galena and sphalerite (zinc blende), respectively. These minerals were probably emplaced in the host rocks by later hydrothermal solutions. The United States is a major producer of lead, with most coming from the Virburnum Trend. Recycled lead is also very important in the United States, accounting for more production than mining. Other major lead-producing countries are Australia, Canada, Peru, and Mexico. Canada, Australia, Peru, Mexico, and the United States are among the world's leading producers of zinc.

FERROALLOYS

The ferroalloys are a group of metals whose chief economic use is for alloying with iron in the production of carbon and various specialty steels. Manganese is the most important of the ferroalloys, and it occurs principally in sandstone deposits. Chemically precipitated nodules on the ocean floor contain up to 20 percent manganese and represent an important potential resource.

The elements chromium, nickel, titanium, vanadium, and cobalt most commonly occur in basic igneous rocks such as gabbro. Examples of such deposits are the Stillwater complex of Montana and the Bushveld complex of South Africa. The Bushveld is the largest such body in the world, and it holds most of the world's chromium reserves. Nickel ranks second in importance to manganese among the ferroalloys. The large igneous deposits at Sudbury, Ontario, Canada, have been the world's major supplier. The most important economic occurrences of titanium are in secondary placer deposits, notably the rutile beach sands of Australia and the ilmenite sands of northern Florida. In addition to its occurrence in dark igneous rocks, vanadium is found as a weathering product in uranium-bearing sandstones such as those of the Colorado Plateau region of the United States. Vanadium is also produced from the residues of petroleum refining and from the processing of phosphate rock. While the primary occurrence of cobalt is in dark igneous rocks, it also is produced from laterites and as a by-product of the sedimentary copper deposits of the African copper belt.

Molybdenum, tungsten, niobium, and tantalum occur commonly in quartz veins and granite pegmatites. Most molybdenum production comes from the very large, but low-grade, porphyry deposits at Climax, Colorado, and as a by-product of the porphyry copper deposits at Bingham Canyon, Utah. Production of the minerals scheelite and wolframite from quartz veins accounts for more than one-half of the world's production of tungsten. Because of their rarity, there is no mining operation solely for the production of either niobium (also known as columbium) or tantalum. Both are always produced as by-products of the mining of other metals.

The United States is self-sufficient and a net exporter only with respect to molybdenum. It is almost totally dependent on foreign sources for manganese, chromium, nickel, cobalt, niobium, and tantalum, and to a lesser degree for vanadium, tungsten, and titanium.

MINOR METALS

There are a number of metallic elements that are rare in nature. Most of these have a primary origin in hydrothermal veins or granite pegmatites and are produced either from such deposits (antimony, cadmium, beryllium, bismuth, arsenic, lithium, cesium, rubidium, and mercury) or from placer deposits that resulted from the weathering and erosion of the pegmatites (zirconium, hafnium, thorium, and the rare-earth elements). The sole exception is magnesium, which is produced mostly from natural brines in wells and lakes, as at Great Salt Lake in Utah, or from seawater.

Most elements in this group are produced exclusively as by-products of the mining and processing of other metals. Antimony, in the mineral stibnite, is closely associated with lead ores in Mississippi Valley-type deposits, and most antimony is produced as a by-product of the mining and processing of lead, copper, and silver ores. China is the world's leading producer of antimony. Cadmium is a trace element that is similar to zinc and therefore substitutes for zinc in some minerals. All cadmium is produced from the mineral sphalerite as a by-product of zinc production. Large but un-

tapped potential resources of cadmium exist in the zinc-bearing coal deposits of the central United States. There are no specific ore minerals of bismuth, and virtually all production is from the processing of residues from lead smelting. Mercury, or quicksilver, is the only metal that is a liquid at ordinary temperatures. It is produced from the sulfide mineral cinnabar and is marketed in steel "flasks," each flask containing 76 pounds of liquid mercury. The mines at Almaden, Spain, have been the leading producers. Magnesium is the lightest metal and is especially strong for its weight. It is an abundant metal, both in the Earth's crust and in seawater. Lithium, cesium, and rubidium are often classified as the rare alkali metals, the abundant alkali metals being sodium and potassium. Like magnesium, lithium is abundant enough in brines and evaporite deposits to be processed economically from these sources. The rare-earths, or lanthanide elements, are a group of chemically similar elements, of which cerium is the most abundant and widely used. Thorium is a heavy metal, the parent element of a series of radioactive decay products that end in a stable isotope of lead. It is not an abundant element, but it is widespread, occurring in veins, placers, and sedimentary rocks.

The United States is well endowed with some of these minor metals, including beryllium, lithium, magnesium, and the rare-earths. On the other hand, it must import virtually all of its supplies of antimony, cadmium, bismuth, mercury, arsenic, and rubidium, and one-half of its annual consumption of zirconium and thorium ores and products.

EXPLORATION FOR METALS

It is a fair assumption that all the large and easily accessible metallic ore bodies have been discovered. Emphasis in the years ahead will be on the detection of deposits that are in more remote localities, are concealed, or are of lower grade. While research provides more sophisticated tools, the basic exploration approach remains the same. Four main prospecting techniques are employed: geological, geochemical, geophysical, and direct. Geological exploration generally involves plotting the locations of rock types, faults, folds, fractures, and areas of mineralization on base maps. Examples of deposits that have been discovered by simple surface exploration and mapping are chromite, which is resistant to weathering and "crops out"; manganese-bearing minerals, which oxidize to a black color; and molybdenite with its characteristic silver color. Because ore minerals are known to be associated with rocks formed in certain geologic environments, the focus of study today is on understanding such associations as clues to locating mineral deposits.

Geochemical exploration consists of chemically analyzing soil, rock, stream, and vegetation samples. Concentrations of metallic elements in the surface environment are assumed to be representative of similar concentrations in the rocks below. Areas of low mineral potential can be eliminated, while targets for further study and testing can be outlined. In some instances, it is necessary to trace metals back through the surface environment to their points of origin. Research is being done not only on the movement and concentration of the economically important metallic elements, but also on other elements that are often associated

CONCENTRATIONS OF METALS IN IGNEOUS ROCKS

Element	Percentage
Aluminum	8.13
Antimony	0.0001
Arsenic	0.0005
Chromium	0.02
Cobalt	0.0023
Copper	0.005
Gold	0.0000005
Iron	5.00
Lead	0.0016
Magnesium	2.09
Manganese	0.10
Mercury	0.00005
Molybdenum	0.00025
Nickel	0.008
Platinum	0.0000005
Silver	0.00001
Tin	0.004
Titanium	0.44
Tungsten	0.00015
Uranium	0.0004
Vanadium	0.015
Zinc	0.011

with them but that, for various reasons, can be more easily moved and concentrated by nature. Cobalt, for example, is mobile, moving through rocks and sediments easily. It may be possible to trace this element back to its source and in the process locate other associated metals.

Geophysical techniques range from large-scale reconnaissance surveys to detailed local analysis. These techniques detect contrasts in physical properties between the ore bodies and the surrounding host rocks. Airborne magnetic and electromagnetic surveys are useful for rapid coverage of remote and inaccessible terrain and of areas where ore bodies are covered by glacial sediments. Radiometric surveys can be used to detect concentrations of radioisotopes and have been effective in locating deposits of thorium, zirconium, and vanadium-bearing minerals. Other applications of geophysical techniques include seismic and gravity studies to determine the thickness of overburden, airborne infrared imagery to detect residual heat in igneous deposits, light reflectance of vegetation, side-looking radar, and aerial and satellite photography. In general, all airborne reconnaissance techniques are followed by detailed geological, geochemical, or geophysical ground surveys.

The final stage in the exploration process is the direct stage. Here, a prospect is directly sampled by drill, pit, trench, or mine to determine its potential. This step is the most expensive. It is the purpose of geological, geochemical, and geophysical prospecting to narrow the possibilities, lower the odds, and thereby reduce the final cost by selecting the most likely prospects for direct sampling.

USES OF MAJOR AND MINOR METALS

The great value of copper derives from its high thermal and electrical conductivities, its corrosion resistance, its ductility, and its strength. It alloys easily, especially with zinc to form brass and with tin and zinc to form bronze. Principal uses of tin are plating on cans and containers, in solder, in bronze, and with nickel in superconducting alloys. Restrictions on the use of lead in pipes and solder should cause an increase in tin consumption as a lead replacement. Bottle and can deposit laws, enacted in a number of states, will increase the use of scrap (recycled) tin. The principal use of lead is in automobile batteries, but cable sheathing, type metal, and ammunition are other important applications. Zinc uses include galvanizing for iron and steel, die castings, and brass and bronze.

The minor metals have a variety of applications in industry, both as metals and in chemical compounds. Antimony is used primarily as a fire retardant, but it is also alloyed with lead for corrosion resistance and as a hardening agent. This last characteristic is important for military ordnance and cable sheathing. Cadmium is used in the electroplating of steel for corrosion resistance, in solar cells, and as an orange pigment. Beryllium is alloyed with aluminum and copper to provide strength and fatigue resistance and is widely used in the aerospace industry. It is a "nonsparking" metal that can be used in electrical equipment. Oxides of beryllium are found in lasers, as refractories, in ceramics, and as insulators. The principal use of bismuth is in the pharmaceutical industry. It soothes digestive disorders and heals wounds. Salts of bismuth are widely used in cosmetics because of their smoothness. Bismuth metal lowers the melting points of alloys so that they will melt in a hot room. This allows them to be used in automatic water sprinkler systems and in safety plugs and fuses.

Mercury, because it is a liquid at room temperature, has applications in thermometers, electrical switches, and, with rubidium, in vapor lamps. It is also used in insecticides and fungicides and has medical and dental applications. Magnesium is alloyed in aircraft with aluminum to reduce weight and at the same time provide high rigidity and greater strength. The metal also burns at low temperature, which makes it suitable for flash bulbs, fireworks, flares, and incendiary bombs. The largest use of magnesium is in the oxide magnesia for refractory bricks. Arsenic is mostly used in chemical compounds as a wood preservative and in insecticides and herbicides. Lithium-based greases have wide application in aircraft, the military, and the marine environment, as they retain their lubricating properties over a wide temperature range and are resistant to water and hardening. Metallic lithium, as well as magnesium, is alloyed with aluminum for aerospace applications. Cesium is used in magnetohydrodynamic electric power generators, in photoelectric cells (automatic door openers), and in solar voltaic cells. Zirconium is used as a refractory in crucibles and brick, as an abrasive and polisher, and for jewelry (cubic zirconium). The rare-earth cerium has applications in photography and as a "colorizer" in television tubes. Most of the thorium consumption in the United States is by the nuclear fuel industry.

USES OF FERROALLOYS

As a group, the economic life of the ferroalloys is closely tied to that of steel. These elements impart to steel such properties as strength at high temperatures (especially titanium and molybdenum), making them vital for aerospace applications; hardness (tungsten and manganese), for use in armor plate, structural steel, and high-speed cutting tools; and resistance to corrosion (chromium and nickel), for plating. The most important of these elements is manganese, for which there is no substitute. It acts as a scavenger during smelting to remove oxygen and sulfur and it is alloyed with the steel for hardness.

The ferroalloys also find important applications in industry beyond their use in steel. Manganese is widely used in dry batteries, pigments, and fertilizers. Chromium is used in refractory brick for high-temperature furnaces. Molybdenum, nickel, and vanadium compounds are catalysts in a number of chemical processes, and molybdenum is used in industrial lubricants. Tungsten carbide is the hardest cutting and polishing agent after diamond, and metallic tungsten is commonly the filament in electric light bulbs. Titanium oxide is an important pigment. It is opaque and forms the whitest of all paints. Nickel is used in coinage, batteries, and insecticides as well as for plating and catalysts. Cobalt is used in magnetic alloys and as a blue pigment in glass and ceramics. The radioisotope cobalt 60 has a number of medical applications. The rare ferroalloy tantalum is used for capacitors and rectifiers in the electronics industry.

Donald J. Thompson

CROSS-REFERENCES

Aluminum Deposits, 1539; Batholiths, 1280; Building Stone, 1545; Carbonatites, 1287; Cement, 1550; Clays and Clay Minerals, 1187; Coal, 1555; Diamonds, 1561; Dolomite, 1567; Fertilizers, 1573; Fluid Inclusions, 394; Fractionation Processes, 400; Geysers and Hot Springs, 694; Gold and Silver, 1578; Granitic Rocks, 1292; Heat Sources and

BIBLIOGRAPHY

Alloway, B. J. *Heavy Metals in Soils*. 2d ed. London: Blackie Academic and Professional, 1995. Alloway's college-level textbook looks at methods used to analyze the heavy metal content in soils, as well as the origin, behavior, and effects of those deposits on the soil and environment. Illustrations, bibliography, and index.

Evans, Anthony M. *An Introduction to Economic Geology and its Environmental Impact*. Oxford: Blackwell Scientific Publications, 1997. Intended for undergraduate students, this book provides an excellent introduction into the field of economic geology. Emphasis is on types of deposits, their environments of formation, and their economic value, along with the impact those deposits have on their environments. Well illustrated and includes an extensive bibliography.

Hutchison, Charles S. *Economic Deposits and Their Tectonic Setting*. New York: John Wiley & Sons, 1983. This text covers the origin and occurrence of all major types of ore deposits, but from the standpoint of their formation within specific plate tectonic environments (such as oceanic spreading centers and subduction zones). Well illustrated, with numerous examples and an extensive bibliography. In addition, there is a brief summary of each of the important mineral resources. Can be read with understanding by a geologic novice.

Jensen, Mead L., and Alan M. Bateman. *Economic Mineral Deposits*. 3d rev. ed. New York: John Wiley & Sons, 1981. Although older, this text is probably the most complete and at the same time the most understandable. Contains good introductory chapters on the origins of ore minerals, followed by well-written descriptions of each of the individual metallic and nonmetallic resources. An excellent reference and one that can easily be used by a beginner.

Jones, Adrian P., Frances Wall, and C. Terry Williams, eds. *Rare Earth Minerals: Chemistry, Origin, and Ore Deposits*. New York: Chapman and Hall, 1996. This book contains extensive information on rare minerals, mineralogy, and mineralogical chemistry. It also focuses on ore deposits, mining practices, and the economic benefits associated with different ore deposits and minerals.

Lamey, C. A. *Metallic and Industrial Mineral Deposits*. New York: McGraw-Hill, 1966. An older, but standard, reference in the field, which may still be used profitably. Includes an initial summary of geologic origins of mineral deposits, followed by a discussion of each individual mineral resource. This style of organization makes the book easier to use if the reader is interested in information about a particular commodity. Suitable for the reader with a limited geological background.

Salomons, Willem, Ulrich Feorstner, and Pavel Mader, eds. *Heavy Metals: Problems and Solutions*. New York: Springer-Verlag, 1995. This collection of case studies and papers looks at the environmental benefits and problems of heavy metals. Several essays focus on the pollution problems associated with heavy metals and the actions that are being taken to correct them.

U.S. Bureau of Mines. *Mineral Commodity Summaries, 1987*. Washington, D.C.: Government Printing Office, 1987. This volume contains up-to-date summaries for eighty-eight non-fuel mineral commodities. For each, information is presented on use; domestic production; price; import quantities, sources, and tariffs; depletion allowances (if any); recycling; stockpiles; world production and reserve base; world resources; significant trends and events; and possible industrial substitutes. Most information is given in tabular summaries. An excellent source of information for research at any level.

_____. *Minerals Yearbook, 1986.* Vol. 1, *Metals and Minerals.* Washington, D.C.: Government Printing Office, 1988. The latest in a series of yearly summaries of the various metallic and nonmetallic resources. Information on each commodity includes uses, consumers, domestic and world production figures, import and export data, prices, and technological advances. The book includes voluminous tables and charts and is probably the finest source of data for economic mineral study at any level.

U.S. Geological Survey. *United States Mineral Resources.* Professional Paper 820. Washington, D.C.: Government Printing Office, 1973. This volume includes an excellent summary for each mineral resource. Some of the data are obsolete, but the summaries are much more complete than those of the Bureau of Mines publications. For each resource, an extensive discussion includes uses, environmental problems, history of exploitation, ore minerals and geologic occurrence, reserves and resources, prospecting techniques, and research methods. An extensive bibliography is included for each resource. Some tables and graphs are included, but no illustrations or diagrams. An excellent source of information, especially for those with some background in geology.

INDUSTRIAL NONMETALS

The nonmetallic Earth resources consist of fertilizer minerals, raw materials for the chemical industry, abrasives, gemstones, and building materials for the construction industry. They therefore provide the necessary base for a technological society.

PRINCIPAL TERMS

ASBESTOSIS: deterioration of the lungs caused by the inhalation of very fine particles of asbestos dust

CATALYST: a chemical substance that speeds up a chemical reaction without being permanently affected by that reaction

GUANO: fossilized bird excrement, found in great abundance on some coasts or islands

METAL: a shiny element or alloy that conducts heat and/or electricity; metals are both malleable and ductile

PROVEN RESERVE: a reserve supply of a valuable mineral substance that can be exploited at a future time

SEDIMENTARY ROCK: rock formed from the accumulation of fragments weathered from pre-existing rocks or by the precipitation of dissolved materials in water

STAR SAPPHIRE/RUBY: a gem that has a starlike effect when viewed in reflected light because of fibrous structure within the mineral

STRATEGIC RESOURCE: an Earth resource, such as manganese or oil, that would be essential to a nation's defense in wartime

DEFINITION OF NONMETALLIC RESOURCES

The definition of a nonmetal is problematic. To the chemist, a nonmetal is any element not having the character of a metal, including solid elements, such as carbon and sulfur, and gaseous elements, such as nitrogen and oxygen. That is the definition found in most dictionaries. To the economic geologist, however, a nonmetal is any solid material extracted from the Earth that is neither a metal nor a source of energy. It is valued because of the nonmetallic chemical elements that it contains or because it has some highly desirable physical or chemical characteristic.

Economic geologists consider the following to be major nonmetallic Earth resources: fertilizers, raw materials for the chemical industry, abrasives, gemstones, and building materials. As can readily be seen, with the exception of the gemstones, the nonmetallic Earth resources lack the glamour associated with metals such as gold and platinum or energy sources such as oil and uranium. Nevertheless, nonmetallic Earth resources play an essential role in the world economy.

Except for gemstones, the nonmetallic Earth resources have certain common characteristics. First, they tend to be more abundant in the Earth's crust than the metals and are therefore lower in price. Nevertheless, most of them are needed in much larger quantities, so the total value of the substances produced is considerable. That is particularly true of the building materials needed for the construction industry. Second, nonmetallic Earth materials tend to be taken from local sources. Most of them are needed in such large quantities that transportation costs would be excessive if they were brought long distances. As a result, regional variations occur in the types of rock used for building stone and crushed rock. Third, problems of supply are not generally associated with nonmetallic Earth materials. Most of them are fairly abundant at the Earth's surface, and few major industrial nations are without deposits of each of them. None of these resources is classified as a strategic material by the United States, for example. Finally, nonmetallic Earth materials tend to require very little processing before being sent to market. In fact, most of them are used in the raw state.

FERTILIZERS

The first category of major nonmetallic Earth resources is the fertilizers. These substances are absolutely essential to a nation's agriculture and therefore to its food supply. The three most important elements for plant growth are nitrogen, potassium, and phosphorus. For years, nitrogen was obtained from nitrogen-rich Earth materials—either from the famous guano deposits in Peru, which were built up by accumulated bird droppings on coastal islands, or from the nitrate deposits in Chile, which cover the floor of a desert. In 1900, however, a German chemist discovered a way of manufacturing nitrates synthetically, using nitrogen extracted directly from the atmosphere. Today, the synthetic nitrate industry provides 99.8 percent of the world's nitrogen needs.

The potassium required for fertilizers was originally obtained from wood ashes, and in many underdeveloped nations, it still is. In 1857, however, potassium-bearing salt beds were discovered in Germany. These had formed as a result of the evaporation of lakes in an arid climate, and they were the world's major source of potassium until 1915, when Germany placed a wartime embargo on their shipment. This embargo forced other countries to explore for replacement deposits, and similar salt beds were eventually found in the former Soviet Union, Canada, and the western United States.

The third element required for fertilizers, phosphorus, was originally obtained from guano or from bones. These sources have been replaced by natural phosphate rock deposits, which are widely distributed around the world. Most of these deposits occur in marine sedimentary rocks, and it is believed that deposition of the phosphate resulted when cool, phosphorus-bearing waters upwelled from the sea floor and were carried into shallow environments, where the phosphorus was precipitated. The largest U.S. phosphate deposits are found in Florida and North Carolina.

RAW MATERIALS FOR CHEMICAL INDUSTRY

The second major category of nonmetallic Earth resources is that made up of the raw materials for the chemical industry. In terms of total production, the most important of these is salt, for which the mineral name is halite. In addition to its use as a dietary ingredient, salt is the raw material from which a number of important chemicals are made, including chlorine gas, hydrochloric acid, and lye. In colder climates, salt is also used for snow and ice control on roads. Salt is produced by the evaporation of seawater and by the mining of underground salt deposits. Although underground salt commonly occurs as deeply buried layers, nature has an interesting way of bringing the salt to the surface. Since salt is lighter in weight than the overlying rocks and is capable of plastic flow, it rises through the surrounding rocks as a salt plug with a circular cross section, known as a salt dome. Salt domes are particularly common along the Louisiana and Texas Gulf Coast.

Another important raw material for the chemical industry is sulfur, a soft yellow substance that burns with a blue flame. The major industrial use for sulfur is in the production of sulfuric acid. Large quantities of sulfuric acid are used in converting phosphate rock to fertilizer. Sulfur also has important uses in the manufacture of insecticides. Most of the sulfur used in the United States comes from the salt domes of the Louisana and Texas coast, where it is found in the upper part of the dome. Superheated steam is pumped down through drill holes to melt the sulfur, and the liquid sulfur is then brought to the surface by the pressure of compressed air. Additional raw materials for the chemical industry include several that are obtained from the beds of dry desert lakes: sodium carbonate and sodium sulfate, which are used in the manufacture of glass, soaps, dyes, and paper; and borax, which is used in making detergents. Another product is sodium bicarbonate, the familiar baking soda.

ABRASIVES

The third major category of nonmetallic Earth resources is the abrasives, which are materials used for grinding, cleaning, polishing, and removing solid material from other substances. Most abrasives are very hard, but those used for cleaning porcelain sinks and silverware need to be fairly soft, so as not to scratch. Abrasives can be either natural or man-made. The natural abrasives are rock and mineral substances that have been extracted from the Earth and that are then either used in the raw state, such as a block of pumice, or pulverized and bonded into sandpapers, wheels, saws, drill bits, and the like. Artificially made abrasives, however, are gradually coming to dominate the market.

The most common abrasive is diamond, which has a hardness of 10 on a scale of 1 to 10 and is the hardest known natural substance. Most natural diamonds are unsuitable for use as gems, however, so about 80 percent of them are used as abrasives. In 1955, General Electric developed a process to make industrial diamonds synthetically, and by 1986 two-thirds of the world's industrial diamonds were produced synthetically. Other important natural abrasives include corundum, the second hardest natural substance, with a hardness of 9; emery, a gray-to-black mixture of corundum and the iron mineral known as magnetite; and garnet, a reddish-brown mineral, with a hardness of approximately 7, that is commonly used in sandpaper. Ninety-five percent of the world's garnet comes from the Adirondack Mountains in New York State.

GEMSTONES

The fourth major category of nonmetallic Earth resources is the gemstones. These are used primarily for adornment and decoration. Unlike the other nonmetals, they are generally not abundant, have a moderate-to-high value, come in small quantities only, are rarely of local origin, and are often in short supply. Desirable properties in a gem are color, brilliance, transparency, hardness, and rarity. Gems are categorized into two principal groups. The precious gems are diamond, ruby, sapphire, emerald, and pearl. All of these can be produced synthetically, except for high-quality gem diamonds. Ruby and sapphire are varieties of corundum and may exhibit "stars." Emerald is a variety of beryl wth a hardness of 7½ to 8. Pearls have a hardness of 3 and are technically not true minerals, because they are produced by a living organism. The semiprecious gems include some one hundred different substances. Most are minerals, except for amber (hardened resin from a pine tree), jet (a dense variety of coal), and black coral (a substance produced by a living organism).

BUILDING MATERIALS

The fifth major category of nonmetallic Earth resources is building materials for the construction industry. They include the familiar building stones obtained from quarries, such as granite, sandstone, limestone, marble, and slate. There is also a high demand for crushed rock, which is used in highway roadbeds, and for concrete aggregate. Sand and gravel are also used in making con-

Asbestos is obtained from flame-resistant flexible mineral fibers and was used to create fire-retardant materials until it was discovered that its inhalation can provoke various cancers. (U.S. Geological Survey)

crete. In addition, many useful products are prepared from Earth materials, such as cement, which is made from a mixture of limestone and clay; plaster, which comes from the mineral gypsum; brick and ceramics, which use clay as their raw material; glass, which is made from very pure sand or sandstone rock; and asbestos, which is obtained from flame-resistant mineral fibers that can be woven into fireproof cloth or mixed with other substances to make fireproof roofing shingles and floor tiles. Concerns about health hazards related to the use of asbestos arose in the 1970's. The very fine particles of asbestos dust can lodge in the lungs, causing asbestosis and lung cancer. As a result, the U.S. consumption of asbestos has declined markedly since the 1970's.

ASSURING SUPPLIES OF NONMETALLIC RESOURCES

An important way in which nonmetallic Earth resources are studied is to analyze proven reserves. Proven reserves are supplies of a mineral substance that still remain in the ground and are available to be taken out at some future time. When experts compare the present rates of production of various nonmetallic Earth resources with the proven reserves of these substances, they can predict which resources may someday be in short supply. In the case of the phosphate rock used in making fertilizers, for example, analysts have found that the United States' phosphate reserves will be exhausted around the year 2010. In 1985, the United States was the world's largest exporter of phosphate; by 2010, it will be the world's largest importer. Clearly, a major exploration program is necessary.

Economic geologists also study ways to assure that adequate supplies of nonmetallic Earth resources will be available for future needs. In the case of phosphate, for example, the need for such studies is critical. Such a program of exploration for new phosphate supplies, or for any other non-metallic resource, must begin with a full understanding of the ways in which the mineral resource originates. Only when geologists understand the conditions under which valuable concentrations of mineral substances form can they successfully search for them.

In the case of phosphate, careful scientific study has shown that the cold waters found in the deep ocean contain thirty times as much dissolved phosphorus as do the warm shallow waters. This observation suggests that when the cold, deep waters are brought to the surface and warmed, by upwelling from the ocean floor and flowing across shallow submarine banks into coastal zones, the warming effect makes the phosphorus less soluble, and it precipitates. As a result, scientists are exploring for phosphates along present or former continental margins, where such upwelling might have taken place.

Another way in which adequate future supplies of nonmetallic Earth resources can be assured is by creating nonmetals synthetically in the laboratory. A good example of this process was the successful synthesis of industrial diamonds by General Electric in 1955. Before then, the United States had no industrial diamond production or reserves and was totally dependent on supplies purchased in the world market, which was controlled by the De Beers group of companies in South Africa. General Electric was able to create diamonds synthetically by subjecting the mineral

graphite—which is composed of pure carbon, as is diamond—to incredibly high temperatures and pressures in a special sealed vessel, using molten nickel as a catalyst. By 1986, two-thirds of the world's industrial diamonds were being produced by this process, and the De Beers monopoly on industrial diamond production was broken.

CONTINUED IMPORTANCE

The United States is fortunate in having a plentiful food supply, and most Americans rarely think about the importance of fertilizers. They are essential, however, for successful agricultural operations. Plant growth requires ample mineral matter, partly decomposed organic matter (humus), water, air, and sunlight. Of these, mineral matter is crucial, because it provides the nitrates and phosphates essential for healthy plant growth. These substances are quickly used or washed away, and they must be replenished regularly so that the soil does not become worn out and infertile. Worn-out soils are frequently encountered in developing nations, where farmers are often too poor to buy fertilizers.

Industrial chemicals such as salt, sulfur, and borax appear on grocery shelves in their pure state, but they are also present as ingredients in products where one would never suspect their existence. In addition, they are frequently needed to manufacture everyday products, such as drinking glasses or writing paper. Salt is a good example. In addition to its use as table salt, it is an ingredient in almost every prepared food item on the grocery shelf: soup, nuts, bread, crackers, canned meats, bottled olives, frozen dinners. . . . The list goes on almost indefinitely. Abrasives, too, are common on grocery shelves, although the word abrasive may not be written on the package. They are in toothpaste, silver polish, bathroom cleanser, pumice stones, sandpaper, and emery boards.

Nonmetals are used in many common construction materials. The beautiful white buildings in Washington, D.C., are made of pure white Vermont marble. Granite, on the other hand, is preferred for tombstones, because it resists weathering better. It is also used for curbstones in northern cities, because it holds up best under the repeated impacts from snowplow blades. The days of buildings faced with cut stone, however, are on the wane; production and transportation costs are simply too high. Today's private dwelling is more likely to be built of cinder blocks, manufactured at a plant outside the city, and downtown office towers are sheathed with walls of glass and prefabricated concrete.

Donald W. Lovejoy

CROSS-REFERENCES

BIBLIOGRAPHY

Bartholomew, John C., ed. *The Times Atlas of the World*. New York: Times Books, 1980. An overview entitled "Resources of the World" is followed by a large world minerals map printed in eight colors. It shows the world distribution of diamonds, chemical and fertilizer minerals, asbestos, clay, magnesite, and talc. The relative importance of each mineral deposit is indicated. Suitable for high school students.

Bramwell, M., ed. *The Rand McNally Atlas of the Oceans*. Skokie, Ill.: Rand McNally, 1977. A well-written and beautifully illustrated atlas. The section entitled "The Great Resource" has an excellent discussion of nonmetallic resources that can be obtained from the sea, such as phosphorite, salt, sulfur, diamond, shell sands, sand, and gravel. There are color photographs, maps, and line drawings. Suitable for high school students.

Constantopoulos, James T. *Earth Resources Laboratory Investigations*. Upper Saddle River, N.J.: Prentice Hall, 1997. An excellent guidebook for any laboratory work involving minerals, deposits, and natural resources. Discusses the protocal and procedures used in working with Earth resources in the laboratory setting.

Craig, J. R., D. J. Vaughan, and B. J. Skinner. *Resources of the Earth*. Englewood Cliffs, N.J.: Prentice-Hall, 1988. A well-illustrated text with numerous black-and-white photographs, color plates, tables, charts, maps, and line drawings. It covers the major categories of nonmetallic Earth resources: fertilizer minerals, chemical minerals, abrasives, gemstones, and building materials. Suitable for college-level readers or the interested layperson.

Evans, Anthony M. *An Introduction to Economic Geology and its Environmental Impact*. Oxford: Blackwell Scientific Publications, 1997. Intended for undergraduate students, this book provides an excellent introduction into the field of economic geology. Emphasis is on types of deposits, their environments of formation, and their economic value, along with the impact those deposits have on their environments. Well illustrated and includes an extensive bibliography.

Fisher, P. J. *The Science of Gems*. New York: Charles Scribner's Sons, 1966. This book has excellent color photographs of the well-known gems. Topics covered include the history of gems, their origins, their characteristics, gem cutting, and gem identification. There is also a glossary of terms and a useful appendix giving detailed information relating to each type of gem. Written for general audiences.

Jensen, M. L., and A. M. Bateman. *Economic Mineral Deposits*. 3d ed. New York: John Wiley & Sons, 1979. This economic geology text provides detailed information on the different metallic and nonmetallic mineral deposits and their modes of formation. There are excellent sections on the history of mineral use and the exploration and development of mineral properties. Cross sections of individual deposits are provided. Suitable for college-level readers.

Skinner, B. J. *Earth Resources*. 3d ed. Englewood Cliffs, N.J.: Prentice-Hall, 1986. An excellent overview of the Earth's nonmetallic resources. It is well written and contains helpful line drawings, maps, tables, and charts, although photographs are few. There are suggestions for further reading and a list of the principal nonmetallic minerals and their production figures for 1982. For college-level readers.

Tennissen, A. C. *The Nature of Earth Materials.* 2d ed. Englewood Cliffs, N.J.: Prentice-Hall, 1983. This useful reference book contains detailed descriptions of 110 common minerals, with a black-and-white photograph of each. Chapter 7, entitled "Utility of Earth Materials," contains helpful sections on the distribution of mineral deposits and the utilization of various Earth materals. Suitable for the layperson.

Youngquist, Walter Lewellyn. *GeoDestinies: The Inevitable Control of the Earth Resources Over Nations and Individuals.* Portland, Oreg.: National Book Company, 1997. This book looks at the Earth's natural, mineral, and power resources, with an emphasis on social and environmental problems that accompany the recovery of such resources. A good book for the layperson. Illustrations and maps.

IRON DEPOSITS

Local exploitation of iron deposits made possible both the Industrial Revolution of Europe in the eighteenth century and the rapid industrial growth of the northern United States at the start of the twentieth century. The mining of iron deposits has moved from these areas during the past half-century to become concentrated within thirty latitudinal degrees of the equator.

PRINCIPAL TERMS

GOETHITE: the most common iron-hydroxide mineral

HEMATITE: a mineral composed of oxygen and fully oxidized iron but no hydrogen

IRON FORMATION: a layered rock deposit that consists mostly of ironstone

IRON HYDROXIDE: a mineral that contains oxygen, hydrogen, and fully oxidized iron

IRON SILICATE: a mineral that contains silicon, oxygen, hydrogen, and abundant iron

IRONSTONE: a chemically precipitated sedimentary rock which contains more than 15 percent iron

MAGNETITE: a mineral composed of oxygen and iron but no hydrogen; the iron occurs in two oxidation states: oxidized and reduced

OOID: a small round grain that is layered internally, like an onion; it typically has a diameter comparable to that of a mineral grain on a sandy beach

REGOLITH: the weathered surface of the Earth; soil is the upper portion of the regolith and includes plant roots

SIDERITE: a mineral composed of iron, carbon, and oxygen

SILICA: silicon with two oxygen atoms; essentially, naturally precipitated glass

ANCIENT IRON DEPOSITS

Most young iron deposits contain marine fossils. Iron formations, therefore, usually are attributed to precipitation from ancient seawater, but the nature of the iron-concentrating process remains unclear. Iron is extremely insoluble within both oxidized and chemically reduced seawater. In a billion parts of seawater, there are only about three parts of iron. The only modern body of seawater with any appreciable dissolved iron is a deep, salty body under the northern Gulf of Mexico, called the Orca basin, which contains 1.6 parts of iron in a million parts of seawater. Given the lack of iron in all other seawater, it is not surprising that modern oceans lack significant iron deposits on the sea floor. In fact, it is extremely difficult to conceive of any reasonable mechanism to produce the enormous ancient iron deposits from seawater. Nevertheless, the very size of the ancient iron formations seems to record precipitation on an oceanic sea floor. Moreover, many ancient iron formations are closely associated with voluminous limestone, which also apparently precipitated from seawater.

The largest iron deposits (iron formations) occur in rocks which are older than 600 million years. Many of the largest iron deposits formed during the Early Proterozoic era (between 2,500 and 2,000 million years ago) and are therefore about half as old as the Earth itself. Rocks that are older than 2,000 million years constitute less than 5 percent of the surface area of the Earth, because subsequent geologic processes either have destroyed the rocks by erosion or melting or have buried them under a thick cover of younger rocks. Deeply buried rocks cannot be mined profitably. Modern industrial society, therefore, depends on exposed remnants of the early Earth that have survived an extremely long time.

No iron is being concentrated on the modern Earth into a significant deposit (a large iron formation), and almost nothing is being produced currently that resembles the ironstone (iron-rich rock) in large iron formations. Ancient iron deposits accumulated within large water bodies, almost certainly oceans, but little is known about the nature of these ancient oceans, as they contained no fish or other animal life which would provide fossil records indicating environmental conditions. There is therefore little agreement among geologists as to the origin of ancient iron deposits.

Iron deposits can be traced laterally for as much as 1,000 kilometers through ancient rock sequences. An individual iron formation may exceed 100 meters in thickness. Large iron formations accumulated on extensive platforms that were not receiving any sediment from rivers. Otherwise, they would not be such pure chemical precipitates. The largest iron formations contain little other than iron, silicon, oxygen, and carbon. Sediment supplied by rivers always contains a wide variety of other chemical elements.

Some of the platforms on which iron formations have accumulated probably lie offshore from a continent, just as the modern Bahama platforms lie off the southeastern United States. The Bahama platform that lies closest to North America is separated from Florida by a deep-water channel, which prevents any continental sediment from reaching the shallow marine platform. The only sediments that are accumulating on the platform are precipitating from calcium and carbon-oxygen molecules dissolved in seawater. Ancient iron formations presumably also precipitated from chemical elements dissolved in seawater.

BANDING

The most voluminous iron formations always contain as much chemically precipitated silica (similar to glass) as they do iron minerals. This silica is concentrated into thin layers that alternate with layers of iron-bearing minerals. The thinnest of these layers commonly ranges from 1 millimeter to 1 centimeter in thickness. In unweathered iron deposits, a common iron mineral is siderite, the iron-carbon-oxygen mineral. However, this mineral is particularly susceptible to weathering and readily becomes oxidized to an iron-oxygen-hydrogen mineral—for example, goethite. Within a well-drained, elevated plateau, weathering of chemically reduced minerals to goethite may occur to great depths, and a large volume of rock may become weathered without being removed by erosion. The weathered deposit subsequently may become deeply buried if the plateau subsides under the ocean, and a thick mantle of marine sediment may then accumulate over the area. If goethite becomes deeply buried, it may become transformed to hematite or magnetite under the great heat and pressure which characterize the deep Earth. Some of the richest iron deposits appear to have experienced such ancient deep weathering and burial prior to becoming exposed to modern weathering.

An interbanded iron formation at the Carol deposits, Wabush Lake area, Iron Ore Company of Canada. (Geological Survey of Canada)

Geologists who are interested in the origin of iron formations study only the unweathered iron formations. Unweathered iron formations are particularly well exposed around the western end of Lake Superior, in both Canada and the United States. There geologists have found fossils of some of the oldest known living cells, enclosed within layers of silica. The iron formations that contain these cells also exhibit large growth forms of ancient photosynthesizing bacteria, or cyanobacteria. These cyanobacteria produced mats that were folded upward from the shallow sea floor, like a crumpled carpet, stretching toward the ancient sunlight. Similarly folded mats are extremely rare on the modern sea floor.

The great bulk of the thin layering (also called banding) in voluminous iron formations is not crumpled but flat. The origin of the alternation between iron-rich and iron-poor (silica-rich) layers has been controversial, because only one other type of chemical sedimentary rock persistently displays such a systematic alternation of layers. This other rock type is calcium-bearing salt that has accumulated under deep water. Salt typically precipitates because of evaporation of seawater, but it is difficult to imagine how a chemical element that is as inherently insoluble as iron could behave like the highly soluble elements that precipitate in salt beds—for example, sodium and calcium.

YOUNG IRON DEPOSITS

Small iron deposits that are younger than 600-million years old typically do not exhibit layering. One explanation is that burrowing animals, which have proliferated on the sea floor during the past 600 million years, have disrupted any layering. Indeed, many young iron deposits contain remnants of animal burrows. The animals that constructed these burrows presumably required oxygen to breathe, but the young iron deposits contain chemically reduced, iron-bearing minerals that could precipitate only in the complete absence of molecular oxygen. The origin of young iron deposits is therefore no better understood than that of the larger old deposits.

Virtually all young iron deposits that are of ore grade exhibit a characteristic texture, which consists of tiny, onion-shaped spheres that are called ooids. Ooids are so beautifully formed that they were thought to be fossil fish eggs by ancient Greek investigators. Although geologists now know that they are not eggs, the origin of iron-rich ooids remains a mystery. Iron-rich ooids are extremely scarce in modern sediment, and the few modern localities where they occur have not yet been thoroughly studied.

Besides the scarcity of silica in young iron deposits, they also differ from old iron deposits in that they typically contain more aluminum. This aluminum occurs within an iron-rich clay mineral. Clay minerals are less abundant in old iron deposits, and the clay minerals that do exist in ancient deposits generally contain no aluminum. Phosphorus is abundant in young iron deposits, which contain an average of 0.5 percent phosphorus, whereas only about one-tenth of this concentration occurs in the old deposits. Old iron formations in Finland are exceptional in that they contain as much phosphorus as do typical young iron deposits. Iron deposits of all ages contain amounts of manganese, but all other metals—for example, copper—characteristically are scarce.

In addition to aluminum-rich silicate minerals, young iron deposits generally contain either goethite, a hydrogen-bearing oxide of iron, or hematite, an iron oxide that lacks hydrogen. It is easy to tell which mineral predominates because goethite imparts a yellow-brown color, whereas hematite is bright red. All iron-bearing silicates are green, regardless of whether they are aluminum-rich.

Young iron deposits in the United States are best known near Clinton, New York, and Birmingham, Alabama. The Clinton deposits are about 430 million years old. In northern Europe, young deposits range from about 150 to 200 million years old. Production of young deposits appears to have been somewhat episodic, as was production of old iron deposits. Although the most voluminous iron deposits formed on the ancient Earth, there have been long stretches of ancient Earth history when production of iron deposits was relatively depressed—for example, between 2,000 and 1,000 million years ago. An eventual understanding of the cyclicity of iron sedimentation probably will provide fundamental information about the global evolution of the Earth.

STUDYING IRON DEPOSITS

Iron deposits are studied both from an economic viewpoint and from an academic viewpoint.

The academic potential of an iron formation is related to its state of preservation and the diversity of features it may exhibit. The prime economic criteria for mining of an iron deposit are ore composition, mining costs, and distance from market. Good ore contains not only a high percentage of iron but also a low percentage of certain chemical elements that would interfere with steel production. Most academic studies of iron formations focus on the information that they may provide about ancient Earth environments, such as the temperature and composition of the Earth's ancient atmosphere and oceans. Iron formations that are suitable for these studies have not been changed much by heat or pressure deep in the Earth and have not experienced much weathering.

The initial step in studying an iron formation is to make thin slices of the various rock types. Each slice is abraded to be so thin that light will pass through the minerals when the slice is placed under a high-power microscope. Microscopic study allows for the identification of all minerals and any fossils of ancient life-forms. Identification of life-forms is facilitated by using an electron-beam microscope, which can achieve higher magnification than does a visible-light microscope.

A complete chemical analysis of an iron formation reveals the degree of concentration of the chemically precipitated elements relative to the composition of average rocks in the Earth's crust. Some chemical elements—for example, carbon and oxygen—occur in more than one variety, called isotopes, and the ratio of these isotopes may indicate something about the chemical process that precipitated these elements. It is already well known that the ratios of oxygen and carbon isotopes in iron formations generally are distinct from those in common chemical sedimentary rocks—for example, limestone—of the same age as the iron formations, although a clear explanation of this distinction remains elusive to geologists.

THE IRON AGE

Voluminous silica-rich iron formations are the only major type of rock body that has not formed since complex life appeared on Earth between 600 and 800 million years ago. An understanding of the origin of these ancient iron formations, therefore, may explain why the ancient Earth did not support complex life. Although modern society depends on voluminous iron formations as a source of iron ore, modern accumulation of a voluminous iron formation may require the chemical composition of the atmosphere and oceans to become toxic to human life.

The present Iron Age, the portion of human history during which tools have been made of iron, started about 1000 B.C.E. Exploitation of iron deposits has increased roughly exponentially since that time. The fact that civilization continues to expand exponentially partly results from the continuing availability of both voluminous iron deposits and the fuels which are needed to convert iron ore into steel. Reserves of iron ore are becoming depleted less rapidly than the known reserves of fuel.

Michael M. Kimberley

CROSS-REFERENCES

Aluminum Deposits, 1539; Building Stone, 1545; Cement, 1550; Coal, 1555; Diamonds, 1561; Dolomite, 1567; Earth Resources, 1741; Elemental Distribution, 379; Evaporites, 2330; Fertilizers, 1573; Future Resources, 1764; Gem Minerals, 1199; Gold and Silver, 1578; Hydrothermal Mineral Deposits, 1584; Hydrothermal Mineralization, 1205; Industrial Metals, 1589; Industrial Nonmetals, 1596; Manganese Nodules, 1608; Metamorphic Mineral Deposits, 1614; Pegmatites, 1620; Platinum Group Metals, 1626; Salt Domes and Salt Diapirs, 1632; Sand, 2363; Sedimentary Mineral Deposits, 1629; Strategic Resources, 1796; Uranium Deposits, 1643; Weathering and Erosion, 2380.

BIBLIOGRAPHY

Chauvel, J. J., et al., eds. *Ancient Banded Iron Formations: Regional Presentations.* Athens, Greece: Theophrastus Publications, 1990. A thorough look at iron ore and the iron mining industry, with an emphasis on what the study of ores and mines can teach people about the past. Illustrations and bibliographical references.

Holland, Heinrich D. *The Chemical Evolution of the Atmosphere and Oceans.* Princeton, N.J.: Princeton University Press, 1984. Only a small portion of this book (pages 374-407) is devoted to a discussion of the origin of iron deposits. Nevertheless, Holland provides a particularly lucid discussion about the importance of understanding iron-concentrating processes if geologists are to achieve a general understanding of the evolution of environmental conditions on Earth.

James, H. L. "Chemistry of the Iron-Rich Sedimentary Rocks." In *Data of Geochemistry*, edited by M. Fleischer. Professional Paper 440-W. Denver, Colo.: U.S. Geological Survey, 1966. This classic publication remains valuable because it includes unchanging data on iron-rich minerals and chemical compositions of iron deposits.

James, H. L., and P. K. Sims, eds. *Economic Geology* 68, no. 7 (1973): 913-1179. This issue of the foremost journal of economic geology is entirely devoted to hypotheses about the origin of the largest iron deposits. A dozen renowned geologists present remarkably diverse explanations of the same rocks. The diversity itself is an indication of how difficult it is to explain the origin of iron deposits.

Kimberley, M. M. "Paleoenvironmental Classification of Iron Formations." *Economic Geology* 73 (1978): 215-229. Prior to this paper, nomenclature for iron deposits had become complicated by the local usage of poorly defined terms. This paper proposes simple definitions of the two most basic terms, "ironstone" and "iron formation," and a global classification scheme for iron deposits. The evolution of iron formations through Earth history is apparent in a chronological listing of the classified types of iron deposits.

Maynard, J. B. *Geochemistry of Sedimentary Ore Deposits*. New York: Springer-Verlag, 1983. Although the largest iron deposits are older than 600 million years, several small iron deposits have formed more recently, as reviewed in this textbook.

Maynard, J. B., E. R. Force, J. J. Eidel, eds. *Sedimentary and Diagenetic Mineral Deposits: A Basin Analysis Approach to Exploration*. Chelsea, Mich.: Society of Economic Geologists, 1991. This book discusses the geochemistry of most major types of metallic sedimentary mineral deposits in a geological context and discusses the strengths and weaknesses of various models that have been proposed to explain their formation. Suitable for any level of reader with some knowledge of both chemistry and geology.

Mel'nik, Yu. P. *Precambrian Banded Iron Formations: Physicochemical Conditions of Formation*. New York: Elsevier, 1982. The rise of the former Soviet Union as an industrial power depended partly on exploitation of its iron deposits. This translation from the Russian describes some of the major iron deposits of the former Soviet Union as well as some of the Russian theories about the processes that formed them.

Morris, R. C. "Genesis of Iron Ore in Banded Iron-Formation by Supergene and Supergene-Metamorphic Processes: A Conceptual Model." In *Handbook of Strata-Bound and Stratiform Ore Deposits*. Vol. 13, edited by K. H. Wolf. New York: Elsevier, 1985. A thorough discussion of the chemical alteration of iron formations. All voluminous iron formations contain as much silicon as iron when unaltered, but alteration (weathering under humid tropical conditions), preferentially removes the silicon, leaving iron within a porous regolith that may be mined readily.

Symons, Martyn C. R., and J. M. C. Gutteridge. *Free Radicals and Iron: Chemistry, Biology, and Medicine*. New York: Oxford Univeristy Press, 1998. This college-level text looks at the geochemical and biochemical make-up of iron and uses its finding to examine the physiological and pathophysiological effects in medicine and in the environment. Intended for readers with some prior knowledge about the subject matter.

Trendall, A. F. and R. C. Morris, eds. *Iron-Formation: Facts and Problems*. New York: Elsevier, 1983. The papers collected in this volume provide a good review of iron formations in western Australia and North America. Provides a description of how, over the past three decades, exploitation of iron deposits in western Australia has expanded greatly, while exploitation in North America has decreased correspondingly.

United Nations Educational, Scientific, and Cultural Organization. *Genesis of Precambrian Iron and Managanese Deposits.* Paris: Author, 1973. The thirty-eight papers in this book record the most international of all major symposia on the origin of large iron deposits, a symposium which occurred in Kiev in 1970. Offers the most comprehensive descriptions of iron deposits from around the globe.

MANGANESE NODULES

Manganese nodules are growths of metal-yielding elements that accumulate in the oceans. Associated with active rift areas, the nodules are formed by precipitation from seawater and by interactions with biological organisms. Their economic potential for development as strategic metal reserves has increased the importance of the nodules to humans.

PRINCIPAL TERMS

ACCRETION: the growth of materials by the addition of new material from a surrounding fluid

NODULE: a spherically shaped, concentrically layered hard mass found on the sea bottom, composed of metallic ions accumulated on seed material

OXIDIZING CONDITIONS: environmental situations in which elements react quickly to the availability of oxygen or other electron-rich atoms

PRECIPITATION: the settling out of ions, elements, or chemical compounds dissolved in a solution as a result of changes in the environment, such as temperature, pressure, or chemical concentrations

SCANNING ELECTRON MICROSCOPY: a technique for resolving extremely small details and for watching crystals grow; electrons are passed over an active surface, then are focused by magnets to give a visual image

VERNADITE and TODOROKITE: typical rare minerals tying up copper, nickel, and cobalt in nodules, derived from continental ore deposits

WHITE and BLACK SMOKERS: vents near active undersea spreading centers, from which large amounts of hot fluids and dissolved substances escape from deeper layers inside the Earth

DISCOVERY OF NODULES

One of the numerous discoveries of the expedition of HMS *Challenger* of 1873-1876 was that of unusual and plentiful sorts of rocks referred to as manganese nodules, retrieved from the ocean by bottom dredging. These nodules, as analyzed first by Sir John Murray, were found to be surprisingly high in manganese localized on red clay, with additional traces of limestone and accretions of diverse heavy metals such as iron, titanium, and chromium. Upon examination, the manganese nodules, measuring 1 to 15 centimeters in diameter, were found to be essentially a conglomeration of heavy metals and other chemical compounds formed around a basic particle acting as a catalyst, or seed nucleus, for crystallization. This centerpiece, represented quite often by a shark's tooth, a piece of bone, or other small, round objects, acts as a condensation center, allowing a surface out onto which the dissolved elements and molecules in solution can precipitate from the surrounding ocean waters. Larger nodules are the result of the gradual growth of such layers on the seed nucleus over long periods of time.

The phenomenon of manganese nodules, however, was not new to geology. Stony concretions of manganese and iron had been used in Sweden for a long time, more than ten thousand tons being mined from Swedish lakes in the year 1860 alone. Professional chemical and mineralogical studies there indicated that the nodules regrew within a cycle of thirty to fifty years, totally replacing the dredged materials. Swedish scientists found that, in the normal oxidizing conditions found in freshwater lakes, with very low rates of sedimentation from surrounding materials, if center seeds such as spores, bark, or clay particles were present, iron and manganese would precipitate as the water met the relevant conditions of temperature and pressure. Growth was continuous and often accelerated by currents sweeping the lake floor, keeping the nodules available at the surface for the further deposition of ions. Such growth ceased when sediments were able to accumulate rapidly on the lake floor.

DISTRIBUTION

In the seas, manganese is found in numerous places and in several forms. In red clays found on the ocean bottom, the element occurs in higher concentration than it does normally in igneous rocks. In other sediments, it occurs principally as manganese dioxide in fine grains, in coatings, and as matrices for other rocks in the process of formation. Although some nodules are large in size, the bulk of manganese is distributed in red clays, igneous rocks, and limestone. The dark colors of the red clays of the Indian and Pacific oceans, for example, have been attributed to the manganese content.

The nodules themselves, with an average content of 29 percent manganese dioxide and 21.5 percent iron oxide, grow by accretion. Slices through the rock show laminations of different shades and textures, indicating differing precipitation rates and chemical compositions of the surrounding waters over time. Found worldwide, manganese nodules, particularly those with the highest metallic concentration, occur in bands running from the north equatorial Pacific to southeast Hawaii to Baja California. Scientists estimate that in the Pacific area alone, one and a half trillion tons of nodules are present, with a rate of formation of some ten million tons per year. A comparable amount is found in the Indian Ocean, with those of high nickel-copper counts discovered adjacent to the equator. Atlantic nodules have much lower strategic metal concentrations. Such distribution and abundance of elements—in addition to manganese—suggest that the water and overlying sediments may control the ultimate composition of the nodules. In the Pacific, diatomaceous and radiolarian siliceous oozes underlie the rocks. In more northern areas, which demonstrate less biological productivity in the surface waters, the nodules have higher iron and lower nickel-copper concentrations where the formed concretions are resting

over pelagic red clays. Indeed, analysis indicates that chemical differences exist between the top water-contacted layer of the nodule and the bottom sediment-immersed layer. The bottom layers test higher in manganese, nickel, and copper; the top layers, in iron and cobalt. Fluctuations in concentration through sectionalized nodules can be accounted for by the episodic rolling over the seafloor.

Manganese nodules contain small, poorly crystalline minerals, the types of which depend on their location in the ocean. The material vernadite, with trivalent cobalt, is found in nodules from seamounts and shallow seas; todorokite, with divalent nickel and copper, is found in rocks from siliceous oozes. Divalent manganese-bearing todorokites come from continental ore deposits.

GROWTH AND FORMATION

The addition of ions causes the nodules to grow, apparently at a very slow rate. Radium, via the mechanism of absorption by manganese peroxide, is included in the nodules; in the concentric layers it falls off rapidly with increasing depth below the nodule surface, indicating duration of formation for each layer. Other isotopes used for determining growth rates include thorium 230, beryllium 10, and potassium 40. Isotopic measurements indicate a growth rate of a few millimeters

A manganese (pyrolusite) nodule from the ocean bottom. (© William E. Ferguson)

per million years. The mystery of this growth process, when the burying red clay sediments accumulate at a thousandfold faster rate, was solved by time-lapse sea-floor photography and sedimentological studies. Both methods showed that the periodic rolling of the nodules, keeping them at the sediment-seawater interface, was accomplished by the high populations of organisms burrowing in the underlying sediments.

Several hypotheses have been proposed for the formation of manganese nodules, the earliest dating back to the time of Murray (1891). These hypotheses include slow precipitation from seawater, deposition from submarine volcanic disturbances, and diagenesis (remobilization) from sediments. Chemical analysis has shown that, compared to normal volcanic rock concentration of manganese, a five- to tenfold concentration of manganese would be required to give the amount found in Pacific red clays. Manganese is leached out of solid materials when it is in a reduced state in acid solutions; the opposite, an alkaline solution in oxidizing conditions, tends to precipitate manganese out as insoluble manganese dioxide. Nodules form, without redissolving into the water, in areas of seamounts and in oceanic deposits where there are even slight amounts of limestone with high alkalinity. Small nodules form this way, as slow precipitates.

In the 1960's, the investigation of the mid-oceanic ridge and of enclosed ocean basins yielded data about the formation of metals. Hot brines above metal-bearing muds were found, particularly near the East Pacific Rise. Water running through those ridge-axis hot springs and flank hydrothermal areas becomes supersaturated with manganese and other metals it dissolves by virtue of a temperature of 400+ degrees Celsius. Deposits of materials dropping out of solution cluster in basins near active spreading centers, known now as black (lacking manganese) or white smokers. Basic metal ions derive from those fluids emitted at the seafloor. White smokers cause the material to be deposited within the rock as cold water percolates downward. Manganese precipitates out when the water cools below white-smoker temperatures; floating in the water column, it then reprecipitates on the seafloor. In general, as the cold seawater travels down into the basaltic crust, it heats, dissolves more metals, and rises to the surface. The leached-out metals are then contributed to the nodule concentrations of iron, cobalt, copper, nickel, and manganese as sulfides and oxide ions.

INVESTIGATIVE METHODS

Numerous investigative techniques have been used to understand both the chemical makeup and the formation process of manganese nodules. Rock samples have been collected by bottom dredging, either by use of simple dredging sacks or by use of submersibles with grappling arms for retrieving specimens.

Once the samples are in the laboratory, a diverse range of analytical techniques are used to determine the chemistry of the nodules. Microchemical analysis starts on sectioned specimens. The chemical composition can be determined using diverse techniques, such as atomic absorption spectroscopy, in which a small sample is burned and the emitted light is analyzed to give concentrations, or ultraviolet spectrophotometry, using light of ultraviolet frequencies that passes through a solution of the nodule in fluid. Chemical analysis for metallic ions can be achieved by affixing concentrations on the order of parts per million or, in some cases, parts per billion. Such chemical work has revealed differences between nodules as well as within nodules themselves, providing information on the spacing of the nodules and on their movement in regard to the water-sediment interface. Such analyses provide fluctuating concentration values for manganese, iron, nickel, copper, cobalt, and other metals, as well as information on the matrix of the red clays and other components, thus furnishing clues to the origin of the rocks.

The minerals found in the nodules, such as todorokite or feroxyhyte, are best studied by use of the geologic polarizing microscope for identification. This instrument, using polarizing lenses that can be aligned at diverse angles to each other, can be used to identify crystals by their color changes and indices of refraction when viewed in polarized (single-plane) light. Each type of mineral responds differently to the light; the various changes as the light's plane of polarization is turned indicate a particular mineral type.

The growth and fine detail of minerals in the nodules can be investigated more precisely by using the scanning electron microscope. In this technique, high-speed electrons are bounced off the surface to be viewed, the returning particles focused by the action of magnetic fields to give a detailed picture, at ultrafine resolution, of the crystal's growing surface. Such observations point to a biological source for the metals in the nodules and to remobilization of the nodules by the underlying organisms in the sediments.

Geophysical, chemical, and physical studies are done to investigate the formation process. Radioactive decay counters measure rates of nuclear breakdown in radioisotopes to find the rate of growth of the nodules. Time-lapse photography, both in the laboratory and on the seafloor, shows how organisms move the nodules, causing them to roll periodically and thus not be buried by accumulating sediments. Such studies of the rates of siliceous ooze or of oceanic red clay sediment deposition give vital information on the sediment-seawater interface and how it changes over time. Chemical and geological laboratory experiments, using conventional chemical tools, show how metals are adsorbed onto the surface of clays or are organically complexed in biological matter settling to the seafloor. In simple chemical experiments using reducing environments, manganese and iron oxides oxidize organic debris to give soluble, positively charged ions such as divalent manganese, iron, nickel, and copper in the pore water of the sediment, making them available for accumulation in the nodules. Further oxidation studies show how manganese and iron are produced in insoluble hydrated forms that accumulate on the nodules themselves. Detailed chemical studies of minerals found in the white smokers on the sea bottom, particularly those minerals left behind in the cracks, and studies of the microscopic "bubbles" associated with the original fluid left trapped in the crystals give a detailed look at the possible source for ores to be found inside the nodules.

COMMERCIAL POTENTIAL

Modern oceanographic science has been able to detail accurate maps indicating where the nodule abundance and associated metal concentrations are highest and where the mining of such minerals would be most economical. It has been suggested that mining of the nodules in deep water could be achieved by giant deep-sea vacuum cleaners, a technique entirely feasible from an engineering standpoint. During the summer of 1970, several companies tested such processes, which allow nodules to be recovered at costs low enough to make a profit. Deposits on land, however, have higher manganese and iron concentrations, complicating the retrieval picture. Three basic systems have been designed for the future: an airlift, a hydraulic lift without air, and a mechanical lift called the continuous-line bucket system. Theoretically, these extended hoses, fitted with spread nozzles, would pass over the sea bottom, using simple suction to remove the supersurface nodules.

While manganese nodules are potentially economically important, as of the end of the twentieth century there was no commercial mining of them anywhere in the world. The two key reasons for this are that there are ample supplies of land-based manganese ore, and recovering the nodule from the seafloor is very expensive. Should either of these factors change, the nodules will be available. Formed by accretion around spherical objects, the nodules, besides containing 20 percent manganese and at least 15 percent iron, have significant enrichments in nickel, copper, cobalt, zinc, molybdenum, and other elements considered important for strategic reasons. Through various chemical and physical techniques, the basic mechanisms of formation seem to be understood, including the dissolving and redeposition of terrestrially derived manganese and the emission as a hot-water-dissolved substance from white smokers found in the undersea volcanic rift areas. Besides the manganese oxide phases accumulating on the surface of the nodules, there are numerous other ions that are bound up in the rock and mineral phases as essential constituents in the crystal structures forming within the nodules.

By understanding the distribution and chemical concentration of manganese nodules, geochemists achieve a more detailed picture of the makeup of the Earth, particularly below the surface, and of the process of mineral enrichment. This knowledge provides vital clues in the search for elements and compounds considered essential for the survival of technological civilization. That information also sheds light on such diverse topics as the origin of the Earth, the formation of the continents, and the planetary evolution over time.

Arthur L. Alt

BIBLIOGRAPHY

Anderson, Roger. *Marine Geology: An Adventure into the Unknown.* New York: John Wiley & Sons, 1986. An introduction to the features of marine geological sciences, this work emphasizes modern knowledge built on plate tectonics. Excellently written, the book provides a detailed account of the black and white smokers of the sea bottom, including a discussion of the strange organisms found there. Contains numerous charts and pictures and an extensive bibliography. For general audiences.

Burns, G., ed. *Marine Minerals.* Washington, D.C.: Mineralogical Society of America, 1979. A vastly detailed account of the variety of rocks and minerals found on the seafloor, this work furnishes an in-depth analysis of chemical elemental distribution. Numerous graphs and tables facilitate data handling. Not for beginners, but a wide range of topics and information provide guidelines for further research.

Chester, Roy. *Marine Geochemistry.* London: Unwin Hyman, 1990. An excellent and thorough introduction to marine geochemistry, chemical oceanography, and marine sedimentology. Well illustrated and clearly written, making it very useful to the beginner. Index and bibliography.

DiToro, Dominic, et al., eds. *Sediment Flux Model for Maganese and Iron.* Vicksburg, Miss.: U.S Army Engineer Waterways Experiment Stations, 1998. This U.S. Army Engineer Waterways publication is filled with mathematical models used to determine the levels of marine, estuarine, and lake sediments. It also includes information on models used to evaluate the manganese and iron content in soils. This highly technical piece requires some background in environmental engineering.

Glasby, G. P., ed. *Marine Manganese Deposits.* New York: Elsevier, 1977. A collection of papers on the types and quantities of manganese deposits found under various conditions. Manganese nodules—their origin, relationship to continents and organisms, growth rates, and chemical compositions—are discussed. Provides extensive tables and charts as well as references. Detailed reading, but a good starting point for further research on numerous topics.

Ingmanson, Dale, and William J. Wallace. *Oceanography: An Introduction.* 3d ed. Belmont, Calif.: Wadsworth, 1979. A general source of information on the oceans, with several chapters specifically about the geology of the seafloor. The formation of features, such as nodules, is related to current theories of seafloor spreading and the evolution of the oceans. Excellent photographs and diagrams. Well written for the layperson.

Seibold, E., and W. Berger. *The Sea Floor: An Introduction to Marine Biology.* New York: Springer-Verlag, 1982. An excellent introduction to marine geology, detailing the types of materials, geologic features, and complex evolution of the ocean bottom. The interaction of organisms, weather, sediments, and humans is explained lucidly. Well written, with numerous charts and pictures. Additional references are listed.

Smith, Peter J., ed. *The Earth.* New York: Macmillan, 1986. A delightful book, filled with pictures and explanatory diagrams, this work describes the structure of the Earth and its evolution. Manganese nodules are discussed both as a typical deposit and as a possible resource for human development. Additional references and a glossary are provided.

Summerhayes, C. P., Warren L. Prell, and K. C. Emeis, et al., eds. *Upwelling Systems: Evolution Since the Early Miocene.* London: Geological Society, 1992. A nice overview of oceanography and marine ecology. This book provides a clear analysis of marine sediments and deposits. Intended for the college reader, although someone without much background in the subject area will also find this book of some use.

Trask, Parker Davies, ed. *Recent Marine Sediments.* Mineola, N.Y.: Dover, 1965. This work records a symposium on the origins and features of sea sediments. Extremely detailed, the section on special sediment features describes the formation of nodules and other deposits found in the world's oceans. An excellent section on methods of study, including mineral analysis, X-ray methods, and bottom-sampling apparatus, is provided. Extensive references. Reading may be difficult for the layperson.

METAMORPHIC MINERAL DEPOSITS

Igneous rock is formed deep in the Earth when magma cools and crystallizes; it yields massive mineral deposits and is an important commodity in the world market. Contact metamorphism is a process that alters preexisting rock through heat; resulting mineral deposits are usually small but contain valuable and rare minerals.

PRINCIPAL TERMS

AUREOLE: a ring-shaped zone around an igneous intrusion; a zone of alteration

GANGUE: the worthless rock or vein matter in which valuable metals or minerals occur

IGNEOUS ROCKS: rocks formed by solidification of molten magma

METAMORPHIC ROCKS: rocks that are alteration products of preexisting rock forms that were exposed to unusual circumstances and changed as a result

METASOMATISM: metamorphism that involves changes in the chemical composition as well as in the texture of the rock

MINERAL: a solid homogenous crystalline chemical element or compound that results from the inorganic processes of nature

PYROCLASTIC: formed by or involving fragmentation as a result of volcanic or igneous action

VESICLE: a small cavity in a mineral or rock

CLASSIFICATION

A mineral deposit can be defined as valuable concentrations or accumulations of one or more useful minerals; if a miner thinks the deposit can be recovered at a profit or a metallurgist thinks it can be treated for a profit, then it is an ore. There are about fourteen hundred mineral species, of which roughly two hundred are considered economic minerals. To determine if a rock is an economic mineral, the miner or metallurgist examines its market value and physical and chemical properties. Although minerals exist to an extent in all rocks, the quantity and quality are not always valuable to the miner or metallurgist.

There are three broad classifications of rocks: igneous, metamorphic, and sedimentary. Metamorphic rocks are a link between the divergent igneous and sedimentary rock forms because they are alteration products of preexisting rock forms that were exposed to unusual circumstances and changed as a result. There are three types of metamorphism; only one of them, contact metamorphism, uses thermal energy exclusively as the catalyst of change. Igneous rock formation also uses thermal energy as the prime cause of formation, only on a much more intense scale than contact metamorphism.

IGNEOUS ROCK FORMATION

Originating at depths of 200 kilometers within the Earth and formed by cooling and recrystallizing magma, igneous rocks have a unique composition of silicate minerals and gaseous elements trapped by rock pressure. Igneous rocks are classified into groups based on mode of origin (formation conditions), mineral constituents (composition), and size and arrangement of mineral crystals (texture). The first group of igneous rocks is granite: large crystalline, quartz-bearing, light-colored rocks. The next group includes basalt and diabase, grouped together because of occurrences close in proximity and similar in composition; quartz-free, dark, and heavy, these rocks are at the opposite end of the igneous spectrum from granite. The third type of rock is pumice and pumicite (volcanic ash): vesicular, glassy, pyroclastic rocks. The fourth and last type of rock, perlite, is sometimes classified with pumice and pumicite because of its similar glassy and vesicular features; however, it has a unique composition that is partly primary (original) and partly an alteration product of rhyolite or obsidian.

Plutons are structures resulting from the emplacement of igneous materials at depth, and they can be studied only after uplifting and erosion, millions or hundreds of millions of years after formation. Plutons are associated with volcanic activity and can be either tabular or massive, discordant or concordant (cutting across existing structures or parallel to existing structures). Dikes are tabular discordant masses that occur where magma is injected into fractures that most likely were vertical pathways followed by molten rock to the surface. Sills—tabular plutons occurring where magma is injected, usually, but not always, horizontally along sedimentary bedding surfaces—can be a mixture of concordant and discordant. Sills force the overlying rock up to a height equal to the sill's thickness. Because of the amount of force needed to push an overlying structure upward, sills form at shallow depths and are easy to spot owing to the deformation of the surrounding surface area. When magma is intruded between layers of sedimentary rock in a near-surface environment, laccoliths form. Like sills, they are easily spotted by the trained eye because of the dome created on the surface by the lens-shaped mass of magma. Batholiths are the largest intrusive igneous bodies to be found; linear formations, they are sometimes the cores of extinct mountain systems.

The formation of a mineral deposit is dependent on the chemical composition of the magma, the time of formation, heat pressure, and the rate of cooling. Experimental data on the factors present during formation were sought in the nineteenth century by N. L. Bowen; his attempt was to determine melting temperatures of rocks and silicate phases. The product of his work was a reaction chart of mineral crystallization behavior. His results indicate the order of minerals formed under specific conditions. Olivine is the first mineral to form, and it remains in the melt, reacting with free atoms to become pyroxene (augite), which becomes amphibole (hornblende), which becomes biotite mica, the last mineral to form in this sequence, named the "discontinuous series" because each mineral has a different crystal structure. The "continuous series" begins with the formation of calcium-rich feldspar crystals, which react with sodium ions in the melt until sodium-rich feldspar is produced; quick cooling prohibits a complete transformation to sodium-rich feldspar and results instead in calcium-rich interiors surrounded by zones progressively richer in so-

dium. Both series occur simultaneously. In the last stages of crystallization, most of the magma is solid, and the remaining melt forms the minerals quartz, muscovite (mica), and potassium feldspar. Basaltic rocks contain the first minerals to crystallize: calcium feldspar, pyroxene, and olivine. Basalts are high in iron, magnesium, and calcium but are low in silicon. Granitic rocks are predominantly composed of the last minerals to crystallize: potassium feldspar and quartz. Andesitic rocks are composed of minerals near the middle of Bowen's table: amphibole and intermediate plagioclase feldspars. Gradations exist in each general mineral composition that distinguish one deposit from the other.

CONTACT METAMORPHISM

Another rock produced from high temperatures is contact metamorphic. Although contact metamorphism involves high temperatures, complete melting never occurs, for that is the realm of igneous rocks. Contact, or thermal, metamorphism is the intrusion of molten magmas into the Earth's crust, accompanied by the exuding of heat and gases that profoundly alter the invaded rocks and give rise to an aureole. The further away the aureole is from the magma body, the lower the grade of its mineral. The size of the altered zone appears to be governed by the size of the rock formation. Contact metamorphism is distinguishable only when it occurs at the surface or in a near-surface environment where the temperature difference between the host rock and environment is great.

There are two kinds of contact metamorphism. Normal contact metamorphism involves recrystallization and reconstruction of the original constituents of the invaded rock: in this process, no additional material is added to the host rock. The second type of contact metamorphism, pneumatolytic, is the one of interest to miners and metallurgists because it involves the same processes of recrystallization and reconstruction as does normal contact metamorphism, in addition to a gaseous transfer of materials from the invading magma. Unlike normal contact metamorphism, the pneumatolytic type involves the formation of new minerals, combining the old constituents with new ones. Rock products yielded are fine-grained, dense, tough rocks of various chemical compositions. Pneumatolytic contact metamorphism occasionally results in economic mineral deposits, which are characterized by an assemblage of very distinctive high-temperature minerals.

The actual mineral deposits are distinguished by the unusual mixture of ores and gangue. Perhaps one of the most unusual features about this comparatively rare phenomenon is the distinctive, high-temperature gangue minerals found with the ores. The deposits usually consist of two or more disconnected bodies, on the average of 100-400 feet, yielding tens of thousands to hundreds of thousands of tons; in a very few cases, the yield has been in the millions of tons. The small size of the deposits also makes them hard to find, necessitating costly exploration and development. The exploitation of the deposits can be annoying because of the size, the random distribution of minerals in the aureole, and the sudden termination of the mineral vein. Occurrence of deposits is within the contact aureole, most often within 100 yards or so of the intrusive contact. The deposits are irregular in outline, taking any shape, and are fairly equidimensional. Their texture is coarse, commonly containing large crystals or clusters of crystals, although very few of the minerals show a crystal outline except garnet, which appears in sugarlike masses of irregular outline.

FIELD AND LABORATORY STUDY

The methods of studying rocks range across several fields of Earth science. For example, a physicist might use spectroscopy, a method of radiation measurement, where a chemist might use laboratory techniques to determine mineral or elemental constituents of a sample rock. Other methods of study range from laboratory techniques to field examination; a partial list includes assaying, chemical analysis, microscopic examination, X-ray examination, infrared scanning, and physical tests. An economic geologist's area of interest, however, would lie in the geology of rocks and the value of mineral deposits within a rock as a commodity to be exploited at the least cost and the most profit. The geologist identifies rocks primarily by noting places of origin, formation, distinctive features, and rock associations; keeping in mind the fact that igneous and contact metamorphic rocks formed hidden from view in the Earth's crust thousands and millions of years in the past, these observable details help the geologist to piece together the mysterious history of the rock's formation and to hypothesize which minerals the rocks contain.

The best way to study rocks is to examine the conditions under which they were formed: temperature, pressure, volatile content, and volatile composition. One can study these components in two ways. Relative conditions compare the formation of one rock to that of another by using petrology and microscopic examination to determine their characteristic positions in the field. Absolute conditions—exact temperature and pressure—of masses of minerals can be calculated to a degree using experimental synthesis, experimental determination of the conditions of reactions between groups of minerals, and the thermodynamic stability field of a mineral; the last method would be of particular interest in the study of contact metamorphism.

Igneous rocks also are studied by the conditions under which they formed. The heat and the composition of the magma involved are key factors to the diversity of mineral deposits. Laboratory experiments have shown that different minerals form under different temperatures; Bowen's reaction chart is one of the many possible igneous rock formation patterns. Magma is a hot liquid that consists, in most cases, of eight elements that are the primary constituents of silicate minerals. The ions in magma are free and unordered until cooling prompts crystallization. Atoms are reformed and arranged in an orderly pattern when cooling restricts rotational and vibrational movement, forcing ions to pack together and chemically rebond. As the magma cools, it does not solidify at once; instead, embryo crystals form and become centers of growth, capturing free ions. The result of the process is a mass of interlocking crystals. The crystal size is directly affected by the rate of cooling; slow cooling produces large crystals with a crystalline network, and rapid cooling results in very small, intergrown crystals with randomly distributed ions and no crystalline network.

TEXTURE

Texture is extremely useful as a key element in learning the order in which the minerals of metamorphic rocks crystallized and the sequence of events that occurred during formation. The texture of a rock refers to shape, size, orientation, and arrangement of mineral grains in a rock. Grain sizes vary from phaneritic (large) to aphanitic (small). The rate of cooling is a primary factor in the development of a rock's texture. A few textures that indicate planar or linear elements in the crystal grains in relation to one another are foliation, a parallel arrangement or distribution of minerals, rather like layering; schistosity, a parallel arrangement of mica or other platy minerals such as slate; and gneissosity, a pattern of alternating light and dark layers.

Three textures that denote equidimensional grains are hornfelsic, nondirectional, and fine-grained (containing dark spots consisting of clusters of grains in common); granoblastic, a mosaic of coarser grains of equal size; and polygonal, equal-sized, interlocking grains of a single mineral that form triple junctions of 120-degree angles. Rocks that contain larger grains are denoted by five textures: euhedral porphyroblast (similar to phenocrysts in igneous rocks, a large crystal surrounded by smaller grains of other minerals); idioblast, a porphyroblast with well-developed crystal faces; xenoblast, an anhedral (irregularly formed) porphyroblast; porphyroclast, a broken fragment of a phenocryst or porphyroblast in a fine-grained matrix; and augen, feldspar porphyroclast with an eyelike shape in a fine-grained gneissic matrix. Other textures of importance include ones that indicate planar features (which deform schistosity), inclusions within or rims on a porphyroblast, concentric features, and the fragmental nature of the whole rock.

INDUSTRIAL USES

An economic mineral deposit is one that can be exploited profitably and one in which the ores are of industrial value. The value of the ore is determined by utilizing standards specific to the ore, such as hardness, cleavage, rift, and grain. Because mineral deposits of the contact metamorphic kind occur rarely, the resulting minerals are often valuable; two examples are garnet, a high-temperature mineral, and crystalline calcite in marble, an alteration mineral. Igneous rocks are more common and yield larger deposits of minerals that are most often used in construction and as abrasives; one common igneous rock is granite, a heavy, sturdy rock that is a very popular building material.

The commercial classification of granite refers to any light-colored, coarse-grained, heavy rock. True granite is a product of the solidification of liquid magma, but some granite, such as granite gneiss, indicates a formation process called metasomatism, which produces a nonigneous granitic rock through pervasively soaking preexisting solid rock via plutonic solutions. The resulting rock is formed without the liquid stage associated with true igneous rocks and, therefore, is a hybrid rock. In industry, granite is desired in one of two forms: crushed or dimension.

The important industrial properties of crushed stone granite are resistance to impact (toughness), to abrasion (hardness), and to disintegration from repeated freezing and thawing (soundness). Crushed granite is used most often for road metal and railroad ballast and as concrete aggregate. One of the minor uses of crushed granite is as poultry grit. Dimension granite has more rigorous properties than does crushed granite. Some important ones are hardness; soundness; compressive strength, the load per unit of area under which the rock fails by shearing or splitting; color, which must be uniform throughout a deposit; texture, in which grain size and distribution of minerals must be uniform; and scaling, or flaking, which occurs on all granite blocks. Dimensional granite is used most often in monuments, memorials, and buildings and as curbstones and paving blocks.

The uses of the other types of igneous rocks are varied according to the class. Basalt and diabase are used in the form of crushed stone for abrasives and aggregates; high-density aggregate is used in nuclear reactors. Some diabase is polished for use as dimension stone and is called black granite. Pumice and pumicite are used primarily as abrasives and lightweight aggregates, as is perlite. Perlite has a unique property that increases its commercial value: the ability to suddenly expand when rapidly heated. The use of contact metamorphic rocks is also varied; two examples are garnet and marble. Garnet, a gemstone, is used in industry as a cutting tool and for decorative purposes. Marble is used in place of other, more common, less attractive stones in the structure of buildings or as decoration. The numerous other products of igneous and contact metamorphic mineral deposits have varied industrial use and range in value, demand, and rarity.

While metamorphic mineral deposits are less profitable and yield fewer tons of product, they contribute to the industrial field in a proportion nearly equal to that of igneous rock deposits.

Earl G. Hoover

Cross-References

Aluminum Deposits, 1539; Biogenic Sedimentary Rocks, 1435; Building Stone, 1545; Carbonates, 1182; Cement, 1550; Chemical Precipitates, 1440; Coal, 1555; Diamonds, 1561; Dolomite, 1567; Earth Resources, 1741; Earth's Age, 511; Elemental Distribution, 379; Evaporites, 2330; Fertilizers, 1573; Future Resources, 1764; Gold and Silver, 1578; Hydrothermal Mineral Deposits, 1584; Hydrothermal Mineralization, 1205; Industrial Metals, 1589; Industrial Nonmetals, 1596; Iron Deposits, 1602; Manganese Nodules, 1608; Mining Processes, 1780; Ocean Drilling Program, 365; Ocean-Floor Drilling Programs, 365; Oceanic Crust, 675; Oxides, 1249; Pegmatites, 1620; Petrographic Microscopes, 493; Platinum Group Metals, 1626; Prokaryotes, 1071; Salt Domes and Salt Diapirs, 1632; Sedimentary Mineral Deposits, 1629; Silica Minerals, 1354; Strategic Resources, 1796; Stratigraphic Correlation, 1153; Uranium Deposits, 1643; Weathering and Erosion, 2380.

Bibliography

Bateman, Alan M. *Economic Mineral Deposits.* New York: John Wiley & Sons, 1942. Although the text is outdated, the field of economic geology is one in which there is little change in fundamentals. Bateman's book is a good introduction, in very understandable terms, to different types of industrial, or economic, rocks. Not only does he discuss the value of the rock to the industries in which it is used, but he also describes the properties of the rock for which professionals look, as well as the origins and occurrences of the rock.

Bates, Robert L. *Industrial Minerals: Geology and World Deposits.* Mineola, London: Industrial Minerals Division, 1990. This text is meant for use by the professional geologist more than the layperson; however, it is written in terms that are easy to understand. Any technical terms can be found in a standard dictionary. This book dovetails nicely with Bateman's work; both are formatted in a very similar fashion, though the tones and stresses of the works are very different.

Best, Myron G. *Igneous and Metamorphic Petrology.* 2d ed. Cambridge, Mass.: Blackwell, 1995. A popular university text for undergraduate majors in geology. A well-illustrated and fairly detailed treatment of the origin, distribution, and characteristics of igneous and metamorphic rocks.

Deer, W. A., R. A. Howie, and J. Zussman. *An Introduction to Rock-Forming Minerals.* 2d ed. New York: Longman, 1992. Standard references on mineralogy for advanced college students and above. Each chapter contains detailed descriptions of chemistry and crystal structure, usually with chemical analyses. Discussions of chemical variations in minerals are extensive. *An Introduction to Rock-Forming Minerals* is a condensation of a five-volume set originally published in the 1960's.

Gillson, Joseph L. *Industrial Minerals and Rocks (Nonmetallics Other Than Fuels).* 3d ed. New York: American Institute of Mining, Metallurgical, and Petroleum Engineers, 1960. An excellent reference written by professionals and giving a brief but thorough synopsis of each mineral's composition, origin, uses, quarrying, and even industrial properties. Although it is a professional publication, it can supply nonprofessionals with general but crucial facts.

Hyndman, Donald W. *Petrology of Igneous and Metamorphic Rocks.* New York: McGraw-Hill, 1985. Hyndman's work is the best source for general study of igneous and metamorphic rocks. Discusses the latest theories and examines the evolution of the two rocks. Includes some technical, scientific information, with the most technical data being appended. Much information is poured into each section of the book, and graphs and illustrations are included when needed.

Lutgens, Frederick K., and Edward J. Tarbuck. *Earth: An Introduction to Physical Geology.* Upper Saddle River, N.J.: Prentice Hall, 1999. 6th ed. An introductory-level college text that contains cursory information on several items of interest: magma and its rock forms; igneous rock origins, occurrences, and classifications; and the three kinds of metamorphism.

Perch, L. I., ed. *Progress in Metamorphic and Magmatic Petrology*. New York: Cambridge University Press, 1991. Although intended for the advanced reader, several of the essays in this multiauthored volume will serve to familiarize the new student with the study of igneous and metamorphic rocks. In addition, the bibliography will lead the reader to other useful material.

United States. Bureau of Mines. *Minerals Yearbook.* Vol. 2, *Metals and Minerals.* Washington, D.C.: Government Printing Office, 1986. An economic evaluation of minerals as commodities, including domestic, foreign, and world production; price; value; supply; demand; technology; uses; and comparisons to earlier years in terms of increase and decrease in percentage. A public document published annually.

PEGMATITES

Pegmatites host the world's major supply of the rare metals lithium, beryllium, rubidium, cesium, niobium, and tantalum. They are also major sources of tin, uranium, thorium, boron, rare-earth elements, and certain types of gems.

PRINCIPAL TERMS

APLITE: a light-colored, sugary-textured granitic rock generally found as small, late-stage veins in granites of normal texture; in pegmatites, aplites usually form thin marginal selvages against the country rock but may also occur as major lenses in the pegmatite interior

CRYSTAL-LIQUID FRACTIONATION: physical separation of crystals, precipitated from cooling magma, from the coexisting melt, enriching the melt in elements excluded from the crystals; this separation, or fractionation, leads to extreme concentration of incompatible elements in the case of pegmatite magma

EXSOLVE: the process whereby an originally homogeneous solid solution separates into two or more minerals (or substances) of distinct composition upon cooling

FLUID INCLUSIONS: microscopic drops of parental fluid trapped in a crystal during growth; inclusions persist indefinitely unless the host crystal is disturbed by deformation or recrystallization

INCOMPATIBLE ELEMENTS: chemical elements characterized by odd ionic properties (size, charge, electronegativity) that tend to exclude them from the structures of common minerals during magmatic crystallization

SOLIDUS/LIQUIDUS TEMPERATURE: the liquidus temperature marks the beginning of crystallization in magmas, and the solidus temperature marks the end; crystals and melt coexist only within the liquidus-solidus temperature interval

VISCOSITY: a property of fluids that measures their internal resistance to flowage; the inverse of fluidity or mobility

SIMPLE PEGMATITES

Pegmatites are relatively small rock bodies of igneous appearance that are easily distinguished from all other rock types by their enormous range of grain size and textural variations. Typically, fine-grained margins of aplite are abruptly succeeded by discontinuous interior layers of coarse, inward-projecting crystals, which, in turn, give way to zones of graphically intergrown quartz and alkali feldspar. Most pegmatites contain numerous isolated pockets, rafts, and radial clusters of abnormally large, even giant, crystals and are cut by fractures filled with late-stage products. The spatial inhomogeneity of mineral distribution, rock texture, and chemical elements exhibited in pegmatites is unequaled in any other igneous product.

The majority of pegmatites are narrow lens-shaped, tabular, or podiform masses measuring from a few meters to several tens of meters in length. Even large, commercially exploited pegmatites rarely exceed 1 kilometer in length and 50 meters in width. Nearly all pegmatites have a bulk composition approximating that of granite. They are composed predominantly of quartz, alkali feldspar, and muscovite, with minor quantities of tourmaline and garnet. Those containing only these minerals are termed simple pegmatites, the word "simple" referring to their mineralogy, chemistry, and internal structural features collectively.

RARE-MINERAL PEGMATITES

Complex pegmatites are composed of the same major mineral assemblage as are simple pegmatites but, in addition, contain a great variety of exotic accessory minerals which host rare metals such as lithium, rubidium, cesium, beryllium, niobium, tantalum, and tin. These pegmatites are also called rare-element or rare-mineral pegmatites, and they are of major economic and strategic importance. In contrast to simple types, they are complex in terms of mineralogy, chemistry, and internal structure. Beyond their economic value, rare-element pegmatites are of interest because they crystallize from the most evolved, or fractionated, granitic magmas in nature. They are the extreme product of extended crystal fractionation and therefore occupy a unique position between normal igneous rocks and hydrothermal vein deposits. The remainder of this article deals exclusively with this important class of pegmatites.

Major rare-mineral pegmatites may exhibit systematic internal variations in mineralogy and texture that are termed zoning. The usual zonal pattern begins with a thin selvage of aplite, a few centimeters thick at most, which typically grades into the country rock. The aplite selvage is abruptly succeeded by a coarse-grained muscovite-rich zone which, by decrease in muscovite abundance, passes into a zone dominated by quartz and feldspar that may also carry abundant spodumene. The innermost zone is composed mainly or exclusively of quartz. Ideally, the zones are crudely parallel structures symmetrically disposed with respect to the center line of the host pegmatite. In a real pegmatite, however, it is usually found that individual zones vary in terms of width, continuity, and position with respect to the center line of the body. The zonal pattern records the inward growth of a pegmatite from the enclosing walls of country rock. The bulk chemical composition of each individual zone is an approximate indication of the composition of the parental magma at the time the zone formed. It follows that the inward zone sequence is an approximate record of compositional changes that occurred in the evolving parent magma as the pegmatite formed. The existence of a common zonal pattern must mean that most pegmatite magmas evolve in a broadly similar fashion.

COMPOSITION OF RARE-MINERAL PEGMATITES

In spite of the general similarity in zone pattern shown by rare-mineral pegmatites as a class, no two bodies are ever exactly alike in terms of the size, shape, and spatial distribution of their zones. The mineral assemblages and textures that compose individual zones are, however, surprisingly consistent. Excluding the selvage, most zoned pegmatites are composed of five distinct mineral assemblages: coarse microcline with or without pollucite; quartz with subordinate lithium aluminosilicate minerals and with amblygonite-montebrasite; massive quartz (virtually monomineralic bodies); albite (sodium feldspar) or fluorine-rich mica bodies that contain abundant tourmaline, apatite, beryl, iron- and manganese-bearing phosphates, rare-earth element phosphates, zircon, and niobium-tantalum-tin oxides; and mixed zones of sequential deposition (composed of very coarse-grained microcline and spodumene or petalite within a finer-grained matrix of albite).

Gross chemical inhomogeneity and remarkable concentrations of exotic minerals have earned zoned rare-mineral pegmatites a reputation for complex chemistry. In fact, the overall bulk composition of most such rock bodies is a rather simple one, dominated by the common elements oxygen, silicon, aluminum, sodium, and potassium, which compose the major pegmatite mineral phases of quartz, albite, microcline, and muscovite. Water and lithium oxide are next in importance, water forming generally 0.5-1.0 weight percent and lithium oxide being present in amounts up to roughly 1.5 weight percent. Oxides of phosphorus and iron usually constitute several tenths of a percent each. The elements calcium, magnesium, and manganese are present in trace amounts only. The total concentration of rare elements (beryllium, boron, cesium, rubidium, tantalum, niobium, tin, tungsten, uranium, thorium, fluorine, and rare-earths), for which pegmatites are famous, is somewhat less than 1 percent by weight in almost every case. This composition, except for the high lithium component, is essentially that of normal alkali granite. The lithium concentration in normal granites, however, is only approximately one-seventieth of that in typical lithium-bearing pegmatites. The bulk composition of a pegmatite body is often assumed to be approximately equivalent to its parent magma. That does not apply to the water content, because most pegmatite magmas exsolve water during crystallization, which then diffuses outward into the surrounding country rock.

FORMATION OF PEGMATITES

Experiments have firmly established that water, as a dissolved constituent in magma, lowers both the liquidus and solidus temperatures significantly. Water also lowers the viscosity of the residual melt, which is a major factor in promoting the growth of the large crystals that distinguish pegmatites from normal igneous rocks. Traditionally, pegmatites have been considered to be "post-boiling" phenomena produced by complex but unspecified interactions between water-saturated silicate melt of low viscosity, a separate coexisting aqueous fluid, and the enclosing rocks. Several important experiments, however, have shown that water vapor in equilibrium with coexisting granite melt has relatively low solvent power with respect to certain major and trace elements concentrated in rare-mineral pegmatites. Further, fluid inclusion data from pegmatite minerals indicate that the trapped fluids are relatively dense and thus intermediate in character between silicate melt and simple aqueous solutions. These data are important because they imply that pegmatite minerals crystallize from highly evolved silicate magma rather than from any aqueous solution that may be present. If that is true, it is likely that one or several incompatible elements in the magma are responsible for the extreme concentration of rare metals in pegmatite magma.

Several lines of evidence suggest that boron, phosphorus, and fluorine play key roles in this regard. On account of their incompatible nature, these elements will be highly concentrated, along with water, in a residual melt prior to "first boiling." Theoretical considerations suggest that high levels of these elements will have two major effects in silicate magmas. First, they will delay water saturation, or "boiling," by increasing the solubility of water in the residual melt, and second, they will promote concentration of Group I elements (lithium, sodium, potassium, rubidium, cesium) in that melt. Thus, the result will be a water-rich, sodium-aluminosilicate-rich, late-stage melt from which albite, tourmaline, phosphate minerals, fluorine-rich micas, beryl, zircon, and niobium-tantalum-tin oxides can crystallize prior to boiling. Such a melt would possess the required low viscosity and low solidus temperature (below 500 degrees Celsius) required to enable pegmatite magma to migrate significant distances from the parent pluton prior to crystallization. High concentrations of boron, phosphorus, and fluorine in the late-stage melt can produce the additional important effect known as immiscibility. At some critical concentration, the single parent melt can split into two mutually insoluble "partner melts" with drastically contrasting chemical compositions. One partner melt will be very rich in silica and could form massive quartz zones enriched in lithium minerals. The remaining partner melt would be a strongly alkaline silicate melt capable of producing albite-dominated zones. The eventual boiling of the boron-phosphorus-fluorine-enriched late-stage melt is not excluded in this theoretical sequence of events, but it does not play the pivotal role in forming pegmatite that has long been attributed to it. The appearance of a mobile aqueous fluid, capable of transporting excess potassium, sodium, fluorine, and boron into the surrounding host rocks, is required for those pegmatites enclosed by intense metasomatic aureoles. At the present stage of knowledge, however, aqueous solutions do not appear to be chemically capable of concentrating major quantities of lithium, rubidium, cesium, beryllium, zirconium, niobium, and other metals that must, therefore, remain in the silicate melt. The fine-grained aplite border selvage displayed by most pegmatites suggests that the initial injection of pegmatite magma occurred while it was still in a water-unsaturated state. The abrupt inward textural transition to coarse-grained interior zones may be an indication that water saturation occurs during (not before) the emplacement process. The crystallization of tourmaline has been shown to be a very effective means of concentrating water in the residual melt and has been suggested to be a "triggering process" to produce rapid water saturation.

MAPPING PEGMATITES

In the past, studies of pegmatites have mainly concerned either the evaluation of their economic potential or their exotic mineralogy. Surprisingly few efforts have focused on pegmatites in a coherent and systematic fashion, and as a result, satisfying theories for their origin are poorly developed relative to other areas of igneous petrology. A comprehensive study of pegmatites in a particular region would include not only the pegmatites but also the country rocks that enclose them and any bodies of exposed granitic rock. Study begins with preparation of a geologic map showing the distribution of pegmatite bodies, by type, throughout the area of interest. The map is compiled from firsthand, detailed observations of rock outcroppings obtained by numerous foot-traverses across the area. Additional useful information may be obtained from earlier geologic maps, aerial photographs, satellite imagery, and mine records. In most cases, the geologic map will show numerous, small, discontinuous pegmatite veins cutting older, high-grade metamorphic country rocks. These are known as external pegmatites. If any granite plutons are present in the area, they must be examined closely, as they are potential parent bodies of the pegmatites. Such granites may themselves host so-called internal pegmatites. External and internal pegmatites, even if derived from the same pegmatite magma (cogenetic pegmatites), generally will differ in mineralogy and zoning traits. It is a considerable achievement if a study can demonstrate that groups of external and internal pegmatites derive from the same parent body of granite. In such a case, the geologist has the rare opportunity to study the chemical and temporal relationships between parent rock, host rock, and an evolving, mobile pegmatite magma. This happy state of affairs is seldom realized because of the all-important "level of erosion." Since external pegmatites form by migration of pegmatite magma upward and away from the parent body along fracture pathways, they will normally be destroyed by erosion before their deeper parent and internal "relatives" are exposed.

By examining a large pegmatite-bearing area, rather than a single pegmatite body (as is often done), it is possible to determine if systematic differences exist in the exposed pegmatites relative to host-rock type, local or regional fracture patterns in the host rock, or distance from igneous plutons. Individual pegmatites can also be compared in terms of shape, size, orientation, mineralogy, and zone characteristics. Systematic variations of these parameters on the scale of a geological map constitute "regional zoning." Increased recognition of such effects is needed to provide greater insight into the operation of the pegmatite-forming process.

After regional geologic relationships are determined, individual pegmatites are mapped in the greatest detail possible (often at scales as large as 1 inch per 10 feet). The objective is to determine the size, shape, and zone sequence of each pegmatite body, which, in turn, will provide a basis for sampling and determining the composition of the pegmatite as a whole. Every effort is made to establish the correct sequence of zone crystallization, because the bulk compositions and fluid inclusion data from each member of the zone sequence can then be used to trace chemical changes in the pegmatite magma during the emplacement process.

COMMERCIAL VALUE

In the past, large and favorably located simple pegmatites were often worked for mica, quartz, and feldspar by the glass and ceramic industries, but production of this type has all but ceased. The post-World War II electronics revolution and other technological advances, such as lithium batteries, have fueled ever-increasing demands for the rare elements found in complex pegmatites.

Rare-element pegmatites with commercial grades and reserves that are sufficiently close to the surface for low-cost open-pit mining are distinctly uncommon. The few that do exist are the only sources for the elements lithium, beryllium, cesium, rubidium, tantalum, and niobium; they are important sources for boron, tin, tungsten, uranium, thorium, fluorine, and rare-earth elements (lanthanides) as well. Because many of the pegmatites in current commercial production are located in Third World countries, some industrialized countries (the United States among them) stockpile rare-element commodities of strategic importance in case of national emergencies or supply disruptions.

Complex pegmatites are valued for reasons other than commercial production of rare elements. From ancient times, they have been mined for precious and semiprecious gems. The intermediate zones of certain pegmatites are the major source of gem-quality topaz, tourmaline, and beryl (including morganite, aquamarine, and emerald). Each year, hundreds of small, commercially subeconomic rare-element pegmatites are prospected by amateur and professional collectors of rare minerals and fine crystals. These pegmatites are, in fact, supporting a small but vigorously growing industry. Many, perhaps even most, of the spectacular crystal specimens displayed in museums were obtained from subeconomic pegmatites by amateur collectors.

Gary R. Lowell

CROSS-REFERENCES

BIBLIOGRAPHY

Best, Myron G. *Igneous and Metamorphic Petrology.* 2d ed. Cambridge, Mass.: Blackwell, 1995. A popular university text for undergraduate majors in geology. A well-illustrated and fairly detailed treatment of the origin, distribution, and characteristics of igneous and metamorphic rocks. Chapter 4 treats pegmatites briefly; chapter 8 discusses late-stage magmatic processes pertinent to pegmatite formation.

Cameron, Evan N., R. H. Jahns, A. H. McNair, and L. R. Page. *Internal Structure of Granitic Pegmatites.* Urbana, Ill.: Economic Geology Publishing, 1949. This classic study of North American pegmatites during wartime illustrates the strategic nature of pegmatite deposits. The detailed maps and descriptions of American pegmatites are the basis for most textbook treatments of pegmatites. Valuable descriptions are given of bodies subsequently removed by mining. Aimed at professionals but can be understood by college-level readers with some background in geology. Will be found in most university libraries.

Desautels, Paul E. *The Mineral Kingdom.* New York: Grosset & Dunlap, 1972. One of many superbly illustrated books devoted to minerals and gems for hobbyists and serious collectors. A skillful blend of art, science, and history of minerals, with a useful index. Very suitable for high school readers and those interested in the aesthetic aspects of gems and crystals.

Guilbert, John M., and Charles F. Park, Jr. *The Geology of Ore Deposits.* 4th ed. New York: W. H. Freeman, 1985. A splendid new edition of a traditional college text for undergraduate geology majors. Pegmatites are treated on pages 487-507. The influence of the early work by Cameron (1949) is striking. Excellent photographs of pegmatite textures and outcrops and a comprehensive review of traditional American perspectives on pegmatites.

Hall, Anthony. *Igneous Petrology.* 2d ed. New York: John Wiley & Sons, 1996. A standard text discussing the general aspects of the occurrences, composition, and evolution of igneous rocks and magmas. Each chapter discusses a different rock group. The role of water in magmas is emphasized in the sections on granites. Text is college level, but the illustrations are useful to the general reader.

Hutchison, Charles S. *Economic Deposits and Their Tectonic Setting.* New York: John Wiley & Sons, 1983. An economic geology text for undergraduate majors in geology. The strength of this book is its modern international perspectives (in contrast to the Guilbert and Park text) on various classes of ore bodies, including pegmatites (pages 125-129). Figure 6-10 presents the Varlamoff spatial classification of pegmatites, which is widely known outside the United States, as well as an excellent cross section of the famed Bikita pegmatite.

Klein, Cornelis, and Cornelius S. Hurlbut, Jr. *Manual of Mineralogy.* 21st ed. New York: John Wiley & Sons, 1999. This edition of a standard mineralogy text for college students with some geology background provides an excellent index of mineral names and tabulates chemical formulas, crystal systems, and physical properties of all the common mineral species and many of the rare ones encountered in pegmatites. This basic mineralogy text is essential for anyone embarking on a study of rare-mineral pegmatites. A brief pegmatite summary is provided.

Norton, James J. "Sequence of Minerals Assemblages in Differentiated Granitic Pegmatites." *Economic Geology* 78 (August, 1983): 854-874. This article updates the classic zone sequence of Cameron (1949) by constructing one that takes into account the Bikita and Tanco pegmatites (the world's largest). The account is technical but understandable by college-level readers with some background in geology. The data tables and bibliography will be useful to anyone with a serious interest in pegmatites.

Press, Frank, and Raymond Siever. *Understanding Earth*. 2d ed. New York: W. H. Freeman, 1998. This comprehensive physical geology text covers the formation and development of the Earth. Readable by high school students, as well as by general readers. Includes an index and a glossary of terms.

PLATINUM GROUP METALS

Although platinum group metals are among the rarest elements known, their chemical inertness, high melting points, and extraordinary catalytic properties make them an indispensable resource for modern industrial society.

PRINCIPAL TERMS

CATALYST: a substance that facilitates a chemical reaction but is not consumed in that reaction

IMMISCIBLE LIQUIDS: liquids not capable of being mixed or mingled

LAYERED IGNEOUS COMPLEX: a large and diverse body of igneous rock formed by intrusion of magma into the crust; it consists of layers of different mineral compositions

MAFIC/ULTRAMAFIC: compositional terms referring to igneous rocks rich (mafic) and very rich (ultramafic) in magnesium- and iron-bearing minerals

MAGMA: molten rock material that solidifies to produce igneous rocks

PLACER: a mineral deposit formed by the concentration of heavy mineral grains such as gold or platinum during stream transport

REEF: a provincial ore deposit term referring to a metalliferous mineral deposit, commonly of gold or platinum, which is usually in the form of a layer

SPECIFIC GRAVITY: the ratio of the weight of any volume of a substance to the weight of an equal volume of water

TROY OUNCE: unit of weight equal to 31.1 grams used in the United States for precious metals and gems

EARLY DISCOVERY OF PLATINUM

The average crustal abundance of the platinum group metals is not known exactly but is comparable to that of gold. Platinum, like gold, can be found as a pure metal in stream placer deposits, where its density and resistance to corrosion have resulted in concentration of the metal during stream transport. Platinum appears to have been recovered from placer deposits since earliest times. A hieroglyphic character forged from a grain of platinum has been dated from the seventh century B.C.E., and ancient South American metalsmiths used platinum as an alloy to improve the hardness of gold as early as the first millennium B.C.E.

The name "platinum" is derived from the Spanish word for silver, to which it was originally considered to be inferior. Spanish conquistadors called the metal *platina del Pinto* (little silver of the Pinto) after its discovery in placer gold deposits of the Rio Pinto. Although the metal looked like silver, it proved to be more difficult to shape. Also, with a density similar to that of gold, platinum in small quantities mixed with gold was difficult to

detect. Fear of its being used to degrade gold and silver led to a temporary ban against the importation of platinum to Europe. The metal was eventually brought to Europe and was described by Sir William Watson in 1750.

LATER DISCOVERY OF OTHER GROUP METALS

The platinum group metals consist of six elements so named because they occur with and have similar properties to platinum. In addition to platinum, they include osmium, iridium, rhodium, palladium, and ruthenium. These five metals were discovered in the first half of the nineteenth century by scientists who examined the residue left when crude platinum was dissolved in aqua regia, a mixture of hydrochloric and nitric acids. Because of their resistance to corrosion, all six platinum group metals and gold are referred to by chemists as the noble metals. Economic geologists classify these metals together with silver as the precious metals. The weight of precious metals, like that of gems, is given in troy ounces.

Four of the platinum group metals were discovered by British scientists in 1803. Smithson Ten-

1626

nant discovered osmium and iridium. The name "osmium" was taken from the Greek word *osme* (meaning "smell"), because the metal exudes a distinctive odor, actually toxic osmium tetroxide, when it is powdered. Osmium is bluish white and extremely hard. Its melting point of about 3,045 degrees Celsius is the highest of the platinum group. The specific gravity of osmium has been measured at 22.57, making it the heaviest known element.

Iridium was named by Tennant for the Latin word *iris*, a rainbow, because of the variety of colors produced when iridium is dissolved in hydrochloric acid. Iridium is white with a slight yellowish cast. Like osmium, it is very hard, brittle, and dense. It has a melting point of 2,410 degrees Celsius and is the most corrosion-resistant metal known.

Rhodium and palladium were first discovered by William Hyde Wollaston, who named rhodium for the Greek word *rhodon* (a rose) because of the rose color produced by dilute solutions of rhodium salts. The metal is actually silvery white, has a melting point of about 1,966 degrees Celsius, exhibits a low electrical resistance, and is highly resistant to corrosion. Wollaston named palladium for Pallas, a recently discovered asteroid named for the Greek goddess of wisdom. Palladium is steel-white, does not tarnish in air, and has the lowest specific gravity (12.02) and melting point (1,554 degrees Celsius) of the platinum group metals. Like platinum, it has the unusual property of absorbing enormous volumes of hydrogen.

The existence of ruthenium was proposed in 1828 but not established until 1844 by Russian chemist Karl Karlovich Klaus. He retained the previously suggested name ruthenium in honor of Ruthenia, the Latinized name for his adopted country of Russia. Ruthenium is a hard, white, nonreactive metal with a specific gravity slightly greater than that of palladium (12.41) and a melting point of 2,310 degrees Celsius.

BUSHVELD COMPLEX

Important deposits of platinum group metals are found in the Bushveld complex, Republic of South Africa; in the Stillwater complex, Montana; and at Sudbury, Ontario. A large deposit similar to that at Sudbury exists at Norilsk in Siberia.

The Bushveld complex holds a special place as

history's greatest source of platinum, and it still contains the world's largest reserves of platinum group elements. Bushveld is a large layered igneous complex. It is located north of the town of Pretoria in the northeast corner of South Africa, and it covers an area roughly the size of the state of Maine. It formed 1.95 billion years ago when an enormous intrusion of mafic magma, the largest known mafic igneous intrusion, was injected and slowly cooled in the Earth's crust. As cooling and solidification occurred, the denser, mafic minerals became concentrated downward in the magma chamber, and the igneous rock became stratified with ultramafic layers at greater depth and layers of increasingly less mafic rocks upward.

Placer platinum was discovered in South Africa in 1924 and was subsequently traced by Hans Merensky to its source, a distinctive igneous layer which became known as the Merensky Reef. The reef is located in the lower part of the Bushveld complex, about one-third of the distance from the base to the top. Although commonly less than 1 meter in thickness, it has been traced for 250 kilometers around the circumference of the complex, and nearly one-half of the world's historic production of platinum group metals has come from this remarkable layer.

The average metal content in the layer is about one-third of a troy ounce per ton of rock, or about 1 part platinum group metals in 100,000 parts rock. Platinum is the most abundant metal extracted from the reef. Other platinum group minerals in order of abundance are palladium (27 percent), ruthenium (5 percent), rhodium (2.7 percent), iridium (0.7 percent), and osmium (0.6 percent). Also produced are significant quantities of gold, nickel, and copper. Mining of such a narrow layer is so labor-intensive that each South African miner produces only about 30 ounces of platinum group metals per year.

STILLWATER COMPLEX

The Stillwater complex is a large-layered, mafic to ultramafic igneous complex remarkably similar to Bushveld. It is exposed for about 45 kilometers along the north side of the Beartooth Mountains in southwest Montana. The Stillwater area has long been famous for its large but low-grade chromium-rich layers, and platinum was discovered there in the 1920's. Serious exploration for economic concentrations of platinum, however, was initiated in 1967 by the Johns-Manville Corporation. This led, in the 1970's, to identification of the J-M Reef, a palladium- and platinum-rich horizon between 1 and 3 meters thick, which, like the Merensky Reef, can be traced through most of the complex.

The Stillwater complex formed 2.7 billion years ago. Like Bushveld, the complex is layered from ultramafic igneous rocks at the base to mafic rocks upward. The J-M Reef lies slightly above the ultramafic zone. It has an average ore grade of 0.8 ounce of platinum group metals per ton of rock with a 3:1 ratio of palladium to platinum. Mining of the J-M Reef commenced in 1987. The ore is concentrated at the mine site and then shipped to Antwerp, Belgium, for refining. The J-M Reef is the only significant source of platinum in the United States, and the Stillwater mine is projected to be in production until about 2020.

SUDBURY COMPLEX

The Sudbury complex, just north of Lake Huron in southeast Ontario, Canada, is similar in many ways to Bushveld and Stillwater, but it is not conspicuously layered. Also, nickel and copper are the main products, with platinum group metals being produced as a by-product. Nickel was discovered in the Sudbury area in 1856. At that time, the region was largely wilderness, and government survey parties were engaged in running base, meridian, and range lines in preparation for a general survey and subdivision of northeastern Ontario. Considerable local magnetic attraction and the presence of iron were noted during the survey. An analysis of the rock showed that it contained copper and nickel as well. The Sudbury magma was intruded 1.85 billion years ago and now appears as an elliptical ring of mafic igneous rock 60 kilometers long by 27 kilometers wide. At depth, the intrusion is believed to have the shape of a funnel, and some fifty ore deposits are found along and just outside its outer edge. The origin of the ore and of the complex itself continues to be the subject of spirited debate.

RESEARCH INTO ORIGIN OF DEPOSITS

Recent industry demands for platinum have stimulated research into the origin of platinum group metal deposits. Much of this research has been directed toward understanding the world's great mafic igneous complexes. It has long been recognized that the origin of the Merensky and J-M reefs is tied to the formation of the layering within these mafic igneous complexes. Geologists have firmly established that the mafic magmas originate in the Earth's mantle and that they derive trace amounts of the platinum group metals from their mantle source rocks. It is also well known that as these magmas crystallize, the various minerals are precipitated from the magma in a fairly well-established sequence. Geologists have long believed that the layered mafic igneous complexes represent the settling of precipitated mineral grains into layers according to their densities. Repetitions and modifications in the layering are considered to be the result of currents churning within the hot magma. As crystallization proceeds, volatile elements such as water, carbon dioxide, and sulfur gradually become concentrated in the remaining magma. Sulfur has the ability to scavenge many metals, including iron, copper, nickel, and platinum group metals. Laboratory studies have shown that if the sulfur concentration is high enough, metallic sulfide droplets can form a separate, immiscible liquid. Like water in oil, the denser sulfide magma droplets sink and accumulate toward the base of the intrusion. Many geologists believe the layers and masses of metallic sulfide ore found in the large mafic igneous complexes formed in this manner.

Detailed studies of the chemical composition of the Stillwater complex, however, suggest that the crystallization sequence was interrupted at about the level of the J-M Reef by an influx of new, somewhat different magma. The evidence suggests that the magma was sulfur-saturated, and its influx is believed to have triggered the precipitation of the platinum minerals. Research on the Bushveld complex has also suggested multiple episodes of magma injection, with the Merensky Reef forming at the base of one of the magma pulses. It should be emphasized, however, that even after a century of investigation, the origin of the ore at the Bushveld, Stillwater, and Sudbury complexes is still a subject of considerable debate.

The hypotheses for ore formation at Sudbury include the separation of droplets of immiscible sulfide liquid from a mafic magma, but a lively debate exists as to the mechanism by which the mafic magma was produced. It has long been noted that a distinctive zone of broken and shattered rock many kilometers wide underlies and surrounds the Sudbury igneous complex. Overlying the complex is a thick sequence of fragmentary rocks, originally interpreted as being volcanic in origin. In 1964, Robert Dietz suggested that Sudbury was the site of a tremendous meteorite impact that formed a large crater and shattered the surrounding rock. It was later proposed that the impact caused the melting that produced the mafic igneous rock and that the supposed volcanic rock was actually material that had been ejected during the meteorite impact and had fallen back into the crater. This theory continues to cause controversy. The evidence for a meteorite impact is strong, but some geologists consider any impact to be unrelated to the Sudbury deposit. Others not only believe the impact theory but also suggest that the Bushveld magma was triggered by a meteorite impact. The debate continues, and its outcome has implications for the presence or absence of metallic ore deposits beneath the large lunar craters.

While field and laboratory work on the great platinum deposits of the world continues, so does experimental work aimed at understanding the conditions under which these deposits formed. Laboratory scientists are duplicating conditions found in nature in order to increase their understanding of the behavior of platinum group elements during crystallization from magmas, their behavior during formation of immiscible liquids, and their mobility at submagmatic temperatures in water-rich solutions.

INDUSTRIAL VALUE

The platinum group metals are used extensively in modern industrial society because of their chemical inertness, high melting points, and extraordinary catalytic properties. Platinum group metals are important parts of the automotive, chemical, petroleum, glass, and electrical industries. Other important uses are found in dentistry, medicine, pollution control, and jewelry. The automotive industry is the single largest consumer of platinum group metals. Since 1974, platinum-palladium catalysts have been used in the United States to reduce emission of pollutants from automobiles and light-duty trucks. A typical catalytic converter contains 0.057 ounce of platinum, 0.015 ounce of palladium, and 0.006 ounce of rhodium. In the European Economic Community, all cars with engines larger than 2 liters produced after October, 1988, must have converters.

The electrical industry is the second largest consumer of platinum group metals. Palladium is used in low-voltage electrical contacts, and platinum electrical contacts protect ships' hulls from the corrosive activity of seawater. The dental and medical professions utilize nearly as much platinum group metals as the electrical industry. Palladium is alloyed with silver, gold, and copper to produce hard, tarnish-resistant dental crowns and bridges. Other medical uses include treatment for arthritis and some forms of cancer. Platinum group metals are also used internally in cardiac pacemakers and in a variety of pin, plate, and hinge devices used for securing human bones.

The chemical industry uses platinum and palladium as catalysts for a variety of reactions involving hydrogen and oxygen. Molecules of either of these gases are readily adsorbed onto the surface of the metals, where they dissociate into a layer of reactive atoms. Oxygen atoms on platinum, for example, increase the rate at which sulfur dioxide, a common industrial pollutant, is converted into sulfur trioxide, a component of sulfuric acid, the most widely used industrial chemical. Other pollution-control devices include the control of ozone levels in the cabins of commercial jet airplanes and oxidation of noxious organic fumes from factories and sewage treatment plants. Platinum group catalysts are also used in the production of insecticides, some plastics, paint, adhesives, polyester and nylon fibers, pharmaceuticals, fertilizers, and explosives. In the petroleum industry, platinum group metals are used by refineries both to increase the gasoline yield from crude oil and to upgrade its octane level. Because material used as a catalyst is not consumed in the chemical reactions (although small amounts are lost), the many important chemical uses actually consume only a small amount of the platinum group metals.

Platinum group metals' ability to withstand high temperatures and corrosive environments has led to their use in the ceramics and glass industry. Thin strands of glass are extruded through platinum sieves to make glass fibers for insulation, textiles, and fiber-reinforced plastics. High-quality optical glass for television picture tubes and eyeglasses is also melted in pots lined with nonreactive platinum alloys. Crystals for computer memory devices and solid-state lasers are grown in platinum and iridium crucibles. As ingots and bars, platinum group metals are sold to investors, and platinum and palladium alloys are commonly used for jewelry. Brilliant rhodium is electroplated on silver or white gold to increase whiteness, wear, and resistance to tarnishing.

SUPPLY VS. DEMAND

The world's reserves of platinum group metals are large, but distribution is concentrated in a relatively few locations. South Africa is the largest producer of platinum. Japan is the largest consumer nation, and the United States is second.

Historically, U.S. production has been extremely small and has consisted almost entirely of platinum and palladium extracted during the refining of copper. Stillwater's J-M Reef is a significant discovery, but it is expected to supply only about 7 percent of the nation's projected needs. The U.S. State Department has therefore added the platinum group metals to a list of strategic materials that are considered essential for the economy and for the defense of the United States and that are unavailable in adequate quantities from reliable and secure suppliers.

Consequently the search continues for new sources of platinum group metals. A potential source may be in the incrustations of iron and manganese found on the submerged slopes of islands and seamounts throughout the world's oceans. These metallic crusts and nodules are believed to have formed by extremely slow precipitation from seawater. Although they are composed mostly of iron and manganese, they contain many metals, including those of the platinum group, and the volume of these deposits is staggering. While commercial development is unlikely before the early part of the twenty-first century, these ferromanganese crusts are considered to be an attractive, long-term resource.

Eric R. Swanson

CROSS-REFERENCES

Aluminum Deposits, 1539; Building Stone, 1545; Cement, 1550; Coal, 1555; Diamonds, 1561; Dolomite, 1567; Earth Resources, 1741; Fertilizers, 1573; Gold and Silver, 1578; Hydrothermal Mineral Deposits, 1584; Hydrothermal Mineralization, 1205; Industrial Metals, 1589; Industrial Nonmetals, 1596; Iron Deposits, 1602; Magmas, 1326; Manganese Nodules, 1608; Metamorphic Mineral Deposits, 1614; Pegmatites, 1620; Salt Domes and Salt Diapirs, 1632; Sedimentary Mineral Deposits, 1629; Strategic Resources, 1796; Uranium Deposits, 1643.

BIBLIOGRAPHY

Evans, Anthony M. *An Introduction to Economic Geology and Its Environmental Impact.* Oxford: Blackwell Scientific Publications, 1997. Intended for undergraduate students, this book provides an excellent introduction into the field of economic geology. Emphasis is on types of deposits, their environments of formation, and their economic value, along with the impact those deposits have on their environments. Well illustrated and includes an extensive bibliography.

Guilbert, John M., and Charles F. Park, Jr. *The Geology of Ore Deposits.* New York: W. H. Freeman, 1985. The best of the available college-level texts on ore deposits. The geology of the Sudbury and Bushveld complexes is cov-ered in the chapter on deposits related to mafic igneous rocks. The book is widely available in university and large public libraries.

Hartley, Frank R., ed. *Chemistry of the Platinum Group Metals: Recent Developments.* New York: Elsevier, 1991. This book provides information on the procedures used to determine the geochemical makeup and properties of platinum group metals.

Loebenstein, Roger J. "Platinum-Group Metals." In *Minerals Yearbook.* Washington, D.C.: Department of the Interior. *Minerals Yearbook* is published annually in three volumes by the United States Department of the Interior. Volume 1 contains general information on nonmetallic resources and all metals, including platinum. Volume 2 has information on U.S. resources by state, and volume 3 has comparable information for other countries of the world. These volumes are widely available in university and large public libraries.

Mertic, J. B., Jr. *Economic Geology of the Platinum Metals.* U.S. Geological Survey Professional Paper 630. Washington, D.C.: Government Printing Office, 1969. A comprehensive survey and extensive bibliography of all platinum deposits, widely available in university and public libraries.

Peterson, Jocelyn A. *Platinum-Group Elements in Sedimentary Environments in the Conterminous United States.* Washington, D.C.: Government Printing Office, 1994. This brief government pamphlet contains detailed information on platinum ores and deposits, as well as sedimentary basins. The publication pamphlet is filled with color and black-and-white maps throughout, and also includes a fold-out map.

St. John, Jeffrey. *Noble Metals.* Alexandria, Va.: Time-Life Books, 1984. This book is part of Time-Life's Planet Earth series. It is an available source of good information on plati-

num and is written for a general audience.

Vermaak, C. Frank. *The Platinum-Group Metals: A Global Perspective.* Randburg, South Africa: Mintek, 1995. Vermaak offers a look into platinum ores and mines, platinum mining, and the platinum group industry. Color illustrations and bibliographical references.

Weast, Robert C., ed. *CRC Handbook of Chemistry and Physics.* Boca Raton, Fla.: CRC Press, 1988. This widely available reference book is the result of the collaboration of a large number of professional chemists and physicists. It contains in condensed form an immense amount of information from the fields of chemistry and physics. Section B contains an alphabetized description of all the known elements and is written in a manner understandable to those with little scientific background. New editions are published at frequent intervals, ensuring that the information is up to date.

Young, Gordon. "The Miracle Metal: Platinum." *National Geographic* 164 (November, 1983): 686-706. This article has excellent pictures and covers many of the uses of platinum.

SALT DOMES AND SALT DIAPIRS

In addition to their importance in the supply of minerals for the food processing and chemical industries, salt domes and salt diapirs are associated with economic reserves of crude oil, natural gas, and sulfur. They are also utilized as repositories for valuable and strategic materials and, potentially, for nuclear waste.

PRINCIPAL TERMS

CAP ROCK: an impervious rock unit, generally composed of anhydrite, gypsum, and occasionally sulfur, overlying or capping any salt dome or salt diapir

EVAPORITE MINERALS: that group of minerals produced by evaporation from a saline solution; for example, rock salt

PRECIPITATION: a process whereby a solid substance is separated from a solution

RIM SYNCLINE: a circular depression found at the base of many salt domes and diapirs

SEDIMENTARY ROCK: any rock formed by the accumulation and consolidation of loose sediment, as in the precipitation of salt crystals from seawater

SHALE SHEATH: a variable thickness of ground-up sedimentary rock found along the flanks of many salt domes and diapirs

BEDDED SALT DEPOSITS

Salt domes and salt diapirs are found throughout the world in association with bedded salt, an evaporative sedimentary rock known to be part of the rock record of every geologic time period. While salt deposits underlie at least twenty-six of the United States, salt domes and diapirs in North America are confined to the Gulf coast basin, extending from central Mississippi and Louisiana and coastal Texas south into the Isthmus of Tehuantepec in Mexico. The Gulf coast basin contains more than five hundred salt domes and diapirs, located within at least nine separate subbasins. Other significant concentrations of salt domes or diapirs occur in Europe, in the Zechstein Basin of Germany and Poland and in Denmark, England, Spain, Portugal, France, and Romania; in Asia, in the Donets Basin of Russia and the northern regions of Siberia; in Africa, along the Mediterranean coast of Algeria and Morocco and in Gabon and Senegal; in South America, in Peru; and in the Iran, Aden, and Yemen sections of the Arabian Peninsula.

Bedded salt deposits result from geological processes that have been in operation for several hundred million years. Precipitation of salt deposits occurs within modified marine environments whenever evaporation exceeds surface runoff and rainfall, and wherever restriction of the free circulation of seawater exists. The evaporation of a 300-meter column of seawater will theoretically result in the precipitation of 4.6 meters of evaporate minerals, of which 3.5 meters will be salt (chemically composed of sodium chloride). Bedded salt units are generally confined to a structural basin, a downwarped region of the Earth's crust containing thick sequences of sedimentary rock. In North America, the Louann Salt, of Middle Mesozoic age (approximately 175 million years old), underlies more than 2 million square kilometers of the Gulf coast basin and is several thousand meters thick.

FORMATION OF SALT STRUCTURES

The mechanism whereby bedded salt units are transformed into salt domes and salt diapirs has been generally known for decades. After precipitation from an evaporating body of saline water, pure, bedded rock salt has a specific gravity of 2.16 grams per cubic centimeter. When that salt bed is initially overlain by younger sedimentary rocks composed of sandstone, siltstone, shale, and clastic limestone—normal rock associations found wherever salt domes occur—the salt is in density equilibrium with the younger rocks, as their specific gravities range from 1.7 to 2.0 grams per cubic centimeter. Upon burial to depths of several thousand meters, however, the specific gravity of the overburden will increase to levels ranging from 2.4 to 2.8 grams per cubic centimeter, while the salt, being incompressible, remains at the 2.16 specific gravity level. Under these conditions, an upper force of buoyancy will act upon the salt, encouraging movement toward the surface of the Earth. This potential for movement is accelerated below depths of 3,000 meters by conditions of increasing temperature and pressure, altering the visco-elastic properties of the salt as it changes from a solid to a consistency comparable to that of warm butter. Through this combination of density imbalance and altered visco-elastic condition, bedded salt will begin to arch overlying rocks at those sites marked by slight salt-surface irregularities. As long as the overlying rocks are arched or folded and thus remain mechanically intact, the resulting structure is termed a salt "dome." Should the overlying strata be broken and penetrated by the rising column of salt, the structure becomes a salt "diapir" (meaning "through-piercing"), a form of intrusive sedimentary rock. Studies of the thickness and age of salt basins suggest that the rate of salt movement and penetration, while variable, generally averages less than 2 millimeters per year. During periods of very active sedimentation and crustal folding, domal activity is most active. In the Aquitaine basin of France, the seven mapped phases of salt upwelling relate directly to periods of crustal deformation in the nearby Alps.

Once initiated, upward flow will continue until the causative forces of buoyancy and flow properties are altered by diminished pressure and temperature, or until the base of the salt dome or diapir is cut off from the bedded unit acting as the salt source. In general, the thicker the bedded salt source and the deeper it is buried, the higher the developing salt structure. A salt dome or diapir will usually not reach to the surface of the Earth; occasionally, though, surface penetration occurs, as in the case of the salt domes of the Persian Gulf region. There, 40 percent of the nearly two hundred salt domes rise above the surrounding desert, often displaying salt "glaciers" that flow laterally away from the center of the exposed salt mass. One exposed dome, Kuh-i-Namak, has a surface elevation of 1,310 meters and a diameter of 9.6 kilometers.

INTEGRAL FEATURES

Through the process of arching or penetrating overlying sedimentary rock, all salt domes and diapirs take on certain identifiable characteristics. These features are found within the interior of the salt structure, forming the top of the dome or diapir and associated with the sedimentary rock adjacent to the salt structure.

Studies of salt mines within near-surface diapirs of the Gulf coast basin give direct evidence of vertical flowage. This flowage is demonstrated by stretched, elongated halite crystals and vertically oriented, tightly folded layers of salt accentuated by color changes between alternating pure and impure halite beds. Further evidence of vertical intrusion into overlying rock is seen in the occasional sighting of inclusions of sandstone and shale within salt structures; such foreign rock is identical in composition to adjacent intruded rock.

Topping most shallow salt domes and diapirs is an impervious mantle of minerals, on average 90 to 120 meters thick, termed the "cap rock." While found on domes up to 3,000 meters deep, cap rock thickness is greatest over the center of the dome, generally decreases with dome depth, and is absent along dome flanks. A typical, well-developed cap rock is mineralogically composed of anhydrite grading upward into calcite (calcium carbonate). Often, a transition zone containing varying amounts of barite (barium sulphate), gypsum, and sulfur separates the two principal layers. Some salt domes within the Poland section of the Zechstein basin of Europe have a cap rock composed of clay minerals. It is generally agreed that cap rock is gradually formed by upward-migrating salt structures encountering waters of decreasing salinity contained within the sedimentary rock being penetrated. Incomplete solution of the impure rock salt results in the accumulation of insoluble residues, principally anhydrite, along the leading edge of the salt intrusion. Transition minerals are believed to result from later alteration of the anhydrite.

ADJACENT FEATURES

The most common features external, or adjacent, to salt domes and salt diapirs include a shale sheath, an overlying domed structure accentuated by a central graben, an overall complex pattern of faulting, and a peripheral rim syncline.

Where encountered by drilling in the Gulf coast basin, the shale (clay) sheath covers the flanks of a dome or diapir much like a cap rock covers the top. This sheath is of variable thickness and is composed of finely ground rock formed by the frictional drag of the upward movement of the salt structure against adjacent sedimentary sandstones and shales.

As bedded salt beds become mobile under conditions of deep burial, they form circular structures created by arching of the overlying sedimentary rock. Diapiric movement is accomplished by salt, and developing cap-rock, penetration of the rock cover along a complex system of fractures (faults) that radiate outward from the circular structure. As such movement actually extends (stretches) the overlying rock cover, the structure center is commonly interrupted by a graben, a depressed block bounded by faults. This combination of a regional arched structure interrupted by complex faulting and a localized central graben is what complicates the geologic interpretation of the upper portions of any salt dome or diapir.

The most significant aspect of the exterior base of a salt structure is the presence of a rim syncline, a structural depression that partially or completely encircles most domes and diapirs. This feature is caused by the downward movement of overlying sedimentary rock into the void formed by the lateral flow of bedded salt into the developing dome. Details of rim synclines are little known as a consequence of their association with the base of salt domes and diapirs.

DISCOVERY OF SALT STRUCTURES

Under normal conditions, salt domes and salt diapirs are not exposed at the surface of the Earth and therefore are not accessible for direct study. An exception would include the diapiric salt structures of the Persian Gulf region, which are surface exposed and which, because of arid climate, are preserved for long-term analysis. Salt domes and diapirs can also be directly studied where salt has been extracted from the upper levels of the structure, either by brine-well production, such as at the Bryan Mound Dome in Brazoria County, Texas, or by direct salt mining, as conducted at the Avery Island Dome in Iberia Parish, Louisiana. Analyses of the walls of such mines give evidence pertaining to the layering and folding of salt beds, permitting development of theories of salt mobility and flowage.

The formation of a salt dome or diapir by long-term upward salt movement usually does not man-ifest itself in the development of surface relief effects identifying the presence of the shallow salt structure. Therefore, the discovery of the great majority of intrusive salt structures is accomplished by instrumentation designed to measure contrasting physical characteristics of the salt and adjacent sedimentary rock indirectly. In 1917, a buried salt dome was identified near Hanigsen, Germany, by a torsion balance, an early prospecting instrument designed to measure alterations in the gravitational field of the Earth. Subsequent drilling into this dome verified the accuracy of the torsion-balance method, which was soon being employed worldwide in the discovery of salt structures. By 1935, the cumbersome torsion balance had been successfully modified in the form of the first practical gravimeter, a modern field instrument that directly reads gravity differences with an accuracy to one millionth part of the Earth's total gravitational field. Because of density differences between the salt and adjacent rock material, such accuracy permits easy identification by gravimeter of a salt dome or salt diapir, as well as identification and thickness analyses of the cap rock.

After World War I, the invention of the refraction seismology process, which measures the change in velocity and deflection of an elastic wave moving from one rock formation to another, opened a new and very successful era of salt dome and salt diapir discovery in the Gulf coast basin. By the second half of the twentieth century, improvements in the reflection seismology process using technology related to refraction seismology were being employed worldwide in the continued discovery of intrusive salt structures, especially within marine waters off the Gulf coast of North America and offshore Africa. The seismic reflection process makes it possible to identify the overall geometry of intrusive salt structures with considerable precision. Adaption of three-dimensional reflection seismology during the 1990's, technology that is capable of documenting not only the geology of the base of salt domes and diapirs but also the structure underlying the source salt layer, is yet another major advancement in the identification and scientific study of intrusive salt.

MINERAL AND FOSSIL-FUEL SOURCES

Halite, the mineralogical term for salt, is the principal evaporite mineral in salt domes and diapirs. Halite has been a significant industrial material and article of world trade for thousands of years. In the United States, half of produced salt is employed by the chemical industry in the manufacture of caustic soda, polyvinyl chloride, hydrochloric acid, and household and industrial bleach. Large volumes are used in cold climates for traction control on highways. Other uses include water treatment; insecticide, medicine, aluminum, and steel manufacture; and food preservation and seasoning.

Extraction of halite from salt domes and diapirs is accomplished by two means: by conventional underground mining, as conducted, for example, in the Winnfield Dome in Winn Parish, Louisiana; and by means of brine (salt solution) wells, as practiced at the Barbers Hill Dome in Chambers County, Texas. Drilled wells are also used to extract economic volumes of sulfur from the cap rock of many salt domes. Production of sulfur from domes and diapirs utilizes the Frasch superheated water process to melt and dissolve the sulfur.

The discovery and exploratory drilling of salt domes and diapirs has been a worldwide priority for decades because of their association with crude oil and natural gas. The bulk of these fossil-fuel deposits is found in a circular area up to 1,000 meters away from the salt periphery. With the dual discovery in 1901 of the Spindletop Dome in Jefferson County, Texas, and the presence of oil within its cap rock, the center of oil and gas production in the United States moved from the Ohio River Valley to west of the Mississippi River. More than several hundred million barrels of oil have been produced from the cap rock and sedimentary rock flanking this historic salt structure.

STORAGE UNITS

In addition to their association with hydrocarbon resources, solution-created and mined cavities in salt domes and diapirs are becoming increasingly important as possible depositories of volatile, strategic, toxic, and nuclear material. The Tatum Dome of Lamar County, Mississippi, is employed by the U.S. Atomic Energy Commission to house an explosion as part of the study of peaceful applications of nuclear energy. With the signing of the Energy Policy and Conservation Act of 1972, five salt domes of the Gulf coast basin became part of the Strategic Petroleum Reserve, the United States' emergency plan for the storage of crude oil. As a typical part of this program, two levels of mined rooms approximately 21 meters high by 30 meters square within the Weeks Island Dome in Iberia Parish, Louisiana, held a capacity 72 million barrels of crude oil during the mid-1990's. Elsewhere in Texas and Louisiana, solution-developed and mined dome cavities have been used for the storage of low-pressure liquid gas, high-pressure ethylene, and refined hydrocarbon products ranging from gasoline to fuel oil.

Intrusive salt structures are particularly amenable to the storage of such products because salt is generally impervious to the passage of liquids. Impermeability combined with a relative lack of humidity makes dome and diapir cavities particularly adaptable to the long-term storage of volatile and radioactive waste materials as well as of historical, legal, and financial records vital to industrial and national security.

Albert B. Dickas

CROSS-REFERENCES

BIBLIOGRAPHY

Alsop, G. Ian, Derek John Blundell, and Ian Davison, eds. *Salt Tectonics.* London: Geological Society, 1996. A collection of essays from the Geological Society discussing the composition of evaporites, salt, and diapirs. Illustrations, index, and bibliographic references.

Halbouty, M. T. *Salt Domes: Gulf Region, United States and Mexico.* Houston, Tex.: Gulf Publishing, 1979. This presentation, written by the acknowledged salt-dome expert among American geologists and petroleum engineers, lists pertinent data for more than four hundred salt domes and salt diapirs within the U.S. and Mexican portions of the Gulf coast basin. Numerous full-color figures illustrate salt-dome distribution, configuration, composition, classification, and economic significance. A very readable reference written for the general public.

Jackson, M. P. A., ed. *Salt Diapirs of the Great Kavir, Central Iran.* Boulder, Colo.: The Geological Society of America, 1991. A useful reference for salt domes of the Central Iranian basin, one of the most spectacular and well-exposed salt dome regions of the world. These salt structures are unusual in that they form circular elevations 80 to 100 meters in relief. Intended for college-level readers.

Jackson, M. P. A., D. G. Roberts, and S. Snelson, eds. *Salt Tectonics: A Global Perspective.* Tulsa, Okla.: American Association of Petroleum Geologists, 1996. A comprehensive presentation of the geologic means whereby bedded salt units are transformed into a variety of geometric shapes, including domes and diapirs. Written at the college level for anyone interested in the exploration of salt structures and their common association with economic reserves of sulfur, crude oil, and natural gas.

Kupfer, D. H., ed. *Geology and Technology of Gulf Coast Salt: A Symposium.* Baton Rouge, La.: School of Geoscience, Louisiana State University, 1970. This two-day symposium, conducted by nine scientists actively engaged in Gulf coast basin salt-dome research, explored the origin, composition, and development of salt structures. This volume is of value in that it presents discussion of topics not commonly found in other references, such as cap-rock development, salt movements, and methods of salt intrusion. Written for the college-level reader, with numerous illustrations.

Lefond, S. J. *Handbook of World Salt Resources.* New York: Plenum Press, 1969. While this volume specifically discusses general salt resources of 196 continental and island countries, its documentation of the worldwide distribution of intrusive salt structures is invaluable to the student of salt domes and diapirs. The history of salt-resource utilization of each country is discussed, along with numerous charts showing chemistry, mineralogy, geographic location, and geology. Very suitable for high school and college-level readers.

Lerche, Ian, and Kenneth Peters. *Salt and Sediment Dynamics.* Boca Raton, Fla.: CRC Press, 1995. This book examines the chemical composition and properties of salt, salt domes, and sedimentary basins. Includes bibliographic references and index.

Murray, G. E. *Geology of the Atlantic and Gulf Coastal Province of North America.* New York: Harper & Brothers, 1961. A general reference to the geology of the coastal regions of the eastern and southern United States. Chapter 5, "Salt Structures," contains a thorough presentation of the salt domes and diapirs of the Gulf coast basin. Though intended for the professional geologist, this volume should be intelligible to the general college-level reader.

Usdowski, Eberhard, and Martin Dietzel. *Atlas and Data of Solid-Solution Equilibria of Marine Evaporites.* New York: Springer, 1998. A good look into seawater composition. There are chapters focusing on evaporites, marine deposits, and salt deposits. This book is accompanied by a computer disk that reinforces the concepts discussed. Illustrations, index, and bibliography.

SEDIMENTARY MINERAL DEPOSITS

Sedimentary mineral deposits are accumulations of economically valuable minerals that occur in sedimentary rocks ranging in age from 2.2 billion to less than 2 million years old. Such deposits have been widely exploited in the past and continue to be important sources of ores.

PRINCIPAL TERMS

DEPOSITION: the physical or chemical process by which sedimentary grains come to rest after being eroded and transported

DIAGENESIS: changes that occur in sediments and sedimentary rocks after deposition caused by interaction during burial with water trapped between the sediment grains

EVAPORITE: a mineral formed by direct precipitation from water resulting from supersaturation caused by solar evaporation in an arid setting

HYDROTHERMAL: characterizing any process involving hot groundwater or minerals formed by such processes

MID-OCEAN RIDGE: a large, undersea chain of volcanic mountains encircling the globe, branches of which are found in all the world's oceans

ORE: a concentration of valuable minerals rich enough to be profitably mined

PLACER: an accumulation of valuable minerals formed when grains of the minerals are physically deposited along with other, nonvaluable mineral grains

STRATA: layers of sedimentary rock

PLACERS

In its broadest sense, a sedimentary mineral deposit is an abnormal accumulation of valuable minerals in sedimentary rocks. When such an accumulation is rich enough to warrant mining, it is referred to as a sedimentary ore deposit. Most geologists would agree, however, that further restriction should be applied to the term. In this discussion, the term "sedimentary mineral deposits" refers only to those deposits that formed through sedimentary processes.

Several types of sedimentary processes may lead to the formation of sedimentary mineral deposits. Primary deposits form at the same time as the host sediments are deposited. They can form either as chemical precipitates or as placers. Placers form when ore minerals are eroded and transported along with sand and silt by rivers. When the sand and silt are deposited as sediments, the ore minerals are deposited along with them. Most placers form in riverbeds and tend to be small. Placer accumulations are also found in some sands deposited on beaches.

The best-known placers are those of gold. Although small by comparison to lode deposits (de-posits in bedrock), placer gold deposits were responsible for setting off the many gold rushes in the United States, including those to Colorado, California, and Alaska. Other placer deposits occur as paleoplacer deposits. These deposits are accumulations that were originally deposited as placer deposits millions or billions of years ago and have since been hardened into sedimentary rocks. Of particular interest to geologists are large deposits of quartz pebbles with sand-sized grains of pyrite (iron sulfide) and uraninite (uranium oxide). Pyrite and uraninite are unstable in the presence of oxygen and quickly break down during erosion. They are common, however, as detrital minerals (minerals that have been eroded and transported by streams) in rocks of Archean age (older than about 2.5 billion years). Geologists have deduced that such accumulations could have formed only if oxygen were not abundant as a free gas (not combined with other elements) in the Earth's atmosphere at that time.

CHEMICAL SEDIMENTS

Chemical sediments form when minerals precipitate directly from seawater or saline lakes onto the seafloor or lake bottom. The most common type of chemical sediment forms when sea or lake water is concentrated by solar evaporation. Such deposits are known as evaporites. Large deposits of anhydrite (calcium sulfate), gypsum (hydrated calcium sulfate), and halite (sodium chloride, also known as table salt) have formed in small seas in areas with arid climates. As water is lost from these bodies because of high rates of evaporation, new seawater enters through narrow straits connecting the sea with the open ocean. Dissolved salts in the water do not evaporate. They are concentrated until the brine becomes saturated, at which time minerals begin to precipitate. Gypsum forms first, followed by halite. Anhydrite forms when gypsum deposits are buried by later sediments. Potash (potassium chloride) can also form in this manner if arid, evaporative conditions persist for a long enough time.

Saline lakes also undergo evaporative concentration of dissolved salts, and evaporite deposits may form from these as well. Lake-water chemistry is controlled by the chemical composition of the surrounding bedrock, and lake water is often quite different in chemical composition from seawater. Thus, lacustrine evaporites (evaporites forming in lakes) commonly contain minerals not associated with marine evaporites. Such deposits include various borate (boron oxide) deposits in California and Turkey and trona (hydrated sodium carbonate) deposits of the western United States.

Chemical sediments also form when hydrothermal solutions are expelled onto the seafloor. The hot, mineral-laden hydrothermal water mixes with seawater and is rapidly cooled. The solubility of most metals decreases drastically as the temperature decreases, and various sulfide minerals of copper, lead, zinc, and iron, as well as barite, precipitate. Deposits of this type occur in many regions of the world, including North America (especially Canada), and are forming today in the Red Sea and the Pacific Ocean along the mid-ocean ridge system.

Primary deposits also form when slight changes in the physical or chemical composition of ocean water cause dissolved minerals to precipitate. For example, the mineral apatite (calcium phosphate) is soluble in cold, deep-ocean water that contains only small amounts of dissolved oxygen. Along

continental margins, upwelling of this water to the surface causes the apatite to precipitate out as sediment. If the influx of other types of sediment, such as sand and silt, is very low, rich accumulations of phosphate can result.

SEDIMENTARY IRON FORMATION

Another type of chemical sediment is known as sedimentary iron formation. During the Phanerozoic eon (544 million years ago to recent times), deposits of this type have formed when iron was leached from rocks and transported by groundwater to shallow, restricted seas, where it precipitated as hematite (iron oxide). The iron ores of northern Europe and eastern North America, which were very important during the Industrial Revolution, are of this type.

A unique type of sedimentary iron formation is found in rocks between 1.6 and 2.5 billion years old. These deposits consist of alternating bands of iron minerals and chert (a hard rock composed of silica). These deposits formed in many parts of the world as iron precipitated from seawater. The source of this iron is the subject of much debate, but one thing is clear: The deposits, though widespread, are clearly restricted to the period in the Earth's history when free oxygen was becoming abundant in the atmosphere. It is believed that this change caused the solubility of iron to decrease (iron is soluble in water only if no free oxygen is present), thereby triggering iron dissolved in seawater to precipitate.

Iron is precipitating on the ocean floor today along with manganese to form widespread deposits of manganese nodules. These nodules are found in all the world's oceans and contain considerable amounts of other metals, such as copper, nickel, and cobalt. It is thought that much, if not most, of the metals are derived by hydrothermal activity along mid-ocean ridges. Because the dissolved oxygen content in the deep oceans is low, the metals can travel considerable distances before precipitating from solution.

SECONDARY DEPOSITS

Secondary deposits form after the host sediments have been deposited. When sediments are buried beneath new sediments, water located between the various sediment grains is squeezed out. As the sediments are buried deeper and deeper, they are warmed by heat from the Earth's interior. As the temperature rises, the water being expelled from the compacting sediment mass reacts with the various grains and becomes increasingly saline. Such reactions, involving pore fluids and the enclosing rocks, are collectively known as diagenesis and occur in all sedimentary environments. Brines formed in this manner are well known from oil and gas exploration. As these brines move toward the Earth's surface, the temperature decreases; in many instances, they also mix with other groundwater as they near the surface.

Cross-bedded fossil sand dunes in a sandstone formation at Zion National Park, Utah. (© William E. Ferguson)

Several important types of secondary deposit are known. In northern Europe, Africa, Australia, and elsewhere, thin (up to several meters thick) but widespread (up to 10,000 square kilometers) layers of shale are enriched in sulfides of copper and, in some cases, other metals, such as lead and zinc. Although there is still much controversy regarding the origin of brines, most evidence indicates that brines formed during diagenesis reacted with the shale shortly after it was deposited.

Secondary deposits are found in limestones (rocks made up of calcium carbonate) and dolomites (calcium-magnesium carbonate). Brines that were formed during diagenesis migrated into cavities in the host sediment, and sulfides (minerals which contain sulfur), particularly of lead, zinc, and, in some cases, barite and fluorite, were precipitated. These deposits exhibit many similarities to shale-hosted deposits but differ in that copper is rare or absent in them. They are commonly called Mississippi Valley-type deposits because they are similar to the extensive accumulations in the Mississippi River drainage area.

Secondary deposits also form when oxidized groundwaters flow through sandstones located near the Earth's surface. Some metals, such as uranium, are present in small quantities in the sandstone. As the water percolates downward through the sandstone, the metals dissolve and are concentrated in the water. Eventually, the dissolved oxygen reacts with organic matter in the rocks. This reaction lowers the solubilities of the metals, and they precipitate. Most of the uranium ores of the western United States formed in this manner. Certain ores of copper are thought to have formed in a similar fashion.

FIELD STUDY

Economic geologists (geologists who study ore deposits) have a number of methods available to them for studying sedimentary mineral deposits. Perhaps the most important method is detailed mapping and description of ore deposits exposed in mines. During mining, rock material is continuously removed, thereby exposing new rocks. This exposure allows the mine geologist (a geologist who works with miners to ensure efficient mining) to observe and describe the deposit from a three-dimensional perspective not readily available to geologists conducting other types of studies. (For example, a geologist examining the face of a mountain can only guess at what the rocks beneath the mountain are like; a mine geologist examining the walls of a mine need only wait until these are stripped away to see what lies behind them.) This three-dimensional viewpoint allows the geologist to determine the spatial relationships of the minerals to one another and to the enclosing strata. Drill cores of rock are normally obtained around the periphery of the deposit, and the geologist can make similar observations of the rocks in the cores. These relationships can then be used to interpret how the deposit formed.

Once studies of individual sedimentary mineral deposits are completed, they can be compared with studies of other, nearby sedimentary mineral deposits. From such comparisons, a regional interpretation of how the deposits formed can be made. Models of this type are extremely useful in deciphering the overall geologic history of an area. They are also useful in the search for similar sedimentary mineral deposits.

MICROSCOPIC STUDY

Examination of rock samples using microscopic techniques allows the geologist to observe relationships not visible to the naked eye. Several types of microscopic study are available. Standard petrographic microscopy involves observation of a thin wafer or rock. This wafer, known as a thin section (0.03 millimeter thick), is mounted on a glass microscope slide and ground until it is thin enough for visible light to pass through it. Polarized light is passed through the thin section and the geologist notes how the light is affected by passing through different minerals.

Some minerals, particularly those containing valuable metals and sulfur, are opaque to visible light no matter how thin they are ground. These minerals are studied by reflected light microscopy. The thin section is polished until a very smooth surface is obtained. Then, light from above is reflected through the microscope for observation. Reflected light microscopy is often combined with standard petrographic microscopy. Together, these techniques are useful for examining the boundaries between the various minerals present in the sample and for determining the order in which they formed.

Additional microscopic techniques are often used to supplement these studies. Epifluorescence microscopy involves illuminating the sample with violet or ultraviolet ("black") light. Under these conditions, certain minerals fluoresce, like the colors on a black-light poster, and they can be easily identified through a microscope. In cathode luminescence microscopy, the sample is put in a vacuum chamber and bombarded with an electron beam. The beam excites certain minerals, causing them to emit visible light. Fluid-inclusion microthermometry involves putting thin chips of the sample into a special heating-cooling microscope stage. The sample is then heated with hot nitrogen gas or cooled with liquid nitrogen. Tiny bubbles of water trapped in the minerals are observed as they freeze, thaw, and expand as the temperature changes. This technique yields information about the temperature and chemical composition of the hydrothermal fluids from which the minerals formed.

GEOCHEMICAL STUDY

Geochemistry is the study of the chemical characteristics of rocks. Economic geologists use several geochemical techniques, including the study of isotopes. An isotope is an atom of an element that contains a greater or lesser number of neutrons than are usually present in that element. Determining the amounts of the various isotopes of an element present in the sample can yield significant information about the origin of the deposit. Radioisotopes, those isotopes involved in radioactive decay, can be used to determine the age of a deposit; radioisotopes of lead minerals can be used to determine the source of lead that has been concentrated in the deposit as well. Isotopes of sulfur (present in most sedimentary mineral deposits) can be useful in determining the source of the sulfur and the temperature and pressure under which the deposit formed. Isotopes of hydrogen and oxygen can yield information about the source of water in hydrothermal fluids (seawater, rainwater, and water released by the melting of rocks) as well as the temperature at which the deposit formed.

CONTINUING VALUE

During the last century, exploitation of the Earth's mineral wealth has expanded at unprecedented rates as advances in technology have led to improvements in mining and extraction methods and the development of new and more expansive uses of minerals and other natural resources. Sedimentary mineral deposits have been, and continue to be, the sources of a major portion of the world's mineral commodities. At one time or another, the largest lead, zinc, gold, uranium, barite, boron, and iron mines in the world have exploited deposits in sedimentary rocks. For these reasons, sedimentary mineral deposits have had a profound effect on the life of every person in the developed world. As technology continues to advance, and as the number of people with access to these advances continues to increase, the demands on the Earth's natural resources, including mineral deposits, will continue to increase as well.

Robert A. Horton, Jr.

CROSS-REFERENCES
Aluminum Deposits, 1539; Building Stone, 1545; Cement, 1550; Chemical Precipitates, 1440; Coal, 1555; Diamonds, 1561; Dolomite, 1567; Earth Resources, 1741; Evaporites, 2330; Fertilizers, 1573; Gold and Silver, 1578; Hydrothermal Mineral Deposits, 1584; Industrial Metals, 1589; Industrial Nonmetals, 1596; Iron Deposits, 1602; Manganese Nodules, 1608; Metamorphic Mineral Deposits, 1614; Mining Processes, 1780; Oil and Gas Exploration, 1699; Pegmatites, 1620; Platinum Group Metals, 1626; Precipitation, 2050; Salt Domes and Salt Diapirs, 1632; Seawater Composition, 2166; Seismic Reflection Profiling, 371; Uranium Deposits, 1643.

BIBLIOGRAPHY

Barnes, H. L. *Geochemistry of Hydrothermal Ore Deposits.* New York: John Wiley & Sons, 1979. This book contains chapters on the genesis of ore fluids in sedimentary environments and on the theory behind various geochemical techniques used for the study of sedimentary mineral deposits. Suitable for college-level readers with some knowledge of chemistry.

Blatt, Harvey. *Sedimentary Petrology.* 2d ed. San Francisco: W. H. Freeman, 1992. A very well illustrated work offering complete coverage of the subject of sedimentary rocks. Intended to be a college-level textbook but perfectly accessible to the interested high school student or layperson. The final two chapters cover the design and conduct of research projects.

Blatt, Harvey, and Robert J. Tracy. *Petrology: Igneous, Sedimentary, and Metamorphic.* New York: W. H. Freeman, 1996. Stated to be intended for the college sophomore or junior but perfectly suited to the interested high school student or general reader. The section on sedimentary rocks is clear and very well illustrated. An excellent introduction to the subject of sedimentary rocks in general and classification in particular.

Edwards, Richard, and Keith Atkinson. *Ore Deposit Geology and Its Influence on Mineral Exploration.* London: Chapman and Hall, 1986. An introductory text on the subject of economic geology. Chapters 5, 6, 7, 8, and 9 are devoted to various types of sedimentary mineral deposits. Generally well written; uses British rather than American spellings for many technical terms, particularly mineral names. Suitable for college-level readers with some knowledge of geology.

Kesler, S. E. *Our Finite Mineral Resources.* New York: McGraw-Hill, 1976. This book covers many aspects of mineral exploitation, from geology and exploration through refining and ultimate use. Chapters 5 and 10, along with parts of chapters 7, 8, and 9, deal with sedimentary mineral deposits. Geological aspects are treated in varying detail; however, geological coverage is generally rudimentary. Written for nonscientists at all levels.

Maynard, J. B., E. R. Force, and J. J. Eidel, eds. *Sedimentary and Diagenetic Mineral Deposits: A Basin Analysis Approach to Exploration.* Chelsea, Mich.: Society of Economic Geologists, 1991. This book discusses the geochemistry of most major types of metallic sedimentary mineral deposits in a geological context and discusses the strengths and weaknesses of various models that have been proposed to explain their formation. Suitable for any level of reader with some knowledge of both chemistry and geology.

Ridge, J. D. *Ore Deposits of the United States, 1933-1967.* 2 vols. New York: American Institute of Mining, Metallurgical, and Petroleum Engineers, 1968. This two-volume set describes ore deposits of many types, including major sedimentary mineral deposits in the United States. The articles contain information on the history of discovery and production as well as geological information. Suitable for college-level readers.

Skinner, B. J. *Earth Resources.* Englewood Cliffs, N.J.: Prentice-Hall, 1986. Chapters 5, 6, 7, and 8 contain information on various types of sedimentary mineral deposits, although these chapters are not exclusively devoted to the subject. Intended for a very general audience and suitable for high school or advanced junior high school readers.

_____. *Economic Geology: Seventy-fifth Anniversary Volume.* El Paso, Tex.: Economic Geology Publishing, 1981. Certain papers in this volume—those by Nash et al.; Pretorius, Gustafson, and Williams; and Bjorlykke and Sangster—provide detailed reviews of the geology of some of the most important types of sedimentary mineral deposits. Another article, by Anhaeusser, gives brief accounts of some other types that are not themselves the subjects of detailed reviews in this volume. Suitable for college-level readers with some background in geology.

Tucker, M., ed. *Sedimentary Rocks in the Field.* 2d ed. New York: Wiley and Sons, 1996. A clearly written and well-illustrated introductory text aimed at British college undergraduates, who learn principally through independent study. The greater part of the text is devoted to classification in terms understandable to the general reader.

United States Bureau of Mines. *Minerals Yearbook.* Washington, D.C.: Government Printing Office. This publication, issued annually, contains statistics on many aspects of mining and mineral exploitation.

URANIUM DEPOSITS

Uranium deposits occur in clastic sedimentary rocks, largely sandstone and conglomerate or their metamorphosed equivalents. These deposits provide the raw material that fuels nuclear power plants.

PRINCIPAL TERMS

COFFINITE: a black uranium-silicon-hydrogen mineral

GUMMITE: a brightly colored mixture of minerals that contains uranium, thorium, and lead; it forms by weathering of a uranium deposit

ISOTOPE: a variation of a chemical element; each variant differs in the number of neutrons in the nucleus

PEGMATITE: an igneous rock that is composed of extremely large crystals, typically much larger than 1 centimeter; it usually contains abundant quartz and potassium feldspar

PITCHBLENDE: a poorly crystalline mass of uranium-oxygen minerals; unlike well-crystallized uraninite, it lacks much thorium

RADIOACTIVE DECAY: spontaneous transformation of one type of atom into another

ROLL-TYPE DEPOSIT: a uranium deposit in sandstone that is characterized by oxidized (brightly colored) sandstone on one side of the deposit and chemically reduced (dull-colored) sandstone on the other

THORIUM: a chemical element that contains two fewer protons than does uranium; it is about three times more abundant than uranium in the Earth's crust

THUCHOLITE: a black mixture of hydrocarbon, uraninite, thorium-bearing minerals, and sulfur-bearing minerals

URANINITE: a mineral composed of uranium and oxygen in which there are twice as many atoms of oxygen as there are of uranium

ISOTOPES OF URANIUM

Uranium deposits are rocks that contain enough uranium to make them of interest to mining companies. Most uranium is exploited to become fuel even though only a small part of the uranium in any deposit is useful as fuel. There are two important types, or isotopes, of uranium. Both of these have 90 protons in their nuclei, but they vary in the number of neutrons. The corresponding isotopes are uranium 235, which has 145 neutrons, and uranium 238, which has 148. Uranium 235 constitutes only 0.7 percent of all uranium, but only uranium 235 is useful as fuel because it discharges more neutrons during its radioactive decay than are required to induce fissioning in adjacent uranium 235 nuclei. Rapid and uncontrolled fissioning of uranium from these naturally produced neutrons characterized the atom bomb that was dropped on Hiroshima. Subsequent fission bombs have relied on the same principle but often used the human-made element plutonium. All nuclear power plants rely on the controlled fissioning of uranium 235.

About 2.8 billion years ago in the Oklo region of Gabon, Africa, groundwater began to deposit uranium oxide, eventually forming an ore body several kilometers long. Uranium was unusually concentrated at several sites in this ore body. About 2 billion years ago, the ratio of uranium 235 to uranium 238 was around 3 percent instead of the 0.7 percent of today, since uranium 235 decays more quickly than uranium 238. At this enrichment of uranium 235, normal water can serve as a moderator in a nuclear reactor, and that is exactly what happened. With groundwater acting as a moderator and a neutron reflector, a series of at least fifteen natural nuclear reactors began operation about 2 billion years ago; some operated for several hundred thousand years. They reached temperatures of 150 to 200 degrees Celsius, and they used several tons of uranium, but none of them exploded or did anything else spectacular. It is noteworthy that none of the radioactive wastes seem to have migrated very far from where they were formed.

DISTRIBUTION

Among all types of ore, uranium deposits have one of the strongest affinities for an environment well within a continent, as opposed to a coastal plain or oceanic environment. This affinity has remained constant throughout Earth's history; the nature of uranium deposits, however, has evolved with time, at least among those deposits occurring in sedimentary rocks.

The oldest sedimentary rocks, the Archean rocks, largely accumulated by underwater slumpage on steep submarine slopes. This sedimentary environment has been particularly poor for accumulation of uranium. The proportion of preserved continental sedimentary environments increases markedly in strata that accumulated near the time of the transition from the Archean to the Proterozoic eon. These sand-rich sedimentary strata contain some of the Earth's largest uranium deposits. Some of these uranium deposits coincidentally contain some of Earth's largest gold deposits, the best-preserved voluminous examples occurring in central Canada, north of Lake Huron, and in the Witwatersrand district of South Africa; other deposits occur in Australia, Brazil, Finland, and the United States. The amount of gold varies, from being unconcentrated in the Lake Huron uranium ore, to being extremely concentrated in the Witwatersrand ore. Uraniferous rock in South Africa averages about 10 parts of gold per million parts of rock. This concentration of gold may not seem very high until compared with average crustal rock, which contains only 0.004 part of gold per million parts of rock. The corresponding concentration factor is 10/0.004, or 2,500. This concentration factor is about ten times higher than that in the Lake Huron ore of Canada. The average for the whole crust is about 3.4 parts per million.

HURON-WITWATERSRAND DEPOSITS

Both the Lake Huron and Witwatersrand uranium ores are mostly ancient gravel beds. The uraniferous conglomerate (gravel that becomes rock is called conglomerate) in both Canada and South Africa contains substantial concentrations of pyrite—a dense, iron-sulfur mineral popularly known as fool's gold. The association of a dense mineral such as pyrite with the even denser uranium minerals has convinced most investigators that the uranium minerals were concentrated to ore grade by physical sorting, just as gold was concentrated into placer gravels in California, Alaska, and elsewhere. Both the observed uranium-thorium minerals and the pyrite are susceptible to oxidation (rusting) and do not survive distant transport under modern temperate-to-tropical weathering conditions. A corollary of the placer hypothesis, therefore, is that the Earth's atmosphere was not very effective at oxidation during the transition from the Archean to the Proterozoic eon, roughly 2.5 billion years ago.

Large uranium deposits such as those in the Lake Huron and Witwatersrand districts do not occur in young strata. It is generally agreed that the lack of Huron-Witwatersrand uranium deposits in young strata records some basic evolution of the Earth, but competing models have been proposed. In the oxygen-poor weathering model, this evolution involves a progressive increase in the oxygen content of the atmosphere. In the contemporaneous-mineralization model, the evolution involves either a decrease in production of uranium-thorium-rich fluids deep in the Earth or a decrease in their ability to rise to the Earth's surface.

The Huron-Witwatersrand uranium deposits are peculiarly lacking in dense iron-oxygen minerals (magnetite and ilmenite), which usually accumulate in placer deposits. In the oxygen-poor weathering model, the lack of magnetite and ilmenite is exemplified by conversion of placer magnetite and ilmenite to pyrite by reacting the iron-oxygen minerals with enough sulfur to convert them to the iron-sulfur mineral pyrite. In the contemporaneous-mineralization model, it is not necessary to hypothesize such a reaction, because iron-oxygen minerals need not have been precipitated by the fluids that were hypothetically precipitating both uranium-thorium minerals and pyrite on eroding mountain slopes.

The highest uranium concentrations in the Huron-Witwatersrand (and younger) deposits typically occur in the most hydrocarbon-rich rocks, although hydrocarbon is only locally abundant in these ancient gravels. Carbon generally occurs within dense hydrocarbons in the Huron-Witwatersrand deposits; in some young uranium deposits, the highest uranium concentrations occur in fossil trees that had accumulated within buried river channels. Trees have existed on Earth for less than 400 million years, but the Huron-Witwatersrand uranium deposits are more than 2.2 billion years old. In the case of uranium-enriched fossil trees, the uranium apparently precipitated from groundwater that was flowing through a buried river channel. Uranium may precipitate from solution by reaction with chemically reduced (electron-rich) material such as a fossil tree. A tree is composed of organic-carbon compounds, and all such compounds are chemically reduced.

The genetic relationship of uranium with hydrocarbons found in the Huron-Witwatersrand deposits is not obvious. Published hypotheses include precipitation of groundwater uranium onto fossil accumulations of bacterial carbon, analogous to precipitation onto younger tree fossils; placer concentration of dense uranium minerals onto a living mat of bacteria or lichen; and precipitation of the hydrocarbons onto previously concentrated uranium minerals. Proponents of the living-mat hypothesis compare the mat to the pieces of carpet that placer-gold miners commonly use to trap fine gold particles. Proponents of the hydrocarbon-precipitation hypothesis note that uranium minerals are constantly emitting radiation that tends to polymerize low-density, low-viscosity hydrocarbons into high-density, congealed hydrocarbons. As a result, through-flowing hydrocarbons could become congealed around preexisting uranium minerals within buried sediment.

ROLL-TYPE DEPOSITS

Uranium-enriched trees occur in young uranium deposits that have formed within sandstone. Sandstone uranium deposits are widespread in New Mexico and in some adjacent states. These deposits consistently lack thorium or gold but are associated with other deposits of copper and/or vanadium within sandstone. Sandstone deposits are called "roll-type" deposits because the ore zone is shaped like a rolled notebook, with the convex side pointing in the direction of groundwater flow through the sandstone. Ordinary dull-colored sandstone lies in front of the ore roll (along the direction of groundwater flow), whereas colored sandstone lies behind the ore zone. The colored sandstone (usually pink or yellow) records the passage of the uranium-rich fluids that transported dissolved uranium to form the ore roll.

Ore rolls predominate over fossil trees as the

host for uranium in sandstone. Most of the uranium in ore rolls occurs as a fine-grained uranium-oxygen mineral (uraninite) or a uranium-silicon-oxygen mineral (coffinite). Both are associated with minor pyrite, hydrocarbons, and calcite (calcium carbonate). The ore volume in individual ore rolls is small compared with that in ancient, uranium-rich gravel beds, but dozens of roll-type deposits may occur scattered across tens of kilometers within an ore district. Most investigators attribute the uranium enrichment in the groundwater that has precipitated roll-type deposits to the ordinary hydrologic cycle of the Earth. They note that groundwater that contains dissolved oxygen from the Earth's atmosphere can oxidize and dissolve uranium from minerals that are dispersed through a large volume of ordinary rock. Such a mechanism is analogous to the passage of a vacuum cleaner over a carpet. An alternative origin for roll-type deposits involves the mixing of rising uranium-rich fluids with descending groundwater. Uranium in the rising fluids may have been carried as dissolved species that consisted of uranium, carbon, and oxygen. Upon entering an aquifer, the mixture of rising and descending fluids may have carried uranium until chemical reaction induced precipitation of uraninite, pyrite, and calcite into an ore roll.

VEIN-TYPE AND OTHER DEPOSITS

Sedimentary uranium deposits are confined to particular sedimentary beds. Another type of deposit consists of veins that cut across rocks. Most of these veins probably were nearly vertical at the time of their formation; the rocks that they cut may have been subsequently deformed so that the veins no longer remain vertical. The vein-type uranium deposits may contain important concentrations of other metals—nickel, cobalt, and silver, for example. It is usually possible to determine that uranium was not precipitated at the same time as the coexisting metals. Vein-type deposits, therefore, may represent the lower end of a nearly vertical "plumbing system" in which the relative concentrations of metals varied markedly with time. Rising fluids with different metal concentrations may have risen to characteristically different levels in the crust. Uranium-rich fluids appear to have commonly reached high crustal levels, including the Earth's surface.

Some ancient uranium deposits, especially those in northern Australia, have been so highly deformed that it is not obvious whether they originally resembled the Huron-Witwatersrand deposits or more closely resembled vein-type deposits. Like the Huron-Witwatersrand deposits, these enigmatic deposits may be voluminous and the ore may follow particular strata faithfully. These strata, however, have been so severely metamorphosed and deformed that it is difficult to reconstruct the original process of uranium supply. For example, the initial uranium minerals may have recrystallized as different minerals during metamorphism. The observed uranium-oxygen minerals typically are finely intergrown in a mixture that is termed pitchblende. Another potential complication, especially in northern Australia, is modern tropical weathering of uranium minerals to gummite.

Young surficial deposits of uranium occur in the deserts of western Australia and in northeastern Africa. The host is soil (called calcrete) that is extremely rich in calcite (calcium carbonate). Uranium in these deposits commonly occurs in a mineral called carnotite, which is composed of uranium, potassium, vanadium, oxygen, and hydrogen.

STUDY BY GEOLOGISTS

Uranium deposits are studied both by exploration geologists and by academic geologists. Exploration geologists (mining-company employees, for example) concentrate on those characteristics that enhance successful exploration for additional ore deposits. Academic geologists (university professors or government researchers) concentrate on the source and transport mechanisms of uranium. Exploration geologists concentrate on the identification of "halos" around uranium deposits. A halo is some peculiar characteristic of the surrounding, uranium-poor rock that is related to the uranium mineralization. This could be a peculiar color, texture, grain size, mineralogy, or abundance of an ordinary element such as calcium. The exploration geologist may not be concerned about why a particular halo recurs as long as it consistently helps find additional ore. An academic geologist wants to know why those features recur.

The difference in sandstone color on alternate sides of a roll-type uranium deposit has been a great boon to exploration geologists. Mining-company geologists examine the core from exploratory drill holes, looking for colored sandstone which records the passage of uranium-rich fluids. Wherever they find such a zone, they direct subsequent drilling to be concentrated in the area between that in which the drills have penetrated colored sandstone and the area in which the drills have penetrated ordinary sandstone. Uranium in ancient conglomerate and associated sedimentary rocks typically occurs with abundant pyrite. This pyrite serves as a useful guide to ore. Because pyrite readily rusts, concentrations of the mineral are easily identified by a rusty coating on exposed rocks.

The mineralogy of uranium deposits is determined by X-ray diffraction and their elemental composition by X-ray fluorescence. One common method of elemental analysis, neutron activation analysis, is not recommended because this technique involves placing a sample within a nuclear reactor. Although the neutron activation technique works well with virtually any other type of ore sample, the fissioning of uranium produces a wide range of spurious elements that distort the elemental analysis. The most common regional exploration technique for uranium deposits uses a gamma-ray detector. One of the daughter products of uranium, bismuth 214, emits gamma rays that are readily detected.

MINING OF URANIUM

Uranium deposits provide an important source of fuel for production of electricity. Within a few decades, the proportion of all electricity that is produced from this source grew from essentially nil to 8 percent of world production as of 1989. The mining of uranium deposits expanded correspondingly. Future exploitation of uranium deposits will depend upon several factors: the availability of alternative fuels for generation of electricity, the safety record of nuclear power plants, the development of a plan for disposal of nuclear waste, and the acceptance of environmental disruption around uranium mines.

The first uranium mines extracted uranium from rich veins. Much of the uranium for the first atom bombs came from uranium-rich veins in metamorphic rocks in the Northwest Territories of Canada. Most modern mining, however, occurs in clastic sedimentary rocks, where the uranium minerals are dispersed between quartz grains of sand-to-pebble size. Modern mining methods favor the large volume of sedimentary deposits despite the fact that vein deposits, although spread out, generally contain higher concentrations of uranium.

The volume of known uranium deposits is sufficient to maintain an adequate supply to nuclear power plants for the foreseeable future. Nevertheless, exploration continues for new deposits because it always is possible to find a deposit with a higher uranium concentration or a greater volume of ore than in any existing deposit. Such a rich deposit could be more profitable to mine than any existing deposit, depending on such factors as transportation and labor cost.

Michael M. Kimberley

CROSS-REFERENCES

Aluminum Deposits, 1539; Building Stone, 1545; Cement, 1550; Chemical Precipitates, 1440; Coal, 1555; Deep-Sea Sedimentation, 2308; Diagenesis, 1445; Diamonds, 1561; Dolomite, 1567; Evaporites, 2330; Fertilizers, 1573; Fluid Inclusions, 394; Future Resources, 1764; Gold and Silver, 1578; Hydrothermal Mineral Deposits, 1584; Hydrothermal Mineralization, 1205; Industrial Metals, 1589; Industrial Nonmetals, 1596; Iron Deposits, 1602; Lakes, 2341; Manganese Nodules, 1608; Metamorphic Mineral Deposits, 1614; Pegmatites, 1620; Platinum Group Metals, 1626; Salt Domes and Salt Diapirs, 1632; Sediment Transport and Deposition, 2374; Sedimentary Mineral Deposits, 1629; Water-Rock Interactions, 449; Weathering and Erosion, 2380.

BIBLIOGRAPHY

Albright, David, Frans Berkhout, and William Walker. *Plutonium and Highly Enriched Uranium, 1996: World Inventories, Capabilities, and Policies.* New York: Oxford University Press, 1997. This book focuses on the world's plutonium and uranium deposits and the safety measures put in place to process and maintain those deposits. Illustrations, maps, index, and bibliography.

Dahlkamp, Franz J. *Uranium Ore Deposits.* New York: Springer-Verlag, 1993. A look at international uranium ores, their conditions, and their economic values. A good introductory book for the layperson with an interest in uranium.

Gabelman, John W. *Migration of Uranium and Thorium: Exploration Significance.* Tulsa, Okla.: American Association of Petroleum Geologists, 1977. This author emphasizes the potential supply of uranium from fluids that are rising from deep in the Earth's crust. Most authors propose this type of supply for the uranium in veins and in highly metamorphosed rocks, but Gabelman goes further to invoke this mechanism for sedimentary uranium deposits as well.

Goldman, Benjamin A. *Discounting Human Lives: Uranium and Global Equity.* Brookfield, Vt.: Avebury, 1994. An examination of the practices and protocols of the uranium industry. Goldman's book focuses heavily on the effects of uranium and the uranium industry on the environment and on human life.

International Atomic Energy Agency. Technical Document Series. Vienna: Author, 1984-1987. The International Atomic Energy Agency (IAEA) is a branch of the United Nations. IAEA publications may be obtained at most major scientific libraries. Five volumes on uranium deposits have been produced in the IAEA technical document series: *Proterozoic Unconformity and Stratabound Uranium Deposits* (1984), *Surficial Uranium Deposits* (1984), *Sedimentary Basins and Sandstone-Type Deposits* (1985), *Vein-Type Uranium Deposits* (1985), and *Uranium Deposits in Proterozoic Quartz-Pebble Conglomerates* (1987). Collectively, these five volumes provide the most comprehensive overview of uranium deposits.

Kimberley, M. M., ed. *Uranium Deposits: Their Mineralogy and Origin.* Toronto: Mineralogical Association of Canada, 1979. The twenty-one papers in this book cover all major aspects of uranium deposits. Like the IAEA books, this book is available at many major scientific libraries.

Rich, R. A., H. D. Holland, and U. Petersen. *Hydrothermal Uranium Deposits.* New York: Elsevier, 1977. This book emphasizes the source and precipitation of uranium from the hot, solute-rich waters (hydrothermal solutions) that have risen through the Earth's crust.

Tang, Yu S., and James H. Saling. *Radioactive Waste Management*. New York: Hemisphere, 1990. An exploration of radioactive plants and government policies on radioactive waste disposal. Illustrations and bibliographical references.

7
ENERGY RESOURCES

GEOTHERMAL POWER

Geothermal power, having its source in the Earth's internal heat, offers a form of energy used in areas of the United States and other countries. Although limited by current technology, the Earth's heat as a power source offers immense resources, high versatility, and cost effectiveness.

PRINCIPAL TERMS

BINARY CYCLE: the process whereby hot water in the primary cycle gives up heat in a heat exchanger; a fluid such as isobutane in the secondary cycle absorbs heat, is pressurized, and drives a turbine generator

DIRECT or SINGLE FLASH CYCLE: the process whereby hot water under great pressure is brought to the surface and is allowed to turn, or "flash," to steam driving an electrical turbine generator

DOUBLE FLASH CYCLE: the process whereby two flash vessels are employed in cascade, each operating a turbine to extract more power

HYDRAULIC FRACTURING: the underground splitting of rocks by hydraulic or water pressure

as a means of increasing the permeability of a formation

HYDROSTATIC PRESSURE: pressure within a fluid at rest, exerted at a specific point

HYPERTHERMAL FIELD: a region having a thermal gradient many times greater than that found in nonthermal, or normal, areas

PERMEABLE FORMATION: a rock formation that, through interconnected pore spaces or fractures, is capable of transmitting fluids

THERMAL GRADIENT: the increase of temperature with depth below the Earth's surface, expressed as degrees Celsius per kilometer; the average is 25 to 30 degrees Celsius per kilometer

GEOTHERMAL PHENOMENA

Geothermal power as evidenced by hot springs, steam vents, geysers, and volcanic activity is the Earth's inner heat escaping through faults and weak spots in the crust. Temperatures below the surface vary from one location to another and depend upon the depth and temperature of the heat source and ability of the subsurface rocks to conduct heat. Generally, geothermal water is not sufficiently hot to produce electricity but still can be used in a variety of industrial and agricultural applications.

Most of the heat energy escapes through the crust at lithospheric plate boundaries near young volcanic centers, such as are found in Hawaii, Alaska, and the western United States. In the western United States, most of the geothermal hot spots are in Nevada, which has more than nine hundred hot springs and wells. This region is the Basin and Range geologic province, with a thin continental crust traversed by a system of north-south faults allowing an easy path for hot water to percolate to the surface. By way of comparison,

the greater part of the Earth's surface is nonthermal, having temperature gradients from 10 to 40 degrees Celsius per kilometer of depth. It is well to keep in mind, however, that even in nonthermal areas there is a great amount of heat several kilometers below the surface. The hot rocks at these depths, although not permeable, could be accessible except for the cost of extracting heat of this grade. The presence of a geothermal field is not always marked by the surface thermal activity. Excellent fields have been detected in places completely devoid of any surface thermal manifestations.

GEOTHERMAL ENERGY RESOURCES

Geothermal heat is now commercially exploitable only in regions where hot pore fluids circulate within permeable formations and may be reached by cost-effective drilling. The geothermal energy resources now in use or under experimentation are dry steam, hot water convection systems, geopressured systems, and hot dry rocks. For temperatures of 200 degrees Celsius or hotter,

A geothermal plant in Wairakei, New Zealand, with steam wells and pipes running to the power house. (© William E. Ferguson)

the most economical and easily used source is dry steam, but this comprises only 0.5 percent of all U.S. geothermal resources. Hot water convection systems that make up 10 percent of the world's geothermal resources are heated by magma sources near the surface, thereby transferring heat to the surface rocks. Water that is confined within the rock layers is heated above the boiling point, but pressure from surrounding rocks prevents the water from becoming steam. Water heated by molten magma may reach temperatures of 300 degrees Celsius or more.

Geopressured fields do not fall within the general classification of geothermal fields, as they occur in nonthermal areas. These fields occur at depths up to 6,000 meters, where temperatures range from less than 93 to more than 150 degrees Celsius. The hot water is pressurized in excess of the hydrostatic values encountered at that depth. These very high pressures are thought to be caused by the gradual subsidence along faults of the rock overburden that has trapped pockets of water below and between alternating layers of sandstone and shale. Geopressurized fields produce three types of energy: thermal, as a result of the fluid temperatures; hydraulic, as a result of the high excess fluid pressure; and chemical, as a result of the caloric value of the methane gas dissolved in the water. The immense energy con-

tained in geopressurized fields is not at this time fully realizable because of their great depths below the surface.

All these geothermal systems utilize groundwater circulation whereby drillers merely need to construct a well to reach the water table. If nature does not oblige, drillers have attempted to inject water in the ground and also to create fractures in the rock strata by use of explosives. The fractures, in turn, provide pathways for the flowing water to pick up heat from the surrounding rock. Hot dry rocks are found at moderate depths by proven methods, but the far greater number of them lie at depths beyond which drilling is not cost effective. Many believe that the large-scale extraction of energy from hot dry rock could have significant long-range payoffs. Considering regions of thermal gradients of 40 degrees Celsius or more for each kilometer of depth, it is estimated that these heat sources could provide up to ten times the heat energy of all coal deposits in the United States.

GEOTHERMAL FACILITIES

The direct flash system is used at the Geysers, the world's largest geothermal facility, located 140 kilometers north of San Francisco. In 1999, the Calpine Corporation, which operates the Geysers facility, announced its capacity to be at more than 1,000 megawatts. Commercial geothermal power has been produced at the Geysers since 1960. Steam at these facilities is first allowed to pass through centrifugal separators, which remove rock particles that may damage turbine blades. The steam travels to the turbines by means of insulated pipes.

The United States' largest geothermal resource is probably located in the Imperial Valley of California. The valley itself is flat but covers six known geothermal resources having thermal fluid temperatures ranging from 120 to 330 degrees Celsius, circulating in layers of porous rock between 1,200 and 6,000 meters below the surface. In

1997, fourteen plants with a combined capacity of more than 400 megawatts were in operation in the Imperial Valley.

A 10-megawatt demonstration plant using the direct flash method is located in Brawley, California. Water at temperatures above 200 degrees Celsius is pumped to the surface under pressure and is routed through a series of pathways, which reduces the pressure, allowing some of the liquid to vaporize or "flash boil." The steam, constituting 20 percent of the fluid, is expanded through a standard steam turbine. A problem with the Brawley geothermal fluid is that it contains 15 percent total dissolved solids, which can choke a three-inch pipe diameter to half an inch after 100 hours of operation. Mineral buildup, or scaling, has been a major engineering challenge at the first hot-water power plants.

FACILITIES UTILIZING BINARY SYSTEMS

Located less than 2 kilometers away, another demonstration plant will be the first commercial facility employing a binary cycle system. In the binary system, pressure prevents the water from flashing to steam. Instead, the hot brine is allowed to flow through a heat exchanger that is surrounded by a heat exchanging fluid (commonly isobutane). The fluid expands and under high pressure drives a turbine. By keeping the brine under high pressure when it leaves the turbine at a temperature of 70 degrees Celsius, engineers believe that they can avoid the scaling problems that clog pipes. The brine is pumped to a well injection station 4 kilometers away on the edge of the Heber geothermal reservoir. It is believed that brine returned to the ground will be geologically recycled and available for future use.

The Imperial Magma test facility in the East Mesa, California, geothermal field is a binary system that uses brine at a low to moderate temperature with two hydrocarbon working fluids, each passing through a heat exchanger. Heat from the East Mesa brine first passes hot brine at 180 degrees Celsius to vaporize isobutane that will run a 12-megawatt turbine. The brine cools and moves on to a second-stage heat exchanger, which vaporizes the propane at a lower temperature and rotates a 2-megawatt turbine generator. The second turbine, operating at a considerably lower temperature, increases the efficiency of the system.

The appeal of the binary system is that it gains high efficiency with moderately low-temperature brines. For high-temperature geothermal fields, engineers believe that using working fluids with lower boiling points will permit engineers to extract energy from the fluids as the brine temperature drops over time. This will help extend the production life of some fields, since most must produce for thirty to thirty-five years to be economical. Another appeal is that a well-designed binary power plant can operate with virtually no emissions.

GEOTHERMAL EXPLORATION

Geothermal exploration involves the teamwork of specialists versed in a variety of disciplines. Geothermal explorers in general will have several objectives in mind: They must find likely locations of geothermal fields, decide whether the field located has a sufficient source of heat, and decide whether the field located is steam- or water-dominated. Of the sources of exploitable geothermal energy, the high-temperature fields are the most promising commercially. These fields are almost always located in young orogenic regions or mountain belts where there has been recent volcanism.

The task of the team's geologist is to construct as accurately as possible a model of the geological structure of the thermal region and to predict promising drilling sites. This is accomplished to a great extent by surface mapping and the study of the tilt of rock outcrops. Much of the model is deduced by direct observations, which are not generally possible below the surface. By studying hot-spring deposits, the geologist can estimate approximate subsurface temperatures.

The function of the hydrologist is to work closely with the geologist and to determine the paths that water will follow underground through geologic strata and within the boundaries of the geologist's model. The hydrogeologist studies the gradients, porosities, and permeabilities of the various geological formations and may be able to offer a reasonable explanation for the thermal fluids reaching permeable zones in the field and, from that point, how they escape to the surface.

The geophysicist's task in geothermal exploration is to determine as accurately as possible the physical properties of the subsurface and to detect anomalies. A geothermal field with large volumes of steam and hot water in permeable rocks will likely appear anomalous when compared to surrounding regions. The geophysicist will use the thermometer and its various forms, including thermocouples and thermistors, to deduce temperature gradients, heat-flow rates, and local hot spots. Electrical resistivity measurements may be taken by placing electrodes into the ground and measuring the voltages between them. Differences in resistance of rocks are attributable not only to the physical differences in rocks but also to the presence of steam or electrically conductive thermal waters. Additional techniques employed may be gravity and seismic measurements, which can delineate variations in the densities of rock strata and the presence of faults for the migration of pore fluids.

The function of the geochemist is to analyze the chemistry of natural thermal discharges. If the fluids are hot, chemical equilibrium will be achieved rapidly; the chemical nature of the discharged fluids is a reflection of the temperature achieved at equilibrium. The geochemist will look for the presence of silica and magnesium and at ratios of sodium to potassium as indicators of deep reservoir temperatures. The lower the sodium-potassium ratio, the higher the fluid temperature. The geochemist is also able to detect valuable minerals present in the thermal fluids that are of interest to industrial geothermal developers.

The aim of this preliminary exploration is to choose promising locations for exploratory well drilling. If the exploration team believes that a useful field exists, then drilling is used to locate zones of permeable rocks saturated with hot thermal fluids. Drilling expense and time are saved if the exploration drilling bore samples are of small diameter; larger bores may follow once the potential of the field has been established.

VALUE AS ENERGY SOURCE

Geothermal energy offers significant savings, 25-55 percent of conventional thermal cost and 30-35 percent of nuclear fuel expense. The construction of geothermal power plants can yield significant savings, as the geothermal energy is already present within the Earth and does not need to be generated. When used along with other forms of energy production, geothermal power can help reduce the overall cost of energy. The

use of geothermal power is becoming of increasing importance on a worldwide scale; in fact, in some twenty countries geothermal exploration is actively pursued, with more than sixty countries involved in geothermal power development. Nevertheless, some predictions are that geothermal energy is unlikely ever to meet more than 10 percent of energy needs. The key lies in the extent to which humans can harness the Earth's heat, particularly the deep heat sources that are at the limits of current technology.

The ultimate goal is universal heat mining, that is, utilizing the Earth's heat wherever it is needed, even in the nonthermal areas. Meeting this goal could require drilling 6.5 to 9 kilometers for power generators operating at thermal efficiences of 15-20 percent. Included in heat mining is the direct tapping of magma pockets of active volcanoes; because of the very high temperatures encountered, this is a formidable task for conventional drilling even at shallow depths.

In addition, the fluids from hydrothermal fields accompanied by gases and certain water soluble chemicals can be potentially hazardous to the machinery components of a geothermal facility as well as to the environment. Some deposits, such as silica and iron sulfide, build up in discharge channels on turbine blades and pipes, which reduces efficiency. Carbon dioxide and hydrogen sulfide in solution may form large corrosion pits on the pipe walls. The discharge of hydrogen sulfide into the atmosphere at geothermal power plants may have an adverse effect on crops, river life, and electrical equipment. Fortunately, scrubber systems can clean hydrogen sulfide and other gases from the effluent, and sometimes those gases can even be converted into useful products such as fertilizer. Also, well-designed geothermal plants release only a few percent of the carbon dioxide and sulfur compounds produced by coal-fired plants and oil-fired plants. As stated above, closed binary cycle plants can operate with virtually no emissions.

By the end of the 1990's, the installed capacity of geothermal electrical power plants was about 6,000 megawatts worldwide, including 2,200 megawatts in the United States. The United States uses an additional 1,000 megawatts for direct heating. This places geothermal energy third in the United States among renewable energy sources (0.5 percent of all energy used), following hydro-electricity (4.5 percent) and biomass (4.3 percent), and ahead of solar power (0.1 percent) and wind power (0.05 percent). Most of the people in Iceland and more than 500,000 people in France use geothermal heat in homes and public buildings.

Michael Broyles

CROSS-REFERENCES

BIBLIOGRAPHY

Albu, Marius, David Banks, and Harriet Nash, et al., eds. *Mineral and Thermal Groundwater Resources.* London: Chapman and Hall, 1997. This well-illustrated book looks at groundwater and its relationship to geothermal processes. Suitable for college-level students with some background in the Earth sciences. Contains bibliographical references and index.

Anderson, Greg M., and David A. Crerar. *Thermodynamics in Geochemistry: The Equilibrium Mode.* New York: Oxford University Press, 1993. An exploration of geochemistry and its relationship to thermodynamics and geothermometry. A thorough resource, but it can be somewhat technical at times. Recommended for the person with a background in chemistry and Earth sciences.

Armstead, H. Christopher. *Geothermal Energy.* 2d ed. Bristol, England: Arrowsmith, 1983. A comprehensive general textbook covering many aspects of geothermal energy, including historical applications of Earth's heat, the nature and occurrence of geothermal fields, and exploration techniques. The methods of electric power generation from geothermal energy from specific regions are discussed, along with the comparative costs in each case. Suitable for the general student of geothermal energy.

_____, ed. *Geothermal Energy: Review of Research and Development.* Lanham, Md.: UNIPUB, 1973. Geothermal energy exploration and utilization are presented with discussion appropriate for the general audience. An excellent supplement to the author's 1983 text. Softcover format with many diagrams, especially of power plant machinery, including noise silencer construction, pipeline expansion arrangements, and circulating water systems.

Gregory, Snyder A., Clive R. Neal, and W. Gary Ernst, eds. *Planetary Petrology and Geochemistry.* Columbia, Md.: Geological Society of North America, 1999. A compilation of essays written by scientific experts, this book provides an excellent overview of the field of geochemistry and its principles and applications. The essays can get technical at times and are intended for college students.

Hodgson, Susan F. *A Geysers Album: Five Eras of Geothermal History.* Sacramento: California Department of Conservation, Division of Oil, Gas and Geothermal Resources, 1997. Hodgson's book provides an easy understanding of geothermal history in the United States. It includes useful graphics, illustrations, and maps as well.

Johnson, T. "Hot-Water Power from the Earth." *Popular Science* 222 (January, 1983): 70. *Popular Science* has published several articles on converting geothermal energy into usable electricity (August, 1972; November, 1974; February, 1979; and January, 1982). Here, the direct flash and binary cycle systems are discussed, along with several case histories from the California region.

Kerr, Richard A. "Hot Dry Rock: Problems, Promise." *Science* 238 (November, 1987): 1226-1229. A discussion on tapping the potentially enormous heat reserves in rocks that are too dry to yield steam or hot water. Deals with the latest research on hydraulic fracturing in deep wells as a means of creating pathways for water injected to circulate in deep hot rocks. Case histories in Los Alamos, New Mexico, and in Cornwall, England, are presented. Suitable for the general science reader.

Kruger, Paul, and Carel Otte, eds. *Geothermal Energy: Resources, Production, Stimulation.* Stanford, Calif.: Stanford University Press, 1973. The authors give a good assessment of U.S. resources in terms of producible geothermal energy, and they elaborate on the characteristic problems of utilization. An especially interesting chapter deals with corrosion and scaling of machinery and presents a complete chemical analysis of geothermal waters from several sources. A valuable reference for the geothermal student.

Plate, Erich J., et al., eds. *Buoyant Convection in Geophysical Flows.* Boston: Kluwer Academic Publishers, 1998. Although highly technical at times, this collection of papers provides good descriptions and discussions about heat convection theories, geophysics, and heat flows. The collection also offers illustrations and a bibliography.

Press, Frank, and Raymond Siever. *Understanding Earth.* 2d ed. New York: W. H. Freeman, 1998. An excellent general text on all aspects of geology, including the formation of igneous and metamorphic rocks. Contains some discussion of the structure and composition of the common rock-forming minerals. The relationship of igneous and metamorphic petrology to the general principles that form the basis of modern plate tectonic theory is discussed. Suitable for advanced high school and college students.

HYDROELECTRIC POWER

Hydroelectric power is the most commonly used renewable energy resource. It can be stored in the form of impounded water, and it is relatively nonpolluting.

PRINCIPAL TERMS

ELECTRICITY: a flow of subatomic charged particles called electrons

ENERGY: the capacity for doing work; power (usually measured in kilowatts) multiplied by the duration (usually expressed in hours, sometimes in days)

FLOW RATE: the amount of water that passes a reference point in a specific amount of time (liters per second)

GENERATOR: a machine that converts the mechanical energy of the turbine into electrical energy

HEAD: the vertical height that water falls or the distance between the water level of the reservoir above and the turbine below

KILOWATT: a thousand watts; a unit of measuring electric power

MEGAWATT: a million watts; a unit of measuring electric power

PENSTOCK: the tube that carries water from the reservoir to the turbine

POWER: the rate that energy is transferred or produced

PUMPED HYDRO: a storage technique that utilizes surplus electricity to pump water into an elevated storage pond to be released later when more electricity is needed

TURBINE: a device used to convert the energy of flowing water into the spinning motion of the turbine's shaft; it does this by directing the flowing water against the blades mounted on the rotating shaft

SUPPLY AND TRANSMISSION

Even though running water has been used by people for centuries to turn the wheels of gristmills and sawmills, it was not until the end of the nineteenth century that waterpower began to be employed to generate electricity. Hydroelectric projects can be as small as a waterwheel supplying energy to a single household or as large as a system of dams and storage projects that supply electricity to many cities and millions of people. Electric energy, generated by water-powered turbines, is transported to houses, factories, mills, and other sites of consumption along high-voltage transmission lines that may extend for more than 1,500 kilometers. These transmission lines are either alternating current (AC), the type of electricity used in houses, or direct current (DC), the type of electricity used in batteries, and can deliver hundreds of megawatts of power. In the United States, agreements between states, regions, and Canada have created a network of transmission lines that allows the flow of electricity from one part of North America to another. This ability to transport electricity from one place to another was one of the driving forces behind the relocation of factories and mills from along the rivers to adjacent to sources of raw materials. The mobility of electricity has also allowed for the growth of numerous cities located away from sources of energy.

The easiest way to harness a river for the purpose of generating hydroelectric power is to construct a dam across a river and funnel the water through a turbine that creates electricity. A dam with a large reservoir of water behind it is best for generating electricity, because both the amount of water in the river and the demand for electricity vary throughout the year. For example, in the Columbia River system in the Pacific Northwest, the river reaches peak flow during the spring snowmelt, but demand for electricity is greatest during the late winter (for heating of homes) and summer (for air conditioning). The ability to store large volumes of water behind each of the dams in the system allows the electric utilities to meet summer and winter demands for electricity by "storing" the water until electricity is needed. Water from spring runoff is stored and then released to generate electricity to power air conditioners in the summer and to heat homes in the winter.

Large storage dams also allow a utility to increase or decrease electric generation to match the demand. Electrical demand in the morning is met by releasing extra water, while at night, when the demand is less, water is either kept in the reservoir or passed through the turbines using a process called spinning. Spinning is the method in which water passes through a turbine but no electricity is produced. In a matter of seconds, the spinning turbine can be engaged and electricity produced.

PUMPED HYDRO

Another form of stored hydroelectric power is pumped storage. During the period from midnight to six in the morning, when energy demand is at its lowest, hydroelectric projects must maintain a minimum outflow of water for navigation, agriculture, mining interests, recreational interests, fish breeding, and water quality. A utility may choose to use the electricity produced by minimum stream flow regulations to pump water into a storage pond. Then, during the day, when energy demand increases, the pumped water is released and electricity generated.

During the late 1980's, pumped hydro required 1.3-1.4 kilowatt hours for every kilowatt hour produced. That may not seem economic, but because the water is pumped using surplus energy (energy that cannot be saved), the utility is able to postpone generation until demand is present. Another consideration is the difference in the cost of energy at peak and nonpeak hours. Electricity sold by the utility during peak demand may be several times the cost of electricity sold during nonpeak hours. Some utilities have even pumped water into excavated caverns. Hydroelectric power produced by the pumped-hydro plants amounts to thousands of megawatts.

WATER FLOW RATE AND HEAD

Hydroelectric power is produced by converting the potential power of natural stream flow into energy. That is commonly done by employing falling water to turn turbines, which drive generators and produce electricity. The amount of electric power produced is dependent upon the flow rate of the water and the head.

The flow rate of water is the volume of water that moves past a point during a specific period of time. The quantity of water is commonly measured by first determining the cross-sectional area (width and depth) of the river; second, the speed of the water is measured by defining a reference length of river to monitor, dropping a float at the upper end of the reference length, and recording the length of time the float takes to travel down the reference length of the river. Then, if one calculates the volume of water (length × width × depth) divided by the time the float took to travel down the reference length of the river, the result is volume per time, or flow rate.

The head of a stream is the vertical height through which the water falls. The head measurement of a hydroelectric project is the elevation difference between where the water enters the intake pipe, or penstock, and the turbine below. When waterwheels are employed to produce electricity, the head measurement is the total distance that water falls to the waterwheel. As with all energy conversions, friction results in some loss. The type of turbine or waterwheel utilized can also contribute to greater or lesser losses of energy. The theoretical maximum power available in kilowatts is equal to the head, measured in meters, multiplied by the flow rate, measured in cubic meters per second, multiplied by 9.8. Efficiencies (actual energy produced divided by the amount of energy available in the flowing water) for hydropower plants (turbines, waterwheels) vary from a high of 97 percent, claimed by manufacturers of large turbines, to less than 25 percent for some waterwheels.

WATERWHEELS

Two devices used to convert the potential energy of water to mechanical energy are the waterwheel and the water turbine. The type of waterwheel or turbine used is dependent upon the flow and head. The ideal situation is high head (more than 18 meters) and high flow, but it is feasible to produce electric energy with any combination of high head and low flow or low head and high flow.

Waterwheels are the simplest machines employed to generate hydroelectric power. The central shaft of the waterwheel, which in the past was directly connected to a grindstone, is hooked to a generator to produce electricity. Efficiency has been claimed by waterwheel manufacturers to be around 90 percent but the usual efficiency ranges from 60 percent down to around 20 percent. The most efficient is the overshot wheel, in which water falls onto the top of the wheel and turns it. The wheel is suspended over the tailwater (the water on the downstream side of the wheel) and is not resting in the water but is suspended above the water surface. This type of waterwheel requires at least a 2-meter head of water.

Three other types of waterwheels are able to operate at lower heads than does the overshot wheel, but all three are costly to construct. The first type, the low and high breast wheels, are turned by water striking the wheel at a point one-third to over one-half of the height of the wheel. The "low" or "high" defines the level at which the water enters the wheel. The second type, the undershot wheel, is probably the oldest style presently in use. The wheel is turned by water running under the wheel. Although this type of waterwheel has an efficiency of less than 25 percent, it can operate with less than a third of a meter head. The third low-head waterwheel is the Poncelet wheel, which is an improved undershot that rests just at the water level and depends upon the velocity of the water to turn the wheel. Because this wheel forces the water through narrow openings on the wheel, it is suitable for heads of less than 2 meters, but it is easily damaged by debris carried in the water.

Because the waterwheel rotates at a slow rate, the gear box, which transfers the rotation energy to the turbine, is a very costly, complex collection of gears. This expense is a major disadvantage. Another disadvantage is the large size of a waterwheel. Given the large amount of time and material involved and the low efficiency, the rate of monetary return is low. Overshot wheels, however, have the advantage of being able to operate with fluctuating water flows better than do water turbines. A second advantage is that once set up, an overshot wheel requires little repair and is not damaged by grit or clogged by leaves as the low-head water turbines are.

TURBINES

Two types of hydraulic turbines are in existence: impulse turbines, which utilize water that is exposed to normal atmospheric pressure, and reaction turbines, which use water under pressure to drive them. The Pelton impulse wheel was designed in 1880 and is the crossover from waterwheels to turbines. The Pelton wheel is composed of a disk with buckets attached to the outside of the wheel. This wheel requires a head of at least 18 meters but can operate under low flow rates because the water is forced under its own pressure through a nozzle to strike the buckets. The water striking the buckets causes the wheel to spin. Because operating efficiencies are commonly over 80 percent, this wheel is still a favorite of many small utilities in North America. The turgo impulse wheel represents an improvement over the Pelton. The water jet is aimed at the buckets at a low angle, thus allowing the stream of water to strike several buckets at once. That results in higher effi-

The electric powerhouse at the base of Hoover Dam captures the energy of the water's 726-foot descent. (© William E. Ferguson)

ciencies with smaller wheels and lower flow rates than those needed for the Pelton. The turgo has an efficiency reported over 80 percent and is suited for use with heads greater than 10 meters. The cross-flow turbine is a drum-shaped impulse turbine with blades fixed along its outer edge. The drum design allows water to pass over the blades twice: once from the outside to the inside, then (after entering the drum) back outside again. The net result is up to 88 percent efficiency in large units and the ability to operate with heads as low as 1 meter. The cross-flow turbine is in widespread use around the world. It is simple to operate and largely self-cleaning.

Reaction turbines are normally used in the large hydroelectric projects; a single unit at Grand Coulee Dam can produce 825 megawatts. Reaction turbines work by placing the whole runner (which is what is left of the "wheel" and resembles all blades set into a central shaft) into the flow of water. The water is carried to the turbines from the reservoir by a long tube called a penstock. The penstock can be more than 10 meters in diameter and tens of meters long. A propeller turbine is a reaction type of turbine that resembles a boat propeller in a tube. This type of turbine may be set either horizontally or vertically, depending on the design of the system. The Kaplan turbine is a turbine with adjustable blades on the propeller to allow operation at different flow rates. The water pressure in this system must be constant or the runner will become unbalanced. Very large hydroelectric plants usually install the Francis turbine. This type of turbine is designed to be set up and adjusted for the specific site. It can be used with a head of 2 meters and has an efficiency rating greater than 80 percent. The turbine spins as water is introduced just above the runner and directed onto the blades, causing the blades to rotate. The water then falls through and out a draft tube. A complicated mechanical governor is often used to guide the water around the runner.

COSTS AND BENEFITS

Hydroelectric power is the most developed of the renewable energy resources. It supplied about 20 percent of the world's electricity as of 1999. Where it is available, hydroelectric power is likely to be less expensive than other alternatives. Its primary environmental effects come from flooding

land behind a dam and from blocking a river's flow. Mercury levels in impounded water rise somewhat for about twenty years in a newly flooded area but then return to normal as the mercury is leached from the flooded soil. Carbon dioxide and sulfur dioxide are released to the atmosphere as newly submerged vegetation decays, but even so, hydroelectric plants emit only about 4 percent as many greenhouse gases as natural-gas-fired or oil-fired power plants, and about 3 percent as much as coal-fired power plants.

During the late 1980's, government and private industry studies were still identifying many regions in the world that could be developed with hydroelectric power plants. The available technology allowed for construction of dams that would permit unhindered fish migration, coexistence of fish hatcheries and hydroelectric projects, and maintenance of natural fish and wildlife populations. Concerns about commercial and sport fishing populations have resulted in the close monitoring of hydroelectric power plant operation by biologists to ensure fish survival.

A major obstacle to the development of new hydroelectric power plants is the large amount of paperwork involved. The cost of environmental impact studies can exceed the cost of actual construction of the projects. Dams must be licensed by the federal government and must meet hundreds of county and state regulations. The amount of water allowed to flow downstream is regulated to ensure that agriculture, sport and commercial fisheries, recreation, environmental considerations, and Native American water rights are satisfied. These competing interests for water imply that no single user of the river will determine how much or when water is moved through the dams.

Susan D. Owen

CROSS-REFERENCES

BIBLIOGRAPHY

Alward, Ron, Sherry Elisenbart, and John Volkman. *Micro-Hydro Power: Reviewing an Old Concept.* Butte, Mont.: National Center for Appropriate Technology, 1979. Delineates all the components of a hydroelectric system with detailed but easy to understand pictures. Although the title refers to micro-systems, this publication also includes pictures of turbines used on larger hydroelectric projects. It contains an international list of manufacturers and suppliers of hydroelectric system components as well as a well-written bibliography.

Basson, M. S., et al., eds. *Probabilistic Management of Water Resources and Hydropower Systems.* Highlands Ranch, Colo.: Water Resources Publications, 1994. This book supplies the reader with mathematical models used to assess management of water supply, watersheds, and hydroelectric plants. Includes illustrations and bibliography.

Freeze, R. Allan, and John A. Cherry. *Groundwater.* Englewood Cliffs, N.J.: Prentice-Hall, 1979. The leading groundwater hydrology text. The subject is presented in an interdisciplinary manner with practical sampling methods, and tests are explained. (It is important to understand the relationships between surface and subsurface water systems before understanding hydroelectric systems.)

Jiandong, Tong. *Mini Hydropower.* New York: Wiley, 1997. Complete with illustrations and maps, this book thoroughly examines the Earth's renewable energy resources, water supply management, and the operations of hydroelectric plants. Index and bibliographical references.

McGuigan, Dermott. *Harnessing Water Power for Home Energy.* Pownal, Vt.: Garden Way Publishing, 1978. Explains how to build any type of small to microscale hydroelectric facility. Lists manufacturers in the United States and the United Kingdom as well as the 1989 cost of the equipment.

Majot, Juliette, et al., eds. *Beyond Big Dams: A New Approach to Energy Sector and Watershed Planning.* Berkeley, Calif.: International Rivers Network, 1997. This collection of essays and case studies focuses on the environmental and social aspects of dams, hydroelectric plants, and electric power production, as well as the government policies surrounding these processes. Suitable for the layperson.

Palmer, Tim. *Endangered Rivers and the Conservation Movement.* Berkeley: University of California Press, 1986. Examines the "flip side" of hydroelectric power: the river. Palmer details the conservation battles that people fought to preserve rivers in their natural states. This book is based on hundreds of interviews and was reviewed by a number of environmentalists and politicians. Easy, enjoyable reading.

Sullivan, Charles W. *Small-Hydropower Development: The Process, Pitfalls, and Experience.* 4 vols. Palo Alto, Calif.: Electric Power Research Institute, 1985-1986. This four-volume Electric Power Research Institute (EPRI) work explains hydroelectric power plants, how to determine where to place them, cost, regulations, environmental impact, and a number of other related topics. This study was completed under contract with the U.S. Department of Energy and is complete to the point of suggesting which computer programs one might utilize when organizing data. This complete study is detailed enough for any group wanting to build a hydroelectric plant and simple enough to read and understand for the nontechnical individual.

United States. Bonneville Power Administration. *Columbia River Power for the People: History of Policies of the Bonneville Power Administration.* Portland, Oreg.: Author, 1980. This publication provides a good description of the development of one of the largest hydroelectric systems in the world. Documents the harnessing and development of the Columbia River and the politics involved. Helps the reader to understand the social, economic, and cultural forces that must be pacified in order to create a successful hydroelectric system.

United States Federal Power Commission. *Hy-

droelectric Power Resources of the United States, Developed and Undeveloped. Washington, D.C.: Government Printing Office, 1976. This document reports hydroelectric power resources, both developed and undeveloped, as of 1976. Lists the state, project owner, river, developed and undeveloped generation capacity, and gross static head of all projects that are licensed by the United States federal government.

NUCLEAR POWER

Nuclear power obtained from the fission of uranium and plutonium nuclei represents a significant percentage of world energy resources. Its production in appropriately designed nuclear fission reactors is especially important as a low-pollution supplement to fossil fuels.

PRINCIPAL TERMS

HALF-LIFE: the time required for half of the atoms in a given amount of a radioactive isotope to disintegrate

ISOTOPES: an element's different forms whose atoms have the same number of protons but different numbers of neutrons

MODERATOR: a material used in a nuclear reactor for slowing neutrons to increase their probability of causing fission

NUCLEAR FISSION: the splitting of an atomic nucleus into two lighter nuclei, resulting in the release of neutrons and some of the binding energy that held the nucleus together

NUCLEAR FUSION: the collision and combining of two nuclei to form a single nucleus with less mass than the original nuclei, with a release of energy equivalent to the mass reduction

RADIOACTIVITY: the spontaneous emission from unstable atomic nuclei of alpha particles (helium nuclei), beta particles (electrons), and gamma rays (electromagnetic radiation)

NUCLEAR FISSION

The idea of the atom as a source of energy developed near the beginning of the twentieth century following the discovery of radioactivity by Antoine-Henri Becquerel. The energy of this spontaneous emission, first measured by Pierre and Marie Curie, was found to be far greater than ordinary chemical energies. Nuclear fission was discovered in 1939 after Otto Hahn and Fritz Strassmann had bombarded uranium with neutrons at their laboratory in Berlin, leaving traces of radioactive barium. Their former colleague Lise Meitner and her nephew Otto Frisch calculated the enormous energy—about 200 million electron volts—that would be released in reactions of this type. These results were reported to Niels Bohr and quickly verified in several laboratories in 1939. Soon Bohr developed a theory of fission showing that the rare isotope uranium 235 (uranium with 235 nucleons: 92 protons and 143 neutrons) is far more likely to produce fission, especially with slow neutrons, than the common isotope uranium 238, which makes up 99.3 percent of natural uranium. It was also recognized that if a sufficient number of neutrons were emitted in fission, they could produce new fissions with even more neutrons, resulting in a self-sus-

taining chain reaction. In this process, the fissioning of one gram of uranium 235 would release energy equivalent to burning about three million tons of coal.

The first nuclear reactor to achieve a controlled, self-sustaining chain reaction was developed under the leadership of the Italian physicist Enrico Fermi in 1942 at the University of Chicago. To increase the probability of fission in natural uranium, only 0.7 percent of which is uranium 235, and to prevent any chance of explosion, the neutrons were slowed down by collisions with carbon atoms in a graphite "moderator." It was necessary to assemble a large enough lattice of graphite (385 tons) and uranium (40 tons) to achieve a "critical mass" of fissile material in which the number of neutrons not escaping from the "pile" would be sufficient to sustain a chain reaction. Cadmium "control rods" were inserted to absorb neutrons during construction so that the chain reaction would not begin the instant the critical size was reached.

The uranium 235 isotope is the only natural material that can be used to produce nuclear energy directly. By early 1941, however, it was known that uranium 238 captures fast neutrons to produce the new element plutonium. Plutonium has a 24,000-year half-life and is fissile, so it can be used as a nuclear fuel. Plutonium can be "bred" in a uranium reactor from uranium 238 with excess neutrons from the fissioning of uranium 235. One other fissionable isotope, uranium 233, can be obtained by neutron capture from the thorium isotope thorium 232. Uranium 233 is a possible future nuclear fuel.

THERMAL REACTORS

The two basic types of reactor in use are thermal reactors, which use slow neutrons, and fast breeder reactors, which use fast neutrons to breed plutonium. Plutonium can be separated by chemical methods, but very expensive physical methods are necessary to separate uranium 235 from uranium 238, involving many stages of gaseous diffusion or centrifuge processes that can distinguish between their slightly different masses. Most nuclear reactors use 3.5 percent enriched uranium, but weapons-grade uranium is usually enriched to 93.5 percent. Fast breeder reactor fuel is generally enriched with plutonium.

Thermal reactors for generating useful power consist of a core containing a critical assembly of fissionable fuel elements surrounded by a moderator to slow the neutrons, a coolant to transfer heat, and movable control rods to absorb neutrons and establish the desired fission rate. Reactor fuel elements are made of natural or enriched uranium metal or oxide in the form of thin rods clad with a corrosion-resistant alloy of magnesium, zirconium, or stainless steel. Moderator materials must be low neutron absorbers with atomic mass close to the mass of neutrons so that they can slow them down by repeated collisions. Most reactors use moderators made of graphite, water, or heavy water, which contains the hydrogen isotope deuterium. Ordinary water is low in cost and doubles as a coolant, but it absorbs neutrons about one hundred times more than graphite and about one thousand times more than

heavy water. Coolants such as water, carbon dioxide, and helium transfer heat liberated by fission from the core, producing steam or hot gas to drive a turbine for generating electricity in the conventional manner. Control rods are made of high neutron absorbers, such as cadmium or boron, and can be adjusted for any desired power output.

LIGHT VS. HEAVY WATER REACTORS

Most reactors in the United States are "light water reactors" (LWRs); they use ordinary water as both moderator and coolant and require some fuel enrichment. Some are "pressurized water reactors" (PWRs), and some are "boiling water reactors" (BWRs). Most LWRs use 2 to 3 percent enriched uranium dioxide fuel elements clad in a zirconium alloy, although the PWR was first developed with much higher fuel enrichments for compact shipboard use. In a PWR, water is circulated at high pressure through the reactor core at above 300 degrees Celsius and then through a heat exchanger, where steam is produced in a secondary loop. In a BWR, the water is boiled in the core at about 280 degrees, eliminating the high cost of an external heat exchanger and highly pressurized containment vessel. LWRs have the fail-safe feature: If the temperature increases fast enough to expel water from the core, neutrons will be slowed less effectively, and the fission rate will decrease.

The Rancho Seco, California, nuclear reactor and cooling towers. (© William E. Ferguson)

Canada has specialized in heavy water reactors, since Canada has access to natural uranium with no need for fuel enrichment. In the Canadian deuterium-uranium system (CANDU), the heavy water coolant is circulated past fuel elements inside pressure tubes, which are surrounded by a heavy water moderator in a low-pressure tank. The coolant is pumped through a heat exchanger to boil ordinary water for driving steam turbines. Since 1968, several CANDU plants in the 200 to 700 megawatt range have been built in Canada, Argentina, India, Pakistan, and South Korea. Variants of this system employ light water or gas as a coolant to reduce the high cost of heavy water, but they may require enriched fuel.

FAST BREEDER REACTORS

The main alternative to thermal reactors are fast breeder reactors, which can obtain about fifty times as much energy from natural uranium by producing more plutonium from uranium 238 than the uranium 235 they use. Since neutron capture by uranium 238 requires fast neutrons (about one thousand times faster than thermal neutrons), no moderator can be used, and a 15 to 30 percent fuel enrichment is needed to sustain the chain reaction. A typical breeder core consists of a compact assembly of fuel rods with 20 percent plutonium and 80 percent depleted uranium (most uranium 235 is removed) oxides surrounded by a "blanket" of depleted uranium carbide to absorb neutrons and yield more plutonium. The "liquid metal fast breeder reactor" (LMFBR) uses sodium in liquid form (above 99 degrees Celsius) as a coolant, since water would slow the neutrons. Loss or interruption of sodium can lead to meltdown of the core, so some designs seal the core in a pool of sodium.

The first commercial fast breeder reactor began operating in the Soviet Union in 1972, producing 350 megawatts. France had the most advanced fast breeder reactor program, with its 1,200-megawatt Super Phénix breeder reactor. India has successfully converted thorium into uranium 233 and used it as fuel in the ICGAR fast breeder test reactor. In its pure form, thorium is a silver-white metal similar to uranium, but thorium metal is more stable in air, retaining its luster for months, whereas uranium quickly tarnishes. Since thorium is three to four times more abundant than uranium, the ability to use it in commercial reactors would greatly extend the nuclear fuel supply.

NUCLEAR FUSION

Most of the problems associated with fission power could be eliminated with nuclear fusion reactors. These problems include the handling, storage, and reprocessing of highly radioactive materials such as plutonium, the possible theft of such materials by terrorists, the disposal of radioactive waste products, the dangers of a reactor accident, and the limited availability of fission fuels. The fusion of hydrogen isotopes to produce helium releases energy comparable to fission but requires no critical mass of fuel that might cause meltdown, has many fewer radioactive products with no storage or disposal problems, and uses a fuel of almost unlimited supply. About 0.01 percent of the hydrogen in ocean water is in the form of deuterium. To overcome electrical repulsion and bring deuterium atoms close enough to cause a fusion reaction, an ignition temperature of about 100 million degrees is required. Ignition and isolation of such reactions require some kind of magnetic confinement of a plasma (ionized gas) or inertial confinement of deuterium pellets. Energy would be extracted by nuclear reactions in a surrounding lithium blanket caused by neutrons emitted during fusion. Some progress has been made in achieving these requirements, but a practical source of fusion power is many years away.

Experiments at several laboratories have suggested the possibility of room-temperature fusion by using electrolysis to draw deuterium ions into the crystal lattice of hydrogen-absorbing materials such as palladium or titanium. Because electrical repulsion between charged nuclei increases greatly as they approach each other, it is difficult to understand how this process could bring deuterium ions close enough for fusion to occur. Even if such experiments are confirmed and explained, the development of a reactor to produce electrical power with this technique may be difficult if not impossible. Hydrogen absorption declines sharply with increasing temperature, decreasing by a factor of at least ten as the temperature approaches the boiling point of water. Much higher temperatures would have to be produced for an efficient steam-driven electrical generator.

SAFETY OF NUCLEAR POWER

The study of reactor safety involves estimating the biological effects of radiation and analyzing the risk factors in possible reactor accidents. Information on the effects of large doses of radiation is based on medical X rays, animal experiments, and studies of Japanese atomic bomb survivors. Radiation doses are monitored by photographic film dosimeters and simple ionization chambers. Normal background radiation from radioactivity in the Earth, radon gas, and cosmic radiation is about double the average dose received by a person for medical purposes annually. The radioactivity from normal reactor operation is considerably less than background radiation. The major public concern focuses on accidental releases of large amounts of radioactivity. The Nuclear Regulatory Commission estimates the risk from a reactor accident at less than one death over its service lifetime.

The risks of a nuclear accident can be studied when one actually occurs. The most serious commercial reactor accident in the United States occurred in 1979 at the Three Mile Island power station in Pennsylvania; the loss of some coolant led to the shutdown of one reactor, as designed, but resulted in costly damage to the core. Because of containment structures, including a thick steel vessel around the core and a reinforced concrete building with walls several feet thick, the highest average dose released was about one-tenth the annual background radiation. A much more serious accident occurred at Chernobyl in the Soviet Union in 1986; a loss of coolant in a graphite reactor led to increased power followed by explosions and fire, killing thirty-one men. Approximately 270,000 people were evacuated from areas in the former Soviet Union where the average radiation from Chernobyl fallout ranged from 6 to 60 millisieverts (mSv). While these levels are above the average of 2.2 mSv for natural background radiation, they are no higher than occur naturally in some regions of Brazil, India, and China. According to the National Research Council report on the health effects of low levels of ionizing radiation, although an increased frequency of chromosome aberrations is found, no increase in the frequency of cancer has been documented in populations residing in areas of high natural background radiation.

The disposal of radioactive wastes is another area of concern and continuing study. Methods of solidifying such waste in glass or other materials for confinement in metal canisters and burial are being studied. The solid waste projected through the year 2010 would cover about 40,000 square meters (10 acres). Of several disposal sites under study, the most likely is deep underground burial in formations of salt or rock.

FUTURE OF NUCLEAR POWER

Nuclear power is an important source of low-pollution energy in spite of serious problems that have emerged since 1980. By the end of the 1990's, 19 percent of electrical energy in the United States was generated by 104 nuclear reactors producing nearly 100,000 megawatts of power, while 16 percent of electrical energy worldwide was generated by about 440 nuclear reactors. Using more than 63,000 metric tons of uranium in 1998, the reactors produced more than 350,000 megawatts of power. At this rate, known uranium reserves will last about fifty years. Reprocessing spent fuel and using the military's excess highly enriched uranium (HEU) may extend the supply for many years, and new reserves are still being discovered. The use of fast breeder reactors rather than thermal reactors would extend nuclear fuel supplies by a factor of fifty or sixty, while the use of thorium could add an additional factor of three or four.

The future of conventional nuclear power is uncertain. Since the Three Mile Island accident in 1979, public distrust of nuclear power has increased. The concern about reactor safety has led to new requirements that have raised the cost of nuclear power plants by a factor of five above inflation. Bankers and investors have become increasingly cautious about financing new plant construction. The high rate of government subsidies has complicated the evaluation of real dollar costs of nuclear power, but estimates indicate that the profit potential of conventional nuclear power is about half that of coal power. The total U.S. nuclear power capacity, operational and planned, slipped from 236 reactors in 1976 to 104 in 1999.

One promising approach to the safer production of nuclear fission power is the development of small-scale modular reactors that use tiny ceramic-coated fuel pellets in small enough quantities in their cores that meltdown is impossible. Although initially expensive, such units would not require expensive safety systems and could be built on an assembly line, producing one module at a time to match operating capacity with the demand for power. Using advanced technology, India, Japan, South Korea, Russia, and other nations are expanding their use of nuclear energy; in 1999, thirty-one new reactors were under construction, and sixty-seven more were on order or planned worldwide.

Joseph L. Spradley

CROSS-REFERENCES

Earth Resources, 1741; Future Resources, 1764; Geothermal Power, 1651; Hydroelectric Power, 1657; Ocean Power, 1669; Radioactive Decay, 532; Solar Power, 1674; Uranium Deposits, 1643; Wind Power, 1680.

BIBLIOGRAPHY

Blair, Ian. *Taming the Atom: Facing the Future with Nuclear Power.* Bristol, England: Adam Hilger, 1983. This book gives a very readable account of the development and use of nuclear power from a British perspective. The basic physical principles of nuclear power are covered, along with a good survey of the nuclear industry. An appendix on world nuclear reactors includes diagrams and data.

Byrne, John, and Steven M. Hoffman, eds. *Governing the Atom: The Politics of Risk.* New Brunswick, N.J.: Transaction, 1996. This volume from the Energy and Environmental Policy Series researches the environmental and social aspects of the nuclear industry. Careful attention is paid to government policies that have been implemented to monitor safety and accident prevention measures.

Cameron, I. R. *Nuclear Fission Reactors.* New York: Plenum Press, 1983. The first half of this book is a technical treatment of nuclear fission and reactor theory. The second half is a more readable survey of reactor types and a good discussion of safety and environmental aspects of reactors. The bibliography lists about 150 references to books and technical articles.

Cohen, Bernard L. *Before It's Too Late: A Scientist's Case for Nuclear Energy.* New York: Plenum Press, 1982. Aimed at the general reader, this book discusses problems in the public understanding of nuclear power. It covers the danger of radiation, the possibility and results of a meltdown accident, the problem of radioactive waste, and the assessment of risks. Each chapter lists many references.

Craig, J. R., D. J. Vaughan, and B. J. Skinner. *Resources of the Earth.* Englewood Cliffs, N.J.: Prentice-Hall, 1988. This introductory college textbook contains a chapter on nuclear power, the nuclear fuel cycle, reactor safety, and uranium mining. Other chapters on fossil fuels and environmental problems give useful comparative information.

Inglis, David R. *Nuclear Energy: Its Physics and Its Social Challenge.* Reading, Mass.: Addison-Wesley, 1973. This book is based on an introductory college course for general students. It explains the basic principles of nuclear energy and describes reactors and their radioactive products. Technical details and historical documents are given in several appendices.

Marion, J. B., and M. L. Roush. *Energy in Perspective.* New York: Academic Press, 1982. Written for a college survey course, this book has chapters on energy consumption, energy sources, nuclear power, and the effects of nuclear radiation. Contains many good diagrams, tables, and photographs.

O'Very, David P., Christopher E. Paine, and Dan W. Reicher, eds. *Controlling the Atom in the 21st Century.* Boulder, Colo.: Westview Press, 1994. This collection of essays focuses on the debates surrounding the legislation and licensing procedures involved in the operation of nuclear power plants.

Priest, Joseph. *Energy: Principles, Problems, Alternatives.* Reading, Mass.: Addison-Wesley, 1984. This textbook is designed for a college survey course on energy. It has chapters on nuclear fission power, breeder reactors, and fusion reactors. Contains interesting illustrations and tables, an appendix giving energy and consumption comparisons, and a glossary of terms.

Young, Warren. *Atomic Energy Costing*. Boston: Kluwer Academic Publishers, 1998. An economic analysis of the nuclear industry, focusing on cost effectiveness, costs of operation, and other economic aspects associated with the production of nuclear energy and the operation of nuclear power plants.

OCEAN POWER

Ocean power encompasses several distinctly different approaches to power generation which, if developed properly, promise potentially large amounts of clean, renewable energy. The importance of developing such alternate energy sources for a world largely dependent on fossil fuels possessing the drawbacks of escalating cost, ultimate exhaustibility, and environmental pollution cannot be overstressed.

PRINCIPAL TERMS

MARINE BIOMASS ENERGY CONVERSION: the cultivation of marine plants such as algaes for conversion of the harvest into synthetic natural gas and other end products

MARINE CURRENT ENERGY CONVERSION: power from the transfer of kinetic energy in major ocean currents into usable forms, such as electricity

OCEAN THERMAL ENERGY CONVERSION: power derived from taking advantage of the significant temperature differences found in some tropical seas between the surface and deeper waters

OCEAN WAVE POWER: the use of wind-generated ocean surface waves to propel various mechanical devices incorporated as an electrical generating system

SALINITY GRADIENT ENERGY CONVERSION: power generated by the passage of water masses having different salinities through a special, semipermeable membrane, taking advantage of osmotic pressure to operate turbines

TIDAL FLOW POWER: power from turbines sited in coastal areas to take advantage of the daily rising and ebbing tidal flow

TIDAL FLOW POWER

Attempts at harnessing ocean power involve the use of specialized technologies developed to exploit natural flows of energy within the marine environment. These energy flows are generated by the interaction of the ocean's waters with the effects of the Sun's energy; the gravitational pull of other celestial bodies, such as the Moon; and, to a much lesser extent, such influences as geothermal activity occurring on the sea bottom. Many engineering schemes have been devised to try to tap into each of these natural energy flows. All recognize the fact that the Earth's surface is mostly ocean—71 percent of it is covered by the sea—and that this immense fluid environment is always in motion in response to an interplay of natural processes.

The most ambitious, and at the time of this writing most productive, ocean power schemes have been tidal power projects. The efficacy of these projects is dependent on how well engineered they are to take advantage of the key factors involved. These factors include the character of a tide at a particular coastal locale as determined by local bottom topography, surface coastline geography, and the orientation of the coast to the open sea. Submarine topographic influences can accentuate the rise of an incoming tide, acting like a wedge to lift the oncoming bulge of tidal water. Thus, tides can reach up to 15 meters on coasts having the right tide-enhancing topography and orientation. The maximum rise and fall of a tide as experienced in a particular location is important in tidal power, as it represents the amount of usable head; "head" is a term used in hydraulic engineering to describe the difference in elevation existing between the level at which water can flow by gravity down from an upper to a lower level, thus making itself available to do work. Unfortunately, only some one hundred coastal sites worldwide are classified as having significant head and thus are optimal candidates for tidal power installations. Scientists conservatively estimate that the global, dissipated tidal power amounts to 3 terawatts per year, or 3 trillion watts. Of this amount, perhaps only 0.04 terawatt would ever be exploitable from feasible tidal power sites.

The French government has been the world pioneer in transforming tidal flow power engineering schemes into reality by constructing a large, functioning tidal energy station at the estuary of the La Rance River in Brittany in northern France. At La Rance, a combination of factors produces a useful hydraulic head and has proven itself economically profitable for several decades. Twenty-four 10,000-kilowatt turbine generators operate within conduits inside the tidal dam, turning at the low speed of 94 revolutions per minute. The turbine blades are designed to operate bidirectionally, in response to either an incoming or an outgoing tide. Thus the French plant exploits the free tidal water movement almost continuously.

OCEAN THERMAL ENERGY CONVERSION

Ranking next to tidal flow power in nearness to economic feasibility and actual, implementable technology is the ocean thermal energy conversion, or OTEC, approach. This method could ultimately be a very large-scale global operation, as many more sites exist that are usable for OTEC than are available for tidal power. Projected, theoretical limits to this energy source are in the neighborhood of 1 terawatt per year. OTEC involves either the construction of floating, open-ocean plants or coastal, land-based plants that exploit the temperature differences existing between water masses at varying depths in the tropical seas. Only small pilot plants have thus far been built, and they have never been run for more than short periods.

The optimal conditions for the most efficient OTEC sites have been calculated to be those where an 18-degree Celsius temperature difference exists between the surface and depths in the range of 600 to 1,000 meters. Regions of the ocean that meet or approach these thermal conditions have been identified as being, among others, Puerto Rico and the West Indies, the Gulf of Guinea, the Coral Sea, many of the Polynesian island groups, and the northwest African island groups. The actual process of converting the thermal difference found in such areas involves a system in which a turbine is turned by heat from the warm, surface water layer. The heat is transferred through devices termed heat exchangers, which introduce the thermal energy into a closed system. A working fluid such as ammonia, contained in sealed pipes, propels the system through the process of controlled convection. The warmed ammonia is heated to a vapor in an evaporator unit that drives the turbine. The used vapor is then conducted through a condenser unit where cold water drawn from below cools it down to a liquid state for another usage cycle. The cool water is brought into contact with the system by the deployment of a very long pipe, hundreds of meters in length, which projects down through the thermocline, or boundary, between the upper and lower water masses. Although the efficiency of the system is typically low, only 2-3 percent, the thermal reservoir is immense and is constantly replenished by the Sun. To make such projects attractive economically, they need only be built on a sufficiently large scale.

MARINE CURRENT ENERGY CONVERSION

Similar to OTEC in its use of the major oceanic flows of thermal energy are plans to exploit large-scale currents. One such current, the Gulf Stream—often called the Florida Current—constantly conveys many millions of gallons past a given point. One plan is a direct approach involving placing large turbines within the main flow of the current. Ideally, the generating site would be close to major electrical consumers such as coastal cities.

A good case can be made for implementing marine current energy conversion along the eastern coast of Florida. The city of Miami, a very large consumer of electrical power, is washed on its doorstep by the Florida Current. This current is estimated to carry approximately 30 million cubic meters of water per second past the city at a rate sometimes reaching 2.5 meters per second. One scheme would involve anchoring a large cluster of special, very slow-speed turbines or water windmills to the bottom that would ride midwater in controlled buoyancy. This complex would function at a depth ranging from 30 to 130 meters and stretch some 20 kilometers across the flow of the Florida Current directly adjacent to Miami. Estimates of the power output of this array are on the order of 1,000 megawatts, provided constantly on a 24-hour, year-round basis. It would extract roughly 4 percent of the usable kinetic energy of the Florida Current at this point in its flow, which is calculated to be in the range of 25,000 megawatts. In addition to the Florida Current, other strong currents exist worldwide that may also be good potential sites for future turbine arrays.

OCEAN WAVE POWER

Another way to take advantage of oceanic energy is to utilize the kinetic energy of surface waves directly to power mechanical devices used to generate electricity. This approach capitalizes on the fact that waves raise and lower buoyant objects. If the floating object is also long and perpendicular to the ocean surface, with most of its mass below the waterline, it is inherently stable and less subject to damage. Such a device is embodied in several variations of a wave-powered pump that is beginning to see practical economic applications. In one form, a large, vertical cylinder floats in the waves. Inside, the lower end is open to the lifting and sinking of the water level in unison with the wave motion. Because of its great length (tens of meters or more), it amplifies the wave motion in the column of air resting above the water. This motion propels air up and down through a double-flow electrical air turbine. As long as it remains in sufficiently deep water, it receives very little damage, no matter the magnitude of wave energy. A buoylike device floats at the cylinder top, keeping out wave splash and snugly maintaining inside air pressure. The cylinder can be anchored by flexible moorings to the sea bottom, and electrical cables can feed the power output to shore.

Coastal-based variants have also been built solidly into rocky cliffs; they respond to wave-driven air pressure that enters from a conduit at its base. An example of the coastal-based type is the plant at Tostestallen in Norway. Numerous rocky coasts worldwide could be similarly utilized for wave power generation.

MARINE BIOMASS ENERGY CONVERSION

Yet another form of ocean power currently in use today is marine biomass energy conversion. This method involves the harvesting of large tracts of plants such as marine algaes. Like other plants, algaes use the process of photosynthesis to convert the Sun's radiant energy into usable chemical energy. In the process, they manufacture various plant tissues that can be processed to yield useful compounds either for manufacturing ingredients or as fuel. In the case of algaes, the fuel would be

in the form of methane. Estimates suggest that 10,000 kilowatts of electrical energy could be recoverable as methane from 2.6 square kilometers of a marine alga known as kelp. Up to a million dollars worth of other useful products can be extracted from the same amount of kelp. In addition to kelp farming for fuel, shallow, coastal basins can also be constructed or modified from natural sites to serve as depositories for sewage waste. Seawater could be introduced to the wastes along with special algaes that would use the sewage as food. The end product would be similar, after harvesting, to the kelp-methane fuel. Thus, sewage would become a recycled fuel instead of an eyesore and a marine pollutant.

SALINITY GRADIENT ENERGY CONVERSION

A form of ocean power that has been envisioned but not implemented is the use of salinity gradients within the sea. This idea is still in the theoretical stage, because of the lack of key materials necessary for the effective technologies to work. A major drawback at this point is the lack of appropriately tough and efficient synthetic membranes necessary for this energy to be economically practical. Salinity gradient energy works on the principle of extracting energy from osmotic pressure by the use of semipermeable membranes. Such membranes would most likely be fabricated from some type of plastic possessing the correct, chemical properties and would take advantage of osmotic pressure between water masses having different percentages of dissolved salts. Influenced by osmotic pressure, less salty water will naturally flow through a semipermeable membrane to the side of greater salinity. The membrane would therefore be designed to be permeable to the fresher water but impermeable to the saltier water. Osmotic pressure would therefore propel a controlled, one-way flow that could be employed to propel water-driven turbines for electricity.

A PROMISING ENERGY SOURCE

Global human population growth is still explosively on the rise, especially in the Third World countries. In hand with population growth is increasing worldwide urbanization and industrialization. Because of these trends, consumption of energy to run manufacturing processes, transportation, food production, climate control, and communication systems has escalated dramatically. Coupled with this ravenous energy use are the byproducts of mounting energy consumption: widespread pollution and a general degradation of the world environment. The world industrial society already faces the specter of diminishing energy sources and the eventual, ultimate exhaustion of all fossil fuels. New energy sources, such as the nuclear option, seem to possess, so far, many drawbacks to widespread usage.

Renewable energy sources that are clean and have a low impact on the quality of the environment are the ideal long-term solutions. Some alternate energy forms fitting this description are either technically feasible today or almost so. The various forms of ocean power that have been actually developed or are in theoretical stages are excellent candidates for helping to alleviate some of the world's more pressing energy-related problems. For at least some geographic areas, ocean power is not only an efficient, clean power source but also economically competitive with fossil fuels or even nuclear power. Tidal flow power is an excellent example of the increasing viability of some forms of ocean power. By the end of the twentieth century, tidal power was producing less than 300 megawatts worldwide, although the largest site

(producing 240 megawatts) has been in operation at La Rance River in France since 1967. As of 1995, eight demonstration plants constituted 685 kilowatts of grid-connected wave-generating capacity worldwide. The most successful test OTEC system has been the one at Kailua-Kona, Hawaii, which has produced more than 50 kilowatts net power. Barring further improvements in technology, OTEC systems will not be competitive with fossil fuel until oil reaches a price of about $35 per barrel. However, the ocean is the Earth's largest solar collector, and supporters of ocean power remain optimistic. All forms of ocean power represent a way of living in harmony with the environment that still offers the option of maintaining a high level of technical civilization.

Frederick M. Surowiec

CROSS-REFERENCES

BIBLIOGRAPHY

Anderson, Greg M., and David A. Crerar. *Thermodynamics in Geochemistry: The Equilibrium Mode.* New York: Oxford University Press, 1993. An exploration of geochemistry and its relationship to thermodynamics and geothermometry. A thorough resource, but it can be somewhat technical at times. Recommended for the person with a background in chemistry and Earth sciences.

Bascom, Willard. *Waves and Beaches: The Dynamics of the Ocean Surface.* Garden City, N.Y.: Anchor Press/Doubleday, 1980. An excellent introduction to the subject of oceanography, this book emphasizes the role of wave and beach processes. A very useful chapter devoted to energy from marine sources is included which objectively presents the pros and cons of ocean power. Well illustrated throughout. Suitable for all readers wishing a working knowledge of the physical oceanographic processes involved.

Carr, Donald E. *Energy and the Earth Machine.* New York: W. W. Norton, 1976. A thorough survey of the primary energy sources that power the industrial world, including the fossil fuels. Ocean power sources are dealt with within the scope of water-derived energy sources in general. Suitable for high school students or anyone wishing a general background on the subject.

Constans, Jacques A. *Marine Sources of Energy.* Elmsford, N.Y.: Pergamon Press, 1980. An excellent treatment of the subject of ocean power, this book dwells on each subcategory in an easy-to-read and enlightening manner. Richly provided with explanatory diagrams, tables, drawings, and maps, which help explain the concepts involved. Appropriate for all readers, high school and above, especially those interested in the technical problems of ocean engineering.

Gage, Thomas E., and Richard Merrill, eds. *Energy Primer, Solar, Water, Wind, and Biofuels*. 2d ed. New York: Dell Publishing, 1978. Similar to hands-on books such as *The Whole Earth Catalog*, this source provides a wealth of information for those interested in the actual "nuts and bolts" involved in implementing alternate energy technologies. Suitable for those with an interest in the subject beyond the theoretical aspect.

Gashus, O. K., and T. J. Gray, eds. *Tidal Power*. New York: Plenum Press, 1972. One of the very best books devoted exclusively to the generation of power through tidal ocean flow, it includes details of the design and operating problems of the large-scale facility at La Rance, France, and the then-proposed facilities at the Bay of Fundy in Nova Scotia, Canada. Suitable for readers at the college level or above.

Hamblin, Kenneth W., and Eric H. Christiansen. *Earth's Dynamic Systems*. 8th ed. Upper Saddle River, N.J.: Prentice Hall, 1998. A college textbook, *Earth's Dynamic Systems* introduces the reader to basic concepts such as gravity, rotation, and the Earth's tides. The text is well illustrated and makes complicated processes understandable to those without a background in physical geology.

Lutgens, Frederick K., and Edward J. Tarbuck. *Earth: An Introduction to Physical Geology*. Upper Saddle River, N.J.: Prentice Hall, 1999. 6th ed. This college text provides a clear picture of the Earth's systems and processes that is suitable for the high school or college reader. In addition to its illustrations and graphics, it has an accompanying computer disc that is compatible with either Macintosh or Windows. Bibliography and index.

Meador, Roy. *Future Energy Alternatives*. Ann Arbor, Mich.: Ann Arbor Science Publishers, 1979. A well-written survey of all the major alternatives to fossil fuels, including the ocean power options. The author assumes no technical background and expertly introduces the general reader to each energy alternative in turn. An excellent book for the high school or college student as well as for the post-academic adult with a curiosity about energy alternatives.

Ross, David. *Power from the Waves*. New York: Oxford University Press, 1995. Ross writes of the potential source of power in the ocean, as well as the electric power that is produced from the ocean by power plants. Illustrations, maps, bibliography, and index.

Teller, Edward. *Energy from Heaven and Earth*. San Francisco: W. H. Freeman, 1979. A good comparison of the major forms of energy used by the industrialized world and the social and economic consequences of their use, it also offers extensive discussions of energy-use policies and their bearing on the development of new energy sources. A worthwhile introduction to the overall energy picture for students or interested laypersons.

Wilhelm, Helmut, Walter Zuern, Hans-Georg Wenzel, et al., eds. *Tidal Phenomena*. Berlin: Springer, 1997. A collection of lectures from leaders in the fields of Earth sciences and oceanography, *Tidal Phenomena* examines Earth's tides and atmospheric circulation. Complete with illustrations and bibliographical references, this book can be understood by someone without a strong knowledge of the Earth sciences.

Wilson, Mitchell, and the editors of *Life. Energy*. New York: Time, 1963. A profusely illustrated overview of the subject of energy. An easy-to-read text outlines the history of energy-related physics and the growth of applied technology designed to exploit energy sources. Suitable for any reader at the high school level or beyond.

SOLAR POWER

Solar power is seemingly boundless, but it has yet to establish itself as a major energy resource for modern civilization. A variety of technologies exist or are under development to harness that power. Many of these technologies work very well; the problem with all of them is cost.

PRINCIPAL TERMS

BARREL: a measure of energy consumption, 1 barrel is equal to the energy in an average barrel (42 gallons) of crude oil; about 5.8 million Btu

BTU (BRITISH THERMAL UNIT): the amount of heat necessary to raise the temperature of 1 pound of water 1 degree Fahrenheit; equivalent to about 0.25 Calorie

INSOLATION: the radiation from the Sun received by a surface; generally expressed in terms of power per unit area, such as watts per square meter

KILOJOULE: a unit of electrical energy equivalent to the work done to raise a current of electricity flowing at 1,000 amperes for 1 sec-

ond (1,000 coulombs) by 1 volt; equivalent to 4.184 Calories

KILOWATT: 1,000 watts, or about one and one-third horsepower

MEGAWATT: 1 million watts, or about 1,340 horsepower

PHOTOLYTICS: the technology that makes use of sunlight's ability to alter chemical compounds in ways that can produce energy, fuels, or both

PHOTOVOLTAICS: the technology employed to convert radiant solar energy directly into an electric current, using devices called solar cells

QUAD: 1 quadrillion Btu; equivalent to 8 billion barrels of gasoline

INSOLATION

The Sun's hydrogen fusion reaction (the same as humankind's hydrogen bomb) produces a relatively steady and uninterrupted 380 billion kilowatts of energy. This energy is liberated into space at the solar surface, where scientists measure a power density of over 75,000 horsepower per square meter. Spreading out radially as it travels outward, this power is greatly reduced by distance. By the time it reaches the Earth, 1 square meter delivers less than 1.5 horsepower. Still, the total of the Sun's energy received on the planet is about 85 trillion kilowatts. The portion that falls on the United States amounts to five hundred times more energy than the nation's total need. If the United States could tap that with only 10 percent efficiency, it could meet all of its energy requirements from what falls on merely 2 percent of its surface, and the opportunity is similarly great for most other countries.

Although the human inhabitants of Earth are only beginning to learn how to harness this power, other life-forms have been using it for at least 2

billion years. The solar energy consumed by plants through photosynthesis is enormous. Within the United States alone, vegetation uses solar energy equivalent to at least 20 million barrels of oil per day. Solar energy is also the power that drives the Earth's weather and stirs the oceans' currents; therefore, any utilization of wind power, ocean thermal power, or the energy stored in plants (whether burning firewood or fuels distilled from biomass) is indirectly an application of solar power. For that matter, the fossil fuels represent solar energy stored in plants and animals that lived hundreds of millions of years ago, but such broad interpretations of solar power have little operational meaning; here, solar power will mean insolation, the radiant energy of the Sun as it falls on the Earth's surface.

CHALLENGES TO UTILIZATION

The practical application of solar power to modern civilization's energy requirements presents some difficult challenges. One problem is that solar energy, although abundant and widely

1674

distributed, is not concentrated, whereas humankind's energy demands tend to require a considerable amount of power brought to bear at very confined locations. Virtually all of North America south of the 48th parallel receives more than 12,000 kilojoules of solar energy per square meter of its surface each day, with large areas of the West and South enjoying more than 16,000 kilojoules. Yet, it takes about 55,000 kilojoules to heat the domestic hot water required by a family of four; this power application alone theoretically requires 3-4 square meters of collection device to satisfy one small family's demand for water with which to cook, wash, and bathe. In actuality, it requires closer to 10 square meters, because the collection devices operate at far below 100 percent efficiency.

Another serious problem in utilizing solar power is the pattern of its availability. Solar power at a given site peaks every day as the Sun passes overhead and disappears entirely during the hours of darkness. Seasonal variations in the location of the Sun affect the number of hours of energy received, and as distance from the tropics increases, the incoming sunlight strikes at ever more acute angles, further diminishing the power available per square meter.

Finally, cloud cover seriously interferes with solar energy reception, making overcast days almost as unproductive of energy as nighttime. The average number of cloudy days per year at a given location is a factor as significant as latitude. For example, sunny, equatorial Kenya enjoys almost twice the insolation as cloudy Nigeria, only 6 degrees of latitude away. Relatively sunny Washington, D.C., at 46 degrees north latitude, has more insolation to work with than Nigeria and almost twice the amount available to the city of Tomei, Japan, which, though 14 degrees closer to the equator, is cloudy much of the year.

These problems notwithstanding, it has been demonstrated that solar power can be successfully harnessed over a wide area of the surface of the Earth. Various schemes to collect and concentrate the energy and to manage its utilization and storage to help match availability to demand are already at a practical stage of development. Moreover, public opinion is strongly favorable toward solar power as an alternative to the continued use of fossil fuels or large-scale development of nu-

clear power. Experts caution, however, that widespread utilization of solar power is still in the future.

THERMODYNAMIC AND PHOTOVOLTAIC POTENTIALS

The essence of the energy problem facing the world is not a shortage of energy but the cost of deriving and distributing "high-quality" energy from abundant but "low-quality" renewable sources. Quality, as traditionally applied to energy, means temperature first and foremost. The value of a unit of energy depends greatly on the temperature at which it can be delivered because, under the laws of thermodynamics, a unit at high temperature can do more useful work than a low-temperature unit. Unconcentrated, just as it falls upon the Earth's surface, solar energy is considered a low-quality thermal resource: It is fine for making "sun tea" and heating swimming pools, but it will neither spin the wheels of industry nor light the nights of a modern, industrialized society. The fact that it is pure, clean energy creating no harmful by-products for the environment is of little consequence if it cannot do the work required of it.

Fortunately for the future, solar power can be collected and concentrated to raise its thermal quality. In fact, solar energy's thermodynamic potential (the highest heat at which it can be realistically supplied) is about 5,000 degrees Celsius, which is 3,000 degrees higher than the thermodynamic potential of conventional nuclear power and equal to the theoretical maximum heat available from the complete combustion of pure carbon in pure oxygen. Moreover, solar radiation is able to stimulate physical effects that have nothing to do with heat. The most important of these to modern civilization is its photovoltaic potential, that is, its ability to transfer electrons in various semiconductor materials, thereby creating an electrical current. Sunlight is also capable of splitting water molecules into hydrogen and oxygen, a fact of enormous significance to more advanced power technologies.

THERMAL SYSTEMS

Solar power technologies already available are quite varied but can be grouped generally as thermal, photovoltaic, and photolytic. Thermal schemes are those best known and understood by the average consumer, because they apply solar power to energy needs in traditional ways. Photolytic technology is still largely in the future.

Thermal systems that require no mechanisms are termed passive designs. They rely on the choice of materials and the size and careful placement of building elements to control the flow of heat within the living space. Typically, large, south-facing glass walls admit solar heat in winter but are shadowed by deep roof overhangs in summer. Interior design facilitates the natural convective flow of warm and cold air and the storage of energy. In favorable locations, good passive thermal architecture can provide as much heating and cooling as an efficient building needs.

"Flat plate" collectors used to heat water are an excellent example of active solar thermal technology. Water is pumped through piping inside a collector designed to gather and hold the solar heat that strikes it. As the water circulates through the collector, its temperature may be raised to as much as 100 degrees Celsius. Such systems most commonly supply domestic hot water but may also be used for space heating and cooling of residences and larger buildings. (Refrigerated cooling with solar-heated water involves a process different from mechanical refrigeration but is not a new idea. Absorption refrigeration, as it is called, was used in early refrigerators.) Other active solar thermal systems for space heating and cooling use the flow of water over building surfaces to add or subtract heat as needed.

Another application for active solar thermal power is in so-called heat engines, which use the expansion, contraction, or evaporation of a fluid or gas to obtain mechanical motion. A number of designs exist, dating back to the early eighteenth century, but all have suffered from low efficiency until now, when they are being interfaced with advanced thermal collectors whose much higher

These photovoltaic silicon cells, located at Natural Bridges National Monument, Utah, are used to generate electric energy from sunlight. (© William E. Ferguson)

temperatures provide the efficiency needed, and they promise to be increasingly important in both terrestrial and space applications.

Also requiring much higher temperatures (from 250 to more than 500 degrees Celsius) are schemes that use the Sun to create superheated steam to operate turbine-powered generators. "Solar farms," such as a 194-megawatt plant in the Mojave Desert 225 kilometers northeast of Los Angeles, use parabolically shaped troughs lined with mirror strips to produce temperatures of 375 degrees Celsius. A working fluid is pumped through the collectors and then to a central heat exchanger to create steam for a generating plant. Another approach, first introduced in Italy, involves reflecting the insolation captured by an array of dozens or hundreds of mirrors to a central "power tower," within which the sum of all the reflected insolation produces operating temperatures in excess of 1,000 degrees Celsius.

PHOTOVOLTAICS

Photovoltaics, the direct generation of electricity from sunlight, was pioneered in the 1950's as an offshoot of semiconductor research. As the space age dawned, devices called solar cells became familiar as the power sources for many satellites. Their very high cost and low efficiency were of little consequence in these applications, as alternatives were either even more expensive or nonexistent. Their high cost stemmed partly from the fact that the silicon disks had to be obtained from ultrathin slices of pure silicon crystal, laboriously grown in laboratories to avoid contamination. Even the slicing operation added significantly to the cost, because half of every crystal produced was consumed by the saw cuts. Low efficiencies compounded the problem, as a great many of the expensive cells were needed to provide even minimal current. Yet, persistent research into new materials and techniques has been changing this gloomy picture. In fact, photovoltaic power generation, though not yet economically competitive with traditional energy sources, has dropped so much in cost that a few major utility companies have put photovoltaic generating stations on line to supplement power from other sources and pave the way for more widespread use in years to come. In addition, more than 10,000 rural residences have their own photovoltaic generating plants, and any home located nearly a kilometer from a power line may find the cost of photovoltaic energy is less than the cost of connecting to the power grid.

A more visionary application for photovoltaics is the Satellite Solar Power Station (SSPS), for which designs but no serious implementation plans exist. Such a satellite would be a gigantic platform in geosynchronous orbit, giving large arrays of solar cells twenty-four-hour-a-day exposure to solar energy undiluted by passage through the atmosphere. The power would be beamed back to receiving sites on the Earth as microwaves and there reconverted to electricity.

PHOTOLYTIC TECHNOLOGY

The most challenging of all solar energy technologies may be ultimately the most rewarding. Research has shown that a small percentage of photons from the Sun are energetic enough to split water molecules, from which three atoms—two hydrogen and one oxygen—are liberated. If this process becomes feasible on a large scale, the oxygen and hydrogen could be readily stored and later either burned as fuels or recombined to water in a reaction that releases electricity. Again, the space age has shown the way through development of "fuel cells," which combine hydrogen and oxygen with almost 100 percent efficiency, yielding significant amounts of fresh water and electric-

ity in the process. If photolysis, as it is called, could draw upon the abundant salt water of the Earth's oceans to produce hydrogen, oxygen, and electricity, with fresh water as an added benefit, humanity's quest for acceptable energy sources would move beyond "renewables" to involve environmentally enhancing power.

COMPETITIVENESS WITH OTHER ENERGY SOURCES

The energy crisis precipitated by the Arab oil embargo of 1973 forced the industrialized nations of the world to make major adjustments in energy use and policy. Through the Energy Research and Development Administration (ERDA) and the Solar Energy Research Institute (SERI), both created in 1974, the United States began a program of government-sponsored research designed to help meet an announced goal of 10 quads of energy from solar power by the year 2000, enough to replace about 5 million barrels of oil daily. The oil embargo, however, also stimulated a sharp reduction in the growth of demand for energy. After years of steady 6 percent per year increases, new energy construction almost came to a standstill. On top of this, oil prices fell sharply again in the 1980's, and environmental constraints on the use of fossil fuels were eased. The net effect was a loss of urgency in the quest for solar power, and federal assistance for research and applications was cut to a small fraction of previous levels. Meanwhile, serious problems developed in the fledgling solar power industry as complaints about inflated pricing and exaggerated savings undermined public confidence and legislative support. A large number of the young solar energy ventures failed between 1980 and 1987.

Because solar heating systems have high initial costs (a large active space heating system costs up to 20 percent of the total cost of a residence), because building contractors tend to be conservative, and because fossil fuels remained relatively inexpensive during the 1990's, solar energy has not yet made a significant impact on energy consumption. Since approximately one-quarter of the United States' annual energy consumption is for heating and cooling, solar energy could have a significant impact. In most regions of the United States, solar energy can best be employed for passive space heating and domestic hot water systems. Active space heating systems are feasible only in areas where the winters are cold and sunny, such as Colorado and Utah. As fossil fuels become depleted and their prices continue to rise in the twenty-first century, however, solar power is expected to become an increasingly viable energy resource.

Richard S. Knapp

CROSS-REFERENCES

Atmosphere's Global Circulation, 1823; Beaches and Coastal Processes, 2302; Climate, 1902; Deep Ocean Currents, 2107; Earth Resources, 1741; Future Resources, 1764; Geothermal Power, 1651; Heat Sources and Heat Flow, 49; Hydroelectric Power, 1657; Nuclear Power, 1663; Ocean Power, 1669; Ocean Waves, 2139; Ocean-Atmosphere Interactions, 2123; Sea Level, 2156; Seawater Composition, 2166; Storms, 1956; Surface Ocean Currents, 2171; Wind, 1996; Wind Power, 1680.

BIBLIOGRAPHY

Behrman, Daniel. *Solar Energy: The Awakening Science.* Boston: Little, Brown, 1976. At nearly four hundred pages, a more detailed discussion of the potential for solar power than any of the other references that are cited. It is written, however, in a popular and anecdotal style that includes attention to the personalities of leading solar proponents, as well as their ideas. Illustrated with photographs and contains a good index.

Brinkworth, B. J. *Solar Energy for Man.* New York: John Wiley & Sons, 1973. Knowledge of the basic principles of solar power has been available for a long time. This 250-page book, which preceded the advent of the "energy crisis," is useful and recommended to the extent that it gives a clear explanation of those principles. Well illustrated and suitable for readers with some grounding in general science.

Camacho, E. F., Manuel Berenguel, and Francisco R. Rubio. *Advanced Control of Solar Plants.* New York: Springer, 1997. A thorough examination of solar power plants and their operation, and of safety systems that govern them. Illustrations, index, and bibliographical references.

D'Alessandro, Bill. "Dark Days for Solar." *Sierra Magazine* (July/August, 1987): 34. Details the profound damage done to America's emerging solar power industry in the 1980's by the combination of low-cost oil, elimination of tax credits, loss of government funding for research, and the industry's own bad business practices.

Economic Commission for Europe. *Solar Power Systems.* New York: United Nations, 1993. First presented at the Seminar for Solar Power Systems held in 1991, this collection of technical papers looks at the international use of solar power and the operation of solar power plants worldwide. Intended for the college reader with background knowledge of solar energy.

Glaser, Peter Edward, Frank Paul Davidson, Katinka I. Csigi, et al., eds. *Solar Power Satellites: The Emerging Energy Option.* New York: E. Howard, 1993. This book is one volume in a space science and space technology series. This volume discusses the solar energy option, its possibilities, and future uses. A good introduction to alternative energy sources.

Halacy, D. S. *Earth, Water, Wind, and Sun: Our Energy Alternatives.* New York: Harper & Row, 1977. Readers wishing a nontechnical overview of the possibilities and problems of solar power may find the twenty-five pages devoted to solar energy helpful. Also recommended for high school readers and for those wanting a brief survey of the field of renewable energy sources.

Landsberg, Hans H., et al. *Energy: The Next Twenty Years.* Cambridge, Mass.: Ballinger, 1979. This report of a study group funded by the Ford Foundation devotes one chapter to solar energy, with primary emphasis on policy recommendations intended to advance its development at economically appropriate rates in the short term while laying the groundwork for it to become a fundamental option in the twenty-first century. As such, it is more of a "why to" than a "how to" look at solar power.

Mathews, Jay. "Solar Energy Complex Hailed as Beacon for Utility Innovation." *The Washington Post* (March 2, 1989): A25. This six-column, half-page article details the stunningly successful introduction of a large commercial solar generating plant in California, which sprang up just as most private ventures in solar power were teetering on the brink of disaster.

Patel, Mukund R. *Wind and Solar Power Systems.* Boca Raton, Fla.: CRC Press, 1999. Patel provides a thorough examination and clear explanation of wind and solar energy plants. Chapters discuss their operations, protocol, and applications. Suitable for the reader without prior knowledge of alternative energy sources. Illustrations, maps, index, and bibliography.

WIND POWER

Power-generating devices that use wind as an energy source are the oldest known devices engineered specifically to produce power. From ancient grain mills to today's complex wind turbines, the wind has provided humanity with a power source. As the world's demand for energy increases at an accelerating rate, the use of wind power as a legitimate alternative energy source will also increase.

PRINCIPAL TERMS

AERODYNAMIC LIFT: a measure of the degree of wind force acting on a rotor, which is translated into power to a generator in a wind-power machine

AXIS SYSTEM: the vertical or horizontal orientation of the power shaft of a wind-power machine

POWER FLUX: the amount of energy that can be obtained from a cross-sectional area of the wind at a given velocity

POWER GRID: the mechanical power distribution system of a social structure and/or its independent power distribution systems

ROTOR: the bladed system on a wind-power machine that is set into motion by the wind and is translated into power to a generator

STEP-UP GEARING: a gear system that increases the revolutions of the downstream shaft (generator) over the revolutions of the upstream shaft (rotor)

EARLY USES

A sailboat was the first device designed to use wind power to replace human energy; sailing craft have been in use since prehistoric times. It was obvious long before the development of science and engineering that there was real, usable energy to be captured in the wind. If the wind could push huge sailing craft over the known oceans of the world, then it could perform other tasks as well. A full thousand years after the first sailing craft, some two millennia before the birth of Christ, the first wind systems were planned for use in Babylonia to pump water through an irrigation system, although there is no evidence that they were ever built. The construction techniques to bring such systems about did not fully materialize until about A.D. 1000, when wind power was used widely by Arabic nations.

The first widespread use of the wind for power was to grind grain, a task usually reserved for slaves or harried, overworked domestics. It was discovered that the huge millstones, some weighing more than a thousand kilograms, could be turned by the wind, hence the name "windmill." Attendant to that discovery was the one that the wind was equally good at pumping water, and wind-powered pumps were put to work lifting wa-

ter from wells and irrigation canals and (of some historical notoriety in Holland) to drain fields. Just before the beginning of the Industrial Revolution, wind-powered devices were in use across the world, being employed to pump water, grind grain, saw lumber, and even turn carousels. Yet, the limits of using the wind for power were well known and, typically, its use was vetoed in favor of more reliable sources of energy, when they were available. Water power, for example, which operated with few interruptions, was usually more consistent than was wind power. Water-powered devices were generally smaller and therefore required less investment in equipment. Also, because the wind shifts direction and velocity, windmills needed to be engineered to operate under a wide range of conditions. Nevertheless, even with its limitations, in many cases, wind was the only choice for power, and tens of thousands of wind-powered systems were constructed around the world.

REEVALUATION AS ENERGY SOURCE

Just before the turn of the nineteenth century came the discovery of new and widespread uses for electricity. One of those devices was the electrical generator. Not long after, Moses G. Farmer was issued a patent for a wind-powered generator. After four thousand years of providing power for humankind, the wind had finally become linked with electricity. Yet, with the onset of the Industrial Revolution, steam power became the energy means of choice. Soon thereafter, petroleum, cheaply priced oil, and internal combustion engines all but drove the use of wind power to extinction.

Beginning in the early 1970's, wind power was recognized as a potentially important world power source. The price of the world's energy had increased when the Organization of Petroleum Exporting Countries (OPEC) forced world oil prices up from $3 to $32 a barrel in seven years. Most nations of the West realized how dependent they were on foreign energy sources and, concomitantly, how closely energy prices were linked to the economic health of nations. Indeed, the free and abundant flow of cheap energy was recognized as directly related to the economic vitality and well-being of any country. Therefore, measures were taken to reduce the dependency on foreign energy sources. The most significant step taken was the encouragement of energy conservation, and second to it was the improvement and development of alternate energy sources. The wind was one of those alternate sources, along with solar energy, hydroelectric power, synthetic fuels, and ocean and geothermal energy—the so-called alternative-renewable energy sources.

In the United States, President Jimmy Carter declared the "moral equivalent to war" against the energy crisis in the late 1970's. Congress responded and passed the Wind Energy Systems Act of 1980, which would provide an eight-year, $900-million program to develop cost-effective wind power systems across the United States. California responded immediately and launched a full-fledged crusade to harness the power of the wind. By 1990, wind power contributed 1 percent of California's energy needs. This number represented more than 1,000 megawatts (million watts), about the output of a single average-sized nuclear reactor. That power, which amounted to three-quarters of the world's total wind energy output, could supply 400,000 households and save 4 million barrels of oil each year. California's example is being mirrored by Hawaii, which is harnessing the constant trade winds. Yet the California experience represents only a fraction of the possible wind-generating power capacity in the United States.

Some of the best sites are in North and South Dakota, Montana, and Wyoming. Since they lie far from population centers, power distribution costs would be high, and so it may never be practical to fully develop them for that reason alone. However, if their wind-generating capacity were fully developed, these states alone might produce 20 percent of the electricity needed in the United States.

Because of supportive government policies in several European countries and because of technical improvements, wind energy became the fastest-growing energy source during the latter part of the 1990's. By 1998, the cost of wind-generating capacity had dropped to one-third of what it was in 1981 and thereby became more competitive with other energy sources. In 1995, the worldwide wind energy generating capacity stood at 4,990 megawatts, but by the end of 1998 it had increased to 9,600 megawatts. A 1998 Danish study concluded that wind energy may provide 10 percent of the world's electricity by the 2020's, and the Worldwatch Institute optimistically predicted that wind-generated power would eventually exceed hydroelectric power, which supplied more than 20 percent of the world's electricity in 1999.

WIND-POWERED GENERATORS

The laws of physics determine the capabilities and efficiencies of wind-powered generators. The physical realities of wind power are these: The wind force varies with the square of its velocity; the power, however, varies with the cube of the velocity; the wind's ability to do work is limited by its flux, or the amount of energy that can be created by a cross-sectional area of wind traveling at a given velocity; and a perfect system can extract only 60 percent of the total wind power available. Realistically, after electrical conversion systems and (in some systems) storage, the conversion efficiency drops to between 20 and 50 percent. Therefore, one needs to garner quite a large area of wind to produce significant power. That equates directly into a very large wind turbine required to produce a relative modicum of electrical power.

A typical wind generator sits atop a tower about 35 meters off the ground and sports a blade approximately 60 or 70 meters in length. This device will generate about 300-500 kilowatts of power. Wind farms may occupy hundreds of acres, popu-

A wind farm with giant turbines at Altamira, California. (PhotoDisc)

lated with many (perhaps several dozen) of these wind turbines. Yet, such a farm will produce only a fraction of the power from a typical nuclear or conventional (oil-fired) power plant. By 1990, California's optimal wind generation capacity, consisting of more than a thousand generators over thousands of acres, producing over 75 percent of the world's total wind power, can at best only provide as much energy as a single nuclear reactor such as the one at Three Mile Island, Pennsylvania.

The typical wind generator consists of a tower, a generator, gears, the rotor, the axis system, and speed control. Some wind generators also have batteries for power storage and inserters to convert power states. Wind generators must be placed on towers to enable them to capture optimal wind states. Generators placed too close to the ground will be adversely influenced by ground obstructions to a steady wind flow, such as buildings, hills, and trees. Obviously, the tower must also be high enough to allow clearance for large, efficient rotors. The rotors (also known as propellers, blades,

or turbines) of the modern wind generator are quite unlike the windmills of the past. They also come in many forms, depending on the task they are required to perform. The generator task itself is principally determined by the kind of generator mounted on the tower. Typically, the rotors are based on one of two designs. The most common wind-power-generator design in use is the horizontal axis rotor system. In this system, the shaft of the rotor is aligned horizontally with the ground, with the rotor blades mounted perpendicular to the rotor shaft. This design requires that the generator be mounted at the top of the tower. The disadvantages are obvious: Installation and maintenance must be done high above the ground. These rotors look much like aircraft propellers—in fact, they are designed along some of the same aerodynamic principles. The other type of design is called the vertical axis system. In this design, the shaft is mounted vertically with the rotors placed alongside the shaft. The rotor designs for these wind generators are not propeller types; one design looks like an eggbeater and is called exactly that. In these designs, the wind rotates the rotor and shaft, which are mounted vertically. That allows the generator to be placed at the bottom of the assembly on the ground. Often there are no requirements for a tower to be constructed. Other vertically mounted designs are called paddle vanes and s-rotors.

REFINEMENTS TO GENERATORS

With few exceptions, in wind-power systems, the wind cannot turn the generator shaft as fast as it needs to rotate to generate power. Therefore, gear systems are required to "step up" the system from the turning rotor to the generator. A typical blade rotation speed of 200-400 rotations per minute (RPM) must be geared up to 1,800-3,600 RPM for the generator to deliver adequate power. These gearing systems require more force be delivered to the turning rotor than if there were no gears between the rotor and the generator shaft. Consequently, that limits the minimum amount of wind necessary to turn the rotor and deliver power from the generator.

The wind-power generator must operate in a very wide range of conditions, considering a constantly variable wind speed and direction. To deliver a consistent energy output, the generator shaft should ideally turn at a more or less constant RPM. In most commercial generators, that is accomplished in a variety of ways. The pitch of the rotor is selectively changed along the entire length, or the rotor is selectively pitched at the tip. Some rotors have flaps, much like an aircraft wing, to change the degree of aerodynamic lift provided by the rotor. The degree of aerodynamic lift ultimately determines the power output of the rotor. Some generators also change the angle at which the system is facing into the wind in order to increase or decrease the amount of impinging wind energy. As important as it is to maintain a maximum effective energy output by regulation, it is also necessary to waste energy by the same mechanisms if the wind speed is greater than the system design can tolerate. There are also systemic control mechanisms relating to aspects other than speed control. Among them are braking systems to slow or stop the generator in the event of dangerous winds or even system failure. Wind systems also need to be shut down if the wind speed is too low, a condition that can deliver unacceptably low or inconsistent power to the regulatory mechanisms.

Wind systems have been designed for home use, and there are many homes around the world whose sole or partial power source is wind power. These homes rely on small wind-power-generation systems nearly identical to their commercial counterparts. Typically, they rely on direct current (DC) generators that feed banks of lead acid batteries storing the power from the wind generator.

Wind-turbine electric generators along Interstate 10 near Palm Springs, California. (AP/Wide World Photos)

That allows for a constant supply of power to the home even when the wind is not blowing. These homes use direct current for many purposes, and many have special DC appliances. Some appliances do not operate on DC, so these homes have devices called inverters, which change direct current to alternating current (AC). During the storage and conversion process, however, there is energy loss so that the wind-power system becomes less efficient at each stage. The typical cost of equipping a home with a wind-power and storage system is about that of a new automobile.

DISADVANTAGES AS AN ENERGY SOURCE

The same problems associated with wind-power generation encountered by prehistoric humans still exist. Generating power from the flow of wind is fraught with myriad difficulties. Consistency and direction have already been mentioned. In addition, storage of wind power is necessary for use in windless conditions or at times when the wind is so strong that the generators have to be shut down for their protection. In a large utility system, the power demand varies considerably from hour to hour and from day to day; therefore, the utility must maintain backup generating capacity. Pacific Gas and Electric gets up to 7 percent of its power from California wind turbines and has had no problem adjusting to their varying power output. Environmental concerns also plague the use of modern wind-power devices. Whereas many environmentalists applaud the benign and clean nature of wind-power generation, many complain that the devices themselves are a blight on the landscape. Unlike Holland's beautiful and architecturally pleasing windmills, the monstrous wind generators that populate the wind farms of today are designed, built, and erected for sheerly mechanical purposes and have little to offer in the way of aesthetics.

ADVANTAGES AS AN ENERGY SOURCE

Although wind power will never replace other, higher level energy sources such as petroleum, coal, or nuclear fission or fusion energy, it remains an important energy source. The relatively low level of technology required is somewhat offset by the initial cost of the equipment. Yet, if wind energy is utilized efficiently by public utilities, it clearly offers a great number of social advantages. It is renewable and nonpolluting, and it is available to a variable degree nearly everywhere. The equipment demands few further technical advances in order to be mass-produced, and the cost per unit would then decline dramatically. The additional development of standard home interfaces would effectively decrease domestic use of other, nonrenewable energy resources, although a technological advance in power storage technology would probably be required. A combination of locally available energy such as wind and solar power with maximal use of conservation techniques could result in a decline or stabilization in the energy requirement by the private sector of the economy. As the free flow and abundance of energy is the life's blood of any economy, such developments serve to benefit society as a whole.

Dennis Chamberland

CROSS-REFERENCES

Air Pollution, 1809; Climate, 1902; Coal, 1555; Geothermal Power, 1651; Hydroelectric Power, 1657; Hydrologic Cycle, 2045; Nuclear Power, 1663; Nuclear Waste Disposal, 1791; Ocean Power, 1669; Oil and Gas Distribution, 1694; Oil and Gas Origins, 1704; Solar Power, 1674; Strategic Resources, 1796.

BIBLIOGRAPHY

Clark, Wilson. *Energy for Survival.* Garden City, N.Y.: Doubleday, 1974. This book discusses all energy alternatives at the disposal of the United States. This widely referenced book includes a very detailed discussion of the use of wind energy in its various forms. Illustrated and indexed.

Gipe, Paul. *Wind Energy Comes of Age.* New York: John Wiley, 1995. A look at the evolution of wind power and the procedures and protocol involved in the conversion process. A good introduction for the nonexperienced reader. Illustrations, index, and bibliography.

Marier, Donald. *Wind Power for the Homeowner.* Emmaus, Pa.: Rodale Press, 1981. The definitive text for homeowners wanting to install their own wind-energy systems. Details the practical and technical side of installation of wind-energy systems. Also discusses cost and legal issues. Well illustrated. Includes relevant tables, references, and addresses of equipment manufacturers.

Naisbitt, John. *Megatrends: Ten New Directions Transforming Our Lives.* New York: Warner Books, 1984. This work details ten directions that are transforming the lives of Americans, from the evolving trends in technology to politics and the economy. Wind power, per se, is not addressed in this book. Yet, the significance of energy and power sources is discussed. This book is vital to understanding the dynamics of a constantly shifting social structure and the role that world energy sources play in it.

Patel, Mukund R. *Wind and Solar Power Systems.* Boca Raton, Fla.: CRC Press, 1999. Patel provides a thorough examination and clear explanation of wind and solar energy plants. Chapters discuss their operations, protocol, and applications. Suitable for the reader without prior knowledge of alternative energy sources. Illustrations, maps, index, and bibliography.

Torrey, Volta. *Wind-Catchers: American Windmills of Yesterday and Tomorrow.* Brattleboro, Vt.: Stephen Greene Press, 1976. This book is a detailed, meticulous recounting of the windmill (a broad category in which the author includes wind generators) throughout world history and its ultimate infiltration into the United States. Yet, the author retains a sharp eye for the scientific as well as the more fanciful historical details. Covers futuristic wind generators and contains some rather exciting photos of futuristic Danish wind-generator designs.

Vogel, Shawna. "Wind Power." *Discover* 10 (May, 1989): 46-49. This up-to-date accounting of the current state of wind power in the world provides an excellent sketch of existing wind-power systems. Discusses the current attitude toward wind energy, foreign investments, and the state of world energy in relation to wind-power economics.

Walker, John F., and Nicholas Jenkins. *Wind Energy Technology.* New York: John Wiley, 1997. A volume in the UNESCO energy engineering series, this book deals with the history of wind energy, the applications of the resource, and the potential for converting wind into electricity. Illustrations and index.

Weiner, Jonathan. *Planet Earth.* New York: Bantam Books, 1986. In association with the National Science Foundation, WQED Television-Pittsburgh provided the funding and support for this extraordinary book based on the television series by the same name. Details in one chapter the energy equation on the planet, its balance, and the methods for extracting it. Provides an excellent look at the balance sheet for world energy, including wind power and its place in the future of world power needs.

8

PETROLEUM GEOLOGY AND ENGINEERING

OFFSHORE WELLS

Offshore wells are drilled in favorable oil and gas areas that are inundated with either oceanic (salt) or inland lake (fresh) waters. Most of the world's future hydrocarbon reserves may be discovered in offshore provinces.

PRINCIPAL TERMS

CONTINENTAL MARGIN: the offshore area immediately adjacent to the continent, extending from the shoreline to depths of approximately 4,000 meters

DIRECTIONAL DRILLING: the controlled drilling of a borehole at an angle to the vertical and at an established azimuth

DRILLING FLUIDS: a carefully formulated system of fluids and solids that is used to lubricate, clean, and protect the borehole

GEOPHYSICS: the quantitative evaluation of rocks by electrical, gravitational, magnetic, radioactive, seismological, and other techniques

HYDROCARBONS: naturally occurring organic compounds that in the gaseous state are termed natural gas and in the liquid state are termed crude oil or petroleum

ROTARY DRILLING: a fluid-circulating, rotating process that is the chief method of drilling oil and gas wells

SEDIMENTARY ROCK: rock formed by the deposition and compaction of loose sediment created by the erosion of preexisting rock

SEISMOLOGY: the application of the physics of elastic wave transmission and reflection to subsurface rock geometry

ESTABLISHING OFFSHORE DRILLING LOCATIONS

Offshore drilling is an extension of on-land oil and gas drilling techniques into waters that either cover or lie adjacent to landmasses of the Earth, such as the Great Lakes of North America and the continental margins that surround each continent. The continental margins are the principal arena in which offshore drilling is conducted; they constitute approximately 21 percent of the surface area of the oceans and may contain a majority of the world's future reserves of oil and gas.

As of the last decade of the twentieth century, oil and gas production had been established off the coasts of more than forty countries in the continents of Africa, Australia, South America, North America, Asia, and Europe, and offshore drilling had been completed in the waters of more than half the nations on Earth. Drilling platforms capable of operating in water deeper than 1,800 meters probe the hydrocarbon potential of the outer continental margin. Even farther offshore, self-propelled drill ships capable of drilling more than 2,000 meters into the seafloor within waters as deep as 6,000 meters have been analyzing the deeper portions of the oceans since the early 1960's.

Initially, the specific offshore drilling location must be established. Onshore, that is the responsibility of the geologist, who, by studying the various rock exposures and their structures, decides on the best surface site for the drilling equipment. Offshore, such exposures are submerged, and geophysical methods of determining the geology of the seafloor must be used. Offshore geophysics is an indirect technique of studying the rocks composing the seafloor by measuring their physical properties. A magnetometer towed behind an aircraft flying low over the water can be used to measure the magnetic properties of underlying rocks. With a gravimeter mounted in a slow-moving boat, the gravity field associated with the seafloor can be analyzed. A combination of these methods will help the geologist locate a site that is underlain by rocks that may contain hydrocarbons. By far, the most common offshore geophysical tool is the seismic reflection survey. Seismology depends on the artificial generation of an elastic sound wave and its transmission through the layered, sedimentary rocks underlying the seafloor within the continental margin zone. These waves reflect off sedimentary rocks and are transmitted back to the surface of the water, where a vessel will record the time difference between transmission and reflection. Millions of such combined reflection arrival times are interpreted as a cross-sectional view of the underlying seafloor. A series of intersecting seismic sectional views presents a simple three-dimensional portrayal of the best site for offshore drilling equipment.

DRILLING METHODOLOGY

The sole method used for drilling offshore wells is the rotary method. Rotary motion is supplied by diesel engines to a length of interconnected drill pipe (the drill string), to the bottom of which is attached the drill bit. Drilling bits come in a variety of styles designed to drill through differing types of sedimentary rock. Each type of drill bit contains an arrangement of high-strength alloy teeth that tear through the rock when rotated under pressure. As the hole deepens, new sections of drill pipe are added. Periodically, the inside of the borehole is lined with cemented casing, which prevents the hole from caving in. During drilling, drilling fluid is circulated through the pipe and hole. This circulating fluid, a mixture of water, special clays, and other minerals and chemicals, is necessary to maintain a safe temperature and pressure in the borehole and to clean the hole of newly created rock chips. The drilling fluid can be formulated to stabilize chemically active rock layers; since it is fluid, it supports (partially floats) 10 to 20 percent of the weight of the drill pipe—a significant consideration when drilling a deep well. All drilling activities take place within the derrick, an open steel structure that often is 60 meters tall. The derrick holds the draw works, whereby the drill string can be drawn out of the borehole and disassembled, one length at a time.

Should a drilling operation discover new reserves of oil or gas, production platforms must be constructed on-site after the movable drilling platform is deployed elsewhere. From these fixed production platforms, as many as fifty or more additional wells are drilled to determine the new hydrocarbon field's size and volume. Each of these boreholes is drilled from the same general location on the production platform; however, at a predetermined depth below the seafloor, individual boreholes veer away from one another, allowing different sectors of the field to be economi-

cally developed from one production platform. This process of deviation is termed directional drilling; it requires the assistance of a specialist, as the bottom positions of the boreholes must be very carefully controlled for effective hydrocarbon production. After the final directional hole has been drilled and the limits of the field fully defined, the production derrick is replaced with equipment used to gather the oil and gas from the many active wells flowing into the platform. An offshore pipeline is laid, connecting the producing platform to onshore pipeline and refining systems, and the new field begins its production history.

DRILLING PLATFORMS

The type of platform that will be used to contain the drilling equipment is critical, considering the inhospitable weather periodically encountered offshore. There are four types of platforms: the submersible, semisubmersible, jack-up platforms, and the drill ship. A submersible platform is stabilized by flooding the hollow legs and pontoons of the platform and establishing seabed moorings. Since these platforms are in contact with the seafloor, they cannot be used in excessively deep waters. In such deep waters, a semisubmersible design is employed; the hollow pontoons are only partially flooded, permitting the platform to settle below the surface of the water but not to rest on the bottom. Inherent in this design is buoyancy that, along with anchor moorings, is sufficient to maintain the platform safely over the drilling site. The ratio of semisubmersible to submersible units in use is approximately 4:1.

By far, the most popular offshore unit is the jack-up platform. These movable structures are towed to the drill location and stationed by lowering massive steel legs to the seabed. In essence, the platform is "jacked up" on the legs to a level sufficient to protect the unit from storm waves. The fourth design is the drill ship, a free-floating, usually self-propelled vessel that contains amidship an open-water drilling unit. The drill ship is kept on location by computer-activated propellers that maintain the horizontal and vertical motion of the standing ship within safe limits. Regardless of the basic design, all offshore platforms contain a drilling derrick, storage and machinery housing, living and eating quarters, recreation and basic health facilities, and sometimes a helicopter landing pad. The entire structure must be of a size sufficient to house and maintain a normal working crew of forty to sixty individuals.

MULTIPLE BOREHOLE PLATFORMS

Because of the expense associated with the siting of an offshore drilling platform, more than one borehole is commonly drilled from the same surface location. More than sixty-five boreholes

An Exxon oil platform off the California coast at Santa Barbara. (AP/Wide World Photos)

have been drilled from a single platform. Most are drilled into the seabed at a predetermined angle to vertical, allowing many wells to "bottom" in an oil or gas reservoir rock over a distance as much as 3 kilometers laterally from the platform site. In this manner, the hydrocarbon reservoir can be economically produced with a minimum risk to the aquatic environment.

Such multiple borehole platforms are designed to extend a safe height above sea level. Platforms in use in the Gulf of Mexico, the North Sea, and the Santa Barbara Channel of California are designed to endure the most severe storm that is likely to occur within a one-hundred-year cycle, including hurricane force storms, as well as earthquake tremors. Special submersible drilling platforms are designed for operation in polar waters, where ice-free waters exist only for a short time period during summer. These structures, which must be able to withstand strong currents and floating pack ice, are protected below the waterline by a steel or cement caisson. Often the drilling deck is circular, allowing easy passage of floating ice.

Floating platforms, semisubmersibles, and drill ships move with wind and water currents. Unless compensated for, these motions affect drilling efficiency. When boreholes are drilled in very deep waters, compliant platforms that yield to weather and water currents are employed.

AN INTERNATIONAL PRIORITY

Offshore drilling for oil and gas, even with the very high costs dictated by its architectural, meteorological, engineering, and safety requirements, is an international priority; it is driven by need, economic return, and national security. Except for difficult-to-explore and environmentally sensitive regions, such as the Arctic, the South American interior, and central Africa, many onshore regions of the world have entered a mature stage of hydrocarbon exploration: Most of the accessible large-volume oil and gas fields have already been discovered. The United States is an excellent example, for it imports almost 50 percent of its daily petroleum requirements, leaving it vulnerable to foreign military and political disturbances. Studies have consistently shown that the best opportunities for increasing the reserves of American oil and gas lie in the continued exploration of the continental margins. Although not all offshore areas contain oil and gas, analyses conducted by the U.S. Geological Survey indicate that within the continental margin of the United States, as much as 40 billion barrels (a barrel equals 42 U.S. gallons, or about 159 liters) of oil and more than 5.7 trillion cubic meters of natural gas are yet to be discovered.

The degree and specific locations of offshore exploration and drilling activities depend on political, environmental, and economic factors, all of which affect petroleum products' availability and price. Certain offshore drilling regulations have already been established. In the United States, states' rights generally prevail to approximately 5,500 meters, or 3 nautical miles, offshore. From there out to designated international waters, federal government policies must be followed. The outer limits of American waters and those of other signatory nations are controlled by the ruling of the 1958 United Nations Conference on the Law of the Sea held in Geneva. Because of economic and engineering constraints, these outer limits have not yet been tested with actual drilling. A common cause for the delay in exploration drilling is public concern for the marine environment. An example of a highly publicized and heavily studied American offshore oil spill is one that occurred in the Santa Barbara Channel, off California, in 1969; another is the 1989 spill in Prince William Sound, in the Gulf of Alaska. The long-term effects of these spills are unknown. One study, however, has shown that between 1950 and 1980, the commercial fish-catch weight in the Gulf of Mexico increased fourfold, suggesting that offshore compatibility between the fishing and hydrocarbon industries is possible. Yet, as environmentally fragile continental margin areas, such as the shores of Alaska and Canada, continue to be evaluated, intensified concerns for the environment may delay exploration agendas. Finally, the costs of finding and producing oil and gas offshore will continue to climb as deeper waters are probed.

Albert B. Dickas

CROSS-REFERENCES

Coal, 1555; Earth Resources, 1741; Geothermal Power, 1651; Hydroelectric Power, 1657; Land-Use Planning, 1490; Nuclear Power, 1663; Nuclear Waste Disposal, 1791; Ocean Power, 1669; Oil and Gas Distribution, 1694; Oil and Gas Exploration, 1699; Oil and Gas Origins, 1704; Oil Chemistry, 1711; Oil Shale and Tar Sands, 1717; Onshore Wells, 1723; Petroleum Reservoirs, 1728; Solar Power, 1674; Strategic Resources, 1796; Well Logging, 1733; Wind, 1996; Wind Power 1680.

BIBLIOGRAPHY

Baker, Ron. *A Primer of Offshore Operations.* 2d ed. Austin: University of Texas Press, 1985. This basic text addresses international offshore operations and is written for the general public. Text covers the chemistry and geology of oil and gas as well as the exploration, drilling, production, and transportation aspects of the business. Illustrated with color photographs and easy-to-understand diagrams.

Duxbury, A. C., and A. B. Duxbury. *An Introduction to the World's Oceans.* 2d ed. Dubuque, Iowa: Wm. C. Brown, 1989. A freshman-level review of the marine environment. Sections are devoted to the physical, chemical, biological, and meteorological structure of the continental margins. Color plates review satellite and research submarine technology.

Ellers, F. S. "Advanced Offshore Oil Platforms." *Scientific American* 246 (April, 1982): 39-49. The methods of construction and emplacement of four different offshore oil platforms, all taller than the World Trade Center and designed to withstand 35-meter waves in water 200 meters deep, are presented in language familiar to readers of this popular monthly magazine.

Engel, Leonard. *The Sea.* Boston: Time-Life Books, 1967. A very easy-to-read introduction to the general physical and chemical composition of the typical ocean. Well illustrated.

Gorman, D. G., and June Neilson, eds. *Decommissioning Offshore Structures.* New York: Springer, 1998. This volume focuses on the engineering and environmental aspects of the abandonment of offshore oil-drilling platforms and other structures. Bibliography and index.

Hall, R. Stewart, ed. *Drilling and Producing Offshore.* Tulsa, Okla.: PennWell Books, 1983. A semitechnical text that introduces the reader to every aspect of the offshore drilling business. Covers drilling, platforms, production, and maintenance, including diving and underwater construction.

Maclachlan, Malcolm. *An Introduction to Marine Drilling.* Tulsa, Okla.: PennWell Books, 1986. The purpose of this book is to introduce the new offshore worker to the principal operations of the business. Provides insight into a way of life directly witnessed by only a very small percentage of the general population. Contains a glossary of marine drilling terms.

Myers, A., D. Edmonds, and K. Donegani. *Offshore Information Guide.* Tulsa, Okla.: PennWell Books, 1988. This comprehensive guide may be considered the bible, telephone directory, and database for offshore operations. One section provides some sixteen hundred references to offshore journals, directories, maps, legislation, and conference proceedings.

Segar, Douglas. *An Introduction to Ocean Sciences.* New York: Wadsworth, 1997. Comprehensive coverage of all aspects of the oceans and the oceanic crust. Readable and well illustrated. Suuitable for high school students and above.

OIL AND GAS DISTRIBUTION

Oil and gas have been discovered on every principal landmass on Earth. Their potential in any particular series of rocks is determined by analyses of the original environment of deposition and its subsequent geologic history and not by present climate or location.

PRINCIPAL TERMS

BASIN: a depressed area of the crust of the Earth in which sedimentary rocks have accumulated

FRONTIER: a region of potential hydrocarbon production; little is known of its rock character or geologic history

GEOPHYSICS: the quantitative evaluation of rocks by electrical, gravitational, magnetic, radioactive, and elastic wave transmission and heat-flow techniques

HYDROCARBONS: naturally occurring organic compounds that in the gaseous state are termed natural gas and in the liquid state are termed crude oil or petroleum

PERMEABILITY: the measure of the ability of a porous rock to transmit liquids and gases

POROSITY: the presence of pore space in a sedimentary rock in which hydrocarbons collect

SEDIMENTARY ROCK: rock formed by the deposition and compaction of loose sediment created by the erosion of preexisting rock

STRUCTURE: a physical rearrangement of sedimentary rocks into geometric forms that favor the accumulation and entrapment of hydrocarbons

PRESENCE IN SEDIMENTARY BASINS

Deposits of petroleum and natural gas are found in sedimentary rocks on every principal continent and within the shallow portions of the ocean basins. Sedimentary rocks cover 75 percent of continental land area and extend into the subsurface to maximum depths of 15,000 meters. The occurrence of hydrocarbons varies greatly among differing continents, countries, and individual sedimentary rock basins. Hydrocarbon quantities vary from oil-stained rock and insignificant surface seeps to subsurface reservoirs containing billions of barrels (1 barrel equals 42 U.S. gallons, or about 159 liters) of petroleum or trillions of cubic meters of natural gas. While hydrocarbons have been discovered in rocks ranging in age from 1 million to 1 billion years, generally the younger the sedimentary rock, the greater the chance of commercial hydrocarbon quantities. Approximately 85 percent of crude oil is found within the Gulf-Caribbean and the Persian Gulf regions and is contained in rock deposited less than 250 million years ago.

The presence of hydrocarbons is related to the environment prevailing at the time localized columns of sedimentary rock were deposited and to the geologic history to which those rocks have been subjected after being deposited. Hydrocarbons are formed by the chemical alteration of the remains of marine life that collect in the pore space (porosity) of sedimentary rocks, most commonly siltstone, sandstone, and porous varieties of limestone and dolostone. Large regions of the Earth's land area that have been subjected in the geologic past to extended periods of mountain building and volcanism are not considered target areas for hydrocarbon exploration. These "shield" regions, composed of rock generally in excess of 600 million years of age, constitute the geologic nucleus of every continent.

Regions surrounding the centrally positioned shields are commonly composed of down-warpings, or depressions, in the crust of the Earth. Within these "sedimentary basins" are accumulations of sedimentary rocks thousands of meters thick. In such basins, oil and gas are found in quantities depending on conditions of hydrocarbon source rock, geologic structure, and geologic history. Assuming that rocks infilling any basin contain source units, plus porous and permeable rocks, hydrocarbons will generate in the course of geologic time in the central and deeper portions of the basin. With increasing overburden pressure and temperature, oil and gas will slowly migrate outward and upward toward the basin periphery, where either entrapment will take place within a structure or surface seeps will occur.

Each basin must be individually evaluated as to its potential for containing economic deposits of oil or gas. Basins that have been infilled with sedimentary rock deposited under marine (saltwater) conditions offer the best potential, while those infilled under terrestrial environments offer the least. Should the sedimentary rock column be thin, overburden pressure would be insufficient to generate crude oil; if the column is excessively thick, the resultant high temperatures will alter the oil to natural gas or even burn off the gas. Hydrocarbons are destroyed when a basin is subjected to mountain-building forces and escape to the surface to form uneconomic tar sand deposits.

DISTRIBUTION IN NORTH AND SOUTH AMERICA

In North America, the principal hydrocarbon-producing regions include the east coast of Mexico, the plains of Alberta and Saskatchewan, and the coastal areas of Alaska. In the lower forty-eight states, the most important producing areas are the coastal states of Alabama, Louisiana, and Texas; the Great Plains and the lower Great Lakes regions; and California. Approximately 10 percent of world oil reserves and 9 percent of world gas reserves are found in North America. Important frontier regions on this continent include the Arctic waters of Alaska and Canada and the Atlantic coastal waters of the United States. The development of both these regions will, however, be constrained by economic and environmental conditions.

Venezuela alone accounts for 84 percent of the oil and 63 percent of the natural gas reserves of South America. The Lake Maracaibo oil fields have historically accounted for a majority of this production. The only other important hydrocarbon-producing country on the American continent is Argentina. With large sections of the middle latitudes of South America composed of the Guyana and the Brazilian shields, most of the oil discovered in South America is found in rock less than 250 million years in age and in structures associated with the creation of the Andean mountain ranges. Recent major discoveries in water depths of 900 meters in the Campos basin, offshore Brazil, indicates that significant reserves are yet to be discovered along the Atlantic seaboard of South America.

DISTRIBUTION IN AFRICA AND ASIA

Africa contains 6 percent of the oil and 16 percent of the gas reserves of the world. North of the Sahara desert, those countries with the greatest hydrocarbon reserves are Libya, Algeria, and Egypt; all border on the prolific Mediterranean basin. In southern Africa, Nigeria controls the majority of production and reserves, while Angola, Congo, and Gabon possess reserves of secondary value. Many sections of this continent have been minimally evaluated for hydrocarbon potential and are considered frontier status. As the African Shield covers much of the 30-million-square-kilometer area of this continent, the best prospects for new oil and gas discoveries are confined to peripheral sedimentary basins, some of which extend into the offshore Atlantic area. These peripheral basins are characterized by thick sedimentary sequences and the common presence of salt deposits; the latter is a positive indicator of hydrocarbon potential.

In the Middle East and Asia, the Saudi Arabian peninsula dominates the distribution pattern of oil and gas. Of the proven world reserves of oil, 65 percent are controlled by the countries of the United Arab Emirates, Iran, Iraq, Kuwait, and Saudi Arabia. Saudi Arabia alone possesses 262 billion barrels of proven reserves, 26 percent of the world total. Saudi Arabia has the largest total reserves of any country. These reservoirs are being depleted by only 700 producing wells, with the average well flowing approximately 13,000 barrels of oil per day. In contrast, in the United States, more than 500,000 wells produce 9 million barrels of oil, for an average production rate of 18 barrels per day. By the end of the twentieth century, the average production per well had decreased to 11.4 barrels per day. All hydrocarbons in Saudi Arabia occur within calcium carbonate sedimentary rocks approximately 155 million years in age. The majority of accumulations are contained within anticline structures, convex upward rock folds formed by compressional forces. These flexures, or folds, are located mainly in eastern Arabia, and the longest (including Ghawar, the largest oil field in the world) extends for more than 400 kilometers.

Australasia, extending from New Zealand to Malaysia, contains approximately 1 percent of world hydrocarbon reserves. Historically, the principal producing areas have been Sumatra, Java, and Borneo, but exploration programs are assessing the offshore waters of this vast region.

DISTRIBUTION IN RUSSIA AND EUROPE

Russia's reserves are principally concentrated in the West Siberian basin. The West Siberian basin has a cumulative oil production of more than 25 billion barrels. The old hydrocarbon regions of the Volga-Ural and North Caucasus regions west of the Ural Mountains and Caspian Sea contain hydrocarbon reserves of secondary importance.

West European hydrocarbon reserves are concentrated in the offshore, North Sea extensions of Norway, Denmark, and the United Kingdom. Western Europe contains only 2 percent of world reserves of oil and 3 percent of world reserves of natural gas. The increases in European production over the past several decades have been derived mainly from basins located offshore; onshore basins, historically, have never been important producers.

INITIAL EVALUATION OF POTENTIAL RESERVE

In the evaluation of the oil and gas potential of a sedimentary basin, whether it be a frontier region or the deeper zones of a mature production area, a routine of analyses has been established. As the geographic distribution of oil and gas accumulations is controlled by regional as well as local factors, such factors must be initially evaluated. The most important of regional factors are those of depositional environment and deformational history.

As a result of the commonly accepted belief that hydrocarbons are derived from oceanic organic matter deposited within sedimentary rocks, basins that have been developed under conditions of marine deposition are given favorable initial evaluation. Minority opinion suggests that hydrocarbons are inorganically (or "abiogenically") derived and, thus, may be found in any type of rock that possesses porosity and permeability. In addition to sedimentary types, such rock might include fractured lava and other igneous rock derived from crystallization of magma (molten rock). This inorganic theory has been tested by the drilling of a deep well in the Scandinavian Shield region of Sweden, an area dominated by heavily fractured granite. As geologic reviews indicated only traces of natural gas and little porosity, the organic theory of hydrocarbon origin appears to be vindicated.

If a world geologic map is superimposed on a hydrocarbon distribution map, relationships dependent on deformational history become apparent. Regionally, oil and gas fields are concentrated in areas of the Earth's crust that are downwarped, such as continental margins, coastal plains, and inland plateaus that are low in elevation in comparison to surrounding areas. Such areas generally contain adequate thicknesses of sedimentary rock possessing the organic material, porosity, and permeability necessary to function as source and reservoir rocks.

FURTHER EVALUATION

After a frontier region is evaluated from the above considerations and found to possess potential for oil or gas accumulation, geophysical technology is used to further localize possible hydrocarbon-bearing structures. The most commonly used geophysical techniques are those that employ seismology (the analysis of basin structure by artificially generated elastic wave reflections), gravitational technology (the association of density distributions with rock types), and magnetic technology (the measurement of the Earth's magnetic field at different locations). Geophysics technology is used to define fundamental basin architecture and geologic history.

As the surface area of a typical basin is on the order of many thousands of square kilometers, each basin must be studied from the viewpoint of preferred habitats of known oil and gas accumulations. Studies of oil accumulations indicate that more than half (54 percent) of world reserves occur along a basin hinge belt, that zone separating intense downwarping of a basin center from the modest downwarping of the basin periphery. Within the deep basin, only 11 percent of the oil is found, while the peripheral shelf contains the remaining 35 percent.

In the final evaluation of oil and gas distribution patterns, it must be emphasized that the majority of known hydrocarbon deposits are concentrated in three intercontinental depressions in the crust of the Earth. In the Western Hemisphere, between North and South America, lies the depression forming the Caribbean Sea and the Gulf of Mexico. Here are found the great oil and gas deposits of Venezuela, Colombia, Mexico, and the Gulf Coast. The second of these depressions contains the hydrocarbon reserves bordering the Red, Mediterranean, Caspian, and Black seas and the Persian Gulf and is formed within the corners of Africa, Europe, and Asia. The last depression is found in the southwest Pacific, between the continents of Asia and Australia, and includes the great fields found in Borneo, Sumatra, and Java. A fourth intercontinental depression, in the early stages of evaluation, occupies the northern limits of North America, Europe, and Asia, surrounding the North Pole. While this region offers great promise and is the site of the largest oil field discovered to date in the United States—the Prudhoe Bay field of Alaska—expanded exploration will be constrained by environmental considerations.

Even with the wide array of available geological and geophysical research procedures, worldwide drilling for new oil and gas reserves is a high-risk endeavor. In spite of continued introduction of new exploration methodologies, approximately 30 percent of all wells drilled since 1859 have been declared dry, and more than 70 percent of the wells drilled annually in frontier areas are unsuccessful.

CONTINUED EXPLORATION OF NEW RESERVES

The worldwide hydrocarbon industry was created in the middle of the nineteenth century to fuel the Industrial Revolution. The recognition of oil and gas as a cheap and efficient energy resource marked the end of wood as an important source of fuel and ultimately reduced "King Coal" to secondary status. By the final decade of the twentieth century, however, the burning of oil and gas was being associated with atmospheric pollution and the greenhouse effect, among other environmental problems. Yet, the promises of solar power have not been realized, and nuclear energy, while economically viable, has been generally considered unsafe.

Oil and gas continue to function in the role that they rapidly assumed after the completion of Edwin Drake's well in 1859; they are the wonder fuels, lubricants, and chemical sources of the modern scientific world. It was popular in the 1960's to predict that the end of the hydrocarbon age would be closely associated with the year 2000, but such predictions have proved premature. The continued exploration for new reserves of oil and gas is partially controlled by economics. Actions taken by the Organization of Petroleum Exporting Countries (OPEC) in the early 1970's caused the value of a barrel of petroleum in the United States to increase from $3 to $35 by 1981. Spurred by this incentive, the American hydrocarbon industry in 1974 created the first annual increase in proven oil reserves in the United States since 1960. In that same year, the world consumption of petroleum products approximated 7 billion barrels annually. Thirty-eight years later, the United States alone was consuming 6.8 billion barrels of crude oil (plus natural gas and plant liquids) every year, with some 51 percent of this supply being imported. Worldwide, the demand for oil and gas, in terms of crude oil equivalent energy, was close to 125 million barrels per day in 1997.

Oil and gas will probably continue for decades to be the world's primary energy source, and new reserves will continue to be sought worldwide. The distribution of oil and gas in the future will be determined by the same factors governing its distribution today. Geologic factors will continue to dictate the best locations for exploration, with local political stability and economic considerations entering into the decision when and whether to explore.

Albert B. Dickas

CROSS-REFERENCES

Offshore Wells, 1689; Oil and Gas Exploration, 1699; Oil and Gas Origins, 1704; Oil Chemistry, 1711; Oil Shale and Tar Sands, 1717; Onshore Wells, 1723; Petroleum Reservoirs, 1728; Well Logging, 1733.

BIBLIOGRAPHY

Andersen, Svein S. *The Struggle over North Sea Oil and Gas: Government Strategies in Denmark, Britain, and Norway.* New York: Oxford University Press, 1993. Andersen deals in great detail with the petroleum industry and trade in the strategically important North Sea, focusing on the government policies of Norway, Denmark, and Britain, which are often at odds with one another. Illustrations and bibliography.

Anderson, Robert C., ed. *The Oil and Gas Opportunity on Indian Lands: Exploration Policies and Procedures.* Golden, Colo.: U.S. Department of the Interior, Bureau of Indian Affairs, Division of Energy and Mineral Resources, 1994. This multiauthored text gives readers a useful overview of the problems that can arise when the distribution of oil and gas is at odds with political and social boundaries that have been established by government agencies. Essays deal with various situations on specific Indian reservations in the United States.

Baker, Ron. *Oil and Gas: The Production Story.* Austin: University of Texas Petroleum Extension Service, 1983. The first four sections of this primer are easy-to-read introductions to the origin and accumulation of hydrocarbons, exploration and testing technologies, and production stimulation procedures. Each section contains full-color diagrams.

Ball, Max W. *This Fascinating Oil Business.* New York: Bobbs-Merrill, 1940. Although an older reference, this 444-page book contains a wealth of information found in no other review of the business. It is especially valuable for its history of early global exploration in remote regions ranging from Afghanistan to Manchukuo. Written for general readers.

Gramling, Robert. *Oil in the Gulf: Past Development, Future Prospects.* New Orleans: U.S. Minerals Management Service, 1995. Provides a brief but informative hisorical account of the offshore oil and gas industries, including the trade practices of the various nations in the Gulf of Mexico. Illustration, maps, and bibliography. Intended for the layperson.

Nawwab, Ismail I., Peter C. Speers, and Paulk F. Hoye, eds. *Aramco and Its World: Arabia and the Middle East.* Washington, D.C.: Arabian American Oil, 1980. A beautifully illustrated volume that explores the history of the discovery and development of oil in a region that possesses 63 percent of the world's petroleum reserves.

Pratt, Wallace E., and Dorothy Good, eds. *World Geography of Petroleum.* Princeton, N.J.: Princeton University Press, 1950. A classic comparative study of the petroleum-producing regions of the world. Coedited by an internationally recognized geologist. A college-level text.

Schackne, Stewart, and N. D'Arcy Drake. *Oil for the World.* 2d ed. New York: Harper & Brothers, 1960. A brief introduction to all aspects of an integrated oil and gas company, from discovery through drilling operations, refining, and marketing.

Tiratsoo, Eric N. *Natural Gas.* 3d ed. Houston: Gulf, 1980. A college-level review of natural gas drilling, production reserves, and economics. Eight chapters discuss principal geographic regions in detail. Includes a valuable compilation of natural gas statistics for the former Soviet Union and Eastern Europe.

_____. *Oilfields of the World.* 3d ed. Houston: Gulf, 1985. This easy-to-read companion reference to the above entry describes oilfields in approximately seventy countries. The geology of each field is presented at a college freshman level.

OIL AND GAS EXPLORATION

Geologists seek to understand the geologic features with which oil and gas are associated in order to make exploration for these minerals less risky and more economical.

PRINCIPAL TERMS

DRY HOLE: a well drilled for oil or gas that had no production

LEASE: a permit to explore for oil and gas on specified land

PERMEABILITY: a property of rocks where porosity is interconnected, permitting fluid flow through the rocks

POROSITY: a property of rocks where empty or void spaces are contained within the rock between grains or crystals or within fractures

PROSPECT: a limited geographic area identified as having all the characteristics of an oil or gas field but without a history of production

REGIONAL GEOLOGY: a study of the geologic characteristics of a geographic area

RESERVOIR: a specific rock unit or bed that has porosity and permeability

SEAL: a rock unit or bed that is impermeable and inhibits upward movement of oil or gas from the reservoir

SOURCE ROCK: a rock unit or bed that contains sufficient organic carbon and has the proper thermal history to generate oil or gas

TRAP: a structure in the rocks that will allow petroleum or gas to accumulate rather than flow through the area

DIFFERING INTERPRETATIONS

Exploration for oil and gas involves the application of many geologic principles toward a single goal: the discovery of new reserves of petroleum and natural gas. The exploration geologist may be a specialist in one of the many fields of geology but will also have a strong foundation in other subdisciplines as well as a good background in support sciences such as physics, chemistry, and mathematics. The methodology involved in the search for oil and gas is dependent on human input. Unlike many other procedures in geology, the actual approach used in the exploration for oil and gas will vary with the individual conducting the search. This results in a diversity of opinions as to the best method of oil and gas exploration, dependent upon the diverse geologic characteristics of the region. As no two regions are exactly alike geologically, it would also be expected that no two oil or gas fields are exactly alike. Given this, a diversity of interpretations of a certain area can actually enhance the exploration of this area rather than hinder it. If two geologists look at the same data on a region, they may both develop different theories about potential oil and gas accumulations within the area. Both cannot be correct, but one may be. If only one viewpoint prevailed, it could very easily be the wrong one. The exploration of oil and gas is an optimistic methodology. If a region does not have production, it is considered poorly understood rather than unproductive. This results in a constant reevaluation of regions as ideas in geology change.

Despite the diversity of opinions, ideas generated about the occurrence of oil and gas within a particular region are not merely random thoughts. All theories concerning the potential occurrence of oil and gas within an area must conform to the regional geology, an understanding of the stratigraphy, structure, and depositional and tectonic history of an area. Raw geologic data can be interpreted in many ways, but the valid interpretation is one that will fit within the regional understanding of the geology.

EXISTENCE OF A TRAP

As the search for oil and gas begins, the exploration geologist identifies very specific areas within a geologic region, called prospects. A prospect is a potential oil or gas field. A prospect is not merely the extension of an existing field, as such work is classified as development geology. Rather, a prospect would be a potential new field some distance away from preexisting production. In order to locate prospects, a model of what such potential fields will look like needs to be developed. This model will be based on already discovered fields with well-known characteristics, although it is recognized that there will be differences between the two, and this knowledge will serve as a starting point for exploration. Despite differences between prospects and models, all prospects have certain criteria that must be met to demonstrate that a prospect is viable or, in other words, drillable. If one of the criteria cannot be met, the prospect may be regarded as too risky to drill.

The first criterion that an exploration geologist must satisfy is proving the existence of a trap. Proving that a trap is present involves the use of several tools. First, the exploration geologist must refer to the model. Once the geologist knows what type of trap to look for, the specific tool is selected. The tool used to locate a trap is a map constructed on the basis of subsurface data. The type of map constructed will depend on the type of trap present in the model. If the trap is a structural trap, a subsurface structure map showing folds and faults will be constructed, similar to a topographic map. The contours on this map connect points of equal elevation relative to sea level, except that they are in the subsurface rather than on the surface. When studied in the subsurface, hills on a structural contour map represent anticlines or domes, valleys represent synclines or depressions, and sharp cliffs usually indicate faults. If the trap in the model is of a stratigraphic type, then a stratigraphic map showing thickness of a rock unit or bed, differing rock types (facies), or ancient depositional environments is developed. The data to construct such maps are generally derived from subsurface information about the rock units in question from previously drilled wells in the vicinity. In addition to well information, data from seismic, gravity, and magnetic surveys may add support to a prospect. Surface geologic data are considered but may not accurately reflect the geologic conditions of the subsurface.

QUALITIES OF A RESERVOIR

Once all the subsurface data have been analyzed and a potential trap located, a detailed examination of the rocks present in the prospect area is needed. Specifically, it must be demonstrated that a rock unit or bed within the trap can function as a reservoir for the oil or gas. The qualities that enable a rock to be a reservoir are porosity, pore spaces between grains or crystals or open fractures, and permeability, the interconnection of pore spaces or fractures that will allow fluid flow (in this case, oil or gas). Demonstrating that these properties exist is usually done through a detailed study of the rocks by using special cylinder-shaped rock samples collected from previously drilled wells called cores. This core analysis is a very important aspect of petroleum geology and is the best way to describe and define reservoirs. Geophysical logs that measure certain physical properties of the rocks are also very valuable in defining a reservoir.

PRESENCE OF SOURCE BED AND SEAL

The next thing to be demonstrated is that a source bed for oil or gas exists in the prospect area. This feature is perhaps the most critical, as oil and gas are generated only under very specific chemical and physical conditions. A rock unit or bed must be located in the rocks of the prospect that have the proper chemical consistency and have been subjected to the proper temperatures needed to generate oil or gas. There must also be a pathway or mechanism to allow the oil or gas to migrate from the source bed to the reservoir, as the two will only rarely be the same unit. What is generally needed is a total organic carbon content of a bed to be greater than 1 percent and a temperature history to allow the organic matter to mature into hydrocarbons such as oil or natural gas. The rock type generally involved will be a shale, but some fine-grained limestones and dolomites as well as some cherts will function as source beds. A pathway of migration for the generated oil and gas must be available. This is not a critical problem if the source bed occurs beneath the reservoir bed because of the tendency of upward migration of oil and gas resulting from the lighter density of those fluids compared to the groundwater. If the source bed occurs above the reservoir, it can still function for the prospect if it is dropped down by a fault below the reservoir unit. Direct downward migration of oil and gas is known to occur, but this process would have to be demonstrated as functioning in the prospect area in order to use it to explain the source for a prospect.

If the above criteria for a prospect are met, the next factor to be considered is the presence of a seal. As the tendency is for oil and gas to migrate upward, an impermeable rock unit or bed must be present above the reservoir bed to prevent the oil and gas from flowing through the reservoir unit rather than accumulating in the trap. If oil and gas cannot migrate upward, they will migrate laterally to the highest point (trap) in the last permeable bed (reservoir) they can enter. Seals can be of any impermeable rock type, but common ones are evaporites (rock salt and gypsum) and shales. Fine-grained limestones and dolomites as well as some igneous and metamorphic rocks can also function as seals.

FINAL CRITERIA

The final gelogical criterion that must be satisfied is the time of oil and gas generation and migration and its relationship to the time of trap formation. In order to determine when these events occurred, a detailed understanding of the geologic history must be attained. In general, the trap must have been formed before the generation and migration of oil and gas in the region. If this is not the case, the trap in question will not likely contain oil and gas. Many excellent traps exist that have been drilled and do not produce because of the timing difference.

If any dry holes (previously drilled wells without production) are near a prospect, the reason for their lack of production must be explained. Generally this will be done by showing that the wells were drilled away from the prospect in question or that technical or economic problems were encountered. Geologists do not want a prospect with a dry hole in the middle of it unless that well was dry for a reason that the present model can explain.

OBTAINING A LEASE

If all the above criteria have been met, the geologist may have a prospect, provided that a lease can be obtained. An oil and gas lease gives the holder the right to explore and produce oil and gas on the land in question. These are the only rights granted to the lease holder. This lease is usually for a specified term (five or ten years) and pays the landowner an annual rental fee plus a

royalty interest in any production. If production is achieved on the lease, the term will not expire until production stops. Production cannot be halted to wait for better economic conditions such as higher prices, as this would endanger the lease. The availability of acreage is a nongeologic factor in the exploration for oil and gas, but it is a critical one. A prospect without at least the potential of acquiring a lease is of little use, as it cannot be drilled and developed.

A PROCESS OF IDEAS

The process of oil and gas exploration is a difficult one because of the large degree of uncertainty involved. Advances such as high resolution seismic profiling, computer analysis and management of well data, and the use of satellite photographs of the land surface have all improved the way geologists explore oil and gas deposits. Despite the benefits of improved technology, the exploration for oil and gas remains a process of ideas rather than equipment. This does not mean that old, established ideas will dominate the search for oil and gas. In fact, quite the opposite is true. Ideas can and do develop at a much faster rate than technology. Because of this, the potential for change in an approach is always present. Thus, a region that might be considered to have no oil and gas potential may appear differently to geologists who have a different or new idea about the geology of the area. Some new ideas affect all regions of the Earth and tend to form a revolution of approaches in the oil and gas industry. Ideas over the last three decades have tended to view geology from a global perspective, and this has changed the exploration for oil and gas. The most important idea of this type is the theory of plate tectonics. This proposed mechanism resulted in a widespread reevaluation of the petroleum potentials of regions, resulting in increased discoveries of new fields.

The effect that the exploration of oil and gas has on society is a very clear one. Oil and gas are needed by almost everyone as sources of relatively inexpensive, clean, and efficient energy. The present structure of the oil industry, including oil exploration and production, is sometimes criticized as being wasteful and inefficient. In some nations, oil and gas exploration has been nationalized and is handled by a team of geologists in order to promote efficiency by eliminating competition for the same prospective new fields. This has been suggested as a model for the oil industry in the United States as a way to increase the known petroleum reserves.

Because the competitive system, in which many geologists develop varying working models of an area, is eliminated by a nationalized approach, only one model is present, and the chances of ever finding oil are greatly reduced. While many nationalized companies are very successful, often their success stories are in areas that simply had not been explored before, and the oil accumula-

tions were very obvious. As these regions mature and oil and gas become harder to find, multiple ideas are needed to locate new fields. A competitive system ensures that as long as a profit can be made, exploration for oil and gas will continue.

Richard H. Fluegeman, Jr.

CROSS-REFERENCES

Offshore Wells, 1689; Oil and Gas Distribution, 1694; Oil and Gas Origins, 1704; Oil Chemistry, 1711; Oil Shale and Tar Sands, 1717; Onshore Wells, 1723; Petroleum Reservoirs, 1728; Well Logging, 1733.

BIBLIOGRAPHY

American Petroleum Institute. *Primer of Oil and Gas Production.* 3d ed. Washington, D.C.: Author, 1978. This is a basic introduction to the procedures and techniques of oil and gas production. Chapters on the origin and accumulation of oil and gas and on the properties of reservoirs are included. This book is written as a beginner's text in nontechnical language and contains helpful illustrations.

Baker, Ron. *A Primer of Oil Well Drilling.* 4th ed. Austin: The University of Texas at Austin, 1979. This book is an introduction to the procedures involved in oil and gas well drilling. It includes chapters on exploration and on oil and gas accumulations. The text is well illustrated with many on-site photographs. Suitable for high school or college students.

Hyne, Norman J. *Dictionary of Petroleum Exploration, Drilling, and Production.* Tulsa, Okla.: PennWell, 1991. This is a dictionary of all terms associated with petroleum and the petroleum industry. A great resource for the beginner in this field. Illustrations and maps.

_____. *Nontechnical Guide to Petroleum Geology, Exploration, Drilling, and Production.* Tulsa, Okla.: PennWell, 1995. An excellent introductory handbook for the reader without prior background in petroleum geology. Discusses exploration, refining, and distribution. Illustrations, bibliography, and index.

LeRoy, L. W., and D. O. LeRoy, eds. *Subsurface Geology.* 4th ed. Golden: Colorado School of Mines, 1977. This book is an important reference for all who are interested in the geology of the subsurface. There are many chapters on petroleum geology, but exploration for oil and gas is not the only focus of this book. The chapters on oil and gas are written by experts in the field and are very complete and well illustrated. Although this book is intended as a reference for college-level students and professionals, it is regularly used as a basic reference for the geology of oil and gas by nongeologists involved in the oil business.

Levorsen, A. I. *Geology of Petroleum.* 2d ed. San Francisco, Calif.: W. H. Freeman, 1967. A full textbook on petroleum geology designed for the college student who has taken some basic geology courses. Serves as a reference for people with an interest in some of the more detailed aspects of petroleum exploration. This book was the training textbook for many present-day exploration geologists. It contains a complete bibliography for each chapter.

Owen, E. W. *Trek of the Oil Finders: A History of Exploration for Petroleum.* Tulsa, Okla.: American Association of Petroleum Geologists, 1975. A detailed history of the development of the petroleum industry throughout the world. It is written from a historical perspective, and the language is technical only when necessary. The text is very well indexed by subject, geographic location, and proper names and is organized into chapters by region.

Selley, Richard C. *Elements of Petroleum Geology.*

2d ed. New York: W. H. Freeman, 1998. A well-organized text covering the specifics of oil and gas and their relationship to geology. This book is designed as a college textbook for students near the end of their course-work in geology or for geologists beginning careers in the petroleum industry. Fairly technical, it requires basic understanding of geological concepts. The book contains subject and proper name indexes, useful illustrations, and appendices that include a well classification table, a glossary of oil terms and abbreviations, and a table of conversion factors.

West Texas Geological Society. *Geological Examples in West Texas and South-eastern New Mexico (the Permian Basin) Basic to the Proposed National Energy Act.* Midland: West Texas Geological Society, 1979. This short publication was designed to inform state and federal government members about the procedures and costs of exploration for oil and gas. This paper is based on case histories and maps; diagrams and actual costs are included. Although not intended as a basic text on petroleum geology, it is a valuable reference, written in clear, nontechnical language.

OIL AND GAS ORIGINS

Oil and gas are two of the most important fossil fuels. The formation of oil and gas is dependent on the preservation of organic matter and its subsequent chemical transformation into kerogen and other organic molecules deep within the Earth at high temperatures over long periods of time. As oil and gas are generated from these organic materials, they migrate upward, where they may accumulate in hydrocarbon traps.

PRINCIPAL TERMS

FOSSIL FUEL: a general term used to refer to petroleum, natural gas, and coal

HYDROCARBONS: solid, liquid, or gaseous chemical compounds containing only carbon and hydrogen; oil and natural gas are complex mixtures of hydrocarbons

KEROGEN: fossilized organic material in sedimentary rocks that is insoluble and that generates oil and gas when heated; as a form of organic carbon, it is one thousand times more abundant than coal and petroleum in reservoirs, combined

METHANE: a colorless, odorless gaseous hydrocarbon with the formula CH_4; also called marsh gas

NATURAL GAS: a mixture of several gases used for fuel purposes and consisting primarily of methane, with additional light hydrocarbon gases such as butane, propane, and ethane, with associated carbon dioxide, hydrogen sulfide, and nitrogen

PETROLEUM: crude oil; a naturally occurring complex liquid hydrocarbon, which after distillation yields a range of combustible fuels, petrochemicals, and lubricants

RESERVOIR: a porous and permeable unit of rock below the surface of the Earth that contains oil and gas; common reservoir rocks are sandstones and some carbonate rocks

PRESERVATION OF ORGANIC MATTER

The origin of oil and gas begins with the production of organic matter by plants and plantlike organisms, through a process called photosynthesis. Photosynthesis converts light energy from the Sun into chemical energy and produces organic matter (or carbohydrates, which include sugars, starches, and cellulose) and oxygen. Carbohydrates burn easily and release considerable amounts of energy in the process. It is this property that makes them ideal fuels.

To become a fossil fuel, the organic matter in an organism must be preserved after the organism dies. The preservation of organic matter is a rare event because most of the carbon in organic matter is oxidized and recycled to the atmosphere through the action of aerobic bacteria. (Oxidation is a process through which organic matter combines with oxygen to produce carbon dioxide gas and water.) Less than 1 percent of the organic matter that is produced by photosynthesis escapes from this cycle and is preserved. To be preserved,

the organic matter must be protected from oxidation, which can occur in one of two ways: The organic matter in the dead organism is rapidly buried by sediment, shielding it from oxygen in the environment, or the dead organism is transported into an aquatic environment in which there is no oxygen (that is, an anoxic or anaerobic environment). Most aquatic environments have oxygen in the water, because it diffuses into the water from the atmosphere and is produced by photosynthetic organisms, such as plants and algae. Oxygen is removed from the water by the respiration of aerobic organisms and through the oxidation of decaying organic matter. Oxygen consumption is so high in some aquatic environments that anoxic water is present below the near-surface oxygenated zone. Environments that lack oxygen include such places as deep, isolated bodies of stagnant water (such as the bottom waters of some lakes), some swamps, and the oxygen-minimum zone in the ocean (below the maximum depth to which light penetrates, where no photosynthesis can occur). Large quantities of organic matter may be preserved in these environments.

The major types of organic matter preserved in sediments include plant fragments, algae, and microbial tissue formed by bacteria. Animals (and single-celled animal-like organisms) contribute relatively little organic matter to sediments. The amount of organic matter contained in a sediment or in sedimentary rock is referred to as its total organic content (TOC), and it is typically expressed as a percentage of the weight of the rock. To be able to produce oil, a sediment typically must have a TOC of at least 1 percent by weight. Sediments that are capable of producing oil and gas are referred to as "hydrocarbon source rocks." In general, fine-grained rocks such as shales (which have clay-sized grains) tend to have higher TOC than do coarser-grained rocks.

TRANSFORMATION OF ORGANIC MATTER

The organic matter trapped in sediment must undergo a series of changes to form oil and gas. These changes take place as the sediment is buried to great depths as a result of the deposition of more and more sediment in the environment over long periods of time. Temperature and pressure increase as depth of burial increases, and the organic materials are altered by these high temperatures and pressures.

As sediment is gradually buried to depths reaching hundreds of meters, it undergoes a series of physical and chemical changes, called diagenesis. Diagenesis transforms sediment into sedimentary rock by compaction, cementation, and removal of water. Methane gas commonly forms during the early stages of diagenesis as a result of the activity of methanogenic bacteria. At depths of a few meters to tens of meters, organic compounds such as proteins and carbohydrates are partially or completely broken down, and the individual component parts are converted into carbon dioxide and water or are used to construct geopolymers, or large, complex organic molecules of irregular structure (such as fulvic acid and humic acid and larger geopolymers called humins). During diagenesis, the geopolymers become larger and more complex, and nitrogen and oxygen content decreases. With increasing depth of burial over long periods of time (burial to tens or hundreds of meters over a million or several

million years), continued enlargement of the organic molecules alters the humin into kerogen, an insoluble form of organic matter that yields oil and gas when heated.

As sediment is buried to depths of several kilometers, it undergoes a process called catagenesis. At these depths, the temperature may range from 50 to 150 degrees Celsius, and the pressure may range from 300 to 1,500 bars. The organic matter in the sediment, while in a process called maturation, becomes stable under these conditions. During maturation, a number of small organic molecules are broken off the large kerogen molecules, a phenomenon known as thermal cracking. These small molecules are more mobile than are the kerogen molecules. Sometimes called bitumen, they are the direct precursors of oil and gas. As maturation proceeds, and oil and gas generation continues, the kerogen residue remaining in the source rock gradually becomes depleted in hydrogen and oxygen. In a later stage, wet gas and condensate are formed. ("Condensate" is a term given to hydrocarbons that exist as gas under the high pressures existing deep beneath the surface of the Earth but condense to liquid at the Earth's surface.) Oil is typically generated at temperatures between 60 and 120 degrees Celsius, and gas is generated at somewhat higher temperatures, between about 120 and 220 degrees Celsius. Large quantities of methane are formed during catagenesis and during the subsequent phase, which is called metagenesis.

When sediment is buried to depths of tens of kilometers, it undergoes the processes of metagenesis and metamorphism. Temperatures and pressures are extremely high. Under these conditions, all organic matter and oil are destroyed, being transformed into methane and a carbon residue, or graphite. Temperatures and pressures are so intense at these great depths that some of the minerals in the sedimentary rocks are altered and recrystallized, and metamorphic rocks are formed.

MIGRATION AND TRAPPING

Accumulations of oil and gas are typically found in relatively coarse-grained, porous, permeable rocks, such as sandstones and some carbonate rocks. These oil- and gas-bearing rocks are called reservoirs. Reservoir rocks, however, generally lack the kerogen from which the oil and gas are generated. Instead, kerogen is typically found in abundance only in fine-grained sedimentary rocks such as shales. From these observations, it can be concluded that the place where oil and gas originate is not usually the same as the place where oil and gas are found. Oil and gas migrate or move from the source rocks (their place of origin) into the reservoir rocks, where they accumulate.

Oil and gas that form in organic-rich rocks tend to migrate upward from their place of origin, toward the surface of the Earth. This upward movement of oil occurs because pore spaces in the rocks are filled with water, and oil floats on water because of its lower density. Gas is even less dense than oil and also migrates upward through pore spaces in the rocks. The first phase of the migration process, called primary migration, involves expulsion of hydrocarbons from fine-grained source rocks into adjacent, more porous and permeable layers of sediment. Secondary migration is the movement of oil and gas within the more permeable rocks. Oil and gas may eventually reach the surface of the Earth and be lost to the atmosphere through a seep. Under some circumstances, however, the rising oil and gas may become trapped in the subsurface by an impermeable barrier, called a caprock. These hydrocarbon traps are extremely important because they provide a place for subsurface concentration and accumulation of oil and gas, which can be tapped for energy sources.

There are a variety of settings in which oil and gas may become trapped in the subsurface. Generally, each of these traps involves an upward projection of porous, permeable reservoir rock in combination with an overlying impermeable cap-

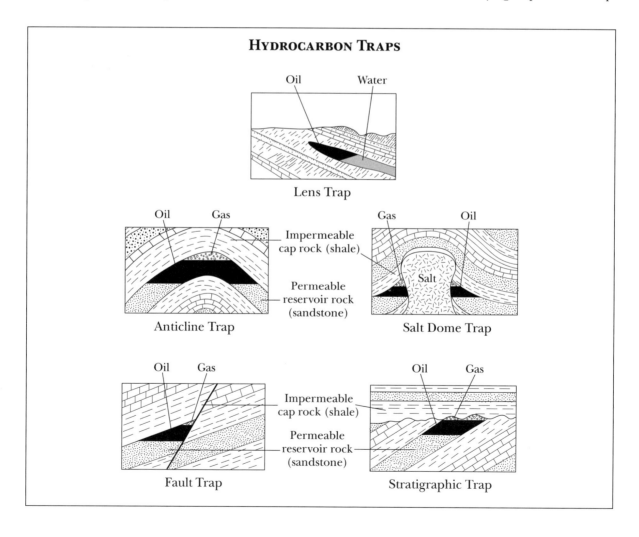

HYDROCARBON TRAPS

Oil   Water

Lens Trap

Oil   Gas

Impermeable cap rock (shale)

Permeable reservoir rock (sandstone)

Anticline Trap

Gas   Oil

Salt

Salt Dome Trap

Oil   Gas

Impermeable cap rock (shale)

Permeable reservoir rock (sandstone)

Fault Trap

Oil   Gas

Impermeable cap rock (shale)

Permeable reservoir rock (sandstone)

Stratigraphic Trap

rock that encloses the reservoir to form a sort of inverted container. Examples of hydrocarbon traps include anticline traps, salt dome traps, fault traps, and stratigraphic traps. There are many types of stratigraphic traps, including porous reef rocks enclosed by dense limestones and shales, sandstone-filled channels, sand bars, or lenses surrounded by shale, or porous, permeable rocks beneath an unconformity. The goal of the exploration geologist is to locate these subsurface hydrocarbon traps. Enormous amounts of geologic information must be obtained, and often many wells must be drilled before accumulations of oil and gas can be located.

PHYSICAL AND CHEMICAL ANALYSES

The origin of oil and gas can be determined using physical and chemical analyses. Petroleum contains compounds that serve as biological markers to demonstrate the origin of petroleum from organic matter. Oil can be analyzed chemically to determine its composition, which can be compared to that of hydrocarbons extracted from source rocks in the lab. Generally, oil is associated with natural gas, most of which probably originated from the alteration of organic material during diagenesis, catagenesis, or metagenesis. In some cases, gas may be of abiogenic (nonorganic) origin. Samples of natural gas can be analyzed using gas chromatography or mass spectrometry and isotope measurements.

Commonly, rocks are analyzed to determine their potential for producing hydrocarbons. It is important to distinguish between various types of kerogen in the rocks because different types of organic matter have different potentials for producing hydrocarbons. In addition, it is important to determine the thermal maturity or evolutionary state of the kerogen to determine whether the rock has the capacity to generate hydrocarbons or hydrocarbons have already been generated.

The quantity of organic matter in a rock, referred to as its TOC, can be measured with a combustion apparatus, such as a Leco carbon analyzer. To analyze for TOC, a rock must be crushed and ground to a powder and its carbonate minerals removed by dissolution in acid. During combustion, the organic carbon is converted into carbon dioxide by heating to high temperatures in the presence of oxygen. The amount of carbon dioxide produced is proportional to the TOC of the rock. The minimum amount of TOC considered adequate for hydrocarbon production is generally considered to be between 0.5 and 1 percent TOC by weight.

INDIRECT METHODS OF ANALYSES

The type of organic matter in a rock can be determined indirectly through study of the physical and chemical characteristics of the kerogen or directly by using pyrolysis (heating) techniques. The indirect methods of analysis include examination of kerogen with a microscope and chemical analysis of kerogen. Microscopic examination can identify different types of kerogen, such as spores, pollen, leaf cuticles, resin globules, and single-celled algae. Kerogen that has been highly altered and is amorphous can be examined using fluorescence techniques to determine whether it is oil-prone (fluorescent) or inert or gas-prone (nonfluorescent). Chemical analysis of kerogen provides data on the proportions of chemical elements, such as carbon, hydrogen, sulfur, oxygen, and nitrogen. A graph of the ratios of hydrogen/carbon (H/C) versus oxygen/carbon (O/C) is used to classify kerogen by origin and is called a van Krevelen diagram. There are three curves on a van Krevelen diagram, labeled I, II, and III, corresponding to three basic types of kerogen. Type I is rich in hydrogen, with high H/C and low O/C ratios, as in some algal deposits; this type of kerogen generally yields the most oil. Type II has relatively high H/C and low O/C ratios and is usually related to marine sediments containing a mixture of phytoplankton, zooplankton, and bacteria; this type of kerogen yields less oil than Type I, but it is the source material for a great number of commercial oil and gas fields. Type III is rich in oxygen, with low H/C and high O/C ratios (aromatic hydrocarbons), as in terrestrial or land plants; this type of kerogen is comparatively less favorable for oil generation but tends to generate large amounts of gas when buried to great depths. As burial depth and temperature increase, the amount of oxygen and hydrogen in the kerogen decreases, and the kerogen approaches 100 percent carbon. Hence, a van Krevelen diagram can be used to determine both the origin of the organic matter and its relative thermal maturity.

The potential that a rock has for producing hydrocarbons can be evaluated through a pyrolysis, or heating, technique, commonly called Rock-Eval. Rock-Eval yields information on the quantity, type, and thermal maturity of organic matter in the rock. The procedure involves the gradual heating (to about 550 degrees Celsius) of a crushed rock sample in an inert atmosphere (nitrogen, helium) in the absence of oxygen. At temperatures approaching 300 degrees Celsius, heating releases free hydrocarbons already present in the rock; the quantity of free hydrocarbons is referred to as S_1. At higher temperatures (300-550 degrees Celsius), additional hydrocarbons and related compounds are generated from thermal cracking of kerogen in the rock; the quantity of these hydrocarbons is referred to as S_2. The temperature at which the maximum amount of S_2 hydrocarbons is generated is called T_{max} and can be used to evaluate the thermal maturity of the organic matter in the rock. In addition, carbon dioxide is generated as the kerogen in the rock is heated; the quantity of CO_2 generated as the rock is heated to 390 degrees Celsius is referred to as S_3. (The temperature is limited to 390 degrees Celsius because at higher temperatures, CO_2 is also formed from the breakdown of inorganic materials, such as carbonate minerals.) These data can be used to determine the hydrocarbon-generating potential of the rock, the quantity and type of organic matter, and the thermal maturity. For example, $S_1 + S_2$, called the genetic potential, is a measure of the total amount of hydrocarbons that can be generated from the rock, expressed in kilograms per ton. If $S_1 + S_2$ is less than 2 kilograms per ton, the rock has little or no potential for oil production, although it has some potential for gas production. If $S_1 + S_2$ is between 2 and 6 kilograms per ton, the rock has moderate potential for oil production. If $S_1 + S_2$ is greater than 6 kilograms per ton, the rock has good potential for oil pro-

duction. The ratio $S_1 / (S_1 + S_2)$, called the production index, indicates the maturation of the organic matter. Pyrolysis data can also be used to determine the type of organic matter present. The oxygen index is S_3/TOC, and the hydrogen index is S_2/TOC. These two indices can be plotted against each other on a graph, comparable to a van Krevelen diagram.

FINITE FOSSIL FUELS

Oil and gas are derived from the alteration of kerogen, an insoluble organic material, under conditions of high temperatures (50-150 degrees Celsius) and pressures (300-1,500 bars). After oil and gas are generated, they migrate upward out of organic-rich source rocks and come to be trapped and accumulate in specific types of geologic settings. The search for oil and gas deposits trapped in the subsurface can be expensive and time-consuming, and it requires trained exploration geologists. Once a promising geologic setting has been located, the only way to determine whether oil and gas deposits are actually present in the subsurface is to drill a well.

Oil and gas are two of the Earth's most important fossil fuels. It is important to understand that a finite amount of these hydrocarbons is present within the Earth. They cannot be manufactured when known reserves are depleted.

Pamela J. W. Gore

CROSS-REFERENCES

BIBLIOGRAPHY

Durand, Bernard, ed. *Kerogen: Insoluble Organic Matter from Sedimentary Rocks*. Paris: Éditions Technip, 1980. This book consists of a series of papers on various aspects of kerogen, ranging from its origin and appearance under the microscope to its chemical composition and structure as determined by a variety of analytical means. The articles are written by specialists; most of them are in English, but a few are in French. Technical but does contain a number of beautiful color plates illustrating the appearance of kerogen-rich rocks and organic microfossils (pollen, spores, acritarchs, dinoflagellates) as seen through the microscope.

Hunt, John Meacham. *Petroleum Geochemistry and Geology.* 2d ed. New York: W. H. Freeman, 1996. Petroleum is a complex organic substance, and its geochemistry is not clearly understood. This book covers many topics, including composition, origin, migration, accumulation, and analysis of petroleum and the application of petroleum geochemistry in petroleum exploration; seep and subsurface prospects; crude oil correlation; and prospect evaluation. While parts of this book are descriptive and explanatory, much of it requires a chemistry background and algebra.

North, F. K. *Petroleum Geology.* Boston: Allen & Unwin, 1985. A long book (607 pages) that covers a wide variety of topics related to petroleum geology, it includes five main parts: introduction; the nature and origin of petroleum; where and how oil and gas accumulate; exploration, exploitation, and forecasting; and distribution of oil and gas. Designed as a college textbook to introduce students to many topics with practical application to exploration and drilling in addition to the basics on the origin of oil and gas. Well illustrated with maps and geologic cross sections representing many oil-producing areas around the world. Suitable for geologists and college students.

Peters, K. E. "Guidelines for Evaluating Petroleum Source Rock Using Programmed Pyrolysis." *The American Association of Petroleum Geologists Bulletin* 70 (March, 1986): 318-329. Although rather technical in nature, this article provides information on Rock-Eval pyrolysis, one of the major analytical techniques for analyzing rocks to determine their hydrocarbon potential. Provides a brief summary of the technique and goes into detail using numerous examples, discussing some of the problems encountered in interpreting samples. Suitable for geologists and advanced college students.

Selley, Richard C. *Elements of Petroleum Geology.* 2d ed. New York: W. H. Freeman, 1998. A well-organized text covering the specifics of oil and gas and their relationship to geology. This book is designed as a college textbook for students near the end of their coursework in geology or for geologists beginning careers in the petroleum industry. Fairly technical, it requires basic understanding of geological concepts. The book contains subject and proper name indexes, useful illustrations, and appendices that include a well classification table, a glossary of oil terms and abbreviations, and a table of conversion factors.

Tissot, Bernard P., Bernard Durand, J. Espitalié, and A. Combaz. "Influence of Nature and Diagenesis of Organic Matter in Formation of Petroleum." *The American Association of Petroleum Geologists Bulletin* 58 (March, 1974): 499-506. This article discusses the generation of hydrocarbons and changes in kerogen that occur during burial. Somewhat technical but well illustrated with graphs. Provides a concise summary of the types of kerogen and depths at which oil and gas are generated.

Tissot, Bernard P., and D. H. Welte. *Petroleum Formation and Occurrence.* 2d ed. New York: Springer-Verlag, 1984. This book is one of the most comprehensive guides to the origin of petroleum and natural gas and should be considered one of the leading references in the field. The book is divided into five parts: the production and accumulation of geologic matter (a geological perspective); the fate of organic matter in sedimentary basins (generation of oil and gas); the migration and accumulation of oil and gas; the composition and classification of crude oils and the influence of geological factors; and oil and gas exploration (application of the principles of petroleum generation and migration). Each part is divided into chapters, which are well written and well illustrated with line drawings and graphs. Easy to read; up-to-date coverage of the field is provided. An indispensable reference for geologists that is suitable for college-level students.

Waples, Douglas W. *Geochemistry in Petroleum Exploration.* Boston: International Human Resources Development Corporation, 1985. This book provides an overview of the origin of oil and gas and should be considered a leading reference in the field. Concise and well illustrated with line drawings and graphs. Easy to read. A good reference for geologists, it is also suitable for college-level students.

Welte, Dietrich H., Brian Horsfield, and Donald R. Baker, eds. *Petroleum and Basin Evolution: Insights from Petroleum Geochemistry, Geology, and Basin Modeling*. New York: Springer, 1997. The authors explore the origins of oil and gas from the perspective of mathematical modeling of sedimentary basins. Somewhat technical, but illustrations and maps help clarify many of the difficult concepts. Bibliography.

OIL CHEMISTRY

Crude oil is fossil organic tissue that has been transformed by geologic processes into a complex mixture of many different chemical compounds called hydrocarbons. Although the composition of oils varies widely, the most abundant hydrocarbons in most oils are the paraffins, naphthenes, aromatics, and compounds with nitrogen, sulfur, and oxygen attached (NSOs). Less abundant compounds, called biomarkers, are true "geochemical fossils" that retain the original molecular structure of the organisms from which the oil is derived.

PRINCIPAL TERMS

AROMATIC HYDROCARBONS: ring-shaped molecules composed of six carbon atoms per ring; the carbon atoms are bonded to one another with alternating single and double bonds

BIOMARKERS: chemicals found in oil with a chemical structure that definitely links their origin with specific organisms; also called geochemical fossils

HYDROCARBONS: natural chemical compounds composed of carbon and hydrogen, usually of organic origin; they make up the bulk of both petroleum and the tissues of organisms (plants and animals)

MOLECULAR WEIGHT: a measure of the mass of the molecule of a chemical compound, as determined by both the total number and the size of atoms in the molecule

NAPHTHENE HYDROCARBONS: hydrocarbon molecules with a ring-shaped structure, in which any number of carbon atoms are all bonded to one another with single bonds

OXIDATION: a very common chemical reaction in which elements are combined with oxygen—for example, the burning of petroleum, wood, and coal; the rusting of metallic iron; and the metabolic respiration of organisms

PARAFFIN HYDROCARBONS: hydrocarbon compounds composed of carbon atoms connected with single bonds into straight chains; also knows as n-alkanes

SATURATED HYDROCARBONS: hydrocarbon compounds whose molecules are chemically stable, with carbon atoms fully bonded to other atoms

PARAFFIN HYDROCARBONS

The various hydrocarbon (organic matter) compounds found in petroleum differ from one another in two fundamental ways: the number of carbon and hydrogen atoms in the hydrocarbon molecule and the shape of the molecule. Hydrocarbon molecules are classified according to the number of carbon atoms they contain. A simple numbering system is used where C_2, for example, refers to a molecule with 2 carbon atoms, and C_3 is a molecule with 3 carbons. In most oils, compounds of very low molecular weight (less than C_5) are dissolved in the oil as natural gas. When the crude oil is pumped out of the ground, these molecules evaporate from the liquid petroleum and are either burned off or collected to be used as fuel.

Carbon atoms can be bonded to one another in straight chains, rings, or combinations of these basic forms. In many hydrocarbon molecules, small molecular fragments, called side chains, are attached like branches of a tree to the main chain or ring of the molecule. The simplest and most abundant petroleum hydrocarbons in crude oils are straight carbon chain molecules, referred to as the paraffin series. The smallest and simplest paraffin hydrocarbon molecule is methane, the most common component of natural gas. The methane molecule is composed of a central carbon atom bonded to four hydrogen atoms in a three-sided pyramid arrangement, or tetrahedron. The structure of paraffin molecules of higher molecular weight (C_2 to C_{30}) is made when additional carbon atoms are attached to the basic methane tetrahedron, making a carbon chain. As these carbon atoms are successively added, a chainlike arrangement of carbon atoms develops, with hydrogen atoms bonded to all the carbon atoms. This series of chainlike molecules is called the n-alkane series, with "n" denoting the number of carbon atoms in the chain. For example, the compound pentane, an abundant component of natural gas, is a five-carbon alkane chain (C_5H_{12}). Octane, the hydrocarbon compound by which gasolines are graded, is an eight-carbon alkane (C_8H_{18}). The highest molecular weight n-alkane found as a liquid in oil is heptadecane ($C_{17}H_{36}$), a hydrocarbon with a boiling point of 303 degrees Celsius.

All carbon atoms have four positions, called bonding sites, at which other atoms can attach themselves to form a molecule. In petroleum hydrocarbons, the carbon atoms are usually bonded to either hydrogen atoms or other carbon atoms. In order for the molecule to be stable so that it does not spontaneously react chemically with other molecules in the oil, all four bonding sites of each carbon atom must be occupied; these stable molecules with filled bonding sites are called saturated hydrocarbons. For any saturated n-alkane in the paraffin series, the number of hydrogen atoms in the molecule (#*H*) can be predicted from the number of carbon atoms (*n*) in the chain by the simple equation #$H = 2n + 2$. Note that for all n-alkane hydrocarbons the carbon-to-hydrogen ratio is always less than 1:2.

NAPHTHENE HYDROCARBONS

If the ends of a paraffin hydrocarbon chain are linked together to form a ring, the result is the shape of the other abundant group of hydrocarbon compounds in crude oils, the naphthene series. Naphthene molecules are composed of carbon atoms bonded together in rings; molecules with this molecular geometry are usually referred to as cyclic hydrocarbons or ring compounds. The simplest naphthene hydrocarbon is cyclopropane (C_3H_6), a three-carbon ring molecule that is a gas dissolved in crude oil. Cyclopropane, like other hydrocarbons of low molecular weight, bubbles out of the oil solution when it reaches the low-pressure conditions of the Earth's surface in the same way that carbon dioxide gas bubbles out of a soft drink when the bottle is opened. In this way, it becomes part of the natural gas that is associated with crude oil production. The liquid naphthene with the lowest molecular weight is cyclopentane (C_5H_{10}); this compound and cyclohexane (C_6H_{12}) are the dominant cyclic hydrocarbons in most oils.

Saturated naphthene hydrocarbon molecules have hydrogen atoms bonded to all the carbon atom bonding sites in a manner similar to that of n-alkanes. In this case, two hydrogen atoms are bonded to each carbon, and the ratio of carbon to hydrogen for these compounds is 1:2.

AROMATIC HYDROCARBONS

Less abundant in oils than the paraffins and naphthenes are the aromatic hydrocarbons. This group of petroleum hydrocarbons, which constitutes from 1 to 10 percent of most crude oils, is so named because many of the compounds have pleasant, sometimes fruity, odors. The aromatics also have a carbon ring structure, but this structure has a different geometry from that of the naphthenes. Aromatic hydrocarbon molecules are formed of one or more six-carbon rings in which the carbon atoms are bonded to one another with alternating single and bouble bonds. For this reason, the aromatic hydrocarbons have a carbon-to-hydrogen ratio of 1:1. The petroleum aromatic with the lowest molecular weight is benzene (C_6H_6), a chemical commonly used as an industrial solvent.

Many of the aromatic hydrocarbons are cancer-causing (carcinogenic) substances; most potent are the high molecular weight, multiringed aromatic molecules referred to as PAHs (poly-aromatic hydrocarbons). The first carcinogen ever discovered is benzopyrene, an aromatic molecule composed of five carbon rings. In the late 1800's, Sir Percival Pott linked this substance with cancer of the scrotum in London chimneysweeps. Their daily exposure to the aromatic hydrocarbons in coal tar and soot, coupled with their poor personal hygiene, was responsible for epidemic proportions of this disease.

NSOs

A small percentage of the hydrocarbons found in oils have distinctive molecular fragments bonded onto basic hydrocarbon structures. These nonhydrocarbon fragments most commonly contain nitrogen, sulfur, and oxygen; hydrocarbon compounds with these attached fragments are called NSOs for this reason. NSO molecules tend to have much higher molecular weights than do the hydrocarbon molecules described earlier. One of the most interesting of the NSO molecular fragments is the amino group; it contains nitrogen and has the formula NH_2. The amino group is the essential component of the amino acids, the building blocks for the many different proteins of which animal organs, muscles, and other tissues are composed. Amino acids are simple molecules that can form by organic or inorganic processes, but their presence in oil is best explained by inheritance from the organic matter source of the oil.

BIOMARKERS

The minor, or trace, hydrocarbon components in crude oil generally make up much less than 1 percent of the total oil; they are probably the most interesting of all petroleum compounds. Many of these chemicals have a chemical composition and molecular structure that definitely links their origin to specific organisms; they are termed geochemical fossils or biological markers (biomarkers) for this reason. Biomarker chemicals were synthesized by the organisms from which the oil originated and have been preserved through the long and complex history of sediment deposition and burial, oil formation (catagenesis), and migration of the oil from its source to the reservoir rock. Biomarkers are generally large molecules, with much higher molecular weights than those of the more abundant oil hydrocarbons, that have a carbon-to-hydrogen ratio greater than 1:2.

Some of the long chain paraffin hydrocarbons are among the best-known biomarkers. One relatively abundant group, the isoprenoid hydrocarbons, have chain-type molecules based on the isoprene group (C_5H_8); isoprene is the primary source of synthetic rubber. Isoprenoids are common in the waxes and chlorophyll of terrestrial green plants and are present in many crude oils and ancient sediments, which indicates that the petroleum isoprenoids are derivatives of chlorophyll and that kerogen from terrestrial plants is a significant source of oils. The most interesting of the plant-derived isoprenoids found in petroleum is pristane, a C_{15} carbon chain with four CH_3 molecular fragments, called methyl groups, attached to the main paraffin chain as side chains. Phytane is an isoprenoid similar to pristane but is composed of a C_{16} carbon chain with four methyl group side chains. The ratio of pristane to phytane is useful to geologists trying to determine the type of organic matter from which the oil was de-

rived. Oil derived mostly from terrestrial plant tissue, for example, has a high pristane-to-phytane ratio (greater than 4:1), while oils from marine animal tissue have much lower ratios.

The study of the biomarker composition of oils has given geologists valuable insight into the origin of petroleum and is used as a valuable tool in oil exploration as well. Every oil has a unique biomarker composition that it inherits from the kerogens that generated the oil and the conditions of catagenesis. Certain biomarker compounds, even when present in exceedingly minute quantities, can be detected with the mass spectrometer. This "geochemical fingerprint" allows petroleum geologists to recognize distinctive chemical similarities between oils and their source sediments. The geochemical fingerprinting technique is routinely used to distinguish different oils from one another, to correlate similar oils from different areas, and to demonstrate a similarity between kerogen-bearing source sediments and the oil that was generated from them. It is also used to increase oil-exploration efficiency by serving as a critical clue to the presence of undiscovered petroleum. The technique has also been used at the sites of oil spills and polluted groundwater and in other instances of oil pollution to identify the source pollutant; it has been successfully used as evidence in courts of law to determine the guilty party and to calculate damages for episodes of oil pollution.

CHROMATOGRAPHY

To determine the chemical composition of an oil, petroleum geochemists employ many different techniques, most of which are modifications of standard techniques of organic chemistry. Crude oils are composed of a vast number of individual chemical compounds that differ from one another in their molecular weight and shape and in the distribution of electric charges on the outer portions of the molecules. These individual compounds are separated from one another by a technique called chromatography. Portions of the oil are slowly passed through a long glass or metal column packed with a chemical substance (usually an organic chemical) that attracts hydrocarbon molecules having certain size or charge characteristics. The attractive chemical inside this "chromatographic column" has a greater affinity for the

PETROLEUM PRODUCTS AND PROCESSING

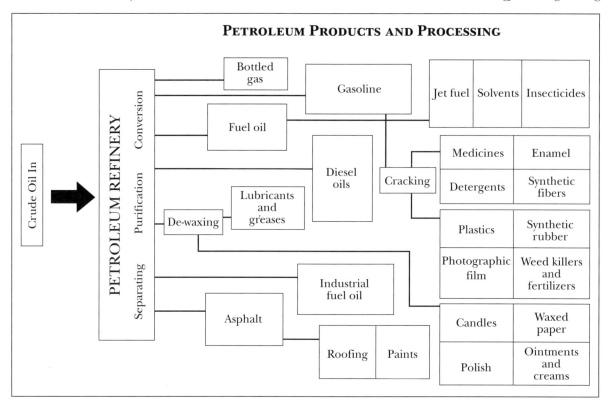

heaviest or most highly charged molecules, so hydrocarbon molecules of different sizes and charges move through the column at different rates and emerge from the end of the column at different times. For example, the n-paraffins are separated from one another by their molecular weight, so the lightest of these molecules emerges first from the column, and the others come out in the order of their carbon number. It should be noted that no column exists that separates all petroleum hydrocarbons, and several different chromatic columns are needed to separate an oil into its constituent chemicals. At the end of the column, a detector that is sensitive to molecular weight, charge, or other characteristics measures the amounts of each hydrocarbon compound as it emerges from the column and graphs the results. As with chromatographic columns, no one detector is able to separate all petroleum hydrocarbon compounds, and several different types are used. One of the most valuable and interesting of these detectors is the mass spectrometer. This detector breaks a hydrocarbon molecule into fragments as it leaves the chromatographic column and mea-

sures the molecular weight of these fragments. Laboratory studies have confirmed the way in which various known hydrocarbon compounds are fragmented in a mass spectrometer; this information is stored in the memory of a computer connected to the detector. During the analysis of an unknown hydrocarbon compound in an oil sample, its fragments are recognized and "reassembled" by the computer to determine the composition of the original hydrocarbon.

PETROLEUM PRODUCTS

Oil is a basic raw material from which many thousands of products are made. The types and quantity of petroleum products that can be obtained by refining crude oil are determined by its bulk chemical composition. For example, paraffin-rich oils with low aromatic content usually yield the highest quantity of hydrocarbon fuels, while highly aromatic oils refine into the best lubricating oils. Oils containing a large percentage of n-alkanes of low molecular weight (C_5-C_{13}) yield high quantities of gasoline, kerosene, jet fuel, and expensive hydrocarbon solvents used in many manufacturing processes; oils rich in n-alkanes of higher molecular weight (C_{14}-C_{40}) yield high quantities of diesel fuel and lubricating oils. Oils dominated by hydrocarbons of high molecular weight (heavy oils) yield asphalt and the hydrocarbons used to make plastics and synthetics.

Oils from different areas of the world can be divided into various types based on their bulk chemistry. Oils composed primarily of paraffins, termed paraffin-based crudes, are the most sought-after of all oil types by the refining industry but represent only a small percentage of the oil being produced worldwide. Most paraffin-based crude oils in North America are of Paleozoic age (about 600-250 million years ago) and are produced from oil fields in the midcontinent region. One of these is the famous Pennsylvania crude, which has historically been the standard against which all oils are compared. Similar oils on other continents are usually much younger. Paraffin-based oils of Mesozoic age (about 250-65 million years ago) are produced in Chile, Brazil, and the Caucasus region. Paraffin-based oils of Cenozoic age (the last 65 million years) are found in Africa, Borneo, and China.

Crude oils dominated by naphthene hydrocarbons, sometimes called asphalt-based oils, are relatively rare. Significant production of naphthenic oils occurs in the Los Angeles-Ventura area of Southern California and some oil provinces of the United States' Gulf coast, the North Sea, and South America. Highly aromatic oils are generally the heaviest and most viscous of all oil types; some are actually solids at surface temperatures. Important deposits of this unusual oil type include the famous black oils of Venezuela, the very large Athabasca tar sand deposits of western Canada, and certain oils from West Africa. Oils of intermediate composition, called mixed-base oils, make up the bulk of worldwide petroleum production.

James L. Sadd

CROSS-REFERENCES

Diagenesis, 1445; Earth Resources, 1741; Offshore Wells, 1689; Oil and Gas Distribution, 1694; Oil and Gas Exploration, 1699; Oil and Gas Origins, 1704; Oil Shale and Tar Sands, 1717; Onshore Wells, 1723; Petroleum Reservoirs, 1728; Unconformities, 1161; Well Logging, 1733.

BIBLIOGRAPHY

Barker, Colin. *Organic Geochemistry in Petroleum Exploration*. Tulsa, Okla.: American Association of Petroleum Geologists, 1979. Published for professional geologists as a short course in petroleum chemistry. The first half of this manual is so well organized and well written that most readers should be able to use it. Not as well illustrated as some of the texts below but an excellent resource on the subject.

Chapman, R. E. *Petroleum Geology*. New York: Elsevier, 1983. Written to emphasize petroleum production and minimize the geological aspects. This book is the best of all listed here for gaining an understanding of petroleum production and aspects of refining oil. Has a good section on basic chemistry.

Hobson, G. D., and E. N. Tiratsoo. *Introduction to Petroleum Geology*. Houston: Gulf Publishing, 1981. An excellent text for the basics of petroleum geology. Accessible to most readers.

Link, Peter K. *Basic Petroleum Geology*. Tulsa, Okla.: Oil and Gas Consultants International, 1987. A concise treatment of those aspects of petroleum chemistry most important to the science and business of oil exploration.

North, F. K. *Petroleum Geology*. Boston: Allen & Unwin, 1985. One of the best general texts on all aspects of petroleum geology. Sections on petroleum chemistry are not as detailed as in some texts but explain basic information very well. Well illustrated.

Selley, Richard C. *Elements of Petroleum Geology*. 2d ed. New York: W. H. Freeman, 1998. A well-organized text covering the specifics of oil and gas and their relationship to geology. This book is designed as a college textbook for students near the end of their coursework in geology or for geologists beginning careers in the petroleum industry. Fairly technical, it requires basic understanding of geological concepts. The book contains subject and proper name indexes, useful illustrations, and appendices that include a well classification table, a glossary of oil terms and abbreviations, and a table of conversion factors.

Taylor, G. H., et al., eds. *Organic Petrology*. Berlin: Gebreuder Borntraeger, 1998. This lengthy, well-illustrated text focuses on the geochemical analysis of coal, peat, and oil shales.

Tissot, B. P., and D. H. Welte. *Petroleum Formation and Occurrence*. New York: Springer-Verlag, 1978. A standard textbook for college-level courses in petroleum geology. A comprehensive text, with a well-written and complete section on petroleum biomarkers. Most readers will find this text informative, but a few of the sections on oil chemistry do require a basic organic chemistry background.

Zamel, Bernard. *Tracers in the Oil Field*. New York: Elsevier, 1995. Zamel covers the use of radioactive tracers in oil and gas exploration and analysis. Illustrations and bibliography.

OIL SHALE AND TAR SANDS

Although oil shales and tar sands are not generally thought of as potential sources of petroleum products, these resources have been developed to possibly replace dwindling oil fields. Tremendous reserves exist worldwide, but the current costs and technology preclude the use of oil shales and tar sands on a full scale.

PRINCIPAL TERMS

BARREL: the standard unit of measure for oil and petroleum products, equal to 42 U.S. gallons or approximately 159 liters

BITUMEN: a generic term for a very thick, natural semisolid; asphalt and tar are classified as bitumens

HYDROCARBON: an organic compound consisting of hydrogen and carbon atoms linked together

KEROGEN: a waxy, insoluble organic hydrocarbon that has a very large molecular structure

OIL SHALE: a sedimentary rock containing sufficient amounts of hydrocarbons that can be extracted by slow distillation to yield oil

RESERVOIR ROCK OR SAND: the storage unit for various hydrocarbons; usually of sedimentary origin

RETORT: a vessel used for the distillation or decomposition of substances using heat

TAR SAND: a natural deposit that contains significant amounts of bitumen; also called oil sand

OIL SHALE RESERVES

Oil shales, or "rocks that burn," are typically fine-grained, stratified sedimentary rock. The term "oil shale" is actually a misnomer because the reservoir rock does not have to be a shale, nor does it contain oil as it is generally known. Organic matter is present in the pores of these rocks in the form of kerogen. Kerogen is produced over a long time as the original organic-rich sediments are transformed into complex hydrocarbons. Unlike oil and natural gas, which move relatively easily in the subsurface, hydrocarbons in oil shale migrate at an almost imperceptible rate or not at all, because the sedimentary parent rock often has a very low porosity and permeability. The richest known oil shales produce between 320 and 475 liters (2-3 barrels) of oil per ton of processed rock. Based on current technology, though, less than 7 percent of these reserves are recoverable.

When examined on a global basis, the United States has large oil shale reserves. Oil shales are located under more than 20 percent of the land area of the United States. Roughly 50 percent of the worldwide reserves are found in the Green River formation, a shale and sandstone unit that formed during the Eocene epoch about 50 million years ago. This formation, the only one in the western United States that has been extensively studied for oil shale potential, covers approximately 42,000 square kilometers of southwest and south-central Wyoming, northeast Utah, and northwest Colorado. The thickest portions of the Green River formation are located in large structural basins that allowed large, shallow lakes to form in the topographically depressed areas. Subsequent deposition of rich organic sediments and later burial and thermal alteration led to the present deposits. Up to 540 billion barrels of oil exist in the rock units having a thickness greater than 10 meters, a thickness that is sufficient to produce enough hydrocarbons to be cost effective using present-day recovery methods. Estimates of the total oil shale reserves for the Green River formation range as high as 2,000 billion barrels of oil. Unfortunately, wide-scale development of these resources will probably not happen, because the extraction process requires large amounts of water, a commodity in small supply in the arid western United States. Approximately 3 liters of water are required to extract 1 liter of oil, so each barrel of oil would require that almost 480 liters of water be used to remove the oil from the reservoir rock. Major projects have been initiated in the Green River formation in western Colorado to set up entire cities to handle the processing of the reservoir rock and its eventual products. Almost all these projects, however, have been terminated or put on indefinite hold because of changes in the worldwide petroleum market.

Although the richest oil shale deposits are in the western United States, another 15-20 percent of worldwide reserves are found in Devonian and Ordovician rocks located from New York to Illinois and into southwestern Missouri. These deposits, however, are not economically usable, because the cost to extract the small amount of oil present far exceeds the value of the oil produced. Several nations attempted using oil shale as a source for petroleum products during World War II, but postwar economic variations shut most of them down within a few years.

TAR SAND RESERVES

Tar sands constitute another major potential source of unconventional oil reserves. These highly viscous deposits—sometimes referred to as tar, asphalt, and bitumen—probably formed as residues from petroleum reservoirs after the lighter, more hydrogen-rich crude oils migrated toward the surface. These porous sands contain asphaltic hydrocarbons, which are extremely viscous. Thus, the hydrocarbons are not bound up as tightly in the reservoir as they are in oil shales. G. Ronald Gray defines these heavy substances as one of the following: bitumen, an oil sand hydrocarbon that cannot be produced using conventional processes; extraheavy oils; and heavy oils (see Bibliography). These three types are usually lumped together when discussing worldwide reserve estimates, which are set at about 5,000 billion barrels.

The seven largest tar sand fields have roughly the same amount of oil as do the three hundred largest conventional oil fields in the world. The three largest tar sand deposits are, in descending order, in northern Alberta, Canada; northeastern Siberia; and along the northern bank of the

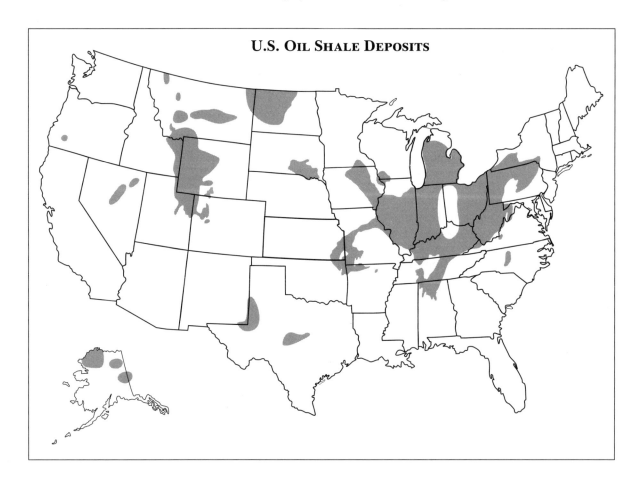

U.S. OIL SHALE DEPOSITS

Orinoco River, Venezuela. The amount of heavy oil in place in the tar sands in the Athabasca deposit of Alberta is essentially equal to all the petroleum reserves found in the Middle East. Tar sand deposits are found in twenty-two states in the United States, the largest deposits being situated in Utah. Given the massive reserves of tar sands throughout the world, they could potentially supply modern society with as much oil as that obtained from conventional flowing wells. The major problem in the future is developing the technology to recover the usable hydrocarbon products in a cost-effective manner.

MINING TECHNIQUES

Once the hydrocarbons are detected in the reservoir rock, they must be extracted. One of two primary mining techniques is used to recover the oil. After the upper soil and rock layers overlying the oil shale are stripped away, the reservoir rock is removed, crushed, and transported to a large retort, where it is heated. This process involves raising the temperature of the rock and hydrocarbons to about 480 degrees Celsius, the temperature at which kerogen vaporizes into volatile hydrocarbons and leaves a carbonaceous residue. When oil shale is heated, the amount of organic matter that is converted to oil increases as the amount of hydrogen in the deposit increases. This vapor is then condensed to form a very viscous oil. After the introduction of hydrogen, the mixture can be refined in a manner similar to that used to refine crude oil, which is drawn from the ground using conventional methods of drilling and pumping. One obvious problem with this method is that a considerable amount of heat (energy) must be expended in order to yield products that themselves are potential energy sources.

The second technique used to recover oil from oil shales involves in-situ heating of the reservoir rock after it has been fractured with explosives or water under pressure. Researchers at the Sandia National Laboratories in New Mexico may have discovered a method to enhance the extraction of kerogen from the reservoir rocks. Underground fracturing of the rock by controlled explosions increases the number and size of passageways for air to pass through. Air is a necessary component to complete the chemically driven thermal reactions that release the hydrocarbons from kerogen.

Heating the rock by pumping superheated water into it loosens the viscous hydrocarbons from the pores and cracks. Hydrocarbons are driven off by the heat, collected, and then pumped to the surface for further distillation and refining.

A key factor in the removal of hydrocarbon compounds from the ground is the carbon-to-hydrogen ratio. Carbon is twelve times as heavy as hydrogen, so carbon-rich heavy crude oils are more dense than conventional oil, which contains a higher percentage of hydrogen. Carbon-rich crude oil yields smaller amounts of the more desirable lighter fuels, such as kerosene and gasoline. Heavy crudes, such as those derived from oil shale and tar sands, can be chemically upgraded by removing carbon or adding hydrogen in the refining process. These procedures, however, are rather complex and certainly add to the already high cost of extraction and refining.

Progress has been made in the recovery rate of hydrocarbons from reservoir rock. Extraction technology has increased the amount recovered from about 12 percent in 1960 to more than 50 percent in some instances by the late 1980's. The recovery rate is dependent on the percentage of complex hydrocarbons present, which varies from formation to formation. One main problem impeding the all-out effort to continue full-scale production is the economics of oil shale mining. When the price of oil, especially that which is imported, is low and the amounts abundant, it is not feasible to consider oil shale as a source.

ENVIRONMENTAL CONCERNS

Several major environmental problems exist with oil shale production. The refining process actually generates more waste than the original amount of rock that is processed. The processed rock increases in volume by about 30 percent, because the hot water and steam used to extract the kerogen from the oil shale enter into the clay molecules present in the rock and cause them to expand. The problem then is what to do with the increased volume of material, as it will overfill the void produced by mining the rock. This expanded material also weathers very rapidly so that it will not remain in place and form a stable terrane. In-situ retorting precludes the need to mine the shale and, thus, circumvents this disposal problem.

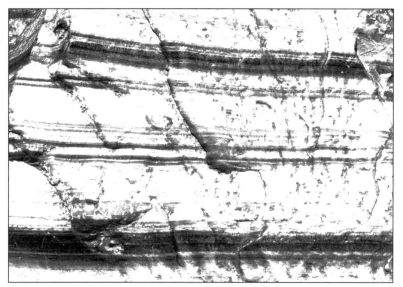

In Debeque, Colorado, the dark strata of these sedimentary layers are oil-bearing shales. (© William E. Ferguson)

Another environmental concern is that of the air quality in the vicinity of the processing plant. Large amounts of dust being thrown into the atmosphere have an adverse effect on the air in the immediate area surrounding the plant. On a large scale, the overall air quality on a global basis can be affected with a marked increase in oil shale production through increases in carbon dioxide. Research has shown that high-temperature retorting methods (those using temperatures exceeding 600 degrees Celsius) may produce more carbon dioxide from the carbonate rocks containing the oil than does the actual burning of the oil produced by the process. This process adds to the amount of carbon dioxide, thus increasing the worldwide greenhouse effect. Additional carbon dioxide is generated through the combustion of free carbon, which is present in the kerogen.

Tar sands are formed in environments different from those in which oil shales are produced. While oil shales appear to form in lake environments, especially those characterized by sandstone and limestone deposits, tar sands are often found in conjunction with deltaic or nonmarine settings. Most of the major deposits of tar sands are in rocks which are of Cretaceous age or younger, whereas oil shales are often associated with older sedimentary formations. The lack of a cement to hold the sand grains together results in high porosities and permeabilities, thus affording the viscous tars an opportunity to flow. Another geologic setting that enhances the formation of large tar deposits includes areas that have an impermeable layer overlying the deposits. The impermeable layer acts as a barrier to prevent the upward movement of the hydrocarbons.

The mining of tar sands must usually be done at or near the deposits, as large amounts of the sands are handled. The actual production of oil from tar sands in the Athabasca field involves large amounts of material being processed. For each 50,000 barrels of oil, almost 33,000 cubic meters of overburden must be removed and about 100,000 tons of tar sand mined and then discarded. Extraction of bitumen and heavy oils from tar sands is relatively difficult because of high levels of viscosity of the hydrocarbons. Bitumen at room temperature is heavier than water and does not flow. The viscosity of these substances can be changed dramatically by applying heat. If tar sands are heated to about 175 degrees Celsius, the bitumens present flow readily and are capable of floating on water. In some cases, the injection of steam into the reservoir sands increases the flow, thus allowing the material to be pumped out of the ground. Hot water can also be used, but this method requires vast volumes of water, something not available in many regions of the west where these sands are presently found. Underground combustion techniques utilize burning the tar sands underground, allowing the resulting heat to warm the bitumens to the point at which they flow and can be pumped to the surface.

PROSPECTS FOR PRODUCTION

The existence of oil shales and tar sands has been known for several centuries. Deposits were used as a source of oil for lamps in Europe and colonial America. Native Americans used them to patch their canoes. Once commercial production of petroleum expanded in the latter half of the nineteenth century, interest in developing and using oil shales and tar sands as a source of oil decreased. Interest increased, however, during World War II and also in the mid-1970's as a result of shortages in petroleum imports. Once the short-term oil crisis was over, though, both the federal government and private industry in the United States dropped most of their research and development associated with extracting oil from oil shale. The Canadian government has moved ahead with its development of extensive tar sand deposits and has been producing substantial amounts of oil from them for many years. Oil can be produced from tar sands for roughly $15 to $30 per barrel, depending upon various factors.

Undoubtedly, interest in developing oil shale reserves in the United States will increase as the national reserves of petroleum diminish and imports become scarcer and more expensive. Widespread production from oil shales will probably not occur until oil reaches a price of $50 to $60 per barrel or a technological breakthrough occurs. Such a breakthrough might include sonic drilling or genetic engineering. Sonic drilling uses a vibrating, rotating drill, and it can be faster and cheaper than other methods. At the end of the twentieth century, however, its use was limited to a depth of about 200 meters. Genetic engineering could involve the use of the many bacteria with the ability to break down hydrocarbons. It may be possible to genetically engineer them to efficiently convert kerogen into liquid or gaseous hydrocarbons. The mining method would then involve drilling holes into the oil shale and fracturing it with explosives below ground. Next, a nutrient solution containing the bacteria would be pumped into the formation; a sufficient time later, oil and gas could be extracted. If the bacteria were made dependent upon the nutrient solution, they would perish when the solution was exhausted. It must be recognized that even the worldwide reserves of oil shale and tar sand are finite and that other energy sources must be developed in the future.

David M. Best

CROSS-REFERENCES

Biogenic Sedimentary Rocks, 1435; Geochemical Cycle, 412; Geothermometry and Geobarometry, 419; Life's Origins, 1032; Mass Spectrometry, 483; Offshore Wells, 1689; Oil and Gas Distribution, 1694; Oil and Gas Exploration, 1699; Oil and Gas Origins, 1704; Oil Chemistry, 1711; Onshore Wells, 1723; Petroleum Reservoirs, 1728; Well Logging, 1733.

BIBLIOGRAPHY

Carrigy, M. A. "New Production Techniques for Alberta Oil Sands." *Science* 234 (December, 1986): 1515-1518. A succinct, fact-filled discussion of tar sand production in Canada. The article is clearly written, so it is suitable for high school readers.

De Nevers, Noel. "Tar Sands and Oil Shales." *Scientific American* 214 (February, 1966): 21-29. A good review of the status of the development of these deposits in the earlier days of production. The author makes some comparisons of the Green River formation oil shale fields and the tar sand sources in the Athabasca tar sand area of Alberta.

Duncan, Donald C., and Vernon E. Swanson. *Organic-Rich Shale of the United States and World Land Area.* U.S. Geological Survey Circular 523. Washington, D.C.: Government Printing Office, 1965. A classic work that serves as the basis for much of the present-day knowledge of worldwide deposits of oil shale, in terms of both locations and reserves. A very readable presentation, it is highly recommended as a key reference. Data presented have since been updated in other works.

Gray, G. Ronald. "Oil Sand." In *McGraw-Hill Encyclopedia of the Geological Sciences*, edited by S. P. Parker. 2d ed. New York: McGraw-Hill, 1988. This article discusses several geological factors that control the environments of formation of tar sands, along with providing specific information on worldwide reserves. Good source of data for those desiring to make comparisons of various areas.

Hughes, Richard V. "Oil Shale." In *The Encyclopedia of Sedimentology*, edited by R. W. Fairbridge and J. Bourgeois. Stroudsburg, Pa.: Dowden, Hutchinson and Ross, 1978. A brief discussion of the environments and characteristics of selected oil shale deposits. More than a dozen technical references are provided for those readers looking for the scientific basis and interpretation of oil shales.

Hyne, Norman J. *Dictionary of Petroleum Exploration, Drilling, and Production.* Tulsa, Okla.: PennWell, 1991. This is a dictionary of all terms associated with petroleum and the petroleum industry. A great resource for the beginner in this field. Illustrations and maps.

_____. *Nontechnical Guide to Petroleum Geology, Exploration, Drilling, and Production.* Tulsa, Okla.: PennWell, 1995. An excellent introductory handbook for the reader without prior background in petroleum geology. Discusses the exploration, refining, and distribution. Illustrations, bibliography, and index.

Production from Fractured Oil Shales. Richardson, Tex.: Society of Petroleum Engineers, 1996. This book covers oil reservoir engineering and oil shale reserves in the United States.

Smith, John W. "Synfuels: Oil Shale and Tar Sands." In *Perspectives on Energy, Issues, Ideas, and Environmental Dilemmas,* edited by L. C. Ruedisili and M. W. Firebaugh. 3d ed. New York: Oxford University Press, 1982. A superb article that discusses the definitions, characteristics, resources, and production of oil shales and tar sands in a well-outlined

presentation. Its very complete reference list provides more sources to readers desiring either general or technical details. This article would serve as an excellent starting point for someone researching the topic. Suitable for high school students.

Smith, John W., and Howard B. Jensen. "Oil Shale." In *McGraw-Hill Encyclopedia of the Geological Sciences,* edited by S. P. Parker. 2d ed. New York: McGraw-Hill, 1988. An excellent summary article, with several figures showing oil shale deposits in the United States. Explains the technical aspects of oil shale properties. An extensive bibliography lists more than a dozen technical references. Suitable for high school readers and beyond.

Snape, Colin. *Composition, Geochemistry, and Conversion of Oil Shales.* Boston: Kluwer Academic Publishers, 1995. This book was published in cooperation with the North Atlantic Treaty Organization (NATO) Scientific Affairs Division. It provides an overview of oil shale properties and production.

Yen, T. F., and G. V. Chilingarian, eds. *Oil Shale.* New York: Elsevier, 1976. A compilation of twelve chapters authored by experts in the various facets of oil shale studies. Some articles are suitable for high school and college students; others are very specific and are suitable for researchers. Numerous references provided with each chapter.

ONSHORE WELLS

Several million wells have been drilled in the exploration for oil and gas within every environment of the Earth. Testing, completion, and evaluation technologies of a completed well are similar whether the borehole is dug by the cable-tool or by the rotary method. Downhole, wire-line analyses have evolved from the simple electric log of the 1920's to a modern array of evaluative services.

PRINCIPAL TERMS

CABLE-TOOL DRILLING: a repetitive, percussion process of secondary use in the boring of relatively shallow oil and gas wells

COMPLETION PROCEDURES: all methods and activities necessary in the preparation of a well for oil and gas production

DOWNHOLE TOOL: a drill bit and motor mounted on the end of the drill string; fluid pumped into the drill string drives the downhole tool, and the drill string is not rotated

DRILL STRING: the length of steel drill pipe and accessory equipment connecting the drill rig with the bottom of the borehole

DRILLING FLUIDS: a carefully formulated system of fluids used to lubricate, clean, and protect the borehole during the rotary drilling process

DRILLING RIG: the collective assembly of equipment, including a derrick, power supply, and draw-works, necessary in cable-tool and rotary drilling

HYDROCARBONS: naturally occurring organic compounds that in the gaseous state are termed natural gas and in the liquid state are termed crude oil or petroleum

ROTARY DRILLING: historically, the principal method of boring a well into the Earth using a fluid-circulating, generally diesel-electric generated, rotating process

WELL LOG: a graphic record of the physical and chemical characteristics of the rock units encountered in a drilled borehole

CABLE DRILLING

The location of the first drilling operation is lost to history, although it is known that the Chinese drilled for brine and water two thousand years ago using crude cable-tool methods. Similar methods of drilling were still being employed in the 1850's. By this process, a well is created by raising and lowering into the borehole a heavy metal bit suspended from a cable or rope. Gradually, the bit will pound its way through the rocks. With the addition of a jar, a mechanical device that imparts a sharp vertical stress to the bit, the process is greatly improved. Surface equipment, contained within a wooden derrick, or rig, was commonly steam driven and repeatedly withdrew the bit from the hole, allowing it to be again dropped to the bottom of the well. As the bottom of the hole fills with rock chips, a bailer is periodically used to remove this debris.

Cable drilling is a slow process: Its greatest advantage is easy identification of oil- and gas-producing rock units. Because minimal drilling fluids are used, uncontrolled surface flows of encountered hydrocarbons (occurrences known as blowouts) are frequent. For this reason, cable drilling is most applicable within depths of 1,000 meters. As late as 1920, cable-tool rigs drilled as many as 85 percent of all wells completed in the United States.

ROTARY DRILLING

Introduced to the industry at Corsicana, Texas, in 1895, the rotary method was used to drill 90 percent of American wells in the 1950's. In rotary drilling, the drill bit is attached to connected sections of steel pipe, or drill string, and lowered into the borehole. Pressure is placed on the bit and the drill pipe is rotated, causing the bit to grind against the bottom of the borehole. In contrast to the cable method, new borehole depths are created by the rock being torn, rather than pounded. When the drill bit becomes dull, the drill string is removed from the borehole, disassembled, and stacked within the tall mast, or derrick. A new bit is attached, and the drill string is reassembled.

The application of a drilling fluid system is a key element of the rotary method. Originally ordinary mud, drilling fluids have become a carefully formulated solution of water, clays, and chemical additives. These fluids are circulated under pressure down the center of the drill string, extruded through the drill bit, and pumped back to the surface through the space between the drill string and the borehole. These fluids serve several important functions: to lubricate and cool the drill bit, to remove rock chips from the borehole, and to protect the borehole from dangerous blowouts. Because of its mechanical advantages, the rotary method is approximately ten times faster than is the cable method in drilling a borehole.

After the borehole is completed, and assuming commercial deposits of oil or gas are discovered, completion procedures are initiated. Because surface instruments cannot detect the presence of subsurface hydrocarbons, the rock units exposed in the wall of a borehole must be evaluated for the presence and quality of contained oil and gas. A preliminary analysis is conducted on the rock chips continuously brought to the surface by the drilling fluid system. Later, instruments lowered into the borehole determine the physical and chemical characteristics of the penetrated rocks and their contained fluids and gases. Should the presence of hydrocarbons be indicated, further testing is conducted to determine the economic value of the discovery. Finally, if economic payout is indicated, the borehole undergoes final completion procedures. Special production tubing systems are installed, and the oil or gas is pumped, or flows under its own pressure, from the rocks up the borehole and into a pipeline or surface storage system.

Rotary drilling procedures vary little with geographic location or climate. In urban areas, the derrick is covered with soundproof material and sometimes even disguised for aesthetic purposes. In sensitive areas, such as Arctic regions and offshore operations, safety and environmental preservation precautions are mandated by state and federal law.

OTHER DRILLING METHODS

Drilling processes other than the cable and rotary have been used. With the turbodrill, or downhole tool, method, the drill string remains stationary while the drill bit rotates under the influence of circulating drilling fluid. This process drills very straight boreholes with minimum mechanical problems, and it excels at directional drilling, especially horizontal drilling. It also greatly reduces wear on the drill stem pipe. The use of a downhole tool is replacing the rotary drilling method. Although a rotary drill rig is still used, its rotary capability is not used. The hammer drill, a combination of slow rotary motion coupled with percussion impact, produces a faster rate of rock penetration, but this method has not been widely accepted in practice. Experiments with vibration and sonic drills have proved unsuccessful or uneconomic.

HISTORY OF A BOREHOLE

After the location for a borehole is determined, rotary drilling equipment is taken to the chosen

A petroleum-well drilling rig. (© William E. Ferguson)

site, an area of about 0.5 to 1 hectare. When the drill rig is assembled, sections of drill pipe, or drill string, are connected within the derrick. The drilling fluid hose is connected to the upper end of the drill string, while a drill bit is attached to the bottom. The rig is now ready to "make hole." The history of a borehole begins with its "spud-in" time, that moment when the ground is broken by the rotating bit.

The rotary table, located in the center of the rig floor and connected to powerful engines, rotates the drill string and attached bit. As the bit rotates, drilling fluid (termed mud) is pumped down the inside of the

An oil-well pump in the Kettleman Hills, California. Hundreds of these pumps dot the landscape in parts of California's southeastern Central Valley. (© William E. Ferguson)

drill string and through openings in the bit. The density of this mud is carefully controlled so that as it exits the bit, it is capable of lifting rock fragments, or cuttings, from the bottom of the borehole, allowing the bit to rotate against a fresh rock surface. The drilling mud, with its contained cuttings, is circulated up the annulus (passage) created between the wall of the borehole and the outside of the drill string. At the surface, cuttings are separated by flowing the drilling mud through a vibrating sieve. Periodically, a sample of cuttings is collected for geologic analysis. Finally, the cleansed mud circulates through the "mud pit," where, after cooling to surface temperature, it is pumped through the drilling fluid hose back into the drill string. While the borehole is being drilled, this mud system is in continuous circulation. Approximately every 9 meters, as the borehole becomes deeper, a new section of drill pipe is added to the drill string, increasing the depth capability of the rig.

At shallow depths, where the bit is penetrating loose soils and poorly consolidated rock formation, the drilling speed is measured in tens to hundreds of meters per day. With increased depths, penetration rates will diminish to as little as a meter per day, depending on rotation pressure and velocity and rock characteristics. At the surface, "conductor pipe" is driven 6-10 meters into the

ground to protect the borehole against collapse. At depths below the conductor pipe, rock units containing fresh water are protected from drilling fluid contamination by cementing "surface casing" through the conductor pipe and into the borehole. At greater depths, progressively smaller radius "intermediate casing" may be cemented into the borehole, keeping the newly drilled hole open while sealing off unusually high or unusually low pressure rock strata. Because each new series of casing must fit into prior cemented casing, the borehole diameter becomes smaller with increased depth. When the borehole reaches programmed total depth (TD), the drilling process is complete. The next phase of activity involves testing for the presence and quantity of oil and gas.

The borehole is protected by cementing "production casing" through the depth of the production zone. Perforating guns, multibarrel firearms designed to fit into the borehole, are lowered to the target production depth and fired electrically. High-velocity bullets penetrate the casing cement and embed in the rock strata, creating pathways through the strata to the wall of the borehole. Oil or gas emitting from the rock through these pathways flows into installed production tubing and to the surface, where the hydrocarbons are either temporarily stored or directed to a nearby pipeline. At this point the well is completed and "on line."

TESTING AND ANALYSES

Drill cuttings are periodically collected from the drilling fluid and analyzed in the field in converted mobile-home vehicles. These field tests determine rock type, contained minerals, density, pore space percentage, and association with either natural gas or crude oil (petroleum). Since drilling is an expensive operation, commonly costing millions of dollars, the majority of boreholes are subjected to additional analyses, termed well logging. Conducted by contracted specialists, well-logging operations involve lowering an elongated instrument called a sonde to the total depth of the borehole. As the sonde is slowly pulled up the hole, it records various characteristics of the rocks exposed within the wall of the borehole and their contained fluids and gases. These characteristics, which include electrical resistivity, conductivity, radioactivity, acoustic properties, and temperature, are transmitted to the surface, where they are recorded and filed for future use. It is common for four or five different logs to be recorded, while on a very important borehole, more than twice this number may be taken.

In the office, individuals trained in geology and engineering study the cuttings analyses and logging data and determine the presence and economic extent of oil or gas by calculating rock porosity, permeability, density, thickness, lateral extent, inclination, and pressure at various depths in the borehole. Should these analyses be pessimistic, the borehole is declared "dry and abandoned" and permanently sealed at several depths by cement plugs. Such is the fate of approximately six out of seven boreholes drilled in frontier (new) geographic regions or to unproven depths; such boreholes are termed wildcat wells. For the one in seven wildcat wells in which logging analyses indicate a chance of success, verification analyses in the form of drill-stem testing (DST) will be conducted.

Drill-stem testing equipment is attached to the base of the drill string and lowered to the rock depth to be tested. After this depth is physically isolated from the rest of the borehole, assuring a valid test, the DST tool is activated, allowing fluids or gases contained within the isolated rocks to flow into the drill string and to the surface. From DST, rock pressures and flow capacities are calculated. When DST verifies positive economic results determined by logging analyses, the commercial quantities of either oil or gas, or both, are declared, and the well is prepared for its final completion phase.

ECONOMIC AND POLITICAL CONSIDERATIONS

The fortunes of the American oil and gas drilling industry are closely tied to the market value of a barrel (42 U.S. gallons, or about 159 liters) of crude oil. As that value increases, so generally does the number of drilling rigs under contract. Adding confusion to this economy-to-rig-use relationship are considerations such as international politics, governmental policies, environmental concern, and marketplace competition for high-risk investment dollars. After a century and a half of drilling wells in the search for new reserves of oil and gas, terrestrial portions of the United States are considered a mature exploration province. The chances of discovering large new reserves of hydrocarbons on land are very small. The future lies in drilling within the offshore provinces (deep water of the Gulf of Mexico and the Atlantic eastern seaboard) and environmentally protected regions (northern Alaska and national parks and forestlands). In order to maintain a hydrocarbon-based energy economy while reducing dependence upon foreign hydrocarbons, oil and gas well-drilling and production programs may have to take place in these frontier exploration regions. Such programs must be governed by consensus regulatory, environmental, and economic policies until either solar, nuclear, or some unforeseen resource assumes the dominant energy position and oil and gas wells no longer need be drilled into the Earth.

Albert B. Dickas

CROSS-REFERENCES

Biogenic Sedimentary Rocks, 1435; Mining Processes, 1780; Mining Wastes, 1786; Offshore Wells, 1689; Oil and Gas Distribution, 1694; Oil and Gas Exploration, 1699; Oil and Gas Origins, 1704; Oil Chemistry, 1711; Oil Shale and Tar Sands, 1717; Petroleum Reservoirs, 1728; Well Logging, 1733.

BIBLIOGRAPHY

Allaud, Louis A., and Maurice H. Martin. *Schlumberger: The History of a Technique.* New York: John Wiley & Sons, 1977. An interesting historical account of the mineral-prospecting methodology invented in 1912 by Conrad Schlumberger and used in modified form in the evaluation of the majority of oil and gas wells drilled throughout the world.

Gray, Forest. *Petroleum Production for the Nontechnical Person.* Tulsa, Okla.: PennWell, 1986. After thirty years in the industry, the author wrote this oil and gas story for an audience that also works daily in the industry but has yet to master the technology and its terminology. Each chapter is accompanied by a series of exercises. A detailed glossary is included.

Hyne, Norman J. *Dictionary of Petroleum Exploration, Drilling, and Production.* Tulsa, Okla.: PennWell, 1991. This is a dictionary of all terms associated with petroleum and the petroleum industry. A great resource for the beginner in this field. Illustrations and maps.

_____. *Nontechnical Guide to Petroleum Geology, Exploration, Drilling, and Production.* Tulsa, Okla.: PennWell, 1995. An excellent introductory handbook for the reader without prior background in petroleum geology. Discusses the exploration, refining, and distribution. Illustrations, bibliography, and index.

Kennedy, John L. *Fundamentals of Drilling.* Tulsa, Okla.: PennWell, 1982. An easy-to-understand basic presentation on oil and gas well drilling. Details the tools and methods used and includes sections on economics and future trends, as well as an introduction to the industry.

Langenkamp, Robert D. *Oil Business Fundamentals.* Tulsa, Okla.: PennWell Books, 1982. A book for individuals interested in learning business perspectives of oil and gas well drilling. In addition to technology, this nontechnical presentation includes chapters on ownership rules, drilling, financing, and the marketing of hydrocarbons.

Welker, Anthony J. *The Oil and Gas Book.* Tulsa, Okla.: PennWell, 1985. Written to bridge the communication gap between the oil and gas industry and members of the general public, such as bankers, investors, and newspersons. Especially valuable because of special-interest topics covering partnerships, joint ventures, promotions, and working interests.

Well Control. Richardson, Tex.: Society of Petroleum Engineers, 1996. This illustrated text covers the dangers of oil well blowouts, as well as safety measures that have been adopted to prevent such hazards. Bibliography.

PETROLEUM RESERVOIRS

A petroleum reservoir is a body of rock that contains crude oil, natural gas, or both, that can be extracted by a well. Two conditions must be met. First, to contain and permit extraction of its fluids, the reservoir rock must be porous and permeable. Second, there must be a reservoir trap, a set of conditions that concentrates the petroleum and prevents its migration to the Earth's surface.

PRINCIPAL TERMS

FIELD: one or more pools; where multiple, the pools are united by some common factor

NATURAL GAS: a flammable vapor found in sedimentary rocks, commonly, but not always, associated with crude oil; it is also known simply as gas or methane

PERMEABILITY: a measure of the rate of flow of fluids through a porous medium

PETROLEUM: a dark green to black, flammable, organic liquid commonly found in sedimentary rocks; it looks like used crankcase oil and is also called crude oil or liquid hydrocarbon

POOL: a continuous body of petroleum-saturated rock within a petroleum reservoir; a pool may be coextensive with a reservoir

POROSITY: the percentage of pore, or void, space in a reservoir

RESERVOIR: a body of porous and permeable rock; petroleum reservoirs contain pools of oil or gas

TRAP: a seal to fluid migration caused by a permeability barrier; traps may be either stratigraphic or structural

WELL LOG: a stripchart with depth along a well borehole plotted on the long axis and a variety of responses plotted along the short axis; there are many varieties, including borehole log, geophysical log, electric log, and wireline log; information may be obtained about lithology, formation fluids, sedimentary structures, and geologic structures

SEDIMENTARY RESERVOIR ROCK

Petroleum (from the Latin) literally means "rock oil." Petroleum reservoirs are volumes of rock that contain or have the potential to contain hydrocarbons (crude oil and natural gas) that can be extracted by wells. A reservoir may contain one or more pools, that is, continuous bodies of oil or gas. One or more related pools form a field. The definition of a pool depends on economic as well as geologic considerations—as the price of oil goes up, the size of a pool goes down. There are four components to a petroleum reservoir: the rock, the pores, the trap, and the fluid.

In the vast majority of cases (99 percent-plus), the reservoir rock is a sedimentary rock. Sedimentary rocks form when unconsolidated sediment, deposited on the Earth's surface by water or wind, becomes a solid rock. Two types of sedimentary rocks contain 95 percent of the world's petroleum: sandstones and carbonates. Sandstone reservoirs contain approximately half the United States' petroleum. A freshly deposited sand is made up of mineral grains, mostly quartz, deposited from water in rivers, along shorelines, and in the ocean in the shallow water adjacent to the continent; less often, it is deposited by wind in sand dunes. Sand-sized grains range from 0.06 to 2.00 millimeters, or smaller than a small pea.

Carbonate rocks include limestones and dolostones. Limestones are made of the mineral calcite (calcium carbonate, $CaCO_3$). Sand-sized grains of calcite form when organisms, such as clams and corals, extract calcite (and aragonite, a related calcium carbonate mineral) from water to build their skeletons. When the animal dies, the skeletons are broken into fragments, which are then sorted by waves and currents and deposited along beaches or on the sea floor. Very fine-grained, mudsized calcite and aragonite may be deposited with the sand-sized grains. This fine carbonate comes from calcareous algae (marine plants). In some cases, corals and algae may combine their activity to construct a wave-resistant mass, a coral reef. Reefs form the best potential reservoir rock. Dolostones, although they form a relatively small portion of the world's total carbonate mass, contain almost 80 percent of the petroleum in carbonate reservoirs in the United States. Dolostones are dominated by the mineral dolomite. Dolomite forms by the reaction of preexisting calcium carbonate minerals with magnesium-rich solutions. How this replacement takes place is not well understood.

POROSITY AND PERMEABILITY

A petroleum reservoir has the same general attributes as a water aquifer; it is a porous and permeable body of rock that yields fluids when penetrated by a borehole. Porosity is a measure of the pore space in a reservoir, the space available for storage of petroleum. It is usually expressed as a percentage. A freshly deposited unconsolidated sand contains about 25 percent intergranular pore space, normally occupied by water. As the sand is buried deeper and deeper, the grains interpenetrate, and the pores are reduced in size and percentage. At the same time, cements will precipitate in the pores. Thus, the pore space decreases, and the sand becomes a solid—a sandstone. To be a petroleum reservoir, a rock needs 10 percent or more porosity. A good reservoir will have 15-20 percent porosity. Sandstone reservoirs seldom exceed 25 percent porosity, but carbonate reservoirs can have up to 50 percent porosity. This greater porosity is a result of very high porosity in the open structure of reef rocks and cave systems of some limestones.

Porosity can be divided into primary and secondary types. Primary porosity is the porosity present at the time the rock was deposited. It depends on several factors, including the roundness and size range of the grains (sorting). Secondary porosity forms after the rock is solidified, when grains and cements are dissolved. Secondary porosity is most common in limestones but is also important in sandstones.

Permeability is a measure of the rate of flow of fluids through a porous medium. In general, the more porous the rock, the more permeable it is likely to be; however, the relation is not simple. For example, Styrofoam (a type of polystyrene plastic) has very high porosity but almost no permeability. Permeability of petroleum reservoirs is seldom less than 1 millidarcy, and a good reservoir needs 10-100 millidarcies.

RESERVOIR TRAPS

A petroleum trap is a geometric situation in which an impermeable layer of rock (a caprock) seals a permeable, petroleum-bearing reservoir rock from contact with the Earth's surface. A common type of caprock is shale, or consolidated, lithified mud. There are two common types of traps, structural and stratigraphic. Structural traps form where the originally horizontal sedimentary rock layers are disrupted by warping (folding) and breaking (faulting). A much sought-after type of fold trap is an anticline. Anticlinal folds have a cross section that is concave down (like the letter *A*). Because oil and gas are lighter than water, they migrate to the high point on the anticline and are prevented from reaching the surface by an impermeable caprock. Fault traps occur where a break in a rock layer brings a reservoir rock into contact with an impermeable rock. Salt-dome traps form where bodies of salt flow upward and pierce overlying rock layers. Salt domes form both anticlinal and fault traps on their crest and margins. They are common on the Gulf coast of the United States. Stratigraphic traps form where the permeability barrier is a result of lateral or vertical changes in the rock, changes as a result of the conditions under which the original sediment was deposited. Stratigraphic traps include buried river channels, beaches, and coral reefs. Geologic methods dominate the search for stratigraphic traps.

Reservoir fluids consist of water, crude oil, and natural gas. When sediments are deposited, they are (or soon become) saturated with water, usually salt water. The oil forms from the alteration (called maturation) of organic material buried in sedimentary rocks. This process takes place in the absence of oxygen at temperatures of 60-150 degrees Celsius when the sediments are buried. Natural gas forms by further thermal alterations of the hydrocarbons (thermogenic gas) or by low temperature alteration of near-surface organic material (biogenic gas, the gas that can be seen bubbling up from the bottom in swamps). After formation, the oil and gas must be expelled from its source rock and migrate into a reservoir rock. As oil is lighter than water, it migrates up the water column to the highest location in the trap. If there is no trap, the oil will seep onto the Earth's surface. Although oil seeps are common, in most cases the oil will be prevented from reaching the surface by a trap. The trapped reservoir fluids will then stratify themselves on the basis of density, lightest at the top (natural gas) and densest on the bottom (water).

AREAS OF CONCENTRATION

Petroleum reservoirs are not uniformly distributed in time or space; certain areas of the world and certain intervals of geologic time have a disproportionate amount of the world's petroleum reserves. One unifying characteristic of these major areas of concentration is the presence of a basin. A basin is an area where rocks thicken from the margin to basin center, rather like a mud-filled saucer. The geologic conditions in basins favor the formation of oil and gas in the basin center and its migration into traps along the basin margin.

Globally, most of the world's oil is in the Middle East: Saudi Arabia (170 billion barrels), Kuwait (67 billion barrels), Iraq (43 billion barrels), and Iran (51 billion barrels). Venezuela (25 billion barrels) and Mexico (48 billion barrels) have the major Latin American reserves, while Africa's largest reserves are in Nigeria (18 billion barrels) and Libya (21 billion barrels). The United States has 27 billion barrels in reserves, of which 85 percent is in four states: Alaska (largely in North Slope fields around Prudhoe Bay), Texas (both along the Gulf coast salt domes and in the Permian Basin of West Texas), California (southwest), and Louisiana (Gulf coast salt domes). While the world's major producing areas are widely separated, they were much closer together 100 million years ago (in the Cretaceous period), and many of them shared a common setting, what geologists call the Tethys Sea.

In terms of time, very little oil is found in rocks older than 500 million years. Rocks of Jurassic and Cretaceous age (about 65-200 million years old) contain 54 percent of the world's oil and those about 35-55 million years old, from the Eocene epoch, contain 32 percent of the world's oil. It is not clear whether this young age (geologically speaking) is related to the origin of oil or results from the fact that deep wells cost more money so that older rocks are less thoroughly drilled.

EXPLORATION FOR RESERVOIRS

The exploration for petroleum reservoirs has two intimately related aspects, one geophysical and the other geological. In geophysical exploration, the scientist, called an exploration geophysicist, uses the physics of the Earth to locate petroleum reservoirs in structural traps. The principal technique is seismic reflection profiling. In this approach, the geophysicist sets off deliberate explosions and uses the energy reflected from subsurface rock layers to interpret the folds, faults, and salt domes that may be present. Variation in the gravity, magnetic, and heatflow characteristics of the Earth may also point to the location of oil and gas pockets in the subsurface.

Geological exploration dominates in the search for stratigraphic traps. Petroleum geologists use facies models to predict the location and extent of petroleum reservoirs. Facies models are based on information about the size, internal characteristics, and large-scale association of modern sediment accumulations. Data based on well logs obtained from previously drilled wells are used to prepare structure maps, which show the "topography" of the reservoir surface, and thickness variation (isopachous) maps. Data from well logs, well cuttings, and cores are used to prepare lithofacies maps, which show the lateral variation in the rocks. These variations influence porosity and permeability trends in the reservoir. Such maps can then be compared to the facies models to make predictions about the size, the location, and the location of the edges of stratigraphic traps in the subsurface. Commonly applied facies models for sandstones include those for rivers, beaches, deltas, and sand dunes. Those for carbonate rocks include beaches, reefs, and dolostones.

A technique that combines geology and geophysics is seismic stratigraphy. In this approach, the data from artificial explosions are interpreted in terms of the depositional system of the rocks; for example, a delta system has a seismic signature that is distinct from that of sediments deposited in the shallow waters of the continental shelf. On the whole, the most favorable "prospect" is the portion of a trap closest to the Earth's surface. Since oil and gas rise to the top of a trap, this location is the most likely volume to contain petroleum.

ECONOMIC CONSIDERATIONS

Petroleum supplies a major portion of the world's fossil energy and, consequently, is an important element in the complex international play of economic forces. In other words, what decides whether a body of rock is a petroleum reservoir is not simply geology but also the reservoir's economic and political setting. The basic problem centers on the fact that petroleum is a nonrenewable resource, is present in finite amounts, and is not randomly distributed in the Earth's subsurface. Millions of oil and gas wells have been drilled in the continental United States, making it the most mature country in the world from the point of view of petroleum exploration. The "easy" oil has been found, and it is becoming harder to find the fewer and fewer undiscovered economically exploitable petroleum pools. On a more pos-

itive note, however, as drilling density has increased, so has knowledge. Better information, better understanding of how and when petroleum enters a reservoir, and better techniques for finding traps have improved the success ratio of oil drilling.

David N. Lumsden

CROSS-REFERENCES

Offshore Wells, 1689; Oil and Gas Distribution, 1694; Oil and Gas Exploration, 1699; Oil and Gas Origins, 1704; Oil Chemistry, 1711; Oil Shale and Tar Sands, 1717; Onshore Wells, 1723; Well Logging, 1733.

BIBLIOGRAPHY

Haun, John D., and L. W. LeRoy, eds. *Subsurface Geology in Petroleum Exploration*. Golden: Colorado School of Mines, 1958. This book contains forty-one short articles covering myriad specific techniques of use to petroleum geologists. Chapters cover the analysis of well cuttings, cores, and fluids; well-logging methods and interpretation; subsurface stratigraphic and structural interpretation; geochemical and geophysical methods; well drilling; formation testing; and well evaluation. Most of the articles assume very little prior knowledge on the part of the reader.

Hunt, John Meacham. *Petroleum Geochemistry and Geology*. 2d ed. New York: W. H. Freeman, 1996. Petroleum is a complex organic substance, and its geochemistry is not clearly understood. This book covers many topics, including composition, origin, migration, accumulation, and analysis of petroleum and the application of petroleum geochemistry in petroleum exploration; seep and subsurface prospects; crude oil correlation; and prospect evaluation. While parts of this book are descriptive and explanatory, much of it requires a chemistry background and algebra.

Hyne, Norman J. *Dictionary of Petroleum Exploration, Drilling, and Production*. Tulsa, Okla.: PennWell, 1991. This is a dictionary of all terms associated with petroleum and the petroleum industry. A great resource for the

beginner in this field. Illustrations and maps.
_____. *Nontechnical Guide to Petroleum Geology, Exploration, Drilling, and Production*. Tulsa, Okla.: PennWell, 1995. An excellent introductory handbook for the reader without prior background in petroleum geology. Discusses the exploration, refining, and distribution. Illustrations, bibliography, and index.

King, Robert E., ed. *Stratigraphic Oil and Gas Fields*. Tulsa, Okla.: American Association of Petroleum Geologists, 1972. Thirteen chapters discuss various exploration techniques and thirty-five chapters discuss case histories of specific reservoirs, pools, and fields. These case histories include information on discovery, development, and trap mechanisms. While written to provide the professional geologist with models upon which to base exploration predictions, some chapters can provide the interested nonspecialist with insights into what a field, pool, and reservoir are.

Levorsen, A. I. *Geology of Petroleum*. 2d ed. San Francisco: W. H. Freeman, 1967. This classic textbook, which was very popular from the mid-1950's to the late 1970's, strongly influenced and unified the way petroleum geologists classify traps and view their discipline. While intended for advanced undergraduate and graduate geology majors in college, it is so well written that its organization, introductory and summary statements on various topics, and specific examples can still serve as a source of general information for the interested nonspecialist. There have been many developments since its publication, however, and its production statistics are out of date.

Moore, Calvin A. *Handbook of Subsurface Geology.* New York: Harper & Row, 1963. Aimed at undergraduate geology students who intend to go into the oil industry, this book has a very pragmatic approach. There is no math or chemistry in it; instead, it stresses the preparation of diagrams of value to the exploration geologist (structure maps, thickness variation maps, lithofacies maps, and cross sections) as well as interpretation of well logs and how to evaluate a formation test.

Selley, Richard C. *Elements of Petroleum Geology.* 2d ed. New York: W. H. Freeman, 1998. This textbook is intended for undergraduate college geology majors. It is more technical than Levorsen's book. Topics covered include the properties of oil and gas, exploration methods, generation and migration of petroleum, reservoir and trap characteristics, and basin classification. Elementary geology, algebra, and chemistry are needed for most of the chapters.

Welte, Dietrich H., Brian Horsfield, and Donald R. Baker, eds. *Petroleum and Basin Evolution: Insights from Petroleum Geochemistry, Geology, and Basin Modeling.* New York: Springer, 1997. The authors explore the origins of oil and gas from the perspective of mathematical modeling of sedimentary basins. Somewhat technical, but illustrations and maps help clarify many of the difficult concepts. Bibliography.

WELL LOGGING

Reservoir rock data obtained by well logging are of vital importance to the petroleum industry. With these data, the production potential of a well can be determined and many problems involving the structure, environment of deposition, and correlation of rock strata can be solved.

PRINCIPAL TERMS

CONDUCTIVITY: the opposite of resistivity, or the ease with which an electric current passes through a rock formation

CORRELATION: the tracing and matching of rock units from one locality to another, usually on the basis of lithologic characteristics

GAMMA RADIATION: electromagnetic wave energy originating in the nucleus of an atom and given off during the spontaneous radioactive decay of the nucleus

HYDROCARBONS: organic compounds consisting predominantly of the elements hydrogen and carbon; mixtures of such compounds form petroleum

LITHOLOGY: the mineralogical composition of a rock unit

NEUTRON: an uncharged, or electrically neutral, particle found in the nucleus of an atom

PERMEABILITY: measured in millidarcies, the capacity of a rock unit to allow the passage of a fluid; rocks are described as permeable or impermeable

PETROLEUM: a natural mixture of hydrocarbon compounds existing in three states: solid (asphalt), liquid (crude oil), and gas (natural gas)

POROSITY: the volume of pore, or open, space present in a rock

RESERVES: the measured amount of petroleum present in a reservoir rock that can be profitably produced

RESERVOIR: any subsurface rock unit that is capable of holding and transmitting oil or natural gas

RESISTIVITY: a measure of the resistance offered by a cubic meter of a rock formation to the passage of an electric current

SONDE: the basic tool used in well logging; a long, slender instrument that is lowered into the borehole on an electrified cable and slowly withdrawn as it measures certain designated rock characteristics

LOGGING PROCEDURE

A well log is a continuous record of any rock characteristic that is measured in a well borehole. The log itself is a long, folded paper strip that contains one or more curves, each of which is the record of some rock property. Since the first "electric log" was run in a well in France for the Pechelbronn Oil Company in 1927, well logs have been the standard method by which well data have been displayed and stored. Well logs and the information they record can be classified in two ways: by type (radioactive, sonic, electric, or temperature) and by purpose (lithology, porosity, or fluid saturation determination).

After a well has been drilled, it is standard procedure to log it. Logging has been compared to taking a picture of the rock formations penetrated by the borehole. The technique consists of lowering the logging tool, or "sonde," to the bottom of the borehole on the end of an electrified cable that is attached to a truck-mounted winch at the surface. The truck also contains the instruments for recording the logged data. The sonde, which is 4.5-6 meters long and has a diameter of 7.5-13 centimeters, is then pulled up the borehole at a constant rate, measuring and recording the data of interest. Measurements are recorded coming "uphole," rather than going "downhole," because it is easier to maintain a constant sonde velocity by pulling it up. On the downward course, the sonde has a tendency to "hang up" on numerous irregular surfaces in the borehole. It is essential to run logs to evaluate petroleum potential before the borehole is lined with steel pipe, because the well completion process is expensive and will be done only if economically justified. Data obtained from the well by logging are used to determine such rock parameters as lithology, porosity, and fluid saturation.

RADIOACTIVE AND SONIC LOGS

Of the logging curves that can be run for lithology identification and correlation, the most useful are the gamma-ray, spontaneous potential, and caliper. The gamma-ray log records the intensity of natural gamma radiation emitted by minerals in the rock formations during radioactive decay. An advance in gamma-ray technology has been the development of the gamma-ray spectrometry tool, a device that measures the wavelengths of the gamma rays and makes possible the identification of individual minerals. The spontaneous potential, or SP, log measures small natural potentials (voltages) caused by the movement of fluids within the formations. These currents largely arise as a result of salinity differences between the pore waters of the formations and the mud in the borehole. Although the borehole is drilled with a bit of a particular size, its diameter is never constant from top to bottom. The caliper log provides a continuous measurement of borehole diameter by means of spring-activated arms on the sonde that are pressed against the wall of the borehole.

Porosity is determined by using singly or in combination the sonic, neutron, and density logs. The standard sonic tool has an arrangement of two transmitters, each with its own signal receiver. The transmitters send out sound waves, which are detected back at the receivers after passing through the rock. The neutron log tool bombards the formation with fast neutrons. These neutrons are slowed by collisions with ions in minerals and fluids. Because a hydrogen ion has approximately the same mass as does a neutron, collisions with hydrogen ions are most effective in slowing the neutron for the same reason that a billiard ball is slowed more by a collision with another billiard ball than it is by a collision with the rail of the table. The slowed neutrons are deflected back to the tool to be counted and recorded. Like the neu-

tron log, the density log is a nuclear log. The density sonde bombards the formations with medium-energy gamma rays. The gamma rays collide with electrons in the formation, causing the gamma-ray beam to be scattered and its intensity reduced before it returns to the detector on the sonde.

ELECTRIC AND TEMPERATURE LOGS

Most logs used to determine water and oil saturations in the formations employ some method of measuring the passage of an electric current through the rock. The electric logs can be subdivided into induction logs and electrode logs. The induction log measures the conductivity of the formation and is the most commonly used device. The induction-logging sonde generates a magnetic field that induces a current deep in the formation. The passage of this current is measured by the logging tool. In the electrode-log system, electrodes on the sonde put current directly into the borehole fluid or the formation. The resistance to the flow of the electric current through the formation is measured as the formation resistivity. The short normal log, microlog, and microlaterolog measure resistivity immediately adjacent to the borehole, while the laterolog and guard log measure resistivity deep in the formation. Deep readings are made by narrowly focusing the electric current beam and directing it straight into the formation rather than letting it diffuse through the mud and into the adjacent formations. The laterolog and microlaterolog are most commonly used when the borehole mud has a saltwater, rather than a freshwater, base.

The temperature log, a nonelectric log, continuously records borehole temperature and can also be used for fluid identification. The dipmeter log is a resistivity device run with three or four electrodes arranged around the perimeter of the sonde. If the rock layers are inclined at any angle to the horizontal, this inclination, or dip, can be detected, because the electrodes will encounter bed boundaries at slightly different times on different sides of the borehole. An on-line computer converts these differences to angle of dip. The cement-bond log is a sonic device that measures the degree to which cement has filled the space between the steel pipe, or casing, that lines the inside of the borehole and the formations behind it (complete filling is desired).

DETERMINATION OF LITHOLOGY

Lithology refers to the mineralogical composition of the rock unit, or formation. Oil and natural gas occur almost exclusively in the sedimentary rocks sandstone, limestone, and dolomite; the latter two are known as carbonate rocks. Shale, the most abundant of all sedimentary rocks, is never a reservoir rock for hydrocarbons, because it is impermeable. An important purpose of well logs is to determine the lithology of the rock formations and, thus, to identify those that possess suitable permeability to serve as reservoir rocks. The principal radioisotopes (thorium, uranium, and potassium), from which most natural gamma radiation emanates, are usually found in minerals in clays and shales. Therefore, the gamma-ray log is used to differentiate shales from sandstones and limestones and to calculate the amount of clay that might be present in some sandstones. Since most fluid movement is in or out of porous and permeable formations, the SP curve may be used to identify such rocks. These rocks are usually sandstones—hence, the identification of lithology. While permeable zones can be located, it is not possible to calculate actual permeability values. Next to permeable formations, the diameter of the borehole is reduced by the buildup of mud cake on the borehole wall. Borehole diameter also changes dramatically through shales, because shale is weak and crumbles, or "caves," thus enlarging the borehole. The caliper log can therefore be used as a lithology log to identify permeable sandstones and "caving" shale.

In addition to assessing rock units for their petroleum content and environmental information, well logs are used extensively for correlation—that is, the matching and tracing of rock units from one locality to another. Since it is lithology, rather than porosity and fluid saturation, that is geologically the most significant factor in rock identification, the lithology logging curves are most commonly used for this purpose. Correlation may be accomplished either by matching log curves "by eye" or by statistical and computer analysis.

DETERMINATION OF POROSITY

Porosity is a measure of the total open space in a rock unit that is available for the storage of hydrocarbons. Such space is normally expressed as a percentage of the total rock volume. Knowledge of formation porosity is necessary to determine the total petroleum reserves in the formation or oil field. The sonic log has historically been the most widely used porosity tool. The time, in microseconds, required for a sound wave to travel through one meter of the rock is continuously plotted on the log. This travel time is the reciprocal of velocity, so a wave that has a high velocity has a short travel time. The use of a dual transmitter-receiver system for modern sonic logs eliminates the effects of changing borehole diameter and deviations of the borehole from the vertical. Formation travel times are functions of lithology and porosity. If the formation lithology is known, the porosity can be calculated. Because a sonic wave travels faster through a solid than through a liquid or gas, increasing porosity causes greater travel times. The presence of shale in the rock formations will also cause unusually high travel times and erroneously high porosity calculations. The neutron tool principally senses the hydrogen ions present in the formation fluids, which, in turn, are found in the pore spaces. This log is affected by lithology, because clays in shales have water within their crystal structure, and the tool senses this water as if it were pore water. The density log measures electron density, which is directly related to the overall, or bulk, formation density. The greater the bulk density, the lower the porosity, because mineral matter is denser than fluid-filled pore space. A related log is the variable-density log, which is used to locate rock zones that are highly fractured and, thus, potential reservoir rocks. The fractures in the rock have the effect of lowering the bulk density.

Because accurate interpretation of the data gathered by the three porosity tools is dependent on a knowledge of lithology, it is common practice in the petroleum industry to run porosity tools in combination, particularly the neutron and density logs. Cross-plotting the readings from the two logs provides both lithology and porosity information. In addition, neutron and density logs respond oppositely to the presence of natural gas in a formation. Density porosity readings increase, whereas neutron porosity readings decrease. Therefore, this log combination will detect the presence of gas-bearing zones by the separation of the two curves.

DETERMINATION OF FLUID SATURATION

Fluid saturations, water and oil, are the most important quantities to be determined from well-log analysis. Because the water and oil saturations together must equal 100 percent, knowing one necessarily determines the other. The significance of these values is clear: If the rock unit of interest is not oil-bearing or if it contains hydrocarbons in quantities that are not economically feasible to produce, the well will be abandoned rather than completed. To understand the quantitative assessment of formation-fluid saturation, one must first understand what occurs within the borehole and in the formations that are penetrated by the well. Because high temperatures are generated by friction as the drill bit grinds its way through solid rock, specially formulated "drilling mud" is continuously circulated down the borehole to cool the bit. In addition, the mud clears the borehole by bringing to the surface the pulverized rock material, or cuttings. The drilling mud is usually a water-based fluid with various mineral additives. Within the borehole, there is a tendency for the fluid portion of the mud to separate from the mineral fraction. The fluid, or mud filtrate, seeps into the permeable rock formations, completely flushing out and replacing the natural formation fluids adjacent to the borehole. This area is the "flushed zone." Some of the filtrate moves deeper into the formation, where it continues to displace the natural fluids, creating a partially flushed area, or "invaded zone." The solid portion of the mud that has separated from the filtrate forms a "mud cake," lining the inside of the borehole on the surfaces of the permeable formations.

SPECIAL-PURPOSE LOGS

Within a reservoir, the rock matrix, fresh water, and hydrocarbons act as electrical insulators. Any electric current that passes through the rock is carried by dissolved ions in salt water in the pore spaces of the formation. Therefore, where current flows readily, the pore fluid is salt water. Where electrical resistance is encountered, it is likely that hydrocarbons occupy the pore spaces. The principal advantage of the arrangement of the induction log is that the current largely bypasses the invaded zone and gives a better picture of the true formation resistivity (the inverse of the conductivity) deep in the formation. Even so, the flushed zone and the invaded zone are still sampled to some extent, and corrections must be made to obtain the true resistivity of the uncontaminated formation. This can be accomplished by running logs that sample the formations only immediately adjacent to the borehole and, therefore, read either the flushed zone or the invaded zone resistivity. In some instances, the short normal log (a curve that is run with most electric log surveys) can be used. In other circumstances, it is preferable to use the microlog or, when saline drilling mud has been used, the microlaterolog. These last two logs are specifically designed to measure the mud cake and flushed zone resistivities.

In addition to corrections for flushed zone and invaded zone resistivities, other corrections must be applied to the log readings to allow for the effects of changing borehole diameter and bed thickness. With a computed true resistivity value and a knowledge of the formation porosity, one can calculate the percentages of water and oil in the formation and make a quantitative determination of the total volume of hydrocarbons. Such formations that contain natural gas rather than oil can be readily identified with the temperature log. As gas moves out of the formation and into the borehole, it is under less pressure and expands. As it expands, it cools, and the borehole temperature opposite the gas-bearing formation is significantly lowered.

Other special-purpose logs are available to the petroleum industry. One of these, the dipmeter log, provides information on the angle of dip of the formations encountered. This information can be used to determine environments of deposition, because rock layers in channel deposits, reefs, offshore bars, and other sedimentary features will have some unique pattern of dip. Increasingly, geologists have been using other well-log curves to refine their environmental interpretations. This is done by examining the curve patterns within the sedimentary rock units and noting whether the log parameters increase or decrease downward, for example, or change gradually or abruptly. The log, in effect, is measuring rock properties, such as particle size, that are controlled by the physical conditions within the site of deposition.

IMPORTANCE TO PETROLEUM INDUSTRY

Well logging is a little-known but necessary part of the petroleum industry. It is the method whereby geologists and petroleum engineers obtain the information on petroleum reservoir rock characteristics that allow them to make decisions about the economic potential of an oil well—that is, whether it can be completed as a "producer" or must be plugged and abandoned as a "dry hole." Such decisions involve millions of dollars and cannot be made without an examination of all the available relevant data. They must usually be correct, or the oil company will not survive financially.

The decade of the 1990's brought increased use of directional drilling, the downhole tool, and a nonrotating drill stem. A different type of well logging called "measure while drilling" (MWD) is well suited to this method of drilling. MWD uses a sonde as part of the drill stem near the tool. The sonde carries the usual logging instruments, along with magnetometers and accelerometers to help in establishing the drill stem's position and the orientation of the hole. High-capacity batteries power the equipment. A transmitter assembly creates a series of pulses in the pressure of the drilling mud by opening and closing the valve. This allows small amounts of mud to pass directly from the drill stem into the borehole, thereby bypassing the drill tool. These pressure pulses are monitored by sensors at the top of the borehole. Computers in the sonde and in the logging truck communicate by means of these pulses and thereby provide logging information continuously while drilling. The logging information is therefore immediately available, and it is easier to interpret than with the older system because the sonde is in a relatively fresh borehole so the drilling mud has had little time to penetrate into the surrounding rock and modify its properties. Based on the logging data, decisions such as whether the well should be completed, if the direction of the borehole should be changed, or which rock zones should be tested for their fluid content and producibility can now be made. As in many business endeavors, time saved in the decision-making process can be turned into money earned.

Donald J. Thompson

CROSS-REFERENCES

Alluvial Systems, 2297; Continental Structures, 590; Earth Resources, 1741; Faults: Thrust, 226; Folds, 624; Gravity Anomalies, 122; Offshore Wells, 1689; Oil and Gas Distribution, 1694; Oil and Gas Exploration, 1699; Oil and Gas Origins, 1704; Oil Chemistry, 1711; Oil Shale and Tar Sands, 1717; Onshore Wells, 1723; Petroleum Reservoirs, 1728; Saltwater Intrusion, 2061; Sediment Transport and Deposition, 2374; Seismic Reflection Profiling, 371; Strategic Resources, 1796; Stratigraphic Correlation, 1153; Transgression and Regression, 1157.

BIBLIOGRAPHY

Asquith, George B., with Charles Gibson. *Basic Well Log Analysis for Geologists.* Tulsa, Okla.: American Association of Petroleum Geologists, 1982. An excellent discussion of the theory of well logging and of the application of each of the major types of log. Each section is complete with description, examples, and problems. Several comprehensive case studies are included at the end. Well illustrated, with a comprehensive bibliography. Intended for the working geologist but would be understood by persons with minimal geologic training.

Berg, Robert R. *Reservoir Sandstones.* Englewood Cliffs, N.J.: Prentice-Hall, 1986. This college-level text includes a good summary of well logs and logging procedures. The various types of log are discussed, with particular emphasis on how they are used in specific formation evaluation and sedimentological problems. While the text as a whole would not be suitable for the novice, the basic discussion of well logging would be.

Brock, Jim. *Analyzing Your Logs.* Vol. 1, *Fundamentals of Open Hole Log Interpretation.* 2d ed. Tyler, Tex.: Petro-Media, 1984. A good, fundamental discussion of the basic characteristics of petroleum reservoir rocks is followed by a description of the theory and use of each of the major types of well log. Fairly easy to understand, with good diagrams but no bibliography. Designed for a short course in well log evaluation. Questions and problems accompany each section. The necessary graphs and correction charts for use with each type of log are included.

Gorbachev, Yury I. *Well Logging: Fundamentals of Methods*. New York: Wiley, 1995. Gorbachev's book is a good introduction to all aspects of geophysical well logging. Includes numerous useful illustrations and a short bibligraphy.

Hyne, Norman J. *Dictionary of Petroleum Exploration, Drilling, and Production*. Tulsa, Okla.: PennWell, 1991. This is a dictionary of all terms associated with petroleum and the petroleum industry. A great resource for the beginner in this field. Illustrations and maps.

_____. *Nontechnical Guide to Petroleum Geology, Exploration, Drilling, and Production*. Tulsa, Okla.: PennWell, 1995. An excellent introductory handbook for the reader without prior background in petroleum geology. Discusses the exploration, refining, and distribution. Illustrations, bibliography, and index.

Pirson, Sylvain J. *Geologic Well Log Analysis*. Houston: Gulf Publishing, 1970. The emphasis in this book is on the use of well logs for sedimentary environment interpretation, structural analysis, facies analysis, and hydrogeology rather than on the evaluation of hydrocarbonate-bearing rock units. Somewhat dated, but the most complete and detailed work of its kind.

Rider, Malcolm H. *The Geological Interpretation of Well Logs*. Houston: Gulf, 1996. Overview of oil well logging and analysis. Illustrations and index.

Selley, Richard C. *Elements of Petroleum Geology*. 2d ed. New York: W. H. Freeman, 1998. An excellent undergraduate text that contains a well-written section on the theory and application of well logging, with a discussion of each of the major types of log. Easily understood by the beginner. Excellent diagrams and an extensive bibliography are included.

9

RESOURCE USE AND ENVIRONMENTAL IMPACTS

EARTH RESOURCES

The Earth's resources of metals, nonmetals, and energy supplies undergird every modern, technological society.

RENEWABLE VS. NONRENEWABLE RESOURCES

All Earth resources can be subdivided into two broad categories. The first category contains the renewable resources. The word "renewable" means that these resources are replenished by nature as rapidly as humans use them up, provided good judgment is used. Renewable resources include the energy of the wind, the timber cut in a forest, or the animals used for food. Each of these Earth resources is constantly being renewed by the energy reaching the Earth's surface from the Sun. As long as the Sun's rays reach the Earth, this pattern of replenishment will continue.

The resources in the second category are known as the nonrenewable resources. These are resources that will not be renewed in a human lifetime. Only limited quantities of these resources are present in the Earth's crust, and they are not replenished by natural processes operating within short periods. Examples of nonrenewable resources include coal, oil, iron, diamonds, and aluminum. While it is true that certain of these resources, such as coal and oil, are being formed within the Earth's crust continuously, the processes by which they are formed are exceedingly slow, being measured in thousands or millions of years.

METALS

The Earth resources of primary interest to the economic geologist can be divided into three categories: metals, nonmetals, and energy sources. The metals are a group of chemical elements that have certain features in common, the most noticeable of which is a high metallic luster, or shine. In addition, they can all be melted, they all conduct heat and electricity, and most of them can be pounded into thin sheets or drawn into thin wires.

Metals can be divided into two classes based on their abundance in the Earth. The first class, which has been called the abundant metals, consists of those metals that individually constitute 0.1 percent or more of the Earth's crust by weight. The metals in this category are iron, aluminum, manganese, magnesium, and titanium. The second class of metals, which are called the scarce metals, consists of those metals that individually constitute less than 0.1 percent of the crust. This class includes such metals as copper, lead, zinc, nickel, mercury, silver, gold, and platinum.

Certain common metals, such as steel, brass, bronze, and solder, are not pure metals but rather alloys, chemical mixtures of two or more metals that have characteristics of strength, durability, or corrosion resistance superior to those of the component metals. Steel, for example, is an alloy in which iron is the main constituent.

Metals are rarely found in the pure state within the Earth's crust. Only the metals gold, silver, copper, platinum, and iron are ever found uncombined. All other metals are found chemically combined with additional elements to form minerals. Geologists use the term "ore deposit" to describe a rock containing metals or metal-bearing minerals from which the pure metal can be profitably extracted. Whether a rock is an ore deposit depends on a variety of factors, including how difficult it is to extract the metal from the metal-bearing mineral, how large the ore deposit is and how accessible, whether valuable by-products can be obtained, and what the current price of the metal is on world markets.

NONMETALS

The second major category of Earth resources is the nonmetals. The term "nonmetal" is widely employed by geologists to describe substances extracted from the Earth that are neither sources of metals nor sources of energy. Nonmetals are mined and processed either because of the nonmetallic elements they contain or because they have some highly desirable physical or chemical characteristic. Some of the Earth's major nonmetallic resources are fertilizers, chemicals, abrasives, gems, and building materials.

Fertilizers contain the elements nitrogen, potassium, and phosphorus. Most of the nitrogen required for fertilizer production is chemically extracted from the air, so the supply is renewable. The potassium and the phosphorus, however, come from rocks dug out of the ground—potash salt layers and phosphate rocks—and the supply of these is nonrenewable.

Several Earth resources provide important raw materials for the chemical industry. They include salt, which is obtained from underground beds of rock salt and from seawater; sulfur, a by-product of oil production; and substances such as borax and soda ash, which are obtained from the beds of dry desert lakes. Abrasives are very hard substances that are used for grinding, polishing, and cleaning. They are obtained from rock and mineral substances dug out of the Earth and then pulverized. Gems are Earth materials that are attractive to the eye. They can be categorized as precious and semiprecious.

Building materials include the stones obtained from quarries, such as granite, sandstone, limestone, marble, and slate. There is also a high demand for crushed rock, which is used as highway roadbeds and for concrete aggregate. Sand and gravel are also used in making concrete. A number of other useful products are prepared from Earth materials, such as cement; plaster, from the mineral gypsum; brick and ceramics, from clay; glass, from very pure sand or sandstone rock; and asbestos, from flame-resistant mineral fibers that can be woven into fireproof cloth or mixed with other substances to make fireproof roofing shingles and floor tiles.

ENERGY SOURCES

The third major category of Earth resources is energy sources. Energy sources are frequently divided into the mineral fuels, which are nonrenewable, and a second, renewable group. The first group contains oil and gas, coal, and uranium.

Crude oil, natural gas, and petroleum provided 48 percent of the United States' energy in 1998. Crude oil is a naturally occurring liquid composed of the elements hydrogen and carbon combined into compounds known as hydrocarbons. Natural gas is a gaseous form of these hydrocarbons. The oil and gas accumulate underground over long periods in source rocks and then migrate into reservoir rocks, where they are trapped. Extraction is accomplished by means of drilling. Two related Earth resources are oil shale, a source rock that still contains oil, and tar sands, reservoir rocks exposed at the surface.

Coal is the third most important energy source in the United States; in 1998, it met 33 percent of the nation's energy needs. Coal originates when partly decayed plant material accumulates on the floor of bogs and swamps and is then buried by overlying sediments that compact the plant material into carbon-rich rocks. The various grades of coal are lignite (brown coal), bituminous coal (soft coal), and anthracite (hard coal). Peat is partly decayed plant material that was never buried at all. Coal is mined at the surface in strip mines or is mined underground. Coal, oil, and gas are sometimes referred to as "fossil fuels."

Uranium is a silver-gray metal used in nuclear reactors to produce electricity. In 1998, nuclear reactors met 10 percent of American energy needs.

Within a reactor, uranium undergoes neutron-induced disintegration, producing heat. This heat is used to drive an electrical generator, just as in conventional power plants. Uranium occurs as veins or grains within a variety of rock types and is mined with standard mining methods.

There are many potential energy sources that are renewable and are mostly underutilized. Foremost among these is hydroelectric power, or water power. This power is generated by means of water falling from a dammed reservoir; the force of the falling water turns the turbine of an electric generator. Barring unforeseen climate changes or silting of the reservoir behind the dam, hydroelectric power can be considered a renewable resource. Sunlight, the wind, the tides, and the steam from geysers are also renewable energy sources, as are living things, in the form of firewood and animal power. Newer applications are gasahol- and methane-fired boilers that use gases fermented from seaweed or cow manure. Renewable resources provided 9.3 percent of the energy used in 1998: 4.5 percent from hydroelectric plants, 4.3 percent from burning biomass, and 0.5 percent from geothermal plants and other sources. The controlled burning of garbage has tremendous potential for the future, as does the use of hydrogen from the ocean to operate "clean" nuclear fusion reactors—a peaceful use of the same chemical reaction that powers the hydrogen bomb.

STUDY OF EARTH SOURCES

Mineral deposits are quite rare in the Earth's crust; either they consist of substances that are uncommon to begin with (gold, for example) or else they are composed of common substances, such as the very pure sand used to make glass, that have been concentrated into workable accumulations. Much study has gone into why such concentrations exist. One valuable tool has been the plate tectonics theory, which proposes that the Earth's surface is divided into a few large plates that are slowly moving with respect to one another. Intense geologic activity occurs at plate boundaries, and many mineral deposits are believed to have been formed by this activity.

It becomes more and more difficult to find mineral deposits, as all the easily discoverable ones have already been found. Aerial prospecting was made possible by the airplane, and it has now been re-

placed by satellite imaging. Most prospecting, however, is based on the search for buried deposits and utilizes indirect methods for detecting favorable underground geologic conditions. A preferred technique is seismic prospecting, in which sound waves are created underground by means of small explosive charges and are then bounced off underground rock layers in order to determine their structure.

Further ways in which Earth resources are studied include calculations of estimated available reserves in view of anticipated future demands; comparison of the fuel values of various energy sources; analysis of the environmental problems and types of pollution caused by the extraction, refining, and utilization of various mineral resources; and investigations of ways to conserve, recycle, or develop substitutes for mineral resources that are in short supply.

ECONOMIC EXTRACTION OF MINERAL RESOURCES

A variety of techniques have been developed for the economical extraction of mineral resources from the Earth. Frequently, extraction costs are the controlling factor in whether a mineral deposit can be profitably worked. In general, extraction techniques can be divided into two groups: surface and underground methods. Surface methods are preferred whenever possible, because they are lower in cost. The traditional surface methods are quarrying, open-pit mining, and strip mining. Strip mining involves the removal of large amounts of worthless overburden so that the mineral deposit can be reached. Underground mining methods include the excavation of shafts, tunnels, and rooms; fluids such as petroleum can be removed by means of drilled wells.

Mineral resources extracted from the Earth are rarely ready to be sent directly to market. Generally, they will require processing to separate undesirable substances from desirable ones. This processing may involve physical separation, as in the case of separating diamonds from rock pebbles of the same size, or chemical separation, such as is required to remove metals from the sulfur with which they are combined in certain ores. Even after the pure mineral substance is obtained, further treatment may be required, as in the smelting of iron to obtain steel or the refining of petroleum to obtain gasoline.

SUPPORT FOR MODERN TECHNOLOGY

The Earth's resources of metals, nonmetals, and energy supplies support all modern technology. Houses and automobiles, televisions and refrigerators, airplanes and roads, jewelry and sandpaper, the electricity that lights a playing field and the gasoline that powers a car—an almost unending list of goods depend on the ability to utilize or harness the resources of the Earth.

A technological society relies on metals. Iron is needed for steel making, aluminum for lightweight aircraft construction, manganese for toughening steel for armor plate, and titanium for making heat-resistant parts in jet engines. Among the scarcer metals, copper is needed for electrical wiring, lead for car batteries and nuclear reactor shielding, zinc for galvanized roofing nails, nickel for stainless steel, and mercury for thermometers and silent electric switches. Silver is used for making photographic film, silverware, jewelry, and coins; gold for coins, jewelry, and dental work; and platinum for jewelry and industrial applications where corrosion resistance is essential.

Among the nonmetallic Earth resources, fertilizers are used for agriculture; salt, sulfur, and soda for the chemical industry; and abrasives for sandpaper and grinding wheels. Cut stone, crushed rock, cement, plaster, brick, glass, and asbestos are all needed by the construction industry.

The energy sources obtained from the Earth enable humans to perform tasks faster than they could manually. Because most machines run on electricity, the output of the energy source often must first be converted into an electric current. Automobiles, however, convert gasoline directly into power by means of the internal combustion engine. Probably the most direct use of an energy source is the powering of a sailboat by the wind. That is a way of using energy that has not changed in the past five thousand years.

Donald W. Lovejoy

CROSS-REFERENCES

Alluvial Systems, 2297; Clays and Clay Minerals, 1187; Diagenesis, 1445; Earth Science and the Environment, 1746; Earth Science Enterprise, 1752; Environmental Health, 1759; Future Resources, 1764; Hazardous Wastes, 1769; Landfills, 1774; Mining Processes, 1780; Mining Wastes, 1786; Nuclear Waste Disposal, 1791; Offshore Wells, 1689; Oil and Gas Exploration, 1699; Oil and Gas Origins, 1704; Oil Chemistry, 1711; Onshore Wells, 1723; Petroleum Reservoirs, 1728; River Bed Forms, 2353; Rocks: Physical Properties, 1348; Sediment Transport and Deposition, 2374; Sedimentary Rock Classification, 1457; Strategic Resources, 1796.

BIBLIOGRAPHY

Craig, J. R., D. J. Vaughan, and B. J. Skinner. *Resources of the Earth*. Englewood Cliffs, N.J.: Prentice-Hall, 1988. A well-written and well-illustrated text with numerous black-and-white photographs, color plates, tables, charts, maps, and line drawings. It provides an excellent overview of the metal, nonmetal, and energy resources of the Earth. There is a useful chapter on Earth resources through history. Suitable for college-level readers or the interested layperson.

Davidson, Jon P., Walter E. Reed, and Paul M. Davis. *Exploring Earth: An Introduction to Physical Geology*. Upper Saddle River, N.J.: Prentice Hall, 1997. An excellent introduction to physical geology. This book explains the composition of the Earth, its history, and its state of constant change. Intended for the layperson, it is filled with colorful illustrations and maps.

Jensen, M. L., and A. M. Bateman. *Economic Mineral Deposits*. 3d ed. New York: John Wiley & Sons, 1979. This economic geology text has detailed information on the different metallic and nonmetallic mineral deposits and their modes of formation. There are excellent sections on the history of mineral use and the exploration and development of mineral properties. Cross sections of individual deposits are provided. For college-level readers.

Lutgens, Frederick K., and Edward J. Tarbuck. *Earth: An Introduction to Physical Geology*. 6th ed. Upper Saddle River, N.J.: Prentice Hall,

1999. This college text provides a clear picture of the Earth's systems and processes that is suitable for the high school or college reader. In addition to its illustrations and graphics, it has an accompanying computer disc that is compatible with either Macintosh or Windows. Bibliography and index.

McGeary, David. *Physical Geology: Earth Revealed.* 3d ed. Boston: WCB/McGraw-Hill, 1998. A thorough introduction to physical geology, this illustrated textbook is accompanieed by a computer laser optical disc that corresponds to the chapters and topics explored. Bibliography and index.

Press, Frank, and Raymond Siever. *Understanding Earth.* 2d ed. New York: W. H. Freeman, 1998. An excellent general text on all aspects of geology. Suitable for advanced high school and college students.

Skinner, B. J. *Earth Resources.* 3d ed. Englewood Cliffs, N.J.: Prentice-Hall, 1986. An overview of all the Earth's resources. It is well written and contains helpful line drawings, maps, tables, and charts, although photographs are few. There are excellent suggestions for further reading and a list of principal ore minerals and production figures for 1982.

Tennissen, Anthony C. *The Nature of Earth Materials.* 2d ed. Englewood Cliffs, N.J.: Prentice-Hall, 1983. This reference book contains detailed descriptions of 110 common minerals, with a black-and-white photograph of each. There are helpful sections on the modes of formation and the classification of igneous, sedimentary, and metamorphic rocks. Chapter 7, "Utility of Earth Materials," is an overview of metallic and nonmetallic mineral resources. Suitable for college-level readers.

Thompson, Graham R. *An Introduction to Physical Geology.* Fort Worth, Tex.: Saunders College Publishing, 1998. This college text provides an easy-to-follow look at what physical geology is and the phases involved in it. Thompson shows the reader each phase of the Earth's geochemical processes. Excellent for high school and college readers. Illustrations, diagrams, and bibliography included.

EARTH SCIENCE AND THE ENVIRONMENT

An understanding of the systems of the Earth, how they interact, and how they are affected by anthropogenic distur-
bances is essential to any study of the environment. As the model of the Earth is refined and improved, it will be possible
to predict the planet's response to anticipated perturbations with greater precision and confidence. This should lead to
better public policy.

PRINCIPAL TERMS

EL NIÑO SOUTHERN OSCILLATION: a reversal in precipitation patterns, ocean upwelling, and thermocline geometry that is accompanied by a weakening of the trade winds; the phenomenon typically recurs every three to seven years

GREENHOUSE GAS: an atmospheric gas capable of absorbing electromagnetic radiation in the infrared part of the spectrum

GYRE: an ocean-scale surface current that moves in a circular pattern

PLATE TECTONICS: a theory that holds that the surface of the Earth is divided into about one dozen rigid plates that move relative to one another, producing earthquakes, volcanoes, mountain belts, trenches, and many other large-scale features of the planet

THERMOHALINE CONVEYOR BELT: a system of oceanic circulation driven by the cooling and sinking of salty surface waters in the Nordic seas

GLOBAL AND SOLID EARTH SYSTEMS

The planet Earth has a large number of interconnected and interacting systems. Although the rates with which these systems respond to perturbations vary greatly, over enough time they all influence one another. Some of these systems are global in scale, while others are local. Many provide feedback that can either amplify or suppress a change.

The systems that have the slowest response times involve the solid Earth. The surface of the Earth is divided into about one dozen tectonic plates. These plates show little interior deformation over hundreds of millions of years, and yet their boundaries are constantly changing. They move across the Earth at rates on the order of centimeters per year and have been doing so for at least several hundred million years. The environmental issues most closely related to this motion are the hazards caused by the earthquakes and volcanoes it produces. However, other, less obvious, connections exist.

Large areas of uplift, such as the Himalayan Mountains and the Tibetan Plateau, dramatically change atmospheric circulation patterns, which in turn affect precipitation and evaporation. Rates at which the tectonic plates are formed along mid-ocean ridges vary, with far-reaching effects. The new material is hot and expanded. The time it takes to cool and contract does not depend on the rate at which it is forming, so faster growth results in larger regions underlain by expanded rock. This raises sea levels. Some scientists have also proposed a connection between volcanic processes acting along the East Pacific Rise and the occurrence of El Niño events.

Although the plate tectonic system interacts with other Earth systems and has long-term effects, it operates at such a slow rate that it is unlikely to cause significant changes to the environment during the course of a few human generations. In one hundred years, a plate may move a few meters. This is more than enough to cause tremendous earthquakes, but it is not likely to produce noticeable climate change, for example.

Continental glaciation causes vertical movements of the Earth's crust, which occur much more rapidly than plate motions. Approximately twenty thousand years ago, a mass of ice about 1.6 kilometers thick covered much of Canada, the northern United States, and Scandinavia. The weight of this ice forced the crust to sink into the mantle; when the ice melted away, the crust began to rise again. As a result, some Viking fishing villages are now well above sea level, and canyons have been cut by meandering rivers into the bedrock of upstate New York.

Some areas are underlain by clays that were originally deposited beneath seawater. As they have been elevated, rainwater has flushed many of the ions from them, leaving them in a metastable state. Such clays are called sensitive clays, or quickclays, and can lose their strength with dramatic effect. This occurred during the 1964 earthquake in Alaska: A large portion of a housing subdivision was destroyed when the sensitive clays beneath it liquefied and collapsed.

As the crust beneath the continental glaciers was being pushed down, uplift occurred just beyond the southern edge of the glacier. (This process can be modeled with a gelatin dessert: Pushing down at some point will cause a bulge to rise around the area being depressed.) This region is now subsiding as mantle material flows back beneath the rebounding crust to the north. This is not noticeable inland, but it adds to the effects of coastal erosion in Maryland, Delaware, and Virginia.

OCEAN SYSTEMS

Much faster than solid rock systems, circulation of the oceans is one of the most significant processes affecting the environment. It is convenient to divide this circulation into surface current circulation and thermohaline (involving the effects of both temperature and salinity) circulation.

Heat from the Sun warms the surface water, which then expands and floats on the cooler water beneath it. This warm body of water has a lens shape, with the center a few meters higher than the edges. As water tries to flow down the slope of the surface, Coriolis force, a result of the Earth's rotation, forces it to move sideways instead. In the Northern Hemisphere, this produces a clockwise movement, and in the Southern Hemisphere, it produces a counterclockwise movement. This ocean-scale circular current is called a gyre. The Gulf Stream is the northwest segment of the North Atlantic Gyre.

The Gulf Stream moves 35 million cubic meters of water past Chesapeake Bay every second. This mass of water is about equivalent to the mass of the entire populations of China and the former Soviet Union. The edge of the Gulf Stream forms an abrupt boundary between cool, nutrient-rich waters near the continents and the warm, nearly lifeless waters of the Sargasso Sea. Although interconnected with weather and climate in a complex way, the operation of the Gulf Stream is not the principal source of heat responsible for the balmy climate of Europe. That heat is given up by water that cools to the north of the Gulf Stream in the Norwegian Sea.

This heat transfer drives the thermohaline circulation. Superimposed on the circular movements of the North Atlantic Gyre is a more gradual movement called the North Atlantic drift. Salty water in the subpolar reaches of the North Atlantic gives up its heat to the cool air above it, cooling and contracting in the process. When sufficiently dense, this water sinks, moving down to the bottom and then to the south, beneath the entire Atlantic Ocean. Moving east off the southern tip of Africa, it splits and enters both the Indian and Pacific Oceans. To replace the water that sank, surface waters gradually drift to the north. These waters have become salty in subtropical regions of high evaporation and so will sink when they cool sufficiently. The importance of this circulation pattern, often called the thermohaline conveyor belt, was not fully appreciated until the 1970's.

This conveyor belt has not always operated, and many scientists are concerned that it might stop again. The surface water needs to be salty enough to sink. About ten thousand years ago, as the glaciers were melting back and the global climate was warming, a layer of very cold fresh water apparently flooded across the North Atlantic and shut down the conveyor belt. Temperatures plummeted. Today the operation of the conveyor belt may depend on a sensitive balance between rates of evaporation and precipitation. Scenarios have been proposed in which anthropogenic disturbances, including carbon dioxide emissions, disrupt this balance, closing down the conveyor belt and triggering another ice age. This system demonstrates the strong link between atmospheric and oceanic systems.

ATMOSPHERIC SYSTEMS

Atmospheric systems constitute the other major heat redistribution scheme on the Earth. Because of the Earth's spherical shape, the equatorial regions receive and absorb far more solar energy than its upper latitudes do.

As this solar radiation reaches the Earth, much of it is absorbed, raising the temperature of the surface. This heats the air above it, which expands and rises into the atmosphere, just like a hot air balloon. It cools as it rises, and the moisture that it contains condenses and precipitates, producing most of the Earth's rain forests. The cooler, drier air moves to the north and south until it eventually descends over belts of latitudes between 20 degrees north and 30 degrees north and between 20 degrees south and 30 degrees south. Descending, it warms up, which increases the amount of moisture it can contain. At the surface, this air evaporates water effectively. Most of the Earth's major deserts are in these latitudinal belts. The convection cell is completed by a return flow across the surface toward the equator. Because of the Coriolis force produced by the Earth's rotation, this flow is deflected. In the Northern Hemisphere, it is deflected to the right, making the winds come out of the northeast. In the Southern Hemisphere, it is deflected to the left, making the winds come out of the southeast. In both hemispheres, these winds are called trade winds.

Another convection cell forms at somewhat higher latitudes, and here the surface winds come out of the southwest in the Northern Hemisphere and out of the northwest in the Southern Hemisphere. These winds, called the westerlies, blow across most of the United States. As they travel over the mountains in the western states, they are forced up into higher elevations where the air loses its water. Deserts lie to the east of these mountains, in what is called the mountains' "rain shadow."

This big picture sketches out the movements of air masses averaged over time. There are interesting variations on this theme, such as the El Niño Southern Oscillation (ENSO) and the North Atlantic Oscillation.

The trade winds of both hemispheres meet along the equator and result in surface winds that tend to push the surface water of the oceans to the west. This surface water is warmer and less dense than the water beneath it, and the boundary between the two is a region of rapidly changing temperature called the thermocline. Because only warm water is pushed to the west, it piles up there, pushing the thermocline lower and lower. Under these conditions, the air just above the surface of the ocean is heated, and moisture is added to it as it travels from east to west. When it reaches the western edge of the Pacific, it rises and its water condenses, resulting in a climate that produces rain forests. Every few years, however, the situation reverses itself. Surface winds weaken, the accumulated warm surface water flows eastward, rain comes to the deserts of Peru, and weather patterns over the entire planet are severely disrupted. The atmospheric effects of such a reversal are called the Southern Oscillation, whereas the oceanic effects are called El Niño. As they both are linked and occur together, it has become common practice to refer to the phenomenon as the El Niño Southern Oscillation (ENSO).

Another oscillation is the North Atlantic Oscillation, which appears to have a cycle time of several decades. Like ENSO, this oscillation seems to occur because of interactions between oceanic and atmospheric thermal processes and reservoirs. Because of its longer period, fewer cycles have been observed since it was detected, and most theories about its origin or history are controversial. Nonetheless, it is thought to have an important influence on the climate of Europe and North America. Many Earth scientists believe that the cool part of its cycle was responsible for a period of cooling from the 1940's through the 1970's. If the warm part of this cycle occurred during the 1980's and 1990's, its effects must be considered in efforts to identify global warming.

GREENHOUSE GASES

Atmospheric geophysicists have long been interested in the interactions between electromagnetic radiation and the molecules that make up air. Most of these molecules are diatomic (consisting of two atoms), including nitrogen and oxygen. Some, such as water vapor, carbon dioxide, methane, and ozone, are not. These molecules have modes of vibration that diatomic molecules do not have, which permits them to absorb electromagnetic radiation in the infrared region of the spectrum. Gases made up of these molecules are called greenhouse gases. This energy absorption is temporary, and they reradiate it later, but it leads to some interesting phenomena.

Most of the energy from the Sun that reaches the Earth is electromagnetic energy in the visible light portion of the spectrum. A cloudless atmosphere lets this pass through easily. Upon reaching the surface of the Earth, some of this energy is absorbed by surface material, which then radiates the energy back, but at a much lower frequency in the infrared region, which humans detect as heat. On its way back to outer space, this energy may be absorbed by a greenhouse gas molecule. When this molecule reradiates the energy, the chances are about equal that it will be directed toward or away from the Earth. In this way, a significant fraction of the infrared radiation is redirected toward the surface of the Earth.

In the end, the temperature of the planet, including its atmosphere, does not change. All the incoming energy is balanced by all the outgoing energy. Because of greenhouse gases, a substantial amount of this energy is radiated from high in the atmosphere, and the temperature at the surface of the planet—where life and the environment that supports it exist—does increase.

The most important greenhouse gas is water vapor. The distribution of water vapor is uneven and subject to wide variations. It is generally assumed that the average relative humidity of the atmosphere remains constant. Because warm air contains more moisture at the same level of relative humidity than cooler air, if global surface temperatures rise, the amount of water in the atmosphere will increase. Thus water vapor is expected to enhance any effects produced by other greenhouse gases.

The greenhouse gas of greatest concern is carbon dioxide. It is well known that although it is present in the atmosphere in only trace amounts, those amounts increased by 12 percent between 1958 and 1998. The amount of coal, oil, and natural gas burned in that time period is also known, and the 12 percent increase represents only about one-half of the carbon dioxide produced by this burning. Where the other half has gone remains a subject of research. Scientists are trying to see if the observed increases in atmospheric carbon dioxide have caused any change in global surface temperatures. It appears that these rose approximately 0.5 degree Celsius between 1880 and 1980, a change well within historical rates of temperature fluctuations, leaving the significance of the increase in atmospheric carbon dioxide open to question.

Earth scientists studying gas bubbles trapped in the ice of Greenland and Antarctica have observed that carbon dioxide concentrations in the atmosphere and global temperatures changed together over the past 200,000 years. Causality has not been established, however, and the situation is complicated because biologic activity, the solubility of carbon dioxide in seawater, and many other factors vary with temperature.

LOCAL SYSTEMS

The environment of planet Earth is maintained by all the interacting global systems. Local areas—such as a flood plain, a watershed, or an earthquake-prone region—also have interconnected systems of environmental significance. Here, too, Earth science provides essential background for policymakers.

Flooding occurs when too much water tries to move through a channel at one time. Earth scientists can determine how much water a channel can hold and how long it will take various volumes of water to move down tributaries. From this information, they can predict how high a particular flood will get and when that crest will occur. By studying geological evidence of previous floods, they can establish how frequently floods of various sizes have occurred in the past. After including the effects produced by agriculture and development, they can then estimate how frequently floods of different magnitudes are likely to occur in the future. This information can be used to determine how much money should be invested for flood control or how restrictive to make zoning within the flood plain.

During floods there is an overabundance of water, but at other times water can be in short supply. In many places, underground aquifers are tapped for drinking water. Earth scientists can study a watershed and evaluate the quantity, quality, and flow patterns of its underground water. They can determine how much water can be safely withdrawn without long-term detrimental effects. If pollution occurs, scientists can determine the likely paths of the pollutants, trace them using remote sensing techniques, and suggest strategies to contain or eliminate the threats they pose.

Threats posed by earthquakes are more difficult to perceive than those posed by pollution or flooding. Large earthquakes are unlikely to strike a region twice within the span of one person's life.

They may leave a record in the soils and structures of a region, and Earth scientists have been able to interpret these to extend knowledge of major earthquakes back several centuries. To detect small earthquakes, which recur frequently, Earth scientists have developed instruments called seismographs. The behavior of geologic units when subjected to prolonged periods of violent shaking can be very different from their behavior under normal conditions. Earth scientists have developed theories to explain such behavior, as well as criteria that can be used to identify those materials that pose the greatest risk. By combining results from these and other areas of research, researchers have been able to estimate the risks posed by earthquakes at various locations. This information can be used to develop building codes and zoning laws to minimize the death and destruction that will accompany the inevitable major earthquakes in the future.

SIGNIFICANCE

The environmentalists' slogan "think globally, act locally" is an acknowledgment of the interconnectedness of many planetary processes and of the importance of making even local decisions based on knowledge that includes some understanding of the potential global consequences of such decisions. Earth science endeavors to develop that understanding.

Earth scientists—by finding and quantifying the many feedback loops with which systems interact with themselves and one another—can identify those natural processes that are most prone to disruption from anthropogenic disturbances. By studying the past behavior of these systems, they can make reasonable estimates of how they might respond to such disturbances. By informing those who are entrusted with establishing public policies, with writing and enforcing laws, and with educating the electorate, Earth scientists can use their knowledge and insight to influence the course taken by the governments of the world.

Otto H. Muller

CROSS-REFERENCES

Aluminum Deposits, 1539; Building Stone, 1545; Coal, 1555; Diamonds, 1561; Earth Resources, 1741; Earth Science Enterprise, 1752; Environmental Health, 1759; Fertilizers, 1573; Future Resources, 1764; Gem Minerals, 1199; Geothermal Power, 1651; Gold and Silver, 1578; Hazardous Wastes, 1769; Hydroelectric Power, 1657; Iron Deposits, 1602; Landfills, 1774; Metamorphic Mineral Deposits, 1614; Mining Processes, 1780; Mining Wastes, 1786; Nuclear Waste Disposal, 1791; Ocean Power, 1669; Oil and Gas Distribution, 1694; Sedimentary Mineral Deposits, 1629; Strategic Resources, 1796; Uranium Deposits, 1643.

BIBLIOGRAPHY

Coch, Nicholas K. *Geohazards: Natural and Human.* Englewood Cliffs, N.J.: Prentice Hall, 1995. A general treatment of environmental problems, this book provides many figures and photographs. A textbook for beginning courses for nonmajors, it puts considerable emphasis on vocabulary.

Keller, Edward A. *Environmental Geology.* 7th ed. Upper Saddle River, N.J.: Prentice Hall, 1996. With an emphasis on pollution but a broad treatment of most environmental issues, this book has been a popular textbook through six previous editions. It includes examples of spectacular successes, and failures, with environmental incidents.

Lundgren, Lawrence W. *Environmental Geology.* 2d ed. Upper Saddle River, N.J.: Prentice Hall, 1999. A comprehensive treatment, with a focus on hazards and hazard mitigation, this book uses case histories to illustrate key points in environmental geology. Most often dealing with issues at a regional or local scale, it provides excellent examples of how knowledge of the Earth sciences can be applied beneficially.

Mackenzie, Fred T. *Our Changing Planet: An Introduction to Earth System Science and Global Environmental Change.* 2d ed. Upper Saddle River, N.J.: Prentice Hall, 1998. This book attempts to provide a fairly thorough scientific foundation for the study of global change. Its treatment varies, sometimes being overly technical and sometimes dismissing important concepts as being too complex to be dealt with.

Murck, Barbara W., Brian J. Skinner, and Stephen C. Porter. *Dangerous Earth: An Introduction to Geologic Hazards.* New York: John Wiley & Sons, 1997. Although a general treatment of environmental issues, this book is unusual because it includes a chapter on meteorite impacts. In addition, it contains a separate chapter on tsunamis.

EARTH SCIENCE ENTERPRISE

The Earth Science Enterprise is an effort to gain a better understanding of the global environment by exploring how Earth's systems of air, land, water, and life interact with one another.

PRINCIPAL TERMS

EL NIÑO: "the boy" in Spanish, referring to the Christ child (since the phenomenon appears at the Christmas season), an irregularly recurring flow of unusually warm surface water in the oceans west of South America; it can strongly affect weather patterns

GLOBAL WARMING: the controversial theory that carbon dioxide released into the atmosphere by humanity's activities is causing the average surface temperature of the Earth to increase

GREENHOUSE GASES: atmospheric gases that trap solar energy and make the Earth warmer; chiefly water vapor, carbon dioxide, methane, nitrous oxide, and chlorofluorocarbons.

LA NIÑA: literally "the girl," the counterpart of El Niño, but involving cold water

OZONE LAYER: an atmospheric layer between the heights of about 20 and 50 kilometers that contains most of the stratospheric ozone

HISTORY OF THE EARTH SYSTEM ENTERPRISE

In 1945 the well-known author of science fiction Arthur C. Clarke predicted the usefulness of geostationary communication satellites, but other common uses of orbiting satellites remained beyond the imagination until much later. The earliest satellites placed in orbit, the Soviet Union's Sputnik 1 on October 4, 1957, and Sputnik 2 on November 3, 1957, and The United States' Explorer 1 on January 31, 1958, were largely for political prestige. The challenge posed by the Soviet satellites led directly to the formation of the National Aeronautics and Space Administration (NASA) on October 1, 1958. As a civilian agency, NASA was charged with developing the necessary technology to explore the atmosphere and space, while military applications were left with the Department of Defense. Eventually NASA's attention turned both outward, to the Moon, planets, and stars, and inward, to Earth as seen from space.

NASA has carried out its exploration of the Earth from space under various programs, including the Mission to Planet Earth enterprise, renamed the Earth Science Enterprise (ESE) in January of 1998. ESE is charged with pioneering the emerging discipline of Earth system science, a discipline that looks at the whole Earth in terms of interacting systems. Using space-, aircraft-, and ground-based measurements, ESE seeks to develop an international ability to assess and forecast the health of the Earth system, to disseminate such information widely, and to enable the productive use of ESE results and related technology by both public and private sectors. ESE is charged with seeking answers to key questions in five areas: land surface cover, near-term climate change, long-term climate change, natural hazards research, and atmospheric ozone.

METEOROLOGICAL SATELLITES

The Television Infrared Observation Satellite, or TIROS, represented NASA's first effort to provide useful observations of Earth from space. The initial goal was to photograph clouds well enough to show the paths of hurricanes and thereby aid the government in making decisions about evacuating the coast. Much lay behind such a simple goal, including the development of the technology and capability to make a satellite and to launch it, to make an infrared television camera that would survive the launch and function in space, to provide the means to point the camera at the Earth, to provide a power source for the satellite, to provide devices to broadcast the pictures from space, and to receive them on the Earth.

The very successful TIROS series was followed

WEATHER SATELLITES

Designation	Launch Dates	Status
Nimbus 1 to 6	1964 to 1974	1983[a]
Nimbus 7/TOMS	1978	1993[a]
ATS 1 to 5	1966 to 1969	1978[a]
ESSA 1 to 9	1966 to 1969	1976[a]
TIROS 1 to 10	1960 to 1967	1968[a]
TIROS-N/NOAA	1978	1981[a]
SMS 1 and 2	1974 and 1975	1982[a]
GOES 1 to 6	1975 to 1983	1993[a]
GOES 7 to 10	1983 to 1997	operational[b]
NOAA 1 to 15 (ITOS)	1970 to 1998	operational[b]

NOTES:
a. Year in which last member of the series was deactivated.
b. At least one member of the series active as of January, 2000.

by the Improved TIROS Operational System (ITOS) series. Once operational, ITOS satellites were renamed National Oceanic and Atmospheric Administration (NOAA) satellites. These satellites were placed in polar orbits, which allowed them eventually to scan most of Earth as it rotated beneath them. They provided improved visible and infrared images of cloud-cover, snow and ice, and the sea surface; temperatures of the cloud-tops and lower atmosphere; and measurements of atmospheric moisture. They collected data broadcast by free-floating weather balloons and ocean buoys worldwide and rebroadcast it to a central processing station. They also carried proton detectors to aid in predicting the arrival of solar storms.

ATMOSPHERIC RESEARCH SATELLITES

The atmospheric research satellites measure various atmospheric constituents along with the structure of the atmosphere. In order to understand the effects of Earth's atmosphere, it is first necessary to know precisely how much solar energy falls on Earth. That is the purpose of ACRIMSAT,

launched in 1999. Active Cavity Radiometer Irradiance Monitor (ACRIM) I and ACRIM II flew aboard previous missions. The active cavity radiometers are pointed directly at the Sun and are uniformly sensitive to radiation from the far ultraviolet to the extreme infrared. They have shown that the total solar irradiance may vary by 0.1 percent, which may account for up to one-fourth of the proposed global warming.

The Total Ozone Mapping Spectrometer (TOMS) has flown on several spacecraft, beginning with Nimbus 7 in 1978. It maps the amount of ozone over various locations and monitors the antarctic ozone hole as well as the smaller arctic ozone hole. This monitoring is necessary to determine if reducing the use of chlorofluorocarbons (CFCs), used in refrigerants and to propel aerosol spray products, is having the desired effect. Since sulfur oxides from volcanoes can affect the ozone, this gas must also be monitored. The Atmospheric Laboratory for Applications and Science (ATLAS) satellites, as

ATMOSPHERE RESEARCH SATELLITES

Designation	Launch Dates	Status
UARS	1991	operational[b]
Meteor 3/TOMS	1991	1994[a]
ADEOS-MIRDORI/TOMS	1996	1997[a]
EP/TOMS	1996	operational[b]
ATLAS 1 to 3	1992 to 1994	1994[c]
ACRIMSAT	1999	operational[b]
QuikTOMS	2000[d]	—
SAGE III/Meteor 3 M	2000[d]	—
SAGE III/ISS	2002[d]	—
SORCE	2002[d]	—
EOS CHEM	2002[d]	—
CloudSat	2003[d]	—
PICASSO-CENA	2003[d]	—

NOTES:
a. Year in which last member of the series was deactivated.
b. At least one member of the series active as of January, 2000.
c. Launches and returns on the space shuttle.
d. Proposed launch date.

well as others, measured the abundance and distribution of many trace gases in the upper atmosphere.

LAGEOS SATELLITES

The Laser Geodynamics (LAGEOS) Satellites are designed to hold to very precise, nearly circular orbits. They are 60-centimeter-diameter aluminum hemispheres bolted onto cylindrical, brass cores. Their high-density-to-surface-area construction guarantees that the solar wind, light pressure, and micrometeoroids have almost no effect on their orbits. Set into wells in the surface of each satellite are 426 corner-cube retroreflectors. Four are made of germanium so that they will reflect infrared light, while the others are made of fused silica glass to reflect visible light. Highway signs are often coated with corner-cube crystals to reflect automobile headlights back toward the driver at night. In the same fashion, these satellites reflect laser light back toward sending stations. Timing laser pulses reflected by these satellites allows ground stations to locate their distances from a satellite to within 1 centimeter. Repeating these measurements over time allows the measurement of land movement associated with earthquake zones and with continental drift. Careful tracking of these satellites also provides information on the exact shape of Earth.

LANDSATS

The launch of Landsat 7 on April 15, 1999, continued a twenty-six-year tradition for the Landsat series. Landsat 7 images a given spot on Earth every sixteen days. Its images are taken through eight different filters spanning the visible and infrared regions of the spectrum. These filters are chosen to enhance details of vegetation, soil type, snow cover, and so forth. Spatial resolution is 30 meters at most wavelengths. Landsat images are used to monitor urban growth, deforestation, erosion, glacial recession, snow accumulation and melt, freshwater reservoir replenishment, vegetation types, agricultural land use, and crop health, among other things. Previously unknown groups of mountains were discovered on Landsat images of southern Victoria Land in Antarctica. Engineers examining Landsat images discovered that a proposed route for an oil pipeline in Bolivia crossed a fault line. Using Landsat data, they were able to avoid the fault and even to find a shorter route. Landsat data have been successfully used in mineral exploration and in monitoring strip mining and strip-mine reclamation.

OCEANOGRAPHY SATELLITES

Seasat and the more advanced TOPEX/Poseidon (Ocean Topography Experiment) are oceanography satellites. TOPEX/Poseidon uses a radar altimeter, a microwave radiometer, and other instruments to measure ocean levels to within a few centimeters, wave heights, wind speeds, and water temperatures. Such measurements are invaluable for tracking El Niño and La Niña phenomena. Jason 1 is the follow-on mission for TOPEX/Poseidon; it is also designed to monitor global ocean circulation and thereby improve global climate predictions.

The SeaWinds satellite uses a new type of radar to look down through clouds to measure ocean winds, currents, and other weather data. SeaWinds demonstrated its all-weather

SURFACE AND GEOPHYSICAL RESEARCH SATELLITES		
Designation	*Launch Dates*	*Status*
Landsat 1 to 7	1972 to 1999	operational[b]
LAGOES 1 and 2	1976 and 1992	operational[b]
Seasat	1978	1978[a]
TOPEX/Poseidon	1992	operational[b]
TRMM	1997	operational[b]
RADARSAT	1995	operational[b]
QuikSCAT/SeaWinds	1999	operational[b]
VCL	2000[c]	—
Jason 1	2000[c]	—
EO 1	2001[c]	—
GRACE	2001[c]	—
Triana	2001[c]	—

NOTES:
a. Year in which last member of the series was deactivated.
b. At least one member of the series active as of January, 2000.
c. Proposed launch date.

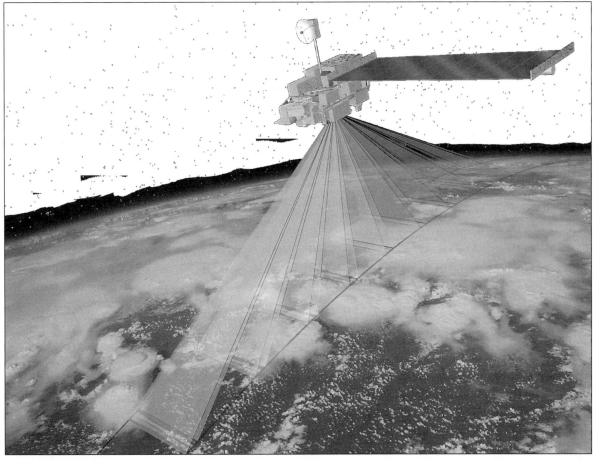

A computer-generated rendering of the Mulit-angle Imaging SpectroRadiometer (MISR), launched aboard the spacecreaft Terra (formerly EOS-AM1) and designed to collect data in four colors at all angles, to provide new information on Earth's climate and atmosphere. Terra is the cornerstone of the Earth Science Enterprise project, and MISR is only one of many projects. (Shigeru Suzuki and Eric M. De Jong, Solar System Visualization Project)

and day-night capabilities by tracking a huge iceberg the size of Rhode Island in July, 1999. Previously, trackers had lost sight of it and had been unable to relocate it. The ocean is huge.

The Tropical Rainfall Measuring Mission (TRMM) measures tropical and subtropical rainfall with microwave and infrared sensors and includes the first spaceborne rain radar. It proved its worth during the 1999 hurricane season when it helped to forecast the tracks and rainfall amounts of hurricanes Dennis and Floyd.

RADARSAT uses cloud-penetrating synthetic aperture radar (SAR) to map the topography of the Earth's surface. It has been especially useful in mapping Arctic sea ice and in producing the most detailed map of Antarctica yet produced.

TRIANA MISSION

The Triana mission is a good example of how ESE is applying available technology. It is named for Rodrigo de Triana, the lookout who first saw the New World from Christopher Columbus's ship. Triana will travel to L1 for an extended look at the Earth. L1 stands for the first Lagrangian point, a place nearly one million miles from the Earth in the direction of the Sun. At L1 the opposing gravitational attractions of the Earth and of the Sun on Triana will provide just the force Triana needs to orbit the Sun in tandem with the Earth. As seen from the Earth, an object at L1 will hoover perpetually over the daytime side, rising, moving across the sky, and setting with the Sun. The effects of other planets would cause Triana to drift away from L1, but making a virtue of necessity, Triana will use rockets to drift slowly about L1 so that it is never directly in line with the Sun. This will greatly aid ground stations in avoiding solar static while receiving data from Triana.

Triana will use ten electromagnetic wavelength channels—from the ultraviolet through the visible and into the near infrared—to image Earth with a spatial resolution of about 8 kilometers. It will provide daily global images of ozone, aerosols, water vapor, cloud cover, cloud height, and vegetation. It will also be able to provide hourly aircraft hazard warnings for volcanic ash. Three cavity radiometers will measure the solar radiation reflected from Earth and the energy emitted into space after being absorbed by Earth. These are all important quantities needed to construct more accurate climate models.

Other instruments on Triana will monitor the solar wind and solar flares. These will provide about an hour's warning before high-energy particles from solar storms reach Earth—enough time to shut down sensitive systems. Streams of such particles can cause damaging currents in some satellites and even in electrical systems on Earth. Triana's managers also see exciting prospects for education. Fresh pictures of the daylight side of Earth could be displayed on the Internet every fifteen minutes. Students could track pollution plumes and watch seasonal changes. They could also follow global changes in ozone, cloud cover, and weather patterns.

GOALS OF THE PROGRAM

Two fundamental and important environmental issues facing humankind today are the loss of stratospheric ozone and the possibility of global warming. Ozone is a chemically active molecule composed of three oxygen atoms joined together. "Bad" ozone near ground level is an irritating pollutant. "Good" ozone is a trace gas in the stratosphere (roughly 20 to 50 kilometers high) that plays an essential role in absorbing harmful ultraviolet light from the Sun. There really is not very much of it. If it were brought down to Earth's surface so that it was at normal atmospheric pressure, it would form a layer only 2 to 3 millimeters thick.

Both natural and manufactured gases can destroy ozone. Most scientists agree that manufactured chlorofluorocarbons (CFCs) gradually make their way into the stratosphere and destroy large quantities of ozone. Governments have mandated that the worst chemicals be phased out of use, even at considerable cost. ESE's proposed ongoing measurements of ozone and other trace gases in the stratosphere are essential, both to our understanding of the chemical processes that destroy ozone and to our ability to monitor progress in halting that destruction.

The situation is far less clear with global warming. It is supposed that carbon dioxide released into the atmosphere by human activities (chiefly by the burning of fossil fuels) enhances the greenhouse effect and produces global warming. Knowledgeable scientists are divided on this issue. Global average surface temperatures are inferred from a limited number of surface stations, but not all scientists are convinced that the averaging has been properly done. These measurements show that surface temperatures increased 0.25 degrees to 0.40 degrees Celsius between 1980 and 2000. Satellite measurements show that the troposphere (the atmospheric layer up to eight kilometers above Earth's surface) has warmed only 0.0 degrees to 0.2 degrees Celsius during the same time period.

It is unclear how to resolve this discrepancy. However, neither this increase in average temperature nor the claimed increase in the numbers and severity of storms is beyond historical deviations—nor is it yet clear that some increase in carbon dioxide increases temperatures. Instead, it may be that increased temperatures produce increased atmospheric carbon dioxide. During the past hundred years, the greatest increase in surface temperature in the United States occurred during the first fifty years, while the greatest increase in carbon dioxide occurred during the last fifty years.

Climate prediction is far more complex than many have supposed. Sunlight, ground cover, oceans, ocean currents, winds, clouds, cloud structure, snow, ice, aerosols, water vapor, a host of trace gases, volcanoes, El Niño and La Niña, and any number of other factors all play roles. So far, climate models are not yet good enough to "predict" what happened in the past under the conditions that scientists believe prevailed then. Until climate models are improved, they cannot be trusted to predict the future accurately. ESE is pursuing a vigorous program to obtain the data necessary to improve climate models so that we may know what course we should take.

Charles W. Rogers

BIBLIOGRAPHY

Bender, David, et al., eds. *Global Warming: Opposing Viewpoints.* San Diego, Calif.: Greenhaven Press, 1997. A collection of short, popular-level articles for or against various aspects of global warming, such as how serious it is, what causes it, what its effects might be, and what should be done about it.

Berner, Elizabeth K., and Robert A. Berner. *Global Environment: Water, Air, and Geochemical Cycles.* Upper Saddle River, N.J.: Prentice Hall, 1996. This book offers a clear and readable introduction into the study of geochemistry and the Earth's systems. Beautiful color illustrations and maps add to the reader's understanding of concepts.

Bruning, David, "Mission to Planet Earth." *Astronomy* (May, 1995): 44-45. An explanation of how the environment can be monitored from space, along with two spectacular images of Lake Taal in the Philippines (the caldera of volcano Taal, currently filled with water) taken by imaging radar aboard the space shuttle *Endeavour.*

Davidson, Jon P., Walter E. Reed, and Paul M. Davis. *Exploring Earth: An Introduction to Physical Geology.* Upper Saddle River, N.J.: Prentice Hall, 1997. An excellent introduction to physical geology and the systems of the Earth. This book explains the composition of the Earth, its history, and its state of constant change. Intended for the layperson, it is filled with colorful illustrations and maps.

Hamblin, Kenneth W., and Eric H. Christiansen. *Earth's Dynamic Systems.* 8th ed. Upper Saddle River, N.J.: Prentice Hall, 1998. A college textbook, *Earth's Dynamic Systems* introduces the reader to basic concepts such as gravity, rotation, and the Earth's tides. The text is well illustrated and makes complicated processes understandable to those without a background in physical geology.

Jones, Laura, ed. *Global Warming: The Science and the Politics.* Vancouver: The Fraser Institute, 1997. A collection of articles showing that several commonly cited global warming predictions have, as yet, insufficient scientific evidence to support their conclusions.

King, Michael D., ed. *EOS: Science Plan.* NP-1998-12-069-GSFC. Washington, D.C.: NASA/Goddard Space Flight Center, 1999. This book is widely available in the government document section of libraries. It is a blueprint for the Earth Observing System within the Earth Science Enterprise program, and it details what is to be measured, why it is to be measured, and how it is to be measured. A good science background is required to understand all of it, but even those with more modest backgrounds will benefit by browsing its colorful charts and figures.

Lutgens, Frederick K., and Edward J. Tarbuck. *Earth: An Introduction to Physical Geology.* Upper Saddle River, N.J.: Prentice Hall, 1999. 6th ed. This college text provides a clear picture of the Earth's systems and processes that is suitable for the high school or college reader. In addition to its illustrations and graphics, it has an accompanying computer disc that is compatible with either Macintosh or Windows. Bibliography and index.

McGeary, David. *Physical Geology: Earth Revealed.* 3d ed. Boston: WCB/McGraw-Hill, 1998. A thorough introduction to physical geology, this illustrated textbook is accompanieed by a computer laser optical disc that corresponds to the chapters and topics explored. Bibliography and index.

National Aeronautics and Space Administration Advisory Council. *Earth System Science Overview.* Washington, D.C.: Government Printing Office, 1986. This exquisite, full-color booklet is the basic document for approaching the subject of Earth system science. Created by the Earth System Sciences Committee of the NASA Advisory Council, the fifty-page document details in easy-to-understand text all the intricate natural mechanisms at work on the planet. Far more important, the booklet describes and pictures the entire Earth system science concept and outlines in depth how the new tools and methods will be brought together to focus on a global data-gathering and archiving information system through international cooperative efforts. The illustrations are excellent examples of highly professional photography and artwork. Written for the high school student and the layperson.

Nilsson, Annika. *Ultraviolet Reflections: Life Under a Thinning Ozone Layer.* New York: John Wiley & Sons, 1996. This is a well-written and easily read book about the formation of the ozone hole: what is being done about it, what ef-fects (if any) it may have already caused, and effects it is likely to have if the hole gets much worse. The book includes an extended discussion on the effects of suntanning.

Parkinson, Claire L. *Earth from Above: Using Color-Coded Satellite Images to Examine the Global Environment.* Sausalito, Calif.: University Science Books, 1997. A heavily illustrated introduction to the use of satellite images to examine atmospheric ozone, sea ice, sea surface temperatures, vegetation, snow cover, and volcanoes. Important environmental topics include the ozone hole, El Niño, deforestation, the effects on climate of sea ice, snow cover, and volcanoes, as well as the missing carbon dilemma.

Press, Frank, and Raymond Siever. *Understanding Earth.* 2d ed. New York: W. H. Freeman, 1998. An excellent general text on all aspects of geology. Suitable for advanced high school and college students.

Thompson, Graham R. *An Introduction to Physical Geology.* Fort Worth, Tex.: Saunders College Publishing, 1998. This college text provides an easy-to-follow look at what physical geology is, and the phases involved in it. Thompson shows the reader each phase of the Earth's geochemical processes. Excellent for high school and college readers. Illustrations, diagrams, and bibliography included.

ENVIRONMENTAL HEALTH

Environmental factors have definite effects on the health of organisms, including humans. For example, trace elements at certain concentrations are required in the diet of organisms for optimum health; environmental exposure to higher or lower levels results in adverse health effects, including disease and death. Natural geologic processes and human activities both act to determine the environmental exposure of organisms to trace elements.

PRINCIPAL TERMS

ALKALINE: the chemical state of a substance that causes it to react with hydrogen ions; acids are the opposite of alkaline substances in that they contain excess hydrogen ions

BIOCONCENTRATION: a general term for the metabolic processes by which organisms accumulate trace elements in their tissues to concentrations greater than that of their living environment

CYCLES: a sequence of naturally recurring events and processes; most cycles consume energy to move a substance through the environment

ENVIRONMENTAL HEALTH: the study of the relationships between an organism's health and aspects of its living environment

POLLUTION: the artificial release into the environment, by human activities, of a substance, resulting in greater environmental concentrations than would occur from natural processes alone

TRACE ELEMENTS: naturally occurring elements present at a very low concentration, usually defined as less than 1 percent

EARTH CYCLES

Several aspects of geology pose potential threats to human health. Management and disposal of hazardous and nuclear wastes, landfill disposal of solid waste, and the potential alteration of the Earth's atmosphere (the "greenhouse effect," ozone depletion) are examples of threats arising from the by-products of human activities. Urban and coastal land-use planning, radon gas, earthquakes, volcanic eruptions, soil liquefaction and expansion, and landslides are areas that involve hazards associated with natural geologic processes. Interactions between Earth materials and geologic processes define the ways in which the Earth operates; they also define the environment. Recognition of the many links between environmental factors and human health has spawned a new scientific discipline: environmental health. Over 4.6 billion years of Earth history, these processes have continuously changed the characteristics of the Earth. Indeed, the primitive Earth underwent a physical and chemical evolution that eventually spawned organic inhabitants. Geologic processes, whether operating at the Earth's surface or within its interior, are often a sequence of recurring events or cycles.

This cyclic view of Earth change was first recognized by eighteenth century Scottish intellect and "father of geology" James Hutton. Hutton recognized that the Earth is fundamentally dynamic and continuously changing in response to myriad cyclic phenomena of Earth components intimately and intricately interacting with one another. He recognized the "rock cycle" as the most fundamental of these. Earth materials are continuously formed into rocks by igneous, sedimentary, or metamorphic processes; the rocks are exposed to processes of weathering and erosion and are eventually destroyed; new rocks are created as the Earth materials are recycled in igneous, sedimentary, and metamorphic processes. The theory of plate tectonics demonstrates that this rock cycle operates on a very large scale involving the lithosphere and asthenosphere of the Earth. Recent discoveries of large-scale tectonic features on some of the other planets and moons in the solar system indicates that the rock cycle is a fundamental characteristic of planetary evolution.

The more scientists learn about Earth cycles, the more they realize that all living things are the way they are because the Earth operates the way it does. No matter how simple, all organisms are functioning components in many geologic cycles; indeed, human lives are inextricably linked to some of the processes responsible for the formation and destruction of rocks and the landscape. Scientists are also beginning to realize that everyday human activities can have an impact on some Earth cycles, often altering the natural rates of change. Examples of human activities that can disrupt environmental health include some mining and natural resource extraction, use of fossil fuels and radioactive substances, the physical alteration of the landscape to suit human needs, many methods of waste disposal, and the destruction or forced extinction of wildlife.

GLOBAL WARMING

The example of carbon dioxide production and the greenhouse effect is less clear. The so-called greenhouse gases allow sunlight to reach the Earth but partially block infrared radiation traveling from the Earth back into space. This makes the planet's surface warmer than it would be without the gases. The primary greenhouse gas in the Earth's atmosphere is water vapor, while carbon dioxide is the second most important. The carbon dioxide cycle is complex. The gas is emitted by animals, decaying vegetation, burning fossil fuel, warming water, and volcanoes. It resides in the atmosphere and is absorbed by plants and dissolves in water—there is far more carbon dioxide in the ocean than in the air. Dissolved carbon dioxide can combine with calcium to form limestone (calcium carbonate) deposits, a process that is greatly aided by life-forms possessing shells; however, it can also take place without such life-forms. Almost all of the Earth's carbon dioxide has been converted to limestone. Over long spans of geological time, limestone sinks deep beneath the Earth's surface, after which its carbon dioxide content is emitted back into the atmosphere during volcanic eruptions, thereby completing the cycle.

There has been more carbon dioxide in the atmosphere at times in the past, and the Earth has been warmer. Measurements show that the amount of carbon dioxide in the atmosphere has been slowly increasing since the nineteenth century, a result that can be traced back to the burning of fossil fuels. The effect that this has, or will have, is con-

troversial. There is no consensus among scientists on whether the Earth is warming. Global ground measurements are difficult, and natural processes such as El Niño easily mask small effects.

If the Earth does warm, warming water will drive dissolved carbon dioxide back into the air, compounding the greenhouse effect. However, it also causes more water to evaporate, thus producing more clouds, which will reflect more sunlight back into space. This should cool the planet. More rainfall should erode rock faster and transfer more calcium into the ocean, leading to the formation of more calcium carbonate and allowing more carbon dioxide to dissolve in the ocean, thus removing it from the air. These natural processes may limit global warming. Unfortunately, climate models are not yet good enough to reproduce the past or the future, so the real effect of increasing carbon dioxide is not yet known. However, prudence dictates that carbon dioxide emissions be reduced until scientists achieve a better understanding of the effects of increased carbon dioxide in the atmosphere.

TRACE-ELEMENT CONCENTRATIONS

All Earth materials are chemical compounds of the ninety-two naturally occurring elements. Of these ninety-two elements, only a few are abundant in the Earth's crust; the eight most abundant elements (oxygen, silicon, aluminum, iron, calcium, sodium, magnesium, and potassium) make up more than 99 percent of all Earth crustal rocks. The other elements are present in rocks in very small quantities (less than 1 percent) and are referred to as trace elements for this reason. Trace elements are sometimes present in greater concentrations in the tissues of organisms. For example, the eight most abundant crustal elements make up less than 70 percent of human tissue; carbon, hydrogen, and nitrogen, all trace elements in the Earth's crust, comprise more than 30 percent of the human body. Although not abundant, most of the trace elements are necessary; that is, they are used in the body to perform many of the metabolic functions and biochemical reactions necessary for life and are continuously provided to the organism in its diet. A given trace element has a range of possible effects on the body. Too little of the trace element in the food and water that are consumed results in poor health or disease because of a deficiency. Conversely, an excess concentration can also cause health to suffer by disease or, ultimately, death by poisoning. The maximum health benefit is attained within a well-defined optimum range of trace-element concentration.

The trace-element concentrations present in the tissues of organisms are generally controlled by the concentration of these same elements in their living environment; for plants, this is the soil. Soils will sometimes accumulate certain trace elements to very high concentrations when the elements are bonded tightly to molecules in the soil or when the chemistry of the soil keeps the trace element from being dissolved and removed by water draining the soil. Plants living in such a soil will naturally contain higher-than-normal amounts of these elements in their tissues; animals feeding on these plants are exposed to these high concentrations. In some cases, the high level of trace elements and nutrients will not harm the health of the plants but may cause poisoning of animals. Certain soils contain such high levels of trace metals that only a few plants can tolerate them. One example is the copperwort, a mosslike plant that grows only in soils with very high copper concentrations. The copperwort has been used historically as an indicator of ore-grade copper to prospectors in Africa and South America; because few animals can tolerate the high levels of copper in this plant, the copperwort has few predators. The trace-element content of many other plant species is used as a tool for metals prospecting.

BIOCONCENTRATION

Organisms like the copperwort tend to accumulate certain trace elements (especially metals) in their tissues at concentrations much higher than in the soil itself by incorporating these elements into very stable body tissue. Termed "bioconcentration," this process results in plant and animal tissues with very high, and sometimes toxic, levels of some elements. Bioconcentration is sometimes responsible for human disease, usually when food organisms that have bioconcentrated high levels of a toxic substance are eaten. A particularly dramatic example occurred with the trace element cadmium. Cadmium atoms tend to mimic calcium atoms in the body and will replace calcium atoms in the molecules of some body tis-

sues. If a person's diet contains a relatively high cadmium content, the cadmium will accumulate in calcium-rich body tissues, usually the bone. During the World War II era in Japan, waste sludge from the lead mining and smelting industry was dumped into the Jintzu River, resulting in high cadmium levels (as well as lead and zinc) in the adjacent rice fields. The dietary cadmium (about 1 part per million in rice) accumulated in the bones of people in the area over a period of five to thirty years. The high skeletal cadmium levels weakened the bone, resulting in skeletal deformities, spontaneous bone fractures, and extreme pain; this unusual and debilitating disease is called *itai-itai*, the Japanese phrase "ouch-ouch." Adverse health effects resulting from cadmium exposure in persons living near lead smelters have been documented in the United States. Cigarette tobacco contains about 1 part per million of cadmium.

Lead is another element that mimics calcium in the body and accumulates in bone; it, too, can cause painful bone weakening at high levels of environmental exposure. Lead also affects the central nervous system, and advanced lead bioaccumulation is believed to have influenced the fall of the ancient Roman civilization. Wine cups, pots, cosmetics, and medicines from Roman ruins contain anomalously high lead concentrations, and the Roman mining industry is known to have produced 55,000 metric tons of lead annually. The wealthy had water delivered to their homes in lead water pipes. Lead poisoning over generations is argued to have caused mental retardation (lead dementia) and eventually death in the ruling class of Rome. Archaeologists have sampled the remains of Roman citizens and documented bone with high lead concentrations and lead accumulation in the skeletal joints, corroborating this hypothesis.

RECOGNITION OF ENVIRONMENTAL HEALTH ISSUES

Interest in and anxiety about human health is a pressing issue. Awareness of the many links between environmental factors and human health has sensitized public consciousness to issues of environmental health. This is especially true in the developed nations of the world. The moral, intellectual, and political turbulence of the 1960's had much to do with a change in the attitude of the U.S. public, for one; that was when it began to see its capacity for changing the natural world and realized that the health of the human race depended, in large part, on the health of the living environment. This awareness was recorded with the passage of benchmark federal legislation (all have been subsequently amended), including the Water and Pollution Control Act (1956), the Clean Air Act (1963), and the Motor Vehicle Act (1965). The philosophy of the U.S. government toward environmental issues was defined by the National Environmental Policy Act of 1969, which required all federally funded actions "to formally consider environmental impact." The birth of the environmental movement attained political maturity on Earth Day, 1970, a mass demonstration attended by average citizens and members of Congress. The environmental movement is still very much alive. National polls have shown that the overwhelming majority of U.S. citizens would support environmental health legislation and regulation, even if it would raise taxes, reduce the number of jobs, or lower the standard of living.

Recognition of environmental health issues is increasing in the Third World as well. Responsible for this increase have been the realization that environmental problems extend across political boundaries, the growth of international exchange of scientific information and expertise, worldwide improvements in literacy and education, and the expanding global exchange of information available to the public. Another factor has been the continuing global assault of human industry on the environment, which grows with worldwide population. International awareness of environmental health issues has had tangible positive results since the mid-1980's. Changing political priorities have resulted in unprecedented international agreements on health and the environment.

James L. Sadd

CROSS-REFERENCES

Air Pollution, 1809; Climate, 1902; Desertification, 1473; Earth Resources, 1741; Earth Science and the Environment, 1746; Earth Science Enterprise, 1752; Future Resources, 1764; Greenhouse Effect, 1867; Hazardous Wastes, 1769; Land Management, 1484; Land-Use Planning, 1490; Landfills, 1774; Mining Processes, 1780; Mining Wastes, 1786; Nuclear Waste Disposal, 1791; Ozone Depletion and Ozone Holes, 1879; Plate Tectonics, 86; Strategic Resources, 1796.

BIBLIOGRAPHY

Cannon, Helen H., and Howard C. Hopps. *Environmental Geochemistry in Health and Disease.* Boulder, Colo.: Geological Society of America, 1971. A collection of articles by various scientists, illustrating known and suspected links between geologic processes and Earth materials (especially trace metals) and human disease and mortality. Some of the articles are of a technical nature, but all have valuable, easily understood maps, graphs, and tables of data that are as interesting to study and as informative as is the text.

Chisolm, J. Julian, Jr. "Lead Poisoning." *Scientific American* 224 (February, 1971): 15-23. A comprehensive nontechnical examination of the toxic effects of lead and routes of lead exposure. Includes many examples.

Foster, Harold D. *Health, Disease, and the Environment.* Boca Raton, Fla.: CRC Press, 1992. Foster looks at the environment for clues and possible connections to the causes of chronic diseases such as cancer, heart disease, and stroke. This is an interesting and easy-to-understand introductory-level book for anyone interested in the subject matter.

Goldwater, Leonard J. "Mercury in the Environment." *Scientific American* 224 (May, 1971): 15-21. An excellent treatment of the mercury cycling through the environment and the accompanying health effects. Includes an excellent explanation of mercury poisoning at Minamata Bay, Japan, a classic example of the interrelationship of trace metals and environmental health.

Keller, Edward A. *Environmental Geology.* 4th rev. ed. Columbus, Ohio: Charles E. Merrill, 1985. A widely used, standard textbook for many college-level introductory courses in environmental geology. Has an excellent chapter on environmental health and geology that includes several different illustrative examples of interactions between geology and environmental health.

Listorti, James A. *Environmental Health Components for Water Supply, Sanitation, and Urban Projects*. Washington, D.C.: World Bank, 1990. A technical paper focusing on health aspects of the water supplies of urban areas. The paper looks at water content and sanitation engineering procedures that have been implemented.

National Academy of Sciences. *Geochemistry and the Environment*. Vol. 1. Washington, D.C.: Author, 1974. Available at most libraries, this volume deals with the relationship of selected trace elements (including selenium) to health and disease, devoting a full chapter to the most significant elements. A technical publication, but it is extremely informative and can be understood and used by most readers.

Noji, Eric K. *The Public Health Consequences of Disasters*. New York: Oxford University Press, 1997. Noji studies the widespread outbreaks of disease and epidemics that often follow natural disasters. He also explores prevention methods and emergency plans to handle disease and contamination. A good examination of how environmental health policies can be implemented in specific situations.

Tanji, Kenneth L'auchli, and Jewell Meyer. "Selenium in the San Joaquin Valley." *Environment* 28 (July/August, 1986): 6-11. A highly readable account of the events and factors involved in selenium poisoning in the Kesterson Wildlife Refuge of central California. Describes the geochemical processes of cycling of selenium between organisms and their environment. Available at most school libraries.

Wilson, Mary E., Richard Levins, and Andrew Spielman, eds. *Disease in Evolution: Global Changes and Emergence of Infectious Diseases*. New York: New York Academy of Sciences, 1994. This collection of papers originally presented at a 1993 workshop examines the environment's role in the evolution of communicable diseases, chronic diseases, and epidemics.

Young, Keith. *Geology: The Paradox of Earth and Man*. Boston: Houghton Mifflin, 1975. A good introductory text with significant information and examples of the effects of geological elements on health.

FUTURE RESOURCES

During the twenty-first century, new resources may well be required to build and energize civilization on other planets and in space as well as to enhance accelerated change in a dynamic Earth-based society.

PRINCIPAL TERMS

CERAMICS: nonmetal compounds, such as silicates and clays, produced by firing the materials at high temperatures

FIBER-REINFORCED COMPOSITES: materials produced by drawing fibers of various types through a material being cast to produce a high weight-to-strength ratio

HYDROLYSIS: the breakdown of water by energy into its constituent elements of water and hydrogen

OFF-PLANET: pertaining to regions off the Earth in orbital or planetary space

PERMAFROST: a layer of soil and water ice frozen together

PHOTOVOLTAIC CELL: a device made commonly of layered silicon that produces electrical current in the presence of light; also, a solar cell

REGOLITH: that layer of soil and rock fragments just above the planetary crust

SUPERCONDUCTORS: materials that pass electrical current without exhibiting any electrical resistance

CLASSIFICATION OF RESOURCES

In the largest possible sense, there will be two broad categories of resources during the twenty-first century: earth-based resources and off-planet resources. The reason for the distinction is an economic one, based on where the materials will be used, and is driven by consideration of the gravitational field of the Earth. The Earth's gravitational field is very strong in comparison with extraterrestrial space, where there is little gravitational influence. It is also strong in relation to the Moon and Mars, where the gravitational field is much weaker than the Earth's. Hence, materials required in space will be dramatically more economical if mined from small bodies in space (such as asteroids or comets) or the Moon. The same resources, if shipped into space from the Earth, would cost many times more. This example is provided as an insight into the distinctive ways of looking at future resources in the developing economy of the twenty-first century. In a more systematic approach to categorization, one could fundamentally classify tomorrow's most basic resources in the same broad classes as are used currently: agricultural products, chemicals (including petroleum and derivatives), metals, ceramics, energy, wood (and derivatives), and power.

IMPACT OF SCIENTIFIC DISCOVERY

Though the same broad categories of resources will be required in the future as are required today, many of their individual identities and uses will be much different. Scientists are beginning to foresee new types of materials that will have a vital use. Such an example was afforded in 1989 when two scientists announced that they thought they had discovered a revolutionary power source. One determinant element in their design was a rare-earth metal called palladium. Until their announcement, there had been few other uses for the metal; after their announcement, the price of palladium temporarily increased many times as a result of its presumed importance. As new discoveries are made, new uses for existing materials such as metals will cause the demand for the resource to change as dictated by availability.

Scientific discovery and technological development are often the key to the advancement and change of society and the resources that drive it. Examples from the past are abundant: Tungsten presently used for light filaments had few uses before the invention of the incandescent light bulb; silicon (the key element of common beach sand) assumed vital importance with the discovery of microelectronics. Probably no other single resource has had a more far-reaching impact on humankind or planet Earth than petroleum and its derivatives. The science of superconductivity produces materials that conduct electrical current without resistance. Late in 1986, this branch of physics took a revolutionary turn when it was discovered that synthetically produced materials could become superconductive at temperatures much higher than had been thought possible. The synthetic materials were described as "artificial rocks" and were made by mixing together various metal oxides under very specific conditions. As the ultimate goal of room-temperature superconductivity is approached, the materials used in producing the superconductors are likely to become highly sought-after and valuable resources.

CERAMICS AND FIBER-REINFORCED COMPOSITES

The use of ceramic materials increased dramatically in the 1980's. Ceramic materials are nonmetal compounds produced by firing at high temperatures. Ceramics can be lightweight, resilient, and heat-resistant. Materials used in the production of ceramics are clays, silicates, and calcium carbonates. Uses of ceramics include refractories (heat-shielding or heat-absorbing materials, such as those used to construct the space shuttle's heat-shielding tiles) and electrical components. They have also been used in tests as automobile engine blocks. The resources that produce such ceramics are projected to have even broader applications in the future.

Another type of product that has become quite important and is expected to play a vital role in the future is fiber-reinforced composites. Such composites are made by drawing fibers through a material being cast. When the material hardens, the fibers cast inside it make even very light products, such as aircraft wings and space vehicle fuselages, much stronger. The materials from which these composites are made (fiberglass, boron, tungsten, aluminum oxide, and carbon) will assume new importance with any increase of the use of fiber-reinforced composites.

ECONOMICS OF OFF-PLANET RESOURCES

Because the exploitation and possible settlement of space will probably be the most significant development of the twenty-first century, resources that will drive that revolution will become highly

valuable in the future. The value of these space materials will be based on their origin. It will always be expensive to ship resources into space from the surface of the Earth. A kilogram of aluminum mined and processed on the Moon and shipped to Earth could one day cost a tenth of the amount of the same aluminum shipped to orbit from the Earth because of launch-energy costs. Even under the best of circumstances, in a well-developed launch system, a liter of water launched from the Earth into space will always cost thousands of times more than will a liter of water on the surface. One must always keep the gravitational interfaces in mind when considering the economics of space resources. Many studies have been conducted concerning the utilization of off-planet resources, their availability and uses. Such studies have been conducted concerning the construction of Moon and Mars bases and colonies as well as the construction of enormous space colonies.

OFF-PLANET WATER AND POWER RESOURCES

The most valuable resources in space will be the two resources of survival, water and energy; the use of one is dependent on the other. Water and energy yield two other vital resources: oxygen and hydrogen. Water is the most basic of all resources necessary for human survival. Aside from the obvious purposes of direct consumption and hygiene, water will be necessary for cleaning, cooling, and producing food. Raw water can be broken down by a process called hydrolysis into elemental hydrogen and oxygen. The hydrogen may be used for energy production and the oxygen for breathing.

The mass of water is great enough that launching volumes of the substance from the Earth that are sufficient to meet the needs of space resources will play a significant role in determining the use of space. Much attention has been paid to the likely sources of water off-planet, especially in the most exploitable deposits. The Moon may contain deposits of water ice, locked up beneath the regolith. The Clementine spacecraft bounced radar signals from the lunar surface during its 1994 exploration of the Moon. Radio telescopes on Earth monitoring those signals discovered that regions near the lunar poles were reflective in a way that was suggestive of the presence of water ice.

The Lunar Prospector spacecraft carried a neutron spectrometer capable of detecting the energy of neutrons blasted from the lunar surface by the solar wind. It was announced on March 5, 1998, that neutrons from the polar regions carried the energy signature of having interacted with hydrogen. The most likely reservoir of hydrogen would be as a constituent of the water molecule in water ice. There may be as much as 300 million metric tons of ice particles mixed with the lunar soil of the permanently shadowed craters at each pole. Presumably, the icy particles have been delivered to the Moon by comet impacts over billions of years.

There is considerable evidence that Mars may contain vast amounts of water locked up beneath its regolith as permafrost (water ice mixed with soil). Third, some asteroids may contain water ice, and comets are generally considered to be mostly water ice. It is speculated that these resources may be captured and relocated to an orbit suitable for mining. Finally, pure water is produced as a waste product of fuel cells that react hydrogen and oxygen to produce power. Such cells, if used extensively for power, may become an important source of water off-planet.

Another source of energy is direct conversion, by photovoltaic (solar) cells, of sunlight into electricity. Such cells may be produced off-planet by using lunar materials such as silicon, a significant constituent of lunar soil. Future space colonists may want to locate reserves of radioisotopes for nuclear power. Such reserves, if found, could be exploited with an excellent promise of safety, especially in high Earth, lunar, or solar orbit, as the isolation of the radioactive contaminants and power plant itself would be assured. Questions concerning other power sources, such as high-temperature fusion and unconventional fusion techniques, remain unresolved. Such devices, when perfected, would overcome nearly every known deterrent, allowing the fullest exploitation of space.

OFF-PLANET BUILDING MATERIAL RESOURCES

Aside from the most basic resources of survival off-planet (water, power, hydrogen, and oxygen), space colonists will require massive building-material resources off-planet. Launching millions of tons of raw building materials into space is as in-

feasible as hauling volumes of water there. A Princeton University physicist, Gerard K. O'Neill, has addressed the question of producing such quantities of building materials for construction of off-planet colonies. O'Neill and his Princeton-based Space Studies Institute have designed a mass driver, a device that could catapult "buckets" of lunar soil into lunar or Earth orbit for processing—in automated space factories—into sheets of aluminum, magnesium, titanium, glass, and other materials for use in construction of lunar or orbiting colonies. Lunar soil has been found to be rich enough in the necessary materials to produce such materials in space. Current studies on the Moon rocks brought back from the Apollo program indicate that roughly 50,000 metric tons of aluminum sheets per year could be produced from 900,000 metric tons per year of Moon ores catapulted into space by the mass drivers.

SURVEYS OF OFF-PLANET RESOURCES

Space probes have been utilized to recover information on the location of resources in space and their amounts. The Apollo mission astronauts returned samples from the lunar surface that were analyzed extensively and used in subsequent detailed engineering and economic studies to reveal how the lunar material might be processed into building materials. One such study was performed by Gerald W. Driggers of the Southern Research Institute of Birmingham, Alabama. In Driggers' account, he estimated the cost of building an orbiting metals factory ($20 billion) and its power requirements (300 megawatts). He also estimated the final cost of the aluminum sheeting ($22 per kilogram) and compared it to the cost of the earth-produced material ($2.2 per kilogram); he then compared it to the cost of the earth-based material launched into space ($88 per kilogram). Such economic studies will drive the capital outlays that will be necessary to fund and justify such substantial expenditures.

The resources of Mars were surveyed at two landing sites in 1976 by the United States' Viking landers. The Viking landers were not designed to return samples directly from Mars but were sent with a device called an X-ray fluorescence spectrometer, which would analyze the Martian soil for its inorganic materials. No probe has ever taken a direct observation of an asteroid, although many scientists are convinced that the Martian moons Phobos and Deimos are captured asteroids. Flyby studies of these bodies indicate that their densities are low enough that they could be made up of some water ice in their interiors. In 1986, several spacecraft flew past Halley's comet as it approached the Sun, returning spectacular, closeup photos of the body and taking measurements of its density. These flybys have supported earlier speculation that comets are largely water ices covered over by exceptionally dark, carbonaceous material.

DEVELOPMENT OF ENERGY PRODUCTION

The survivability and quality of every human's life is directly affected by the resources available to each individual. From water and power to food and building materials, humankind has been in a constant struggle to improve the availability and quality of resources while reducing the magnitude of the struggle to obtain them. The historical propensity of civilization, if not its fundamental purpose, is to ease that struggle and increase the availability of the resource base while constantly improving its superiority and basic usefulness.

The resources of the future will follow this trend, with momentous advances being made in the most elemental domain: power, or energy. Energy production directly affects the acquisition of all other resources. The lives of all humans should benefit from this resource development. Improved ways of generating power will directly influence food production for all people but, most profoundly, those populations in regions fixed on the edge of continual famine. High-temperature superconductivity would change the basis of electronic devices, resulting in more efficient and cheaper electrical power production, storage, transmission, and use. The exploitation of this science would enable supercomputers that operate at unprecedented speed. Transportation on Earth and in space would be revolutionized.

Dennis Chamberland

CROSS-REFERENCES

Desertification, 1473; Earth Resources, 1741; Earth Science and the Environment, 1746; Earth Science Enterprise, 1752; Environmental Health, 1759; Freshwater Chemistry, 405; Geochemical Cycle, 412; Greenhouse Effect, 1867; Groundwater Pollution and Remediation, 2037; Hawaiian Islands, 701; Hazardous Wastes, 1769; Hydrologic Cycle, 2045; Land-Use Planning, 1490; Land-Use Planning in Coastal Zones, 1495; Landfills, 1774; Mining Processes, 1780; Mining Wastes, 1786; Nuclear Waste Disposal, 1791; Radon Gas, 1886; Soil Chemistry, 1509; Soil Formation, 1519; Soil Types, 1531; Strategic Resources, 1796; Surface Water, 2066.

BIBLIOGRAPHY

Arnopoulos, Paris. *Cosmopolitics: Public Policy of Outer Space.* Toronto: Guernica, 1998. This book offers a thorough discription of space exploration and industrialization, and the laws and legislation surrounding these fields.

Cadogan, Peter. *The Moon: Our Sister Planet.* New York: Cambridge University Press, 1981. This work discusses the discoveries arising from the exploration of the Moon during the decade of the 1960's. Details the resources discovered from the close analysis of the Apollo Moon rocks. Also speculates about water on the Moon and other possible Moon-based resources. Well illustrated, although its index is somewhat limited.

Cooper, Henry S. F., Jr. *The Search for Life on Mars: Evolution of an Idea.* New York: Holt, Rinehart and Winston, 1980. Although this book is a superlative accounting of the search for life on Mars, it relates the instruments used in all the Viking lander science, including a discussion of the device that measured the constituents of Martian soil. An excellent work for understanding how scientists comprehend the details of another planet's resources through automated robotics and how the inferences to planetary resources are thus enabled. Although not illustrated, a well-written scientific narrative.

Gulkis, Samuel, D. S. Stetson, Ellen Renee Stofan, et al., eds. *Mission to the Solar System: Exploration and Discovery.* Pasadena, Calif.: National

Aeronautics and Space Administration, Jet Propulsion Laboratory, 1998. This collection of essays written by experts in their respective fields deals with space exploration and the possibility of resource exploration on other planets. It provides a look into current practices, as well as future potential in space exploration.

Hazen, Robert M. *The Breakthrough: The Race for the Superconductor.* New York: Summit Books, 1988. A story that features the discovery of high-temperature superconductors in 1987. Relates what resources were used in the construction of the "artificial rocks" that were made superconductive. Also describes the revolutionary changes that such discoveries would bring.

Heppenheimer, T. A. *Colonies in Space.* Harrisburg, Pa.: Stackpole Books, 1977. This detailed, illustrated book depicts what resources will be required to settle space en masse. It names and enumerates studies that have independently addressed the issue in sufficient particulars that the reader will understand the complexity and degree of effort required to make this next great "leap" of humankind.

Miles, Frank, and Nicholas Booth. *Race to Mars: The Harper and Row Mars Flight Atlas.* New York: Harper & Row, 1988. This book specifies the details for a flight to Mars and its ultimate settlement. Discusses the possibilities of Mars' resources, such as water, availability of building materials, and power sources. An excellent accounting of the difficulties in finding sufficient resources to set up colonies off-planet and the efforts required to establish a foothold on new frontiers.

O'Neill, Gerard K. *2081: A Hopeful View of the Human Future.* New York: Simon & Schuster, 1981. O'Neill discusses the facts as they will be encountered in delivering raw materials not only to the space and planetary colonies but also to the Earth from space. He discusses the "capture" of asteroids and the use of Moon-based mass drivers to deliver raw materials for use in space. Illustrated; written for all readers.

Schmidt, Victor. *Planet Earth and the New Geoscience.* 2d ed. Dubuque, Iowa: Kendall/Hunt, 1994. Produced through the University External Studies Program at the University of Pittsburgh in coordination with the television series *Planet Earth*, this college-level textbook provides an overview of the Earth sciences, mineral resources, and astronomy, and their future possibilities.

HAZARDOUS WASTES

Hazardous wastes include a broad spectrum of chemicals, produced in large quantities by industrial societies, that are dangerous to the health of the environment. These wastes often reach the environment, where reactions in soils and groundwater may work to reduce the hazard.

PRINCIPAL TERMS

BIODEGRADATION: biological processes that result in the breakdown of a complex chemical into simpler building blocks

HALOGEN: one of a group of chemical elements including chlorine, fluorine, and bromine

HEAVY METAL: one of a group of chemical elements including mercury, zinc, lead, and cadmium

HYDROCARBONS: chemicals composed chiefly of the elements hydrogen and carbon; the term is usually applied to crude oil or natural gas and their by-products

XENOBIOTIC: refers to a chemical that is foreign to the natural environment; an artificial chemical that may not be biodegradable

---

HAZARDOUS BY-PRODUCTS OF MANUFACTURING PROCESSES

Broadly defined, hazardous wastes include chemical, biological, or radioactive substances that pose a threat to the environment. Small amounts of radioactive wastes are rendered harmless by dilution, while larger amounts are held in isolation. Biological wastes may be incinerated, and some chemical wastes, such as nerve gas, are turned into harmless by-products by incinerating them at sufficiently high temperatures. The largest volume of hazardous wastes are the raw materials, the by-products, or the end-products of modern manufacturing processes. In that context, the elements and chemicals are all useful and necessary. When they end up in places where they do not belong—such as soils, groundwater, lakes, rivers, the ocean, and the air—and in higher than acceptable concentrations, then they become hazardous wastes. For example, heavy metals, such as lead and chromium, are used in many chemical processes. They may enter the environment from the time they are mined from the ground until they are used in a chemical process. They become hazardous wastes when their concentrations in the environment are sufficient to harm plants and animals.

Many hazardous waste chemicals are derived from petroleum. Petroleum is a complex mixture of chemicals, called hydrocarbons, composed mainly of carbon and hydrogen atoms. During refining and processing, the various petroleum hydrocarbons in the mix are separated; the carbon and hydrogen atoms are rearranged; and new elements, such as chlorine, bromine, and nitrogen, are introduced into the hydrocarbon molecules. In this way, naturally occurring chemicals, such as those that make up gasoline, are separated and concentrated, and new, synthetic chemicals are produced. These synthetic chemicals, and products derived from them, enable the production of several types of plastics (such as PVC pipes and polyethylene bags), nonstick surfaces, synthetic fibers in clothes and carpets, and many pharmaceuticals. Many of the synthetic chemicals, however, that are used to make the useful products—and the end-products themselves—may be hazardous wastes in the environment. For example, polychlorinated biphenyls (PCBs) are synthetic halogenated hydrocarbons, which are useful as heat-insulating materials in electrical transformers. When the transformer is scrapped, the PCBs become a waste that must be discarded with care.

HAZARDOUS WASTE RELEASES

In time, most environments can cleanse themselves of accidental spills or discharges of natural hazardous wastes. Several metals will be removed from water by attaching to sediment or to soil mineral particles. Although the metals are still in the environment, they are in a form that makes them less mobile and less likely to be absorbed or eaten by plants and animals.

Hydrocarbon spills and discharges—from crude oil blown out of a damaged offshore oil rig to leaks of refined gasoline from the underground storage tanks at filling stations—are probably the most common type of hazardous waste release. Hydrocarbon spills are also a formidable cleanup task for the environment. A major factor in the natural cleansing process is that the hydrocarbon molecules represent sources of energy. The stored energy is tapped in each molecule by burning the molecules in combustion engines. Bacteria and fungi in the ocean, lakes, and soils can also tap the energy in hydrocarbon molecules and use them as food. Some of the molecules in the complex mixtures are easier to use than others but, given enough time and the right conditions (such as a plentiful supply of oxygen, as the primary biological process is respiration), most of the hydrocarbons can be broken down, or biodegraded, to carbon dioxide and water.

Synthetic chemical spills are much more difficult for the environment to handle without help. The chemicals, being man-made, are new to the environment, and the natural mechanisms to biodegrade them into simple, harmless chemicals are not immediately available. For this reason, synthetic chemicals may remain in the environment for longer periods of time. As a result, there is more time for contact between the synthetic chemicals and plants and animals. There is also more opportunity for the plant or animal to absorb or eat harmful quantities of the chemical. Once inside an organism, a xenobiotic chemical will tend to accumulate. Very few biochemical mechanisms are available to the organism to break down the molecule so, as more molecules are ingested, the concentration of the chemical increases in the organism's tissue.

TRANSPORTATION OF HAZARDOUS WASTES

When a scientist is studying hazardous wastes in the environment, there are three important questions to answer. First, how is the waste transported? Second, what is the concentration of the waste during transport? Finally, can processes in the natural environment reduce the distance the waste is transported?

Most hazardous wastes are transported either in air or in water. Some wastes can be transported both ways. Since plants and animals use water and air in their daily lives, the wastes are taken into an organism as part of the normal life process. To determine how much of a given hazardous waste will be transported in solution (dissolved in water) or in air, a scientist determines the solubility or volatility of the waste. Solubility is a measure of how much of a chemical will dissolve in water. It is said that oil and water do not mix, suggesting that substances such as crude oil are not soluble in water. Crude oil, however, is a complex mixture of chemicals, some of which can dissolve in water. Once dissolved, they may be carried away from the site of a spill in solution to a river or to a drinking-water well.

CONCENTRATION OF WASTES DURING TRANSPORT

Just as some of the chemicals in a crude oil spill are soluble in water, some of the chemicals are volatile. The smell of gasoline (a refined product of crude oil) vapors at a filling station indicates that some chemicals volatilize and are transported in the atmosphere. In the open atmosphere, air is moving so rapidly that under most normal weather conditions, volatile hazardous chemical concentrations never get large enough to cause any concern. When weather conditions stagnate over large open areas, such as the Los Angeles basin, or in small areas, such as a cellar, the concentrations may become large enough to cause illness and even death.

In an area where a hazardous waste spill is suspected, scientists will take samples of the local water and air. The samples will be analyzed for the soluble and volatile components of the waste. Surface water samples may be collected from lakes or streams. Groundwater samples may be collected from existing water wells or from specially constructed monitoring wells. Air samples for volatile chemical analysis may be collected in restricted areas, such as basements, by using a vacuum pump to draw a certain amount of air through an activated carbon filter. The hazardous chemicals are absorbed by the carbon, and the sample is taken to a laboratory for analysis. Sometimes volatile hazardous chemicals can be identified in the soil atmosphere (the air filling the spaces between soil par-

Lowry Landfill in Arapahoe County, Colorado, 1978: Garbage and hazardous wastes mix with groundwater. (U.S. Geological Survey, photo by N. Gaggiani)

ticles above the water table). Investigators pound a hollow pipe into the ground and, using a similar vacuum pump and carbon filter, draw soil air up through the pipe and over the absorbent carbon.

Sorption and Biodegradation

Natural processes in soils and groundwater often act to destroy the chemicals in a hazardous waste. For example, a waste's rate of movement through the environment can be slowed by natural means. Some of the chemicals and elements in hazardous wastes have low solubility in water. These are called hydrophobic ("afraid of water"), because they attach themselves to the surface of a soil mineral grain or piece of dead plant material rather than stay in solution. Sorption, a natural process involving the removal of chemicals from solution by attachment to solid surfaces, slows a waste's rate of migration from the spill site. Some of the heavy metals and many of the hydrocarbons are hydrophobic. Sorption studies in the laboratory with soil materials and pure chemicals help to explain this complex natural phenomenon.

Microscopic organisms, such as bacteria and fungi, living in the soil and groundwater can limit the spread of a hazardous waste spill by eating some of the chemicals. Hydrocarbons, common chemicals in a hazardous waste spill, can be an energy source for organisms. Biodegradation of hazardous chemicals is an important natural mechanism for limit-

ing the transport of the chemical and ultimately cleaning up the wastes. Unfortunately, the microorganisms that consume the most hazardous chemicals are not common in most environments. The more common microbes, when they are in the environment, are often overwhelmed by the sudden influx of food represented by the spill. Other chemicals, such as oxygen gas, necessary for them to destroy the hazardous chemicals are quickly depleted, and the rate of destruction rapidly diminishes.

Scientists hope to exploit the natural biodegradation of hazardous chemicals in two ways. One way is to stimulate microbial activity in the area of a spill. This usually means providing a fresh supply of oxygen and other nutrients necessary for the organisms to work at full speed. The other method is to grow microorganisms in the laboratory. When a spill occurs, the microorganism or group of microorganisms is brought from the lab to the spill site. Once there, it may be injected into the ground, or the water contaminated with waste may be pumped to the surface and into tanks containing the microbes. The biologically treated water is filtered before being discharged, to keep the microorganisms in the treatment tanks. The latter technique seems to be more effective because it is easier for scientists to provide the optimum conditions for biodegradation in the controlled treatment tanks.

Economic Impact

The chemicals that comprise hazardous wastes can be found in kitchens, workshops, garages, or garden sheds. Hazardous chemicals are present in household cleaners, gasoline and oil, paints, varnishes, thinners, herbicides, and insecticides. Handling hazardous chemicals has become part of the way of life. Through carelessness and ignorance, hazardous chemicals become hazardous wastes when they are released into the environment. When they are released, their movement can become uncontrolled, and their concentration in air or water sufficiently high to damage the environment. Most of the chemicals comprising hazardous wastes at one time were useful. It is only when they are accidentally spilled or disposed of improperly that they become hazardous wastes.

The government spends a significant amount of money to clean up hazardous waste sites and to prevent future spills. Billions of dollars will be spent as part of the Comprehensive Environmental Response, Compensation and Liability Act (CERCLA), also known as Superfund, to find and clean up uncontrolled waste sites. More is being spent through the Resource Conservation and Recovery Act (RCRA) to limit or eliminate hazardous chemical releases to the environment, particularly from the factories that use and produce large quantities of the chemicals.

Banks and savings and loans, which hold the mortgages on all types of properties—from large industrial complexes to private homes—are now concerned about hazardous wastes. If buried or spilled hazardous wastes are found on a piece of property, the expense of a cleanup may be so high that the property occupant will declare bankruptcy. The holder of the property mortgage may then be liable for the cleanup. Before committing themselves to finance a land purchase, banks now require a preliminary hazard assessment of the property. Such an assessment usually includes an examination of historical records, which indicate the uses of the land by previous owners, and a site survey to identify any obvious signs of problems.

Richard W. Arnseth

CROSS-REFERENCES

BIBLIOGRAPHY

Brown, Michael. *Laying Waste: The Poisoning of America by Toxic Chemicals*. New York: Pantheon, 1980. The author was a reporter who closely followed the events surrounding the discovery of hazardous wastes at Love Canal. The book opens with a recounting of that story, followed by case histories of hazardous waste's impacts on communities around the country. Having been written by a newspaper reporter, the text is accessible to the general reader. Contains a good index but no bibliography to guide further reading.

Byrne, John, and Steven M. Hoffman, eds. *Governing the Atom: The Politics of Risk*. New Brunswick, N.J.: Transaction, 1996. This volume from the Energy and Environmental Policy Series researches the environmental and social aspects of the nuclear industry. Careful attention is paid to government policies that have been implemented to monitor safety and accident prevention measures.

Dadd, Debra L. *The Nontoxic Home: Protecting Yourself and Your Family from Everyday Toxics and Health Hazards*. Los Angeles: Jeremy P. Tarcher, 1986. The author discusses the toxic chemicals found in common household products and their potential health hazard. The book does not specifically deal with disposal of these substances but gives insight to the wide range of common products that become hazardous wastes when discarded. The author also provides information on nontoxic alternative products. Suitable for the general reader.

Gay, Kathlyn. *Silent Killers: Radon and Other Hazards*. New York: Franklin Watts, 1988. A short book (130 pages) with many news photo illustrations. The text covers a range of hazardous chemicals from Agent Orange to lead-based paint to radioactive soil used as fill for home construction. Small bits of technical description are inserted between larger discussions of the political, economic, and social impacts of hazardous wastes. Written in a nontechnical style with a bibliography of books and magazine articles. Recommended for the general reader.

Hawkes, Nigel. *Toxic Waste and Recycling.* New York: Glouster, 1988. A short (32-page), profusely illustrated introduction to toxic waste problems for young readers. As the title indicates, the book emphasizes recycling as a solution. A good introduction to the scope of the problem and potential solutions for junior high students.

O'Leary, Philip R., Patrick W. Walsh, and Robert K. Ham. "Managing Solid Waste." *Scientific American*, December, 1988: 36. In this short article, the authors discuss the mounting solid waste disposal problem related to dwindling landfill space and the restriction on siting new landfills. They discuss alternatives to landfill disposal, for example, recycling, volume reduction techniques such as incineration, landfill design, and energy from garbage. Suitable for the general reader.

Parker, Sybill P., ed. *McGraw-Hill Encyclopedia of Environmental Science.* 2d ed. New York: McGraw-Hill, 1980. A number of sections in this specialized encyclopedia are relevant to hazardous wastes. Sections covering industrial wastes, soil, groundwater, and solid waste management are especially pertinent. Each section is written in a nontechnical style, with a few illustrations and tables of facts, and is followed by a short bibliography of sources for more technical reading.

Turnberg, Wayne L. *Biohazardous Waste: Risk Assessment, Policy, and Management.* New York: Wiley, 1996. Turnberg has written a complete account of the policies, laws, and legislation surrounding medical, infectious, and hazardous waste management. He also provides an analysis of the risks involved.

Wagner, Kathryn D., et al., eds. *Environmental Management in Healthcare Facilities.* Philadelphia: W. B. Saunders, 1998. This book provides an examination of the procedures by which health-care facilities abide in regard to the disposal of medical, infectious, and hazardous waste, and the precautions they take to prevent environmental pollution. Comes with a computer laser optical disc that complements the essays.

LANDFILLS

Humankind has found it convenient throughout history to dump unwanted wastes into nearby ravines, swamps, and pits. With increased emphasis on sanitation, the open dumps have been replaced with landfills in which wastes are placed into excavations and covered with soil. Landfilling is the most common method of disposing of garbage and other unwanted material generated by cities and industry.

PRINCIPAL TERMS

AQUIFER: a porous, water-bearing zone beneath the surface of the Earth that can be pumped for drinking water

CLAY: a term with three meanings: a particle size (less than 2 microns), a mineral type (including kaolin and illite), and a fine-grained soil that is puttylike when damp

GEOMEMBRANE: a synthetic sheet (plastic) with very low permeability used as a liner in landfills to prevent leakage from the excavation

GROUNDWATER: water found below the land surface

LEACHATE: water that has seeped down through the landfill refuse and has become polluted

PERMEABILITY: the ability of a soil or rock to allow water to flow through it; sands and other materials with large pores have high permeabilities, whereas clays have very low permeabilities

POLLUTION: a condition of air, soil, or water in which it contains substances that make it hazardous for human use

SATURATED ZONE: that zone beneath the land surface where all the pores in the soil or rock are filled with water rather than with air

VECTOR: a term used in waste disposal when referring to rats, flies, mosquitoes, and other disease-carrying insects and animals that infest dumps

WATER TABLE: the upper surface of the saturated zone; above the water table, the pores in the soil and rock contain both air and water

HAZARDOUS VS. NONHAZARDOUS WASTES

All human activities produce unwanted by-products called wastes. For normal household living, these unwanted wastes are garbage and trash. Stores, factories, gas stations, and all other businesses also create large amounts of waste materials. Waste materials are classified as hazardous and nonhazardous: Hazardous wastes contain chemicals and other constituents that, if inhaled, eaten, or absorbed by humans and other life-forms, are detrimental to health; nonhazardous wastes may contain small amounts of toxic or hazardous ingredients. They present no threat, however, to the welfare of society if disposed of correctly.

Because of the severe environmental restrictions and legal liabilities associated with burying hazardous materials in the ground, recycling and incineration have become the principal disposal options for hazardous wastes. Sites at which toxic wastes are buried are tightly controlled and carefully monitored to ensure that no leakage occurs that could pollute the groundwaters.

LAND DISPOSAL

One of the primary methods of getting rid of wastes is by land disposal. Land disposal may be in the form of placing the unwanted material directly on the land surface, especially in low, swampy areas, ravines, and old gravel or other mined-out pits. These types of land disposal are known as dumps, or open dumps, because the unwanted wastes literally are haphazardly dumped at the site, and the waste piles often are not covered until some other use is made of the surface. The term "landfill," although sometimes used to describe dumps, refers to land disposal of wastes in which the disposal site is designed and operated to a specific plan. Modern landfills are designed to receive a certain amount of refuse over a specified period of time, such as twenty years. At the end of that period, the site is reclaimed and converted for a different land use.

In ancient times, when the Earth's population was small, waste products were discarded on the ground surface or thrown into a stream at the campsite. The amount was small enough at that time so that there was no significant pollution or adverse effects on the environment. As the population grew and more people began to live in towns, trash and garbage often were thrown in ravines and in low, swampy areas as a way of filling in the land. This practice of open dumping has existed in the United States and other Western nations up to the present.

The Earth's population has soared to more than 6 billion people, nearly 300 million of them living in the United States. The population concentration has shifted from rural to urban areas. To supply the demands of the public, industry has developed and produced a wide variety of materials and chemicals, such as plastics, that end up as waste products after use. More than 10 billion metric tons of solid waste are generated each year in the United States: Some 200 million metric tons are municipality generated rubbish and garbage, which amounts to 730 kilograms of nonindustrial waste per person per year.

As the amount of municipal wastes increased, more and more space was needed for open dumps.

Many serious problems arose as the use of dumps conflicted with land use, sanitation, and aesthetics in the surrounding communities. Pollution of rivers and the groundwater was directly traced to uncontrolled dumping of toxic wastes in dumps and pits. The federal government, therefore, enacted a series of laws over the years to protect the environment and especially the surface and groundwaters from pollution. The result of the legislation was to define and separate wastes that were hazardous to humans from the municipal wastes in land disposal. Special restrictions were placed on the disposal of hazardous wastes in the Earth. Although originally no special restrictions were placed on land disposal of municipal wastes, the protection of rivers and the groundwater from pollution required states to establish a permit process that set limits on where dumps and landfills could be located. Thus, disposal of municipal and industrial wastes has become expensive.

LANDFILL FACILITIES AND FEATURES

There are more than 16,000 active nonhazardous waste landfills of various types in the United States. More than 6,000 of them are classified as municipal, which means that most of the waste placed in the landfill is from households or is general community refuse. Modern landfills cover many acres. Some are more than 100 acres in size.

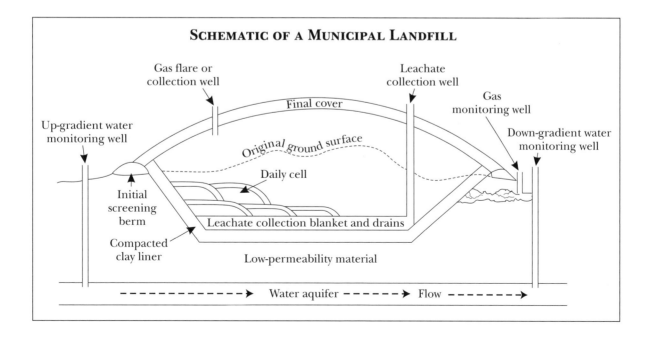

SCHEMATIC OF A MUNICIPAL LANDFILL

Although the actual size and shape of each landfill depend upon the site geology and the amount of waste that is to go in it, landfill facilities generally are excavated into the ground to form a pit. The refuse is placed in the pit and gradually builds up throughout the life of the landfill to a predetermined vertical height above the original ground level. Once the design height is reached, a soil cover is added, and the ground surface is reclaimed to form a moundlike hill. The depth of excavation to form the pit will depend upon the thickness of low-permeability clay soils over a buried water-bearing deposit, called an aquifer, that supplies drinking water to households and towns. Enough clay soil must be left to prevent seepage of any leachate that may escape from the landfill into the aquifer and pollute it. (Leachate is rainwater that has seeped down through the landfill refuse and has become polluted.) Monitoring wells are placed around the landfill into the aquifer to ensure that no pollution occurs.

After the excavation has been dug to its intended depth, a protective liner is placed over the bottom and sides of the pit. The liner may be of clay compacted by a construction roller, or it may be a plastic sheet called a geomembrane. Waste material that is brought to the landfill daily is placed in specified layers called cells. At the close of each day, the exposed waste is covered with soil to prevent odors, blowing debris, and infestation by vectors. "Vector" is a term used to collectively refer to any disease-carrying insects or animals, such as rats, flies, and birds, that would infest the waste material. The daily cells are stacked one on top of another until the final height is reached. Leachate and gas collection pipes are installed throughout the landfill to collect and dispose of dangerous gases and liquids. These perforated pipes are covered with a layer of gravel through which the leachate can easily flow from the landfill waste to the collection network.

SITE EVALUATION

Before a landfill can be constructed and operated at a proposed location, the site must be evaluated to ensure that it satisfies all local, state, and federal regulations relating to protecting the environment and the health and welfare of the citizens. The landfill operator must collect and assess much information before applying to a state pollution control board or similar governing commission for a permit. The information must verify that the landfill will present no danger to the public or to the environment. A large number of factors must be evaluated and those evaluations presented to the permit board.

By far the most important factor for determining whether a site is satisfactory for a landfill is the geology. The local geology must contain thick and continuous clay soils or their rock equivalent, shale. These deposits do not allow water or contaminated water from landfills (leachate) to flow rapidly through them; their low permeability will prevent any leachate that may escape from the pit from flowing into surface streams or to underlying aquifers that supply water to the surrounding communities. It is very important to protect these aquifers; once they become polluted, it is extremely difficult (and costly) to clean them up for public use in a short period of time. Compared to water in surface streams, groundwater in porous aquifers moves exceedingly slowly. Also, because the aquifer is hidden beneath the surface, it is difficult to trace and clean up the polluted water flow. To evaluate the subsurface geology of the site and the surrounding region, the company proposing the landfill will drill holes around the area and take soil samples so that a geologist can identify the different soils and rocks and construct a cross section illustrating the thicknesses, types, and relationships of the different materials.

Besides the geology, other major factors that must be assessed are climate and weather, flooding, ecology, historical landmarks, nearness to airports, traffic, and land-use and zoning restrictions. The climate and weather describe the rainfall and winds to be expected. All landfills must be above the 100-year flood height or have suitable flood protection. Special emphasis is placed on the ecology: A landfill cannot be built in protected wetlands or destroy the habitat of endangered species of plants and animals. Also, landfills are not allowed to destroy historical landmarks and archaeological sites. Municipal landfills attract birds; therefore, landfills must be more than a mile away from airports. Information on the road system and traffic volumes must be gathered to assess the impact of waste trucks adding both more traffic and greater loads to the roads. Landfills are not the most desirable land use and, therefore, such

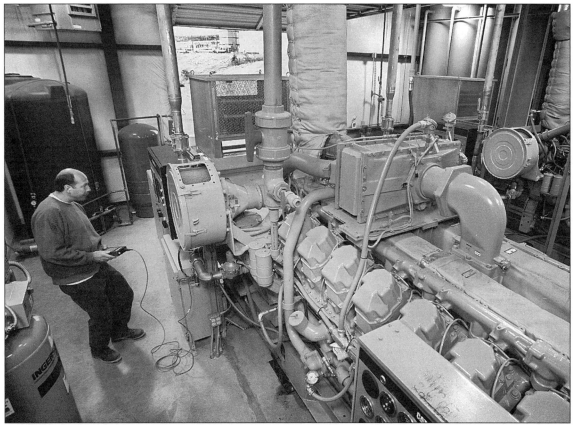

Landfills that contain garbage and other organics will generate methane gas that must be collected and either flared into the atmosphere or piped away for fuel. Some landfills use the gas at the site for commercial and light industrial energy. Here, an operator monitors an engine that turns methane gas from the landfill into electricity. (AP/Wide World Photos)

developments must prove they are compatible with the surrounding land use. Many communities have zoning restrictions that must be met. Thus, evaluating a proposed site for a landfill is a complex and time-consuming job.

LANDFILL DESIGN

A company must supply to the state permit board, in addition to information and assessment on the suitability of a proposed site, data and plans outlining the overall design, construction, operation, and reclamation when the landfill is closed. The design information must include drawings showing the depth and size of the pit area. The plans must give details on how the leachate is prevented from escaping from the pit and where monitoring wells will be drilled around the site to detect any leakage that may occur. Landfills that contain garbage and other organics

will generate methane gas that must be collected and either flared into the atmosphere or piped away for fuel. Some landfills use the gas at the site for commercial and light industrial energy.

The final aspect in a landfill design is the closing of the operation and the reclaiming of the site for a different land use. A final soil cover is placed over all the waste to isolate it from the public. Five to ten feet in thickness, this layer of soil controls vectors, prevents odors, and beautifies the landfill's surface. Periodic checks are made by the owner after the landfill has been reclaimed to sample the monitoring wells and to repair any erosion features.

PUBLIC POLICY

The federal government, the states, and communities have had to reevaluate their policies of uncontrolled dumping of municipal and industrial wastes because of serious pollution to the water supplies. With enactment of laws to protect the environment and to prevent pollution, land disposal of wastes is now controlled and monitored. Hazardous wastes are a severe threat to the health and welfare of communities. As a result, with few exceptions, land disposal of these wastes is no longer the primary disposal option.

Land disposal remains the primary method of disposing of nonhazardous wastes. Liquid nonhazardous wastes usually are disposed of in lagoons. Solid wastes, such as garbage and trash, are placed in landfills. For cities and other urban areas, large acreage of open land is scarce, and citizens generally object to having a landfill constructed nearby. Thus, cities and adjacent communities have been forced to cooperate on large regional landfills. It is therefore necessary for companies wanting to site new landfills to hunt for undeveloped land of a hundred acres or more where the geology can prevent pollution of the public water supplies. Even though municipal landfills, which often are called sanitary landfills, are considered nonhazardous, they will contain from 5 to 10 percent toxic material. If the landfill is small, such as in rural communities, the amount of hazardous leachate that may escape from the pit site is small and will thus have no appreciable detrimental effects on the environment and public health. The large landfills serving cities, however, do pose a threat to the public welfare and health if major amounts of leachate escape from the site. It is extremely difficult and costly to purify groundwater used for public water supplies if an aquifer becomes polluted. When households and communities lose their groundwater supply to pollution, the hardships to the citizens are both severe and costly. Some communities have had to transport water from other areas and to impose tight controls on how the water is used.

Waste disposal is a very important part of the lifestyle of modern society; with the quest for more and more conveniences and services, the unwanted by-products of civilization continue to increase and must be disposed of in some way. In addition, the natural environment has deteriorated to a point of serious concern. Therefore, landfilling must be done with the utmost care and planning in order to guard against pollution.

N. B. Aughenbaugh

CROSS-REFERENCES

BIBLIOGRAPHY

Bagchi, Amalendu. *Design, Construction, and Monitoring of Landfills.* 2d ed. New York: Wiley, 1994. Bagchi discusses the elements of design and the procedures involved in the construction and upkeep of sanitary landfills. Includes twenty-five pages of bibliographical references and an index.

California Integrated Waste Management Board, Permitting and Enforcement Division. *Active Landfills.* Publication Number 251-96-001. Sacramento: Author, 1998. A state examination of the protocol and maintanance of active California landfills. Includes a review of sanitation policies and safety measures.

Cook, James. "Not in Anybody's Back Yard." *Forbes* 142 (November 28, 1988): 172-177. One of several articles that *Forbes* magazine has printed on waste disposal and pollution of the environment.

Foreman, T. L., and N. L. Ziemba. "Cleanup on a Large Scale." *Civil Engineering,* August, 1987. An example of a large number of articles published in technical journals on waste disposal and pollution. Although they can contain technical data and scientific details, the bulk of the texts are written in a very understandable way and provide an important source of information on landfills.

Mulamoottil, George, Edward A. McBean, Frank Rovers, et al., eds. *Constructed Wetland for the Treatment of Landfill Leachates*. Boca Raton, Fla.: Lewis Publishers, 1999. A collection of essays that examine the use of the wetlands for sanitary landfills. Provides an extensive look at the construction and waste treatment procedures involved.

U.S. Environmental Protection Agency. *Subtitle D Study*. Washington, D.C.: Government Printing Office, 1986, 1988. This study on nonhazardous wastes is one of many reports by the U.S. Environmental Protection Agency on all aspects of waste disposal and pollution control. The agency is an excellent source of both general information and specific data. Interested readers can write to the U.S. Environmental Protection Agency, Office of Solid Waste (WH-562), 401 M Street S.W., Washington, D.C. 20460.

U.S. Geological Survey. *Toxic Waste—Groundwater Contamination*. Water Fact Sheet. Washington, D.C.: Government Printing Office, 1983. An informational brochure, or fact sheet, for the general public on groundwater supplies and pollution. Interested readers should write to the Hydrologic Information Unit, U.S. Geological Survey, 419 National Center, Reston, Virginia 22092.

MINING PROCESSES

Mining techniques are methods whereby ore is extracted from the Earth. Surface mining techniques are employed when ore is found close to Earth's surface. Whenever valuable solid minerals occur at depths beneath the land surface too great to be recovered at a profit using other techniques, underground methods are employed.

PRINCIPAL TERMS

DRAGLINE: a large excavating machine that casts a rope-hung bucket, collects the dug material by dragging the bucket toward itself, elevates the bucket, and dumps the material on a spoil bank, or pile

GRADE: the classification of an ore according to the desired or worthless material in it or according to value

LEVEL: all connected horizontal mine openings at a given elevation; generally, levels are 30-60 meters apart and are designated by their vertical distance below the top of the shaft

ORE: any rock or material that can be mined at a profit

OVERBURDEN: the material overlying the ore in a surface mine

PANEL: an area of underground coal excavation for production rather than development; the coal mine equivalent of a stope

PILLAR: ore, coal, rock, or waste left in place underground to support the wall or roof of a mined opening

RAISE: vertical or steeply inclined excavation of narrow dimensions compared to length connecting subsurface levels; unlike a winze, it is bored upward rather than sunk

SCRAPER: a digging, hauling, and grading machine having a cutting edge, a carrying bowl, a movable front wall, and a dumping or ejecting mechanism

SHAFT OR WINZE: vertical or steeply inclined excavation of narrow dimensions compared to length; shafts are sunk from the surface, and winzes are sunk from one subsurface level to another

STOPE: an excavation underground to remove ore, other than for development work; the outlines of a stope are determined either by the limits of the ore body or by raises and levels

VEIN: a well-defined, tabular mineralized zone, which may or may not contain ore bodies

SURFACE MINING METHODS

There are eleven varieties of surface mining methods: open-pit, mountaintop removal, conventional contour, boxcut contour, longwall strip, multiseam, blockcut, area, block-area, multiseam scraper, and terrace mining. Open-pit mining involves removing overburden and ore in a series of benches from near surface to pit floor. Expansion can occur outward and downward from the initial dig-in point. Many metallic deposits and industrial minerals are mined by this method; an example is the Bingham Canyon mine in Utah. The chief disadvantage of this method is the near impossibility of reclaiming large pits such as at Bingham, where nearly all the extracted material has been processed or transported away.

Mountaintop removal mining was popularized in eastern Kentucky. It involves removing—in boxcut fashion spoiling (dumping the overburden) in valleys on the first cut—an entire mountaintop, leaving a fairly flat pit floor upon which to reclaim overburden from succeeding cuts. A postmining surface of gently rolling terrain will replace the rugged premine surface. Disadvantages of this method are limited valley fill material, increased capital costs for additional equipment, and a need for extensive mine planning so as to ensure maximum productivity at minimum cost. It is, however, more effective than is contour mining.

In conventional contour mining, material is spoiled downslope from the active pit, often resulting in toxic material resting on native ground and causing erosion, poor vegetative reestablishment, and potential acid mine drainage. This type of mining results in a long striplike band running around the hillside and has been discouraged by stricter environmental laws in Kentucky and West Virginia. It is generally favored by smaller operators because expenses are lower and premine and reclamation planning efforts are reduced, but the method disturbs much more acreage than do other steep slope methods. A method similar to conventional contour mining, but with somewhat improved environmental impacts, is boxcut contour mining. Spoil segregation and terrace regrading help to inhibit toxic downslope material and to aid revegetation. The low wall of the boxcut helps to support most of the spoil.

Longwall strip mining has been attempted in some parts of West Virginia. This innovative method involves removing a small pit by conventional stripping methods, followed by the setup of a continuous miner (or excavator), chocks (blocks or wedges), and a conveyor. After a pass by the continuous miner, the chocks are advanced, allowing the roof to fail, similar in fashion to underground longwall mining. Although production costs are higher, land may be more easily restored. This benefit may extend to groundwater (preventing acid mine drainage) when geology is favorable.

Multiseam mining is done in many locations throughout the world. Frequently, a bottom and top split are separated by about 2 to 6 meters or more. This distance means that this "interburden" material must also be removed and often drilled and shot as well. Yet, even though the overall stripping ratio may be higher, it may be beneficial and economical to extract both seams. Permitting and assessment costs are lower, as well. Block-cut mining is also suitable in areas of moderate to steep topography. A single box- or block-cut is excavated, and mining progresses outward in two directions from the initial cut. Environmental disturbances are minimized as the first cut is regraded and revegetated during mining of subsequent cuts.

Large areas of flat or gently rolling terrain are

Terrace mining for sand and gravel in a former river bed. (© William E. Ferguson)

ideal for area mining, in which draglines or large stripping shovels remove overburden. Succeeding pits are dug normal or parallel to the strata-bound ore body. This method is the most common type of mining attempted in the major lignite and sub-bituminous coal fields of western North Dakota, Montana, and northeastern Wyoming.

Block-area mining is similar in many respects to both area and block-cut methods. It is designed to recover seams as thin as 0.3 meter. Capital requirements are less than in some methods, and overburden removal can be sequenced with reclamation regrading.

Multiseam scraper mining involves uncovering large blocks of coal (up to 122 meters wide by 305 meters long) by scraper. After the coal is removed, scrapers again remove the parting between succeeding seams until all the coal is mined. This method is ideal for the coal and thin parting sequences found in the upper Midwest.

Terrace mining involves overburden and ore removal in a series of benches, or terraces. These benches may follow natural features, as in stream or lake terraces. This type of mining has been effective in recovery of diamonds and sand and gravel.

COSTS OF SURFACE MINING

Surface mining does not come without its price and, in many cases, is far from easy. Surface mines are frequently located far from populated areas. In some of the Arctic regions, operations may be inhibited for much of the year by unfavorable weather. Some surface mines experience difficult mining conditions year-round.

Ugly scars from surface mining left on the landscape in many developing countries and in some parts of the eastern United States have prompted tougher environmental legislation. In the United States since the 1950's, toxic materials have been specially handled, and surface and groundwater problems are being seriously addressed.

UNDERGROUND MINING METHODS

When the valuable mineral to be extracted occurs at a depth too great to allow surface mining at a profit, underground mining methods are used. The depth at which profitability decreases depends on the value of the mineral deposit to be mined. Thus, some metallic minerals have been surface (open-pit) mined to depths equaling or

exceeding the maximum depths at which coal has been mined by underground methods. A number of physical factors influence the underground mining method to be chosen, including size, shape, continuity, and depth of the mineral deposit; range and pattern of ore quality; strength, hardness, and structural characteristics of the deposit and the surrounding and overlying rock; groundwater conditions; subsurface temperatures (some mines are so hot they must be air-conditioned); local topography and climate; and environmental protection considerations. If solution mining (flooding the pit) is being considered, the chemical composition of the ore mineral also will be an important consideration. Additionally, technologic and economic factors must be considered, such as availability of workers and worker safety; availability of water, power, and transportation; and, most important, a market for the mineral.

Once a decision has been made to develop an underground mining operation, a method is chosen from one of the following four categories: a method involving predominantly self-supporting openings, a method predominantly dependent on artificial supports, a caving method, or a solution mining method. In solution mining, men do not work underground to mine the valuable mineral. Instead, wells are drilled, and liquid flows or is pumped down the wells and through the deposit to dissolve the mineral. The mineral is later recovered by processing the pregnant solution (liquid containing the dissolved mineral) after it has been pumped to the surface.

STOPING METHODS

Methods involving predominantly self-supporting openings are open stoping, sublevel stoping, shrinkage stoping, and room-and-pillar mining. Prior to production of the mineral in marketable quantities, the mine must be developed. Development work begins with gaining access to the mineral deposit by sinking a shaft, driving a horizontal tunnel-like excavation (adit) into a hillside, or driving a sloping tunnel-like excavation (decline) from the surface. Once accessed, portions of the deposit to be mined are blocked out by bounding them with a three-dimensional network of horizontal tunnel-like excavations (levels) and vertical shaftlike excavations (raises or winzes). Each of these blocked out areas defines a stope to be

Surface subsidence pits and troughs above an abandoned coal mine north of Sheridan, Wyoming. (U.S. Geological Survey)

mined by the stoping method selected for the particular mine.

Open stoping is any mining method in which a stope is created by the removal of a valuable mineral without the use of timber, or other artificial supports, as the predominant means of supporting the overburden. Open stopes include both isolated single openings, from which pockets of ore have been extracted, and pillared open stopes, where a deposit of considerable lateral extent has been mined with ore pillars left for support. In sublevel stoping, large blocks of ore between the levels and raises that bound the stope are partitioned into a series of smaller slices or blocks by driving sublevels. The ore is then drilled and blasted by miners in the sublevels in a sequence that causes the sublevels to retreat *en echelon* (in step formation) with each successively lower sublevel being mined slightly ahead of the one immediately above it. This process allows all the broken ore to fall directly to the bottom of the stope, where it passes through funnel-like openings (called draw points) to the haulage level below for transport out of the mine.

In shrinkage stoping, the ore in the stope is mined from the bottom upward, with the broken ore allowed to accumulate above the draw points to provide a floor from which blasting operations can be conducted. The broken ore is periodically drawn

down from below (shrunk) to maintain the proper headroom for blasting. Room-and-pillar mining is a form of pillared open stoping employed where the mineral deposit is relatively thin, flat-lying, and of great lateral extent. It is employed most commonly in the mining of coal but is also used to mine other nonmetals such as rock salt, potash, trona, and limestone.

Stull stoping, cut-and-fill stoping, square-set stoping, longwall mining, and top slicing are among the methods predominantly dependent on artificial support. Stull stoping is a method that can be considered transitionary between those methods involving predominantly self-supporting openings and those involving predominantly artificially supported openings. It is used chiefly in narrow, steeply sloping (dipping) vein deposits. Timbers (called stulls) are placed for support between the lower side wall (called the footwall) and the upper side wall (called the hanging wall) of the vein.

The cut-and-fill stoping method, like shrinkage stoping, has a working level floor of broken rock; however, in this case, the broken rock is waste rather than ore. Instead of draw points, heavily timbered ore chutes are constructed through the waste rock to pass the broken ore to the haulage level for transport from the mine. The solid ore is fragmented by blasting so that it falls on top of the waste rock floor, where it is collected and dumped down the ore chutes. Waste rock, often from development work elsewhere in the mine, is placed in the stope to build up the floor and maintain the desired headroom for blasting.

Square-set stoping is a method employing extensive artificial support. "Square sets" (interlocking cube-shaped wooden frames of approximately 2.5-3 meters on a side) are constructed to form a network resembling a playground's "monkey bars" to provide support for the stope. As mining progresses, the square sets are immediately filled with waste rock—except for those sets kept open for ventilation, ore passes, or passageways for miners (called manways). In the longwall method, a massive system of props or hydraulic supports is used to support the roof over a relatively long, continuous exposure of solid mineral to be mined. Virtually complete extraction results, and the overlying roof behind the line of support collapses into the mined void as the mining advances. A method similar to longwall is top slicing, in which ore is extracted in horizontal timbered slices starting at the top of the ore deposit and working downward. A timbered mat is placed in the first cut, and the overburden is caved. As subsequent cuts are advanced, caving is induced by blasting out props (timbers) behind the face, while working room under the mat is continuously maintained. It differs from longwall mining in that several levels are developed *en echelon* rather than mining progressing on a single level.

CAVING METHODS

Two caving methods are commonly used in underground mining: sublevel caving and block caving. In sublevel caving, work begins in the uppermost sublevel of the stope and progresses downward. As the ore is blasted and collapses onto the sublevel floor for removal, the wall rock of the stope immediately caves behind the ore. The broken ore is removed through the sublevels. In block caving operations, the ore body is induced to cave downward because of gravity for the entire height of the stope (usually more than 30 meters). The process is initiated by undercutting the block of ore and allowing it to collapse. The collapsed ore is removed through draw points at the bottom of the stope.

ADAPTATION TO OTHER USES

The relevance of mining methods to society is not limited simply to the production of useful and valuable minerals. The ideas and concepts developed for underground mining also have been applied and adapted to other uses. Techniques used in underground mining are used in the construction industry, especially on projects involving underground openings for transportation, electric power generation, fuel storage, and military purposes. The construction of highway, railroad, and subway tunnels can be cited as examples of transportation applications. Large hydroelectric dams and similar projects utilize rock tunnels and chambers to channel the falling water used to turn the turbines that generate the electricity. Underground caverns created using underground mining technology are being utilized for petroleum and liquefied natural gas storage, and minelike underground excavations in rock also are being investigated for use as deep underground storage, or disposal, facilities for nuclear wastes. Military applications include underground chambers for housing intercontinental ballistic missiles and for nuclear weaponproof command facilities.

Paul S. Maywood and Dermot M. Winters

CROSS-REFERENCES

BIBLIOGRAPHY

Chironis, N. P., ed. *Coal Age Operating Handbook of Coal Surface Mining and Reclamation.* New York: McGraw-Hill, 1978. An excellent, nontechnical introduction to surface coal mining techniques. Well illustrated, and the photographic reproductions are of high quality. Many site-specific case studies, which make this volume particularly interesting to all levels.

Crickmer, D. F., and D. A. Zegeer, eds. *Elements of Practical Coal Mining.* 2d ed. New York: Society of Mining Engineers of the American Institute of Mining, Metallurgical, and Petroleum Engineers, 1981. This book is intended to be a beginners' book and will give the reader an overview of the fundamentals of coal mining. Appropriate for use in vocational schools, high schools, community colleges, and libraries, and by those persons in or expecting to enter the coal mining industry.

Gayer, Rodney A., and Jierai Peesek, eds. *European Coal Geology and Technology*. London: Geological Society, 1997. A complete examination of the technologies used in European coal mines and the coal mining industry. Illustrations, maps, index, and bibliographical references.

Ghose, Ajoy K., ed. *Mining on a Small and Medium Scale: A Global Perspective*. London: Intermediate Technology, 1997. This volume collates the proceedings of the Global Conference on Small and Medium Scale Mining held in Calcutta, India, in 1996. The volume focuses on the social and environmental aspects of mining in developing countries.

Institution of Mining and Metallurgy. *Surface Mining and Quarrying*. Brookfield, Vt.: IMM/ North American Publications Center, 1983. This symposium volume contains thirty-eight papers on a variety of topics related to surface mining. Many of the papers are illustrated, but some are quite technical. The illustrations and photographs vary somewhat in quality, but it is a good reference overall and very good on specific topics. Good for college-level students.

Peters, W. C. *Exploration and Mining Geology*. 2d ed. New York: John Wiley & Sons, 1987. Chapter 6 of this book, "Approaches to Mining," provides a short, but thorough, summary of surface and underground mining methods. The remainder of the book provides an overview of the geologist's work in mineral discovery and mineral production.

Stefanko, Robert. *Coal Mining Technology Theory and Practice*. New York: Society of Mining Engineers, 1983. Although this book is geared to underground coal mining, there are many excellent sketches illustrating the basics behind most of the different types of surface mining. The language is for the most part nontechnical. Suitable for high-school-level readers.

Stout, K. S. *Mining Methods and Equipment*. New York: McGraw-Hill, 1980. Profusely illustrated with sketches, drawings, and photographs, this book provides a thorough introduction to the fundamental concepts of mining and their applications in producing ore and coal. The technical detail presented makes it suitable as a text for an entry-level course in mining engineering and for persons with more than a casual interest in mining. The number and quality of the illustrations make the technical detail less daunting to the layperson.

Thomas, L. J. *An Introduction to Mining: Exploration, Feasibility, Extraction, Rock Mechanics*. New York: Halsted Press, 1973. An introductory text written for students of mining and mining engineering, this book covers the entire mining process, from initial exploration to mine operation. Chapter 6, entitled "Underground Metalliferous Mining," and chapter 7, entitled "Underground Mining of Bedded Deposits," provide lucid descriptions of the principal underground mining methods.

U. S. Department of Labor, Mine Safety and Health Adminstration. *Coal Mining*. Washington, D.C.: Government Printing Office, 1997. This brief government publication examines the state of coal mines and mining in the United States. There is a strong focus on the health risks associated with coal mining. Illustrations and maps.

MINING WASTES

Mining has produced vast areas of disturbed land and disrupted ecosystems. The magnitude of extraction activities, combined with inadequate responses by mining companies, has inspired reclamation laws and research.

PRINCIPAL TERMS

ECOLOGICAL SUCCESSION: the process of plant and animal changes from simple pioneers such as grasses to stable, mature species such as shrubs or trees

ECOSYSTEM: a self-regulating, natural community of plants and animals interacting with one another and with their nonliving environment

EROSION: the movement of soil and rock by natural agents such as water, wind, and ice, including chemicals carried away in solution

LANDFORMS: surface features formed by natural forces or human activity, normally classified as constructional, erosional, or depositional

ORPHAN LANDS: unreclaimed strip mines created prior to the passage of state or federal reclamation laws

RECLAMATION: all human efforts to improve conditions produced by mining wastes, mainly slope reshaping, revegetation, and erosion control

TOPSOIL: in reclamation, all soil which will support plant growth, but normally the 20-30 centimeters of the organically rich top layer

BY-PRODUCTS OF MINING

Over the years, humankind has found more and more uses for the minerals within the Earth and has been able to develop deposits of lower and lower quality. Miners remove 15 meters or more of overburden (soil and rock) to obtain 30 centimeters of coal; some copper mines process ores as low as 0.5 percent. These endeavors produce literally mountains of waste, often referred to as spoil; all human exploitation of Earth resources produces waste as a by-product, but none more than the mining of coal. Another inevitable by-product is ecosystem disruption—the alteration of plant and animal habitat—and changes in hydrology, which includes all aspects of water. Reclamation is the effort to heal the altered environment through landform modification, revegetation, and erosion control.

Until the mid-twentieth century, a "use it" ethic of land use was dominant. Nature was all too often seen as an enemy to be conquered. Since World War II, however, with the rise of environmental concerns, the United States has recognized the need to restore these disturbed lands. Nevertheless, coal strip mining has increased with the growing demand for energy. Between the 1930's and early 1970's, electrical demand doubled every de-cade. When the 1973 energy crisis hit, the huge deposits of western coal were seen as one pathway to energy independence. Additionally, the abrupt halt in orders for new nuclear power plants meant that almost all new electrical power plants since 1973 have been fueled by coal.

GROWING CONCERN FOR THE ENVIRONMENT

Reclamation efforts in the past have been slight because of apathy, the relatively small areas involved, and the cost. Current efforts are directly linked to the growth of conservation and environmental movements in response to population growth and the development of available lands. In the United States, population and economic growth following World War II resulted in pollution conditions that shocked the nation. The first large-scale environmental conference was held at Princeton University in June, 1955. Entitled "The International Symposium on Man's Role in Changing the Face of the Earth," it set the agenda for the coming national concern for the environment. Steward Udall, Secretary of the Interior, was instrumental in the 1967 publication of *Surface Mining and Our Environment: A Special Report to the Nation.* It was a clear call to attack the growing problem of unreclaimed coal strip mines, described as "orphan" lands.

The creation in 1970 of the Environmental Protection Agency led the way to what came to be called "the environmental decade." The Surface Mining Control and Reclamation Act of 1977, though covering only coal mining, was landmark legislation. It requires revegetating and restoring the land to its approximate original contour; bans mining of prime agricultural land in the West and where owners of surface rights object; requires operators to minimize the impact on local watersheds and water quality; establishes a fee on each ton of coal to help reclaim orphan lands; and delegates enforcement responsibility to the states, except where they fail to act or where federal land is involved. Major efforts are underway to correct the land disruptions associated with coal mining, including research which may apply to other mining operations. Although mining has become more expensive, reclamation gives substantial hope for restoring the land.

WATER QUALITY AND HYDRAULIC IMPACTS

Important research on water quality dates from the 1950's and was focused on acid mine-drainage problems in Appalachia, where it has long been recognized as a problem from underground mines. Acid is produced in any type of Earth movement where iron pyrite and other sulfides are removed from a reducing environment to an oxidizing one. Within the Earth these sulfides are stable, but when exposed to oxygen and water they readily oxidize, producing sulfuric acid, the main ingredient in acid mine drainage. Coal strip mines can produce acid when the sulfide-rich layers above and below the coal (often called fire clay) are scattered on the surface or left exposed in the final cut. A major reclamation need is to identify these acid-producing materials and bury them. Acids speed up the release of dangerous metals such as aluminum and manganese, which can be toxic to vegetation; they also cause deformities in fish.

Hydrologic impacts mainly involve changes in runoff and sediment. A major study was conducted by the U.S. Geological Survey in Beaver Creek basin, Kentucky. This area was selected because mining operations were just beginning there under typical Appalachian conditions. Data collected from 1955 to 1963 revealed large increases in runoff and sediment. By contrast, another study in Indiana demonstrated the tremendous water-holding capacities of the disturbed lands. The surface topography had much to do with the different results because a more level condition existed in Indiana. Another study was conducted at a basin in northeastern Oklahoma while it was being subjected to both contour and area strip mining. After initial increases in runoff and erosion, huge decreases occurred as much of the drainage became internalized. When reclamation began under a new state law, however, erosion rates comparable to unprotected construction sites were measured (up to 13 percent sediment by weight). Clearly, slopes lacking a protective vegetation cover will produce much more runoff, which in turn will carry enormous sediment loads. Nevertheless, in several orphan lands, water and sediment are trapped internally because of the drainage obstacles and depressions created by the mining operation.

A scientist with the U.S. Geological Survey holds samples outside an iron mine near Redding, Calfornia, in March, 2000. The water in the mine is thought to be the most acidic in the world. (AP/Wide World Photos)

USES OF MINING WASTE LANDFORMS

Early research on mining waste involved description of the new landforms created by human activity. The unreclaimed landforms produced by strip mining, still visible in orphan lands, are spoil banks, final-cut canyons (which are often filled with ponds), headwalls (above the final cuts), and transport roads. Mining waste landforms have some interesting uses. Perhaps most important is recreation, especially fishing in final-cut ponds. In some North Dakota mines, natural revegetation has formed refuges for deer and other wildlife. One site in northeastern Oklahoma, given to the local Boy Scouts, now supports a dense population of trees and birds and is a wonderful small wilderness. Other uses include forestry in the East and sanitary landfills near urban areas. Most softwoods, such as pine, prefer more acidic soils, which are common in eastern strip mines. Use in sanitary landfills is aided by the already disturbed condition of the land and by the fact that the clays and shales commonly found in coal strip mines provide relatively impermeable conditions.

The linkage between mining waste landforms, vegetation/soil conditions, and erosion was recognized early. One study in Great Britain found that even in successfully revegetated strip mines, erosion rates were 50-200 times those of undisturbed areas. Revegetation is the key to reclamation; it, in turn, is heavily dependent upon the quality of the soil. Because of the diverse nature of soil and local ecosystems, a specific reclamation plan is never made easily.

EFFECTIVE RECLAMATION PLANNING

Effective reclamation plans require sufficient data on soil pH (a measure of acidity or alkalinity), soil chemistry for fertility and toxicity problems, water retention capability, soil organisms, and useful native plants, especially perennials. The best way to provide the needed soil conditions is to save and rapidly replace the topsoil and to establish a vegetation cover quickly in order to protect the soil from rain and gully erosion. Even thin layers of good topsoil are a major improvement over the deeper subsoils, which tend to be low in organic matter and rich in acid-producing sulfides.

Detailed planning is now an integral part of the reclamation process. When planning is completed before mining begins, some of the impact can be minimized; reclamation can proceed in concert with mining to shorten the interval before replanting. Since erosion is highly destructive to revegetation efforts, erosion control systems are vital. They are designed to direct the flow of water and to dissipate its energy, enhancing soil moisture for vegetation.

Also vital to a good reclamation plan is a thorough understanding of local ecosystems and the potential success or failure of restoration efforts. Revegetation is mainly an exercise in ecological succession. Once an ecosystem is disrupted, it is difficult to reestablish. In some cases, it may be wiser to pursue a different course. Furthermore, because ecological succession is slow, fifty years or more may be needed to gauge success or failure accurately.

NEED FOR CONSERVATION AND RECYCLING

The need for conservation and recycling is supported by the ecological axiom that, in nature, matter cycles but energy flows; by recycling matter, large energy savings result from reusing materials such as steel, aluminum, and glass. Since energy flows, and so cannot be recycled, conservation is the best source of additional energy. These savings reduce the pressure to disturb more land in order to extract energy-producing coal. A key factor in the environmental impact of mining is the adaptability of natural ecosystems. How much stress can they tolerate, and what kind of conditions can be expected from the consequent ecological succession? Ecosystems are remarkably resilient and adaptable. Still, the pace of human activity is so rapid and the recovery of ecosystems so slow that the human race would be wise to err on the side of conservation.

Questions are currently being raised about the impact of coal burning on potential atmospheric heating. Because coal is nearly pure carbon and because carbon dioxide (produced when carbon is burned) is a major greenhouse gas (one that absorbs and thus traps heat radiating from the Earth's surface), the world's increasing dependence on coal for electrical power production could lead to climatic changes that could affect agriculture. Revegetation efforts can help in this area because growing plants remove carbon dioxide from the atmosphere. One suggested replacement for coal, nuclear power, has its own set of mining waste problems, along with a need for long-term storage of radioactive wastes.

Two opposing land-use viewpoints provide a framework for discussion: the historic "use it" ethic and the growing "preserve it" ethic. Many, however, prefer a middle road of scientific conservation or sustainable Earth practices, both of which include reclamation. From a political perspective arise the issues of tolerance levels and affordability. Government has a responsibility to protect the health and welfare of its citizens. Where clear dangers exist, government has a responsibility to act. Many pollution issues, however, are controversial. How much is too much, and at what level does the cost of abatement become so high as to be unaffordable? Human use of the Earth without pollution is impossible, and there are limits to what can be done to abate it. Still, public support remains strong for pollution control, cleanup of past pollution problems, and reclamation of disturbed land.

Nathan H. Meleen

CROSS-REFERENCES

Building Stone, 1545; Coal, 1555; Earth Resources, 1741; Earth Science and the Environment, 1746; Earth Science Enterprise, 1752; Environmental Health, 1759; Evaporites, 2330; Fertilizers, 1573; Future Resources, 1764; Hazardous Wastes, 1769; Hydrothermal Mineralization, 1205; Industrial Metals, 1589; Industrial Nonmetals, 1596; Iron Deposits, 1602; Landfills, 1774; Mining Processes, 1780; Nuclear Waste Disposal, 1791; Oil Shale and Tar Sands, 1717; Sedimentary Mineral Deposits, 1629; Strategic Resources, 1796; Uranium Deposits, 1643.

BIBLIOGRAPHY

Abandoned Mines and Mining Waste. Sacramento: Californian Environmental Protection Agency, 1996. This short state government publication focuses on the environmental and social aspects of mines that are no longer operating. It is accompanied by an easy-to-read fact sheet provided by the U. S. Department of Health and Human Services.

Collier, C. R., et al., eds. *Influences of Strip Mining on the Hydrologic Environment of Parts of Beaver Creek Basin, Kentucky, 1955-66.* U.S. Geological Survey Professional Paper 427-C. Washington, D.C.: Government Printing Office, 1970. One of the first basinwide studies over an extended period of time, it attempted to cover all aspects of hydrologic impacts. A fundamental reference for any serious student of strip mining. Parts 427-A (basin description) and 427-B (covering 1955-59) were published in 1964.

Dalverny, Louis E. *Pyrite Leaching From Coal and Coal Waste.* Pittsburgh, Pa.: U. S. Department of Energy, 1996. This brief government publication looks at coal-mining policies and protocol, as well as the state of disposal procedures used by coal-mine operators. Illustrations and bibliography.

Hanson, David T. *Waste Land: Meditations on a Ravaged Landscape.* New York: Aperture, 1997. This photo essay shows hazardous waste sites and mines, and allows the reader to see the effects of mining on the environment.

Law, Dennis L. *Mined-Land Rehabilitation.* New York: Van Nostrand Reinhold, 1984. A heavily illustrated hardback with an attractive layout and print style. Provides extensive coverage of reclamation issues, which sometimes seems simplistic and sometimes overly complex but is thorough. Heavily footnoted and contains useful references and tables. Intended as an introduction to the topics involved in rehabilitation, sections include mining procedures, environmental impacts, surface rehabilitation, revegetation, and planning.

Miller, G. Tyler, Jr. *Living in the Environment.* 5th ed. Belmont, Calif.: Wadsworth, 1988. The dominant textbook in environmental science for college students, it is also suitable for use by any serious high school student or adult. Its strength is the diversity of topics covered, ranging from ecosystems to politics. Contains a useful glossary, an extensive bibliography, and an index. An excellent primer on ecosystems, land use, and energy. Contains much information about environmental legislation.

Narten, Perry F., et al. *Reclamation of Mined Lands in the Western Coal Region.* U.S. Geological Survey Circular 872. Alexandria, Va.: U.S. Geological Survey, 1983. Free and highly useful, this source has sixty pages filled with photos and detailed information about the current status of reclamation research and knowhow. Well written, informative, and thorough; a must for anyone interested in reclamation, especially revegetation.

Thomas, William L., Jr., ed. *Man's Role in Changing the Face of the Earth.* Chicago: University of Chicago Press, 1956. These are proceedings of the international symposium by the same name, with contributions by many well-known scientists of the period. Divided into sections entitled "Retrospect" (past), "Process" (present), and "Prospect" (future), the book concludes with summary remarks by the conference cochairmen. A classic in environmental literature. Topical coverage is extensive. Few illustrations, but a thorough index.

U.S. Department of the Interior. *Surface Mining and Our Environment: A Special Report to the Nation.* Washington, D.C.: Government Printing Office, 1967. This exquisitely illustrated, full-color, oversized paperback is now a classic for those interested in surface mining. Telling much of its story in vivid photo-essays and striking illustrations, it alerted the nation to the problems of surface mining and appealed for action. Topics include the nature and extent of surface mining, its environmental impact and related problems, past achievements and future goals, existing laws, and recommendations for action.

U.S. Environmental Protection Agency. *Erosion and Sediment Control: Surface Mining in the Eastern U.S.* EPA Technology Transfer Seminar Publication. Washington, D.C.: Government Printing Office, 1976. This two-part paperback manual ("Planning" and "Design") was written for technicians, professionals, and laypersons. Includes a useful glossary, but lacks a bibliography. Provides an understanding of the underlying mechanics and rationale of erosion control and basic information on procedures and design. Illustrated with photos and diagrams.

Vogel, Willis G. *A Guide for Revegetating Coal Minesoils in the Eastern United States.* USDA Forest Service General Technical Report NE-68. Broomall, Pa.: Northeast Forest Experiment Station, 1981. With much useful information about revegetation under humid conditions, this publication is a must for anyone interested in revegetation of mining waste.

NUCLEAR WASTE DISPOSAL

Nuclear waste disposal in a geologic repository is considered the safest and surest way to achieve isolation of the wastes from the surface environment. No one can guarantee that a repository will last forever, but its combined geologic characteristics will minimize damage in the event of failure.

PRINCIPAL TERMS

HALF-LIFE: the time required for one-half the radioactive isotopes in a sample to decay

HIGH-LEVEL WASTES: wastes containing large amounts of dangerous radioactivity

ISOTOPES: atoms of the same element that differ in the number of uncharged neutrons in their nuclei

LEACHATE: water that has come into contact with waste and, as a result, is transporting some of the water-soluble parts of the waste

LOW-LEVEL WASTES: wastes that are much less radioactive than high-level wastes and thus less likely to cause harm

PERMEABILITY: a measure of the ease of flow of a fluid through a porous rock or sediment

POROSITY: a measure of the amount of open spaces capable of holding water or air in a rock or sediment

RADIOACTIVITY: the spontaneous release of energy accompanying the decay of a nucleus

SOLUBILITY: the tendency for a solid to dissolve

SORPTION: the process of removing a chemical from a fluid by either physical or chemical means

TRANSURANIC: an isotope of an element that is heavier than uranium and that is formed in the processing and use of nuclear fuel and plutonium

HIGH- VS. LOW-LEVEL WASTES

Nuclear waste disposal is a necessary evil in a world where radioactive isotopes and the energy produced by their decay are used, among other things, for generating electricity, diagnosing and treating diseases, and making nuclear weapons. Nuclear wastes are not all the same. Some wastes, because of their composition, are more dangerous than others, and some are more dangerous for longer periods of time. When evaluating a geologic site for disposal of nuclear wastes, the characteristics of the site must be identified so that the waste will be isolated from the Earth surface environment for a minimally acceptable period of time. Ensuring the minimal length of isolation is most important for those wastes that are dangerous for a long time and those that contain large concentrations of dangerous isotopes. Such wastes, usually called high-level wastes, consist of isotopes of uranium and plutonium produced in the use and processing of nuclear fuel. The length of isolation is not so important for those wastes that are dangerous for shorter times and are less

concentrated. These wastes are called low-level wastes and may include some long-lasting wastes but in lower concentrations. An example of a low-level waste is slightly contaminated garbage, such as disposable laboratory equipment, disposable medical equipment, and disposable gloves and overalls (which are used in the handling of radioactive materials). The length of time that a particular waste is dangerous depends on which radioactive isotopes are in the waste and on the half-life of each isotope. The half-life is the time required for one-half of the radioactive isotopes in a sample to decay.

During the natural process of radioactive decay, the nucleus of a radioactive isotope, the "parent," is changed into that of another isotope, the "daughter," and energy is released. This radioactive decay from parent to daughter is usually one step in a series of many, as the original parent isotope changes into a nonradioactive, stable isotope. At each step, energy is released in the form of energetic particles or energetic rays. Thus, in a sample of uranium, the parent uranium isotopes are constantly decaying in a series of steps through various daughter isotopes until the sample consists of pure lead, the ultimate daughter isotope in the uranium decay series. It takes a long time for all the uranium to decay completely to lead. The half-life of one uranium isotope, uranium 238, is 4.5 billion years. The long half-life of uranium suggests that the half-lives for some of the daughter isotopes in the decay series are also long.

ISOLATION IN GEOLOGIC REPOSITORY

Exposure to radiation is never without risk. The problem with a long half-life is that the isotope is giving off energetic particles or rays that are dangerous to plants and animals for an extended period. Thus, high-level wastes, where the isotopes are in concentrated form, must be isolated from the surface environment for a long time. Low-level wastes are less dangerous only because the concentrations of radioactive isotopes are lower. While they need to be isolated from the surface environment, they are not as destructive to life as are high-level wastes.

Plants and animals are exposed to radiation through air, water, and food. Naturally occurring radioactive isotopes may be in the air (in the form of radon, for example) and taken into the lungs. Some radioactive isotopes are dissolved in the water we drink (potassium 40) and in the food we eat (carbon 14). To minimize excessive exposure to radiation, a geologic repository must isolate radioactive waste from the surface environment. A geologic repository is any structure in either rock or soil which uses, in part, the natural abilities of these materials to isolate the wastes from the surface environment. A geologic repository may be a shallow trench dug in the soil, partially filled with low-level waste and covered with the excavated soil, or it may be a cavern constructed deep underground in hard rock. Regardless, the purpose of the repository is to isolate the waste in such a way that there is no excess exposure of organisms to radiation as a result of the presence of the repository. In this sense, isolation means that the waste must be kept from contaminating the surrounding air, water, and food.

CHARACTERISTICS OF REPOSITORY

The rocks or soil of a geologic repository must have certain characteristics to isolate wastes prop-

erly. Some of the more important characteristics are porosity, permeability, mineral solubility, and sorption capacity. Porosity and permeability are related but are not the same. Porosity is a measure of the ability of a rock or soil to hold a fluid, either liquid or gas. Expressed as a percentage of the volume of the sample, porosity is the volume of open, or pore, space in the rock or soil. If the soil or rocks of a geologic repository have high porosity, then large amounts of water or air may contact the waste; however, contact with the waste does not necessarily mean that the contaminated water or air will reach the surface environment. If the rock or soil is also permeable—that is, if the pores are interconnected enough so that the water or air can flow from the repository to the surface—then the repository is not likely to provide adequate isolation. Porosity can exist without permeability when many small pores contain a large volume of fluid, but the connections between pores are too small to allow fluid flow. Permeability can exist with small porosity when a few penetrating cracks in an otherwise solid rock allow easy fluid flow.

Mineral solubility is an important characteristic in cases where water may contact the waste. Soluble minerals, those that tend to dissolve in water, may be removed from the surrounding rock or soil by moving water. When minerals dissolve, a space is left behind that adds to the porosity, and may increase the permeability, of the repository.

The sorption capacity of a mineral is the ability of that mineral to remove a dissolved molecule or ion from a fluid. Sorption may occur by either chemical or physical means. A type of chemical sorption is the ion exchange that takes place in a home water-softening unit. Water containing problem ions (in the case of a repository, radioactive ions) flows over the exchanging minerals. The problem ions attach to the mineral and, in the process, force the ion that was attached originally into the solution. The water is thus cleansed of problem ions. In physical sorption, the water and the contaminant are attached to the mineral by the force of friction, which results in a thin layer of water attached to the mineral surface. The attached, or physically sorbed, water and any material dissolved in it are not moving.

All these characteristics of rock and soil are related in some way to the possibility that the waste will escape from the repository and be transported by a fluid to the surface. It is unlikely that any repository will be a perfect candidate to provide isolation when all the characteristics are considered. Some characteristics, however, may be more important than others, and weakness in one may be offset by strength in another.

DESCRIPTION OF REPOSITORY MATERIAL

Shallow burial of low-level wastes and deep burial of high-level wastes are preceded by careful description of the strengths and weaknesses of the geologic material of the repository. In the case of shallow burial, the repository material is usually soil or saprolite (near-surface rock that has been partially turned to soil by the actions of water). To determine the strengths and weaknesses of a particular soil for containing low-level waste, the Earth scientist must study the structure, hydrology, mineralogy, and chemistry of that soil.

When water, in the form of rain or snowmelt, enters a soil, in most cases it does not flow through the soil uniformly. Water tends to flow more easily through the soil along certain permeable paths. In nearly every soil there are preferred flow paths and isolated areas. Sometimes the structure is inherited from the original rock that broke down to soil. Fractures and other cracks contained in the original rock may be preserved as cracks in the soil. Structures inherited from the parent rock are especially evident in saprolite, which exists in a stage between rock and true soil.

Other soil structures are the products of soil processes. Such structures include the cracks developed when a soil dries out, the tunnels formed by burrowing organisms such as worms and ants, and the openings left when the roots of dead plants decay. All these structures are important because they determine how and how fast water will flow through a soil.

The means and speed of water flow through soil are important aspects of its hydrology. The flow of water into a shallow burial trench is usually restricted by engineered, low-permeability barriers such as compacted clay or plastic liners. If the liners fail and water flows through and out of the wastes, the surrounding soil should slow the flow of contaminants in two ways: physically and chemically. Physically impeding, or retarding, the flow of contaminated water from the waste trench means that the soil structure does not provide permeable

flow paths; the water is forced to flow through many small, interconnected pores. As a result, the waste-transporting water contacts more of the mineral material in the soil. Chemical retardation of wastes occurs when the contaminants in the slowly moving fluid interact with the minerals of the soil. Some of those minerals have the capacity to sorb contaminants, thereby further slowing their migration. Other reactions between the minerals and the wastewater result in the chemistry of the water changing in such a way that the contaminants may become insoluble and form new solids in the soil. The effect is the same: The flow of contaminants from the burial trench is slowed or stopped.

STABILITY OF REPOSITORY

Disposal of high-level wastes in a geologic repository requires that many of the same characteristics of the host rock be determined. In addition to having the appropriate hydrological, mineralogical, and chemical characteristics, the host rock is required to be reasonably stable for ten thousand years. For example, the modern hydrological characteristics of the proposed Yucca Mountain repository site in Nevada are ideal. The proposed repository will be located below the desert surface, out of the reach of downward-percolating rainwater and well above the nearest fresh groundwater. In the desert environment, rainfall is so infrequent that it cannot reach the repository from above, nor can the infrequent rainfalls raise the groundwater level to flood the repository. Construction of 8 kilometers of test tunnels in Yucca Mountain began in 1994, and evaluation of the site was to be completed by 2001, with construction to begin in 2005 and emplacement of high-level waste to start in 2010.

The problem is that climates have changed in the past—the desert Southwest of the United States used to be much wetter—and climates will undoubtedly change in the future. In addition, the repository may be disturbed, whether inadvertently or purposely, by human activities. The most likely inadvertent disturbance is that created by future generations, who, in a search for mineral deposits, will puncture the repository with drilling equipment. To avoid this scenario, the repository should be located in an area unlikely to yield mineral wealth. Deeply buried layers of rock salt, thought to be ideal repositories because of their stability and their resistance to water flow, are no longer being considered. Many of the rock salt deposits are already being mined for the salt, and future generations may mine the remaining ones. In addition, some of these deposits are associated with oil and natural gas. Although these considerations make salt beds unsuitable for storing high-level waste, some are suitable for storing low-level waste.

The federal Low-Level Radioactive Waste Policy Act of 1980 delegates the responsibility for the disposal of low-level waste to the state in which that waste is produced. The Waste Isolation Pilot Plant (WIPP) was constructed 42 kilometers east of Carlsbad in the southeastern corner of New Mexico in order to develop and prove the required technology. By law, WIPP can only accept low-level waste from the nation's nuclear weapons program; the first shipment arrived and was placed in storage during 1999. The waste is stored 650 meters below the surface in rooms excavated in an ancient, stable salt formation.

POLITICAL CONSIDERATIONS

Nuclear waste disposal, whether of low-level or high-level wastes, is a problem that will remain. More waste is being created every day. A geologic repository and its host rock or soil must safely isolate the wastes from the surface environment for a specified period of time. Should the repository and host material fail, plants and animals will be exposed to radioactive elements. The risk of exposure to radiation can never be eliminated, but careful design of the repository and careful determination of the host material's strengths and weaknesses can result in a disposal site that minimizes the risk.

Ultimately, the siting of a nuclear waste disposal facility will be a political decision. That political decision must be based on solid technical information and judgments. As nuclear wastes accumulate in temporary holding facilities (for example, spent fuel elements from nuclear reactor power generators are stored in large water-filled pools near the reactor), political pressures build to solve the disposal problem. An informed public, while pressing for a solution, will understand delays necessitated by the need to gather and interpret the data. It must also be understood that risk can be minimized but never eliminated.

Richard W. Arnseth

CROSS-REFERENCES

BIBLIOGRAPHY

Bartlett, Donald L., and James B. Steele. *Forevermore: Nuclear Waste in America.* New York: W. W. Norton, 1985. Written by two reporters, this book looks at the history and future of nuclear waste disposal in the United States. The strength of the book is in tracing the political aspects of decision making in a technical field. Accessible to any reader with an interest in the subject, regardless of technical background.

Burns, Michael E., ed. *Low-Level Radioactive Waste Regulation: Science, Politics, and Fear.* Chelsea, Mich.: Lewis, 1987. One of the few books available for the general reader that deals exclusively with the problems of low-level wastes. Individual articles by groups of authors cover a range of issues from technical aspects of exposure risks to the politics among states when deciding where to locate a repository. Some of the articles may be too technical for the lay reader, but there is a wealth of data contained in tables throughout. Contains extensive references for further reading and an adequate glossary.

Byrne, John, and Steven M. Hoffman, eds. *Governing the Atom: The Politics of Risk.* New Brunswick, N.J.: Transaction, 1996. This volume from the Energy and Environmental Policy

Series researches the environmental and social aspects of the nuclear industry. Careful attention is paid to government policies that have been implemented to monitor safety and accident prevention measures.

Carter, Luther J. *Nuclear Imperatives and Public Trust: Dealing with Radioactive Waste.* Washington, D.C.: Resources for the Future, 1987. Carter provides a detailed but not overly technical history of the radioactive waste repository story. The book details the political and bureaucratic decision-making processes and shows how ill-informed decisions led to early disasters. Contains a good glossary of terms and extensive footnotes.

Gerber, Michele Stenehjem. *On the Home Front: The Cold War Legacy of the Hanford Nuclear Site.* Lincoln: University of Nebraska Press, 1992. A look at the measures that have been used at Hanford Nuclear Weapons Plant to dispose of hazardous waste, and an examination of the effects the disposal has on the environment. Includes illustrations, maps, index, and sixty-three pages of bibliographical references.

Grossman, Dan, and Seth Shulman. "A Nuclear Dump: The Experiment Begins (Beneath Yucca Mountain, Nevada)." *Discover* 10 (March, 1989): 48. A newsy, nontechnical article on the proposed Yucca Mountain, Nevada, high-level waste repository. A very accessible, if brief, discussion of some of the tests needed to characterize the rocks and their history. The authors point out areas of disagreement on technical questions and indicate the complexity of such problems. There are a few good illustrations of what the repository will look like. In general, a good introductory article for anyone with no background in the subject.

League of Women Voters Educational Fund Staff. *The Nuclear Waste Primer: A Handbook for Citizens.* New York: Lyons, 1987. A very compact (ninety-page) introduction to the nuclear waste problem. After a quick introduction to some of the technical jargon used in radiation chemistry, the book describes the magnitude of the problems of dealing with low- and high-level wastes, concluding with suggestions for citizen action. The glossary is brief but comprehensive.

Noyes, Robert. *Nuclear Waste Cleanup Technology and Opportunities.* Park Ridge, N.J.: Noyes Publications, 1995. A complete discription of the legislation enacted to handle the clean-up and disposal of radioactive and hazardous waste, as well as the environmental and social aspects associated with the disposal. Includes illustrations, index, and bibliography.

STRATEGIC RESOURCES

A nation's strategic resources are those resources that are essential for its major industries, military defense, and energy programs. For the United States, these resources include manganese, chromium, cobalt, nickel, platinum, titanium, aluminum, and oil.

PRINCIPAL TERMS

ALLOY: a substance composed of two or more metals or of a metal and certain nonmetals

CATALYST: a chemical substance that speeds up a chemical reaction without being permanently affected by that reaction

MANGANESE NODULES: rounded, concentrically laminated masses of iron and manganese oxide found on the deep-sea floor

OIL RIGHTS: the ownership of the oil and natural gas on another party's land, with the right to drill for and remove them

ORE DEPOSIT: a natural accumulation of mineral matter from which the owner expects to extract a metal at a profit

PROVEN RESERVE: a reserve supply of a valuable mineral substance that can be exploited at a future time

SALT DOME: an underground structure in the shape of a circular plug resulting from the upward movement of salt

WHAT MAKES RESOURCES "STRATEGIC"

Reference is frequently made to strategic resources, but unfortunately, there is no general agreement on what makes resources "strategic." Because the word "strategy" has a military connotation, strategic resources are often considered to be those resources that would be of critical importance in wartime. A somewhat broader definition of strategic resources is that they are those resources that a nation considers essential for its major industries, military defense, and energy programs. Similarly, there is no general agreement as to just which of the many Earth resources are the strategic ones. Several authors have restricted the definition to metals that are in short supply, with some even limiting the definition to the six metals alloyed with iron in the making of steel. Others, taking a broader view, have included among the strategic resources nonmetals such as fertilizers and energy sources such as petroleum.

In listing those Earth resources that are strategically significant, it is important to realize that each nation's list will be different. In other words, a resource that is in critically short supply for one nation may be possessed in abundance by a second. Furthermore, what is considered to be a strategic resource today may not be considered one

tomorrow or a hundred years from now. Before the 1970's, the opening of the large Middle Eastern oil fields had driven the price of oil down to $1.30 per barrel and the price of gasoline at the pump as low as 20 cents per gallon. Excess production capacity continued until the early 1970's, when increasing political tensions in the Middle East finally resulted in a united front on the part of the members of the Organization of Petroleum Exporting Countries (OPEC). First came voluntary production cutbacks, then the Arab oil embargo of 1973-1974. Eventually, the price of oil reached $35.00 per barrel, and gasoline was selling for more than $1.50 per gallon, when it could be obtained at all. A resource which everyone had taken for granted had suddenly become a strategic resource.

PROTECTIONS AGAINST SHORTAGES

To ensure adequate oil supplies for the United States in future emergencies, the U.S. government authorized the establishment of the Strategic Petroleum Reserve. The purpose of this legislation was to purchase 1 billion barrels of oil and to store it in large caverns hollowed out of underground salt domes in coastal Texas and Louisiana. The creation of the Strategic Petroleum Reserve illustrates an important principle. Nations can protect themselves against a possible interruption in supplies of imported strategic resources by stockpiling them during peacetime. Shortages during World War I, for example, caused the U.S. Congress to pass the Strategic Materials Act of 1939 and to begin stockpiling tin, quartz crystals, and chromite in anticipation of the outbreak of another war.

Another way that a major industrial nation protects itself against possible wartime shortages of strategic materials is by arranging for access to supplies of these materials in the event of war. Frequently, this preparation will involve trade agreements with neutral nations or with nations to which a country is bound by political alliances. In extreme cases, it may be necessary to invade a neighboring nation in order to obtain access to strategic resources. The history of Alsace-Lorraine in Europe exemplifies such a situation. This important iron-mining district is traditionally French, but it shares a border with Germany. In both world wars, Germany occupied Alsace-Lorraine in order to assure access to these iron ore deposits.

A third way that a nation can ensure adequate supplies of strategic resources during wartime is to develop substitutes for scarce materials using cheap materials already available. Certain mineral resources, such as mercury and uranium, have unique properties, and for these resources, no satisfactory substitutes can be found. Some of the other metals can be synthesized from related elements but only by prohibitively expensive methods. A number of strategic resources, however, can be synthesized from cheap, readily available materials at comparatively low cost.

U.S. Strategic Resources in Shortest Supply

The four strategic resources that would be in shortest supply in the United States were it to go to war are manganese, chromium, cobalt, and nickel. Manganese is a soft, silver-gray metal that is essential in the making of steel. Up to 7 kilograms of manganese are necessary for the production of each ton of iron or steel, and no satisfactory substitute has ever been found. Manganese removes undesired quantities of oxygen and sulfur from the iron during the steel-making process, yielding a hard, tough product suitable for bridge steel,

projectiles, and armor plating. U.S. manganese deposits are small, low in grade, and expensive to work; 95 percent of its manganese is imported.

Chromium is a hard, silver-gray metal that is also essential in the making of steel and for which no satisfactory substitute has been found. Chromium makes steel resistant to corrosion and appears on the shiny, chromium-plated surfaces on automobiles. The chromium content of stainless steel varies from 12 to 30 percent. Chromium-steel alloys are also used in aircraft engines, military vehicles, and weapons. The United States has large, low-grade deposits of chromium ore in Montana, but they are expensive to process. Consequently, 90 percent of its chromium is imported, primarily from South Africa.

Cobalt is a silver-white metal that is also needed in steel making. The addition of small quantities of cobalt makes steel harder and heat-resistant. Consequently, cobalt steel has important applications in the manufacture of metal-cutting tools, jet engines, and rockets. Cobalt alloys are magnetic, and the magnetism is retained permanently; therefore, these alloys are used for the manufacture of magnets. The United States has a small cobalt production from low-grade ores in Missouri, but 90 percent of its cobalt is imported, primarily from the country of Zaire, in Africa.

Nickel is a nearly white metal that is another important alloying agent in the making of steel. Nickel steels do not corrode or rust, so large quantities of nickel are used in the manufacture of stainless steel. Nickel is also used in plating because of its shine, in coinage (as in the familiar nickel coin), and for alloys that have important applications in the defense industry. The United States has very few high-grade nickel ore deposits, but there is a large, low-grade deposit in Minnesota. About 80 percent of its nickel is imported from Canada, the world's largest nickel producer.

Additional mineral resources that might be in short supply in the United States during wartime include platinum, titanium, and aluminum. Platinum has valuable chemical properties, because it resists corrosion and acts as a catalyst to speed chemical reactions. Titanium is a silver-gray metal used as an alloy. Because it imparts great strength, heat resistance, and resistance to corrosion, it is used in the construction of supersonic aircraft, jet engines, and space capsules. The major use for aluminum in the United States is in beverage cans, 102 billion of which were produced in 1998.

ESTIMATING RESERVES

Strategic resources can be studied in a variety of ways. The first is to identify which of the various Earth resources are in shortest supply. That is done by analyzing production and usage figures for each of a nation's industrially important minerals in order to determine how much the nation relies on imports for each of these materials. When usage exceeds production, it is relying on imports. The greater the reliance on imports, the more strategic the resource becomes.

A second way in which one can study strategic resources is to identify resources which are adequate at present but which may become scarce in the future. That is done by estimating reserves. A reserve is a supply of a mineral substance that still remains in the ground and is available to be extracted at some future time. Two types of reserve can be distinguished: proven reserves and undiscovered reserves. Proven reserves are reserves which have already been outlined by drilling or some other means; there is practically no risk of the desired substance's not being there. Undiscovered reserves, on the other hand, are reserves which are believed to be present on the basis of geologic studies but which are still inadequately explored.

SEARCH FOR EXPANDED SUPPLIES

A third way of studying strategic resources is to begin a search for expanded supplies before current reserves are exhausted or before imports have become unavailable because of wartime conditions. Various methods have been employed for increasing the supply of a strategic resource. They include stockpiling, trade agreements (or territorial annexation, in extreme cases), manufacture of synthetics or the substitution of other substances, and the development of conservation or recycling programs.

Two additional avenues have also become available in the search for expanded supplies of strategic resources. One is the use of the plate tectonics theory. This theory proposes that the Earth's surface is divided into a few large plates that are slowly moving with respect to one another. Intense geologic activity occurs at plate boundaries, and many mineral deposits are believed to have been formed by such activity. Much exploration for new deposits of strategic resources is being concentrated at plate boundaries. A second new avenue is the exploration of the deep-sea floor. Although manganese nodules were discovered on the deep-sea floor by the *Challenger* expedition in the late 1800's, it was not until the advent of manned submersibles and remote-controlled television cameras that it was realized that 15 percent or more of the ocean floor may be covered by such nodules. They have valuable amounts of iron, nickel, manganese, cobalt, and other strategically important substances. As exploration of the deep-sea floor continues, other supplies of strategic resources may be found there, as well.

CONSERVATION MEASURES

Because a nation's strategic resources are those that would be of critical importance in wartime, one would expect not to hear much about them in peacetime. They are substances which must be imported, however, and excessive imports lead to trade deficits. As a result, a government may resort to conservation measures in order to reduce such imports. A good example is the American program for recycling aluminum beverage cans. This program has two objectives: to reduce U.S. imports of costly aluminum ore and to conserve the large amounts of electricity needed to process aluminum ore.

In troubled times, such as during the 1973-1974 Arab oil embargo and the ensuing energy crisis, measures aimed at conserving strategic resources become very noticeable to the general public. To reduce oil consumption, Americans were asked to turn down their thermostats, the nationwide speed limit was reduced to 55 miles per hour, and automobile companies were told to improve the gas mileage of their cars, which led to a whole new generation of midsized cars. Furthermore, ripple effects caused by the oil shortage spread through the economy, triggering a recession, costing people their jobs, and setting off a stock market decline. Overseas, the value of the dollar declined against other currencies, and soon American tourists found themselves paying more for a hotel room or a meal.

In wartime, the need to conserve strategic resources may even result in rationing. This happened in the United States during World War II. Each driver was given a ration card entitling him or her to purchase a certain number of gallons of gasoline each week; drivers were placed in different categories depending on whether they drove for pleasure or to work.

Donald W. Lovejoy

CROSS-REFERENCES

Clays and Clay Minerals, 1187; Earth Resources, 1741; Earth Science and the Environment, 1746; Earth Science Enterprise, 1752; Environmental Health, 1759; Evaporites, 2330; Future Resources, 1764; Geomorphology of Dry Climate Areas, 904; Groundwater Movement, 2030; Groundwater Pollution and Remediation, 2037; Hazardous Wastes, 1769; Hydrologic Cycle, 2045; Landfills, 1774; Mining Processes, 1780; Mining Wastes, 1786; Nuclear Power, 1663; Nuclear Waste Disposal, 1791; Pyroclastic Rocks, 1343; Radioactive Decay, 532.

BIBLIOGRAPHY

Bartholomew, John C., ed. *The Times Atlas of the World.* New York: Times Books, 1980. A well-written overview entitled "Resources of the World" is followed by large, eight-color maps. The world mineral map shows the world distribution of metals and nonmetals and world geology. The world energy map shows the distribution of energy sources and their comparative consumption by nations. Excellent graphs. Suitable for high school readers.

Bramwell, M., ed. *The Rand McNally Atlas of the Oceans.* Skokie, Ill.: Rand McNally, 1977. A beautifully illustrated atlas with subsections on the various energy sources in the ocean, the continental shelves, manganese nodule deposits on the deep-sea floor, and offshore oil. Excellent color photographs, maps, and line drawings. Suitable for high school readers.

Craig, J. R., D. J. Vaughan, and B. J. Skinner. *Resources of the Earth.* Englewood Cliffs, N.J.: Prentice-Hall, 1988. A well-written text with numerous black-and-white photographs, color plates, tables, charts, maps, and line drawings. Chapter 2, entitled "Earth Resources Through History," contains excellent sections on strategic resources, resources and international conflict, and the reliance of the United States on imports. Suitable for college-level readers.

Davidson, Jon P., Walter E. Reed, and Paul M. Davis. *Exploring Earth: An Introduction to Physical Geology.* Upper Saddle River, N.J.: Prentice Hall, 1997. An excellent introduction to physical geology and the systems of the Earth. This book explains the composition of the Earth, its history, and its state of constant change. Intended for the layperson, it is filled with colorful illustrations and maps.

Jensen, M. L., and A. M. Bateman. *Economic Mineral Deposits.* 3d ed. New York: John Wiley & Sons, 1979. This outstanding economic geology text provides detailed information on metallic and nonmetallic mineral deposits and their modes of formation. Discusses the history of mineral use, the exploration and development of mineral properties, and the role of strategic minerals in international relations and war. For a college-level audience.

Klein, Cornelis, and C. S. Hurlbut, Jr. *Manual of Mineralogy.* 21st ed. New York: John Wiley & Sons, 1999. This well-known manual provides detailed descriptions of the various metallic and nonmetallic minerals, including their crystallography, physical properties, and chemical characteristics. There is also helpful information on the mode of formation of the various minerals, where they are found, and why they are useful. Summary tables are included.

Lutgens, Frederick K., and Edward J. Tarbuck. *Earth: An Introduction to Physical Geology.* Upper Saddle River, N.J.: Prentice Hall, 1999. 6th ed. This college text provides a clear picture of the Earth's systems and processes that is suitable for the high school or college reader. In addition to its illustrations and graphics, it has an accompanying computer disc that is compatible with either Macintosh or Windows. Bibliography and index.

McGeary, David. *Physical Geology: Earth Revealed.* Boston: WCB/McGraw-Hill, 1998. 3d ed. A thorough introduction to physical geology, this illustrated textbook is accompanieed by a computer laser optical disc that corresponds to the chapters and topics explored. Bibliography and index.

Press, Frank, and Raymond Siever. *Understanding Earth.* 2d ed. New York: W. H. Freeman, 1998. An excellent general text on all aspects of geology. Suitable for advanced high school and college students.

Skinner, B. J. *Earth Resources.* 3d ed. Englewood Cliffs, N.J.: Prentice-Hall, 1986. This book provides an excellent overview of the Earth's metallic, nonmetallic, and energy resources. Contains helpful line drawings, maps, tables, and charts; few photographs. Includes suggestions for further reading and a list of principal ore minerals and their production figures for 1982. Suitable for the interested layperson.

Tennissen, Anthony C. *The Nature of Earth Materials.* 2d ed. Englewood Cliffs, N.J.: Prentice-Hall, 1983. Contains detailed descriptions of 110 common minerals, with a black-and-white photograph of each. Chapter 7 includes helpful sections on the distribution of mineral deposits and the utilization of various Earth materials. Suitable for college students.

Thompson, Graham R. *An Introduction to Physical Geology.* Fort Worth, Tex.: Saunders College Publishing, 1998. This college text provides an easy-to-follow look at what physical geology is, and the phases involved in it. Thompson shows the reader each phase of the Earth's geochemical processes. Excellent for high school and college readers. Illustrations, diagrams, and bibliography included.

EARTH SCIENCE

Alphabetical List of Contents

Categorized List of Contents